Nondestructive Evaluation of Semiconductor Materials and Devices

NATO ADVANCED STUDY INSTITUTES SERIES

A series of edited volumes comprising multifaceted studies of contemporary scientific issues by some of the best scientific minds in the world, assembled in cooperation with NATO Scientific Affairs Division.

Series B: Physics

RECENT VOLUMES IN THIS SERIES

Volume 39 – Hadron Structure and Lepton–Hadron Interactions – *Cargèse* 1977
edited by Maurice Lévy, Jean-Louis Basdevant, David Speiser,
Jacques Weyers, Raymond Gastmans, and Jean Zinn-Justin

Volume 40 – Kinetics of Ion–Molecule Reactions
edited by Pierre Ausloos

Volume 41 – Fiber and Integrated Optics
edited by D. B. Ostrowsky

Volume 42 – Electrons in Disordered Metals and at Metallic Surfaces
edited by P. Phariseau, B. L. Györffy, and L. Scheire

Volume 43 – Recent Advances in Group Theory and Their Application to Spectroscopy
edited by John C. Donini

Volume 44 – Recent Developments in Gravitation – *Cargèse* 1978
edited by Maurice Lévy and S. Deser

Volume 45 – Common Problems in Low- and Medium-Energy Nuclear Physics
edited by B. Castel, B. Goulard, and F. C. Khanna

Volume 46 – Nondestructive Evaluation of Semiconductor Materials and Devices
edited by Jay N. Zemel

Volume 47 – Site Characterization and Aggregation of Implanted Atoms in Materials
edited by A. Perez and R. Coussement

Volume 48 – Electron and Magnetization Densities in Molecules and Crystals
edited by P. Becker

Volume 49 – New Phenomena in Lepton-Hadron Physics
edited by Dietrich E. C. Fries and Julius Wess

This series is published by an international board of publishers in conjunction with NATO Scientific Affairs Division

A	Life Sciences	Plenum Publishing Corporation
B	Physics	London and New York
C	Mathematical and Physical Sciences	D. Reidel Publishing Company Dordrecht and Boston
D	Behavioral and Social Sciences	Sijthoff International Publishing Company Leiden
E	Applied Sciences	Noordhoff International Publishing Leiden

Nondestructive Evaluation of Semiconductor Materials and Devices

Edited by
Jay N. Zemel
The University of Pennsylvania
Philadelphia, Pennsylvania

SPRINGER SCIENCE+BUSINESS MEDIA, LLC

Library of Congress Cataloging in Publication Data

Nato Advanced Study Institute on Nondestructive Evaluation of Semiconductor Materials and Devices, Villa Tuscolano, Italy, 1978.
  Nondestructive evaluation of semiconductor materials and devices.

  (NATO advanced study institutes series: Series B., Physics; v. 46)
  "Published in cooperation with NATO Scientific Affairs Division."
  Includes index.
  1. Semiconductors—Testing—Congresses. I. Zemel, Jay N. II. North Atlantic Treaty Organization. Division of Scientific Affairs. III. Title. IV. Series.
TK7871.85.N376 1978   621.3815′2′028   79-16499
ISBN 978-1-4757-1354-1   ISBN 978-1-4757-1352-7 (eBook)
DOI 10.1007/978-1-4757-1352-7

Lectures presented at the NATO Advanced Study Institute on
Nondestructive Evaluation of Semiconductor Materials and Devices,
held at the Villa Tuscolano, Italy, September 19–29, 1978

© 1979 Springer Science+Business Media New York
Originally published by Plenum Press, New York in 1979
Softcover reprint of the hardcover 1st edition 1979

All rights reserved

No part of this book may be reproduced, stored in a retrieval system, or transmitted,
in any form or by any means, electronic, mechanical, photocopying, microfilming,
recording, or otherwise, without written permission from the Publisher

Preface

From September 19-29, a NATO Advanced Study Institute on Nondestructive Evaluation of Semiconductor Materials and Devices was held at the Villa Tuscolano in Frascati, Italy. A total of 80 attendees and lecturers participated in the program which covered many of the important topics in this field. The subject matter was divided to emphasize the following different types of problems: electrical measurements; acoustic measurements; scanning techniques; optical methods; backscatter methods; x-ray observations; accelerated life tests.

It would be difficult to give a full discussion of such an Institute without going through the major points of each speaker. Clearly this is the proper task of the eventual readers of these Proceedings. Instead, it would be preferable to stress some general issues. What came through very clearly is that the measurements of the basic scientists in materials and device phenomena are of substantial immediate concern to the device technologies and end users. It was also very clear that there is a monstrous documentation problem confronting large scale producers and users involving field studies of failure. While the accelerated life test and quality control engineer attempts to employ both good science and statistics to decide on the future of a producer line, feedback to the engineers and their support scientists is an essential aspect of closing the informational loop. The surface defects, noise properties and structural flaws turned up by varied measurements will influence the yield and accelerated life test data in a moderately causal fashion. However, it is not clear that the life assurance required by the large scale data processors, the end users, can be adequately met unless these users take the time, bother and resultant expense to chase down the failures to their source.

Even individual measurements that would appear well characterized such as spreading resistance and four point probe methods have enough current problems to suggest that they would be fit topics for many additional Ph.D. dissertations and numerous research papers. More important than the academic consequences of such research is that the possible improvements in electrical measurement could significantly lower the cost of quality assurance. That would be a real fallout from such a program.

Buried in these excellent papers at various depths are numerous scientific and technological jewels that will require different levels of effort to extract. What was most gratifying was the outstanding character of the presentations of the fifteen lecturers and the enthusiastic response of the students. Toward the end of the second week, the students themselves had sent back home for materials to present. This was done in the evening after a full day of lecturers and during free time. That speaks for itself.

The fine work of A. D'Amico in local arragements and G. Spisso in heading the Secretariat needs to be mentioned in closing. A committee consisting of Prof. A. A. Quarante, Modena, Prof. E. DeCastro, Bologna and Prof. E. Gatti, Milano, advised the Director on policy.

I would like to acknowledge the financial aid of STET and FIAT of Italy, the U. S. National Science Foundation for support of student travel and Drs. T. Kester and A. Di Lullo of the NATO Scientific Affairs Office for their wise counsel. The staff of the Laboratorio di Elettronica dello Stato Solido were essential to the orderly operation of the Institute and the cooperation of Professor A. Paoletti was appreciated by the Director.

Finally, I would like to express my appreciation to T. Harris and R. Woldow for their help on the organization and to my daughter-in-law Susan Blackwood Zemel for the typing of the manuscript.

Philadelphia, Pennsylvania
March 1979

Jay N. Zemel

Contents

CHAPTER 1: Two Probe (Spreading Resistance) Measurements for Evaluation of Semiconductor Materials and Devices   1
J. R. Ehrstein

    I. Spreading Resistance as an Empirical Technique
   II. Contact Models
  III. Depth Profiling of Semiconductor Structures
   IV. Applications of Spreading Resistance Measurements
    V. Possible Errors in Spreading Resistance Measurements

CHAPTER 2: Four-Terminal Nondestructive Electrical and Galvanomagnetic Measurements   67
H. H. Weider

     I. Introduction
    II. The Collinear Four-Probe Array
   III. The Square Four-Probe Array
    IV. Reliability of Four-Probe Techniques
     V. Resistivity of Lamellae of Arbitrary Shape
    VI. Error Introduced by Contact Size, Position and Geometry
   VII. Hall Effect Measurements
  VIII. Hall Effect of Arbitrary Contour Specimens
    IX. Apparatus for Hall Effect Measurements
     X. Probe Mapping and Iso-Resistivity Contour Maps

CHAPTER 3: Characterization of Surface States
at the Si–SiO$_2$ Interface   105
G. DeClerck

   I. Introduction
  II. Metal-Insulator-Semiconductor (MIS) Structure
 III. MIS Structure with Surface States
  IV. Fluctuations of Surface Potential
   V. Experimental Techniques Using MOS Capacitors
  VI. Experimental Techniques Using MOS Transistors
 VII. Surface State Measurements by Means of Charge Coupled Devices

CHAPTER 4: Steady-State and Non-Steady-State
Characterization of MOS Devices   149
J. G. Simmons

   I. Relaxation Current Techniques
  II. C-V Techniques
 III. Current Voltage Measurements

CHAPTER 5: Semiconductor Material Evaluation
by Means of Schottky Contacts   201
K. Heime

   I. Scope of the Paper
  II. Introduction
 III. Evaluation of Shallow Levels
  IV. Evaluation of Deep Levels
   V. Conclusions

CHAPTER 6: Noise   257
A. D'Amico

    I. Introduction
   II. Stochastic Processes, Stationarity and Ergodicity
  III. Measurements
   IV. Thermal Noise
    V. $1/f^a$ Noise (Flicker Noise)
   VI. Generation-Recombination (G-R), Shot and Burst
  VII. Noise in Devices
 VIII. Nonlinearity by Third Harmonic Index Method

CONTENTS ix

CHAPTER 7: Optical Characterization of
           Semiconductors 315
   E. D. Palik and R. T. Holm

     I.   Review of Basic Optical Properties
     II.   Infrared Reflection and Transmission
           in Bulk Materials
     III. Infrared Reflection and Transmission
           of a Thin Film on a Substrate
     IV.   More Optical Characterization Techniques
     V.   Characterization at the Band Edge

CHAPTER 8: Use of Photoemission and Related
           Techniques to Study Device Fabrication 397
   W. E. Spicer

     I.   Introduction
     II.   Physics of Photoemission
     III. Synchrotron Radiation Sources
     IV.   Sputter Auger Techniques
     V.   Oxides and Oxygen Adsorption on
           GaAs, IuP and GaSb
     VI.   Sputter Auger Studies of the $Si-SiO_2$
           Interface on Real Device Oxides
     VII. A New Mechanism for Fermi Level Pinning
           of Schottky Barriers
     VIII. Schottky Barriers on Si and II-VI
           Compounds
     IX.   Conclusions

CHAPTER 9. Scanned Photovoltage and Photoemission 457
   T. H. DiStefano

     I.   Introduction
     II.   Internal Photoemission Probe of
           Interfaces
     III. Scanned Electro-Optical Techniques
     IV.   Photoemission Images
     V.   Scanned Photovoltage Analysis of
           Semiconductor Surfaces
     VI.   Scanned Photovoltage Images

CHAPTER 10: SEM Methods for the Characterization
of Semiconductor Materials and Devices   515
  C. J. Varker

    I.   Introduction to Scanning Electron
Microscopy
   II. Properties of Semiconductor Materials
and Devices
  III. Principle SEM Methods and Techniques
   IV. Characterization of Electrically Active
Defects in Silicon Materials and
Devices Using EBIC Microscopy
    V. Electrical Analysis of Integrated
Circuits

CHAPTER 11: Backscattering Spectrometry and
Related Analytic Techniques 581
  M.- A. Nicolet

    I.   Introduction
   II. Backscattering Spectrometry (BS)
  III. Nuclear Reaction and Ion Beam
Induced X-Rays

CHAPTER 12: The Acoustic Microscope: A Tool for
Nondestructive Testing 631
  J. Attal

    I.   Introduction
   II. Scanning Acoustic Microscope Design
In Transmission Operating Mode
  III. Resolution Performance
   IV. Acoustic Images: Nondestructive
Evaluation of Solid State Devices

CHAPTER 13: Nondestructive Tests Used to Insure
the Integrity of Semiconductor
Devices with Emphasis on Passive
Acoustic Techniques 677
  G. G. Harman

    I.   Introduction
   II. Some Current Production Line Assembly
Tests
  III. Passive Acoustic Techniques

CONTENTS   xi

CHAPTER 14: Lifetime Data Analysis   739
      F. H. Reynolds

         I.   Introduction
        II.   Lifetime Functions
       III.   Lifetime Distributions
        IV.   Parameter Sensitivities
         V.   Graphical Parameter Estimation
        VI.   Model Validation
       VII.   Numerical Parameter Estimation
      VIII.   Analysis of Experimental Lifetime Data
        IX.   Alternative Approaches
         X.   Conclusion

Index   771

Chapter 1

TWO-PROBE (SPREADING RESISTANCE) MEASUREMENTS FOR EVALUATION

OF SEMICONDUCTOR MATERIALS AND DEVICES

>James R. Ehrstein
>
>Electron Devices Division
>
>Center for Electronics and Electrical Engineering
>
>National Engineering Laboratory
>
>National Bureau of Standards
>
>Washington, DC 20234

I. SPREADING RESISTANCE AS AN EMPIRICAL TECHNIQUE

A. Introduction

Interest in two-probe resistance, (spreading resistance) measurements dates back perhaps 20 years. It arose at a time when the development of a number of techniques for measuring resistivity and resistivity profiles was being pursued, primarily for germanium and silicon technology. In the early days of its use, diffusions were relatively deep and control of epitaxial resistivity was an important problem. The structures of interest have changed noticeably in the interveninng years. Ion implantation has come into regular use and the dimensional scale of diffusions and epitaxy have been considerably reduced.

Spreading resistance has evolved into a rather mature measurement technique during that time. Although not qualifying in all respects as a perfect measurement, i.e. one which has high resolution, is fast, accurate, contactless and cheap, spreading resistance does have unique capabilities compared to other methods for measuring semiconductor resistivity and its variation. It is the purpose of this paper to give an overview of the state of knowledge of the spreading resistance technique, its control, interpretation and possible limitations.

B. General Capability

The range of specimen values which can be measured is limited primarily by the dynamic range of the electronics employed. Measurements in silicon over the carrier concentration range 10^{12} to $10^{21}/cm^3$ are common with commercial electronics. In contrast to the powerful technique of secondary ion mass spectroscopy, the range is independent of dopant species. Since the measurement has a lateral spatial resolution on the order of 10 µm, it can be used for measuring the lateral dopant uniformity on the surfaces of bulk crystals, slices, and epitaxial layers. The applications of such surface measurements range from studies of crystal growth dynamics to inspection of and quality control over the resistivity uniformity of device starting material.

Depth profiles of resistivity (dopant density) of an arbitrary combination of epitaxial, diffused, or implanted layers can also be measured. To accomplish this it is customary to section the specimen at a shallow angle thus exposing the structure in depth. Using specimen sectioning angles of about 10 millirad (0.5°) and lateral probe step increments of 10 m, measurements can readily be taken which represent incremental steps of about 25 nm (250 Å) of specimen depth. The measurement always averages over a small specimen volume. For a specimen which is nonuniform with depth, the size of this volume is determined by the nature of the resistivity variations with depth. For routine process monitoring and control, it is often possible to utilize raw spreading resistance depth profile data and look for deviations in shape or value from raw profile data taken on similar structures which had been found to give proper device operation. However, to derive quantitative resistivity (carrier concentration) profiles from the measured data it is necessary to use "correction factors" to account for the effect of the structure on the sampling volume. Quantitative resistivity values are important for development or modification of device processing steps or for relating device performance models to material parameters.

C. Basics of the Spreading Resistance Measurement

The measurement is made by placing small radius probes of hardened highly conducting (micro-ohm cm) alloy to the surface of the semiconductor material of interest and applying pressure. The electrical resistance between the probes is measured using a small dc bias. The analysis of the measurement is based on the recognition that there is a component of circuit resistance due to the constriction

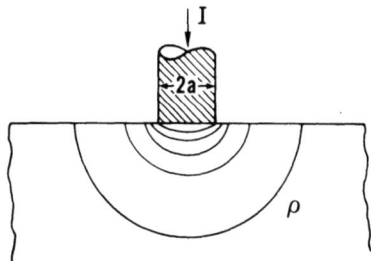

Fig. 1. Representation of potential distribution in a semi-infinite medium due to current constriction through a small contact.

of the current and its subsequent spreading when contact between macroscopic bodies is made by means of microscopic areas. Holm [1] has shown for a single indenting hemispherical contact of radius, a, assuming ohmic behavior and one contact member having a resistivity much larger than the other member, that the resistance, R, due to current constriction (and subsequent spreading) is given by:

$$R = \frac{\rho}{2\pi a} \qquad (1)$$

providing the body of higher resistivity, ρ, is large (semi-infinite) compared to the size of the contact.

For a nonindenting contact with plane circular interface (cylindrical contact model), Holm gives:

$$R = \frac{\rho}{4a} \qquad (2)$$

where a is the circle radius, see Figure 1. Thus, for either contact shape, when the probe is highly conductive, the resistivity of the semiconductor controls the measured resistance.

About 80% of the total potential drop in the horizontal plane due to the spreading phenomenon occurs within a distance of about 5 times the contact radius for both hemispherical or cylindrical contacts. Because the contact damage in silicon is shallow (several thousand angstroms) compared to the contact diameter (2 to 10 microns), see Figure 2, the cylindrical contact model, eq. (2), is the customary starting point for describing spreading resistance measurements. The contact radius, a, is not known absolutely, but as a scale parameter relating specimen resistivity and measured two probe resistance it will recur in all further development of the spreading resistance method. As a point of reference note that for contacts with a radius on the order of 4 µm, a typical value for the spreading resistance measurement on thick slice specimens is about

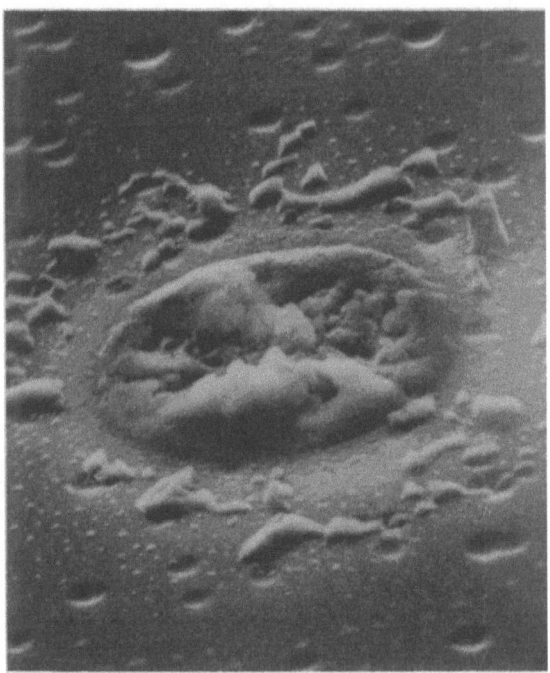

Fig. 2. Scanning electron microscope picture of damage in silicon due to a spreading resistance probe (after Mayer and Schwartzman [49]).

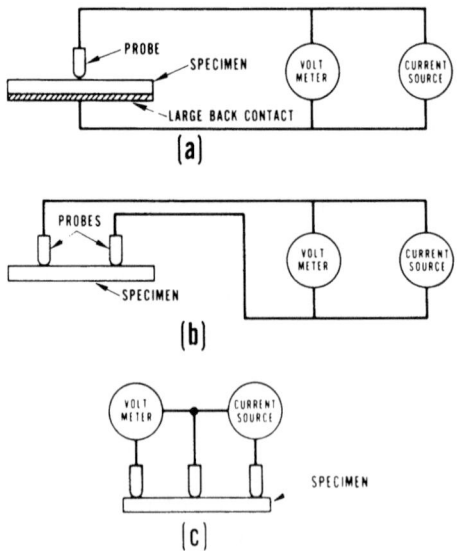

Fig. 3. Schematic arrangement for spreading resistance measurement using one, two, and three probes.

1000 times the resistivity of the specimen.

The spreading resistance measurement can actually be made using one, two, or three probes, see Figure 3. As applied to semiconductors, the case of one probe has little practical importance since it is limited to specimens without underlying junctions and requires an ohmic contact on the back side or elsewhere on the specimen. In the case of two probes, both are assumed to contribute equally to the total spreading resistance, and we note from the foregoing paragraph that for reasonable spacing, typically 20 µm to 1000 µm, the bulk resistance between the probes is negligible compared to the spreading resistance due to either probe. Hence, the net measured resistance to a very good approximation is due to the spreading resistancte of the pair of probes only, or:

$$R = \frac{\rho}{2a} \tag{3}$$

In the case of three probes, one probe serves as the common point to both the current and voltage circuits, and is the only probe contributing to the measured resustance. Hence, as for the single probe, $R = \rho/4a$. This three probe form does not have the application limitations of the single probe form, and is more ideal than the two probe form in that no assumption of exact equivalence of the probes is necessary. However, it is more difficult to fabricate and to keep aligned. Since alignment of the probes parallel to the intersection of the beveled and top surfaces of the specimen is a crucial factor for depth profiling of structures, the three probe form is not commonly used. Except where necessary to refine specific concepts, spreading resistance will be taken to mean the net measured resistance between two probes on a semiconductor. This will admittedly include possible barrier resistance effects and a small bulk resistance component to be added to the true geometrical spreading resistance. However, these components cannot be extracted unambiguously from the measured data in any event.

D. Some Practical Considerations

Since the utility of the spreading resistance measurement hinges on being able to make a series of constant quality contacts, which primarily means that the scale factor "a" does not vary in an uncontrolled fashion, several considerations regarding the probe system arc found to be empirically necessary. The first is that static pressures must be used which are at, or exceed, the elastic limit of the specimen. In silicon this corresponds to a load of about 1000 kg/mm^2 and results in a plastically deformed damage area on the silicon. Such damage apppears necessary in most instances for best reproducibility of measured resistance. Second, the application of the probes to the semiconductor must be well controlled: use of a dash pot to control the rate of descent of the probes to a value on the order of 1 mm/sec, or the use of spring loaded probes is required. Well-designed bearings and supports capable of keeping unwanted lateral motion to virtually undetectable levels (a small fraction of a micrometer, or less) are also required. In the case of undamped or very lightly damped probed descent, the impact momentum generally causes conchoidal fracture of the semiconductor which will result in unstable measurements. Alloy probe tips in which the binder metal has been worn away exposing numerous grains of hardened alloy offer improved contact stability, and the probes must be kept macroscopically, but not atomically, clean. Periodic light abrasion of the probe tips is generally necessary to ensure this condition.

It has also been found that applying constant voltage is preferable to the use of constant current, presumably because of effects due to large high frequency transients generated when the constant current supply is switched on. In practice a constant voltage of 1 to 10 mV is applied between the probes. A voltage in this range minimizes nonohmic properties of the metal semiconductor contact, as well as joule heating.

In the real world, spreading resistance measurements have been basically limited to silicon and germanium. Most compound semiconductors, either because of reduced hardness, wide band-gap difficulty in controlling surface potential, or some combination of these factors, have proven difficult for spreading resistance probing.

E. Spreading Resistance as a Comparison Measurement - Calibration

Even when restricted to measurements on silicon and germanium, the contacts are not ideal. Effects due to the physics of metal-semicondctor contacts are encountered and are complicated by work damage and surface states resulting from specimen preparation. In order to more reasonably describe the measurement response, the simple relation, eq. (3), must be modified. On a gross scale, for homogeneous bulk specimens of fixed conductivity type and crystallographic orientation (within a few degrees), it is found that the relation

$$\log R = m \log \rho + B \qquad (4a)$$

or

$$R = B' \rho^m \qquad (4b)$$

reasonably well describes the measured spreading resistance, R, as a function of resistivity, see Figure 4. It is noted, comparing

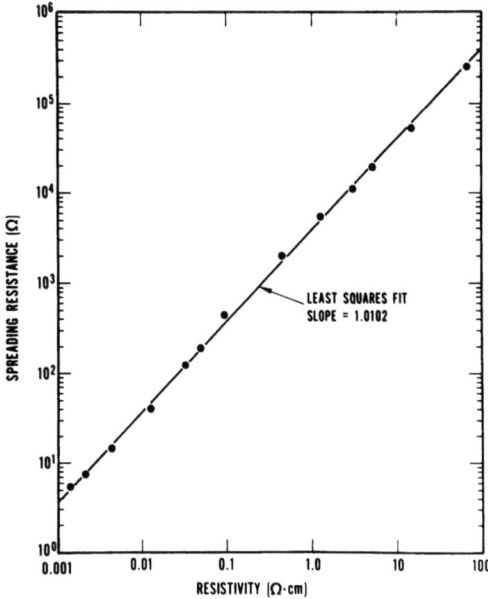

Fig. 4. Typical spreading resistance calibration data: (111) n-type silicon.

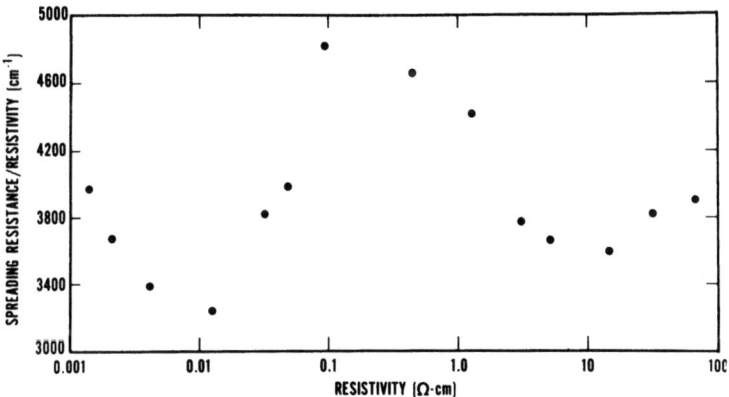

Fig. 5. Typical calibration data with ordinate normalized: (111) n-type silicon.

eq. (3) and eq. (4b) that an estimate of the contact parameter (in centimeters) can be obtained at $\rho = 1$ $\Omega\cdot$cm from the relation

$$R\Big|_{1\ \Omega\cdot\text{cm}} = \frac{1}{2a} = B' \tag{4c}$$

where B' is the estimate of measured resistance at 1 $\Omega\cdot$cm.

The slope, m, and the intercept, B, above are found to be functions of conductivity type, specimen orientation and surface preparation even for a fixed set of probe conditions. Values for m are typically in the range 0.85 to 1.05. Even if this relation fully described the observed data, it would agree with the simple model of Holm only for the case where m = 1.0. However, on closer examination of typical bulk speciemn data, noticeable deviations from the above relation are found, see Fig. 5. The deviations are attributed to metal-semiconductor barrier effects and like the slope and intercept, they depend on conductivity type, specimen orientation, and preparation, as described in Section II.

Many variations in the fine scale relation between measured spreading resistance and specimen resistivity can be found in the literature and are illustrated in part in Figures 13 and 14. However, details of the conditions of measurement for these are generally lacking. A rather extensive set of measurements described in reference [2] illustrates the range of responses obtainable as a result of various specimen preparations and probe types even when care has been taken to eliminate, or minimize by randomization, effects due to extraneous variables. No one of the responses illustrated there should be considered the "true" relation. In general, each is valid for the particular set of conditions under which it was obtained.

Spreading resistance is not an absolute but a relative technique, with a set of probes requiring periodic calibration against a set of specimens of known resistivity. Such a calibration is valid for interpreting measurements on specimens of interest only when using the same probe, probe load, specimen conductivity type, orientation, and surface preparation for which it was generated. Even so, if the calibration was generated with badly worn probes or on contaminated specimen surfaces, its validity may be in doubt.

The amount of calibration information required will in general depend on the intended application. When measurements of the lateral resistivity uniformity or of the average absolute resistivity of "unknown" bulk or thick epitaxial specimens are to be made, extra calibration data is often required in the restricted resistivity range of interest. In this application, the principal source of measurement error results from uncertainty in the calibration. Causes of the uncertainty are data scatter on one or more calibration specimens and inconsistencies between measurements from various calibaration specimens. Such inconsistencies are usually due to variations in surface preparation or to inherent nonuniformities in the chosen calibration specimens themselves. It is advisable in this case to accumulate and measure a larger number of calibration specimens in order to average such variations in the resistivity range of importance.

Since the principal error in quantifying measurements on the surface of specimens relates to the variation of the contact parameter, a, and its determination as a function of resistivity, it is advisable for best accuracy to bypass the simplification of eq. (4). It is common to represent calibration data by a modified form of eq. (3) which allows more detail:

$$R = \frac{\rho}{2a(\rho)} \quad (3a)$$

Here, $a(\rho)$, called an effective contact radius, incorporates any nonlinearity with resistivity and is determined from the measured calibration values over small intervals of resistivity. This is not the only manner of accounting for calibration nonlinearity, as will be shown in Sections II D and III G, but in general it suffices for surface measurements.

When depth profile measurements on thin layers are to be made, a somewhat less detailed knowledge of the calibration response may be sufficient. In the first place, for analysis of depth profile data, an additional term, called a sampling volume correction, is necessary in the relationship between measured resistance and specimen resistivity. This sampling volume correction accounts for the additional effect of the layer boundaries or internal gradients on the potential distribution within the specimen, and hence on the measured resistance. Values for such correction factors may often be two orders of magnitude larger than typical calibration nonlinearity. Moreover, the calculation of such volume factors is very cumbersome in most schemes, as will be discussed in Section III. Often the calculations require several levels of iteration to obtain convergence, even without the complication of a resistivity dependent effective contact parameter. It is common practice, although not necessarily correct, to resort to eq. (4) to represent the calibration relation when interpreting depth profile data. Some practitioners will include the appropriate $a(\rho)$ value determined from calibration data obtained on bulk specimens as a final step when interpreting depth profile data. However, this may not be correct either, as will be discussed in Section III G.

The principal point of this section is that even after certain experimental precautions are taken with respect to apparatus and probes, spreading resistance measurements are related to, but are not a fundamental measure of, specimen resistivity. Hence, for meaningful measurements, spreading resistance should be considered a comparative technique between specimens of known resistivity, the calibration set, and the unknown specimen of interest. Material surface preparation and all other conditions of measurement must be kept equivalent between the sets.

F. Further Considerations for Meaningful Measurements

The foregoing does not mean that all choices of measurement condition are equally good. The acceptable specimen preparation procedures are those which (1) give reproducible response for a given specimen, (2) are stable over a reasonable period (hopefully for days or weeks, but certainly for several hours) and (3) do not mask real structure or add spurious structure. For example, lapped surfaces are rapid to prepare and among the most stable. They also give quite repeatable average measurement response from one preparation to the next. However, due to the coarseness of the surface finish, compared to the probe contact size, rather high data scatter must be tolerated. Further, such surfaces may be too rough for bevel section profiling of thin layers. Polished surfaces, on the other hand, particularly scratch-free chem-mechanically polished surfaces yield the lowest scatter. It has been shown, however, that aqueous polishes whether mechanical or chem-mechanical cause a spurious increase in the spreading resistance measured on (111) p-type specimens, particularly above 1 $\Omega \cdot cm$ [3]. This increase, which has been found to be as large as several hundred percent at 100 $\Omega \cdot cm$, is not quantitatively reproducible. It can, however, be removed by a simple thermal treatment of the specimen at 150°C for 15 minutes. A smaller spurious shift, generally towards reduced measurement values, is often seen on aqueous-polished (111) n-type specimens above about 1 $\Omega \cdot cm$. Here the level shift does not respond well to thermal treatment and nonaqueous polishing may be advisable.

The previous paragraph serves to illustrate that there is no unique answer to choice of specimen preparation. Measurements made in a limited resistivity range or with relaxed requirements on noise may allow choices that would not be wise in other instances. Depth profiling of a thin transistor structure will require a low noise surface with simultaneous optimization of the beveled surface for both p- and n-type layers covering a wide range of resistivity levels; such a specimen presents a severe test of surface preparation optimization.

The probe itself can be a major source of measurement error. A method for rapidly verifying acceptability of a set of probes for the intended application is required. Typical probe alloys such as tungsten-osmium, tungsten-ruthenium, and tungsten-carbide have different hardnesses, grain sizes and abrasion resistances. Control of the metallurgy from batch to batch and sometimes from probe to probe is not ideal. Wear proceeds at various rates, with some probes having useful lives of several thousand measurements. Some will operate satisfactorily for a wide range of probe loads, still others seem incapable of generating repeatable data for any chosen load. In addition, such material as airborne dust may accumulate on the probe tips, temporarily impairing probe quality for even the best of probes.

Methods to qualify probes as being satisfactory are somewhat circular at best. Satisfactory behavior has no meaning except in relation to the type of semiconductor material ultimately to be measured. Yet tests of the probes should be independent of spurious specimen surface effects as might be found on the semiconductor. Several empirical probe qualifications found to be useful are as follows: (1) to ensure reasonable agreement between the actual contact and the cylindrical symmetry models developed for data analysis, the plastically deformed damage region on a silicon specimen should form a closed compact area with a minimum of spurs or fingers; (2) a circle circumscribing the damaged area should have a diameter in reasonable agreement with that predicted by Hertzian stress analysis for the load being used, (see reference [5,6] and section II A; (3) the probe response, in terms of average measurment value as well as point-to-point measurement repeatability, should be checked on a well-stabilized specimen or specimens reserved for this purpose. It is common practice to use an etched and aged 1 Ω·cm (111) p-type specimen for general use, but there is evidence that lower values, i.e., 0.01 Ω·cm p-type silicon, give a more sensitive test of repeatability particularly at the light loads (5 to 20 grams force) used to minimize penetration during depth profiling. Gorey, et. el. suggest a more elaborate probe qualification procedure [4]. The first part requires a light abrading of the surface of a new probe to enable good agreement between the forward and reverse polarity measurements on a 20 Ω·cm p-type specimen (an ohmic backside contact to the p-type specimen is needed to individually check each probe in a two probe system). The second step requires testing the probe load dependence of two separate types of measurements. The first is the ratio of the forward bias to reverse bias resistance for the p-type specimen mentioned (a range of probe loads with a constant value of unity for this ratio, indicating good contact quality, is required before proceeding). The second measurement is the ratio of the resistance of a 2 µm thick, 2 Ω·cm n/n$^+$ epitaxial specimen to that of 2 Ω·cm bulk n-type specimen. This ratio should remain constant for a range of probe loads. It will inevitably begin to decrease with increasing load after some critical value indicating excessive probe penetration through the epilayer. To be fully satisfactory for further uses a pair of probes should exhibit some range of probe forces over which both of the preceding tests have no dependence on probe force. When such a range of forces if found for a pair of probes, the center of the range is used as the operating force for regular measurement use.

We have seen that there is no method for preparing a specimen surface which is best for all specimens in all applications. Similarly, there is no test for quality of operation of a probe which

is fool-proof. Those tests mentioned here are simply examples of
tests which work reasonably well. Nevertheless, it is common
experience that a well-conditioned set of probes can give point-to-
point repeatability on the order of a fraction of 1% for a uniform
specimen. Further, measurement repeatability on the order of 5%
following repreparation of specimens can be obtained for much of
the normal resistivity range.

G. Summary

Spreading resistance measurements are made under less than
ideal conditions. Specimens are not measured in vacuum isolation
nor under the protection of a passivating film, and the use of
preparations to ensure high surface recombination velocity often
cannot be used. Many problems with regard to specimens and probes
as well as some of the empirical solutions have been detailed in
this section. These should serve the experimenter both as a
warning regarding possible measurement pitfalls, and as a guide to
reasonable choices for measurement conditions.

II. CONTACT MODELS

The previously mentioned deviation of the empirical spreading
resistance-resistivity relation from the simple model of Holm or
even from a log-log relation, suggests that a closer look at the
contact is in order.

A. Mechanical Description

The metallurgical contact process is described simply by Mazur
[5] using the Hertzian formula to derive the circular radius of
contact, a, under the application of a force, F, to a member with
hemispherical radius, r, in contact with a flat surface:

$$a = 1.1 \{(Fr/2) \ [(1/E_1 + 1/E_2)]\} \tag{5}$$

where E_1 and E_2 are the Young's moduli of the materials involved
in the contact. Note that probe loads reported for work on silicon
generally fall in the range 5 gf to 45gf.

A more detailed calculation by Kramer and van Ruyven [6] considers that the above equation is valid only within the elastic
limit of the contact members. Hence, they consider a two-region
contact, see Fig. 6. In the central region out to a distance,
X_1, the elastic limit of silicon is initially exceeded, and the
silicon plastically deforms until it provides enough contact area
to support the probe. Under this assumption the central region

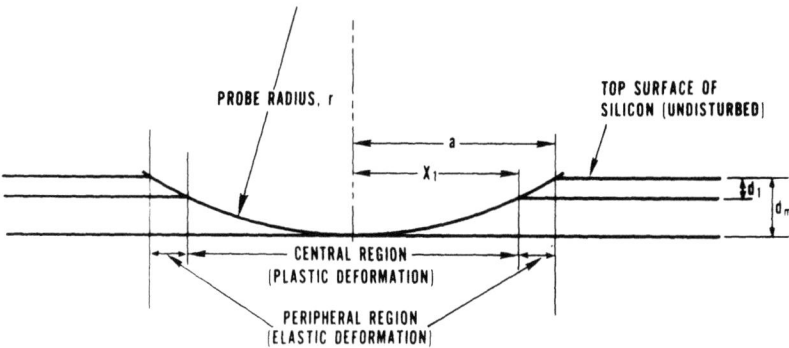

Fig. 6. Schematic of stress distribution under a hemispherical contact, with the yield limit of the silicon being surpassed out to X_1; elastic deformation from X_1 out to a, (after Kramer and van Ruyven [6]).

finally stabilizes when uniformly stressed to the silicon elastic limit. The peripheral part of contact from X_1 out to a is stressed below the elastic limit to a level which decreases with distance away from the central region.

Various properties under the contact are sensitive to pressure: effective bandgap, barrier height, and carrier mobility, as well as the ability to penetrate native SiO_2 layers. Different stress distributions may therefore cause different electrical response from the two contact regions. If the material properties change with resistivity, the relative contributions of the two contact regions may likewise change, with the net effect being a resistivity-dependent contact resistance.

Tong and Popaniak [7] show a definite correlation between resistivity dependence of the Knoop hardness of silicon and the corresponding dependence of the apparent contact radius of a spreading resistance probe on both (111) n-type and (111) p-type specimens. However, the spreading resistance contact radius, obtained from the ratio of spreading resistance to resistivity, shows perhaps five times the variation seen in the Knoop hardness tests.

B. Piezoresistance

Fonash [8] discusses the interpretation of the voltage measured between the probes. He considers both the boundary conditions which can be assumed when measurements are made on small nonuniform structures and the physics of the probe-semiconductor interaction for any structure. In particular he notes the importance of piezoresistance effects. The resistivity, ρ_{eff}, of the stressed layer immediately under the contact is given by a second rank tensor related to the unstressed resistivity, the stress tensor and the piezoresistance tensor. Fonash calls this an effective resistivity and after appropriate tensor algebra writes for contact to a (100) plane of silicon:

$$\rho_{eff} = (1 - p\pi_{11}) \tag{6}$$

where p is the applied pressure, ρ is the resistivity of the unstressed layer and π_{11} is the surviving component of the collapsed fourth rank piezoresistance tensor.

For contact to a (111) plane of silicon

$$\rho_{eff} = \frac{\rho[(1-p/3(\pi_{11}+2\pi_{12}))^3 - 2(p/3\cdot\pi_{44})^3 - 3(p/3\cdot\pi_{44})^2(1-p/3(\pi_{11}+2\pi_{22}))]}{[(1-p/3(\pi_{11}+2\pi_{12})) + (p/3\pi_{44}) + 2(p/3\pi_{44})(1-p/3(\pi_{11}+2\pi_{12}))]} \tag{7}$$

Using $p = 10^{10}$ dynes/cm^2 and values of π_{ij} from the literature [9], he calculates for (100) planes:

$\rho_{eff} \simeq 2\rho$, for n-type Si
$\rho_{eff} \simeq 0.94\rho$, for p-type Si

and for (111) planes

$\rho_{eff} \simeq \rho$, for n-type Si
$\rho_{eff} \simeq 0.1\rho$, for p-type Si

These must be taken as estimates only since data for π_{ij} does not exist at stress levels comparable to the elastic limit of silicon, and further π_{ij} is doping level dependent. In addition, as is seen from the mechanical calculations of Kramer and van Ruyven [6] and the models discussed in the following sections, the stress distribution under the probe is nonuniform and the electrical properties are best modeled by a probe consisting of two regions with differing properties. However, such piezoresistance effects are probably the best explanation of the difference in spreading resistance value between n- and p-type specimens of the same resistivity.

C. A Simple Electrical Model

Keenan et. al. [10] consider the probe contact to consist of two operationally different regions much in the manner of the contact in Fig. 6 (but using different symbols for the two radii). A central region of radius a_1 is ohmic, has pure spreading resistance behavior and perhaps is equivalent to the plastically deformed area of radius X_1 in Fig. 6. The outer region is annular in shape having outer radius a_2 and inner radius a_1. It consists of a surface barrier diode with barrier height V_D in series with an additional spreading resistance dependence in the surface beneath the barrier. The two regions conduct in parallel as represented by the two branches of the equivalent circuit in Fig. 7. The authors consider the resistivity dependence of the points of divergence for the forward and reverse bias current-voltage (I-V) response of a normal probe. Different I-V behavior is measured for (111) n- and p-type silicon and is shown in Fig. 8. They also consider the nonlinearity in the relation between log resistance and log resistivity for (111) n- and p-type silicon. Their model for the barrier primarily addresses n-type material, but observations on its implications for p-type silicon are made.

Both types of data indicate the existence of a nonlinear contact element, whose contribution to the contact response depends on specimen conductivity type and resistivity value. The behavior of this element is postulated to be controlled by specimen surface states. In the nonequilibrium flat band condition, surface states with a density N_{ss} are considered to be filled to a level ϕ_o which is above the valence band, but below the Fermi level, E_F. To establish equilibrium, electrons from the conduction band fill the states to the Fermi level. Following Henisch [11], the resultant band bending eV_D, assuming all donors are ionized in the region where $dV/dx \neq 0$, can be found from:

$$V_D = \frac{E_F - \phi_o}{e} + \frac{KN}{16\pi N_{ss}^2 e^3} - \frac{1}{2}\left\{\frac{KN}{2\pi N_{ss}^2 e^3}\left[\frac{E_F - \phi_o}{e} + \frac{KN}{32 N_{ss}^2 e^3}\right]\right\}^{1/2}$$

(8)

where K is the dielectric constant, e is the electronic charge, N is the carrier concentration, and the barrier height, V_D, is measured from the top of the barrier to the edge of the conduction band.

For ϕ_o below mid-gap, a depletion layer and rectification will result for n-type material. For high resistivity (low carrier concentration) n-type, V_D should be quite large, thus preventing any meaningful conduction through the periphery of the contact. However for high carrier concentration, V_D should become quite small, significant conduction through the outer portion of the contact will occur. This behavior, with the rectification component increasing with the resistivity of n-type material, is consistent with the observed points of divergence of the forward and reverse I-V characteristics for various resistivity n-type specimens, see Fig. 8b. For these specimens the points of divergence occurred for lower voltage values as resistivity increased, with the divergence points following the relation $\log I = (\text{const}) \cdot V^3$.

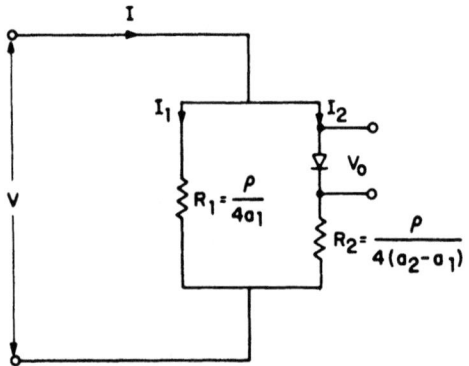

Fig. 7. Equivalent circuit for the two regions of the contact as modeled by Keenan et. al. (after Keenan et. al. [10]).

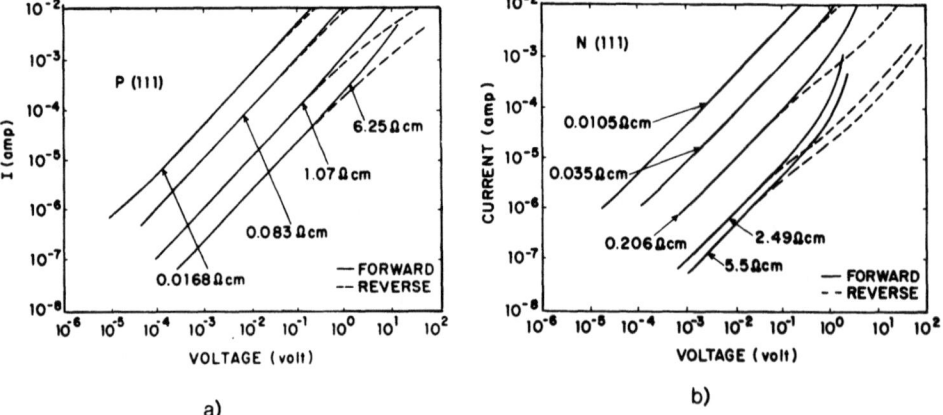

Fig. 8. Current voltage characteristics for a spreading resistance contact on silicon specimens of various resistivities.
(a) (111) p-type silison, (b) (111) n-type silicon (after Keenan et. al. [10]).

Qualitatively, at least, the same model also describes the general results obtained on p-type specimens. For the same surface state conditions on p-type silicon, an enhancement layer is formed if E_F is above ϕ_o, allowing ohmic behavior of the contact periphery. With increased doping concentration, E_F moves through ϕ_o, a depletion layer is formed and rectification should begin. As with n-type silicon, the change in rectification properties due to the barrier in the contact periphery is consistent with the nonlinear relation observed between measured resistance and resistivity of p-type specimens. Further, the predicted increase in rectification

with decreasing resistivity for p-type silicon is consistent with
the observation that the points of divergence between the forward
and reverse I-V curves occur for higher voltages, as specimen resistivity increases.

More extensive calculations were made for n-type silicon which
is always expected to exhibit some amount of rectification. The
Schottky barrier reverse-bias diode equation is used (minority
carrier, tunneling, and image force effects ignored) to describe the
annular part of the contact. This is then used to calculate the
equivalent circuit resistance, see Fig. 7, as a function of resistivity. To accomplish this, the radii of the central and peripheral
annular portions of the contact are estimated respectively from the
high resistivity and low resistivity linear portions of the n-type
calibration relation and values of N_{ss} and ϕ_o are assumed. V_D
can then be calculated from eq. (8) and used to calculate the resistance of the Schottky barrier as a function of resistivity (carrier
concentration). The current density, j, through the Schottky barrier, after Henisch [11], is given by:

$$j = \sigma \left[\frac{(V_D + V_B) 8\pi Ne}{k} \right]^{\frac{1}{2}} \exp\frac{-eV_D}{kT} \frac{1 - \exp(-eV_D/kT)}{1 - \exp(-2e(V_D + V_B)/kT)} \quad (9)$$

where σ is the conductivity and V_B is the applied voltage. The I-V
characteristic of the Schottky portion of the contact can be obtained from this equation. Using $N_{ss} = 5 \times 10^{11}$/eV cm and $\phi_o = 2eV$,
Keenan et. al. obtain good detailed agreement with the measured
spreading resistance resistivity relation [10]. Figure 9 shows this
comparison for the transition region between two linear portions of
the measured relation on (111) n-type silicon. Good agreement is
also obtained between the measured I-V behavior and that predicted
by this model for two n-type specimens. Quantitative results were
not presented for p-type specimens owing to complications in the
apparent transition of the surface from enhancement to depletion
as a function of resistivity.

D. A Model With Barrier Mechanisms

A more comprehensive treatment of the contact has been given
by Kramer and van Ruyven [6]. They begin by assuming that the
central, plastically deformed region of the contact is not, in fact,
simply ohmic but is composed of a pure spreading resistance component in series with a barrier-controlled resistance. This barrier
controlled resistance is modeled as having two mechanisms conducting
in parallel by which charge can cross the barrier: thermal emission
over the barrier and tunneling through it. A component of resistance
is assigned to each mechanism. The peripheral contact is ignored
at this point.

The temperature dependence of spreading resistance was measured on a set of (111) n-type and (111) p-type specimens over the temperature range 100 to 325 K with measurements being taken at 125 Hz. The temperature dependence of the measurements was analyzed according to the temperature dependence of the resistance controlled by the thermal emission and tunneling mechanisms to yield information on the barrier height and the nonlinearity of the spreading resistance-resistivity relation.

According to Kramer and van Ruyven the resistance per unit contact area R_b^* in the thermal emission branch is given by:

$$R_b^* = (k/qA^{**}T)\exp(E_b'/kT) \qquad (10)$$

where q is the electronic charge, A^{**} is the effective Richardson constant and E_b is the effective barrier height. The corresponding normalized resistance in the tunnel-current limited branch is given by:

$$R_t^* = kT\exp(rE_b'(B + 1/kT))/(vZNq^2)(-1 + rE_b'(B + 1/kT) \qquad (11)$$

where v is the striking velocity of electrons against the barrier, B is a combination of material and barrier parameters and Z is the area enclosed by the curve of tunnel current vs. energy level.

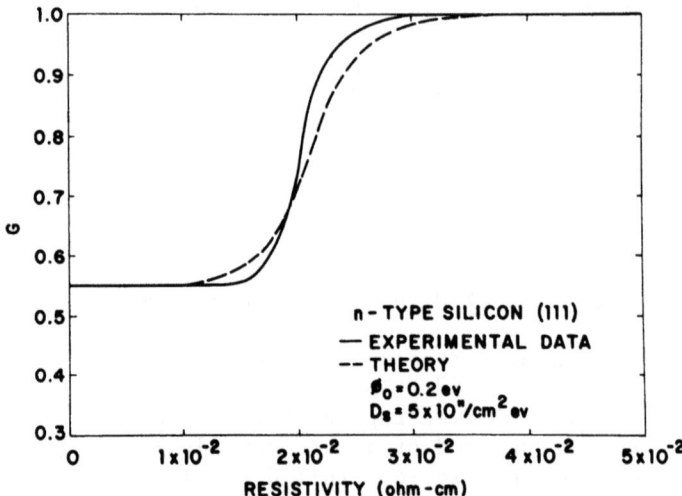

Fig. 9. Experimental and theoretical calibration nonlinearity factor in region of strongest nonlinearity for (111) n-type. Contact model of Keenan et. al. used for theoretical calculation (after Keenan et. al. [10]).

Values of the resistances R_b and R_t are calculated as a function of resistivity based on values of E_b' obtained from fits to the temperature dependence data and from contact areas based on the Hertzian stress calculations. In contrast to assuming that the nonlinear resistivity dependence of spreading resistance can be built into an effective contact radius, $a(\rho)$, as in Section I D the nonlinear dependence of measured resistance is represented here by a factor, $K(\rho) \equiv R_{measured}/R_{spreading}$ with the contact parameter assumed to be constant. While the spreading resistance is normally considered to be the same as the total measured resistance, in this section the term is reserved for that portion of resistance which is due to pure geometric spreading and which is linear in resistivity as in eq. (3). For the model of the contact assumed, the expression for K reduces to:

$$K(\rho) = 1 + 4R_b^* R_t^* / \pi \rho a (R_b^* + R_t^*) \qquad (12)$$

where a is now a constant. Values for $K(\rho)$ can be calculated using values of R_B^* and R_t^* obtained from the previous equations.

Using the resistivity dependent values of E_b obtained from the temperature dependence measurements described above, Kramer and van Ruyven find reasonable agreement with the barrier lowering described by Andrews and Lepselter [12] for all specimens below 10 Ω·cm and limiting values for E_b of 0.32 eV for p-type and 0.25 eV for n-type. They observe that if the interface Fermi level has the same energy with respect to the band edges for p-type and n-type silicon as is found in zero pressure contacts, the total band gap in the central contact region must be about 0.57 eV (0.32 + 0.25 eV). This suggests a band gap narrowing of about -5.4×10^{-4} eV/kg mm^2 obtained from diode characteristics measured at pressures up to 800 kg/mm^2 [13]. Dependence of the components of contact resistance for the central region calculated from the foregoing are shown in Figures 11 and 12.

Values of $K(\rho)$ calculated by Kramer and van Ruyven from eq. (12) are seen to be in good agreement with values from their own room temperature data, but in poorer agreement with $K(\rho)$ values from other authors, see Figures 13 and 14. This discrepancy is probably due to a wide range of measurement conditions used. Experimental values of $K(\rho)$ less than unity are seen for very high and very low resistivities and can be accounted for by a change of effective contact radius. This might be expected if the peripheral contact regions, which are much less stressed and face a higher barrier than the central region already treated, contribute an additional conduction mechanism at extreme resistivities. It is suggested by Kramer and van Ruyven that this might be the case at very low resistivities where tunneling is likely and at very high resistivities where the spreading resistance of the central contact is very large with respect to peripheral barrier resistance.

Since spreading resistance measurements are normally made at about 10 mV, both resistance terms above are given for zero bias. In contrast to Keenan [10], the barrier height is taken here as the difference between the top of the barrier and the Fermi level.

Data for the p-type specimens show, see Fig. 10, exp(1/T) dependence from 325 K down to about 180 K with lesser dependence for lower temperatures. Values of E_b derived from a fit of R_b to data in the range 180 to 325 K range from 0.16 eV for a 0.03 Ω·cm specimen to 0.34 eV for a 40 Ω·cm specimen. The lowest resistivity specimen considered, 0.008 Ω·cm shows virtually no temperature dependence whatever. The n-type specimens, in general, appear to show a more limited correlation with the pure thermal emission, exp(1/T), dependence. For a number of specimens in the range 250 to 325 K, little or no variation of resistance is discernible from the data scatter. Values of E_b' obtained for n-type specimens which do show a temperature dependence over a limited temperature range run from 0.26 eV for a 24 Ω·cm specimen to 0.17 eV for a 0.028 Ω·cm specimen.

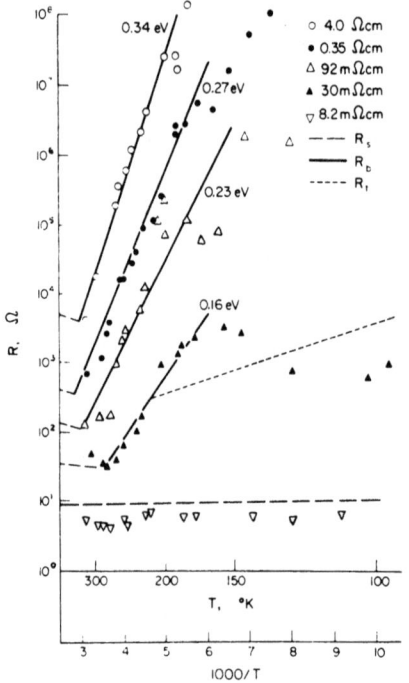

Fig. 10. Contact resistance vs. temperature for a spreading resistance on (111) p-type specimens at five resistivities (after Kramer and van Ruyven [6]).

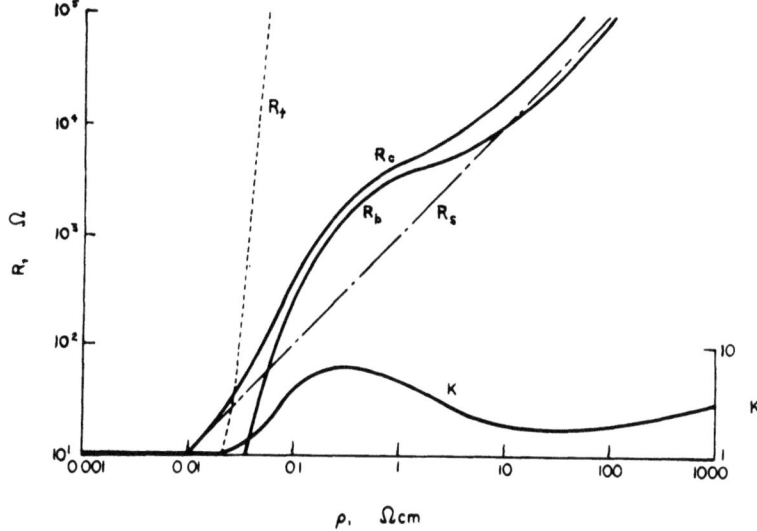

Fig. 11. Theoretical values of the components of contact resistance vs. resistivity as derived from the contact model and temperature dependence data for (111) p-type silicon (after Kramer and van Ruyven [6]).

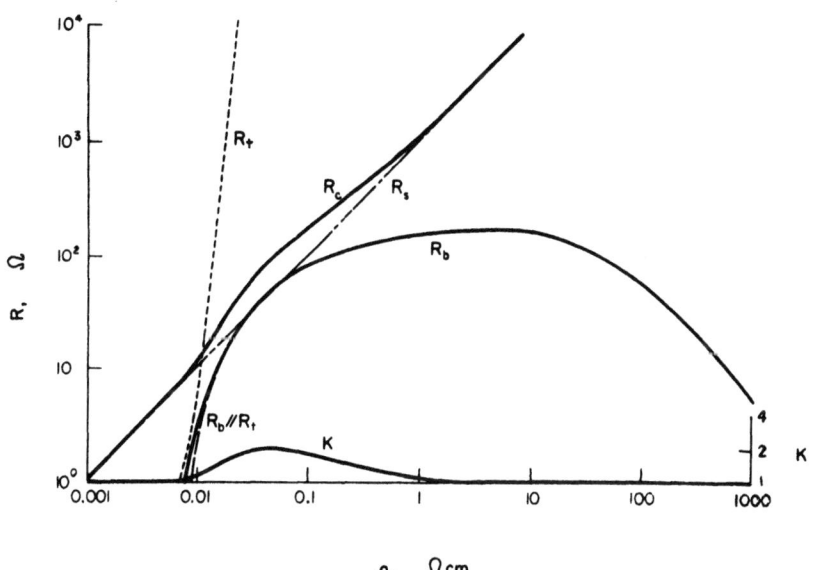

Fig. 12. Theoretical values of the components of contact resistance vs. resistivity as derived from the contact model and temperature dependence data for (111) n-type silicon (after Kramer and van Ruyven [6]).

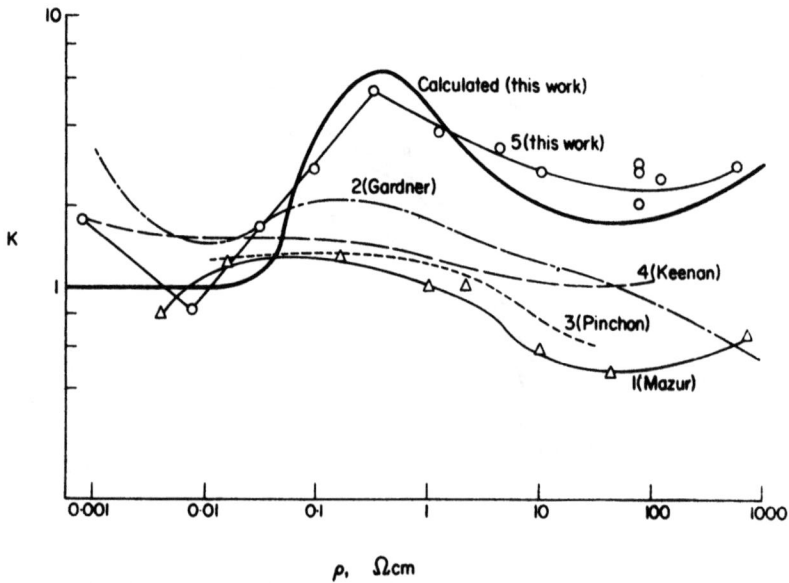

Fig. 13. Nonlinearity factor, K, as a function of resistivity for (111) p-type silicon. Values calculated by Kramer and van Ruyven and empirical values form several authors shown (after Kramer and van Ruyven [6]).

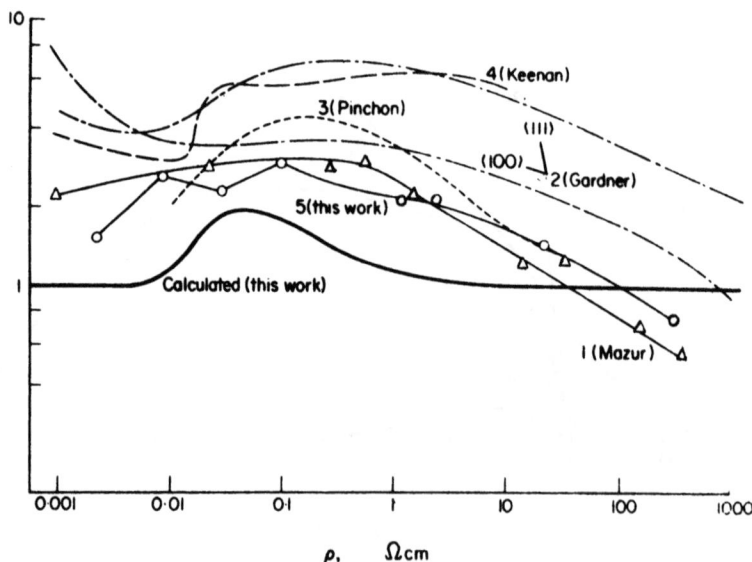

Fig. 14. Nonlinearity factor, K, as a function of resistivity for (111) n-type silicon. Values calculated by Kramer and van Ruyven and empirical values from several authors (after Kramer and van Ruyven [6]).

Several results from the work of Kramer and van Ruyven have general practical application. The first of these is the extreme temperature sensitivity of p-type silicon and the strong, but less extreme temperature, sensitivity for n-type. The second is that the barrier-related nonlinearity factor, $K(\rho)$, as seen from eq. (12) can be reduced by increasing probe load. While such a reduction may be of advantage in the case of calibration and interpretation of measurements, it is done at the price of spatial resolution. The last result from this work, in conjunction with data on the crystal orientation dependence of band-gap narrowing [13,14], is that n-type silicon is expected to be more sensitive to orientation and to the mechanical properties of the probe than is p-type.

E. Models vs. Practice - A Recapitulation

Spreading resistance measurements are seen to be transfer measurements from a set of calibration specimens which have been appropriately chosen and prepared to the match test specimen of interest. Successful transfer is seen to have four aspects: three are experimental and one is mathematical. The experimental aspects have been dealt with at length: (1) use of criterion to readily evaluate the current operating condition of a set of probes; (2) careful replication of measurement conditions between calibration and test specimens; (3) use of a specimen surface preparation which yields a noise level and reproducibility consistent with the intended application. The mathematicl consideration lies in the choice and method of application of an algorithm to account for the effects of resistivity variations in graded or thin layers on depth profile measurements of those layers. Development of any such algorithm requires a model of the entire structure to evaluate the influence of structure boundaries on current potential distribution within the specimen. Section III outlines some of the models used to generate such a mathematical algorithm.

The models for the spreading resistance contact interface described in Sections II A and II D serve to explain the deviation from the linear relation between spreading resistance and resistivity as well as to point out the sensitivities of the measurement to such factors as temperature, crystal orientation, conductivity type, and probe load. However, such models are both inadequate and too cumbersome for direct use in interpreting the measurements on an arbitrary specimen even when homogeneous and semi-infinite. When analyzing graded or multilayer data, the contact models do not even allow a unique answer to the choice of boundary condition under the contact.

III. DEPTH PROFILING OF SEMICONDUCTOR STRUCTURES

A. General Concepts

Clearly, depth profiling is the prime application of spreading resistance measurements. The wide range of dopant concentration to which it responds, (the range being independent of dopant species for shallow-level dopants), the applicability to an arbitrary combination of layer conductivity types and thicknesses, and a depth resolution which can be of the order of a few hundred angstroms all combine to give this technique unparalleled "flexibility" among semiconductor depth profiling techniques. A schematic arrangement for depth profiling of a bevelled specimen is shown in Fig. 15.

The contact parameter, a, is not directly calculable in terms of fundamental contact mechanisms. The difficulty of its determination in the presence the previously mentioned experimental factors which influence its value is one of the limitations to extraction of accurate resistivity depth profiles from spreading resistance data. A second limitation occurs in the determination of the angle at which the specimen is sectioned. Knowledge of the sectioning angle, which is often as small as 0.25° for submicrometer layers, is necessary to relate the horizontal travel of the probe to effective depth below the original surface of the specimen. However, the most significant limitation to accurate profile analysis, is the correction factor which must be used to account for the fact that depth profile measurements are not made on a semi-infinite homogeneous

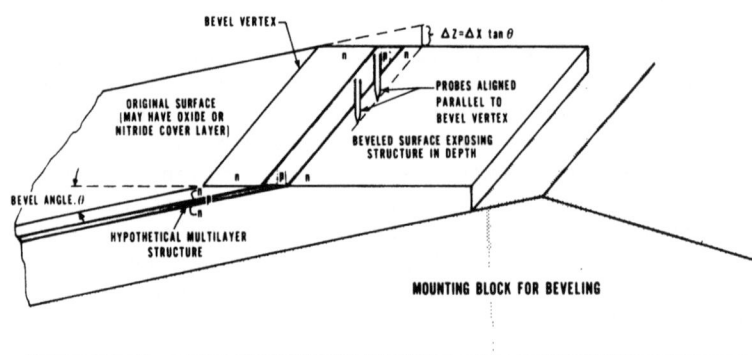

Fig. 15. Schematic of two probe depth profile measurements on an n/p/n structure beveled at angle θ.

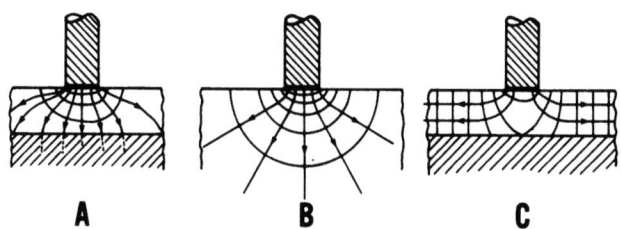

Fig. 16. Hypothesized current distribution under a probe on three limiting case structures. (a) thin layer on highly conducting substrate, (b) semi-infinite, uniform layer, (c) thin layer isolated from substrate.

media. Figure 16 illustrates the effect on current and potential distribution of simple structure boundaries parallel to the surface. Whatever the structure's deviation from uniformity and semi-infinite extent, the result is a modification of the current and potential distributions compared to the case of a uniform semi-infinite structure. The effect on the probes can be pictured as a modification of the sampling volume to which the probes respond, or as a modification at the surface of the potential which is measured by the probes. The end result is a difference in measured resistance between a uniform semi-infinite structure having resistivity ρ_1, and a nonuniform or finite structure having the same resistivity at its surface. A reasonably simple expression for the effect of nonuniform structure on measurements by two probes is given by a modification of eq. (3a):

$$R = \frac{\rho F}{2a(\rho)} \qquad (13)$$

Here, F, called a sampling volume correction factor, accounts for any and all aspects of the structure, such as resistivity gradients or layer boundaries, which are able to modify the response of the probes. Methods for calculating F will be outlined in this section. The contact parameter is written as a function of resistivity for simplicity but its use in this form is subject to limitations which will be discussed.

For each step in the profile, if the probes are aligned parallel to the bevel vertex, both probes may be considered to make contact to material with the same resistivity value. The value of the contact parameter, a, and its effect on the measured "spreading" resistance through a barrier dependent contact resistance term, is considered to be determined by the resistivity at the surface only. However, once current has passed through the contact into the semiconductor, the shape of its distribution and the extent to which spreading occurs, see Fig. 16, is usually determined much more strongly by the structure of the layer itself and its relation to probe contact size and probe spacing, than by any barrier phenomena.

B. Image Model of Non-graded Layers

The earliest modeling of the sampling volume factor, F, is due to Dickey [16] and is based on a capacitor plate analog of the probe contacts. Small but finite capacitor plates with separation, S, are given charges +q and -q and placed in contact with the surface of an isolated uniform resistivity slab. The image charge series due to the charges discs can be summed to evaluate surface potential. Using eq. (13) and the scaling relation between charge on the disc and current through the contact:

$$q = I\rho/4\pi \quad \text{esu} \tag{14}$$

the value for F was found to be:

$$F = 1 + \frac{2a}{\pi t} \{\ln(s/2t) - 0.116\} \tag{15}$$

when the slab thickness, t, is greater than the contact radius, a. This "unilayer" model is limited to junction isolated layers, but because of the assumption of a homogeneous slab it is not useful for highly graded layers as might be found in low power analog or digital transistor structures. However, it serves quite well for thick, slowly graded layers, such as the diffused layers found in thyristors and other power control devices, as well as for most thick epitaxial structures.

In a later publication, [17], Dickey used the capacitance per unit length of a two parallel wire transmission line to set up the capacitive analog of two probes separated by S on an isolated layer with thickness, t < a. From this model:

$$F = \frac{2a}{\pi t} \ln(s/a) \tag{16}$$

In this same work he also derived forms for F based on capacitance models for the cases t < a and t > a where the bottom surface of the slab is now highly conducting. Since all these "unilayer" forms account only for the existence of a sharp boundary to the layer in question and fail to consider effects due to gradients within the layer, they have limited utility for device structures of interest and will be regarded as a point of reference only. However, eq. (16) constitutes the basis for a recent treatment of graded structures which will be discussed in Section III E.

TWO PROBE (SPREADING RESISTANCE) MEASUREMENTS   27

C. Multilayer Boundary Value Approach

To evaluate graded layer structures a completely different approach based on the boundary value problem was taken by Schumann and Gardner (S-G). It will only be outlined here because of the length of the derivation and the number of publications available to those seeking more detail [18,19,20,21,22]. It leads to a cumbersome numerical computation for the factor, F, but because of its generality it has served as the benchmark analytical procedure since its development. Much recent work has been done to improve its computation speed.

In the S-G approach, it is necessary to consider the graded layer to be composed of a vertical stack of lamellae or sublayers of equal thickness, d. Each sublayer is of infinite lateral extent and is assumed to have constant resistivity, ρ_i. The contact between probe and semiconductor is assumed to have cylindrical symmetry. Thus the entire structure as modeled has cylindrical symmetry. The results applied to measurements on shallow beveled specimens although the beveling process destroys the assumed symmetry. A schematic representation is shown in Fig. 17.

The principle assumption in the S-G solution is that Laplace's equation is valid. Charge accumulation or depletion at boundaries must therefore be assumed to be a second order effect. Other assumptions are that the metal-semiconductor contact is assumed ohmic, a sharp transition in resistivity is assumed to exist between each of the sublayers, and the presence of a junction is approximated by transition to an infinitely resistive layer. The following boundary conditions are assumed:

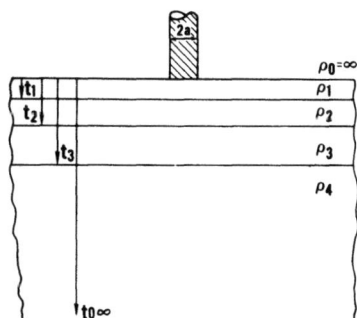

Fig. 17. Multilayer model for probe contacting a graded layer.

1. Current density is zero across the top surface boundaries except at the contact:

$$\left.\frac{\partial V(r,Z)}{\partial Z}\right|_{\substack{Z=0 \\ r>a}} = 0 \qquad (17)$$

2. The potential is continuous across all sublayer interfaces:

$$V_i(r,Z) = V_{i+1}(r,Z), \qquad \text{at } Z = t_i \qquad (18)$$

for all i where the subscript indicates one of the sublayers and t_i is the cumulative thickness from the surface to the ith interface.

3. The current is continuous across the internal interfaces:

$$\frac{1}{\rho_i}\frac{\partial V_i(r,Z)}{\partial Z} = \frac{1}{\rho_{i+1}}\frac{\partial V_{i+1}(r,Z)}{\partial Z} \qquad \text{at } Z = t_i \qquad (19)$$

for all i where ρ_i is the resistivity of the ith layer.

4. The potentials vanish with large distance from the current source:

$$\lim V(r,Z) = 0, \qquad r \to \infty, \; Z \to \infty \qquad (20)$$

5. The current through the probe is given by:

$$\frac{\partial V_1}{\partial Z} = \frac{I\rho_1}{2\pi a(a^2 - r^2)^{1/2}}, \qquad \text{for } r \leq a, \text{ and } Z = 0 \qquad (21)$$

This latter boundary condition is given in terms of current density to avoid mixed boundary conditions for the multilayer problem, which would arise in light of the condition given by Eq. (17). Boundary condition 5 can be derived from the solution of the semi-infinite uniform space problem which does have mixed boundary conditions since the realistic condition under the contact is actually one of constant potential, V_0 and the surface condition away from the contact is one of zero current as in eq. (17). This mixed boundary value problem with cylindrical symmetry can be solved by dual integral equations for the conditions at the surface. We digress to the semi-infinite layer mixed boundary value problem for which the general form of the solution is:

$$V(r,Z) = \int_0^\infty f(\lambda) e^{-\lambda Z} J_0(\lambda r) d\lambda \qquad (22)$$

where $f(\lambda)$ is a function to be determined, J_0 is the Bessel function, and that part of the solution in $e^{+\lambda Z}$ has already been set equal to zero to keep the potential bounded at infinity.

TWO PROBE (SPREADING RESISTANCE) MEASUREMENTS

The boundary conditions at the surface are given by:

$$V(r,0) = \int_0^\infty f(\lambda) J_0(\lambda r) d\lambda = V_0, \qquad 0 < r \le a \qquad (23a)$$

and,

$$-\frac{\partial V(r,0)}{\partial Z} = \int_0^\infty f(\lambda) J_0(\lambda r) d\lambda = 0, \qquad r > a \qquad (23b)$$

It can be shown that

$$f(\lambda) = \frac{2V_0}{\pi} \frac{\sin(\lambda a)}{\lambda} \qquad (24)$$

satisfies these boundary conditions since:

$$\frac{2V_0}{\pi} \int_0^\infty \frac{\sin(\lambda a)}{\lambda} J_0(\lambda r) d\lambda = V_0, \qquad 0 < r \le a \qquad (25a)$$

and

$$\frac{2V_0}{\pi} \int_0^\infty \sin(\lambda a) J_0(\lambda r) d\lambda = 0, \qquad r > a$$

$$= \frac{2V_0}{\pi(a^2 - r^2)^{\frac{1}{2}}} \qquad 0 < r \le a \qquad (25b)$$

Recognizing $V = I\rho_1/4a$, we see that eq. (25b) gives the result used in eq. (21) for the boundary condition in the multilayer problem. Thereupon, the general solution meeting the boundary conditions for the semi-infinite single layer problem is:

$$V(r,Z) = \frac{I\rho_1}{2\pi a} \int_0^\infty - Z \frac{\sin \lambda a}{\lambda} J_0(\lambda r) \frac{d\lambda}{\lambda} \qquad (26)$$

Returning to the multilayer problem, the general solution must contain new terms with the same cylindrical symmetry so as to satisfy the additional internal boundary conditions plus the general solution for the single layer problem above. For the potential at the top of the nth layer of our N layer structure the solution will have the form:

$$V_n(r,Z) = \frac{I\rho_1}{2\pi a} \int_0^\infty e^{-\lambda Z} \sin(\lambda a) J_0(\lambda r) \frac{d\lambda}{\lambda}$$

$$+ \int_0^\infty \theta_n(\lambda) e^{-\lambda Z} \sin(\lambda a) J_0(\lambda r) \frac{d\lambda}{\lambda} + \int_0^\infty \psi_n(\lambda) e^{\lambda Z} \sin(\lambda a) J_0(\lambda r) \frac{d\lambda}{\lambda} \qquad (27)$$

where the 2N functions, $\theta_n(\lambda)$ and $\psi_n(\lambda)$, will be used to satisfy the 2N boundary conditions: current and potential continuity at each of N-1 interfaces, finite potential at infinity, and the same current distribution at the top surface as for the single layer model above.

To keep the potential bounded at infinity, S-G set $\psi_N = 0$. Then from the boundary condition of zero surface current outside the contact requires that: $\theta_1 = \psi_1$.

The boundary condition on continutiy of the potential at the interfaces results in terms of the form, (one equation for each of N-1 interfaces):

$$(1 + \theta_n(\lambda)) e^{-\lambda t_{n+1}} + \psi_n(\lambda) e^{\lambda t_n} = $$
$$(1 + \theta_{n+1}(\lambda)) e^{-\lambda t_{n+1}} + \psi_{n+1}(\lambda) e^{\lambda t_{n+1}} \quad (28)$$

For all interfaces but the last. For this last interface the term of the form $\psi_N e^{\lambda t_N}$ is omitted from the right hand side since ψ_N has already been set to zero to satisfy the condition at infinity.

The boundary condition on current continuity at the interfaces results in terms of the form:

$$\frac{1}{\rho_n} (-) [1 + \theta_n(\lambda)] e^{-\lambda t_n} + \psi_n(\lambda) e^{\lambda t_n} = $$
$$\frac{1}{\rho_{n+1}} (-) [1 + \theta_{n+1}(\lambda)] e^{-\lambda t_{n+1}} + \psi_{n+1}(\lambda) e^{\lambda t_{n+1}} \quad (29)$$

Again, the equation for the last interface has no term in $\psi_N e^{\lambda t_n}$ on the right hand side.

These sets of coupled equations can be solved for the coefficients θ_n and ψ_n using for example the method of determinants. To simplify the resulting equations the following shorthand is adopted:

$$K_n = \frac{\rho_{n+1} - \rho_n}{\rho_{n+1} + \rho_n} \quad (30)$$

and

$$d_n = t_{n+1} - t_n \quad (31)$$

where d_n is the individual thickness of the nth layer, although all layers have a common thickness, d. Since the coupled equations result in each θ_n and ψ_n being functions of the parameters of all the layers, an illustration of the first few coefficients is given for the simple case of a three layer structure:

$$\theta_1 = \psi_1 = \frac{K_1 e^{-2\lambda t_1} + K_2 e^{-2\lambda t_2}}{1 - K_1 e^{-2\lambda t_1} - K_2 e^{-2\lambda t_2} + K_1 K_2 e^{-d\lambda}} \quad (32)$$

$$\theta_2 = \frac{K_1 + K_2 e^{-2\lambda t_1} + K_2 e^{-2\lambda t_2} - K_1 K_2 e^{-2\lambda d}}{1 - K_1 e^{-2\lambda t_1} - K_2 e^{-2\lambda t_2} + K_1 K_2 e^{-2\lambda d}} \quad (33)$$

For a multilayer structure, many additional coefficients could be calculated. However, they, and even θ_2, are immaterial. The reason is that the probe senses only the potential V_1 at the surface and this depends only on θ_1 and ψ_1 ($=\theta_1$), see eq. (27) for $i = 1$. Hence at the surface, ($Z = 0$):

$$V_1 = \frac{I\rho_1}{2\pi a} \int_0^\infty (1 + 2\theta_1(\lambda)) \sin(\lambda a) J_0(\lambda r) \frac{d\lambda}{\lambda} \quad (34)$$

with θ_1 being evaluated by determinant or other methods from the coupled boundary condition equations.

This general expression for the potential at the surface can be used to evaluate the difference in potential measured by a two probe measurement system, $\Delta V = (V_A - V_B)$ at $Z = 0$. The assumption is made that the two probes, A and B, have separation large enough that neither modifies the field due to the other. Therefore the principle of superposition applies and eq. (34) can be used to evaluate the potential difference ΔV. First, it must be recognized that the potential at each probe consists of two terms:

$$V_A = V_A(a,0) - V_A(S,0) \quad (35a)$$

and

$$V_B = V_B(S,0) - V_B(a,0) \quad (35b)$$

where the term with $(a,0)$ dependence is the self potential of either probe, the term with $(S,0)$ dependence is the potential due to the other probe. Opposite signs for the component terms reflect the opposite directions of current through the respective probes. By symmetry, the form of the component potentisl is the same for each probe. The resultant form for $\Delta V = (V_A - V_B)$ is then:

$$V = 2[V(a,0) - V(S,0)] \quad (36)$$

where the subscripts A,B can be dropped. The self potential of either probe is taken from the average potential over the probe:

$$V(a,0) = \frac{\int_0^a \int_0^{2\pi} V_1(r,0) \, r \, dr \, d\theta}{\int_0^a \int_0^{2\pi} r \, dr \, d\theta} \quad (37)$$

Substituting from eq. (34) for V (r,0) and interchanging the order of integration, the self potential of each probe is given by:

$$V(a,0) = \frac{I\rho_1}{\pi a^2} \int_0^\infty [1 + 2\theta_1(\lambda)] \sin(\lambda a) \, J_1(\lambda a) \, \frac{d\lambda}{\lambda^2} \quad (38)$$

The term with (S,0) dependence in eq. (26) can be evaluated directly from eq. (34) with r = S in the Bessel function. Then substituting for V(a,0) and V(S,0) in eq. (36) and dividing by the measurement current, I, we obtain:

(39)

$$\frac{\Delta V}{I} = R = \frac{\rho_1}{2a} \, \frac{4}{\pi} \int_0^\infty [1 + 2\theta_1(\lambda)] [\frac{J_1(\lambda a)}{\lambda a} - \frac{J_0(\lambda S)}{2}] \, \frac{\sin(\lambda a)}{\lambda} \, d\lambda$$

In this equation, the parameter a is assumed not to depend on resistivity. Recognizing $\rho_1/2a$ as the classical two-probe spreading resistance on a semi-infinite homogeneous structure of resistivity ρ_1, we can identify the term in brackets as the sampling volume correction factor, F. From eq. (13) and eq. (39):

$$F = \frac{4}{\pi} \int_0^\infty (1 + 2\theta_1) \, \frac{J_1(\lambda a)}{\lambda a} - \frac{J_0(\lambda S)}{2} \, \sin(\lambda a) \, \frac{d\lambda}{\lambda} \quad (40)$$

When the correction factor, F, is applied to depth profile data, a value for the sublayer thickness must be used to evaluate θ_1, (see eq. 32). The depth increment between an adjacent pair of measurement positions down the beveled specimen is used for this sublayer thickness. To analyze a full profile, eq. (39) is sequentially applied to the "effective" surface for each measurement location down the beveled specimen. Since θ_1 is always a function of the resistivity values of all sublayers below the measurement location down to the nearest isolating boundary, the analysis must be unfolded for the simplest value of θ_1. This is the value belonging to the lowest point on the profile. In this case, the substrate resistivity is assumed to be known, and θ_1 incorporates only a single unknown resistivity value. Once this resistivity is solved for, the equation for F can be applied to the next higher measurement location, where again only a single new value of resistivity need be evaluated. The resistivity profile of the structure is thus deconvoluted.

Although simple in concept, the procedure as outlined is laborious in practice. Normally, the first step is to invert eq. (13):

$$\rho_1 = \frac{2aR}{F} = \frac{2a\Delta V}{FI} \tag{41}$$

where R is the measured resistance or spreading resistance between the two probes. The equation is transcendental, with ρ_1 occurring both as the unknown on the left and as a parameter folded into F (through θ_1) on the right.

An iterative procedure is therefore required to converge on the value of ρ_1 appropriate to each layer. Discussion of the use of simple iterative, Newton-Raphson, and bracketing techniques is contained in the work of Lee [22].

As each new layer (data point) is added, θ_1 becomes dependent on one more layer. It is then necessary to evaluate the new form for θ_1 prior to trying to converge on the new layer's resistivity value. Moreover, as presently constructed, the integral for F in Eq. (40) involves three periodic functions, and must be reevaluated for each new value of θ_1. Again a useful summary of the numerical evaluation of the integral is contained in the work of Lee [22]. Although not including the probe self potential in his evaluation, Lee notes that nearly 23,000 values of λ are required for an accurate evaluation of the integral. When there are a large number of layers, the determinant solution for θ_1 is usually done numerically and not with prestored analytic forms. The result is that the full determinant solution for θ_1 must be performed for each value of λ occurring in the numerical integration. The integration and evaluation of θ_1 must be performed for each step in the iterative procedure to determine the new value of ρ_1, with 3 to 20 iterations on ρ_1 typically being required per layer. It should be rather apparent at this point that using an $a(\rho)$ dependence in Eq. (41) adds an undesirable complication to this iteration procedure.

D. Computational Improvements For The Boundary Value Solution

Hu [21] notes that an early implementation of the Schumann and Gardner procedures as outlined by Yeh and Khokhani [20] required 10 min. of CPU time on an IBM 360 Model 85 to perform the calculations for a 25 point profile.

Improvements by Hu include a study of the integration factor, $1 + 2\theta(\lambda)$, noting that it can be approximated by a simple power law as a function of λ with certain limitations. Use of this power law enables interpolation of θ_1 between select values of the integration parameter, λ, at which θ_1 is evaluated exactly. His second operational improvement lies in the development of a spatial grid concept in which a reduced number of sublayers is selected with evaluation of the matrix elements occurring only for this reduced number of sublayers. He also includes a discussion of the accuracy loss encountered (less than 5%) and notes a computational time reduction by a factor of 30.

Choo [25] notes that based on similar equations developed by Sunde [26] for geoelectric measurements, great simplification is possible for evaluating the θ_1 terms for the successive sublayers. A recurrence formula can be used to relate the successive sublayer values of the function $(1 + 2\theta)$. Let A_i represent $(1 + 2\theta_1)$ for the ith layer, then:

$$A_i = \frac{1 - b_i e^{-2\lambda d_i}}{1 + b_i e^{-2\lambda d_i}} \qquad (42)$$

where λ and d_i are as previously defined, and the coefficients b_i are given by:

$$b_i = \frac{\rho_i - \rho_{i-1} A_{i-1}}{\rho_i + \rho_{i-1} A_{i-1}} \qquad (43)$$

Then from eq. (42) with $d_1 = 0$:

$$A_1 = 1$$

Using $T_i = \rho_i A_i$, a more convenient form results:

$$T_i = \frac{W_i \rho_i + T_{i-1}}{1 + (W_i T_{i-1})/\rho_i} \qquad (45)$$

where

$$W_i = \frac{1 - e^{-2\lambda d_i}}{1 + e^{-2\lambda d_i}} \qquad (46)$$

and

$$T_1 = \rho_1 \qquad (47)$$

It should be noted that the use of the index, i, here is opposite to that previously used for the description of the Schumann-Gardner model. Schumann and Gardner assigned $i = 1$ to the top layer for development of the multilayer model. The set of equations (42) to (47) are being developed to simplify calculation of correction factor resulting from the model. When calculating the correction factor and applying it the analysis of data, the process begins not at the top layer, but at the bottom of the structure where the structure, and hence the calculation is simplest. It is appropriate, in this case, to assign $i = 1$ to the lowest layer, with the index increasing toward the surface. At any given stage of profile analysis, the index will have the value, n, the number of sublayers (data points) below the measurement point being analyzed.

With each d_n having the common value of the effective depth increment between data points, the exponentials in the W_i terms need to be evaluated just once. The problem of iterating to find the correct value of the top sublayer resistivity at each step no longer requires repeated inversion of a 2n x 2n matrix. Rather, the recurrence relation for $T_i = \rho_i A_i = \rho_i(1 + 2\theta_1)$ at any measurement point involves only the value of T_{i-1} and this will already have been evaluated. Therefore, only the remaining parameter, ρ_i, the resistivity of interest, needs to be iterated to obtain convergence.

Use of this recurrence relation not only noticeably reduces computation time, it allows many more sublayers to be handled in a computer of given CPU size. Choo was able to take advantage of this ability to handle a large number of sublayers in the calculation. His results show, for example, that for an assumed exponential profile, a 10 sublayer approximation to the profile gives an error of 20% and a 20 sublayer model gives an error of 10%, both with respect to the value calculated for 300 sublayers. While 300 sublayers or measurement steps is impractical for actual profiles measurements of a single layer, this calculation points out the error incurred by representing a smooth continuous resitivity profile by a sequence of discrete stair steps, unless the number of steps is reasonably large.

The most recent step in the development of computation simplification for the boundary value based calculation is due to D'Avonzo et. al. [27,28]. It is based on the recognition by Hu [21] that the integration factor, $1 + 2\theta_1$, is a well behaved function of the integration parameter, and can be interpolated between judiciously chosen points. D'Avonzo [27,29], chose to use cubic spline functions dependent on λ, the integration variable, rather than a simple power law as used by Hu. D'Avonzo states that the error from spline interpolation is always less than that from power law interpolation. When the calculation is handled in this manner the resistivity dependence of the integration factor is absorbed in the polynominal coefficients of the spline functions. Since these are constant within each interval of the integration, they can be removed from the integration. One is then left with integrals over each λ interval which depend only on the radius of and separation between the probes. If one operates with a fixed probe spacing and uses two estimates of probe radius, one each for n-type and p-type material, the integrals over each interval can be precalculated and stored. These simplifications, like those of Choo [25], not only speed the computation but reduce the size of the CPU required to make the calculation. A simple form of the correction factor resulting from a two-layer evaluation of the boundary value procedure is shown in Fig. 18.

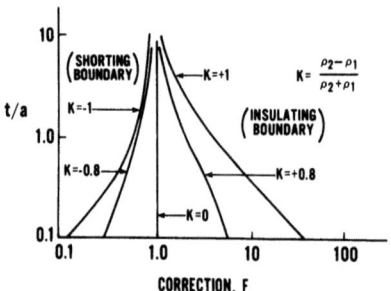

Fig. 18. Sampling volume correction factor of Schumann and Gardner for a two layer structure as a function of layer resistivity ratio.

E. The Local Slope Approximation

A completely different approach by Dickey [23] leads to another simplified calculational procedure. It is based on consideration of two limiting cases for graded thin layers, one where the thin layer is over a nonconducting substrate (or is isolated by a junction from the substrate), and the other where the thin layer is over a highly conducting substrate. Here, a thin layer is one with thickness less than the contact parameter, or probe radius, a. The first case has as its foundation, Dickey's transmission line calculation for a thin uniform layer over an insulating substrate [17], from which eq. (16) can be combined with eq. (13) to give:

$$R = \frac{\rho}{\pi t} (\ln S/a) \tag{48}$$

Note hat if this equation is simply applied to a layer with nonuniform resistivity, ρ/t has no unique value, but it does have the dimension of sheet resistance, R_s. Making this tentative identification:

$$R = \frac{R_s}{\pi} (\ln s - \ln a) \tag{49}$$

This suggests that if two probe measurements are taken for various probe spacings on a thin isolated layer and plotted vs. logarithm of probe spacings, the slope will be the sheet resistance of the layer and the X-axis intercept will be the logarithm of the contact parameter a. This was confirmed by Dickey [23] for a number of p-type and n-type diffusions, for which the sheet resistance was

TWO PROBE (SPREADING RESISTANCE) MEASUREMENTS

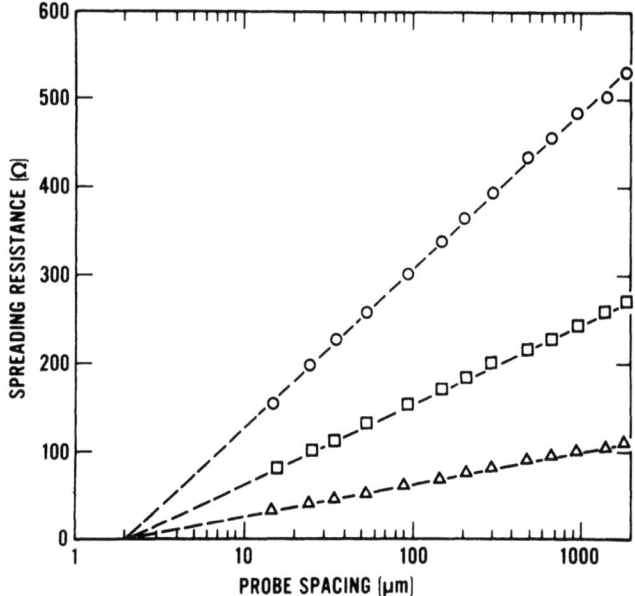

Fig. 19. Probe spacing dependence of two probe data of three junction isolated diffused layers with different sheet resistances. Resistance values: O - 52 Ω; □ - 126 Ω, Δ - 250 Ω (after Dickey [23]).

measured independently by a four-point probe. Results for three p-type diffusions are shown in Fig. 19.

This confirmation that two probe spreading resistance values on thin isolated layers are proportional to the sheet resistance of the layers has the following implication for depth profiling: a spreading resistance value measured on a beveled specimen should be related to the sheet resistance value appropriate to the layer remaining beneath the measurement location. For this reason, analysis of spreading resistance profile measurements on thin isolated layers should proceed like differential sheet resistance profiling using four probe resistivity measurements in conjunction with anodic removal of silicon layers [29,30]. In the differential sheet resistance method, resistivity values are extracted from sheet resistance values measured before and after removal of a thin layer of the specimen. Although layer removal is simulated by beveling the structure for the case of spreading resistance measurements, the similarity in measurement analysis should remain.

Formally, Dickey notes that for the thin isolated graded layer the problem is best handled in terms of conductance since the sublayers should conduct in parallel between the probes. That is, the lines of current are almost exclusively parallel to the surface as illustrated in Fig. 16(c). The conductance between the probes for any measurement point, n, on the bevel sectioned specimen is given by:

$$G_n = \frac{\pi}{\ln(s/a)} \sum_{i=1}^{n} \sigma_i \Delta t \tag{50}$$

where the n sublayers have conductivity values, σ_i, and common thickness Δt. The summation is over the number of sublayers below the point of measurement and the scale factor, $\pi/\ln(S/a)$ comes from eq. (48). Writing a similar equation for an adjacent measurement position, the conductivity of the sublayer between two measurement locations can be obtained from the difference of the two equations:

$$\sigma_n = \frac{\ln(s/a)}{\pi} \frac{G_n - G_{n-1}}{\Delta t} = \frac{\ln(s/a)}{\pi} \frac{\Delta G}{\Delta t} \tag{51a}$$

A schematic representation of this layer partitioning is shown in Fig. 20. Similar analysis for a thicker isolated slab results in an additional term on the right involving the second derivative of the sampling volume correction factor. It will not be discussed here but can be found in reference [23].

For the case of a thin graded layer over a highly conducting substrate, the substrate tends to "short-circuit" the top layer. Lateral conduction between the probes should occur primarily in the substrate with the current in the graded layer being normal to

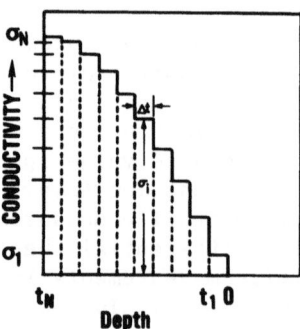

Fig. 20. Schematic representation of sublayer construction for parallel superposition model of a thin isolated layer. (after Dickey [23]).

TWO PROBE (SPREADING RESISTANCE) MEASUREMENTS

the surface in the form of cylinders, see Fig. 16(a). The resistance measured at the surface is postulated to be that of a series of cylindrical resistances due to the sublayers under the probes in addition to the spreading resistance between virtual contacts at the substrate:

$$R_1 = 2 \sum_{i=1}^{n} \rho_i \frac{t_i - t_{i-1}}{\pi a^2} + \frac{\rho_s}{2a} \tag{52}$$

where a is the contact parameter, ρ_s is the substrate resistivity, and the summation is over all sublayers between the contacts and the substrate.

By writing a similar equation for measured resistance, R_2, one sublayer deeper into the structure, and then subtracting the two equations the resistivity of the intervening sublayer can be obtained from the difference of the two equations:

$$\rho_1 = \frac{\pi a^2}{2\Delta t} [R_1 - R_2] = \frac{\pi a^2}{2} \frac{\Delta R}{\Delta t} \tag{53a}$$

As with the case of thin isolated layers, the resistivity of the sublayer between any successive pair of measurement points on thin layers over conducting substrates appears to be extractable on a simple fashion from the difference in measurement values at the two locations.

A simple transformation can be used to give eq. (51a) and eq. (53a) compatible form. This form also is compatible with the manner in which the commercial spreading resistance amplifier acquires data; that is, the logarithm of resistance. The transformation is based on equations (51a) and (53a) containing terms of the type, $\Delta Y/\Delta t$, which in the differential limit can be replaced by $Y\Delta \ln(Y)/\Delta t$.

The results of the transformation are:

$$\sigma = \frac{1}{\rho} = \frac{\ln(s/a)}{\pi} \frac{G\Delta \ln G}{\Delta t} \tag{51b}$$

and

$$\rho = \frac{\pi a^2}{2} R \frac{\Delta \ln R}{\Delta t} \tag{53b}$$

Then the conductance scale in eq. (51b) is changed to a resistance scale:

$$\frac{1}{\rho} = \frac{\ln(s/a)}{\pi} \frac{1}{R} [\frac{-\Delta \ln R}{\Delta t}] \tag{51c}$$

Combining the relation for F, from eq. (13) with eq. (51c) for thin isolated layers gives:

$$F = \frac{-2a\ln(s/a)}{\pi} \frac{\Delta \ln R}{\Delta t} \tag{54a}$$

and from eqs. (13) and (53b) for thin layers over conducting substrate:

$$F = \frac{4}{\pi a} \frac{\Delta t}{\Delta \ln R} \tag{55a}$$

These two limiting case equations can be related to the data, as acquired, by noting that $\Delta \ln R/\Delta t$ is 2.303 times the local slope, $m = \pm d\log R/dt$, of the measured resistance values, with the result

$$F = K_1 m \tag{54b}$$

for thin isolated layers, and

$$F = -k_2/m \tag{55b}$$

for layers over a conducting substrate, where

$$K_1 = \frac{(2.3)2a\ln(s/a)}{\pi} \tag{56}$$

and

$$K_2 = \frac{(2.3)4}{\pi a} \tag{57}$$

respectively.

These two limiting forms can be combined in a single equation for application to structures more general than simple single layers:

$$F = K_1 m/2 + \frac{K_1 m^2}{2} + K_1 K_2 \tag{58a}$$

This hypothesized form for F reduces to the previous values for the cases of large positive slope (junction isolated) and large negative slope (shorting substrate). Figure 21 represents the two limiting regimes where the local slope model is expected to be correct, and the regime of moderate slopes where the hypothesized form must be tested for correctness.

While giving the correct form for large values of slope, the results of eq. (58a) can be seen clearly to be in error for the trivial, but definable, case of depth profiling of a thick uniform piece of bulk material. The spreading resistance data for this case should have a constant value, the slope should be zero and the volume correction factor F should be unity, yet eq. (58a) gives F = 1.4 for $s/a = 10$ and F = 1.9 for $s/a = 100$.

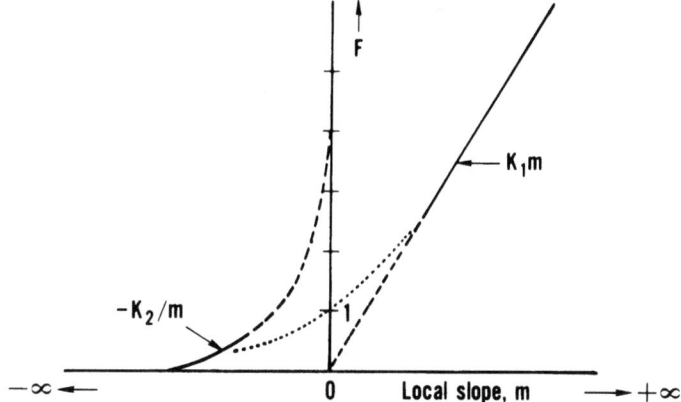

Fig. 21. Schematic representation of the volume correction factor based on the local slope model showing the regions of large negative and positive slope where the model is quite accurate and the intermediate region of slope where a bridging function is required (after Dickey [23]).

An empirical modification of eq. (58a) is suggested to give more reasonable values for F for the case of data with small slope, whether positive or negative [23]. It is suggested that the term K_1K_2 under the square root sign in eq. (58a) be replaced by the function

$$K_3 = 1 + (K_1K_2 - 1)(2/\pi)\tan^{-1}[\log(R_o/R_s)]$$

where R_o is the resistance at the point being measured and R_s is the resistance measured approximately one contact parameter deeper into the structure. The suggested form for K is admittedly arbitrary. However, the resulting equation

$$F = \frac{K_1m}{2} + \sqrt{\frac{K_1m}{2}^2 + 1 + (K_1K_2 - 1)\,2/\pi\tan^{-1}[\log R_o/R_s]} \quad (58b)$$

gives the forms of Eqs. (54b) and (55b) for large positive and large negative values of local slope of the measured data, and gives unity when the slope is zero. The arctan function is slowly varying in resistivity. The deeper layer "look-ahead" term in R_o/R_s is intended to account for the effects for layers thicker than the limiting cases discussed and is shown by Dickey to contribute to the F factor in the same fashion as the second derivative contribution to the junction isolated case mentioned subsequent to eq. (51a).

Although the result is seemingly derived in an overly simplitic fashion, more detailed consideration of the assumptions and the results have been give by Dickey, including extensive comparison with results from the Schumann and Gardner approach [18,19]. Over a wide range of conditions the comparison is quite favorable.

F. Depth Profiling - Other Considerations

It is suggested in conjunction with the multilayer boundary value calculation that the value of the contact parameter, a, appearing in eq. (39), (40, and (41) be considered to be independent of resistivity. The amount of numerical evaluation inevitably required to extract resistivity values from these equations even with a = constant makes this a desirable simplification. However, incorporation of additional resistivity dependence may well result in improved accuracy of the final profile. This incorporation can be accomplished by taking an initial value for "a" and solving for values of resistivity and F factor which satisfy eq. (41). Then, calibration data can be used to find a better estimate of the contact parameter for that resistivity. With this new contact parameter, F must be recalculated and brought to convergence with a new estimate of resistivity. This two level convergence loop, one for the volume correction factor, the other for contact parameter via empirical calibration are successively performed to obtain the best estimate for resistivity.

The local slope formulation of Dickey is most simply implemented using a single value of the contact parameter when using eq. (58) to calculate the factor F. This estimate of "a" can be obtained from a simplified interpretation of calibration data using eq. (4c) or from eq. (49) after performing a series of measurements on a thin isolated layer as a function of probe spacing. Having then evaluated the factor F and having applied to the measured resistance to calculate resistivity, a better estimate of a for that resistivity can be obtained from calibration data using eq. (3a). If it is significantly different from the value of a initially used to calculate F, a reevaluation of resistivity from the raw data and this new "a" value can be performed, and the process repeated as necessary.

When applied to data taken as a function of probe spacing, eq. (48) gives sheet resistance values in very good agreement with values measured by four-probe (generally within 5%). Therefore, it is judged to be a reasonably accurate description of two probes measuring a thin isolated layer. However, the value of contact parameter found from this equation often differs by as much as a factor of two from the value obtained on a bulk specimen whose resistivity is the same as that found at the surface of the thin film. Further, as suggested in Fig. 16, the effect of nearby boundaries is to change the current distribution immediately under the probes themselves. These considerations suggest that a calibration using bulk specimens to obtain a spectrum of $a(\rho)$ values, may not be a proper description of the nature of the probes response when measuring thin layers. The question of the best method to evaluate the contact parameter, a, and its possible resistivity dependence for use in depth profile analysis cannot be answered at this time. The related question of proper mathematical boundary conditions under the probes is discussed briefly in section V D.

Although the proper method of evaluating the contact parameter and incorporating it into depth profile interpretation is not fully understood, a somewhat different procedure than any of the proceeding is plausible. The procedure, discussed by Mazur [31], Pinchon [32], and Gruber [33], considers the volume correction factor to be applicable only to the true geometrical spreading resistance. In this procedure, the measured resistance is not longer considered identical to spreading resistance. The measured resistance, now written R_m is taken to be composed of a contact resistance R_c and a true spreading resistance R_{SR}. For bulk specimens, where F = 1:

$$R_m = R_c + R_{SR} = R_c + \frac{\rho}{2a} \equiv \frac{K(\rho) \cdot \rho}{2a} \qquad (59)$$

where R_c is given by

$$R_c = \frac{(K(\rho) - 1)}{2a} \qquad (60)$$

Here all the nonlinearity with resistivity found from calibration is incorporated not into an effective contact paramter $a(\rho)$ but rather into a separate term, $K(\rho)$. Such a $K(\rho)$ dependence was also used in eq. (12) for the barrier modeling of Kramer and van Ruyven [6]. When this construction is applied to depth profiles, where $F \neq 1$, the equation corresponding to eq. (59) is

$$R_m = R_c + FR_{SR} = \frac{(K(\rho) - 1 + F) \rho}{2a} \qquad (61)$$

where the volume correction factor is applied only to the true geometrical spreading resistance. From this relation:

$$\rho = \frac{2aR_m}{(K(\rho) - 1 + F)} \qquad (62)$$

In this form, where a = constant, iteration on an effective contact parameter, $a(\rho)$ is no longer necessary. Once the volume correction F has been obtained using eq. (41) or eq. (58), incorporation of calibration nonlinearity requires a single additional step.

G. Auxiliary Two Probe Measurements

None of the mathematical procedures or algorithms for evaluating the volume factor used in depth profiles is capable of working "blind". That is, no procedure has been devised to allow the conductivity type and dimensional extent of any given layer to be recognized from raw data only. For junction isolated diffusions or implants, the junction is generally quite evident from a large peak in resistance. For junction isolated epitaxy, however, this feature is usually weak or absent. Various other layer combinations also make a unique interpretation of the exact extent of any one layer difficult to interpret from raw spreading resistance data alone. The expected layer structure, with conductivity types and approximate layer thicknesses is generally known a priori and can be used when analyzing data. However, sometimes for new or modified processes, layer type conversion is never reached, unintended layers are formed by uncontrolled dopant sources, or junction location cannot be determine without auxiliary information.

In such cases, two auxiliary measurements are possible. In the first, one of the two probes can be heated by the use of a small fractional watt resistor mounted near the tip of one of the two probe arms and applying a dc current. It is then possible using a high impedance amplifier instead of the spreading resistance electronics to monitor the Seebeck voltage generated between the two probes by the temperature difference. The probes can be stepped along a beveled specimen and the Seebeck voltage which differs in sign for p- and n-type material can be plotted as a function of position. Figure 22 illustrates use of the Seebeck testing procedure.

While a very useful technique for determining the conductivity type of a layer, the Seebeck method has been found to be somewhat uncertain regarding the exact point of transition in conductivity type between layers. For the purpose of junction location it is possible to use a photovoltage instead of a Seebeck measurement. In this case, one probe is offset slightly, about 10 μm, from the other in the direction of motion of the probes when traversing the specimen. In this manner one probe crosses the junction before the other. If the area around the probes is now illuminated with a microscope lamp, the resulting photovoltage between the probes may be measured with a high impedance amplifier. When both probes are on the same layer of the structure, the photovoltage will arise from photogenerated electron-hole pairs being separated by a field due to local variation of carrier concentration. This signal may be on the order of millivolts. However, when the leading probe first crosses the junction so that one probe is on each side of the junction, a much larger junction photovoltage is measured. This drastic change in photovoltage is a clear indication of junction location.

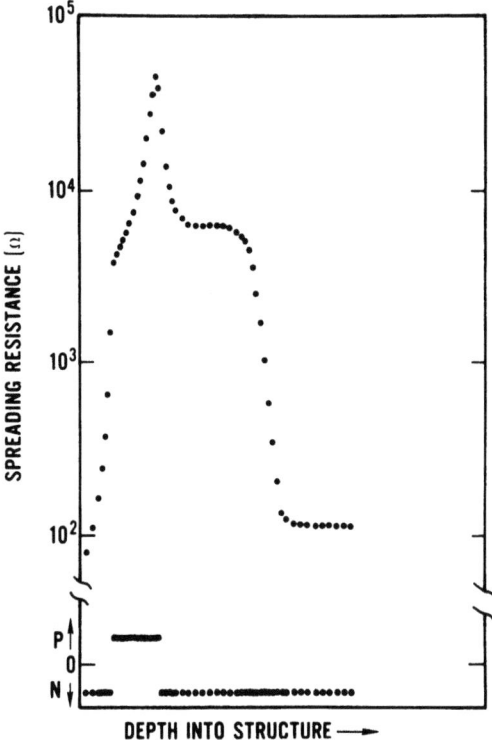

Fig. 22. Spreading resistance data for a bipolar structure with two probe Seebeck data superimposed showing layer conductivity type determination.

IV. APPLICATIONS OF SPREADING RESISTANCE MEASUREMENTS

Applications fall into two categories: (a) measurements of lateral resistivity profiles on the surface of a slice or longitudinal section of a crystal and (b) depth resistivity profiles through one or more layers fabricated into a semiconductor substrate.

A. Lateral Resistivity Profiling

Papers by Mazur [34] and Chu [35] presented demonstrations of resistivity fine-structure due to microsegregation of dopant during crystal growth for the cases of silicon and germanium respectively. Such variations primarily constitute a problem for large area devices such as those used for power control applications. Edwards and Nigh [36] show the correlation of similar microvariations to the operation of wafers fabricated for videophone use. Voltmer [37] shows that a Fourier analysis of resistivity variations or striations can be related to the crystal growth conditions: pull

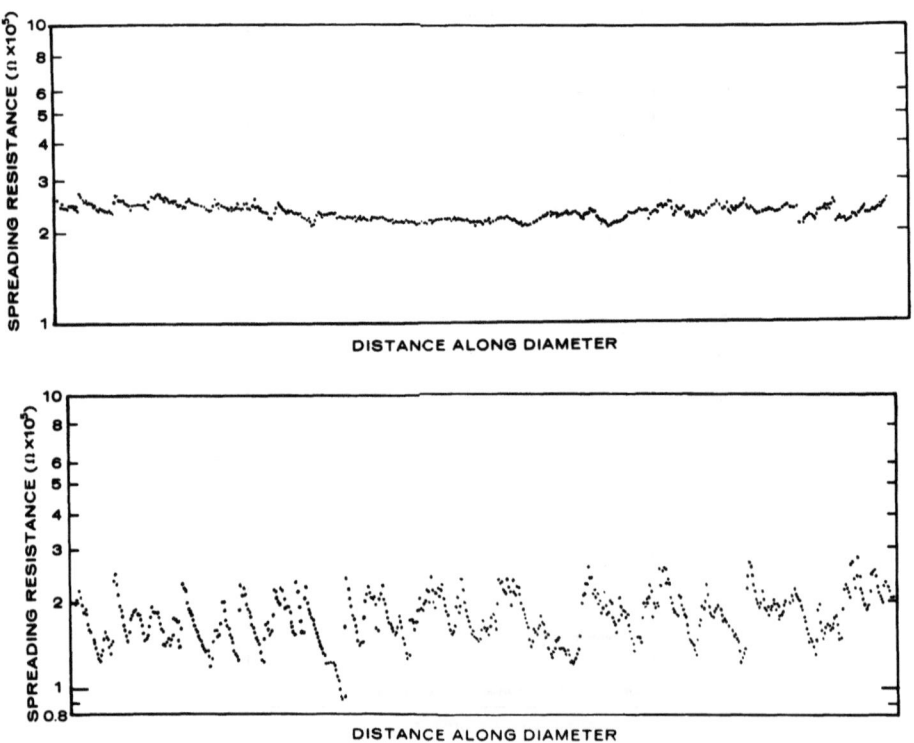

Fig. 23. Spreading resistance determination of radial resistivity variations of two (111) n-type slices about 65 ohm·cm. Top trace: Neutron transmutation doped silicon; bottom trace: Float zone grown. Step size: 25 μm.

rate, and melt and crystal rotation rates. Murgai et. al. [38] also demonstrate use of the technique to gain an understanding of crystal growth dynamics. Vieweg-Gutberlet [39] found the effects of local variations in oxygen distribution. Figure 23 illustrates the level of detail in resistivity variation obtainable for any of the above mentioned applications.

B. Depth Profiling

Depth profiling of layered structures can be subdivided into relative or absolute measurements. As a relative measurement, often done for process control purposes, direct use is made of spreading resistance data, without reduction to a carrier concentration profile. As described by Karstaedt [40] for the production of

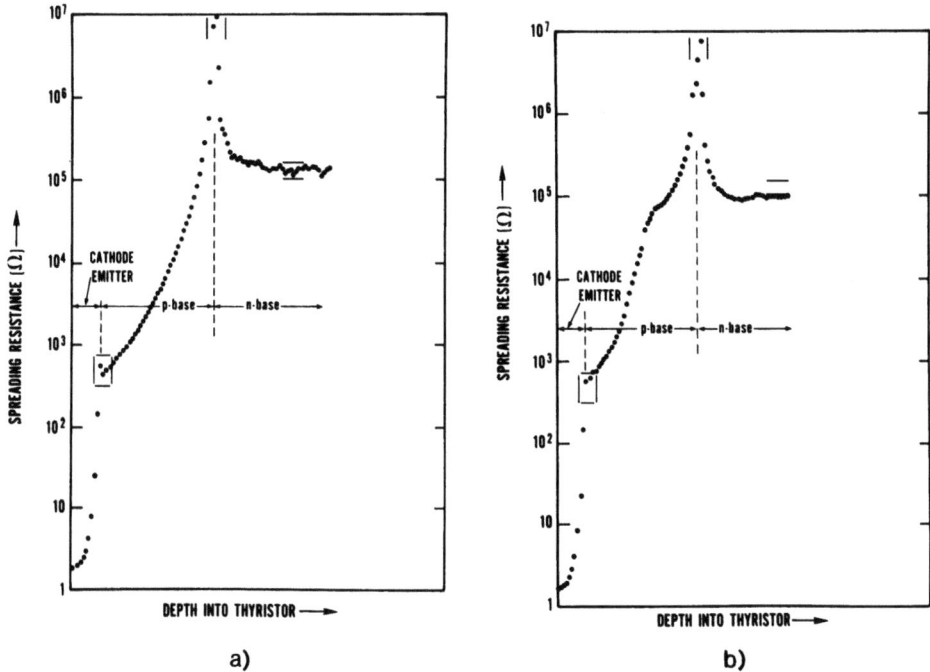

Fig. 24. Typical depth profile data for top three layers of thyristor structure showing typical allowed limits on critical parameters. (a) properly prepared specimen surface, (b) improperly prepared specimen surface showing spurious data in p-base region.

thyristors, high and low limits can be determined for the spreading resistance values of key device features. To obtain such limits for a given type of device, similar devices which were tested and found to operate up to design specifications are sectioned and measured by spreading resistance. The range of spreading resistance values for key parameters of these devices is then used to define the limits for process control measurements. An example of such measurements is shown in Figures 24(a) and (b). As another application of relative depth profiles, Assour [41] shows their use to study the effect of heat treatments on gold and oxygen in power devices. This is illustrated in Figs. 25 and 26.

Generally, absolute depth profile measurements are the most significant application of the measurement since it is here that the full advantage of the measurement's feedback information capability is realized. In this application, two levels of conversion of

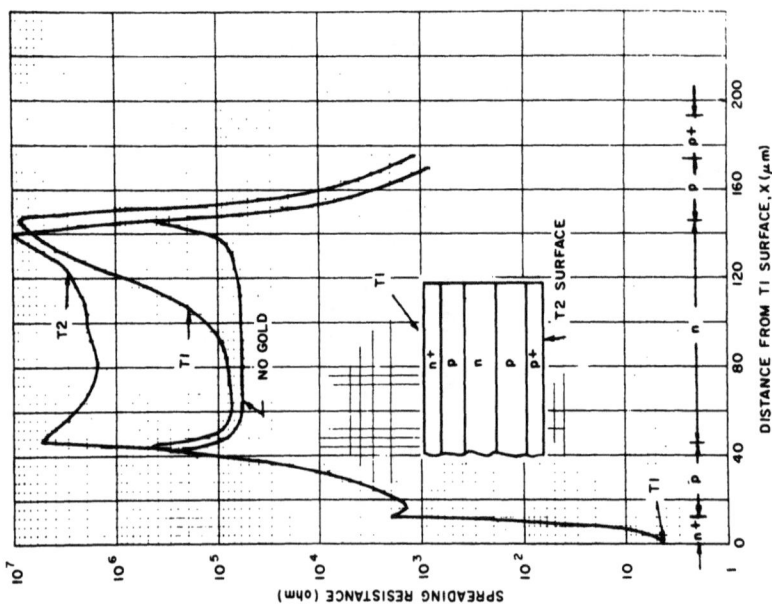

Fig. 26. Relative depth profiles by spreading resistance showin the effect of thermal treatments interacting with gold, used as a life-time control, to alter the resistivity profile as well (after Assour [41]).

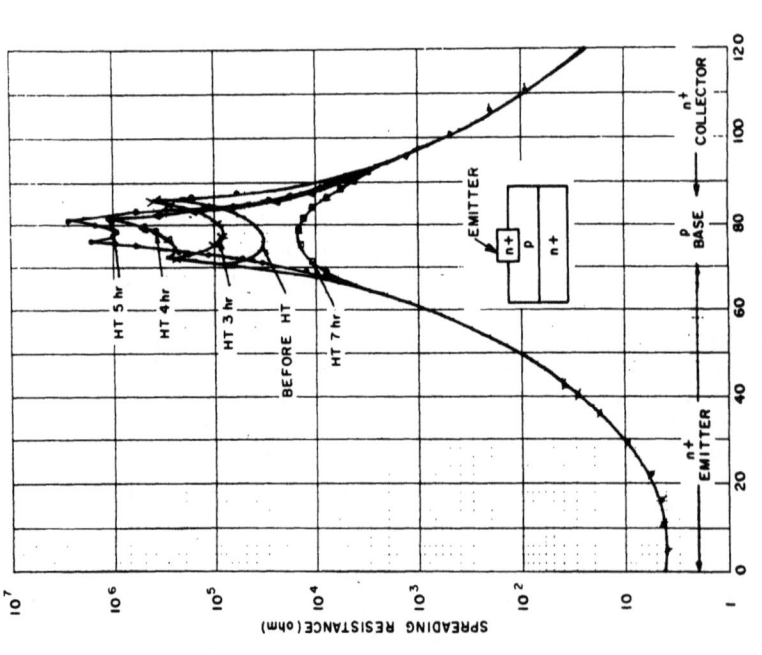

Fig. 25. Relative depth profiles by spreading resistance showing the effect of various thermal cycles on oxygen in silicon in compensating the p-region (after Assour [41]).

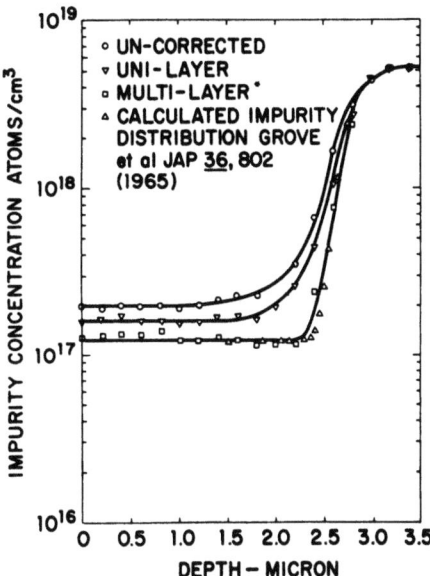

Fig. 27. Epitaxial n/n⁺ structure profile determined by spreading resistance and as predicted by a model (after Yeh [20]).

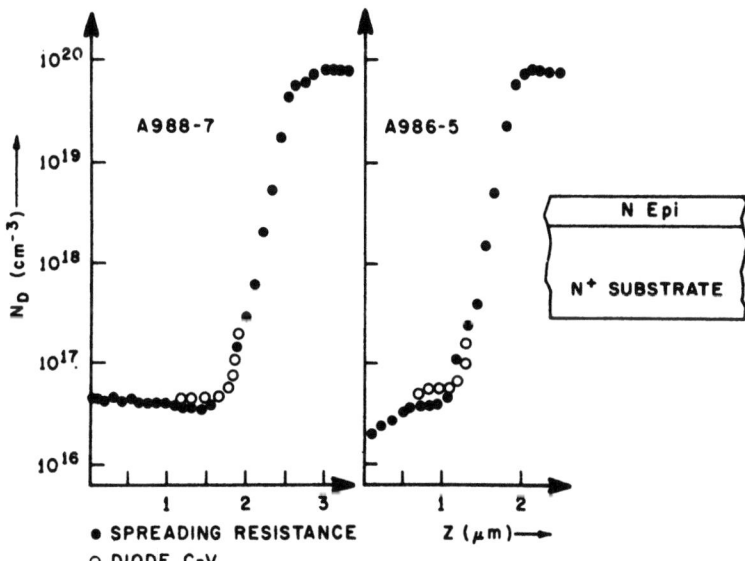

Fig. 28. Comparison of spreading resistance and capacitance voltage determinations of the doping profile of an n/n⁺ epitaxial structure (after Morris [42]).

data is generally required. The first is from resistance profile to resistivity profile. The second is from resistivity to dopant concentration profile, this latter conversion being necessary for relation to process control and modeling.

The number of examples of quantitative depth profiles in the literature is unfortunately not large and there are only a few examples of critical investigations of the quality of such profiles. The latter situation is probably due to the limited range of applicability of such alternate measurements as capacitance-voltage (C-V) and incremental sheet resistance, the relatively early stage of development of secondary ion mass spectroscopy and the proprietary nature of many profiles of interest.

Yeh [20], as shown in Fig. 27 finds good agreement of an n/n^+ epitaxial profile corrected using the S-G multilayer procedure with a profile calculated from a model. Limited comparisons between spreading resistance profiles and C-V profiles of epitaxial structures are also given by Morris [42] as illustrated in Fig. 28 and by Iida [43]. The quantitative agreement is within perhaps 10%, however, the range over which comparison can be made is clearly limited by the range over which C-V measurements are applicable.

A paper by Schroen [44] utilizes psuedo data based on typical experience to illustrate the relation between profile features derived from spreading resistance measurements and from other common techniques as they might be used to monitor a bipolar process, see Fig. 29.

Schroen et. al. [45] show a comparison between diode C-V, mercury probe C-V and spreading resistance profiles of an epitaxial layer on a conducting substrate, see Fig. 30. Surface concentrations are in very good agreement between the techniques for this structure, but the spreading resistance analysis appears to introduce a nonreal dip in the profile just prior to the transition to the interface. The authors attribute the dip to "over correction" resulting from the correction algorithm. It is apparently influenced by the value used for the contact radius as shown in section V C. Schroen [45] also shows a diffused layer profile as determined by spreading resistance and by anodic oxidation-incremental sheet resistance profiling, see Fig. 31. Here, although both techniques can be used to profile the entire structure, and agreement is seen to be excellent over the deepest two-thirds of the diffused layer, strong discrepancies exist in the two values for carrier concentration near the surface. A plausible cause for such discrepancies is found in the work of D'Avonzo et. al. [27,28], and is discussed in section V A, below.

Comparison between spreading resistance profiles and the predictions of LSS (Lindhard, Scharff and Schratt) theory for boron

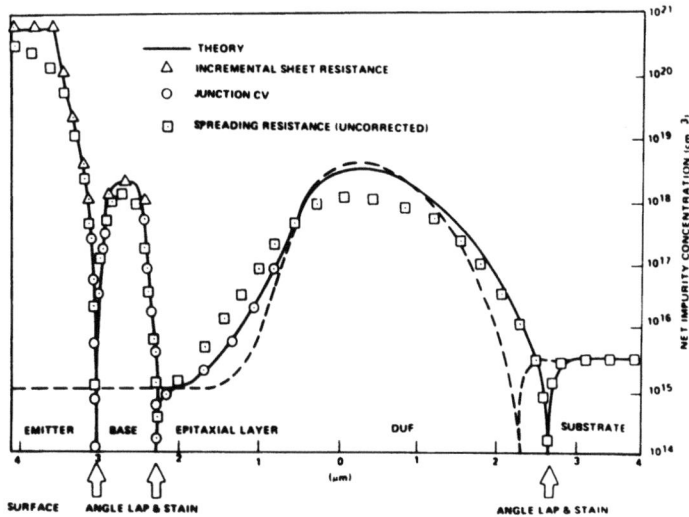

Fig. 29. Representation of the comparative measuring capabilities of several profiling techniques on a typical bipolar structure (after Schroen et. al. [45]).

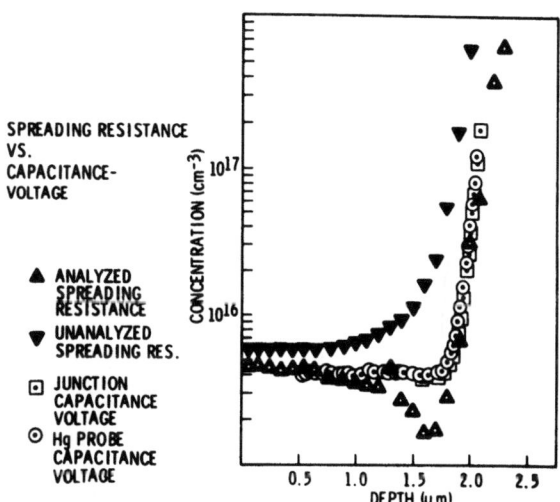

Fig. 30. Comparison of depth profiles on an epitaxial n/n$^+$ structure as determined by spreading resistance and two forms of capacitance-voltage measurement (after Schroen et. al. [45]).

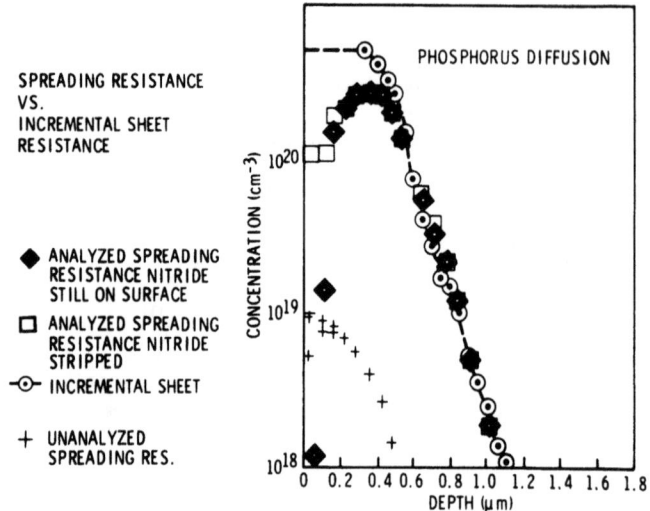

Fig. 31. Comparison of profiles for a phosphorous diffused layer derived from spreading resistance and incremental sheet resistance (four-probe) measurements (after Schroen [44]).

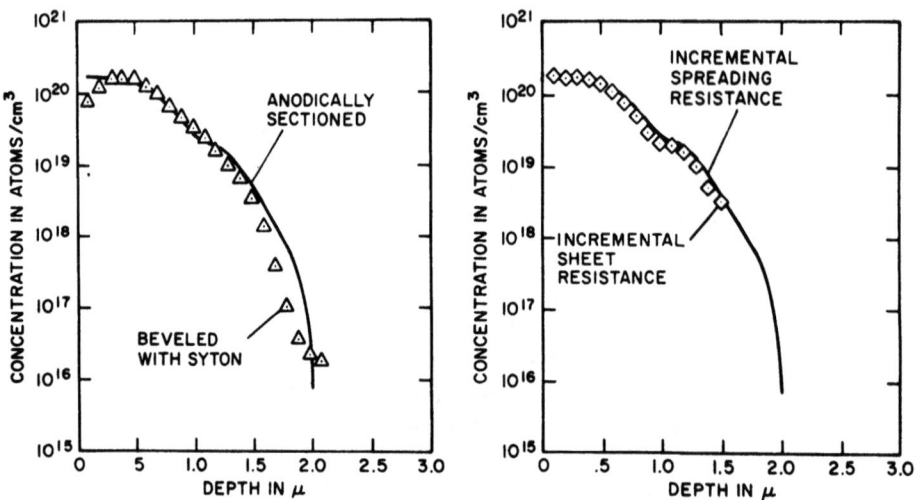

Fig. 32. Comparison of profiles of a phosphorous diffusion as measured by spreading resistance and by incremental sheet resistance, (a) beveled specimen spreading resistance, (b) anodically sectioned spreading resistance (after D'Avonzo et. al. [27]).

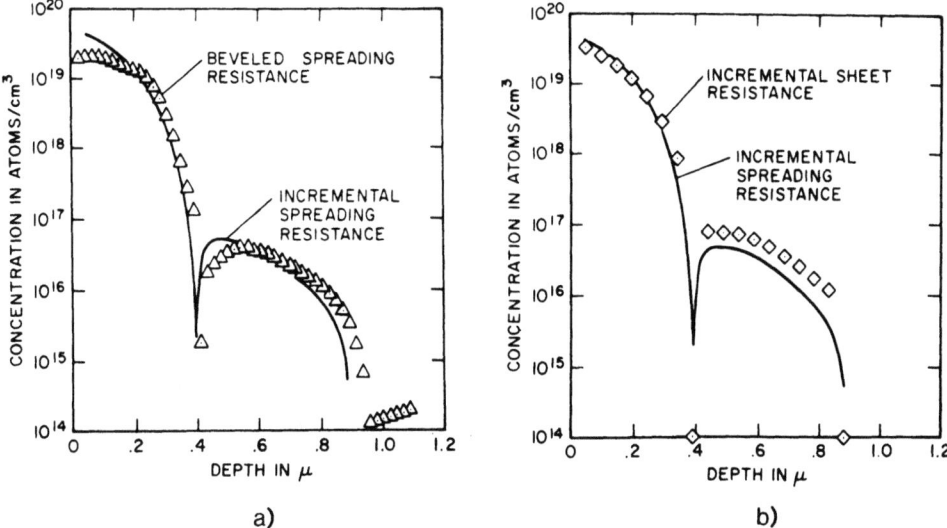

Fig. 33. Comparison of profiles of the course-channel region of a VMOS/DMOS transistor as measured by spreading resistance and by incremental sheet resistance, (a) beveled specimen spreading resistance, (b) anodically sectioned specimen spreading resistance (after D'Avonzo et. al. [27]).

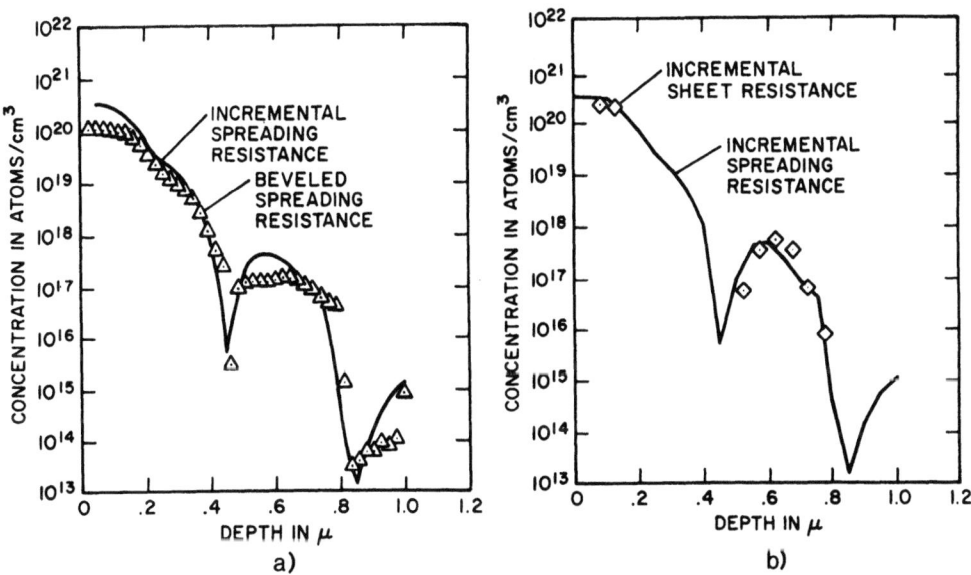

Fig. 34. Comparison of profiles of the emitter-base region of a bipolar transistor as measured by spreading resistance and by incremental sheet resistance, (a) beveled specimen spreading resistance, (b) anodically sectioned spreading resistance (after D'Avonzo et. al. [27]).

implants covering nearly two orders of magnitude of ion flux density are made by Kudoh [46]. Although LSS theory is not adequate to describe the complete profile shape, the measured and LSS-predicted values for depth of maximum concentration are in excellent agreement. The magnitude of the peak concentration for all but the lowest dose implant, (about $1 \times 10^{13}/cm^3$) is also in excellent agreement with LSS predictions [47,48]. In addition, for a shallow boron implant Kudoh shows reasonable agreement between spreading resistance and incremental sheet resistance profiles.

Comparisons of several forms of electrical profile measurements are found in the work of D'Avonzo et. al. [27,28]. Figures 32(a), 33(a), and 34(a) show comparisons of profiles on junction isolated structures as taken by spreading resistance on a beveled structure and by oxidation-incremental sheet resistance. The difference in apparent depth scales seen in Fig. 33(a) may be due to difficulties in the calibration of the anodic stripping process. The low value of dopant concentration at the surface as derived from beveled specimen spreading resistance measurements is much more noticeable, since it is apparent in all three profiles. This feature was already illustrated in the work by Schroen (Fig. 31). Further experiments by D'Avonzo, made on companion specimens to the three already shown are illustrated in Figs. 32(b), 33(b) and 34(b) respectively and suggest a reason for the near surface profile discrepancy. For these latter specimens, measurements of spreading resistance and sheet resistance by four-probe were taken on the planar top surface of the structure. Depth profiling was accomplished by growing and stripping anodic oxide layers of silicon between successive measurements. The near surface discrepancy between spreading resistance and four-probe sheet resistance disappears in each case. These results strongly suggest that the use of beveled specimens is responsible for the underestimate of near surface concentration demonstrated by Schroen and D'Avonzo. Specimen beveling removes the lateral symmetry assumed in all the sampling volume correction factor models.

V. POSSIBLE ERRORS IN SPREADING RESISTANCE MEASUREMENT

A. Errors Related to Beveled Specimens

Loss of lateral symmetry in the specimen due to beveling has been shown to cause errors at least near the bevel edge. The importance of this loss of symmetry is found to depend on the structure being profiled and is conjectured to be due to lateral and vertical variations in the structure competing in their ability to influence the current distribution between the probes. In thin high surface concentration diffused or implanted layers, the parallel superposition model of Dickey [23] shows that the near surface sublayers provide the bulk of the contribution to the conductance of the sheet resistor between the probes. When a portion of this sheet conductance contribution is removed by beveling, the measured resistance between the probes is increased for measurement positions near the bevel edge in the manner exemplified in Fig. 35. The size of the increase is related to the importance of the near surface region in determining the sheet resistance of the layer. No correction factor exists for this lateral effect except in the case of proximity to a sharp discontinuity in the specimen that is, a "90° bevel". When the beveled structure is not highly graded as in the case of an epitaxial structure, or of a deep diffusion, the effect of proximity to the bevel edge is usually not severe. Applying the parallel superposition concept to the probe response for these cases, the sheet conductance between probes is less dominated by the near surface sublayers, and the effect of removing a portion of the contribution of these layers by beveling has less overall effect on the data taken near the surface. However, for all structures, sensitivity to bevel edge effects increases with increasing probe spacing. The bevel is not believed to be the cause of the low concentrations seen in the channel region of the structure in figure 33(a) and in the base region in Figure 34(a). It was suggested by D'Avonzo that these discrepancies are due to depletion resulting respectively from the presence of source and emitter regions.

In addition to effects due to bevel edge proximity, three other possible errors due to the use of bevelled specimens can be identified. The first is that the use of beveled specimens

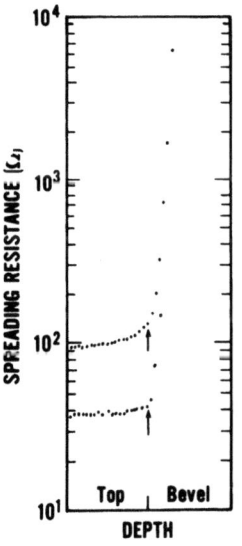

Fig. 35. Examples of heightened resistance measured by two probes in proximity to the bevel edge on the TOP surface of a beveled specimen, where the top surface has uniform resistivity. Upper trace: probe spacing = 200 μm; lower trace: probe spacing = 12 μm (after Dickey [23]).

in addition to violating the lateral symmetry assumed in the models, conceivably allows lateral "uphill" conduction by sublayers which are more highly conducting than those existing below the point of contact. Such an effect cannot be accounted for by the existing sampling volume corrections which are downward looking only. The remaining effects relate to the accuracy of bevel angle measurement and to the physical quality of the bevel. Bevel angles are generally measured by one of four methods; microscope depth of focus, stylus profilometer, laser spot reflection [49], and image splitting viewed in a microscope [50]. Multiple beam interferometry is also possible but not often used. Whatever method is used, each has accuracy limitations even for perfectly prepared bevels. An error in bevel angle measurement results in an error in the specimen depth scale proportional to the tangent of the angle. Bevel errors may also result from the specimen itself. It is not uncommon to generate specimen bevels, particularly at the shallower angles, which are nonuniform in one of two ways. Bevel edge rounding, as shown in Fig. 36 is the most serious. Since most bevel angle measurements give an average bevel angle, this defect results in inaccurate depth scale for points near the surface. The other bevel fabrication error is

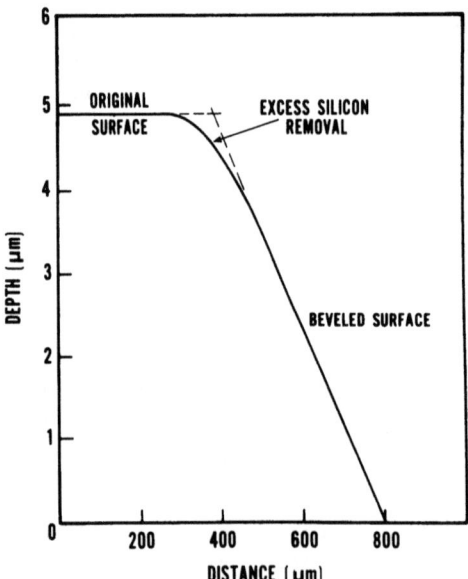

Fig. 36. Profile view of a specimen showing the bevel rounding defect.

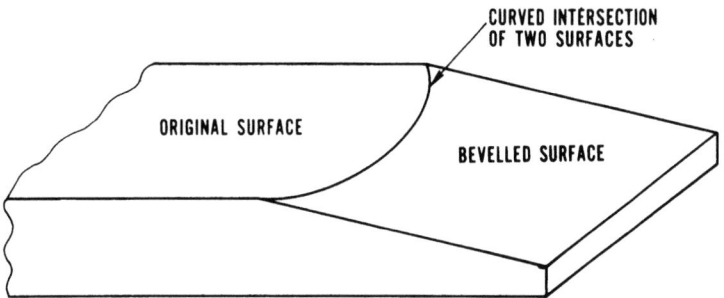

Fig. 37. Representation of the bevel vertex curvature defect.

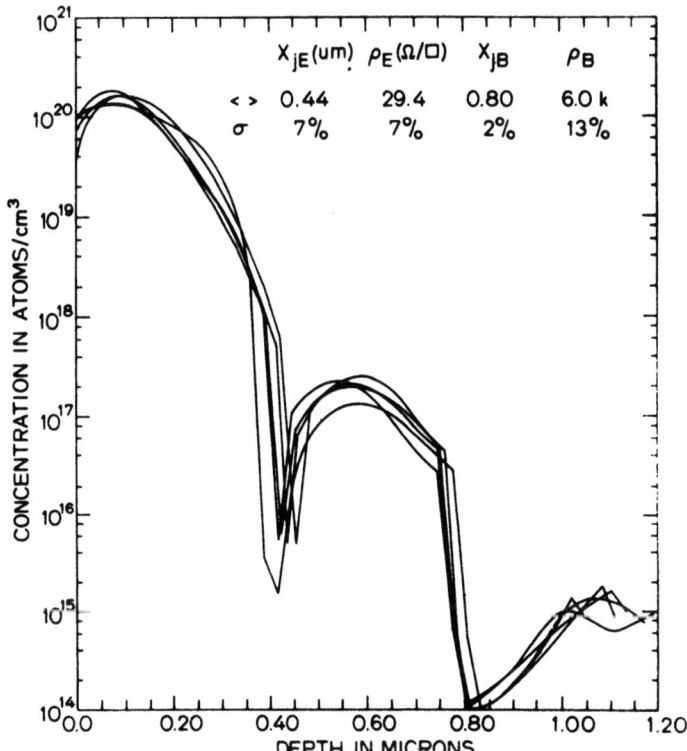

Fig. 38. Illustration of the reproducibility of profile determination obtainable on separately prepared sections of the same specimen (after D'Avonzo et. al. [28]).

that of bevel curvature illustrated in Fig. 37. This defect appears to cause problems predominantely when using the image splitting procedure [50] to measure bevel angles.

B. Surface Preparation, Calibration, and Contact Parameter Determination

Control of specimen surface preparation and replication of surface effects on calibration and test specimens has been discussed at length. Figure 38 illustrates the level of reproducibility obtainable on separately prepared sections of the same specimens for a generally well controlled process.

Probe calibration attempts to account for such effects as barrier potential and piezoresistance. The effects of such nonideal contact behavior on current distribtuion under the probes may not be identical on bulk specimens normally used for calibration and on thin layers being profiled. The contact parameter derived from one may not be strictly applicable to the other. A choice must be made regarding incorporating resistvity dependence of calibration into $a(\rho)$ or $K(\rho)$ or simply using a = constant with no further resistivity dependence. Each choice will change the final quantitative profile to some extent. Figure 39, taken from the work of Lee [22], shows a noticeable effect of the assumed contact radius, 0.5 µm, noticeably smaller than the value, 1.556 µm, determined from calibration data is able to remove the "over-correction" cusp evidenced at the epi-substrate transition. However it also noticeably shifts the calculated profile of the entire structure.

C. Data Acquisition - Electronics and Probes

Errors due to drift or nonlinearity in the data acquisition electronics may exist, but generally can be readily quantified. More difficult to quantify are differences encountered due to choice of probe configuration. It might be expected that decreased probe spacing would increase resolution of fine substructure by tending to localize the current distribution. Figure 40 from the work of Deines [49] shows that decreased probe spacing also can change the profile values even on a reasonably simple structure. Choice of probe load can also have an important effect on analysis of structures particularly if thin. Figures 41 and 42 also from the work of Deines show the effect of probe load on interpretation of epi-substrate interface location and on near surface concentration in a more complicated $p^-/p^+/n^-$ structure. In both cases, excessive probe penetration is expected to be the source of problems at the higher load.

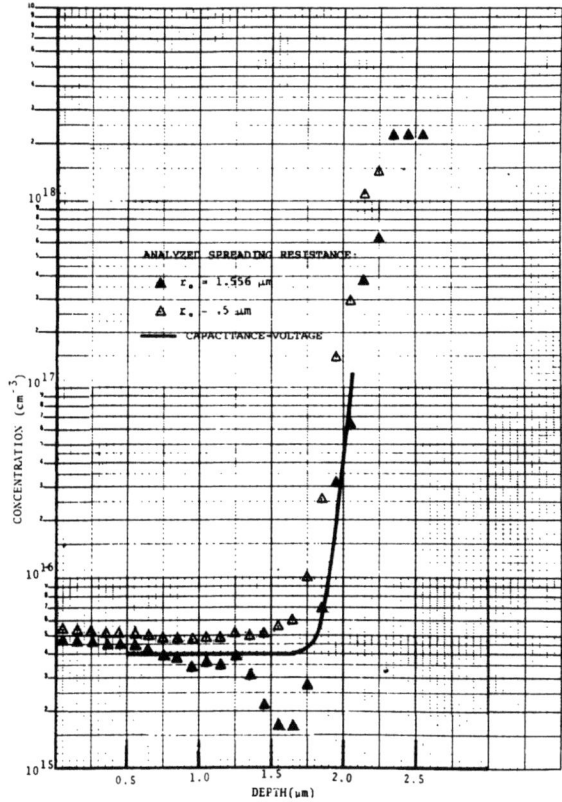

Fig. 39. Effect of assumed contact radius on the final profile after analysis, case of an n/n+ epitaxial structure (after Lee [22]).

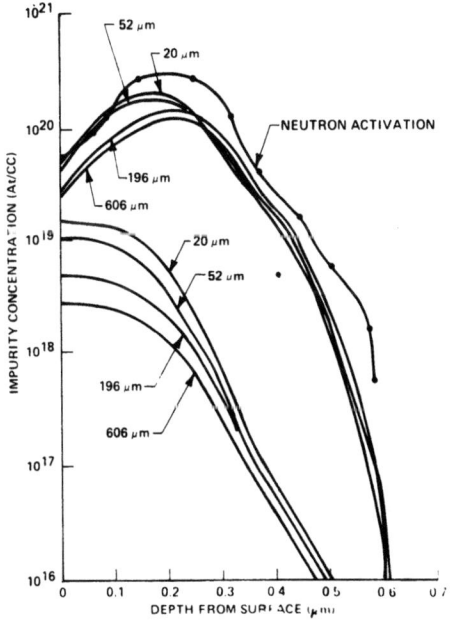

Fig. 40. Effect of probe spacing on an uncorrected and corrected profile of an ion implant. The uncorrected profile arises from conversion of raw resistance data to resistivity, then to dopant density without use of the sampling volume factor, F (After Deines et. al. [51]).

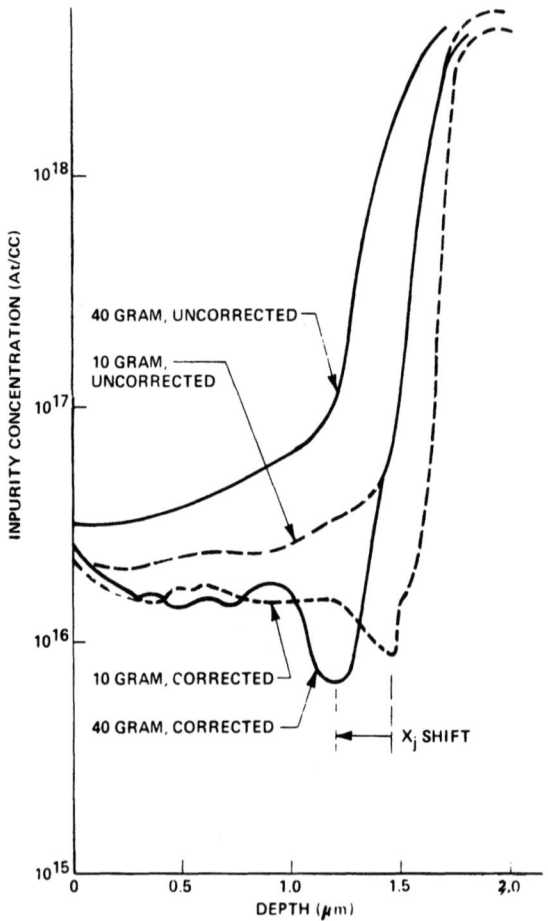

Fig. 41. Effect of probe loading on the uncorrected and corrected profiles of an n/n^+ epitaxial structure (after Deines et. al. [51]).

A price must be paid for reducing probe load, however. Equation (12) from the work of Kramer [6], predicts greater calibration nonlinearity at lighter load as one consequence. More important, however, is the increased scatter typically encountered in both calibration and specimen data as load is decreased. Figure 43, also from Lee [22], shows that a small amount of noise in the original data can be substantially amplified in the final processed profile. While not always causing an error in the sense of an offset in interpreted position or dopant density, such noise always leaves open the possibility of missed structure and diminished confidence in absolute values calculated. Presmoothing of data is often needed prior to analysis.

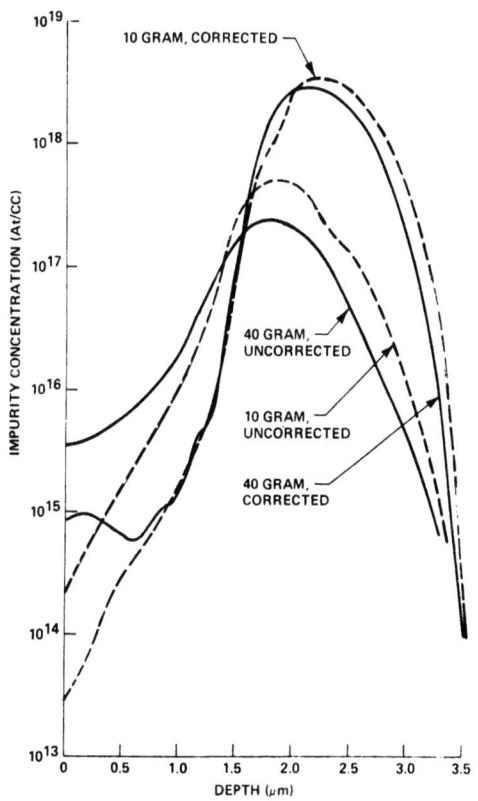

Fig. 42. Effect of probe loading on the uncorrected and corrected profiles of a $p^-/p^+/n^-$ structure (after Deines et. al. [51]).

D. Sampling Volume Correction Factor and Data Reduction

The approach to deconvoluting data to account for probe sampling volume which has received the most attention is the boundary value problem solution as elaborated by S-G and others. A number of the assumptions in that treatment have been considered in recent publications. Severin [15], Dickey [23], and Leong et. al. [24] discuss the effect of various boundary conditions and current distributions under the probes other than that used by S-G. Reasons for the altered boundary conditions include contact resistance and the modification of the current distribution by physical composition of the structure being probed. In general none of these calculations give volume correction factors which differ by more than 10% from the S-G values.

Similar, but less estensive testing of the local slope algorithm is reported by Dickey [23]. This testing generally shows very favorable agreement (10% discrepancy or less) with results based on the S-G method for select simple structures. Further

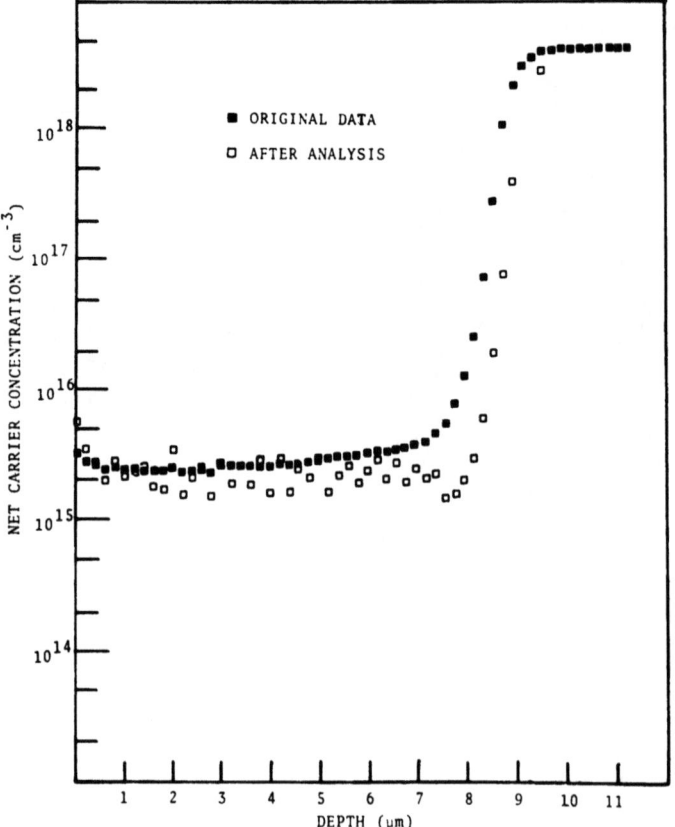

Fig. 43. Example of the amplification of small scatter in uncorrected profile (raw data) when processed by the sampling volume correction algorithm (after Lee [22]).

testing of the local slope algorithm is best done by comparing actual profile data analyzed by it and also by one of the other algorithms such as that of S-G. More extensive comparison of profiles derived from spreading resistance with profiles resulting from other methods of measurement, is needed for all spreading resistance correction factor algorithms.

Regardless of choice of sampling volume correction, care must be taken when comparing spreading resistance derived profiles with profiles derived from most other methods. Most methods other than spreading resistance and incremental sheet resistance respond to net electrical dopant density as in MOS and diode C-V profiling or to total majority carrier density as in the case of secondary ion mass spectroscopy (SIMS) profiling and the resulting resistivity and dopant density scales can only be compared by resorting to an empirically derived relation. Much recent information has been developed by Li [52], Thurber [53], and others [54,55,56] on the nature of this relation. Most of this work is in reasonable agreement and differences with the earlier work of Irvin [57] for the $10^{16} - 10^{18}$ /cm^3 range of boron are clearly seen. However, some differences do exist among the recent authors and this uncertainty must be added to the measurement errors of the other techniques which might be used for profile measurements to compare with those from spreading resistance.

VI. SUMMARY

Among profiling techniques, spreading resistance has no parallel for universality of application in terms of structure types and dynamics range of response. Using specially tailored probes [51], spacing between probes on the order of 10 µm have been reported. Using the technique of taking step sizes smaller than the probe induced damage area, as little as 0.25 µm, and calibrating under similar conditions, profile data can be taken on layers with total thickness less than 0.1 µm (1000 Å). In this manner, spatial resolution is comparable to or better than attainable with alternate profiling techniques.

Many considerations are clearly involved in taking and interpreting profiles by spreading resistance. Qualification of probes, controlled preparation of specimens, and empirical determination of the spreading resistance-resistivity relation via calibration, all affect the quality of data obtained. The quality of the data interpretation can be affected by the following: (1) the quality of the model for, and the method of application of a sample volume correction factor, (2) use of beveled rather than the planar structures which are generally modeled, (3) failure to account for semiconductor phenomena such as charge accumulation and depletion, and (4) selection and use of a mobility-resistivity relation to generate profiles in terms of dopant density.

These considerations suggest that accurate determination of absolute resistivity values from spreading resistance measurements is unlikely. It has been shown previously that spreading resistance is very sensitive to changes of resistivity. It might appear then that spreading resistance should be reserved as a tool for high resolution relative resistivity measurements only in cases where changes of, or differences in resistivity are important, thereby leaving absolute profile determination to other techniques. Relative profile measurements might be satisfactory, in a study of the relative effects of various alternatives process steps on the electrically active dopant distribution for fixed ion implantation conditions. Spreading resistance analysis is, in fact, particularly applicable for this application. However, while other profiling techniques, subject to their own inherent limits of applicability, may indeed be preferred for quantitative analysis of certain structures, the emerging body of results indicates highly reliable quantitative profiles can be obtained from spreading resistance in most instances.

ACKNOWLEDGEMENT

I would like to thank those authors who granted kind permission to use figures as noted in the figure captions and in particular, the publishers of Solid State Electronics and the Journal of the Electrochemical Society for the following:

Figures 7, 8, and 9 were originally presented at the Fall 134th Meeting of the Electrochemical Society Inc. held in Montreal, Canada.

Figures 27, 32, 33 and 34 are reprinted by permission of the publisher, The Electrochemical Society, Inc.

Figures 6, 11, 12, 13 and 14 are reprinted with permission from "Solid State Electronics", Volume 15, P. Kramer and L. Van Ruyven, The Influence of Temperature on Spreading Resistance Measurement, Copyright 1972, Pergamon Press, Ltd.

REFERENCE

1. R. Holm, Electric Contacts Theory and Application (Springer Verlag, New York 1967).
2. J. R. Ehrstein, Semiconductor Silicon 1977, Proceedings Volume 77-2, The Electrochemical Society, 377 (1977).
3. J. R. Ehrstein, Semiconductor Measurement Technology: Spreading Resistance Symposium, NBS Special Publication 400-10 (December 1974).
4. E. F., C. P. Schneider and M. R. Poponiak, J. Electrochem. Soc. 117, 721 (1970).
5. R.G. Mazur and D.H. Dickey, J. Electrochem. Soc. 113, 255 (1966).
6. P. Kramer and L. J. Van Ruyven, Solid State Electronics 15, 757, (1972).
7. A. H. Tong and M. Poponiak, IBM Technical Report 22-507 (1967).
8. S.J. Fonash, Spreading Resistance Symposium, op. cit., 17.
9. C.S. Smith, Solid State Physics, (F. Seitz and D. Turnbull, Eds.), 6, 175, Academic Press, (1958).
10. W.A. Keenan, P.A. Schumann, A.H. Tong, R.P. Phillips, Ohmic Contacts to Semiconductors, (The Electrochemical Society, 1969), 263.
11. H. K. Henisch, Rectifying Contacts to Semiconductors, 179, Oxford Univ. Press, (1957).
12. J. M. Andrews, and M.P. Lepselter, Solid State Elec. 13, 1011 (1970).
13. K. Bulthuis, Philips Res. Report 23, 25 (1968).

14. I. Balslev, Phys. Rev. 143, 636 (1966).
15. P.J. Severin, Spreading Resistance Symposium, op. cit., 45.
16. D.H. Dickey, Abstract #57, Pittsburgh Meeting, The Electrochemical Society (1963).
17. D.H. Dickey, Spreading Resistance Symposium, op. cit., 45.
18. P.A. Schumann and E.E. Gardner, J. Electrochem. Soc. 116, 87 (1969).
19. E.E. Gardner and P.A. Schumann, Solid State Electronics 12, 371 (1969).
20. T.H. Yeh and K.H. Khokani, J. Electrochem. Soc. 116, 1461 (1969).
21. S.M. Hu, Solid State Electronics 15, 809 (1972).
22. G.A., Lee Spreading Resistance Symp., op. cit., 75.
23. D.H. Dickey and J.R. Ehrstein, Semiconductor Measurement Technology: Spreading Resistance Analysis for Silicon Layers with Non-Uniform Resistivity, NBS Spec. Pub. 400-48, to be published.
24. M.S. Leong, S.C. Choo and L.S. Tau, Solid State Elec. 21, 933 (1978).
25. S.C. Choo, M.S. Leong and K.L. Kuan, Solid State Electronics 19 561 (1976).
26. E.D. Sunde, Earth Conduction Effects in Transmission Systems, 54-57, Dover (1968).
27. D.C. D'Avonzo, R.D. Rung, R.W. Dutton, Tech. Rep. 5013-2, Stanford Elec. Lab. Stanford Univ. (1977).
28. D.C. D'Avonzo, R.D. Rung, A. Gat, R.W. Dutton, J. Electrochem. Soc. 125, 1170 (1977).
29. E. Tannenbaum, Solid State Elec. 2, 123 (1961).
30. R.P. Donavan, R.A. Evans, NBS Spec. Pub. 337, (C.A. Marsden, Ed.) 123-131 (1970).
31. R.G. Mazur, Operators Manual for ASR-100 Spreading Resistance Instrument.
32. P.M. Punchon, Spreading Resistance Symposium, op. cit. 209.
33. G.A. Gruber, Spreading Resistance Symposium op. cit. 209.
34. R.G. Mazur, J. Electrochem. Soc. 114, 255 (1967).
35. T.L. Chu and R.L. Ray. Solid State Tech. 37-40, (Sept. 1971).
36. J.R. Edwards and H.E. Nigh, Spreading Resistance Symposium, op. cit., 179.
37. F.W. Voltmer, and H.J. Ruiz, Spreading Resistance Symp., op. cit. 191.
38. A. Murgai, H.C. Gatos, A.F. With, J. Electrochem. Soc. 123, 224 (1976).
39. F. Vieweg-Gutberlet, Spreading Resistance Symp., op. cit., 185.
40. W.H. Karstaedt and K.S. Tarnega, Proc. of the Annual Mtg. IEEE Inst. and App. Soc., 882, IEEE (1977).
41. J. Assour, Spreading Resistance Symp., op. cit. 201.
42. B.L. Morris and P.H. Langer, SRPROF, Spreading Resistance Symp., op. cit. 63.
43. Y. Iida, H. Abe, and M. Kondo, J. Electrochem. Soc. 124, 1118 (1977).
44. W.H. Schroen, Spreading Resistance Symp. op. cit. 235.

45. W.H. Schroen, G.A. Lee and R.W. Voltmer, Spreading Resistance Symp. op. cit. 155.
46. O. Kudoh, K. Uda, Y. Ikushima and M. Kamoshida, J. Electrochem. Soc. 123, 1751 (1976).
47. J. Lindhard, J. Scharff, and H.E. Schiott, Mat. Fys. Medd. Dan. Vid. Selsk 33, 14 (1963).
48. J. Gibbons, W. Johnson and S. Mylroie, Projected Range Statistics, (Dowden, Hutchinson and Ross 1975).
49. F. Mayer and S. Schwarzman, Spreading Resistance Symp., op. cit., 123.
50. A.H. Tong, E. Gorey, and C.P. Schneider, Rev. Scient. Instrum. 43, 320 (1972).
51. J.L. Deines, E.F. Gorey, A.E. Michel and M.R. Poponiak, Spreading Resistance Symp. op. cit. 169.
52. S.S. Li and W.R. Thurber, Solid State Elec. 20, 609 (1977).
53. W.R. Thurber, R.L. Mattis and Y.M. Liu, Proc. of the Topical Cong. on Char. Tech. for Sem. Mat. and Dev., (P. Barnes and G. Rozgonyi, Eds.), 81 (Electrochem. Soc. 1978).
54. S. Wagner, J. Electrochem. Soc. 119, 1570 (1972).
55. F. Mousty, P. Ostoja, and L. Passari, J. Appl. Phys. 45, 4576 (1974).
56. G. Baccarani and P. Ostoja, Solid State Elec. 18, 579 (1975).
57. J.C. Irvin, Bell System Tech. J. 41, 387 (1962).

Chapter 2

FOUR-TERMINAL NONDESTRUCTIVE ELECTRICAL AND GALVANOMAGNETIC MEASUREMENTS

H. H. Weider

Electronic Material Sciences Division

Naval Ocean Systems Center

San Diego, CA 92152

I. INTRODUCTION

One of the well-established procedures used for a quantitative evaluation of the electrical properties of semiconductors is the measurement of their resistivity. Modern technological processes, particularly those concerned with manufacturing of semiconductor devices and integrated circuits require that such measurements be made not only of the spatial average of macroscopic specimens but also of localized microscopic fluctuations in their resistivity. Such fluctuations may be present in virgin semiconducting wafers cut from a single crystal boule; they may be due to a radial and axial microsegregation of impurities during growth or to defects generated by sawing, slicing or other forms of plastic deformation to which a wafer may be subjected during preliminary handling before it is accepted for further processing. Changes in resistivity and its spatial distribution may also occur in a wafer during any one or more of the various stages of device or integrated circuit fabrication procedures. These process-induced changes can serve as diagnostic tools for monitoring and quality control of a fabrication process. Complementary Hall effect measurement, used in conjunction with resistivity measurements yield more complete and better data. They can provide the spatial average and the fluctuations in free carrier concentration and mobility as a function of parametric variations of a technological process. The object of these lectures is to present a description of those methods and techniques used for nondestructive quantitative four-electrode characterization of the resistivity, Hall coefficient, carrier concentration and mobility of electronic materials.

II. THE COLLINEAR FOUR-PROBE ARRAY

A collinear four-probe (CFP) array probe can be used to measure the resistivity of specimens having a wide variety of shapes including those with irregular boundaries as well as the resisitivity of smaller regions included in a matrix with different electrical properties. Collinear four-probe measurements are used extensively in the silicon device and integrated circuit technology for the evaluation of Si wafers, epitaxial layers, ion implanted regions and impurity diffusion processes. Figure 1 shows schematically a representation of a CFP array. It consists of four pointed equispaced probes in contact with a plane surface of a uniform isotropic material to be measured. The probes are considered to be far from any of the other boundaries of the specimen so that the latter may represent an essentially semi-infinite volume of uniform resistivity. Further constraints applicable are:

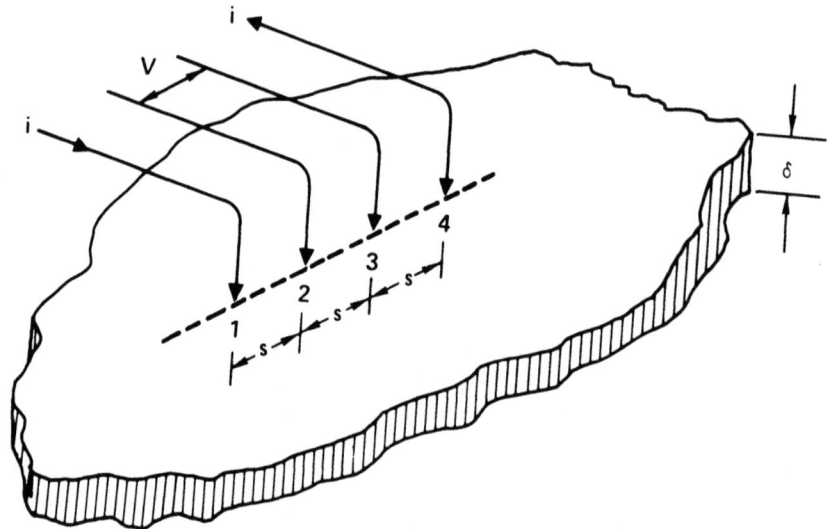

Fig. 1. Collinear four-probe array on a semi-infinite sheet of thickness, δ; outer probes 1 and 4 are the current input and output probes, inner probes 2 and 3 measure the potential difference, V, between them.

(a) the diameter of the contact between each of the probes and the specimen should be small compared to the distance, s, between the probes,

(b) the surfaces should have a high recombination rate so that any electrons or holes injected by the probes recombine in their immediate vicinity and their effect on the conductivity is negligible, and

(c) the boundary between the current-carrying electrodes and the bulk material is small in diameter and hemispherical in shape. Let a current, i, be impressed on the two outer probes (1 and 4) while the potential difference, V, is measured between the inner probes (2 and 3). The current streamlines are considered to have radial symmetry. The potential, V_f, at a distance, r, from a current-carrying electrode in a specimen whose thickness $\delta \gg s$ and whose boundaries are at infinity is

$$V_f = \rho i/(2\pi r) \tag{1}$$

Since there are two current-carrying electrodes, $V_f = (\rho i/2\pi)(1/r_1 - 1/r_4)$ where r is the radial distance from probe 1 and r is the distance from probe 4. The floating potentials at proves 2 and 3 are

$$V_{2f} = (\rho i/2\pi)(1/s - 1/2s) \tag{2}$$

$$V_{3f} = (\rho i/2\pi)(1/2s - 1/s)$$

The potential difference between the probes, $(V_{2f} - V_{3f}) = V$ and the current i are used to calculate the resistivity,

$$\rho_o = 2\pi s \cdot (V/i) \tag{3}$$

This relation and another, more complex one developed for a collinear array having unequal spacing between its proves were derived by Valdes [1], who also calculated by the method of images the correction factors $F(\ell/s)$ applicable to the resistivity, $\rho = \rho_o \cdot F(\ell/s)$ under the additional constraint that one of the exterior probes of the array is at a distance, ℓ, from either a parallel or a perpendicular specimen boundary. Uhlir [2] extended these calculations to include correction factors for a finite specimen thickness. If Equation (3) is used to calculate the resistivity of a specimen whose thickness, $\delta \geq 10s$ with the four-probe array oriented perpendicular to and at a distance, ℓ, from a nonconducting boundary, then the error in ρ is equal to or less than 2.5% provided that $\ell/s \geq 2.5$. If the array is parallel to a nonconducting boundary, then this error $\leq 2.5\%$ if $\ell/s \geq 2.5$; it is the same for a parallel conducting boundary if $\ell/s \geq 1.5$. The specimen or layer thickness is a significant parameter in the evaluation of the actual resistivity if the interprobe spacing, s, is comparable to the thickness, δ, shown in Figure 1. In the limit if the thickness $\delta \ll s$, the specimen may be considered as essentially two-dimensional, an infinite sheet having an infinitesimal thickness, δ. Since the current streamlines have radial symmetry in the homogenous isotropic sheet, Ohm's law is $(\partial V/\partial r) = -\bar{J}$ and the current density $\bar{J}$ and potential difference, $V = V_2 - V_3$, between probes 2 and 3 are respectively,

$$\bar{J} = \frac{ir}{2\pi\delta r} - \frac{ir'}{2\pi\delta r'}, \quad V = -\int_{2s}^{s} J \, dr \qquad (4)$$

where the unit vectors $r = r'$ and $r' = 3s - r$; therefore,

$$V = \frac{i\rho}{2\pi\delta} \int_{s}^{2s} \frac{1}{r} + \frac{1}{3s - r} \, dr \qquad (5)$$

The solution of this integral then leads to the resistivity,

$$\rho = \frac{V}{i} \frac{\pi\delta}{\ln 2} = 4.5324 \delta R \qquad (6)$$

Equation (6) is adequate for determining the sheet resistivity, $\rho_s = \rho/\delta$, of an essentially infinite conducting sheet of infinitesimal thickness in comparison with the interprobe spacing, s.

Smits [3] has employed the method of images to calculate a correction factor, C, for such sheet resistivities, measured by means of a centered four-probe array placed on a line of symmetry such as the diamter, d of a circular specimen or on the bisectrix, d/2, of a rectangular $\ell \times d$ specimen. The correction factor was found to be dependent on both the ratio d/s and ℓ/d; thus $\rho_s = RC(d/s, \ell/d)$. For small d/s the sheet resistivity $\rho_s = Rd/s$. Table 1 shows the calculated dependence of the correction factor C on d/s and ℓ/d.

III. THE SQUARE FOUR-PROBE ARRAY

A square four-probe (SFP) array, such as shown in Figure 2, provides some advantages over a collinear array if the resistivity of small surface area specimens are to be measured. For a thin semi-infinite conducting sheet whose thickness δ is much smaller than the interprobe spacing, the resistivity determined in terms of the current i applied to the current probes and the potential difference V measured between the potential probes in Figure 2 is [2]

FOUR-TERMINAL NONDESTRUCTIVE MEASUREMENTS

TABLE 1

d/s	Circle C for dia. d	Square C for $\ell/d = 1$	Rectangle C for $\ell/d = 2$	Rectangle C for $\ell/d = 3$	Rectangle C for $\ell/d \geq 4$
1.0				0.9988	0.9994
1.5			1.4788	1.4893	1.4893
2.0			1.9454	1.9475	1.9475
3.0	2.2662	2.4575	2.700	2.7005	2.7005
4.0	2.9289	3.1137	3.2246	3.2248	3.2248
5.0	3.3625	3.5098	3.5749	3.5750	3.5750
10.0	4.1716	4.2209	4.2357	4.2357	4.2357
20.0	4.4364	4.4516	4.4553	4.4553	4.4553
∞	4.5324	4.5324	4.5324	4.5324	4.5324

$$\rho = 2\pi\delta \, (V/I)/\ln 2 \tag{7}$$

Further corrections are required if the specimen is of specific finite shape. If the SFP is centered on a circular wafer of diameter d and thickness δ then R = V/I the resistivity,

$$\rho = \frac{\pi\delta R}{\ln 2} F(d,s) \tag{8}$$

where the geometrical correction factor F(d,s) is presented as a function of d/s in Table 2.

TABLE 2

d/s	2	2.0	3.0	4.0	5.0	10.0	20.0	∞
F(d,s)	1.000	1.082	1.325	1.518	1.645	1.892	1.971	2.000

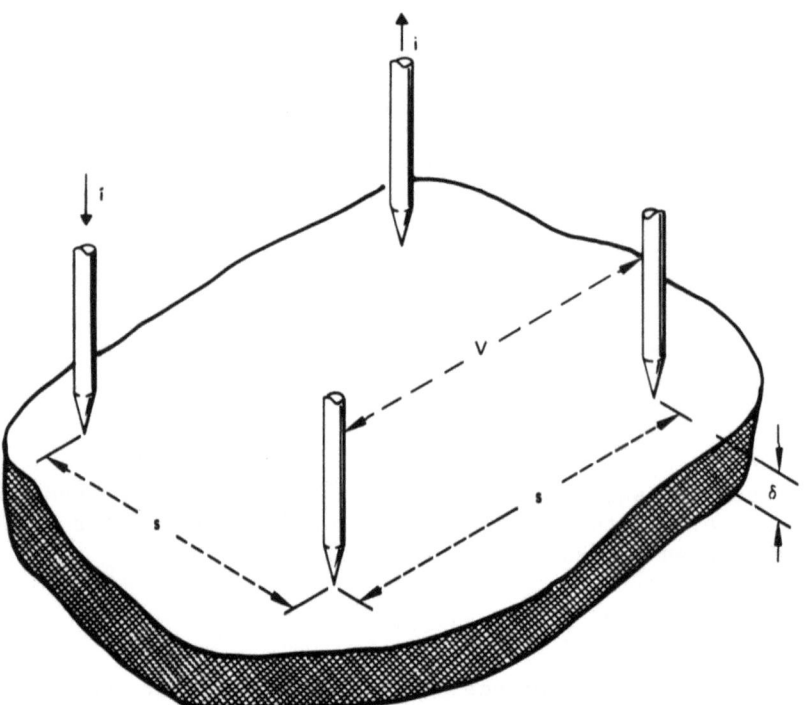

Fig. 2. Square four-probe array on a semi-infinite surface of thickness, δ; current input and output points constitute current dipole; interpole spacing is s.

The geometrical correction factor for a SFP on a square specimen of side d was determined by Mircea [4] in terms of the relation

$$\rho = RK(s,d) \tag{9}$$

The dependence of the geometrical correction factor K(s,d) on the ratio (s,d) and on the relative angle between the array and the edges of the square specimen are shown in Figure 3.

Mircea [5] has calculated the geometric correction factors by the application of Green's function to a parallelipiped and Stephens et.al. [6] have extended these calculations to include: (a) non-equidistant point probes located symmetrically on the top face, and (b) point-probe potential contacts located symmetrically on the top face and point-probe current contacts centered on the end faces of the parallelipiped. Their results lead to the following generalizations for case a:

(1) The apparent, measured resistivity is always greater than or equal to the correct value.

(2) Longer samples and thinner samples give apparent resistivities that are more nearly correct.

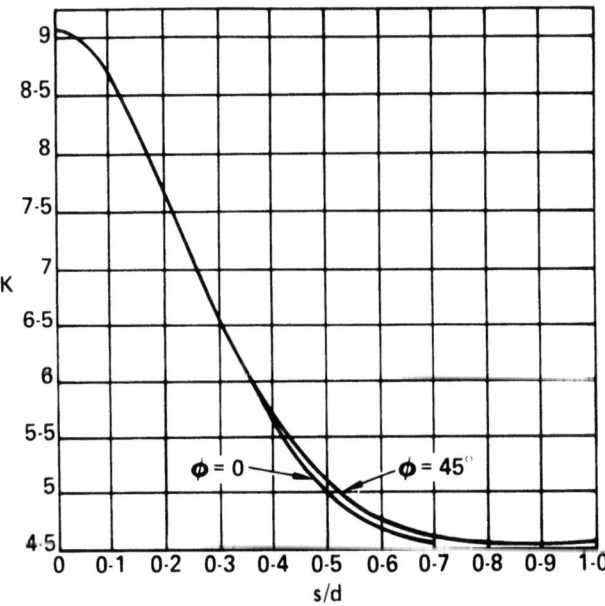

Fig. 3. Dependence of correction factor K on ratio s/d and on angle ϕ between array and side of square specimen (after A. Mircea, J. Sci. Instr., 41 (1967) 679).

(3) Smaller potential contact separations give apparent resistivities that are more nearly correct.

  (4) The longer and thinner the sample, the closer the potential contacts can be placed to the current contacts while still measuring a resistivity equal to the correct resistivity.

Specifically, if the current contacts are located one-tenth the length of the sample from the ends, then the actual resistivity can be calculated from the relation, $\rho = 2\pi Rs$, given in Equation (3) provided the sample is at least 4 times as long as it is wide, the sample is at least as thin as half its width, and the voltage-contact separation is not greater than two-thirds the current-contact separation. For case (b) the applicable generalizations are:

  (1) Longer samples give apparent resistivities that are more nearly correct.

  (2) Except for very short samples, smaller potential contact separations give apparent resistivities that are more nearly correct.

  (3) The longer the sample, the closer the voltage contacts can be placed to the ends of the sample while still obtaining apparent resistivity equal to correct resistivity.

  (4) For large potential contact separation, samples thinner than half their width give apparent resistivities greater than the correct resistivity, while samples thicker than their width give apparent resistivities less than the correct value.

The actual resistivity of a specimen using this probe configuration can be calculated by means of Equation (3) and $\rho = 2\pi Rs$ provided that the sample is at least 3 times as long as it is wide, is not thicker than it is wide, is not thinner than half its width, and provided the potential contact separation is not greater than half the length of the sample.

Figure 4 is a block diagram of auxiliary apparatus used for four-point probe measurements. It includes a constant current source equipped with a voltage limiting control rheostat. The latter is used to protect the thin film or epitaxial layer specimens from voltage breakdown which might otherwise occur following initial contact between probe and specimen. A "calibrate-measure" switch is used to connect an external precision ammeter for precise setting of the input current. The current polarity switch is used for identifying unipolar conduction and the operational amplifier provides for the amplification of the output of the voltage probes when measuring low resistivity specimens with small input currents.

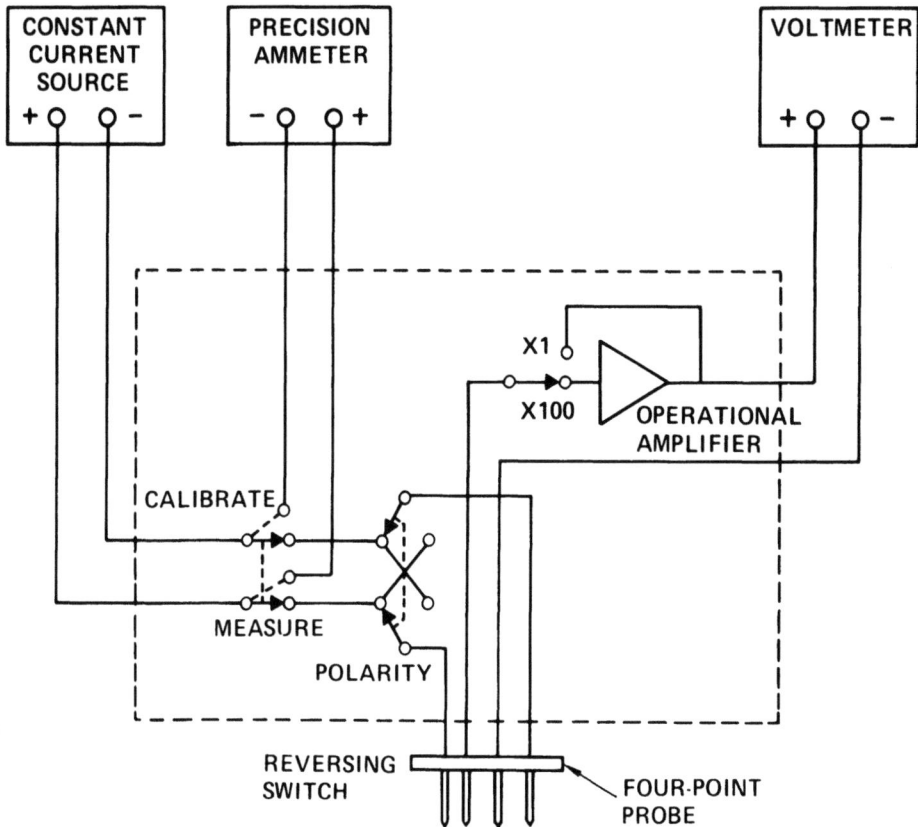

Fig. 4. Apparatus used for precise four-point probe measurements includes a precision current monitor used for calibration, current polarity reversing switch and operational amplifier for measuring low resistivities.

IV. RELIABILITY OF FOUR-PROBE TECHNIQUES

Four-probe techniques are subject to a variety of systematic and random errors. The former appear to be more significant for materials which have a higher resistivity. The current passing through a specimen during the measurement can affect the resistivity: (a) by an apparent increase in the resistivity produced by heating; (b) by sweeping minority carriers injected at the forward-biased contact to the vicinity of the potential probes where they decrease the resistivity [7]. On higher resistivity materials (> 10 ohm-cm) the increase of resistivity due to heating is more than offset by its decrease caused by minority carrier injection. The measurements are also subject to random errors which become evident if the probe head is moved between measurements. These errors can be attributed to two specific mechanisms: surface current leakage and probe wander. Surface leakage currents can be suppressed to a large extent by enclosing the entire voltage measuring circuit in a screen which is brought to the potential of the voltage probes before each measurement by means of an auxiliary potentiometer. The variation in interprobe spacing is a random error; when a probe makes contact with a wafer then any displacement directed along the line of the probe has a pronounced effect on the error of measurements.

Consider the sheet resistivity of a wafer which may be assumed to be semi-infinite; from Equation (3) $\rho_s = 2\pi s R$ if the measurement is made by means of a collinear four-point probe. If the displacements of the four probes are, respectively, x_1, x_2, x_3 and x_4 then the error in the measured resistivity 7 ,

$$\frac{d\rho}{\rho} = \frac{dV}{V} = \frac{1}{4s}(3x_1 - 5x_2 + 5x_3 - 3x_4) \tag{10}$$

If the probe displacements are random with a standard deviation then the error in the measurement of the resistivity,

$$\frac{d\rho}{\rho} = \frac{\sqrt{17}\alpha}{2s} = 2.06\ \alpha/s \tag{11}$$

For essentially two-dimensional thin layer specimens where Equation (6) applies the error in the resistivity due to a random displacement α of the equispaced probes [8] is

$$\frac{d\rho}{\rho} = \frac{\sqrt{5}\alpha}{2s\pi n 2} = 1.613\alpha/s \tag{12}$$

The precision can be improved by averaging the result of several independent measurements with the probe head lifted and replaced on the specimen between successive measurements.

V. RESISTIVITY OF LAMELLAE OF ARBITRARY SHAPE

Often, measurements need to be made on specimens having an irregular contour or on lamellae of arbitrary shape whose contacts are on their periphery. The theoretical foundation of such measurements is based on the conformal representaiton developed by van der Pauw [9,10] for mapping a specimen of arbitrary shape on a semi-infinite half-plane.

FOUR-TERMINAL NONDESTRUCTIVE MEASUREMENTS

Consider a continuous single-valued function of the form $z = f(t) = x(u,v) + jy(u,v)$ which defines a complex-variable, $z = x + jy$ as a function of another variable $t = u + jv$. Let a particular value, t', represent a point in the complex plane of t. Then in the complex plane of z some particular value (or values), z', corresponds to t'. There is a similar correspondence for other points on the respective complex planes of t and z so that a curve on the t-plane is transformed by "conformal mapping" from the t-plane to the z-plane. Conformal mapping is an analytical technique used by van der Pauw to demonstrate a procedure for measuring the resistivity of a specimen having an arbitrary shape and thickness with four, not necessarily equispaced, line-electrodes on its periphery. Consider the semi-infinite plane of thickness, δ, shown in Fig. 5. Line-electrodes 1, 2, 3, and 4 on the plane boundary are separated from each other by the dimensions a, b, and c. Let a current, i, be introduced at electrode 1 and removed at electrode 2. The appropriate current density, J, electric field, E, and the potential difference developed between electrodes 3 and 4, $V_{3,4}$, for the current i, flowing into the half-plane is

$$V_4 - V_3 = -\left(\frac{\rho i}{\pi \delta}\right) \int_{a+b}^{a+b+c} \frac{dr}{r} \tag{13}$$

Solving the integral in Equation (13) leads to

$$V_{3,4} = \frac{\rho i}{\pi \delta} \int_b^{b+c} \frac{dr}{r} = \frac{\rho i}{\pi \delta} \ln\left(\frac{a+b+c}{b+c}\right) \tag{14}$$

The potential $V_{3,4}$ developed by the current, i, flowing <u>out</u> of the half plane, between electrodes 3 and 4 is

$$V_{3,4} = \frac{\rho i}{\pi \delta} \int_b^{b+c} \frac{dr}{r} = \frac{\rho i}{\pi \delta} \ln\left(\frac{b+c}{b}\right) \tag{15}$$

The potential difference between Equations (14) and (15) is, therefore,

$$V_4 - V_3 = V_{3,4} = \frac{\rho i}{\pi \delta} \ln \frac{(a+b)(b+c)}{b(a+b+c)} \tag{16}$$

and the resistance determined from Equation (16) is

$$R_{12,34} = \frac{V_{3,4}}{i} = \left(\frac{\rho}{\pi \delta}\right) \ln \frac{(a+b)(b+c)}{b(a+b+c)}$$

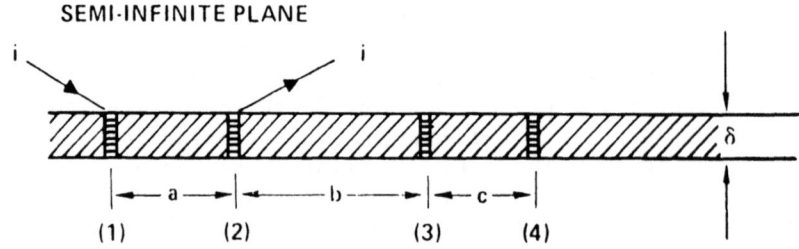

Fig. 5. Representation of a semi-infinite half-plane of thickness, δ, with line-electrodes 1 and 2 for current terminals and 3 and 4 used for voltage probes.

If the currents and voltage electrodes are commutated so that 2 and 3 are current electrodes and 1 and 4 are potential probes, then, solving the analog equations for the resistance leads to

$$\exp(-\frac{\pi\delta}{\rho} R_{12,34}) + \exp(-\frac{\pi\delta}{\rho} R_{23,41}) = 1 \qquad (18)$$

and the resistivity can be expressed as

$$\rho = (\frac{\pi\delta}{\ln 2}) (\frac{R_{12,34} + R_{23,41}}{2}) f \qquad (19)$$

The function f can be evaluated from

$$\frac{R_{12,34} - R_{23,41}}{R_{12,34} + R_{23,41}} = (\frac{f}{\ln 2}) \text{ arc cosh } \frac{\exp(\ln 2/f)}{2} \qquad (20)$$

The function, f, on the right hand-side of this equation depends only on the ratio $R_{12,34}/R_{23,41}$ (i.e., on the commutated current and measured voltages). Figure 6 presents, in graphical form, the dependence of f on the ratio $R_{12,34}/R_{23,41}$.

Van der Pauw has shown that a lamina of arbitrary shape can be mapped by a conformal transformation onto the semi-infinite plane of Figure 5 provided that: $\nabla \cdot \bar{J}=0$ and $\nabla \times \bar{J}=0$, that the lamina is simply-connected, that it is homogenous and isotropic, that it is uniform in thickness, its four "ohmic" line-electrodes are on its periphery, and the line-electrode projections on the surface of the lamina are point-contacts. The resistivity of a lamina or wafer of arbitrary shape can be determined by applying a current, i, to

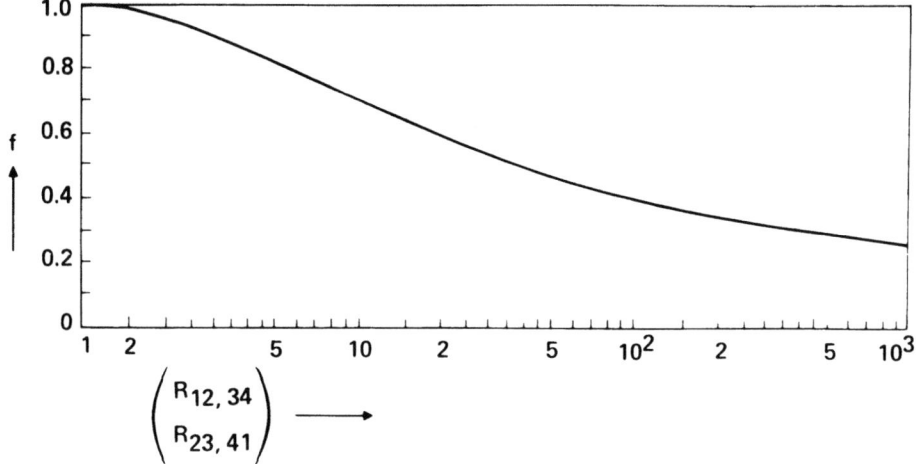

Fig. 6. The function f used to determine the specific resistivity of a specimen with an arbitrary contour as a function of the resistance ratio $R_{12,34}/R_{23,41}$. (after L. J. van der Pauw, Phillips Res. Repts., 13 (1958) 1).

line-electrodes 1 and 2 along its thickness dimension and measuring the potential difference between electrodes 3 and 4; this leads to the resistance $R_{12,34}$. In a similar manner, after commutating the current and voltage electrodes, $R_{23,41}$ is obtained. Figure 6 is then used for the calculation of the appropriate $f(R_{12,34}/R_{23,41})$ and, thereafter, of the resistivity.

Note that if the specimen used were a perfect circle with the electrodes along orthogonal diameters, then $R_{12,34} = R_{23,41}$. Therefore a solution of Equation (20) leads to f = 1. In consequence, the sheet resistivity, $\rho_s = (\pi/\ln 2)R_{12,34}$, is the same expression as that of Equation (6), the idealized equispaced four-point contact probe touching an infinite sheet. Rymaszewsky [11] has shown that for a collinear equispaced four-point probe used on a thin film, the sheet resistivity can be determined by means of a correction factor based on conformal representation similar to that used by van der Pauw. For two commutated measurements such that

$$R_1 = V_{32}/i_{41}, \qquad R_2 = V_{34}/i_{21} \qquad (21)$$

related by

$$\exp(-2\pi R_1/\rho_s) + \exp(-2\pi R_2/\rho_s) \qquad (22)$$

ρ_s can be expressed in the form

$$\rho_s = \frac{\pi R_1}{\ln 2}\left[1 + \frac{R_2}{R_1}\right] f\left(\frac{R_1}{R_2}\right) \qquad (23)$$

Thus Equation 6 is multiplied by a correction factor which includes van der Pauw's f-function and the function of the commutated resistance, R_2/R_1. He suggested that this method can be applied to a broad class of point-contact probes disposed either in line or along the circumference of a specimen.

VI. ERROR INTRODUCED BY CONTACT SIZE, POSITION AND GEOMETRY

The finite size of the electrodes applied to the periphery of arbitrary contour specimens affects the measurement of their resistivity made by means of van der Pauw's method [9,10]. The errors introduced by finite electrodes or by the displacement of an electrode from the specimen periphery were evaluated by van der Pauw [10] for a circular specimen of diameter D, whose contacts are placed symmetrically at 90° intervals; only one of the contacts is assumed to deviate from the ideal, as shown in Figure 7; the relative fractional errors in resistivity for the three cases illustrated here are:

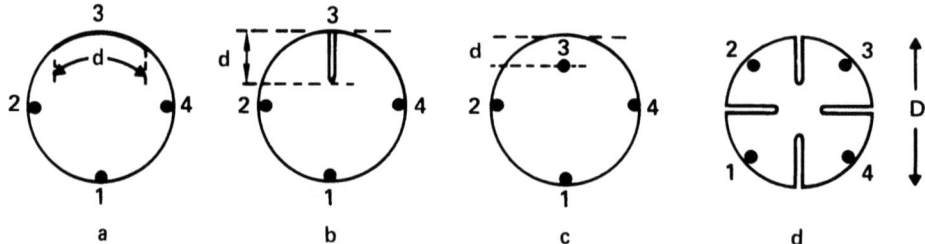

Fig. 7. Error in contact size and position: (a) one of the contacts has a length, d, along the periphery; (b) one of the contacts has a length, d, perpendicular to the periphery; (c) one of the point contacts is at a distance, d, from the periphery; (d) cloverleaf shaped specimen.

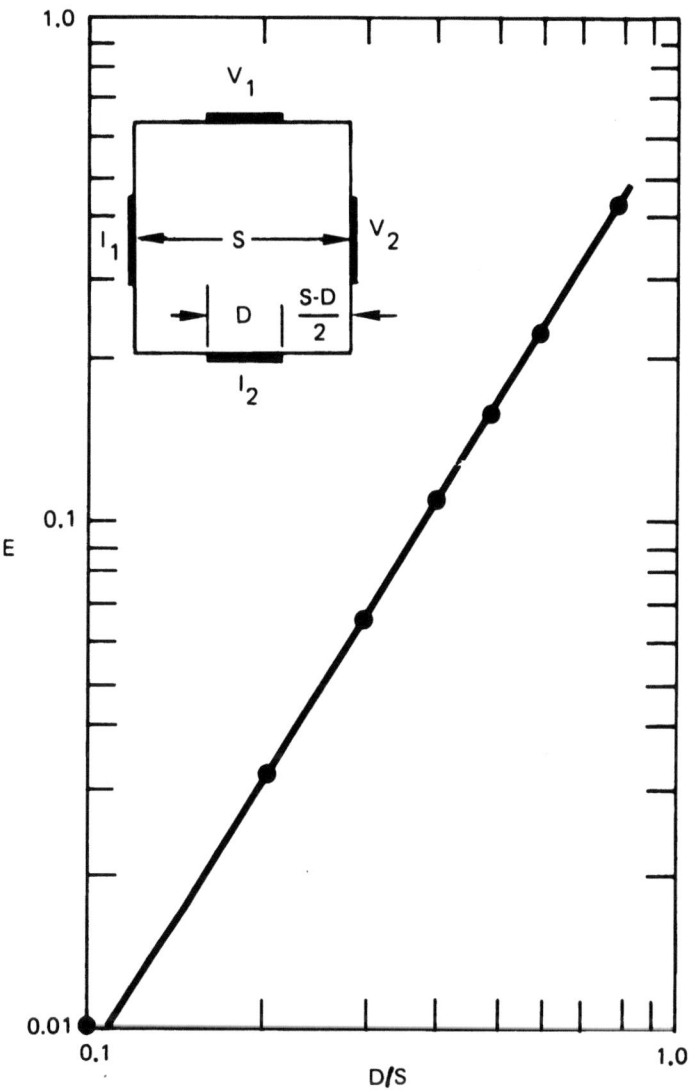

Fig. 8. Normalized sheet resistance error, E, with contacts centered on each side of a square specimen with dimensions as shown in the insert as a function the ratio of contact length to side of the square. (after M. G. Buehler and J. M. David, Natl. Bureau of Stds., Special Publ. 400-29 (1967) 64).

(a) $\Delta\rho/\rho \simeq -d^2 (16d^2 \ln 2)^{-1}$

(b) $\Delta\rho/\rho \simeq -d^2 (4D^2 \ln 2)^{-1}$ (24)

(c) $\Delta\rho/\rho \simeq -d^2 (2D^2 \ln 2)^{-1}$

If none of the contacts are ideal then, to a first approximation, the total error is the sum of all the errors. The influence of the contact and the errors associated with them can be eliminated by the use of a clover-leaf shaped specimen such as that shown in Figure 7(d). A clover-leaf configuration, applied to thin films or epitaxial layers, requires complex photolithographic processing which may alter their surfaces; such a structure usually occupies too much to the useful area of a specimen removing that portion from consideration for other applications. For this reason, simple circular or square structures with contacts which can be applied directly to the corners or edges of an epitaxial layer or thin film are considered desirable.

In the silicon integrated circuit technology there is a need for measurements to be made of the sheet resistivity of microstructures placed on an integrated circuit chip; the contact dimension of such a structure can be an appreciable fraction of the device perimeter. Consider the square configuraiton shown in the insert of Figure 8 with side-centered contacts. Let the idealized van der Pauw sheet resistivity for this structure be $\rho_s = \pi R/\ln 2$ and let the measured sheet resisitivity be ρ_{sm}. Figure 8 shows the dependence of E on the ratio of contact length to the side dimension of the square specimen [12]; the error of measurements fits the empirical relation

$$E \simeq (0.595 \pm 0.015)(D/S)^{1.82 \pm 0.03}$$ (25)

If D/S < 0.1 then the geometrical error associated with this structure is less than 1%. A different configuration also subject to a van der Pauw-type of analysis is the Greek cross shown in the insert of Figure 9. Except for very short cross-arm lengths such that A/S < 0.1 the error of measurement fits the empirical relation

$$E = (0.59 \pm 0.006) \exp[(-6.23 \pm 0.02) A/S]$$ (26)

If A/S > 0.6 then the geometrical error is less than 1% and if A/S > 1.02, a condition achieved easily in practice, then the corresponding error is less than 0.1%. The technological implication of the test structure, shown in Figure 9 for planar integrated circuit process control, is its use for measuring accurately the sheet resistivity of a wafer; and yet it can be made to very small sizes because the width of the arms of the Greek cross is limited only by the minimum linewidths which can be obtained by conventional photolithography. This type of geometrical configuration is, as we

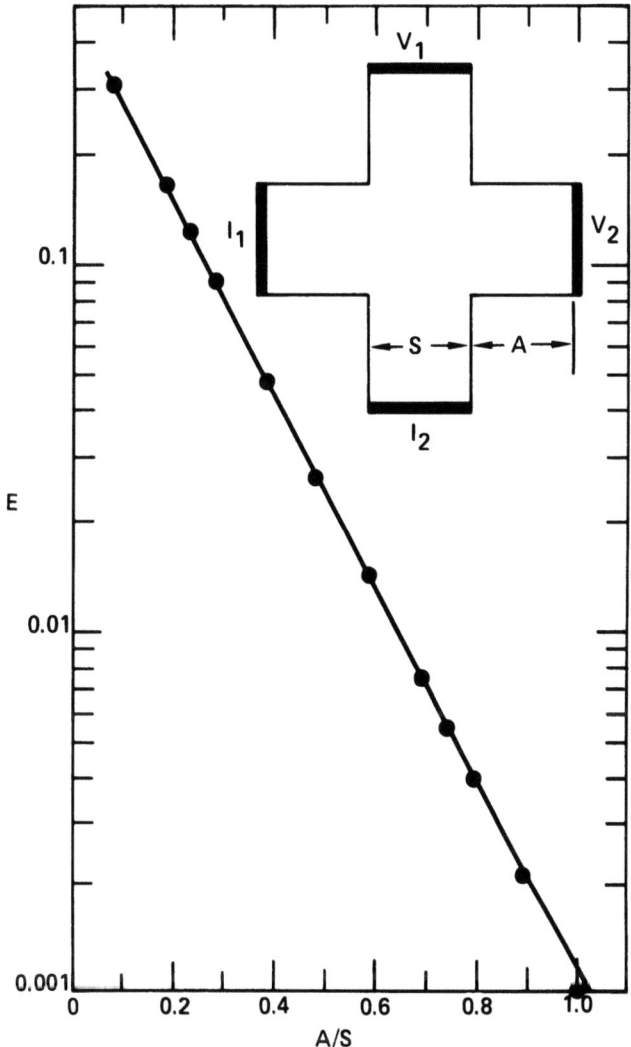

Fig. 9. Normalized sheet resistance error resulting from contacts applied to the cross-arm terminals of a Greek cross as a function of the length to width ratio of a symmetrical cross-arm as shown in insert. (after M. G. Buehler and J. M. David, Natl. Bureau of Std's Special Publ. 400-29 (1976) 65)

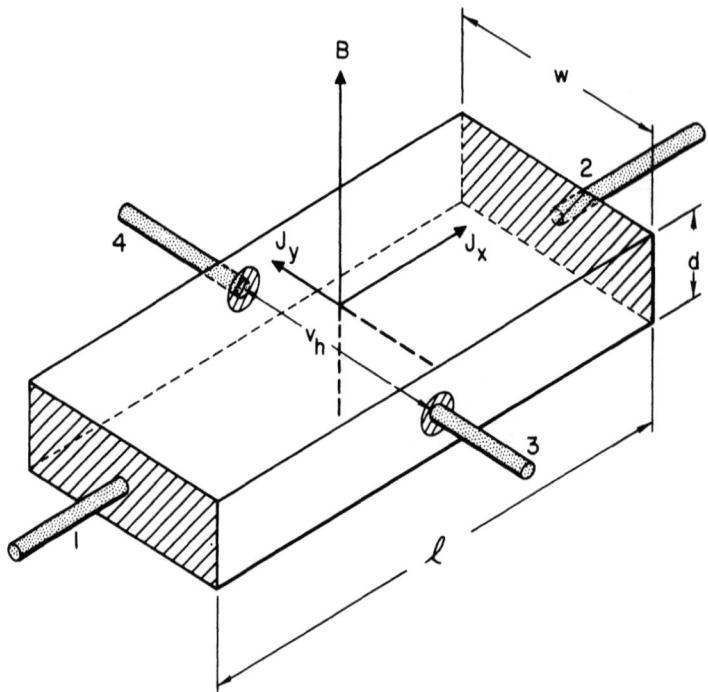

Fig. 10. Schematic representation of wafer in which electrons are the dominant charge carriers. Broad area faces, wd are current contacts designated by numerals 1 and 2. Hall voltage contacts transverse to magnetic field B and current density J_x are designated 3 and 4.

shall see in the following section, a particularly well adapted structure for both Hall effect and resistivity measurements and can be used for rapid automatic measurements of the specific parameters of large area wafers.

VII. HALL EFFECT MEASUREMENTS

Hall effect measurements made on rectangular wafers such as shown in Figure 10 require that the Hall voltage v_h be determined as a function of the transverse magnetic field B and DC input current i in accordance with

$$v_h = \int_0^w E_y \, dy = -R_h (iB/d) \tag{27}$$

where R_h is the Hall coefficient and E_y is the Hall field. The amplitude and polarity of v_h depends on the magnitude and polarity of i and on the direction of B relative to the coordinate axis. If terminal 1 is the positive polarity for a DC current i_+ and the direction of the steady-state magnetic field shown in Figure 10 is defined as B_+ then the Hall voltage, measured between terminals 3 and 4 is such that terminal 3 is negative for an n-type material (dominant electronic conduction) and positive for a p-type material (dominant hole conduction). In an ideal situation, these Hall electrodes are point contacts placed on an equipotential plane so the $v_h = 0$ for $B = 0$; in practice such conditions can be approximated only; there is always a finite potential difference between them in $B = 0$ defined as a misalignment potential v_0. The potential measured between the Hall electrodes for $B \neq 0$ is therefore, the algebraic sum of the misalignment potential and of the Hall voltage, $v = v_h + v_0$. The misalignment potential can be eliminated by a permutation of the current polarity and orientations of the magnetic field while keeping their amplitudes of i and B constant.

If the current polarity and magnetic field orientations shown in Figure 10 are indexed as i_+ and B_+, respectively, and the potential difference measured between Hall coefficient can be described in terms of Equation (27) as

$$R_h (C/cm) = \frac{2.5 \times 10\ d}{B} [\frac{v(B_+, i_+)}{i_+} + \frac{v(B_+, i_-)}{i_-} - \frac{v(B_-, i_-)}{i_-} - \frac{v(B_-, i_+)}{i_+}]$$

(28)

The choice of a rectangular wafer with needle-like point-contacts is, in most cases, not a practical test structure for Hall effect measurements because of contact resistance fluctuations with time, temperature and other variables. The bridge-shaped test structures shown in Figure 11 have some specific advantages for both Hall effect and resistivity measurements. Such bridge-shaped specimens can be made by photolithographic processing of epitaxial layers deposited on insulating or semi-insulating substrates or by cutting a similar contour ultrasonically from an appropriately oriented wafer and then making ohmic contacts to all of the terminals. The companion asymmetric Hall bridge shown in Figure 11 provides some advantages where space is at a premium. It can be used as a test structure on a portion of a wafer where it does not interfere with other devices and components. Furthermore, its misalignment voltage can be compensated by means of the circuit shown in Figure 12(b).

Compensation of the misalignment voltage of Hall test structures may be desirable for high-resolution Hall measurements, in particular for the case where the Hall voltage is an order of magnitude smaller

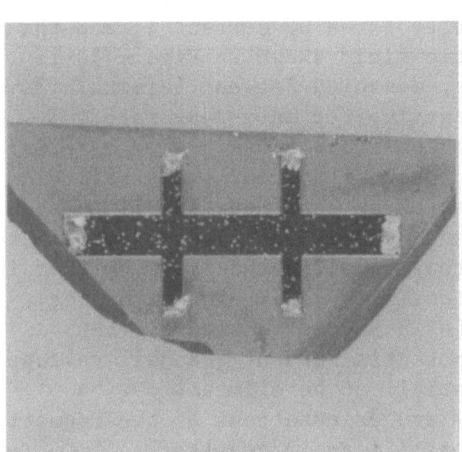 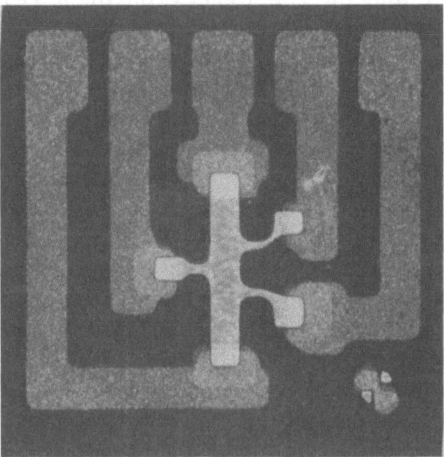

Fig. 11. Photolithographically-processed epitaxial GaAs layers made into bridge-shaped Hall effect structures. The device on the left is magnified 10X, that on the right, by 72X. Contacts shown are Au:Ge eutectic alloy covering edges of Hall structures and extending over semi-insulating substrates.

than the uncompensated misalignment voltage. The compensation circuits shown in Figure 12 provide a temperature-independent compensation provided that the input current i_1 is constant, that the branch current $i_c \ll i_1$, and that $i_c = i_1(r_c/r)R_i$. Here the resistances r and r_c are indicated in Figure 12 and the internal resistance R_i is that present between contacts 1 and 2 for (a); between 3 and 3' for (b); and between 1 and 3 for (c). Only the compensation circuit shown in Figure 12(b) is useful for large variations in magnetic field. The reason for this is that any magnetoresistance effects which influence R_i are the same as those which influence the test structure between contacts 1 and 2 at least to first order. For the other compensation schemes the misalignment voltage is itself a function of the magnetic field and, therefore, a potential source of magnetic field-dependent errors.

The essential components of apparatus for Hall effect measurements are: (a) a DC constant current source with provision for reversing its polarity; (b) an electrometer, galvanometer of high-impedance high-resolution digital or analog voltmeter with provision for reversal of its input polarity without affecting its accuracy or precision; (c) an electromagnet capable of providing an adjustable flux density from ~ 0.05 T to ~ 1.00 T with a homogeneity over

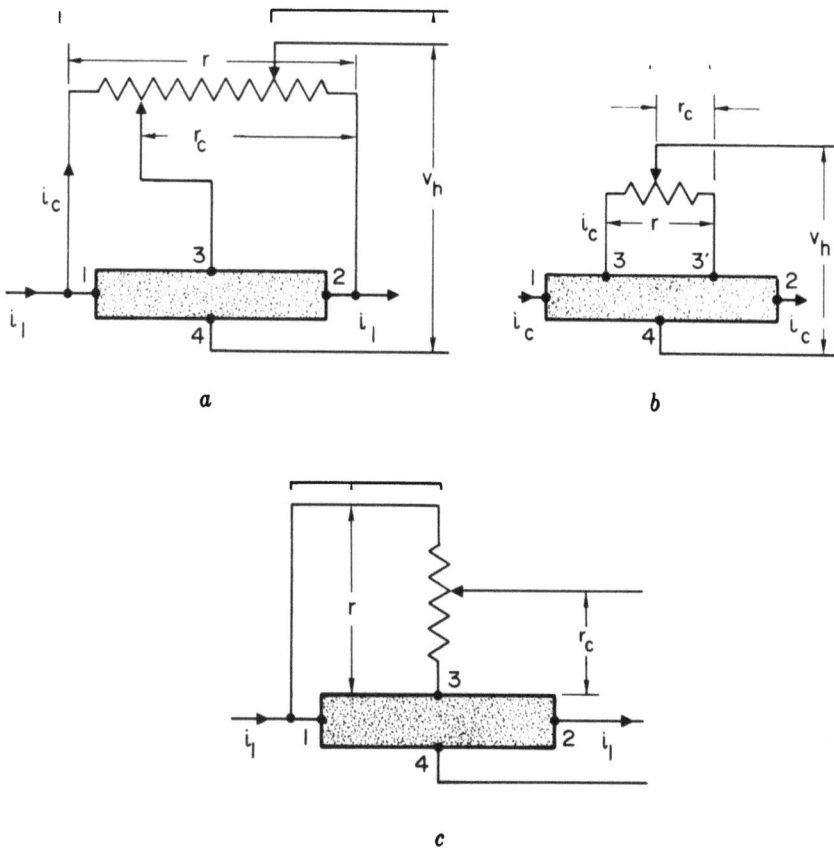

Fig. 12. External compensation circuits for eliminating the misalignment potential between Hall electrodes.

the specimen volume of ±1% and a precision and long-term stability of ±1%. A nuclear resonance magnetic field probe or a precision-calibrated Hall generator should be available to monitor the magnetic field between the magnet poles in the course of the measurements. Adequate provision should be made for reversing the direction of the magnetic field. If a permanent magnet is used to provide the magnetic field then provision should be made for rotating the specimen 180° about its axis parallel to the direction of current flow.

In order to determine the Hall coefficient from such Hall measurements, the thickness d of a wafer must be determined by means of mechanical probe measurements, by the use of a calibrated microscope equipped with a filar eyepiece, or a precision vernier caliper. An optimum specimen thickness d_{opt} might be chosen [13] in order to reduce the root mean square error in v_h and the rms error in d,

$$d_{opt} = (\Delta d/\Delta v_h)^{\frac{1}{2}} (R_h Bi) \tag{29}$$

For a given R_h, an optimum thickness exists because a decrease in d increases v_h thus increasing the accuracy with which v_h can be determined; however, by decreasing d the accuracy of the thickness measurement is reduced.

A minor source of systematic error is the inhomogeneity of the magnet. It was investigated analytically by de Mey [14] by an integral equation technique. As a rough approximation he found that if the magnetic field has a homogeneity better than x% then the error in the Hall voltage is less than 0.5x%. Koppe and Bryan [15] have considered an extreme case in which the magnetic field B, oriented as shown in Figure 10 had a distribution, B(x), which decayed to half its peak value at a distance along the x-direction equal to half the width of the specimen; for such a case, the ratio of the Hall voltage v_{hi} to that measured in a homogenous magnetic field is $v_{hi}/v_h \sim 0.8$. The fractional error in the value of the Hall coefficient can be expressed in the form

$$\frac{\Delta R_h}{R_h} = \frac{\Delta v_h}{v_h} + \frac{\Delta d}{d} + \frac{\Delta i}{i} + \frac{\Delta B}{B} \tag{30}$$

For a specimen chosen to have the optimum thickness by means of Equation (29) the fractional error in R_h is

$$\frac{\Delta R_h}{R_h} = \frac{\Delta i}{i} + \frac{\Delta B}{B} + 2\left(\frac{\Delta d \cdot \Delta v}{R_h iB}\right)^{\frac{1}{2}} \tag{31}$$

If the length of a rectangular wafer is not much greater than its width, then a correction factor must be included in Equation (27) to account for the reduction in Hall voltage produced by the electrostatic shorting of the Hall field by the current electrodes. Various analytical techniques [16-21] have been used to calculate the perturbation of the equipotentials and current streamlines in a rectangular plate produced as a function of its length to width, ℓ/w ratio, the extent of its contacts, and the Hall angle θ,

$$v_h = R_h \cdot (Bi/d) \cdot F(\theta, \ell/w) \tag{32}$$

The geometrical correction factor, $F(\theta, \ell/w)$ shown in Figure 13, is normalized in terms of v_h for $\ell/w \to \infty$. Results obtained experimentally and justified theoretically by Drabble et. al. [22] indicate

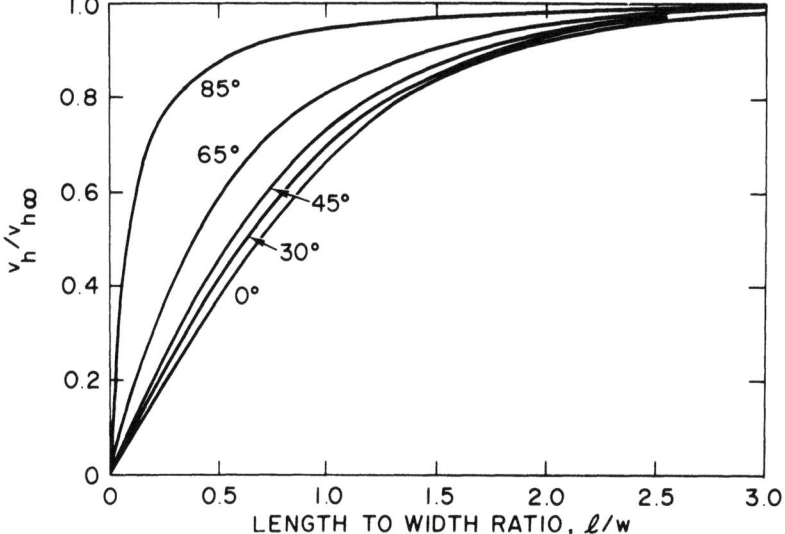

Fig. 13. Hall voltage v_h normalized in terms of v_{h_∞}, that of an infinite length rectangular wafer as a function of its length to width ratio with the angle θ as a running parameter.

that it is possible to obtain Hall voltages comparable (within 5%) to rectangular specimens with $\ell/w \gg 1$ provided that the metal contacts of specimens with $\ell/w < 1$ are replaced by a material whose resistivity is much larger than that of the specimens and that these contacts have lengths that are at least equal to their widths.

VIII. HALL EFFECT OF ARBITRARY CONTOUR SPECIMENS

The Hall coefficients of a lamella with an arbitrary contour and a thickness δ can be determined by the method of van der Pauw [9,10] provided that: (a) the lamella is simply connected and its thickness is uniform; (b) the contacts are sufficiently small to be considered as point contacts; (c) the magnetic field is perpendicular to the surface of the lamella. Figure 14 illustrates schematically such a lamella and suggests the distribution of the current streamlines considered to be invariant in a magnetic field. A current i is introduced at point contact M and extracted at contact O. The potential difference v is measured in B = 0 between point contacts N and P and the resistance is then calculated, $R_{MO,NP} = v/i$. If a transverse magnetic field is now applied to the lamina then $R_{MO,NP}$

changes by an amount $R_{MO,NP}$ and the Hall coefficient is

$$R_h = (\delta/B) \cdot \Delta R_{MO,NP} \tag{33}$$

The magnetic field produces a rotation of the equipotential planes in the lamella. The electric field and current density vecotrs are no longer collinear and the change in potential difference between P and N in Figure 14, $\Delta(V_P-V_N)$ can be obtained by integrating the Hall field E_h between P and N', across the lamella, along a path s orthogonal to the current streamlines and then along a streamline between N" and N. The latter makes no contribution to the integral; therefore

$$\Delta(V_P-V_N) = \int_P^{N'} E_h ds = R_h B \int_P^{N'} J ds = R_h B i_{MO}/\delta \tag{34}$$

which leads directly to Equation (33).

The distribution of current streamlines and equipotentials in a disc of thickness δ with peripheral point contacts placed on orthogonal diameters is shown schematically in Figure 15. For such a structure the Hall coefficient R_{h1} can be determined by one measurement in which the current and voltage contacts are opposite each other. Thereafter, a Hall coefficient R_{h2} can be determined with the Hall contacts and current contacts interchanged. The magnetic field must be small enough to keep any magnetoresistance-induced nonlinearities between v_h and B to a negligible value. If R_{h1} is ±10% of R_{h2} then the average Hall coefficient is $<R_h> = (1/2)(R_{h1} + R_{h2})$; if the discrepancy between R_{h1} and R_{h2} is greater than 10% then the specimen may be too inhomogenous for further investigation.

If the contacts are of finite size or if one or more of the point contacts are not located on the circumference of the disc then errors are introduced in the Hall measurements. These have been evaluated by van der Pauw [9] in terms of Figure 7; the fractional error in R_h for (a) $\Delta R_h/R_h \simeq 2d\,(\pi^2 D)^{-1}$; for (b) $\Delta R_h/R_h \simeq -4d(\pi D)^{-1}$, and for (c) $\Delta R_h/R_h \simeq -2d(\pi D)^{-1}$. If none of the contacts are ideal then, to a first approximation, the total error is the sum of the errors of each contact. The use of the clover leaf-shaped specimen, shown in Figure 7, reduces these errors. It provides a relatively larger Hall voltage than that of bridge-shaped specimen for the same amount of heat dissipation. It has greater mechanical strength and is easier to make than bridge-shaped specimens. Jacobs and de Mey [23] have investigated the dependence of the Hall voltage measured on a clover leaf-shaped structure, such as shown in Figure 16, as a function of contact

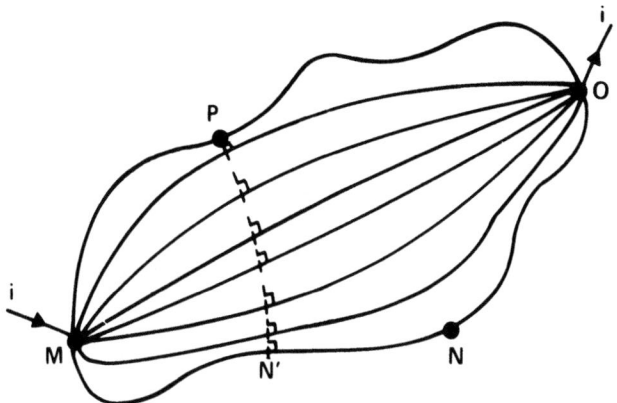

Fig. 14. A lamella of arbitrary contour with peripheral point contacts. A current i introduced and extracted at points M and O, respectively. The current streamlines are shown schematically and dashed line between P and N' is an equipotential.

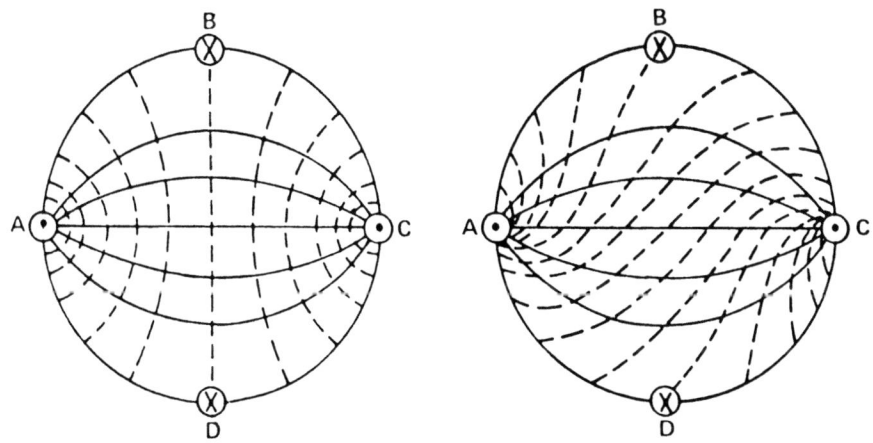

Fig. 15. Current streamlines (continuous lines) and equipotentials (dashed lines) in a disc; current electrodes are identified by dots, voltage probes by crosses. On the left B=0, on the right B is normal to plane of page.

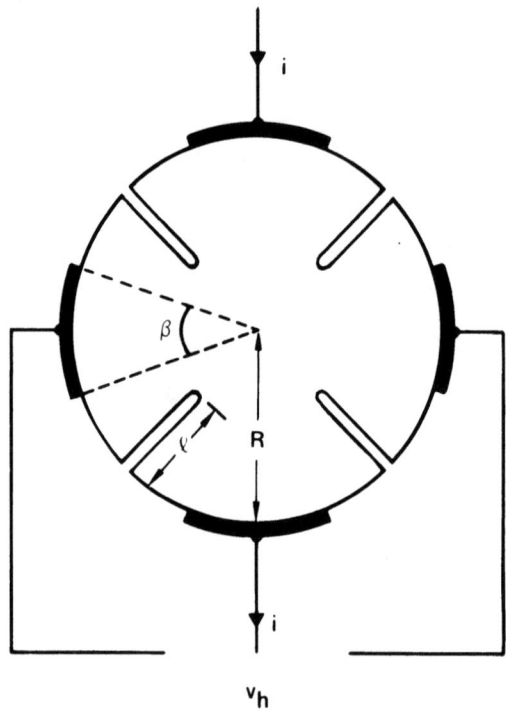

Fig. 16. Clover leaf-shaped structure for Hall effect measurements shows slots of length ℓ cut into disc of radius R. Contacts subtend angle β (after E. Jacobs and G. de Mey, Electron. Lett., 12 (1976) 4).

size and geometry by solving Laplace's equation by an expansion technique. With the potential ψ expressed as a power series in the Hall tangent, μB, the Hall voltage contains only odd terms in μB,

$$v_h = v_{h_1} \cdot \mu B + v_{h_3} (\mu B)^3 + \ldots \quad (35)$$

The first term of Equation (35) is the conventional Hall effect and the higher order terms represent the nonlinear dependence of v_h on B. The calculations of Jacobs and de Mey [23] indicate that the size of the contacts has a strong effect on the geometrical correction factor of the linear Hall effect and the nonlinearity of the measured Hall voltage of a clover leaf-shaped specimen is also a function of the size of the contacts and of the angle β in Figure 16. For small contacts the slots in Figure 16 provide an improvement only if β→R; however they do provide a decrease of the nonlinearity of v_h vs. B inherent in such measurements.

A symmetrical peripherally-contacted cross such as shown in Figure 9 can be considered to be a special case of a clover-leaf structure. The analytical and experimental investigations of Haeusler and Lippmann [24] have shown that for a ratio of S/A = 1 (refer to Figure 2) the error in the linearity of v_h with respect to B is $\sim$0.3% for B=1T and $\sim$0.8% for B=2T for large Hall angles. Further investigations made by de Mey [25] indicate that v_h is essentially independent of B for Hall tangents < 0.10 and that error in R_h is primarily of geometrical origin except in very high magnetic fields where magnetoresistance contributions can no longer be neglected [26]. The symmetrical cross shown in Figure 9 provides several advantages for low systematic error resistivity and Hall measurements. These include the simplification of lead attachment to the larger contacts of this structure, smaller current densities compared against the same current applied to rectangular or circular configurations, a reduction in Joule heating due to the better thermal conduction through the contacts of the cross compared to bridge-shaped specimens, smaller thermally-induced error voltages due to a reduction in thermal gradients and a reduction in contact-related noise.

IX. APPARATUS FOR HALL EFFECT MEASUREMENTS

The Hall effect measurements described thus far have been concerned, implicitly, with DC measurements. These require a stable current source capable of providing currents in the range from $\sim 5 \times 10^{-10}$A to 5×10^{-3}A with a stability of at least 0.01% for resistive loads which may vary from a few ohms to 10^8 ohms.

Potentiometric methods may be used for the measurement of small Hall voltages. A potentiometer can be used to produce a potential which matches the Hall voltage. The difference between them is then amplified for ease of observation. Similar apparatus with a DC output can be coupled either to a mechanical chopper amplifier or a photoelectric galvanometer amplifier with a precision of $\sim 10^{-9}$V. Clark and Ficket [27] have described apparatus suitable for the detection of low level Hall voltages with a precision of $\sim$25 pV.

Superconductive galvanometers or superconductive parametric amplifiers [28] can be used to achieve Hall voltage sensitivities of 10^{-14}V for low resistance specimens such as metals measured at cryogenic temperatures. Even higher sensitivities can be obtained [29] with galvanometers using Josephson tunneling junctions.

Hall effect measurements can also be made, on occasion, more advantageously than with DC by using a synchronous time variation of the measurement parameters. Such measurements can be made by means of alternating current supplied to a specimen with a static magnetic field, or, as an alternative, DC supplied to the specimen and a transverse time-variable magnetic field or a double AC technique in which both the specimen current and magnetic field are time variables. Techniques based on AC specimen currents and static magnetic fields were investigated and used by Donoghue and Eatherly [30] and Lavine [31] with a Hall voltage resolution of one part in 10^5. The use of AC provides some advantages in terms of the apparatus used for the detection, amplification, and for the use of noise rejection techniques such as sampling, synchronous detection and "lock-in" techniques. However, if the misalignment voltage of a specimen is too large then the AC amplifier may be driven into saturation. It is necessary to reduce the misalignment voltage to a minimum by external circuit compensation. A temperature-dependent drift of the misalignment voltage can complicate the compensation procedure. Mechanical vibration of the specimen or its leads can induce spurious voltages in the Hall circuit; in order to minimize these it is essential that the specimen, its mounting and leads be mechanically stable.

One of the principal advantages of the alternative method, that of a DC input to the specimen and an AC-driven magnetic field, is that the misalignment voltage for this situation is a DC parameter while the Hall voltage appears at the frequency of the magnetic field. Narrow bandwidth detection or synchronous detection of the Hall voltage can be used to eliminate error signals thus improving the resolution of the apparatus. However, the AC magnetic field is coupled inductively to the leads at the current input and Hall voltage output of a specimen. For a magnetic field of $\sim 0.5T$, a frequency of 1 Hz and a Hall voltage loop ~ 0.3 mm^2 an error signal of $\sim T\mu V$ is obtained in the Hall voltage circuit. The elimination of such inductive pick-up is, as a rule, a complex process which requires attention to detail in design of the experiment.

A time-variable magnetic field and DC specimen input current were used by Lomas et. al. [32] for Hall effect measurements. The magnetic field was provided by an electromagnet capable of producing magnetic fields up to 0.5T in a 2.54 cm air gap. It was driven by a variable 1 to 3 sec/Hz symmetrical square wave current with a polarity reversing switch so arranged that the magnet armature is short-circuited before the power supply polarity is reversed. To prevent any damage to the microvoltmeter used to measure the Hall voltage, due to switching transients, arrangements were made to short-circuit it during the switching cycle by a relay triggered from a pulse derived from the switching circuit; after polarity reversal is completed the relay contacts are opened and Hall voltage is recorded. Inductive pick-up in the current loop and Hall voltage circuit is circumvented by gating the Hall sensor circuit in synchronism with the magnetic field and drift of the misalignment voltage is not significant due to the synchronous time-variable character of the Hall voltage. This apparatus [32] was reported to be capable of resolving Hall voltages of the order of 10^{-7}V.

In the double AC method, a sinusoidal current with an angular frequency ω_1 is used to drive the specimen and a different sinusoidal current of angular frequency ω_2 is used to provide the magnetic field. In consequence, the Hall voltage may be expressed by

$$V_h = R_h \frac{i_{max} B_{max}}{2\delta} [\cos(\omega_1-\omega_2)t - \cos(\omega_1+\omega_2)t] \qquad (36)$$

where the specimen current is $i = i_{max}\sin\omega_1 t$ and the magnetic field $B = B_{max}\sin\omega_2 t$ with δ the thickness of the specimen. The Hall specimen behaves as a suppressed carrier modulator and the Hall voltage has two sidebands: $(\omega_1-\omega_2)$ and $(\omega_1+\omega_2)$ with each of them representing half the amplitude of the equivalent DC Hall voltage.

Russell and Wahlig [33] used a 70 Hz Hall current and a 60 Hz magnetic field to perform double AC Hall measurements with the Hall voltage detected at $\nu = 10$ Hz, the difference frequency. It is desirable to have the two sidebands far apart in the frequency domain in order to minimize intermodulation errors. The frequency used for the magnetic field is limited by the rise of the impedance of the magnet armature with frequency. This, in turn makes the peak attainable magnetic field a function of frequency. Lupu et. al. [34] developed a variation of the double AC Hall measurement method in which sequential double phase-sensitive detectors are used. These are synchronized with the input current frequency and also with the frequency of the magnetic field. The relative phase of the Hall voltage can be determined with respect to either of the two excitation signals. This permits a differentiation between spurious components and the actual Hall voltage. The latter appears at the frequency of the input current. With the magnetic field varied at 2 Hz and the input current with a frequency of 510 Hz and the signal is then phase-detected against a reference signal of 2 Hz thus extracting the Hall voltage. A detailed analysis of the spurious signals which can arise in this type of double AC Hall measurement have been give by McKinzie and Tannhauser [35]. They also provide a repertory of diagnostic techniques for distinguishing between the real Hall voltage and the various error signals.

X. PROBE MAPPING AND ISO-RESISTIVITY CONTOUR MAPS

The spatial homogeneity of the resistivity of virgin semiconductor wafers and the perturbations in their resistivity introduced subsequently by fabrication procedures such as oxidation, diffusion, ion implantation and metal deposition affect the performance reliability and ultimate cost of the devices and integrated circuits made of them. One of the primary methods used for the characterization of spatial inhomogeneities is the mapping of the localized sheet resistivity by means of a collinear four-probe (CFP) array used in conjunction with an automatic probe-positioning apparatus. The sheet resistivity of a circular wafer whose radius is much greater than the interprobe spacing of a centered CFP is $\rho_s = k(\Delta V/i)$ where k is a geometrical correction factor. If the probe is displaced from the center of the wafer then k is a function of the position of the probe and of its orientation which may be considered to either perpendicular or along the wafer radius [36-38]. For $(\ell/r) \ll 1$ an arbitrarily oriented CFP located at a distance a from the nonconducting edge of a thin circular sample will have the same correction factor as that of a similarly oriented CFP of probe separation $s' = s(1-a/2r)^{-1}$ located a distance a from the nonconducting edge of a thin semi-infinite sample [38]. These correction factors have to be taken into account in producing maps of the distribution of the sheet resistivity over a circular wafer.

A conformal transformation of a rectangle into a semi-infinite plane was used by Perloff [39] to calculate the geometrical correction factors used to determine the sheet resistivity of a square specimen of side a measured by means of a square array of side s centered or displaced by Δ_x, Δ_y from the center of the specimen. The correction factor was calculated in terms of the maximum displacement $\Delta^* = \Delta^*_x = \Delta^*_y$ with the current dipole oriented either parallel to or at 45° with respect to the side of the specimen. The latter is shown in Figure 17. It represents the calculated dependence of $k(\Delta_x, \Delta_y)$ on $a \cdot (2s)^{-\frac{1}{2}}$ with the nature of the displacement as a running parameter. The $k(0, 0)$ curve corresponds to that of the centered array, the $k(\pm\Delta^*, 0)$ curve corresponds to the displacement of the array to either of the extrema along the x-axis only, etc. If the electrodes are at the midpoints of the sides, $a \cdot (2s)^{-1} = 1$ of the square sample, the $k(0,0) = \pi/\ell n2$ and if $a \gg s$ then the correction factor in Figure 17, $k(0,0) \to 2\pi/\ell n2$, the limiting case of a square array on an infinite conducting sheet. It is evident from Figure 17 that the correction factor $k(\Delta_x, \Delta_y)$ can differ substantially from the value $K(0,0)$ for a given probe spacing and sample size due to the influence of the nonconducting boundaries on the potential distribution within the square sample. Similar calculations were made by Perloff for a parallel probe and specimen orientation and for a collinear four-probe array displaced from the center of a square specimen. Figure 18 shows that the ratio $(k_{max} \cdot k_{min})$, considered to represent the positioning error of the square array is a function of the ratio $a \cdot s(\cos\theta + \sin\theta)^{-1}$, where θ is the angle between the current dipole and the side of the specimen; it shows that the sensitivity of the geometrical correction factor to positional errors may be minimized by orienting the array at $\theta = \pi/4$, as shown in Figure 17, rather than in the parallel configuration, $\theta = 0$, particularly for a $s(\cos\theta + \sin\theta)^{-1} \geq 1.2$.

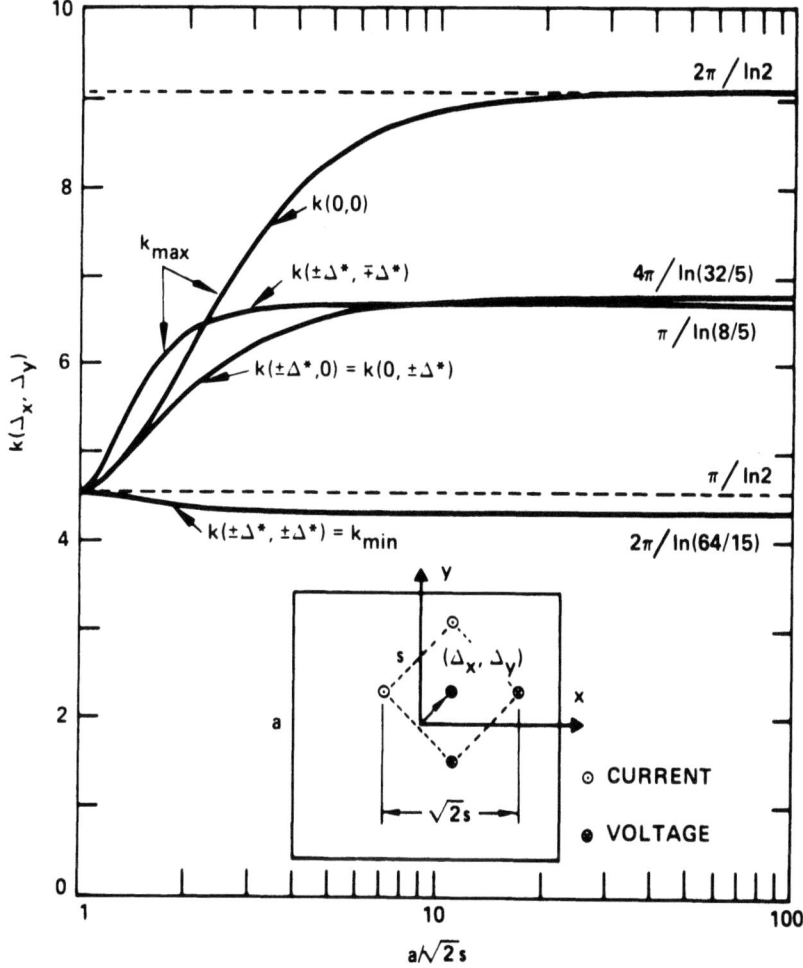

Fig. 17. Correction factors $k(\Delta_x, \Delta_y)$ for various locations of a square four-probe array oriented at $\Theta = 45°$ with respect to the side of a square specimen of side a. The quantity $\Delta^* = (a - \sqrt{2}s)/2$ is the maximum displacement of the center of the probe. (after D. S. Perloff, Solid State Electron., 20 (1977) 681)

Furthermore, if the displacement error of a collinear and square four-probe array are compared by differentiating the relevant expressions for their geometrical correction factors, then it appears that the probable displacement error of a collinear array is approximately 4 times larger than that of the square array.

Collinear or square four-probe array mapping measurements can be made over the surface of a wafer and the resultant data can be

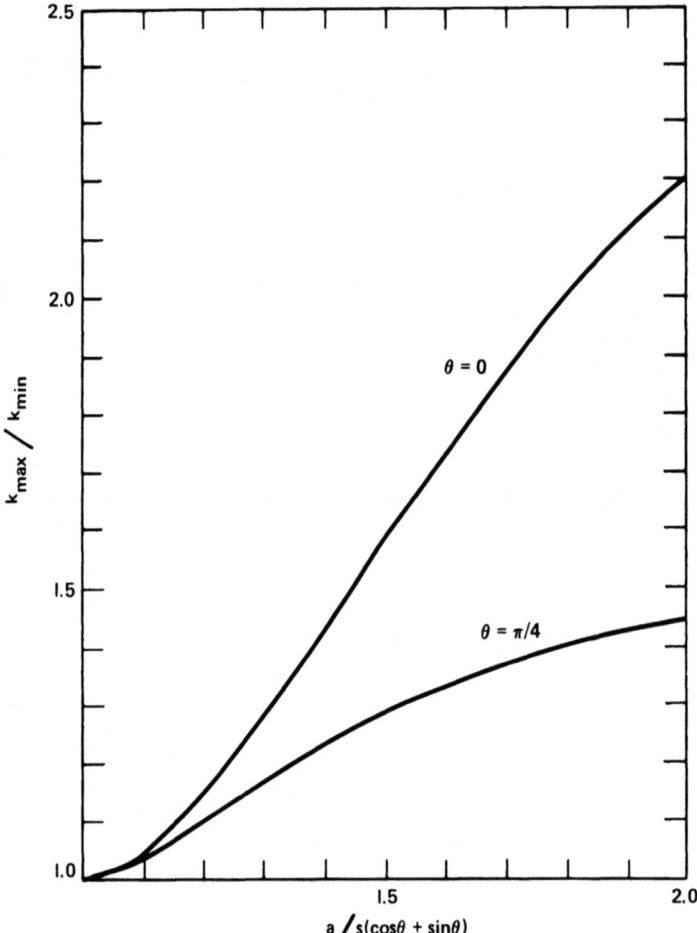

Fig. 18. Positioning error of geometrical correction factor k_{max}/k_{min}, as a function of the angular orientation Θ of the square array on a specimen of side a. (after D. S. Perloff, Solid-State Electron., 20 (1977) 681).

presented in various formats. One-dimensional profiles provide detailed information concerning point-to-point variations produced by doping or ion implantation. Two-dimensional maps may be made by dividing a wafer into a grid pattern with an arbitrary mesh of n test sites and measuring the mean sheet resistivity $\bar{\rho}_s$ and the standard deviation σ_d given by

$$\bar{\rho}_s = \frac{1}{n} \sum_{i=1}^{n} (\rho_s)_i \qquad \sigma_d^2 = \frac{1}{n-1} \sum_{i=1}^{n} [(\rho_s)_i - \bar{\rho}_s]^2 \qquad (37)$$

Figure 19 (a) is an example [40] of such a map determined on a 7.62 cm diameter phosphorous-diffused p-type silicon wafer. Values of $(\rho_s)_i$ obtained from CFP measurements were used to compute deviations from the wafer mean resistivity according to the relation

$$\Delta_i = [(\rho_s)_i/\bar{\rho}_s - 1] \times 100\% \qquad (38)$$

A format such as that shown in Figure 19(a) may be preferable in some instances since it presents deviations above and below the mean by (+) and (−) signs. It requires relatively little computation and can be implemented by means of a line printer. An alternative method of presentation of the data is in the form of two-dimensional histograms which associate measured values and interpolations with one of a number of intervals into which the range between $(\rho_s)_{max}$ and $(\rho_s)_{min}$ is divided. These intervals, each of which contains an equal number of values are identified by circles of various weight (thickness) and the symbols (−) and (+) in Figure 19(b) identify values above or below the median.

A different representation of the data may be made in terms of contour lines which represent equal values of sheet resistivity. Figure 19(c) shows such a contour map; the contour lines differ from each other by increments of 0.01 $\bar{\rho}_s$ and the symbols (+) and (−) represents experimental data greater or less than $\bar{\rho}_s$. The darker line in Fig. 19(c) represents the contour which corresponds to the mean sheet resistivity $\bar{\rho}$. Data interpolation and extrapolation has been used [40] to provide for the continuity of the contour lines and their extension to the periphery of the wafer.

Perloff et. al., [41] have used CFP probes either on unprocessed wafers or in conjunction with photolithographically defined rectangular test patterns for the characterization of diffused or ion-implanted layers. The test system for the acquisition of sheet resistivity data consisted of an automatic x-y probe, recording and sequencing devices operated by a minicomputer. Display of the data was made in the form of contour maps. a <111> oriented n-type wafer was found to have a radial resistivity gradient. An n-type epitaxial layer grown on p-type silicon substrate had a tear drop-shaped resistivity contour pattern which reflected the spatial distribution of the doping gases during the growth process.

The resistivity contour maps of silicon wafers which had been subjected to an open-tube flow gas-phase diffusion of impurities

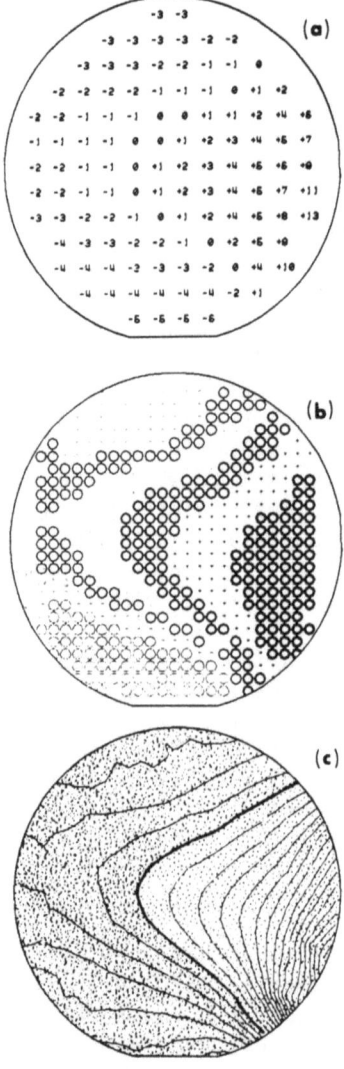

Fig. 19. Method of displaying spatial sheet resistivity variations obtained by means of collinear four-point probe measurements: (a) percent deviation from mean; (b) histogram; (c) iso-sheet resistivity. (after D. S. Perloff, F. W. Wahl and J. T. Kerr, Electron and Ion Beam Sci. & Tech., (1976) 464).

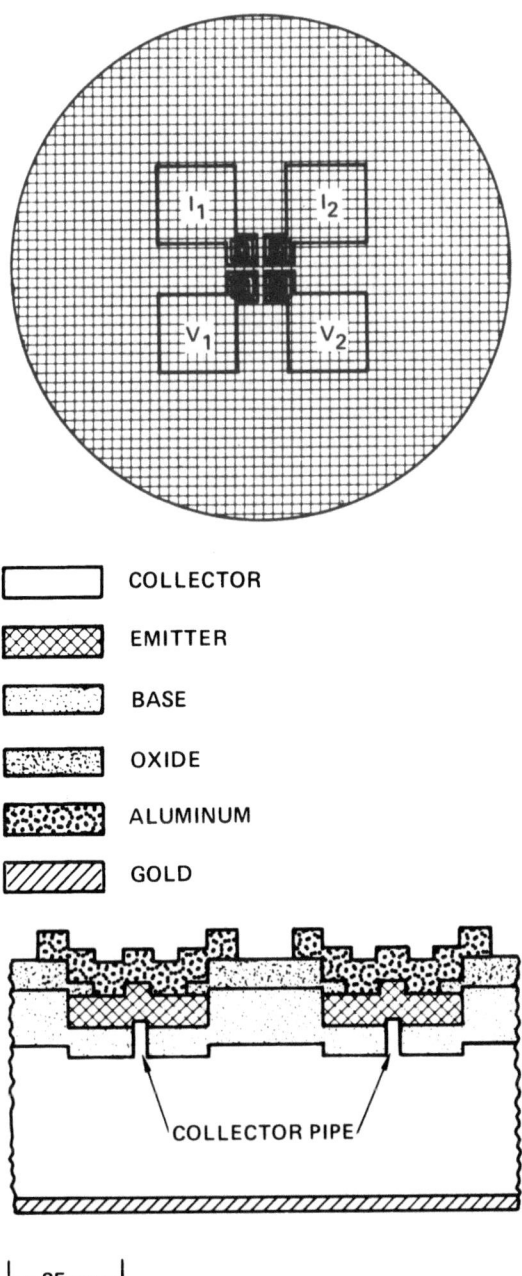

Fig. 20. (a) Planar square array four-probe test structure, I_1 and I_2 are current contacts, V_1 and V_2 are potential probes; (b) cross-sectional view (after M. G. Buehler and R. Thurber, IEEE Trans. Electron Dev. ED-23(1976) 968).

reflect the perturbations in flow of the gas through the reactor in contrast with gas phase diffusion in a sealed quartz ampoule; such wafers were found to have a uniform resistivity over the entire wafer within 1%. Ion implantation offers a degree of doping uniformity which is superior to diffusion. However, geometric effects in electrostatic scanning and displacement velocity effects in mechanical scanning limit the uniformity attainable with ion implantation [42].

The use of CFP or SFP arrays is not always advantageous for mapping fluctuations in resistivity induced by processes such as those used for the fabrication of bipolar transistors. A small planar four probe test structure which is compatible with such fabrication processes was developed by Buehler and Thurber [43]. Analogous to a mechanical square four-point array the planar four-probe array consists of a large area base diffusion broken at four points through which contact is made to the undiffused collector material as shown in Figure 20. Including bonding pads such a structure occupies a square region 318 μm on a side. A probe spacing of 57.2 μm allows the resistivity of the silicon wafer to be measured with good spatial resolution. During fabrication a base is diffused into 0.013 to 12 ohm-cm n-type silicon except at the four-probe positions which are protected from the base diffusion by four base-oxide islands. These islands define four collector pipes which make contact with the collector material. These are square pipes 6.4 μm on a side; they must be large enough to prevent lateral base diffusion from closing the collector pipes but, on the other had they have to be small enough so that the collector pipes are point-contacts. The situation can be improved by increasing the probe spacing; however if the probe spacing approaches the thickness of the wafer then geometrical correction factors, such as discussed in section 1 have to be used.

Crossley and Ham [44] have described a variety of test structures used to obtain parametric data for characterizing the processing steps in the fabrication of silicon-on-sapphire integrated circuits. The circuit elements used were MOS capacitors which can provide information about the oxide thickness, charge, doping density, inversion voltage, minority carrier lifetime and maximum depletion metal-oxide-semiconductor transistors used as planar test structures, can provide information on junction properties, the field-effect mobility of majority carriers and related properties. Gated van der Pauw-type resistivity and Hall effect structures biased to flatband were used to make accurate resistivity measurements. Extensive data logging and computer-based analysis and correlation of the raw data was found necessary for a characterization of the capabilities and limitations of integrated circuit fabrication processes. The data was presented in the form of histograms, two-dimensional correlation plots or quasi-three-dimensional plots of parameter values or combinations of parameters as a function of wafer position.

REFERENCES

1. L. B. Valdes, Proc. IRE, 42 (1954) 420.
2. A. Uhlir, Jr., Bell Syst. Tech. J., 34(1955)105.
3. F. M. Smits, Bell Syst. Tech. J., 37(1958)711.
4. A. Mircea, J. Sci. Instr., 41(1964)679.
5. A. Mircea, Solid-State Electron., 6(1963)459.
6. A. E. Stephens, J. J. Mackey and J. R. Sybert, J. Appl. Phys. 42(1971)2592.
7. J. K. Hargreaves and D. Millard, Brit. J. Appl. Phys., 13(1962) 231.
8. R. Hall, J. Sci. Instr., 44(1967)53.
9. L. J. van der Pauw, Philips Res. Repts., 13(1958)1.
10. L. J. van der Pauw, Philips Tech. Rev., 20(1958)220.
11. R. Rymaszewsky, Electron. Lett., 3(1967)57.
12. J. M. David and M. G. Buehler, Solid-State Electron., 20(1977)593.
13. C. M. Hurd, J. Sci. Instr., 42(1965)465.
14. G. de Mey, Solid-Stae Electron., 20(1977)139.
15. H. Koppe and J. M. Bryan, Can. J. Phys., 29(1951)274.
16. H. Weiss, Structure and Applications of Galvanomagnetic Devices, Pergamon Press, New York (1969) 16.
17. I. Isenberg, B. R. Russel, R. F. Greene, Rev. Sci. Instr., 19(1948)685.
18. R. F. Wick, J. Appl. Phys., 25(1954)741.
19. H. J. Lippmann and F. Kuhrt, Z. Naturforsch., 13a(1958)474.
20. J. R. Drabble and R. Wolfe, J. Electron. Contr., 3(1957)259.
21. J. Haeusler, Arch. Electrotech, 52(1968)11.
22. J. R. Drabble, M. M. Kaila and H.J. Goldsmid, J. Phys. D. Appl. Phys., 8(1975)790.
23. B. Jacobs and G. de Mey, Electron. Lett. 12(1976)4.
24. J. Haeusler and H. J. Lippmann, Solid-State Electron., 11 (1968)173.
25. G. de Mey, Arch. Elektron. Uebertragungstech., 27(1973)309.
26. G. de Mey, Y. Burvenich and M. de Molder, Phys. Stat. Sol., a23(1974)K45.
27. A. F. Clark and F. R. Ficket, Rev. Sci. Instr., 40(1969)465.
28. R. P. Ries and C. B. Scatterthwaite, Rev. Sci. Instr., 38(1967)1203.
29. J. Clarke, Phil. Mag., Ser. 8, 13(1966)15.
30. J. J. Donoghue and W. P. Eatherly, Rev. Sci. Instr., 22(1951)513.
31. J. Lavine, Rev. Sci. Instr., 29(1958)970.
32. R. A. Lomas, M. J. Hampshire and R. D. Tomlinson, J. Phys. E. Sci. Instr., 5(1972)819.
33. B. R. Russel and C. Wahlig, Rev. Sci. Instr., 21(1950)1028.
34. N. Z. Lupu, N. M. Tallan and D. S. Tannhauser, Rev. Sci. Instr., 38(1967)1658.
35. H. L. McKinzie and D. S. Tannhauser, J. Appl. Phys., 40(1969)4954.
36. M. A. Logan, Bell Syst. Tech. J., 40(1961)885.
37. L. J. Swarzendruber, Solid-State Electron., 7(1964)413
38. D. S. Perloff, J. Electrochem. Soc., 123(1976)1745.

39. D. S. Perloff, Solid-State Electron., 20(1977)681.
40. D. S. Perloff, F. E. Wahl and J. T. Kerr, Proc. Electron and Ion Beam Science and Tech., 7th Internat. Conf., Electrochem. Soc. (1976)464.
41. D. S. Perloff, F. E. Wahl and J. D. Reimer, Solid State Tech., 20(1977)311
42. D. S. Perloff, F. E. Wahl and J. Conrogan, J. Electrochem. Soc., 124(1977)582.
43. M. G. Buehler and W. Thurber, IEEE Trans Electon Dev., ED-23 (1976)968.
44. P. A. Crossley and W. E. Ham, J. Electron. Mat., 2(1973)465.

Chapter 3

CHARACTERIZATION OF SURFACE STATES AT THE Si-SiO$_2$ INTERFACE

G. DeClerck*

Katholieke Universiteit Leuven

Departement Elektrotechniek

Afdeling E.S.A.T.

Kardinaal Mercierlaan 94

B - 3030 Heverlee, Belgium

I. INTRODUCTION

The present knowledge of thermally grown silicon dioxide layers is extremely broad. A lot of effort has been given in the past to understanding the Si-SiO$_2$ system, giving rise to the modern fabrication techniques for high quality LSI-circuits. Nevertheless, the exact physical nature of the Si-SiO$_2$ interface is still unresolved and many device characteristics still depend on poorly known properties of the thin interfacial layer between the silicon bulk and the amorphous silicon dioxide film. Surface generated leakage currents, 1/f-noise in MOSTransistors, carrier trapping and information losses in surface channel charge coupled devices are ascribed to so called "surface states" or "interface states".

Surface states are allowed energy levels in the forbidden energy gap located at the Si-SiO$_2$ interface. They can be donor or acceptor like. Occupied donor states are neutral, unoccupied donor states are positive, while acceptor states are negative when occupied and neutral when unoccupied. In the Si-SiO$_2$ system the

---

*Bevoegd Verklaard Navorser NFWO (Fellow National Research Foundation)

interface states are mostly encountered as fast states, with time
constants varying from a few nsec to tens of msec. The considerable
dispersion in time constants at a fixed gate voltage can be explained
by two different models. The tunneling model proposed by Preier
accounts for the dispersion by a depth distribution over about 10
Angstroms yielding a broad range of tunneling time constants [1].
On the other hand, the surface potential fluctuation model of
Nicollian and Goetzberger is based on oxide charge fluctuations [2].

The physical origin of surface states is located within a
disordered region of a few atomic layers at the interface [3]. The
defects responsible for the energy levels can be of an intrinsic
nature, such as structural defects with dangling bonds [4],
deficiency or excess of silicon or oxygen (Cheng [5], Johanessen
and Spicer [6]). Extrinsic surface states are due to chemical
impurities such as sodium or charged centers situated in the vicinity
of the interface [7].

This paper describes some of the experimental techniques to
assess the important electrical properties of interface states such
as density, capture cross section and energy distribution. The
classical high frequency C-V, low frequency C-V and conductance
technique will be briefly reviewed. Charge pumping, deep level
transient spectroscopy (DLTS), will be discussed. The weak inversion technique based on drain current measurements of MOS transistors in the weakly inverted region will be explained. Surface
states characterization by means of 1/f-noise measurements of MOS
transistors and by transfer loss experiments on surface channel
charge coupled devices will also be discussed.

II. Metal-Insulator-Semiconductor Structure

For a detailed treatment on this subject we refer to the
excellent handbooks of S. M. Sze [8], A. S. Grove [9], P. Cobbold
[10]. The present review will be limited to those features of the
C-V or G-V curves needed for the subsequent discussion on the
experimental techniques. Although the derivation is valid for
other kinds of MIS-systems, we refer to the Si-SiO$_2$ system.

Fig. 1 shows a MOS capacitor, consisting of a thin insulating
SiO$_2$-layer on top of a p-type silicon substrate. V_G is the voltage
applied to the field plate and t_{ox} is the thickness of the insulator which ranges from a few nanometers to several microns. A
typical value for surface characterization is 100 nanometers.

The energy band diagram of the p-type semiconductor is shown
in Fig. 2. Energy increases in a positive direction following
the standard convention, the energy increases positively in Fig. 2

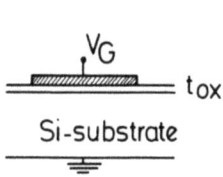

Fig. 1. MOS Capacitor

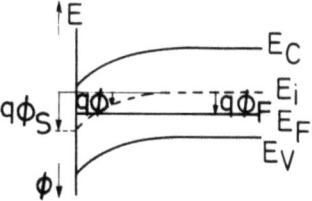

Fig. 2. Energy band diagram

while the potential increases negatively. The potential, ϕ, is measured with respect to the intrinsic level, E_i, and is zero in the bulk of the semiconductor. At the semiconductor surface ϕ takes the value ϕ_S. ϕ_F is the potential difference between the intrinsic bulk level and the Fermi level.

The relation between V_G, ϕ_S and the semiconductor space charge Q_{si} is found by solving the one-dimensional Poisson equation:

$$\frac{\partial^2 \phi}{\partial x} = - \frac{\Gamma(x)}{\varepsilon_{Si}} \tag{1}$$

where ε_{Si} is the permittivity of the silicon and $\Gamma(x)$ is the total space charge density given by:

$$\Gamma(x) = q(N_D^+ - N_A^- + p - n) \tag{2}$$

N_D^+ and N_A^- are the densities of ionized donors or acceptors. Solution of the Poisson equation gives the electric field at the surface E_S as a function of the surface potential:

$$E_S = \pm \frac{2kT}{qL_D} F(\phi) \tag{3}$$

with positive sign for $\phi_S > 0$ and negative sign for $\phi_S < 0$. k is the Boltzmann constant and L_D is the extrinsic Debye length defined for a p-type substrate by:

$$L_D = \left[\frac{2kT \, \varepsilon_{Si}}{N_A \, q^2} \right]^{1/2} \tag{4}$$

and $F(\phi_s)$ is given by:

$$F(\phi_s) = [(\exp^{-\beta\phi_s} + \beta\phi_s - 1) + \frac{n_0}{p_0}(\exp^{\beta\phi_s} - \beta\phi_s - 1)]^{\frac{1}{2}} \quad (5)$$

n_0 and p_0 are the bulk equilibrium concentrations of electrons and holes respectively.

The space charge in the silicon per unit area is equal to:

$$Q_{Si} = -\varepsilon_{Si} E_s = \frac{2\varepsilon_{Si} kT}{qL_D} F(\phi) \quad (6)$$

with negative sign for $\phi_s > 0$ and positive sign for $\phi_s < 0$.

The differential capacitance of the silicon space-charge region is defined by:

$$C_{Si} = -\frac{dQ_{Si}}{d\phi_s} = -\frac{\varepsilon_{Si}}{L_D} \frac{1 - \exp^{-\beta\phi_s} + \frac{n_0}{p_0}(\exp^{\beta\phi_s} - 1)}{F(\phi_s)} \quad (7)$$

At flat-band condition, i.e. $\phi_s = 0$, C_{Si} can be obtained by series expansion of the exponential terms in (7), yielding:

$$C_{Si\,(\phi_s=0)} = \frac{\sqrt{2}\,\varepsilon_{Si}}{L_D} \quad (8)$$

V_G is divided into a voltage drop V_{ox} across the oxide and ϕ_s

$$V_G = V_{ox} + \phi_s \quad (9)$$

This also means that the total differential capacitance C of the MIS capacitor is a series of combination of the insulator capacitance C_{ox} and the silicon space charge capacitance C_{Si}.

$$C = \frac{C_{ox} \cdot C_{Si}}{C_{ox} + C_{Si}} \quad (10)$$

with $$C_{ox} = \frac{\varepsilon_{ox}}{t_{ox}} \quad (11)$$

V_{ox} can be written as $V_{ox} = t_{ox} E_{ox} = -\frac{Q_{Si}}{C_{ox}} \quad (12)$

Equation (9) becomes then

$$V_G = \phi_s - \frac{Q_{Si}}{C_{ox}} \quad (13)$$

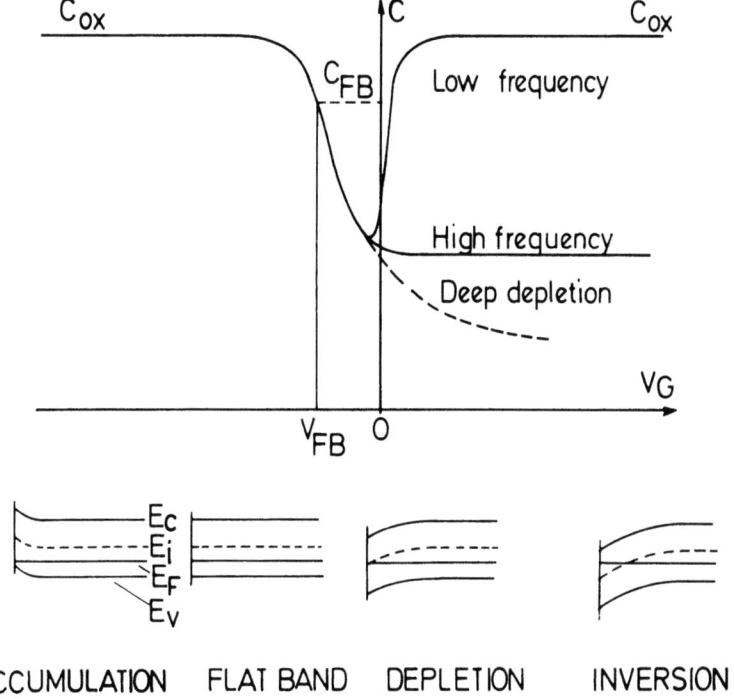

Fig. 3. MOS capacitance - voltage curve; the energy band bending in the four regions is also shown.

The work function difference ϕ_{MS} between the gate and the semiconductor and oxide charge S, Q_{ox}, can be taken into account by introducing a flat band voltage V_{FB}.

$$V_{FB} = \phi_{MS} - \frac{Q_{ox}}{C_{ox}} \tag{14}$$

The relation between gate voltage, surface potential and silicon space charge finally becomes:

$$V_G - V_{FB} = \phi_s - \frac{Q_{Si}}{C_{ox}} \tag{15}$$

Four different regions can be distinguished for the surface potential depending on the carrier concentrations at the surface. Figure 3 shows the different ranges in the case of a p-type substrate and the associated capacitances.

(a) $\phi_s < 0$ $(V_G - V_{FB} < 0)$. Q_{Si} is positive.

The energy bands are bent upwards. Those holes are accumulated at the surface, resulting in a large C_{Si}. C in equation (10) tends towards the C_{ox}.

(b) $\phi_s = 0$ $(V_G - V_{FB} = 0)$. Q_{Si} is zero.

The energy bands are flat; therefore this situation is called the "flat band" condition. The capacitance C_{Si} is given by (8).

(c) $0 < \phi_s < 2\phi_F$ $(V_G - V_{FB} > 0)$. Q_{Si} is negative.

The energy bands are bent downwards. The holes are repelled from the surface and the semiconductor charge in this depletion layer consists of the negatively ionized acceptor ions.

In the depletion approximation the width X_d of the depletion layer is given by:

$$X_d = \left(\frac{2\varepsilon_{Si} \phi_s}{qN_A}\right)^{1/2} \tag{16}$$

and the depletion capacitance is then equal to:

$$C = C_D = \frac{\varepsilon_{Si}}{X_d} = \left(\frac{qN_A \varepsilon_{Si}}{2\phi_s}\right)^{1/2} \tag{17}$$

However, a more rigorous treatment of expression (7) yields:

$$C_D = \frac{\varepsilon_{Si}}{L_D} \frac{1}{(\beta\phi_s - 1)^{1/2}} = \left(\frac{qN_A \varepsilon_{Si}}{2(\phi_s - kT/q)}\right)^{1/2} \tag{18}$$

At $\phi_F < \phi_s < 2\phi_F$ the onset of <u>weak inversion</u> occurs. The negative charge Q_{Si} still consists mainly of the depletion layer charge. The capacitance is given by (18)

(d) $2\phi_F \leq \phi_s$ <u>Strong inversion region</u>.

Here Q_{Si} is composed of ionized acceptors and electrons from the inversion layer. The electron concentration at the surface is larger than the hole concentration in the bulk. When V_G is made more positive, a small increase in band bending results in a very large increase of the electron density within the inversion layer. This also means that the depletion layer width reaches constant value which is found by inserting $\phi_s = 2\phi_F$ in expression (16).

In this region a distinction has to be made between the low frequency and the high frequency behavior of the semiconductor capacitance. The inversion layer electrons have to be provided for by the generation-recombination mechanisms of the minority carriers in the depletion layer and at the semiconductor surface. In a MOS transistor or a charge coupled device the arrival of the inversion charge can be enhanced by the presence of an external source of minority carriers. In these cases (low frequency and/or the presence of an external source) the electrons can follow the a.c.-signal. The capacitance is then still given by expression (7) where the $\exp\beta\phi$ - terms will account for the high density of inversion charges.

If the inversion charges cannot be supplied or withdrawn fast enough, the equilibrium with the a.c.-signal is no longer maintained. A high frequency capacitance is then obtained, which as a first approximation is given by (17) or (18) with $\phi_S = 2\phi_F$.

The frequency below which the MOS capacitance corresponds to the theoretical low frequency characteristic depends strongly on the minority carrier generation lifetime and on the surface generation currents are minimized so that frequencies much lower than 1 Hz are needed to obtain a low frequency response.

When the gate voltage applied to a MOS capacitor is pulsed from accumulation into strong inversion, the inversion charge cannot be built up instantaneously and a "deep depletion" situation occurs. The depletion layer is larger than predicted by the theory of thermal equilibrium and the depletion layer capacitance is smaller. Total capacitance is then smaller than the theoretical high frequency capacitance. The transition from the non-equilibrium deep depletion capacitance towards the equilibrium high frequency capacitance is governed by the generation mechanisms and has been described by many authors (Zerbst [11], Heiman [12], Hofstein [13], Schroder [14]).

III. THE MOS-STRUCTURE WITH SURFACE STATES

Let us consider a MOS capacitor made on a n-type substrate and biased in depletion. Surface states are assumed to be present lying at one single energy level E_T. The density of these surface states is N_S (units per cm^2) and their capture cross section for electrons and holes respectively σ_n and σ_p. The occupancy of the states is governed by Fermi-Dirac statistics expressed by:

$$f_o = \frac{1}{1 + g_A \exp\frac{E_T - E_F}{kT}} \quad (19)$$

where g_A is the spin degeneration factor which is equal to 2 for accpetor states and ½ for donor states.

The capture and emission of electrons or holes by surface states is not infinitely fast but associated with a time delay. Application of the Hall-Schockley-Read generation-recombination statistics and using a small signal approximation [2], the admittance Y_s of single-level surface states interacting with the conduction band is found to be

$$Y_s = j\omega \frac{q^2}{kT} \frac{N_{ss}f_o(1-f_o)}{(1+\frac{j\omega f_o}{c_n n_{so}})} \tag{20}$$

where $j = (-1)^{1/2}$ is the imaginary unit, ω is the angular frequency of the small sinusoidal, n_{so} is the free electron density at the <u>interface</u> corresponding to the applied bias, c_n is the electron capture rate given by $\sigma_n v$ with v the thermal velocity of the electrons. Only capture and emission of electrons by the interface states are considered in deriving (20). This is valid in most practical cases when the capacitor is biased in depletion. For a more complete model however where interactions with both conduction and valence band are taken into account, we refer to Lehovec-Slobodskoy [15] and Cooper-Schwarz [16]. If a p-type substrate is used, $c_n n_{so}$ should be replaced by $c_p p_{so}$ and expression (20) is valid in depletion with the same restrictions as explained above.

Equation (20) gives the admittance of a series RC network with a capacitance C_s, a resistor R_{ss} and a time constant τ:

$$C_s = q^2 N_{ss} f_o (1 - f_o)/kT \tag{21}$$

$$\tau = \frac{f_o}{c_n n_{so}} = \frac{f_o}{c_n n_o} \exp \frac{-q\phi_s}{kT} = C_s R_{ss} \tag{22}$$

This series RC-network should be located in parallel with the depletion capacitance C_D. In practice, however, a single level surface state distribution seldom occurs but a continuous distribution of interface states across the energy gap must be considered. The density N_{ss} of interface states is now expressed per cm^2 eV. Integration of (20) over the bandgap yields:

$$Y_{ss} = j\omega \frac{q^2}{kT} \int_{E_V}^{E_c} \frac{N_{ss} f_o (1-f_o) \, dE}{1+j\omega f_o/c_n n_{so}} \tag{23}$$

If the density N_{ss} and the capture cross section σ_n are only slightly dependent on E_T, the integral can be solved analytically and is:

$$Y_{SS} = \frac{qN_{SS}}{2\tau_m} \ln(1+\omega^2\tau_m^2) + jq\frac{N_{SS}}{\tau_m} \arctan(\omega\tau m) \qquad (24)$$

where

$$\tau_m = \frac{1}{c_n n_{SO}} = \frac{1}{v\sigma_n n_{SO}} \qquad (25)$$

The equivalent parallel conductance G_p caused by a continuous distribution of interface states is given by:

$$\frac{G_p}{\omega} = \frac{qN_{SS}}{2\omega\tau_m} \ln(1+\omega^2\tau_m^2) \qquad (26)$$

The surface state capacitance C_p, and given by the last part of equation (26), and conductance G_p are located in parallel with C_{Si} (Fig. 4(a)).

Experimental G_p/ω -ω curves are smeared out over a much broader frequency range than predicted by (26). They cannot be described by a single time constant τ_m but by a dispersion of time constants which has been attributed by Nicollian-Goetzberger to statistical fluctuations of surface potential. This will be discussed in next section.

IV. FLUCTUATIONS OF SURFACE POTENTIAL

To understand the experimental G_p/ω -ω behavior, the theory of the nonuniform surface potential has to be considered. This nonuniformity is mainly caused by statistical fluctuations of interface charges, either fixed oxide or surface state charges. Two models for these interface charge inhomogeneities will be described and their relative importance will be discussed.

A. Patchwork Model

In their original theoretical analysis [2], Nicollian and Goetzberger introduce the patchwork model, also referred to as "quasi-uniform" model. The MOS structure with interface charge inhomogeneities is approximated by an array of equal-area patches with characteristic area α. Each elementary device of the array has a uniformly distributed interface charge and hence uniform surface potential, and satisfies the one-dimensional Poisson equation for a uniform interface charge distribution.

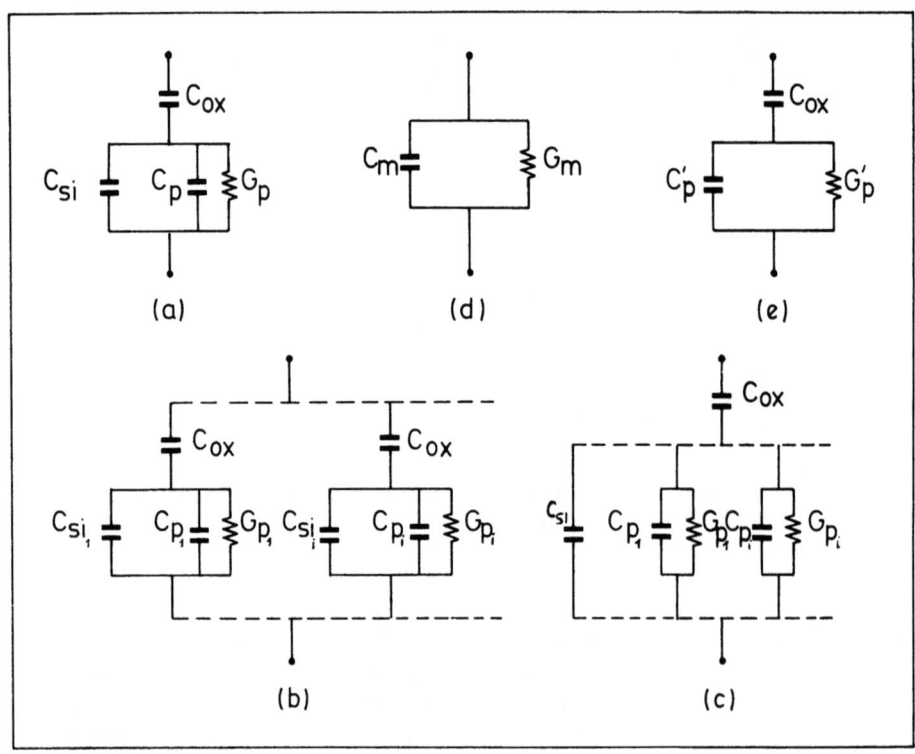

Fig. 4. Equivalent circuit of a MOS capacitor with a continuum of surface states. (a) Equivalent circuit of a microcapacitor with homogeneous interface charge. (b) Equivalent circuit of the total capacitor when short-wavelength interface charge fluctuations are absent. (c) Equivalent circuit of the total capacitor, valid for small-amplitude charge fluctuations of arbitrary wavelength. (d) Transformation of equivalent circuit a- to a pure parallel circuit. (e) Transformation to a series-parallel circuit with separated oxide capacitance. (Figure taken from Ref. 39)

The number N of interface charges ($N = Q_{ox}\, \alpha/q$) in a characteristic area is Poisson distributed over the patches. When the mean number $\bar{N}$ is large, the Poisson distribution can be approximated by a Gaussian distribution with average $\bar{N}$ and variance also equal to $\bar{N}$. The probability for having an oxide charge Q_{ox} in an elementary area is

$$P(Q_{ox}) = \frac{1}{(2\pi\sigma_q^2)^{1/2}} \exp[-(Q_{ox} - \bar{Q}_{ox})^2/2\sigma_q^2] \tag{27}$$

SURFACE STATES AT THE Si–SiO$_2$ INTERFACE

The standard deviation σ_q is expressed in C cm^{-2} and is equal to

$$\sigma_q = \left(\frac{q\bar{Q}_{ox}}{\alpha}\right)^{1/2} \tag{28}$$

For small deviations of the interface charge from the average charge, the resulting potential fluctuations are assumed to be linearly related to these charge fluctuations. As a result of this linearization, the surface potential also becomes Gaussian distributed, with mean value $\bar{U}_S$ and standard deviation σ_s, (both normalized in kT/q-units). The relation between σ_s and σ_q, after including the damping effect of the surface states [17,18] through their low frequency capacitance $C_{SS} = N_{SS}$ is given by

$$\sigma_s = \frac{q}{kT} \frac{\sigma_q}{C_{ox} + C_{Si} + C_{SS}} \tag{29}$$

A rigourous treatment of the patchwork model leads to the equivalent circuit of Fig. 4(b), which is valid in depletion and weak inversion. Every micro-capacitor is described by the series-parallel circuit as discussed earlier (Fig. 4(a)).

However, for simplicity of numerical calculations this equivalent circuit has been transformed by Nicollian-Goetzberger into the equivalent circuit of Fig. 4(c). The oxide capacitances are all lumped together and arranged in series with the distributed surface state admittance. The oddity of this equivalent circuit is that it connects all the parallel surface state branches together to a common equipotential although the model presupposes that surface potentials vary from one surface state branch to another. However, the surface state branches still differ in their d.c. quiescent values of surface potential.

Equation (26) is replaced by:

$$\frac{G_p}{\omega} = \frac{1}{(2\pi\sigma_s^2)^{1/2}} \int_{-\infty}^{\infty} \exp\left[-(U_s-\bar{U}_s)^2/2\sigma_s^2\right]$$

$$\cdot \frac{qN_{SS}}{2\omega\tau_m} \ln(1+\omega^2\tau_m^2) \, dU_s \tag{30}$$

Deuling et. al. [17] showed, within the linearization approximation, that both the equivalent circuit of Fig. 4(c) and the more direct patchwork-circuit of Fig. 4(b) are in close agreement for small fluctuations and low surface state densities.

B. The Random Model

Brews [18] rejects the patchwork model because of edge effects and interactions of the patches, due to short-wavelength charge variations. Instead he proposes a small-fluctuation mathematical model for a random distribution of the interface charge. Brews does not break up the MOS capacitor into discrete elementary capacitors, but solves the three-dimensional Poisson equation for the entire device. He finds the variance of the surface potential dependent on the minimal wavelength λ of the charge variations:

$$\sigma_s^2 = \{\frac{q}{kT(\varepsilon_{Si}+\varepsilon_{ox})}\}^2 \frac{q\bar{Q}_{ox}}{4\pi} \ln\{1+(\frac{\varepsilon_{Si}+\varepsilon_{ox}}{C_{ox}+C_{Si}+C_{SS}})\}\frac{1}{\lambda^2} \quad (31)$$

λ may also be thought of as the correlation length of the charge distribution. In this expression, $\bar{Q}_{ox}$ is the mean interface charge per unit area.

For large wavelengths λ satisfying the condition

$$\frac{\varepsilon_{Si}+\varepsilon_{ox}}{C_{ox}+C_{Si}+C_{SS}} < \lambda \quad (32)$$

equation (31) reduces to

$$\sigma_s^2 = \frac{q\bar{Q}_{ox}}{4\pi\lambda^2}(\frac{q}{kT})^2\frac{1}{(C_{ox}+C_{Si}+C_{SS})^2} \quad (33)$$

which is seen to be similar to the original Nicollian-Goetzberger result for the patchwork model (29) if the characteristic area is chosen as:

$$\alpha = 4\pi\lambda^2 \quad (34)$$

For thin oxides, heavy doping or large surface state densities, condition (32) can be satisfied and the random model is equivalent to the patchwork model. If these assumptions are not fulfilled, the patchwork model is only valid if the interface charge inhomogeneities have long wavelengths. In his random distribution model Brews [18] assumes $5 < \lambda < 100$ Å and if some realistic numerical values are used in (32) ($C_{ox} = 2.5 \times 10^{-8}$ F cm^{-2}, $C_{Si} = 3.0 \times 10^{-8}$ F cm^{-2} and $C_{SS} = 3.0 \times 10^{-9}$ F cm^{-2}) this leads to $\lambda > 2410$ Å for the patchwork model to be valid.

It is very important to point out here that, based on his mathematical analysis, Brews [19], proves that Fig. 4(c) gives the correct equivalent circuit for a random interface charge distribution for small fluctuations in the depletion and weak inversion range of biases. He also indicates that Fig. 4(b) is indeed adequate for a patchwork distribution if the patch size exceeds the depletion width. It should be noticed here that Nicollian et. al., by starting from a patchwork distribution, and making some approximations (linearization, lumping of the oxide capacitances) accidentally come to the right circuit for a random charge distribution.

Several attempts to determine whether the random model or the patchwork model prevails can be found in literature [20,21,22,23], but not enough data are available to date to give a clear answer to the question how the interface charge is distributed. It can also be expected that this distribution is considerably influenced by the oxide preparation technique.

As a conclusion, one should keep in mind that the surface potential of a MOS structure is not uniform but statistically distributed. This will influence the interpretation of several experimental techniques discussed in next section.

V. EXPERIMENTAL TECHNIQUES USING MOS CAPACITORS

A. High-Frequency C-V Method

The method of Terman [24] compares an experimental C-V curve with a theoretical one calculated for the same oxide thickness and substrate doping. The frequency of the a.c.-signal superimposed on the d.c.-bias is supposed to be sufficiently high so that surface states cannot follow the a.c.-test signal. However they will still contribute to an interface charge variation when their occupancy is changed due to a variation of bias voltage. Although this technique is certainly useful for the detection of large surface state densities (high 10^{10} cm^{-2}eV^{-1} and higher) which show up as gross irregularities of the measured C-V curve, the method suffers from the following disadvantages making it useless for lower densities (low 10^{10} cm^{-2} ev^{-1} and 10^{9} cm^{-2}eV^{-1}).

1. The redistribution of impurity atoms at the surface during high temperature treatments must be taken into account by looking for the exact value of the doping concentration at the surface. This problem was extensively investigated by Baccarani and Severi [25] and by Margalit et. al. [26].

2. A severe limitation of the high-frequency C-V method is the need for numerical or graphical differentiation in order to obtain the N_{SS}-ϕ_S curve. The voltage shift, ΔV, between the theoretical and experimental C-V plots is measured as a function of

surface potential ϕ_s (known from their high-frequency capacitance value) and N_{SS} is found from

$$N_{SS} = \frac{C_{ox}}{q} \frac{\partial \Delta V}{\partial \phi_s} \tag{35}$$

3. Experimental C-V curves are slightly distorted by surface potential fluctuations. It has been shown by Castagne and Vapaille [27] and by Declerck et. al. [28] that these statistical fluctuations cause fictitious surface state peaks close to flat band. The magnitude of these fictitious peaks are in the 10^{10} cm^{-2} eV^{-1} as illustrated in Figure 5.

4. The condition that the surface states cannot follow the a.c.-signal is not always fulfilled for surface states lying around flat band when a standard measuring frequency of 1 MHz is used. This can also lead to some fictitious structure in the N_{SS}-ϕ_s curve obtained by means of this method [28].

B. Low-Frequency C-V Method

The basic theory for the low frequency C-V technique was developed by Berglund in 1966 [29]. Because of the extremely low generation and recombination rates obtained by modern technologies, a.c. test frequencies much lower than 1 Hz are needed to obtain a low frequency C-V plot. As a more practical alternative technique, Castagne [30], Kerr [31] and Kuhn [32] introduced the quasistatic or slow-ramp method, illustrated in Fig. 6. A slow triangular voltage sweep is applied to the MOS capacitor and the displacement current is measured by means of a high input impedance opamp or electrometer. The voltage output is linearly related to the differential capacitance of the MOS-structure

$$V_{out} = - R \, C(V) \, \frac{dv}{dt} = -\alpha RC \tag{36}$$

The rise and fall time of the voltage sweep must be slow enough so that both the inversion layer charges and the charges trapped in the surface states always correspond to thermal equilibrium. For present state of the art devices, typical sweep rates are between 10 and 100 mV/sec. Distortions of the obtained C-V plots due to non-equilibrium effects have been discused by Kuhn and Nicollian [33].

The relation between ϕ_s and V_G is found by expressing equation (9) in a differential form:

$$dV_G = dV_{ox} + d\phi_s \tag{37}$$

SURFACE STATES AT THE Si–SiO$_2$ INTERFACE

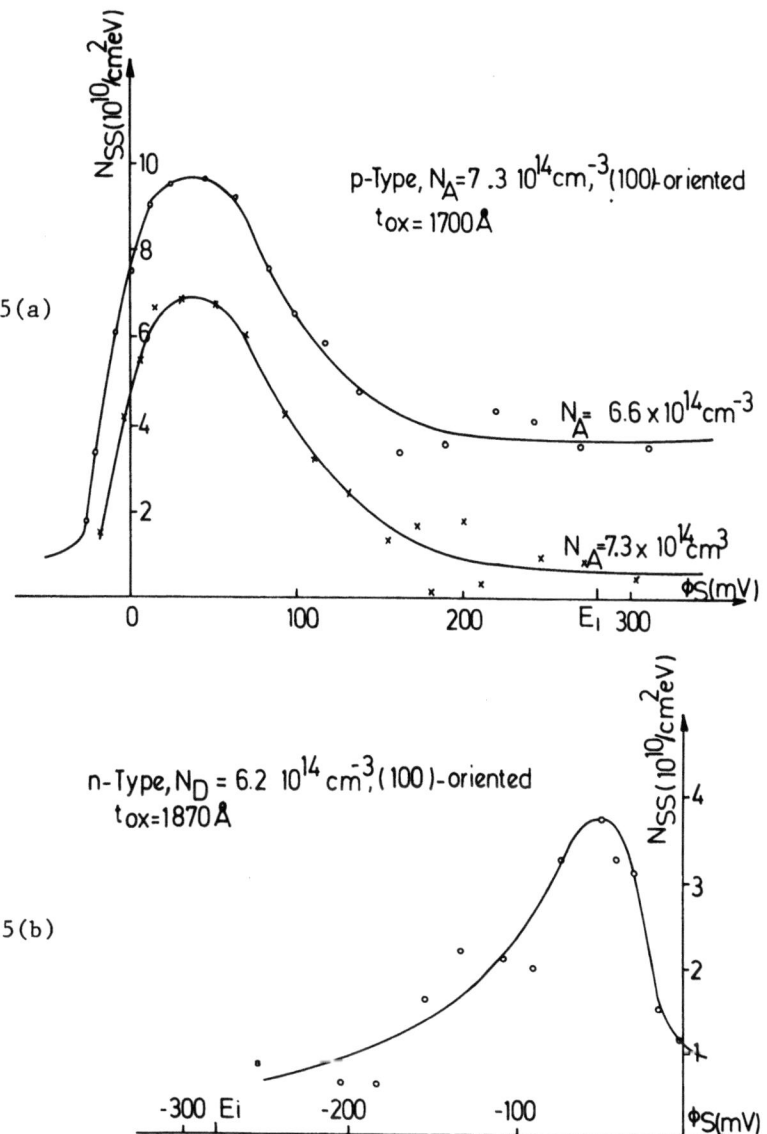

Fig. 5. (a) Experimental density of fictitious surface states N_{SS} vs. surface potential ϕ_s, obtained by the high frequency C-V method on a p-type substrate. The influence of substrate doping is also shown.

(b) n-type substrate.

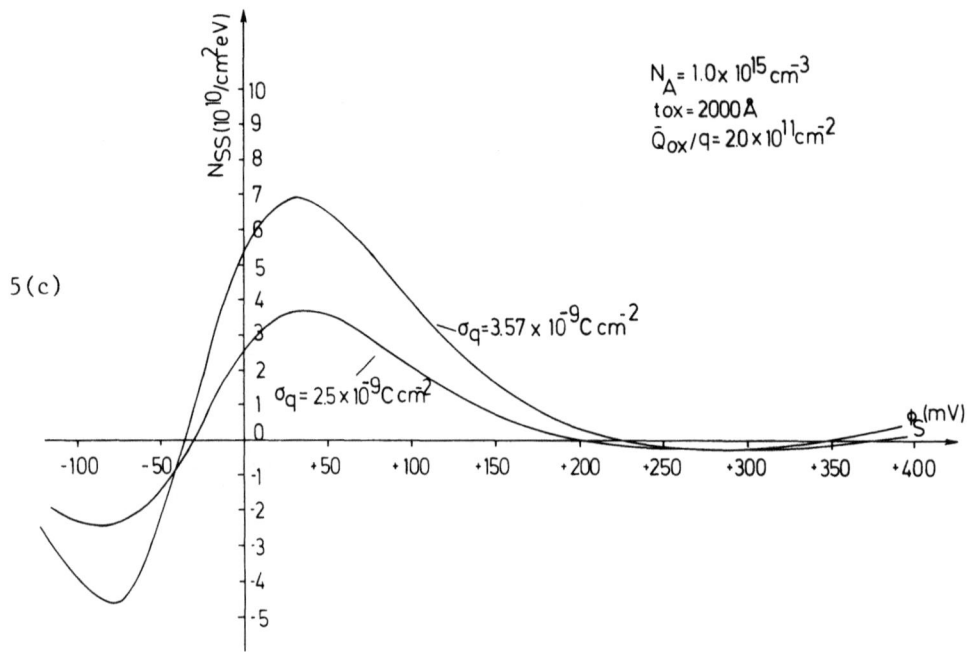

Fig. 5. (c) Fictitious N_{SS} calculated from a simulated C-V plot with τ_q = 0.25 x 10^{-8}C cm^{-2} and 0.375 x 10^{-8} C cm^{-2} respectively.

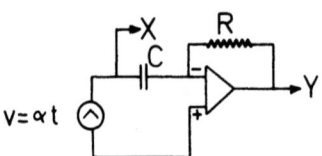

Fig. 6. Schematic set-up for quasi-static C-V measurements.

Reordering of (37) and using

$$dV_{ox} = \frac{dQ_M}{C_{ox}}$$

and $dQ_M = C_{LF} \, dV_G$ with Q_M the charge on the metal plate, leads to the so-called Berglund integral:

$$\phi(V_G) = \int_{V_{G1}}^{V_G} (1 - \frac{C_{LF}}{C_{ox}}) \, dV_G + \Delta_B \tag{38}$$

The integration constant Δ_B is given by the surface potential at V_{G1}. As explained by Kuhn [32], Δ_B is obtained by fitting theoretical and experimental $C_{LF} - \phi_s$ curves in accumulation and strong inversion.

Another technique to find the integration constant Δ_B is based on equation (18) giving the depletion capacitance C_D as a function of surface potential ϕ_s. This can be used when a high frequency C-V curve is available. In this case, the intercept of the $C_D^{-2} - \phi_s$ plot with the ϕ_s - axis yields the integration constant according to:

$$C_D^{-2} = \frac{2}{qN_A \varepsilon_{Si}} (\phi_s - \frac{kT}{q}) \tag{39}$$

The surface state capacitance C_{SS}, which appears in parallel with the semiconductor capacitance C_{Si}, and the surface state density N_{SS} are derived from

$$qN_{SS} = C_{SS} = \frac{C_{LF}}{1 - \frac{C_{LF}}{C_{ox}}} - C_{Si} \tag{40}$$

C_{Si} can be calculated from theory or can be obtained in depletion and accumulation from an experimental high-frequency C-V plot since in these ranges, inversion charge response may be neglected. The ultimate accuracy obtainable with the quasi-static technique is about 1×10^{10} cm^{-2} eV^{-1}, depending on device parameters, and is limited by several factors such as

(a) The determination of the integration constant Δ_B which is critical. A small error of only a few mV will greatly influence the N_{SS}-values in accumulation and strong inversion (Figure 7).

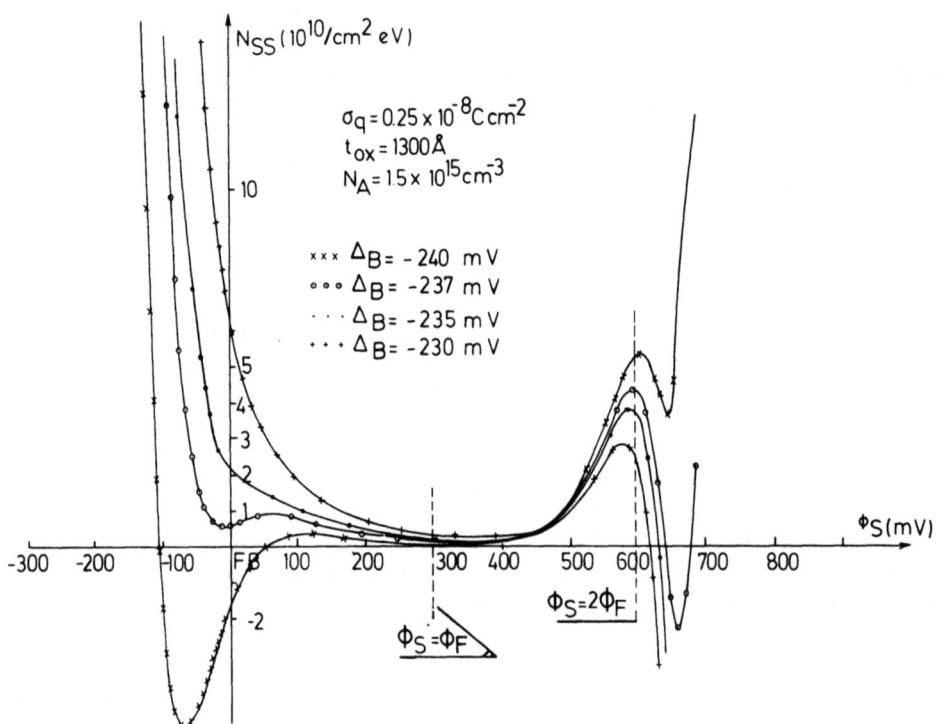

Fig. 7. Fictitious density of surface states calculated from a low frequency C-V plot with $\tau_q = 0.25 \; 10^{-8}$ C cm^{-2}. Δ_B is the Berglund integration constant. (The Figure is taken from Ref. 28.)

(b) Small experimental errors in C_{ox}, C_{LF} and C_{HF}, substrate doping and oxide thickness contribute significantly to erroneous N_{SS}-curves.

(c) The non-uniform surface potential which also influences the technique and results in a fictitious surface state peak at $\phi_s = 2\phi_F$ when experimental low frequency plots are compared to calculated ideal low frequency curves (Fig. 7). This can be partially eliminated by comparing the experimental low frequency plot with an experimental high frequency plot as surface potential fluctuations will influence both curves in approximately the same manner.

It can be concluded that application of the quasi-static C-V technique in the low 10^{10} cm^{-2} eV^{-1} N_{SS} range requires an accurate determination of all measured parameters and a careful interpretation of the obtained N_{SS} results. The technique however is extremely useful for a quick evaluation of test capacitors. For this purpose a graphical method has been developed which is based on the measurement of the minimum of the low-frequency C-V plot [34]. The N_{SS} value for this particular applied bias can then be determined from graphs when oxide thickness and substrate doping are known (Fig. 8).

C. Gray-Brown Technique

In this technique, which was introduced by Gray and Brown [35] in 1966, high-frequency C-V curves are plotted at various temperatures and the flatband shift is measured as a function of temperature. The surface state density can be obtained from the temperature behaviour of the flatband shift since at lower temperatures, the Fermi level moves towards the majority carrier band edge and the interface charge changes due to the varying occupancy of the surface states. Using this technique at a measuring frequency of 150 kHz, Gray and Brown found the well known peaks in surface state distribution near the valence and the conduction band edge.

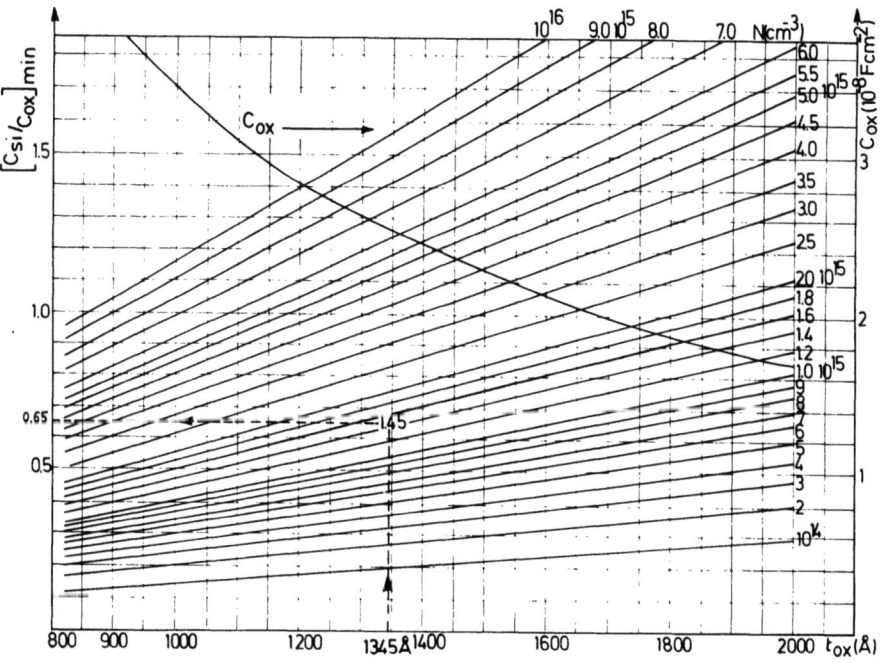

Fig. 8 (a) $(C_{Si}/C_{ox})_{min}$ is found as a function of measured N and t_{ox} values. The oxide capacitance C_{ox} is also presented.

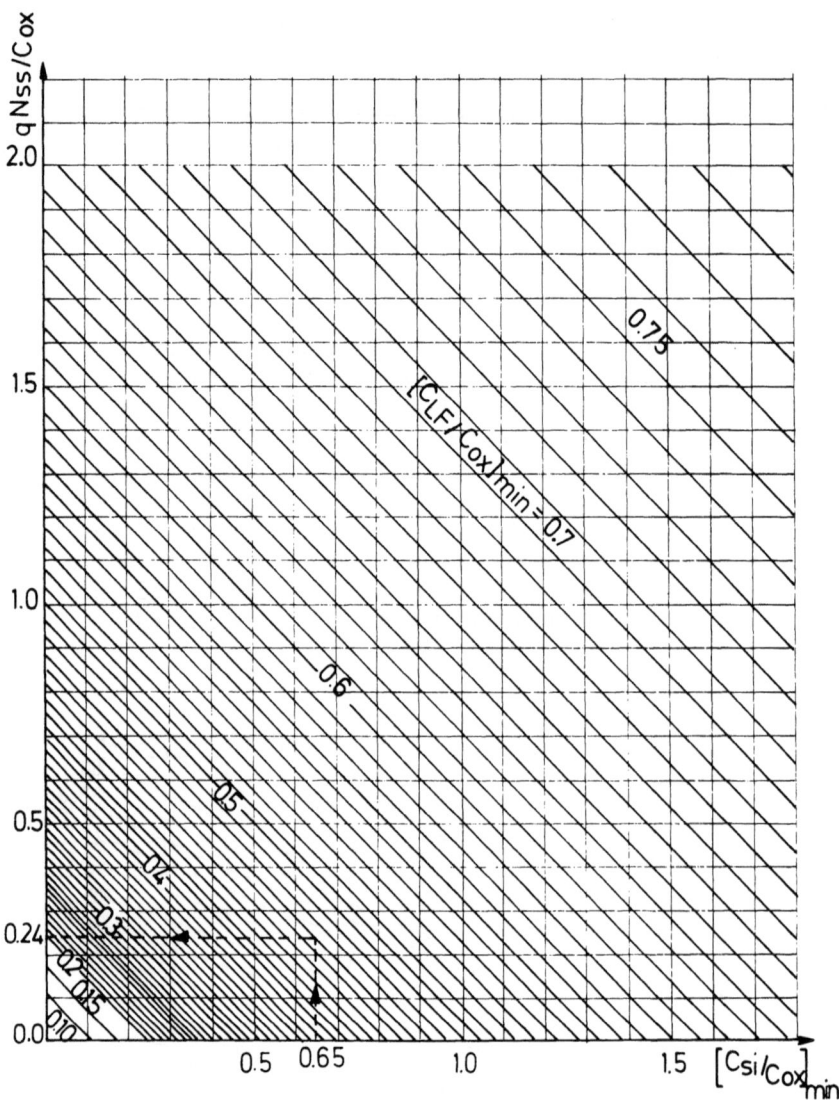

Fig. 8. (b) Determination of qN_{SS}/C_{ox} as a function of experimental $(C_{LF}/C_{ox})_{min}$ and of the $(C_{Si}/C_{ox})_{min}$ found in Figure 8(a). (These figures are taken from Ref. 34.)

e.g. $t_{ox} = 1345$ Å $\qquad N = 1.45 \times 10^{15} cm^{-3}$

Fig. 8(a) gives $(\frac{C_{Si}}{C_{ox}})_{min} = 0.65 \qquad C_{ox} = 2.52 \times 10^{-8} F\ cm^{-2}$

$(\frac{C_{LF}}{C_{ox}})_{min}$

Fig. 8(b) we find $\dfrac{qN_{SS}}{C_{ox}} = 0.24$

or $N_{SS} = 3.8 \times 10^{10}$ cm^{-2} eV^{-1}

In 1973 Boudry [36] made computer simulations using similar experimental parameters and taking into account a decrease of the capture cross section near the band edge. He showed that fictitious peaks are indeed found in the same energy range when the a.c. test frequency is too low. According to his results, frequencies near 200 MHz must be used to maintain high frequency conditions near the band edge. It is clear that the Gray-Brown results are erroneous and that all results obtained with this technique are questionable. There is also uncertainty about the temperature dependence of other parameters such as work function and electron affinity.

D. Conductance Technique

The MOS a.c. conductance technique proposed by Nicollian-Goetzberger in 1967, is still considered as one of the most complete and most reliable techniques to characterize surface state properties. This is due to the fact that MOS-capacitor losses, measured by means of an impedance bridge, are directly related to surface state trapping phenomena, at least in the depletion region of bias. On the other hand, in the C-V techniques the surface state effects show up as relatively small distortions of the ideal C-V curves. The complete conductance technique yields information about N_{SS}, σ_p or σ_n and the standard deviation of the surface potential fluctuations. Cooper and Schwartz [16] have recently extended the conductance technique to the weak inversion region by taking into account the interactions with both majority and minority carriers. It should also be noted that the experimentally observed broadening of the $G_p/\omega-\omega$ plots has been attributed by Preier [1] to a tunneling mechanism into oxide traps concentrated within about 10 Å from the interface. Eaton-Sah [37] and Ushirokawa et al. [38] explained their experimental $G_p/\omega-\omega$ plots by a combined effect of interface charge fluctuations and tunneling into traps located in the oxide at a small distance from the interface.

In principle the conductance technique consists of the measurement of the parallel capacitance C_m and the parallel conductance G_m of the equivalent circuit of Fig. 4(d) versus bias voltage at different frequencies. The measured capacitance and conductance values are then converted into the equivalent circuit for the patchwork model (Fig. 4(b)) or for the random model (Fig. 4(c) and 4(e)). If necessary a series resistance R_S, due to the finite conductance of the bulk material and to the contact resistances, should be taken into account. For a discussion about the origin and the validity of these models we refer to section IV. It has been proven however by Muls et. al. [39] that for realistic σ_s-values both circuits lead to the same estimate for N_{SS} and σ_s, whereas the capture cross section strongly depends upon the choice of the interface charge distribution model. For ease of calculations, the circuit of Fig. 4(c) is preferred. It is noticed again that this circuit corresponds to a random charge distribution with small amplitude fluctuations but that Nicollian and Goetzberger came to the same equivalent circuit starting from their patchwork model and lumping the oxide capacitances together.

ϕ_S can be determined by means of a Berglund integration of the low-frequency C-V curve [38] or by using the experimental value for the depletion capacitance C_D and the relation $C_D(\phi_S)$ [39].

Several techniques have been described in the literature for analyzing the experimental $G_p/\omega-\omega$ curves and will be briefly reviewed.

A. Direct Fitting Technique

A direct method can be based on fitting an experimental $G_p/\omega-\omega$ curve with a theoretical one calculated with formula (30). Three parameters have to be determined: N_{SS}, σ_s and σ_n for n-type wafers or σ_p for p-type substrates.

An example illustrating the accuracy of this technique is given in Fig. 9.

B. Computer Technique

The direct fitting technique is probably the most accurate but is very time consuming and not practical for evaluation of a large number of MOS-devices. Therefore computer programs have been developed in order to facilitate the interpretation of the measured C-V and G-V data. For a discussion of these computer programs we refer to the literature [28,40]. These techniques allow the determination of N_{SS} and σ_n or σ_p over a surface potential range of a few hundred mV between flatband and midgap. Its disadvantage is that C-V and G-V curves have to be measured at different test frequencies.

C. Quick Evaluation Procedure

Based on the analytical properties of the $G_p/\omega-\omega$ relation, graphs have been calculated by several authors [7,40,41] enabling a

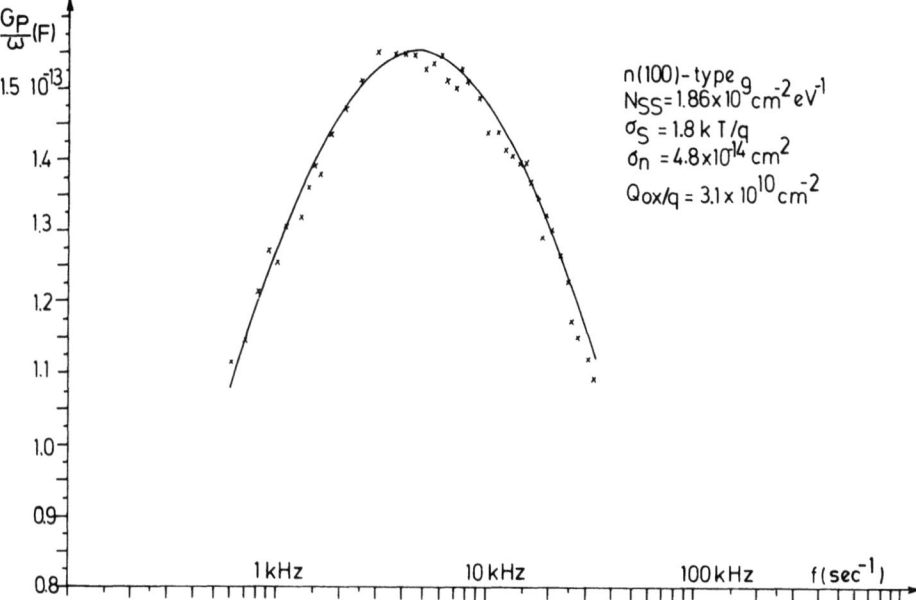

Fig. 9. (a) G_p/ω vs. ω for a MOS-structure on a n-(100)-type substrate with $N_D = 5.25 \times 10^{15}$ cm^{-3} and $Q_{ox} = 1295$ Å, for $V_G = -1.65$ V, $\phi_s = -313$ mV. The direct Nicollian-Goetzberger fitting of the experimental curve with the theoretical one yields $N_{SS} = 1.86 \times 10^9$ cm^{-2} eV^{-1}, $\sigma_s = 1.8$ kT/q, $\sigma_n \sim 4.8 \times 10^{-14}$ cm^2. The area of the MOS-structure is 2.07×10^{-3} cm.

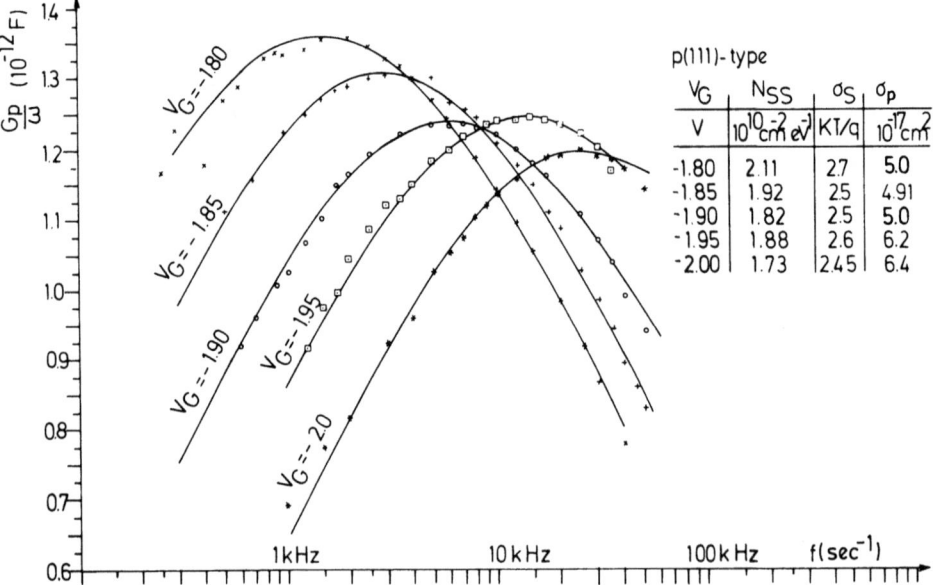

Fig. 9. (b) G_p/ω vs. ω for a MOS-device on a p-(111)-type substrate with $N_A = 2.24 \times 10^{15}$ cm^{-3} and $Q_{ox} = 1360$ Å, for $V_G = -1.80, -1.85, -1.90, -1.95$ and -2.00, after a negative bias-temperature stress of 10 min at 150° C with $V_G = -26$ V. The area of the MOS-structure is 2.07×10^{-3} cm^2. (Figures are taken from Ref. 28)

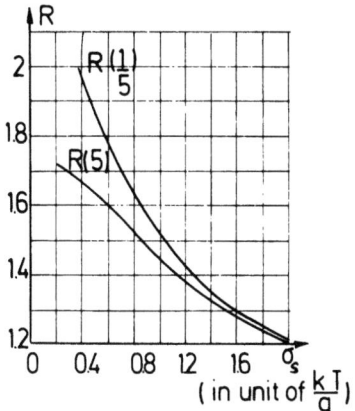

Fig. 10. R(1/5) is the ratio of the parallel conductance G_p/ω for $\omega = \omega_{max}$ and $= \omega_{max}/5$, plotted vs. standard deviation σ_s in units of kT/q; R(5) is the same ratio for $\omega = \omega_{max}$ and $\omega = 5\omega_{max}$, vs. σ.

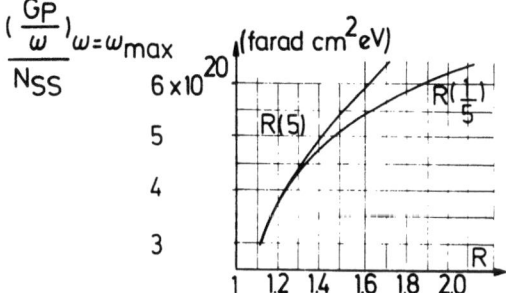

Fig. 11. The ratio $(G_p/\omega)/N_{SS}$ for $\omega = \omega_{max}$ is plotted vs. R(1/5) and R(5). (Figures 10 and 11 are taken from Reference 41.)

quick evaluation of experimental $G_p/\omega - \omega$ data. After converting the measured C_m and G_m-values to the parameters of Fig. 4(e), the following procedure is normally taken:

(a) The standard deviation σ_s is obtained from a first graph by measuring the ratio R of the maximum G_p/ω at ω_{max} to the G_p/ω at angular frequencies which are respectively equal to $\omega_{max}/5$ or $5\ \omega_{max}$ (Fig. 10).

(b) A second graph allows the direct evaluation of N_{SS} from the measured maximum G_p/ω when the above mentioned R-values are known (Fig. 11).

VI. EXPERIMENTAL TECHNIQUES USING MOS TRANSISTORS

A. Charge Pumping Technique

This technique, which is suitable for production control on a large number of devices, was introduced by Brugler and Jespers in 1969 [42] and examined in greater depth by Elliot in 1976 [43]. The basic experiment is illustrated in Figure 12. The source and drain of a p-channel MOS transistor are reverse biased with respect to the substrate. When a sufficiently large negative pulse is applied to the gate of the device, the surface becomes deeply depleted and holes will flow from the source and the drain regions into the channel as soon as the threshold voltage V_T is reached. The inversion layer will be established very quickly, however some holes will be captured by surface states as the surface state capture time constant is very small due to the high density of minority carriers in the inversion layer. The surface state occupancy

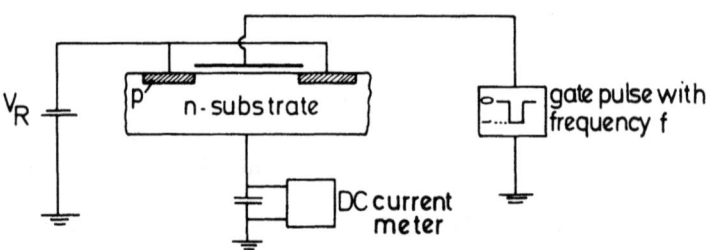

Fig. 12. Basic set-up for charge pumping measurements.

is now governed by the quasi-Fermi level of the minority carriers. When the gate pulse is turned off the mobile charge moves back to the source and drain regions under the influence of the reverse bias voltage but some charge remains in the surface states as the emission rate to the valence band is much slower and given by the emission time constant τ_e

$$\tau_e = \frac{1}{\sigma_p v N_v} \exp[(E_T - E_V)/kT] \tag{41}$$

where N_v is the number of states in the valence band, E_T the trap level and E_V the valence band edge. A similar expression can be written for the interaction of electrons with the conduction band when a n-channel device is used (eq. 44).

As soon as the gate pulse is turned off, the holes trapped by the surface states will recombine with electrons from the n-type substrate which are now amply present at the surface. This recombination gives rise to a net flow of positive charge to the substrate.

A repetitive application of negative gate pulses results in a "pumped current" of holes from the source and drain diodes to the substrate which is in the opposite direction of the diode leackage current after subtraction of the leakage current is given by the equation:

$$I_b = A_G f\, (qN_{SS}\,|\Delta\phi_S| + \alpha C_{ox}\,|V_G - V_T|) \tag{42}$$

A_G is the area of the transistor channel, f is the frequency of the gate pulse and $\Delta\phi_S$ gives the change of surface state occupancy under application of the gate pulse. The second term in the expression is called the "geometrical component" and accounts for a fractional charge loss after switching off the clock pulse which is not due to surface state effects. This extra loss occurs when the inversion charge can not be removed quickly enough e.g., when channel length is too large or fall time of the pulses too small. Under appropriate geometrical and electrical conditions it can be eliminated so that the pumped charge will only be caused by surface states trapping phenomena. Typical charge pumping characteristics are shown in Figure 13.

Elliot extended the charge pumping technique and proved its usefulness as a quick evaluation tool giving the integrated number of surface states $N_{SS}\Delta\phi_S$ on standard MOS transistors. It is important to notice that the technique remains applicable for small geometry transistors (10 x 10 μm) where C-V or G-V techniques fail because of experimental inaccuracies. This charge pumping technique is also related to N_{SS} measurements from transfer loss studies on charge coupled devices (Section VII).

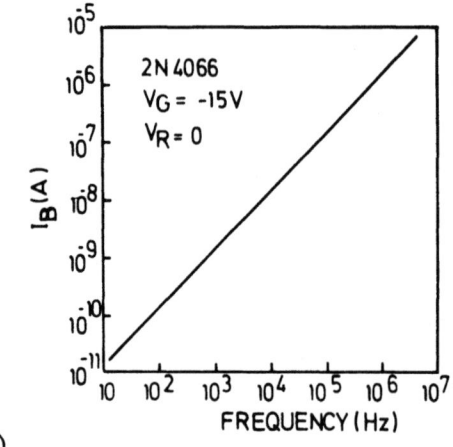

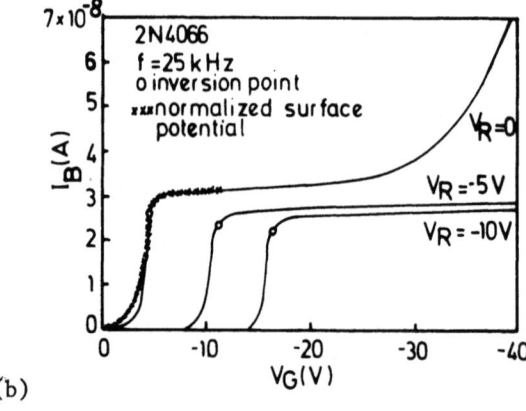

Fig. 13. Typical charge pumping characteristics (from Brugler and Jespers, Ref. 42) (a) d.c. substrate current versus gate pulse frequency.
(b) d.c. substrate current versus gate voltage.

B. Deep Level Transient Spectroscopy Technique (DLTS-Technique)

Surface state properties can also be studied by measuring capacitance transients resulting from electrons and hole emission from these states to the conduction and the valence bands. The fundamental theory for this capacitance transient technique has been developed by Sah et. al. [44] and by Lang [45,46] who introduced the DLTS-technique as a valuable tool for analyzing deep traps in bulk semiconductors using p-n junction diodes. Later, Wang et. al, [47,48] adapted the technique for characterization of semiconductor surfaces by means of MOS capacitors or MOS transistors. When the latter is used, more information can be gathered from the experiments.

The experimental set-up is given in Figure 14. The source and drain of the p-channel MOS transistors are reverse biased and the gate is held at a negative voltage causing a deep depletion situation at the Si-SiO$_2$ interface. The steady state capacitance is measured by means of a capacitance bridge. The band bending at the silicon surface is depicted in Figure 15(a) showing that the surface states are not occupied by electrons due to the position of the quasi-Fermi level of electrons. It should be noticed that the exact position of the quasi-Fermi level of holes for this "gate controlled diode structure" is determined by the flow of the charge generated at the surface and in the depletion layer as demonstrated by Pierret [49].

A positive pulse will now be applied to the gate which leads to accumulation of the surface with electrons as shown in Fig. 15(b) When the gate pulse is removed the surface condition returns to deep depletion but the electrons captured at surface or bulk traps can only be released at a finite rate. Indeed they will be emitted to the conduction band according to the Hall-Schockley-Read process [50,51]. The instantaneous width of the depletion layer will be larger than the steady state condition to compensate for the excess negative charge of the trapped electrons. This modulation of depletion width results in the capacitance transient observed, the capacitance initially being smaller than the steady state value but tending towards it with a time constant determined by the release of trapped charges. If the emission process is dominated by one single level at energy level E_T, having an emission time constant τ_e, the capacitance transient can be written as:

$$\Delta C(t) = \Delta C_i \exp[-t/\tau_e] \tag{43}$$

ΔC_i is the initial value of the transient capacitance which is related to the number of traps. The emission constant τ_e is given by

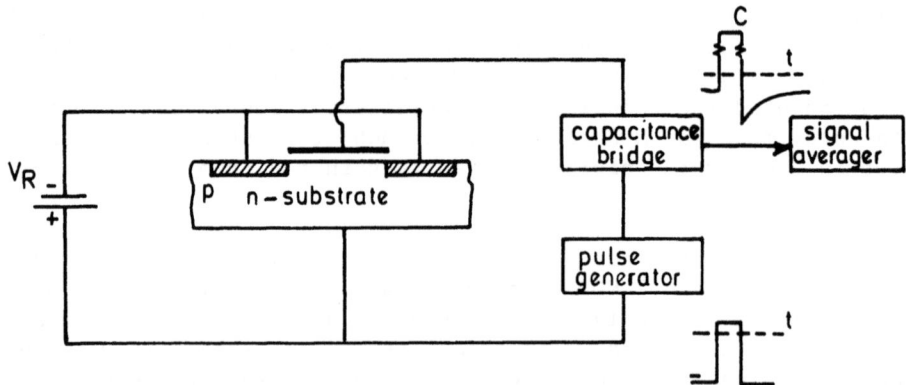

Fig. 14. Experimental set-up for DLSTS-technique (See also Reference 47).

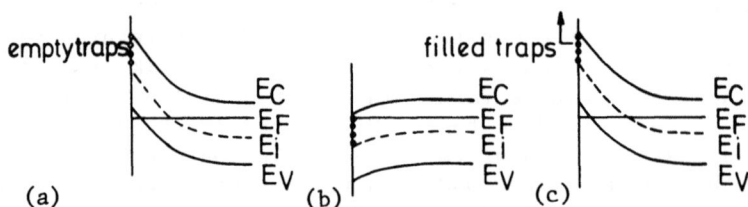

Fig. 15. Schematic representation of DLTS-technique.
(a) Deep depletion after reaching steady state
(b) Accumulation: the traps are filled
(c) Deep depletion: the traps are emitting the electrons to the condition band.

$$\tau_e = \frac{1}{\sigma_n V N_C} \exp\left[\frac{E_C - E_T}{kT}\right] \qquad (44)$$

where N_C is the density of states and E_C the edge of the conduction band.

Because of the strong temperature dependence of the emission time constant τ_e, the characteristic capacitance transient from different energy levels may be detected by measuring the transient as a function of temperature. A two channel signal averager is normally used for enhanced accuracy. For a detailed discussion of this experimental technique we refer to the papers by Lang [45], Wang [47] and to the recent work by Johnson et. al. [52].

It should be emphasized again that the DLTS-technique has only recently been applied to the study of the MOS strucutres. Its sensitivity is proven to be at least as good as for the conductance technique [48]; it offers the possibility of studying capture cross sections and it covers a wide energy range. However the results can be disturbed by bulk trapping phenomena making the interpretation somewhat difficult and the experimental set up needs a fast response, high frequency capacitance bridge and a high quality signal averager to detect the small transient capacitances.

C. Determination of N_{SS} From 1/f-Noise Experiments

It is widely accepted that low frequency 1/f-noise or flicker noise in MOS transistors is related to the presence of surface states at the Si-SiO$_2$ interface. Trapping of free carriers in states located at the interface or at a small distance from the interface (2-10 Å) (in which case trapping occurs through a tunneling mechanism), is believed to be responsible for the broad spectrum of flicker noise observed in MOS transistors [53,54]. The dependence of the equivalent input noise power on geometrical parameters of the device such as gate area and oxide thickness, and on applied voltages or channel current is not yet fully established as different models have been developed [55]. However there is some consensus about the fact that the relation between noise power and surface state density can be expressed as:

$$V_{eq}^2 \sim \frac{N_{SS}}{A_G f} \qquad (45)$$

A_G is the gate area and f is the a.c.-test frequency. It is obvious that N_{SS} here is assumed to be constant or only slowly varying over the energy range of interest. This linear relation between V_{eq}^2 and N_{SS} was demonstrated first by Abowitz et. al. [56], after whom it was extended to lower N_{SS}-values by Klaassen [54] and Broux et. a.l. [57]. These data, corrected for geometrical factors

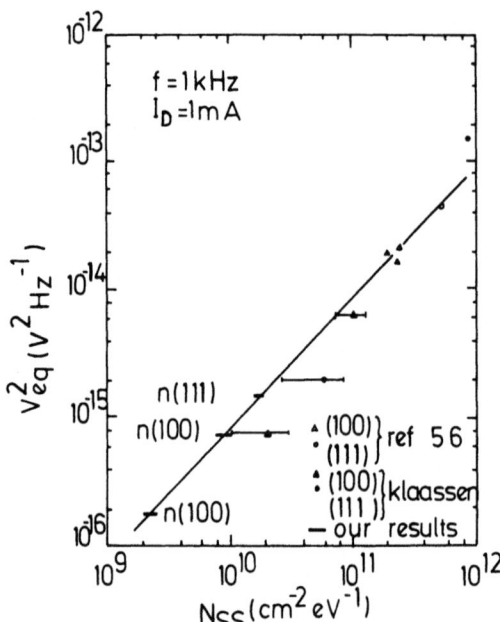

Fig. 16. Relation between MOS transistors 1/f noise and surface state density N_{SS} (figure taken from Reference 57).

and shown in Figure 16, illustrate the possibility of controlling N_{SS} over a large number of wafers by monitoring the 1/f-noise on a small area test device at a fixed drain current and test frequency. This kind of measurement is quickly performed on standard test-transistors by means of easy to operate noise measuring equipment.

D. Weak Inversion Technique

MOS transistors are normally operated in the strong inversion mode, i.e., for an n-channel device with source to substrate voltage equal to zero, ϕ_S in the channel region is larger than $2\phi_F$. However when the gate voltage is equal to or even slightly smaller than the classically defined threshold voltage, the channel current does not become zero but a considerable electron current still flows from source to drain. This is the exponential tail of the experimental I_D - V_G characteristic, illustrated in Fig. 17 on a linear and a logarithmic scale respectively.

It should be noticed from Fig. 17 that when the oxide thickness is scaled down from 1000 Å to 200 Å, the slope of the weak inversion current does not change considerably. This is one of the most

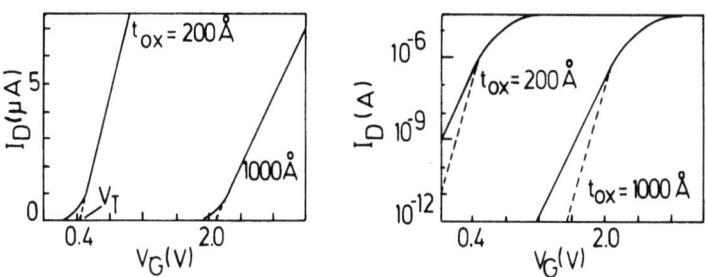

Fig. 17. Drain current versus gate voltage characteristics of a MOS transistor illustrating the weak inversion operation (Figure taken from Ref. 62).

severe limitations on the developement of small geometry devices since this weak inversion current forms an enhanced leakage current leading e.g. to a decrease in the refresh time of memory-type circuits.

The channel current in weak inversion was first calculated by Guzev et. al. [58], as a drift current due to a lateral field. Swanson and Meindl [59] used a weak inversion model for the study of low-voltage complementary MOS circuits. Stuart and Eccleston [60] explained the weak inversion channel current as a diffusion current from source to drain. Barron [61] started from the general expression for the channel current including diffusion and drift component and developed a model that adequately describes the influence of the drain and gate voltage on the weak inversion drain current. A detailed discussion on the theory and applications of the weak inversion operation is given by Muls et. al. [62].

It can be shown [62] that the weak inversion drain current is

$$I_D = \frac{W}{L} \mu C_D(\phi_{SO}) \frac{N}{M} (\frac{kT}{q}) \exp[\beta(\phi_{SO} - 2\phi_F)]$$
$$+ (\frac{V_G - V'_T}{N}) \cdot (1 - \exp[-\beta\frac{M}{N}V_D]) \quad (46)$$

For a fixed gate voltage

$$I_D = I_{DMAX} (1 - \exp[-\beta\frac{M}{N}V_D]) \quad (47)$$

with I_{DMAX} the maximal drain current at a given gate voltage when V_D is infinite in equation (46). V_D is the drain to substrate voltage; the source to substrate voltage is assumed to be zero. W and L are the channel width and length, μ is the mobility of the carriers in the channel. The influence of oxide charge fluctuations on this mobility will be discussed later. $C_D(\phi_{SO})$ is the depletion capacitance calculated for the surface potential ϕ_{SO} at the source.

$$N(\phi_{SO}) = \frac{C_{ox} + C_D(\phi_{SO}) + C_{SS}}{C_{ox}} \tag{48}$$

$$M(\phi_{SO}) = \frac{C_{ox} + C_D(\phi_{SO})}{C_{ox}} \tag{49}$$

V'_T is a new threshold voltage which is slightly different from the classical V_T [62].

It is clear from equation (47) that in weak inversion the drain current versus drain voltage characteristics is governed by an exponential law depending on C_{SS} through the exponent M/N given by:

$$\frac{M}{N} = \frac{C_{ox} + C_D(\phi_{SO})}{C_{ox} + C_D(\phi_{SO}) + C_{SS}} \tag{50}$$

The M/N value is determined by the magnitude of C_{SS} compared to $C_{ox} + C_D$. To find N_{SS} by means of the weak inversion technique it will be necessary to obtain M/N from an experimental $I_D - V_D$ plot and to calculate N_{SS} from:

$$N_{SS} = \frac{1}{q} \frac{N}{M} [C_{ox} + C_D(\phi_{SO})] (1 - \frac{M}{N}) \tag{51}$$

More insight into the physical interpretation of equations (46) to (51) is obtained by looking at Fig. 18. It shows the band bending at the surface and the occupancy of the surface states at flat band condition ($\phi_S = 0$) and in weak inversion ($\phi_F < \phi_S < 2\phi_F$). For the latter case both an equilibrium (Fig. 18(b)) and a quasi-equilibrium (Fig. 18(c)) situation are represented. This quasi-equilibrium situation is established by the application of a reverse voltage at the drain diode of the MOS transistor shifting the electron quasi-Fermi level apart from the majority Fermi level over a potential difference ϕ_c. The quasi-Fermi level difference ϕ_c is zero at the source, is very small in weak inversion throughout most of the channel and becomes equal to V_D at the drain junction.

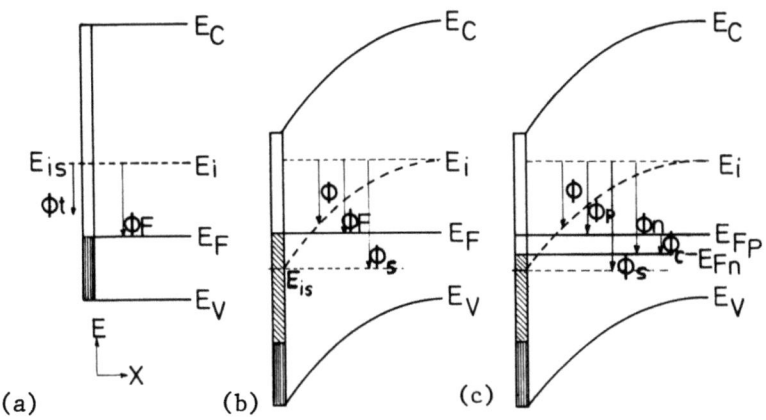

Fig. 18. Band bending and occupancy of the surface states
(a) flat band condition
(b) weak inversion, equilibrium
(c) weak inversion, quasi-equilibrium.

The occupancy of the surface states in weak inversion is governed by the quasi-Fermi level of the minority carriers if capture cross sections for holes and electrons are of the same order of magnitude. Due to the variation of ϕ_c from zero at the source to V_D at the drain, the occupancy of the surface states will also vary from source to drain. The interface charge density and, consequently, the surface potential and inversion charge will be affected in this way by the presence of surface states. This explains why the drain current becomes dependent on the surface state density leading to expressions (46) to (51).

It can be shown that the N_{SS} value obtained by the weak inversion technique corresponds to the surface state density at an energy position $q(\phi_{SO} - \phi_F)$ above E_i when an n-channel device is used and at $q|\phi_F - \phi_{SO}|$ below E_i in the case of a p-channel transistor [62]. Computer simulations demonstrate that for practical values of oxide thickness (1000 Å) and substrate doping ($1 \times 10^{15} cm^{-3}$) N_{SS} can be measured with an accuracy of better than $1 \times 10^{10} cm^{-2} eV^{-1}$ for surface potentials satisfying:

$$3/2\phi_F - 50 \text{ mV} \leq \phi_{SO} \leq 3/2\phi_F + 100 \text{ mV} \tag{52}$$

For a p-type substrate with $N_A = 1 \times 10^{15} cm^{-3}$ this goes from 100 meV to 250 meV above midgap at room temperature. The accuracy of the technique is illustrated in Fig. 19. The weak inversion

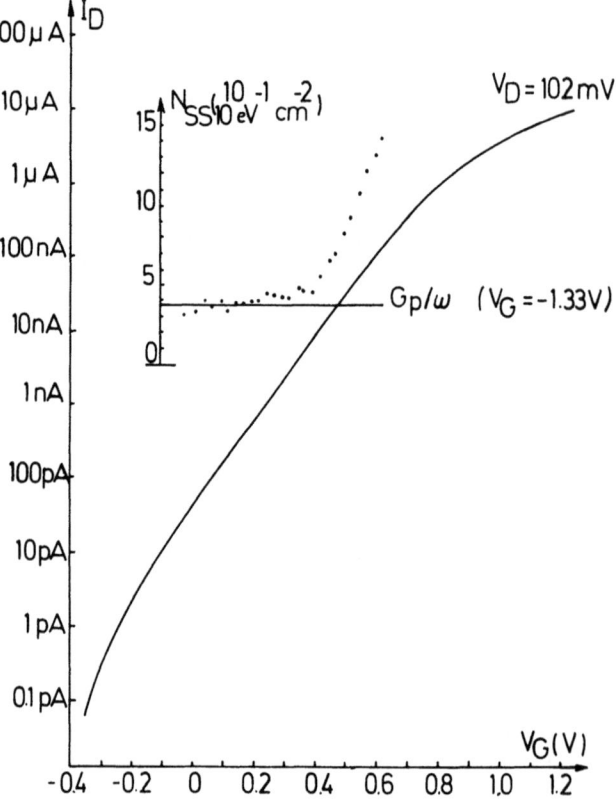

Fig. 19. Experimental in I_D - V_G curve, of an ion-implanted n-channel MOSFET. The N_{SS} values resulting from I_D - V_D measurements at different gate voltages are shown and compared with the N_{SS} value found by the method of Simonne (The figure is taken from Reference 62).

drain current is plotted versus gate voltage and the N_{SS} values obtained from equation (51) are shown in the insert at the corresponding gate voltages. The C_{ox} and C_D-data are obtained from three terminal capacitance measurements [63]. This example is particularly interesting since this n-channel transistor received a boron ion implantation to shift the threshold voltage. It proves that the technique can be applied in the case of nonuniform substrate doping. Also indicated in the figure is the N_{SS} density of $3.8 \times 10^{10} cm^{-2} eV^{-1}$ obtained in depletion by means of the $G_p/\omega-\omega$ conductance technique. The agreement between both measuring techniques, although applied at different bias ranges, has been observed during all experiments. This is typical of structures with low N_{SS} whose distribution is

SURFACE STATES AT THE Si-SiO₂ INTERFACE

considered to be rather uniform through most of the energy gap.

It should be remarked that equation (46) assumes a constant channel length L. To avoid channel shortening effects relatively long channels and low drain voltages (< 1V) must be used in order to facilitate the experimental determination of I_{DMAX} from equation (47). The most accurate technique consists of a least square fit of the measured I_D - V_D curve to (46). A quick evaluation technique can be developed which is based on the definition of M/N (equation (50)) and on the fundamental I_D - V_D relation (46). N_{SS} can then be expressed as:

$$N_{SS} = \frac{C_{ox} + C_D(\phi_{SO})}{q} \left\{ \frac{-\beta V_D}{\ln(1 - \frac{I_D}{I_{DMAX}})} - 1 \right\} \tag{53}$$

Inserting the drain voltage V_D corresponding to a ratio I_D/I_{DMAX} in (53) is sufficient to find N_{SS}. It can be proven however that the best accuracy is obtained when N_{SS} is calculated with the I_D/I_{DMAX} value for which

$$\ln(1 - \frac{I_D}{I_{DMAX}}) = -1 \tag{54}$$

This leads to

$$\frac{I_D}{I_{DMAX}} = 0.632 \tag{55}$$

equation (53) can then be written as

$$N_{SS} = \frac{C_{ox} + C_D(\phi_S)}{q} \left(\frac{qV_D}{kT} - 1 \right) \tag{56}$$

In summary, the surface states density can be found from the value of V_D which has to be applied in order to obtain a drain current equal to 63.2% of the maximal current for the considered gate voltage.

The technique is illustrated by Fig. 20, showing the theoretical I_D/I_{DMAX} versus V_D curve for $N_{SS} = 0$, together with a set of experimental curves measured on a same device but at different gate voltages.

The value of V_D at the intersection of each curve with the line $I_D/I_{DMAX} = 0.632$ has to be inserted in (56). If $C_D(\phi_{SO})$ is determined

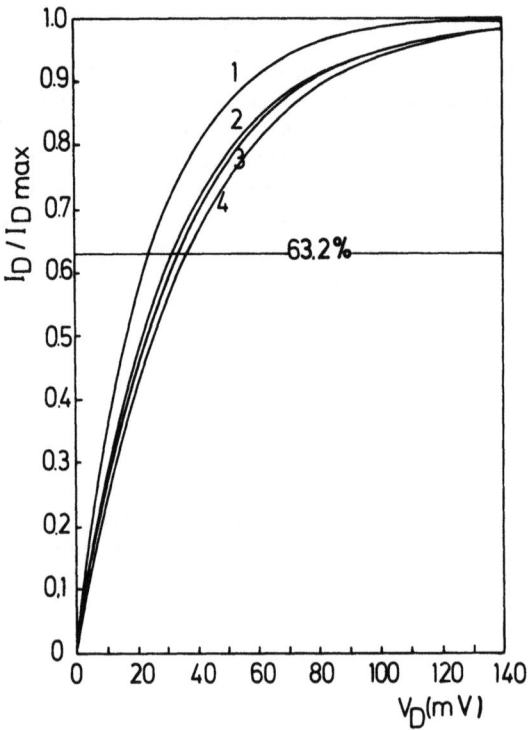

Fig. 20. Curve 1 is the theoretical I_D/I_{DMAX} versus V_D for $N_{SS} = 0$.
Curves 2-4 are experimentally obtained characteristics for a transistor with $t_{ox} = 2000$ Å, $N_A = 1.11 \times 10^{15} cm^{-3}$ and $N_{SS} = 5.9 \times 10^{10} eV^{-1} cm^{-2}$
(2) $\phi_{SO} = 404$ mV
(3) $\phi_{SO} = 485$ mV
(4) $\phi_{SO} = 567$ mV
(The figure is taken from Reference 62.)

for every value of ϕ_{SO}, N_{SS} can be calculated. This results in a density of about $6 \times 10^{10} cm^{-2} eV^{-1}$ which agrees very well with the value of $5.9 \times 10^{10} cm^{-2} eV^{-1}$ found independently with the conductance technique of Nicollian-Goetzberger.

The need to know C_{ox} and C_D at ϕ_{SO} complicates this technique somewhat. C_{ox} is easily found, C_D can be determined experimentally by three terminal capacitance measurements [63] or from a theoretical calculation when the substrate doping and surface potential are known. ϕ_{SO} can be found from a Berglund integration of low frequency C-V plot or from fitting the measured I_D - V_G curve to theory.

It should be emphasized that the weak inversion technique as described here is based on the study of the $I_D - V_D$ characteristic and not of the slope of the $\ln I_D - V_G$ plot (59). The slope of this curve cannot be used for determination of N_{SS}, since it is strongly affected by oxide charge fluctuations [64]. It has been demonstrated theoretically by Brews [65] that the influence of oxide charge fluctuations on the channel current of the MOS transistor can be expressed by a mobility variation:

$$\mu = \mu_M \left(1 - \frac{<\sigma_s>^2}{2}\right) \tag{57}$$

with μ the MOSFET conductance mobility where μ_M is the surface mobility [66], not influenced by fluctuations and $<\sigma_s>^2$ is the variance of the relative carrier-density fluctuations, averaged over the depth of the inversion layer [65]. In most practical cases this can be interpreted as the variance of the surface potential fluctuations.

A mobility behavior similar to equation (57) was experimentally observed by Fang and Fowler [67], and has been verified by Chen and Muller [68] and by Muls et. al. [62].

VII. SURFACE STATE MEASUREMENT BY MEANS OF CHARGE COUPLED DEVICES

Charge coupled devices consist of a series of closely spaced MOS capacitors connected together to form a periodic set of electrodes [69]. By the application of clock pulses, the capacitors are driven into deep depletion and potential wells are created in the silicon. A schematic representation of a three-phase n-channel charge coupled device is shown in Fig. 21. An electrical or optical input structure provides charge packets which are stored in the potential wells and transferred from one well to the next one by using an appropriate clock sequence. At the output section of the CCD the charge packets are detected and converted into an electrical voltage or current signal. In surface channel CCD's the charge packets are transferred along the silicon surface, whereas in a buried channel device the charge transfer occurs in the bulk at a small distance from the silicon surface.

The transfer of charge packets in a CCD is not perfect but limited by several mechanisms, resulting in a loss of charge at each charge transport. This transfer loss or transfer inefficiency ε is typically between 10^{-5} and 10^{-3} per transfer. At high clock frequencies, when the carrier transit time is comparable to the clock period, the fundamental physics of the transfer mechanism itself will limit the efficiency of the transport. At lower clock frequencies, interactions of the mobile carriers with interface

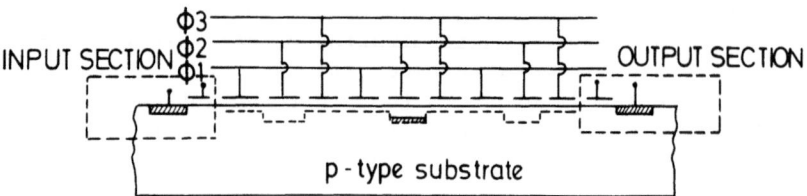

Fig. 21. Basic structure of a three phase n-channel charge coupled device.

traps in a surface channel CCD or bulk traps in a buried channel CCD is the main cause of charge transfer inefficiencies.

The capture of mobile charges by surface states in a CCD is very similar to the previously discussed charge pumping effect (VIa). As soon as a charge packet moves into an empty potential well, the surface states will fill quickly up to the quasi-Fermi level of the minority carriers. However, when the charge packet moves to the next potential well, the charges trapped by the surface states cannot be released instantaneously but will be emitted from the surface states with a time constant given by equation (44) for an n-channel CCD.

When a time t_e is available for the emission process, the surface states will empty to the energy level:

$$(E_c - E_T)_{t_e} = kT \ln(t_e \sigma_n v N_c) \tag{58}$$

It is possible to derive information about the surface state density N_{SS} by examination of the charge lost from a first "one" packet after a string of "zero's". Assume the number of zero's, n_{zero}, the number of clock phases m and the clock frequency f_c. The total charge loss from the first one can then be found by evaluating (58) for $t_{e_1} = 1/mf_c$ and for $t_{e_2} = (mn_{zero} + 1)/mf_c$ respectively. If the charge emitted during t_{e_1} is considered to catchup with the "one" packet, the charge loss can be written as a function of N_{SS} as:

$$Q_{loss} = qN_{SS} \left[(E_C - E_T)_{t_{e_2}} - (E_C - E_T)_{t_{e_1}} \right] \tag{59}$$

N_{SS} is expressed per eV and is assumed to be constant over the energy range of interest. Using (58) it becomes:

$$Q_{loss} = qN_{SS} \, kT \, \ln(mn_{zero} + 1) \tag{60}$$

The fractional charge loss ε_{10} of a first "one" consisting of N_{SIGN} electrons is given by:

$$\varepsilon_{10} = \frac{N_{SS}}{N_{SIGN}} \, kT\ln(m \, n_{zero} + 1) \tag{61}$$

This formula has been derived first by Carnes and Kosonocky [70]. A detailed analysis of the effect of surface state trapping on CCD performance has been given by Tompsett [71], Mohsen et. al. [72], Heald [73] and by Klar et. al. [74]. Recently Kriegler et. al. [75] have presented a new technique, based on the influence of the fat zero level on the transfer inefficiency, which allows the determination of N_{SS} and capture cross section.

The technique discussed above is illustrated in Fig. 22. The transfer inefficiency ε of two n-channel CCD's is plotted versus the number of zero's between the ones [76]. From equation (61) a surface state density of respectively 2×10^{10} cm^{-2} eV^{-1} and 7×10^{9} cm^{-2} eV^{-1} is found. It is clear from the figure that an excellent agreement between experiments and theoretical curve is obtained.

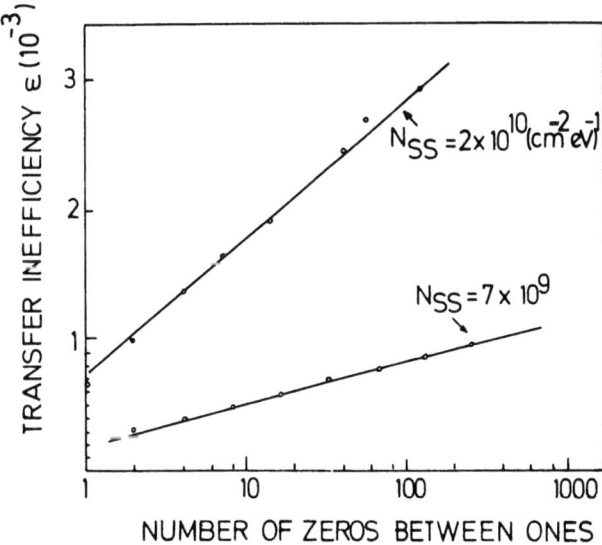

Fig. 22. Transfer inefficiency versus the number of zeroes between the ones. (Figure taken from Reference 75)

REFERENCES

1. M. Preier, Appl. Phys. Lett., Vol. 10, 361 (1967).
2. E. Nicollian and A. Geotzberger, Bell Syst. Techn. J., Vol. 46 1055 (1967).
3. B. Deal, J. of the Electrochem. Soc., Vol. 121, 198 (1974).
4. E. Kooi, Philips Res. Rept., Vol. 21, 477 (1966).
5. Y. Cheng, Surface Science, Vol. 23, 432, (1970).
6. J. Johanssen and W. Spicer, J. Appl. Phys., Vol. 47, 3028 (1976).
7. A. Goetzberger, E. Klausmann and M. Schulz, CRC critical reviews in solid state sciences, Vol. 6, 1 (1976).
8. S. Sze, "Physics of Semiconductor Devices", Wiley (1969).
9. A. Groove, "Physics and Technology of Semiconductor Devices", Wiley (1967).
10. R. Cobbold, "Theory and Applications of Field-Effect Transistors", Wiley (1967).
11. M. Zerbst, Z. Angew. Phys., Vol. 22,30 (1966).
12. F. Heiman, IEEE Trans. Electron Devices, Vol. 14, 781 (1967).
13. S. Hofstein, IEEE Trans. Electron Devices, Vol. 14, 785 (1967)
14. D. Schroder, IEEE Trans. Electron Devices, Vol. 19, 1018 (1972).
15. K. Lehovec and A. Slobodskoy, Solid-State Electronics, Vol. 7, 59 (1964).
16. J. Cooper and R. Schwartz, Solid-State Electronics, Vol. 17, 641 (1974).
17. H. Deuling, E. Klausmann and A. Goetzberger, "Solid-State Electronics, Vol. 15, 559 (1972).
18. J. Brews, J. Appl. Phys., Vol. 43, 2306 (1972).
19. J. Brews, J. Appl. Phys., Vol. 43, 3451 (1972).
20. K. Ziegler and E. Klausmann, Appl. Phys. Lett., Vol. 28, 678 (1976).
21. G. Baccarani et. al., Appl. Phys. Lett., Vol. 23, 265 (1973).
22. M. McNatt and C. Sah, J. Appl. Phys., Vol. 45, 3916 (1974).
23. G. Declerck et. al., J. Appl. Phys., Vol. 45, 2593 (1974).
24. L. Terman, Solid-State Electronics, Vol. 5, 285 (1962).
25. G. Baccarani et. al., Alta Frequenza, Vol. 11, 310 E (1971).
26. S. Margalit et. al., IEEE Trans. Electron Devices, Vol. 7, 861 (1972).
27. R. Castagne and A. Vapaille, Surface Science, Vol. 28, 157 (1971).
28. G. Declerck et. al., Solid State Electronics, Vol. 16, 1451 (1973).
29. C. Berglund, IEEE Trans. Electron Devices, Vol. 13, 701 (1966).
30. R. Castagne, C. R. Acad. Sci., Paris, Vol. 267, 866 (1968).
31. D. Kerr, International Conference on the Properties and use of MIS strucutres, Grenoble, 303 (1969).
32. M. Kuhn, Solid-State Electron., Vol. 13, 873 (1970).
33. M. Kuhn and E. Nicollian, J. Electrochem. Soc., Vol. 118, 370 (1970).
34. R. Van Overstraeten et. al., J. Electrochem. Soc., Vol. 120, 1785 (1973).
35. P. Gray and D. Brown, Appl. Phys. Lett., Vol. 8, 31 (1966).

36. M. Boudry, Appl. Phys. Lett., Vol. 22, 530 (1973).
37. D. Eaton and C. Sah, Phys. Stat. Sol. (a), Vol. 12, 95 (1972).
38. A. Ushirokawa et. al., Jap. J. of Appl. Phys., Vol. 12, 388 (1978).
39. P. Muls et. al., Solid State Electron, Vol. 20, 911 (1977).
40. E. Nicollian et. al., Solid-State Electron., Vol. 12, 937 (1969).
41. J. Simonne, Solid-State Electron., Vol. 16, 121 (1973).
42. J. Brugler and P. Jespers, IEEE Trans. Electron. Devices, Vol. 16, 297 (1969).
43. A. Elliot, Solid-State Electronics, Vol. 19, 241 (1976).
44. C. Sah et. al., Solid-State Electronics, Vol. 13, 759 (1970).
45. D. Lang, J. Appl. Phys., Vol. 45, 3023 (1974).
46. D. Lang, J. Appl. Phys., Vol. 45, 3014 (1974).
47. K. Wang, Semiconductor Silicon 1977, edited by H. Huff and E. Sirtl, 404.
48. K. Wang, Topical Conference on Characterization Techniques for Semiconductors, Materials and Devices, May 1978, Seattle.
49. R. Pierret, Solid-State Electronics, Vol. 17, 1257 (1974).
50. R. Hall, Phys. Rev., Vol. 87, 387 (1952).
51. W. Schockley and W. Read, Phsy. Rev., Vol 87, 835 (1952).
52. N. Johnson et. al., Int. Topical Conference on the Physics of SiO_2 and its Interface, Yorktown Heights, March 1978.
53. A. Van Der Ziel, Solid-State Electronics, Vol. 17, 110 (1974).
54. F. Klaassen, IEEE Trans. Electron. Devices, Vol. 18, 887 (1971).
55. R. Ronen, RCA Review, Vol. 34, 280 (1973).
56. E. Abowitz et. al., IEEE Trans. Electron. Devices, Vol. 14, 775 (1967).
57. G. Broux et. al., Electron. Lett., 97 (March 1975).
58. A. Guzev et. al., Sovjet Physics-Semiconductors, Vol. 4, 1245 (1971).
59. R. Swanson and J. Meindl, IEEE J. Solid-State Circuits, 146 (1972).
60. R. Stuart and W. Eccleston, Electron. Lett., Vol. 8, 223 (1972).
61. M. Barron, Solid-State Electronics, Vol. 15, 293 (1972).
62. P. Muls et. al., to be publishedin Advances in Electronics and Electron Physics.
63. I. Bateman and J. Magowan, Electron. Lett., Vol. 6, 669 (1970).
64. R. Van Overstraeten et. al., IEEE Trans. Electron. Devices, Vol. 20, 1150 (1973).
65. J. Brews, J. Appl. Phys. Vol. 46, 2181 and 2193 (1975).
66. J. R. Schrieffer, Phys. Rev.
67. F. Fang and A. Fowler, Phys. Rev. Vol. 169, 619 (1968).
68. J. Chen and R. Muller, J. Appl. Phys., Vol. 45, 828 (1974).
69. C. Sequin and M. Tompsett, "Charge Transfer Devices", Academic Press (1975).
70. J. Carnes and W. Kosonocky, Appl. Phys. Lett., Vol. 20, 261 (1972).
71. M. Tompsett, IEEE Trans. Electron Devices, Vol. 20, 45 (1973).
72. A. Mohsen et. al., IEEE J. Solid-State Circuits, Vol. 8, 125 (1973).
73. D. Heald, Solid-State Electronics, Vol. 20, 657 (1977).
74. H. Klar et. al., Proc. of ESSCIRC 1977, 101.
75. R. Kriegler et. al., Device Research Conference, Corneel University, (June 1977).

76. G. Declerck et. al., Proc. Intl. Conf. on the Application of Charge-Coupled DEvices, Setp. 1976, Edinburgh, P. 23.

Chapter 4

STEADY-STATE AND NON-STEADY-STATE CHARACTERIZATION OF MOS DEVICES

John G. Simmons

Department of Electrical Engineering

University of Toronto

Toronto, Canada   M5S 1A4

There are many a.c., d.c. and transient techniques available for the study of interfacial [1-20] and bulk trap [21-27] properties of MOS devices, far more than can be covered in these talks. The methods presented here reflects the personal interests of the author, and involve many of the physical underlying principles involved in other techniques. The features of the methods are the relative ease with which the desired information on the trap parameters can be extracted from the experimental data.

I. RELAXATION CURRENT TECHNIQUES

These methods involve the biasing of the device with a constant voltage and monitoring the current flow as a function of time or temperature. Using the appropriate voltage biasing, the technique can be used for the investigation of both the interface [28,30] and bulk [29,30] traps in MOS devices. We begin with a discussion of the study of interface traps.

A. Interfacial Isothermal Relaxation Currents (IIRC)

This method [28] consists of simply applying a step voltage to the MOS system and monitoring the d.c. relaxation current as a function of time as the system relaxes to the equilibrium state, under isothermal conditions.

We will present generalized equations that provide for closed-form analytic solutions of the isothermal I-t decay characteristic of MOS systems whatever the interface trap distribution. More important, however, we will demonstrate, by means of an appropriate artifice, how to extract directly the trap distribution and its capture cross section from the isothermal data.

1. Theory. Consider a metal-oxide-semiconductor* device with its gate electrode positively biased, so that the device is in the accumulation mode. In this case, the interface traps in the upper-half of the band gap will be filled to an energy, say E_1, at which the Fermi level intersects the semiconductor surface (Fig. 1(a)). Suppose now that the gate electrode is negatively biased so that the device is in the inversion mode. In this case, under quasi-equilibrium conditions only those traps are filled that lie in levels below the energy, say E_2, at which the Fermi level intersects the surface in the lower-half of the band gap (Fig. 1(d)). Thus, when the sample is switched from the accumulation to the inversion mode, the interface traps located between the energies E_1 and E_2 lose their electrons. This relaxation process does not occur instantaneously, and during the period it occurs, the device is in the non-steady-state.

Electrons escape from the interface traps in the upper-half of the band gap by thermal excitation into the semiconductor conduction band (Fig. 1(b)); when these traps have emptied, surface or bulk generation of electron-hole pairs becomes significant. In this event, electrons in the traps located in the lower-half of the band gap escape by recombining with the generated holes in the vlaence band (Fig. 1(c)).

The electrons that are emitted to the semiconductor conduction band are immediately swept out of the depletion region by the high field therein, causing a current to flow in the external circuit. Thus the current that initially flows in the circuit is due to the interface-generation process; it is also possible for a bulk-generation current to succeed the interface-generation current, depending on the relative density of the bulk and interface traps. Here we will be concerned solely with the emission current.

The current I flowing in the external circuit is proportional to the rate of emission of electrons from the interface traps. The rate of emission is determined as follows. We define a parameter $n_i(E,t)$, as the energy distribution of electrons per unit energy

---

*Here we will confine ourselves to n-type semiconductors. The case of p-type semicnductors follows directly from the discussion of the n-type system, the principle difference being that in the p-type system the surface states in the lower half, rather than the upper half, of the band gap are studied.

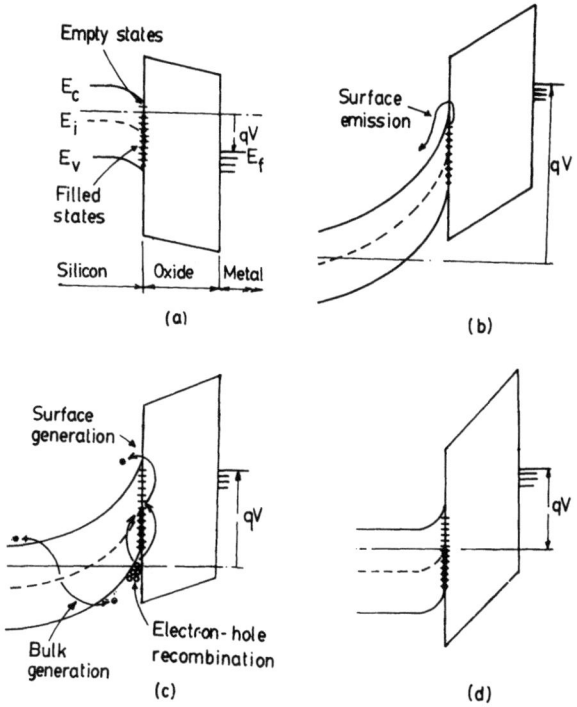

Fig. 1. Energy diagram illustrating: (a) accumulation; (b) emission process; (c) interface and bulk generation of electron-hole pairs and recombination of electron m interface states with holes in the valence band; (c) inversion mode.

per unit area at the interface at any time t. The number of electrons per unit area in an incremental energy range dE at any time t after a negative pulse has been applied to the system is, therefore, equal to $n_i(E,t)dE$. It follows then that the rate of emission of electrons per unit area from the traps located between E and E + ΔE is

$$d(\delta n_i)/dt = -e_n \delta n_i \equiv -e_n n_i(E,t)dE \tag{1}$$

In Eq. (1), e_n is the thermal emission coefficient of the traps, which is given by

$$e_n = v\sigma_n N_c \exp[(E_i - E_c)/kT] \qquad (2)$$

$$\equiv \nu \exp[(E_i - E_c)kT]$$

The solution of (1) for <u>constant</u> temperature is

$$\delta n_i = N_i(E) \exp(-e_n t) dE \qquad (3)$$

In obtaining (3) we used the boundary condition that $\delta n_i = N_i(E)dE$ at $t = 0$, where $N_i(E)$ is the energy distribution per unit area per unit energy of the interface traps; note that this statement implies that in the process of accumulating the device the states below E_1 are completely filled (Fig. 1(a)). Substituting Eq. (3) into Eq. (1) yields

$$\frac{d(\delta n_i)}{dt} = -N_i(E) e_n \exp(-e_n t) dE \qquad (4)$$

From an inspection of Eq. (1) and (4) it will be apparent that

$$n_i(E,t) = N_i(E) \exp(-e_n t) \qquad (5)$$

from which it will be clear that $\exp(-e_n t)$ is the non-steady-state occupancy function. The function $\exp(-e_n t)$ is shown plotted in Fig. 2 as a function of energy with time as a parametric variable,

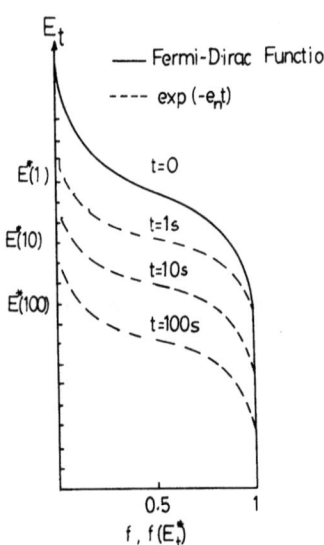

Fig. 2. Non-steady-state distribution as a function of energy, with time as a parameter variable.

and it is seen to have a similar dependence on energy as the Fermi-Dirac function. Thus, we may define a non-steady-state Fermi level E_f^*, give by

$$\exp[-e_n(E_f^*)t] = 1/2$$

or**

$$E_f^* = E_c - kT[\ln(\nu t) + 0.365] \simeq E_c - kT \ln(\nu t) \qquad (6)$$

For energies more than about 2kT above E_f^* the interface traps are essentially empty, while for energies more than about 2kT below E_f^* the interface traps are essentailly full. In essence, then, Eq. (6) gives the energy of the uppermost-filled trap at time t after the device has been switched into the non-steady-state mode. Consequently, we may write

$$n_i = \int_{E_v}^{E_c} N_i(E) \exp(-e_n t) dE \simeq \int_{E_v}^{E_f^*} N_i(E) dE \qquad (7)$$

Hence, it follows that if E_f^* decreases in energy by dE in a time dt, then the rate of change, dn_i/dt, in the number of the occupied traps is

$$\frac{dn_i}{dt} = -N_i(E_f^*) \frac{dE_f^*}{dt}$$

and, thus, the rate of increase of positive charge dQ_i/dt† at the interface is

$$\frac{dQ_i}{dt} = -q \frac{dn_i}{dt} = qN_i(E_f^*) \frac{dE_f^*}{dt} \qquad (8)$$

Differentiating Eq. (6) with respect to time and substituting the result in Eq. (8) yields

$$\frac{dQ_i}{dt} = \frac{qN_i(E_f^*)kT}{t} \qquad (9)$$

From the charge neutrality requirements we have

---

† Note that since the traps are donor type, the net positive charge at the interface is $Q_i = q(\int_{E_v}^{E_c} N_i(E)dE - n_i)$; thus $\dot{Q}_i = -q\dot{n}_i$.

** For typical values of $\nu (\approx 10^{11} \text{sec}^{-1})$ and practical (experimental) values of t ($\geq 10^{-5}$sec) so $\ln \nu t \gg 0.365$.

$$Q_g + Q_i + Q_d + Q_{ox} = 0 \tag{10}$$

where Q_g is the charge per unit area on the gate electrode, Q_i is the interface-trap charge per unit area, Q_d is the charge density in the semiconductor depletion region, and Q_{ox} is the fixed charge per unit area in the oxide. Differentiating Eq. (10) yields

$$\dot{Q}_g + \dot{Q}_i + \dot{Q}_d = 0 \tag{11}$$

Now the current density J circulating in the external circuit is simply the rate of change of the charge on the gate electrode Q_g; hence, substituting $J = \dot{Q}_g$ in Eq. (11) yields

$$J = -\dot{Q}_d - \dot{Q}_i \tag{12}$$

Furthermore,

$$Q_d = qN_d x_d \tag{13}$$

where N_d is the donor density, and x_d is the width of the semiconductor given by [31]

$$x_d = -\frac{\varepsilon_s}{C_{ox}} + \left[\left(\frac{\varepsilon_s}{C_{ox}}\right) - \frac{2\varepsilon_s}{qN_d}\left(\frac{Q_{ox}+Q_i}{C_{ox}} + V_g\right)\right]^{\frac{1}{2}} \tag{14}$$

From Eq. (13) and (14) we obtain

$$\dot{Q}_d = qN_d \dot{x}_d = -\frac{C_d \dot{Q}_i}{C_d + C_{ox}} \tag{15}$$

where $C_d = \varepsilon_s/x_d$ is the capacitance of the semiconductor depletion region. Substituting Eq. (15) into (12) gives

$$J = -\frac{C_{ox}\dot{Q}_i}{C_{ox} + C_d} \tag{16}$$

and substituting $\dot{Q}_i$ from Eq. (9) into (16) yields

$$J(t) = \frac{qC_{ox}}{C_{ox} + C_d} \frac{kT}{t} N_i(E_f^*) \tag{17}$$

Normally $C_d \ll C_{ox}$, so Eq. (17) further reduces to

$$I(t) = \frac{qAkT}{t} N_i(E_f^*) \tag{18}$$

where A is the area and I the isothermal current; hence, the isothermal current at time t is directly proportional to the trap density $N(E_f^*)$ at E_f^* where E_f^* is given by Eq. (6).

2. **I-t Characteristics for Special Trap Distributions.** In order to demonstrate the efficacy and accuracy of Eq. (18), we compare the exact and approximate isothermal current-time characteristics for a few special trap distributions.

(i) Uniform trapping distribution. Substituting $N(E) = N_t =$ constant into (18) yields

$$J = qKTN_1/t \tag{19}$$

In Fig. 3 curves (a) illustrate the exact [28] (numerical) and approximate I-t characteristics for this case. The correlation is seen to be very good.

(ii) Exponential trapping distribtuion. The trapping distribution is assumed to be given by:

$$N(E_i) = N_2 \exp[-\alpha(E_c - E_i)]$$

$$N(E_f^*) = N_2 \exp[-\alpha(E_c - E_f^*)] \tag{20}$$

where N_2 and α are characteristic constants of the distribution. Substituting Eq. (20) into (18) and using (6) we obtain

$$J = qN_2kT(\nu^{-\alpha kT})(t^{-kT+1}) \tag{21}$$

Correlation between the exact and approximate characteristics is again very good, as illustrated by the curves (b) in Fig. 3.

(iii) Gaussian trapping distribution. We assume a trapping distribution given by

$$N(E) = N_3 \exp[-\gamma(E - E_0)^2] \tag{22}$$

that is,

$$N(E_f^*) = N_3 \exp[-\gamma(E_f^* - E_0)^2]$$

where N_3, γ, and E_0 are characteristic constants. Substituting (22) into (18) and using (6) yields

$$\ln J = C_1 + C_2 \ln t + C_3 (\ln t)^2 \tag{23}$$

where

$$C_1 = \ln\{2qkTN_e \exp[-\gamma(E_c - E_0 - kT \ln\nu)^2]\}$$

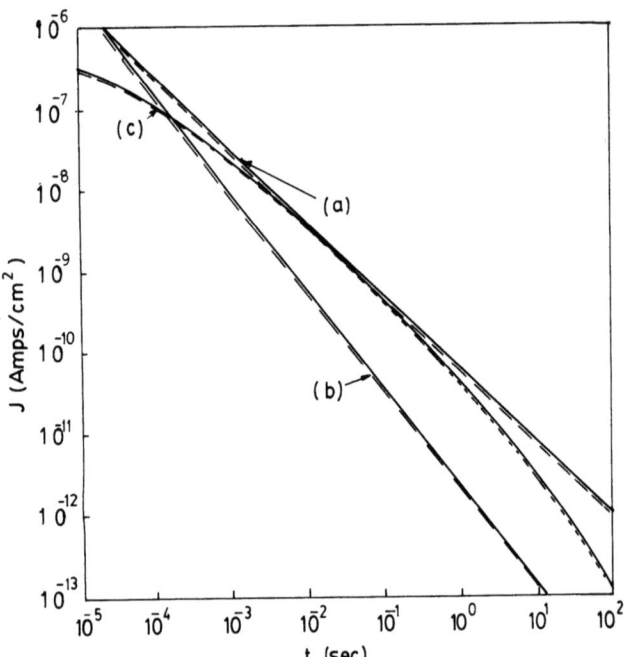

Fig. 3. Approximate and exact J-t characteristics for: (a) a uniform trap distribution, $N_S(E) = 5 \times 10^{12} \text{cm}^{-2}(\text{eV})^{-1}$; (b) an exponential trap distribution, $N_S(E) = 10^{13} \exp[-2(E_c-E)]\text{cm}^2(\text{eV})^{-1}$ and; (c) Gaussian trap distribution, $5 \times 10^{12}\exp[-25(E-0.38)^2]\text{cm}^{-2}(\text{eV})^{-1}$. Other parameters used: $\nu = 10^{11}$ sec and $T = 200°K$.

$$C_2 = 2kT\gamma (E_c - E_o - kT \ln \nu)^{-1}$$

$$C_3 = -\gamma(kT)^2$$

The exact and approximate characteristics are illustrated in Fig. 3 by curves (c), and, again, the correlation between the two characteristics is seen to be extremely good.

3. **Direct Determination of Trap Distribution.** The isothermal I-t characteristics are not particularly fruitful from the point of view of yielding data on the surface trap distribution. That this is so is particularly well demonstrated by the three sets of curves in Fig. 3, for although the trap distributions responsible for the curves are quite disparate, the curves themselves are very similar in nature, and might well have been thought to have been generated by

almost identical trap distributions. Although the I-t characteristics per se do not provide much information on the trap distributions responsible for them, we will now show that the product of the isothermal current and time (It) plotted as a function of $\log_{10} t$ actually yields directly the interface trap distribution.

Equation (18) may be expressed in the form

$$N_i(E_f^*) = \frac{It}{AqkT'} \qquad (24)$$

which relationship shows that the produce of current and time is linearly proportional to the trap density at E_f^*. Also, we know that E_f^* is related to time by

$$E_c - E_f^* = kT \ln \nu t$$

or

$$E_c - E_f^* = 2.303 kT (\log_{10} t + \log_{10} \nu) \qquad (25)$$

hence, using (25) the $\log_{10} t$ axis can be transformed into energy measured with respect to the bottom of the conduction band. Therefore, it will be apparent from an inspection of Eq. (24) and (25), that plotting It as a function of $\log_{10} t$ provides a direct image of the energy distribution of the interface traps in the upper-half of the band gap. In other words, using the transformations provided by (24) and (25), the It-$\log_{10} t$ characteristic may be directly converted into a $N_i(E_f^*)$ vs. $E_c - E_f^*$ characteristic, that is, into an actual plot of the interface trap distribution.

To demonstrate the applicability and efficacy of the method, the I-t curves shown in Fig. 3 are shown replotted in Fig. 4 as It vs. $\log_{10} t$ characteristics; the axes have been transformed to $N_i E(f^*)$ and $E_c - E_f^*$ axes to obtain the energy distributions.

B. Interfacial Thermal Relaxation Current (ITRC)

The basic principle which underlies the thermal relaxation technique [29] can be best understood with reference to the energy band diagrams shown in Fig. 5, which are for an n-type silicon substrate. As in the IIRC technique, the traps in the region of the silicon-oxide interface are filled by applying a positive voltage bias to the metal electrode with respect to the n-type silicon substrate at a low temperature T_o, to drive the device into the accumulation mode as shown in Fig. 5(a). The polarity of the voltage bias is now reversed and the device goes into the deep-depletion mode as illustrated by Fig. 5(b). Now, while the temperature of the device is held constant at T_o, the electrons escape from the interface traps by the process of isothermal emission as described earlier.

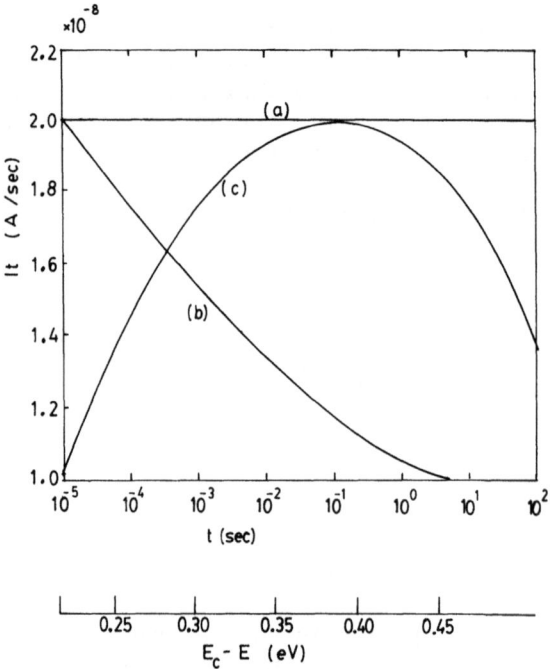

Fig. 4. The curves of Fig. 3 replotted in the form It vs. $\log_{10} t$. The It and $\log_{10} t$ axes are also expressed in terms of $N_s(E_m)$ and $E_c - E_m$, respectively.

Consequently, we can associate a non-steady-state Fermi level E_f^* with the trapped interface charge. However, if the temperature of the device is sufficiently low, the rate of emission of the interface charge and thus the associated emission current after the first two or three seconds will be negligible. In other words, E_f^* is essentially held constant by the prevailing experimental conditions.

In the TDRC technique, the temperature of the device is now raised at a <u>constant</u> rate β:

$$T = \beta t + T_o \qquad (26)$$

Consequently, the emission probability for the trapped electrons now increases rapidly with the result that the rate of release of the electrons trapped at the interface increases, with the shallow traps releasing their charge before the deeper lying levels. It follows, then, that E_f^* now begins to move to lower energies, as shown in Fig. 5(c). Under these conditions it can be shown [29] that E_f^* is

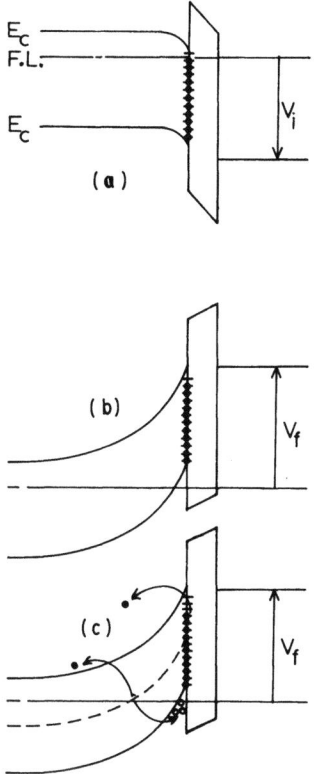

Fig. 5. Energy band diagrams illustrating (a) the filling of interfacial and nitride traps in accumulation; (b) non-steady-state deep-depletion mode; and (c) thermal emission from interfacial traps.

related to temperature via the relationship

$$E_c - E_f^* = 10^{-4} \times T[1.92 \log_{10}(\nu/\beta) + 3.20] - 0.0155 \quad (27)$$

This expression is the analog of Eq. (6) for the IIRC technique, except that for the case in hand, since β is chosen to be constant, E_f^* is a linear function of temperature, whereas in the IIRC method E_f^* is a linear function of time.

Following the arguments given in the previous section, the rate of emission of charge $\dot{Q}_i$ from the interface states may be written in the form

$$\dot{Q}_i = -qN_s(E_f^*) \frac{dE_f^*}{dt} = -qN_s(E_f^*) \frac{dE_f^*}{dt} \frac{dT}{dt} \qquad (28)$$

$$= q\beta N_s(E_f^*) \frac{dE_f^*}{dt}$$

Differentiating (27) with respect to time and substituting the result in Eq. (28) yields

$$\dot{Q}_i = q\beta N_s(E_f^*) \times 10^{-4}[1.92 \log_{10}(\nu/\beta) + 3.2]. \qquad (29)$$

Finally, following the steps leading from Eq. (10) to (18), we obtain

$$J(T) = q\beta n_s(E_f^*) \times 10^{-4}[1.92 \log_{10}(\nu/\beta) + 3.2] \qquad (30)$$

The important feature of (30) is that it states that the current $J(T)$ measured at temperature T is directly proportional to the interface trap density at E_f^*, i.e.,

$$N_s(E_f^*) = \frac{J(T)}{q\beta \times 10^{-4}[1.92 \log_{10}(\nu/\beta) + 3.2]} \qquad (31)$$

However, Eq. (31) is of no value unless E_f^* is known at temperature T; well, indeed, (27) supplies this relationship:

$$E_c - E_f^* = 10^{-4}T[1.92 \log(\nu/\beta) + 3.2] - 0.0155 \qquad (32)$$

Consequently, Eq. (30) and (32) which show $N(E_f^*)$ and $(E_c - E_f^*)$ as being a linear function of $J(T)$ and T, respectively, tell us that the ITRC plot, $J(T)$ vs. T, is in fact a <u>direct</u> image of the interfacial trap distribution. The transformation from the J vs. T curve to the N(E) vs. E curve is provided by Eq. (31) and (32) respectively, as shown schematically in Fig. 6.

C. Typical Interfacial Results

One of the features of the relaxation methods is the simplicity of the experimental technique. We will describe measurements obtained from a Metal-Nitride-Oxide-Silicon (MNOS) sample using both the ITRC and IIRC methods and compare the resulting trap distributions obtained by the two methods. These structures are used in the fabrication of electrically programmable non-volatile memory devices.

The MNOS devices used were capacitor structures in which the oxide thickness was nominally 18 Å and the nitride layer 850 Å. The silicon substrate was n-type and approximately 9 Ωcm resistivity.

CHARACTERIZATION OF MOS DEVICES

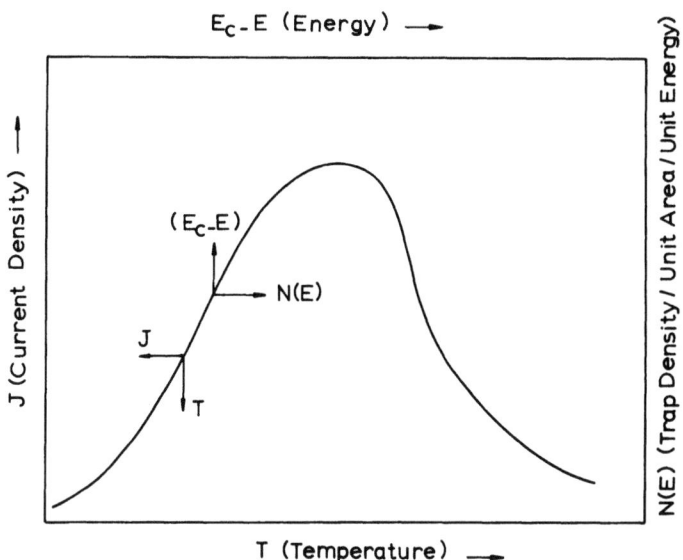

Fig. 6. Graph illustrating relationship between I-T and N(E) - (E_c-E) curves.

1. <u>ITRC Results [30]</u>. As mentioned earlier, the ITRC method involves first biasing the device into the accumulation mode in order to fill the interface states. Then the temperature is lowered, and at the appropriate low temperature T_o the device is biased into deep-depletion. The lower the starting temperature, the closer to the band edge are the traps that may be probed.

Figure 7(a) illustrates the I vs. T characteristics as a consequence of raising the temperature of the device at two different constant heating rates, $\beta = 0.1°K$ and $0.01°K$. Figure 7(b) is the corresponding interface trap distribution obtained using the transformation Equations (31) and (32). The resulting shape of trap distribution curve obtained from the two I-T curves are seen to be very similar. Adjusting ν, the attempt to escape frequency, to obtain the correlation between these two curves resulted in a value $\nu = 5.6 \times 10^{10} sec^{-1}$. More will be said about this value after we have discussed the IIRC results.

2. <u>IIRC Method [30]</u>. How close to the conduction band edge a study of the interface states may be made using this technique depends on the time-resolution of the measuring apparatus; the faster

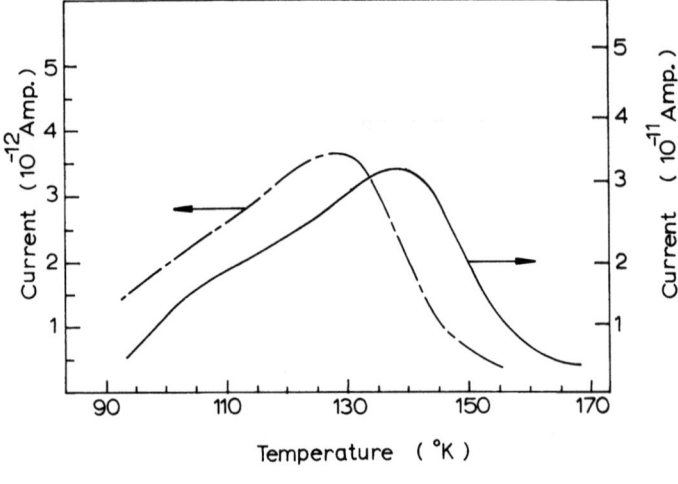

Fig. 7(a)

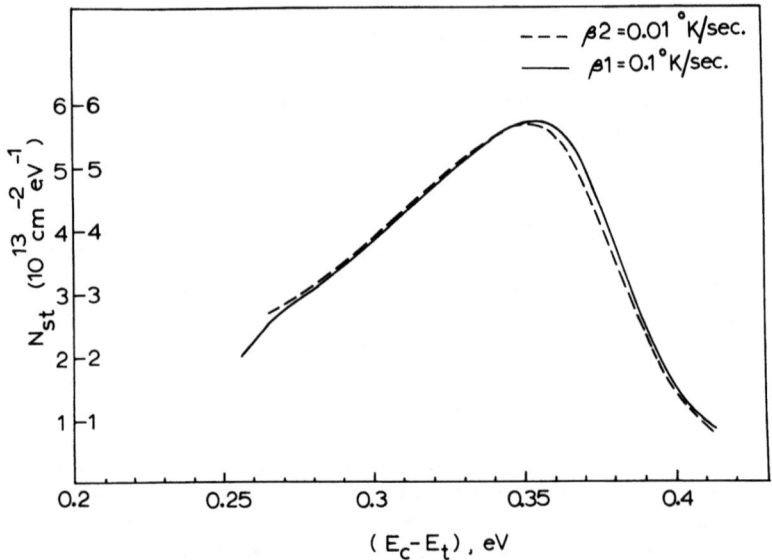

Fig. 7(b)

Fig. 7. Thermal Relaxation: (a) I vs. T for two heating rates, β = 0.1K/sec and 0.01K/sec; (b) Interface trap distribution obtained from (a).

its time response, the closer to the bottom of conduction band the interface states can be studied. It will be apparent from Eq. (6) that the effective energy resolution of the measuring equipment can be enhanced by cooling the sample to low temperatures. For example if the maximum response time of the apparatus is 10^{-6} sec and $\nu = 10^{11} \text{sec}^{-1}$, then at 300°K one is limited to studying surface states within about 0.28 eV of the bottom of the conduction band. However, at liquid-nitrogen temperature (77°K), the surface states within about 0.07 eV of the conduction band edge can be studied, and at liquid-helium temperature (4°K) within about 0.0035 eV.

Although we have been concerned with the interface states density throughout the upper half of the band gap, by judicious initial and final biasing, various energy ranges may be investigated. The whole of the upper half may be investigated in this manner and the distribution constructed piecemeal from the accumulated data. Thus, in the construction of the trap distribution curve it is normally necessary to have to make several transient current measurements at different temperatures. Consequently, this method tends to be more tedious than the ITRC method, which normally requires only two measurements to be made. However, the experimental technique is somewhat simpler with the IIRC method in that it is easy to hold the sample at a constant temperature than to ramp it at a constant rate.

With these points in mind, Fig. 8(a) illustrates the transient IIRC obtained at various constant temperatures, resulting from biasing the device into the deep depletion mode, after first having biasing it in the accumulation mode, to fill the interface states.

From these curves, $N_i(E_f^*)$ was obtained using (24), and E_f^* was obtained from Eq. (6). The value of the attempt-to-escape frequency used was $5.6 \times 10^{10} \text{sec}^{-1}$, as obtained in section I C-1, above. The resulting $N_s - E_f^*$ (i.e. $N(E)$ vs. $E_i - E$) characteristic is displayed as the dashed line of Fig. 8(b).

A comparison of the interface trap distribution obtained by the isothermal and the thermal methods is shown in Fig. 8(b). The good correlation between the two curves is provides for confidence in both techniques. It should be noted that the value of the attempt-to-escape frequency used influences significantly the relative magnitude of the two curves. Thus, the fact that the two curves require exactly the same value of the attempt-to-escape frequency in order for them to correlate gives further confidence in the validity of the techniques.

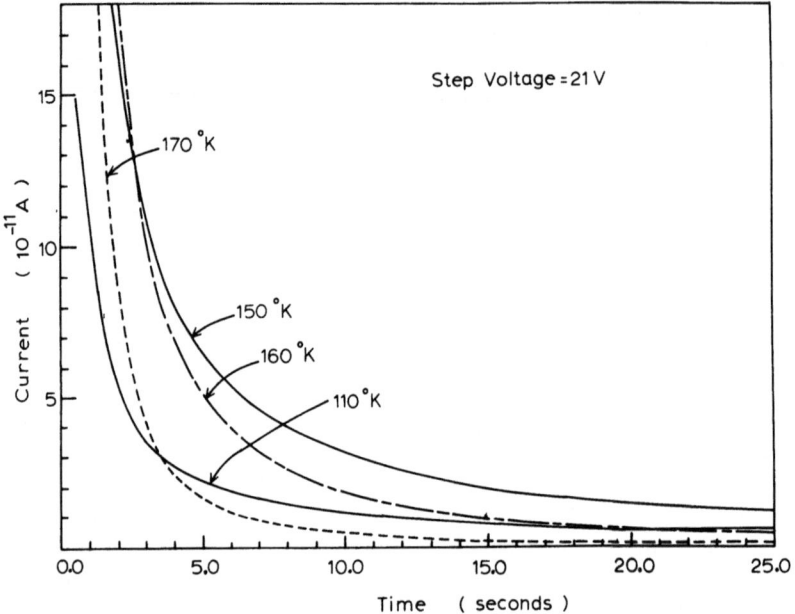

Fig. 8(a)

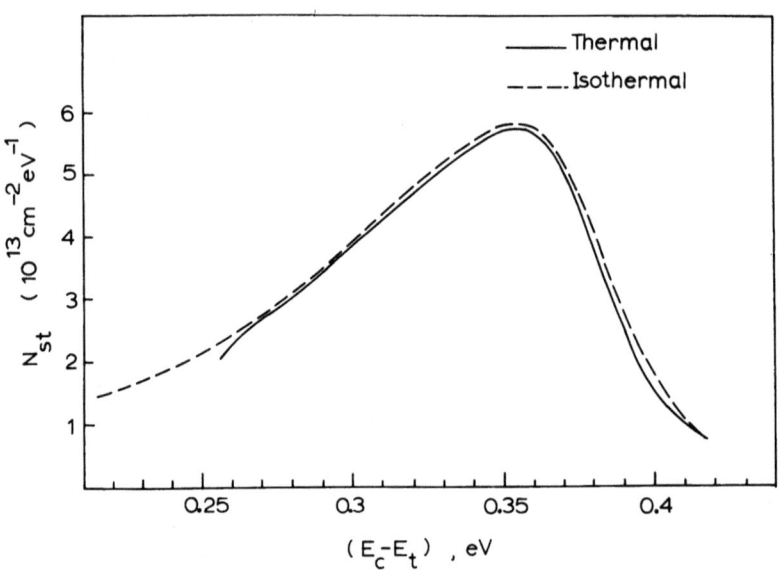

Fig. 8(b)

Fig. 8. Transient Isothermal Method: (a) Transient currents at various temperatures, (b) Trap distribution according to isothermal method (dotted line) and thermal method.

D. Bulk Thermal Relaxation Currents

The thermal relaxation current method [32] used for studying interface traps can also be used for the study of the bulk traps in the silicon substrate of MOS devices. Interface emission effects are distinguished from bulk generation effects by appropriate initial biasing of the device.

1. **Physical Principles.** The procedure used to obtain the non-steady-state bulk-generation I-T curve for a capacitor structure involves first applying an initial large reverse-voltage V_i to the device at room temperature so that the surface of the silicon is strongly inverted (Fig. 9(a)). In this mode, the contribution of the interface-generation current is negligible. The device is then next cooled in the strong inversion mode to some low temperature, T_o, and then the reverse voltage on the device is further increased to some value, V_f. The increase in the reverse voltage on the device is now increased of a uniform rate (β) and the non-steady-state generation current that flows in the external circuit is plotted as a function of temperature. The generation current can be shown [32] to be related to the temperature by

$$I_g = qN_b A e_n(E_{tb})[L_o - (\frac{L_o - L_f}{L_o})CT^2 \exp -E_{tb}/kT] \tag{33}$$

where

$$C = \frac{v\sigma_n N_b \varepsilon_s N_v k}{\beta(C_{ox} + C_d)N_d E_{tb}} \tag{34}$$

L_o is the width of the depletion-layer at T_o, and C_{ox} and C_d are the oxide and depletion-layer capacitances per unit area, respectively (in practical devices $C_{ox} \gg C_d$). The I_g-T curve manifests a broad peak and the temperature, T_m, of the maximum is related to the bulk trap energy, E_{tb}, and carrier lifetime $\tau = (v\sigma_n N_t)^{-1}$ through the following relationship

$$E_{tb} = kT_m \log_e [\frac{\varepsilon_s N_v kT_m (2L_o^2 - L_f)}{\beta\tau(C_{ox} + C_d)N_d L_o^2 E_{tb}}] \tag{35}$$

The constant in the argument of the logarithm can be determined by performing the experiment at two heating rates β_1 and β_2, and by substituting these values and the corresponding values of T_m into Eq. (35); here the same initial conditions are assumed for the two experiments. Hence, we obtain

$$\log_e B = \frac{T_{m1} \log_e(\frac{T_{m1}^2}{\beta_1}) - T_{m2} \log_e(\frac{T_{m2}^2}{\beta_2})}{(T_{m2} - T_{m1})} \tag{36}$$

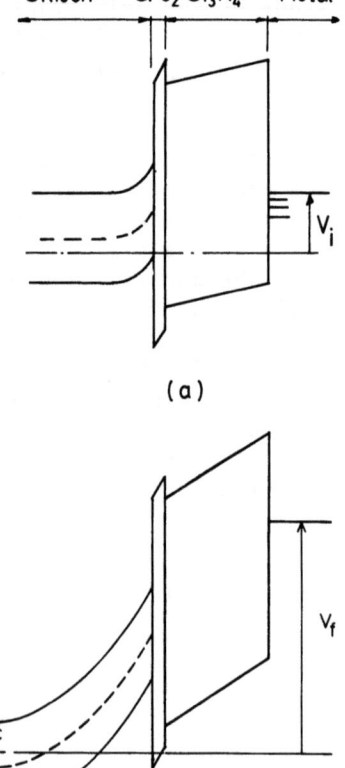

Fig. 9. Energy-band diagrams of the MNOS structure for n-type silicon illustrating: (a) the strong inversion mode at room temperature and, (b) the deep-depletion mode at the initial low temperature.

where

$$B = \frac{\varepsilon_s k N_v (2L_o - L_f)}{\tau(C_{ox} + C_d) N_d L_o^2 E_{tb}} \tag{37}$$

Thus, having determined the value of B, the trap energy may now be obtained from (35).

Another important trap parameter that is readily determined from the data is the so-called carrier lifetime, τ. From a knowledge of the oxide capacitance and the initial width of the depletion-layer, the carrier lifetime is now determined by making use of Eq. (37).

CHARACTERIZATION OF MOS DEVICES

2. *Effect of Reverse-Voltage Bias.* Figure 10 shows a set four I_g-T plots. The voltages V_f applied during the heating are shown alongside the curves, and the initial voltage applied during cooling from room temperature was in all cases $V_i = -6V$; the rate of increase in temperature was $0.08°K/sec$. It is observed that the temperature at which the maximum of each peak occurs is relatively independent of the bias voltage.

Figure 11 illustrates a log I vs. 1/T plot obtained from the leading edge of the -18V curve of Fig. 10. From this curve and Eq. (33) an activation energy of 0.55 eV was obtained.

3. *Effect of Heating Rate.* Figure 12 illustrates curves obtained at two different heating rates $0.02°K/sec$ and $0.06°K/sec$. In both cases $V_i = -6V$ and $V_f = -12V$. It is observed that the maximum of each of the peaks occurs at significantly different temperatures. From the temperature T_m at which the two maxima occur and Eq. (37) we find that the carrier lifetime $(= v\sigma N_t)^{-1} = 4$ μsec. Using this value in Eq. (37) we obtain $E_{tb} = 0.54$ eV, which is in excellent agreement with that obtained from Fig. 11. Using these computed values in Eq. (33) we generated the theoretical I-T curves corresponding to the experimental curves in Fig. 10. The theoretical characteristics are shown by the dotted lines, and it is seen that the correlation between the two sets of curves is extremely good.

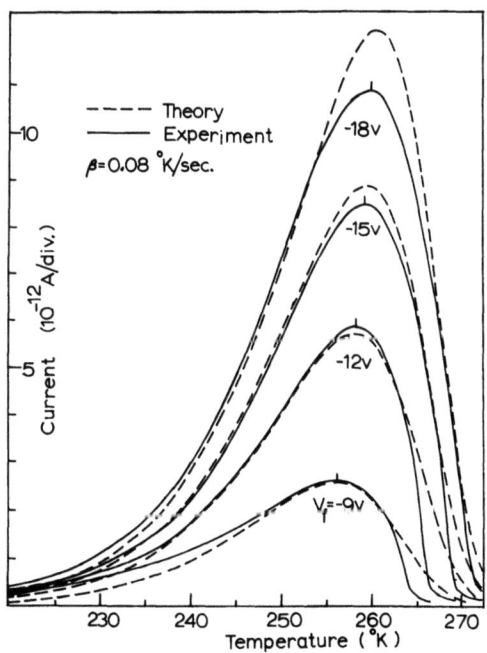

Fig. 10. Experimental and theoretical I-T curves for $V_f - V_i$ ($V_i = -6V$).

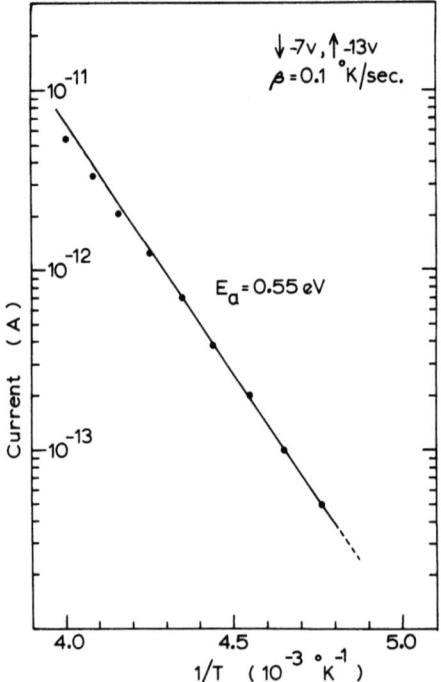

Fig. 11. Plot of log I vs. 1/T from -18V curve of Fig. 10.

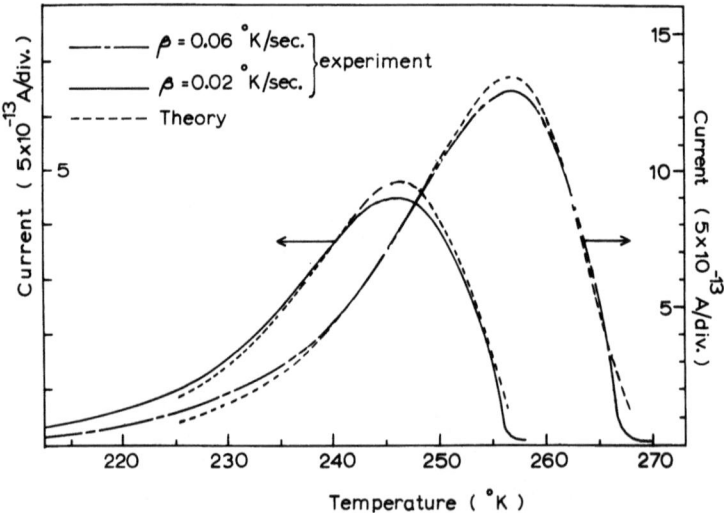

Fig. 12. I-T curves for two heating rates.

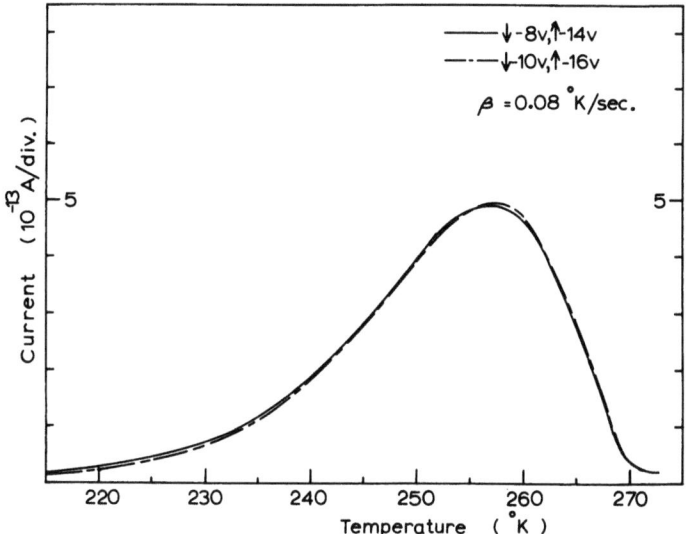

Fig. 13. I-T characteristics demonstrating that bulk-generation in dominant process.

4. **Effect of $(V_f - V_i)$.** The curves shown in Fig. 13 were obtained using the same difference $(V_f - V_i)$ in the final (V_f) and initial (V_i) voltages but with different V_i and V_g, and using a heating rate of 0.08°K/sec. It is observed that the curves are almost identical in shape, indicating that the I-T characteristic depends only on the difference in initial and final voltage rather than their absolute values. This fact is borne out by the theory.

In the case of interface-generation, it can be shown that for the same difference in inital and final voltages, the I-T curves are quite disparate. Thus, the method serves as a means of distinguishing between bulk and interface-generation.

II. C-V TECHNIQUES

The MOS structure is in essence a voltage variable capacitor, the capacitance being dependent in both oxide thickness and silicon doping concentration and trapping properties of the system. Furthermore, the system capacitance is a very sensitive function of the trap parameters of the system. Consequently, capacitance methods have been used widely in the study of the trap properties of MOS systems.

A few of these methods, in particular, those methods in which the principles pertain to Part III of this chapter, will be presented here.

A. The System Capacitance

Charge neutrality requires that

$$Q_g + Q_i + Q_s = 0 \tag{38}$$

where Q_s is the total charge (depletion plus inversion charge) associated with the silicon depletion (space charge) region [34]:

$$Q_s(\phi) = \sqrt{2\varepsilon_s kTN_d} \; [(e^{-\lambda\phi} + \lambda\phi - 1) + (n_i/N_d)^2(e^{\lambda\phi} - \lambda\phi - 1)]^{\frac{1}{2}} \tag{39}$$

where ϕ and Q_s are both taken as negative in accumulation and positive in depletion and inversion. We note that the first term in the square brackets is due to the contribution of the depletion charge and the second to the contribution of the inversion charge.

We also have for the voltage distribtuion

$$V_g = -V_{ox} - \phi \tag{40}$$

$$V_g = \frac{Q_g}{C_{ox}} - \phi \tag{41}$$

From Eq. (38) we have

$$\frac{dQ_g}{dV_g} = -\left(\frac{dQ_i}{d\phi} + \frac{dQ_s}{d\phi}\right)\frac{d\phi}{dV_g}$$

By definition

$$\frac{dQ_g}{dV_g} = C_T \tag{43a}$$

$$\frac{dQ_i}{d\phi} = C_i \tag{43b}$$

and

$$\frac{dQ_s}{d\phi} = C_s \tag{44}$$

so

$$C_T = -(C_i + C_s) \frac{d\phi}{dV_g} \tag{45}$$

Differentiating Eq. (41) with respect to the gate voltage we obtain

$$1 = \frac{1}{C_{ox}} \frac{dQ_g}{dV_g} - \frac{d\phi}{dV_g}$$

or

$$-\frac{d\phi}{dV_g} = 1 - \frac{C_T}{C_{ox}} \tag{46}$$

Substituting Eq. (46) into (45) yields

$$\frac{1}{C_T} = \frac{1}{C_{ox}} + \frac{1}{C_i + C_s} \tag{47}$$

The equivalent circuit according to Eq. (47) is used in Fig. 14.

B. High-Frequency C-V Characteristic

The definition of high-frequency as defined here is a frequency above which the interface charge and inversion charge cannot respond to the small signal. Thus, $C_i = 0$ so (47) reduces to

$$\frac{1}{C_T} = \frac{1}{C_{ox}} + \frac{1}{C_s} \tag{48}$$

and the second (inversion) term in Eq. (39) is ignored in determining C_s:

$$C_s = \frac{dQ}{d\phi} = \sqrt{q\phi_s \lambda N_d} \frac{(1 - e^{-\lambda\phi})}{(e^{-\lambda\phi} + \lambda\phi_s - 1)^{\frac{1}{2}}} \tag{49a}$$

where the + sign is used for depletion and inversion and the − sign for accumulation. The flat-band capacitance of the silicon, that is when $\phi_s = 0$, may be obtained from (49a) by expanding the exponential to yield

$$C_s(\text{Flat band}) = \sqrt{q t_s \lambda N_d} \tag{49b}$$

It is also noted that when ϕ is positive by more than about 2 kT/q, that is when the device is in the depletion or inversion mode, $\lambda\phi_s \gg 1$ and Eq. (49a) reduces to the form

$$C_s = \sqrt{\frac{q\varepsilon_s N_d}{2\phi}}, \quad \phi > 2kT \tag{49c}$$

which is called the <u>depletion approximation</u>.

Although the inversion and interface charges cannot respond to the high-frequency signal, these charges will respond to the large-signal biasing voltage V_g, unless the applied voltage is applied at a high rate. In this section we will assume V_g is applied sufficiently slowly so that the system is in quasi-equilibrium with the large signal:

$$V_g = \frac{Q_g}{C_{ox}} - \phi$$

and using Eq. (38)

$$V_g = -\frac{(Q_s + Q_i)}{C_{ox}} - \phi \tag{50}$$

1. <u>Ideal High-Frequency C-V Characteristic</u>. In generating the ideal C_T-V_g curve it is assumed that interface states don't exist, so $C_i = 0$ and $Q_i = 0$. Thus, Eq. (47) reduces to

$$\frac{1}{C_T} = \frac{1}{C_{ox}} + \frac{1}{C_s(\phi)} \tag{51}$$

and Eq. (50) to

$$V_g = -\frac{Q_s(\phi)}{C_{ox}} - \phi \tag{52}$$

where $C_s(\phi)$ is given by (49) and $Q_s(\phi)$ by (39).

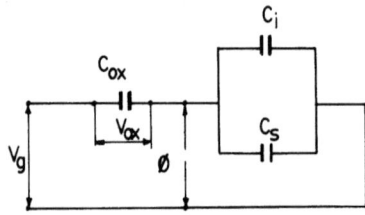

Fig. 14. Equivalent circuit for MOS capacitor.

CHARACTERIZATION OF MOS DEVICES

Equations (51) and (52) are a pair of parametric equations in ϕ. A typical ideal high-frequency curve is shown in Fig. 15. From Eq. (49a) it is seen that for large negative values of ϕ (corresponding to large positive values of V_g, C_s reduces to

$$C_s = \sqrt{\frac{\varepsilon_s q N_d}{2kT}} \; e^{-\lambda\phi/2}, \quad \phi < 0 \tag{53}$$

Consequently C_s becomes very large, and when $C_s \gg C_{ox}$ then we see from Eq. (53) that

$$C_T = C_{ox} \tag{54}$$

It follows, then, that we may determine the oxide capacitance adn, hence, the oxide thickness from the strongly ($V_g \gg 0$) accumulation section of the curve.

It will be noted that for sufficiently large <u>positive</u> values of ϕ,

$$Q_s \propto e^{\beta\phi/2}$$

so Q_s increase rapidly with small changes in ϕ. This condition prevails at the onset of strong inversion; that is, when

$$\phi = 2\phi_n = 2(E_{Fn} - E_m)/q$$

$$= \frac{E_g}{2} - \frac{kT}{q} \ln\left(\frac{N_c}{N_D}\right) \tag{55}$$

Consequently, from Eq. (52) we see that in the strong inversion mode large changes in V_g result in only small changes in ϕ. In fact, to a good approximation we may say that ϕ becomes essentially constant in the strong-inversion mode. Consequently, the width, X_d, of the depletion region and, hence, C_s becomes essentially constant and equal to

$$X_d = \left(\frac{2\varepsilon_s \phi}{qN_d}\right)^{1/2} = \left(\frac{4\varepsilon_s \phi_n}{qN_d}\right)^{1/2} \tag{56}$$

and

$$C_s = \frac{\varepsilon_s}{X_d} = \left(\frac{\varepsilon_s q N_d}{4\phi_n}\right)^{1/2} \tag{57}$$

Since C_s may be considered constant in the strong inversion mode, then it follows from (51) that C_T becomes constant:

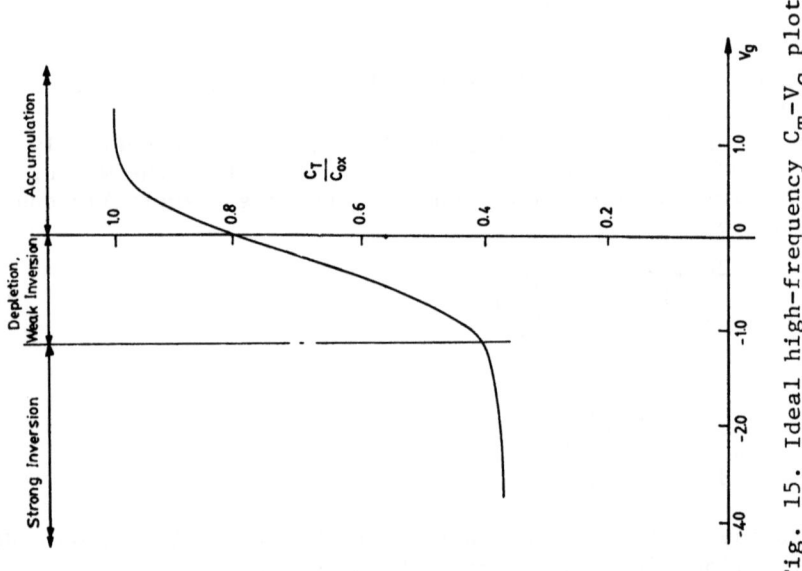

Fig. 15. Ideal high-frequency C_T-V_G plot.

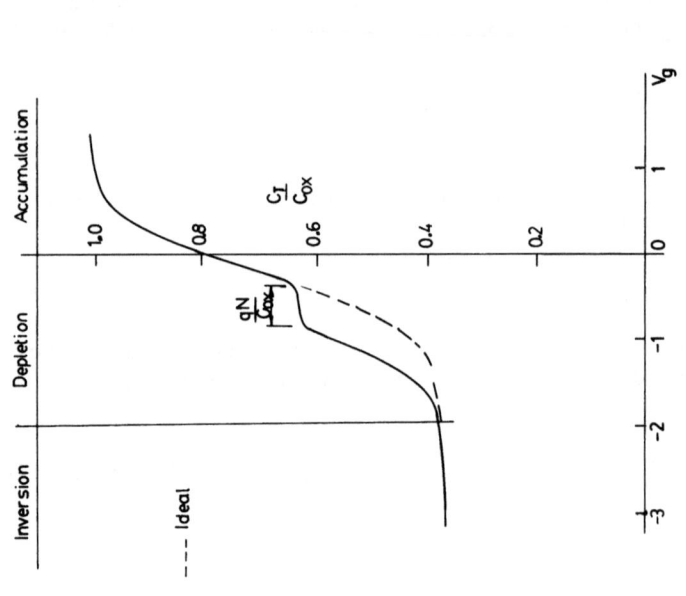

Fig. 16. High frequency C_T-V_g curve for MOS capacitor with a discrete trap level.

$$\frac{1}{C_T} = \frac{1}{C_{ox}} + \left(\frac{4\phi_n}{\varepsilon_s q N_d}\right)^{1/2} \tag{58}$$

This, then, is the reason that the C-V curve flattens out at high negative values of V_g, as seen in Fig. 15. Consequently, it follows from Eq. (58) that the doping concentration, N_d, may be determined by measuring the capacitance of the system in the strong-inversion mode.

2. *Effect of Discrete Traps.* Even when interface states exist in MOS systems, within the definition of high-frequency C-V curves, C_T is still given by (51). However, since the interface states respond to the large (biasing) signal, then

$$V_G = \frac{Q_g}{C_{ox}} - \phi = \frac{-(Q_s + Q_i)}{C_{ox}} - \phi \tag{59}$$

Hence, from a comparison of Eq. (52) and (59), it is apparent that the effect of the interface charge is to displace the C-V curve along the negative V_g axis relative to the ideal curve by an amount $-Q_i/C_{ox}$.

Figure 16 illustrates the effect of a discrete interface trap of density N_i per unit area located at an energy $E_c - E_{ti}$ below the bottom of the conduction band. Assuming the traps are acceptor-type, that is, neutral when filled and thus positively charged when empty, then as ϕ increases from zero, the flat-band position, the C-V curve follows the ideal characteristic, curve a-b, since the traps are essentially completely filled. When ϕ reaches a value ϕ_i given by

$$\phi_i = E_{ti} - \Delta E_F \tag{60}$$

that is when the Fermi level begins to enter the trap level, the electrons start to escape from the interface traps, since they are in dynamic equilibrium with the silicon conduction band. Thus, further decrease in the gate voltage results in the release of electrons from the interface traps rather than the depletion region growing in size in order that Eq. (38) will be satisfied. The surface potential, ϕ, is thus essentially pinned to the value given by Eq. (60) until such times as the traps are emptied; thereafter Q_s and, hence, ϕ must increase to satisfy Eq. (38). The capacitance is, therefore, practically independent of voltage, as illustrated by the ledge b-c in the C-V curve. The capacitance at which the ledge occurs provides the energy of the trap level, which using Eq. (48), (49c) and (60) is obtained from the expression

$$\frac{1}{C_T} = \frac{1}{C_{ox}} + \left[\frac{q\varepsilon_s N_d}{2E_{ti} - \Delta E_F}\right]^{1/2} \tag{61}$$

It will be noted that the ledge moves to lower capacitance levels with decreasing temperature because E_F moves closer to the bottom of the conduction band.

Just as ϕ approaches ϕ_i, and remembering that the trap is still full, ($Q_i = 0$), then from Eq. (59)

$$V_{g1} = \frac{-\phi_s}{C_{ox}} - \phi_i$$

and just after the trap level has emptied $Q_i = qN_i$, then

$$V_{g2} = -\left(\frac{qN_i + Q_s}{C_{ox}}\right) - \phi_i$$

Hence, the length $\Delta V_g = V_{g1} - V_{g2}$ of the capacitance ledge is

$$\Delta V_g = qN_i/C_{ox}$$

from which the density of the discrete trap can be determined:

$$N_i = C_{ox}\Delta V_g/q \tag{62}$$

Since the traps are now empty, there is no further contribution of the interface charge to the system. Thus, as the gate voltage goes further negative, the C_T vs. V_g curve is identical to the ideal curve but displace from it along the V_g axis by an amount ΔV_g (see section c-d of Fig. 16).

3. <u>Distributed Traps</u>. Again, since the electrons interface states are assumed to be unable to repond to the small-signal, then the system capacitance is again given by Eq. (51). Assuming again that the interface traps are acceptor-type then (see Fig. 17).

$$Q_i = q \int_{E_{Fi}}^{E_c} N_i(t)(1-f)dE \tag{63a}$$

$$\approx q \int_0^{\phi+\Delta E_F} N_i(E_i)dE_i \tag{63b}$$

and ϕ is obtained from (59):

$$V_g = -\frac{q}{C_{ox}} \int_0^{\phi+\Delta E_F} N_i(E_i)dE_i - \frac{Q_d}{C_{ox}} - \phi \tag{64}$$

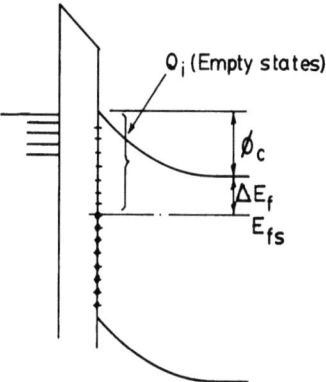

Fig. 17. Energy diagram for MOS capacitor with distributed interface states.

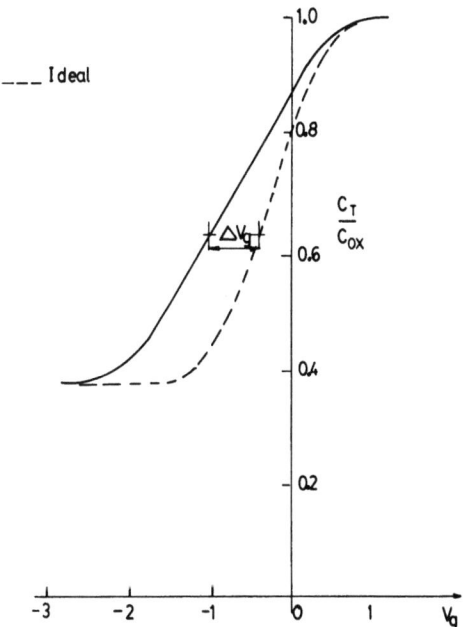

Fig. 18. High frequency C_T-V_g curve for MOS capacitor with distributed interface traps.

For uniformly distributed traps, that is $N_i(E_i)$ = constant = N_i, and assuming the device is in the depletion or weak-inversion mode, so that (49c) may be used, then (64) becomes

$$V_g = -\frac{1}{C_{ox}}[N_i(\phi + E_F) + (2q\epsilon_s N_d \phi)^{1/2}] - \phi$$

Figure 18 illustrates the high-frequency C_T-V_g for a constant trap density. It is seen that increasing values of N_i causes the flat band voltage, V_{FB}, to move towards more negative values:

$$V_{FB} = -\frac{qN_i \Delta E_F}{C_{ox}}$$

Indeed, it follows from Eq. (65) that the voltage during depletion or weak-inversion is displaced by an amount ΔV_f from the ideal curve

$$\Delta V_g = -\frac{1}{C_{ox}} N_i(\phi + E_F)$$

Generally speaking, for any trap distribtuion we have from (64)

$$\Delta V_g = -\frac{q}{C_{ox}} \int_0^{\phi + \Delta E_F} N_i(E_i) dE_i \qquad (65)$$

or

$$\frac{d\Delta V_g}{dE_i} = -\frac{q}{C_{ox}} N_i(E_i) \qquad (66)$$

Since $E_i = \phi + \Delta E_F$, then Eq. (66) may be written in the form

$$\frac{d\Delta V_g}{d\phi} = -\frac{q}{C_{ox}} N_i(\phi) \qquad (67)$$

Now, since N_d and C_{ox} can be determined from the C_T vs. V_g curve in the strong inversion and the accumulation mode, respectively, the corresponding ideal curve can be constructed. Furthermore, since ϕ as a function of C_T is known (see Eq. (51)), then a V_g vs. ϕ curve may be constructed. Hence, in light of Eq. (49), differentiation of this curve yields the trap distribution.

From a practical point of view, however, it is really only possible to construct the trap distribution for the energy range corresponding to depletion and weak inversion, $0 < \phi \leq 2\phi_n$ (that is the energy range $\Delta E_F < (E_g - E_i) < (\Delta E_F + 2\phi_n)$. This is because as $\phi \to 0$, trap levels close to E_c are being probed, and extremely high frequencies are required if they are not to respond to the small signal. On the other hand, for $\phi > 2\phi_n$, ϕ is only weakly dependent

CHARACTERIZATION OF MOS DEVICES

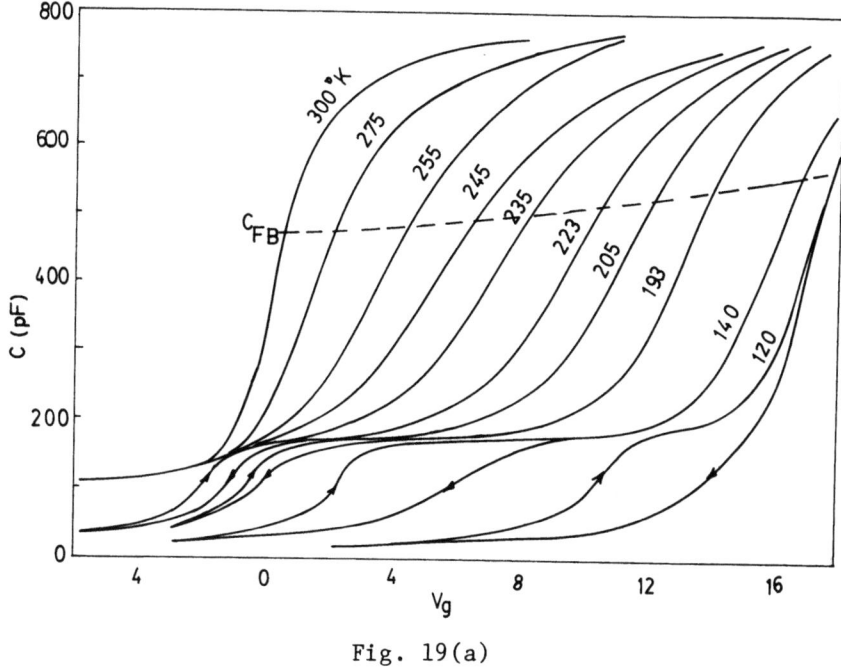

Fig. 19(a)

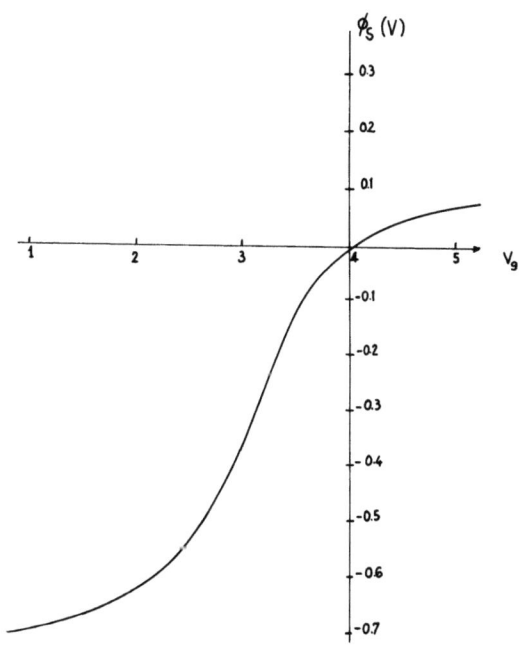

Fig. 19(b)

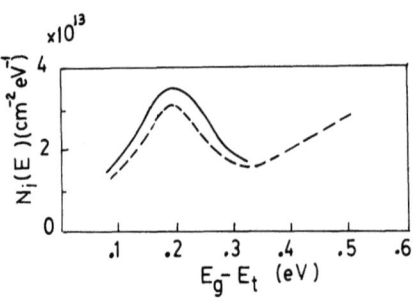

Fig. 19(c)

Fig. 19. Characteristics of an MNOS device: (a) C_T-V_g characteristics; (b) V_g vs. ϕ characteristic; and (c) Computed trap distribution obtained from isothermal method (full line) and temperature method (dotted line).

on V_g and, hence, it is difficult to determine the surface potential and, thus, the trap level being probed, with any accuracy.

As a practical example of technique, we illustrate the family of C_T vs. V_g curves obtained at various constant temperatures from a MNOS non-volatile memory capacitor. Using the T = 245°C curve we obtained the V_g vs. ϕ curve (Fig. 19(b)), and, consequently, the interface trap distribution (Fig. 19(c)), as indicated by the full line.

It is clear from Fig. 19(a) that the C_T-V_g curve is temperature dependent, shifting towards more positive voltage levels. This is because E_F decreases with decreasing temperature. Thus, if the surface potential is held constant with temperature, E_{Fi} actually sweeps through the bandgap, from lower to high energy levels, as the temperature decreases:

$$E_{Fi} = \Delta E_F + q\phi_c = kT \ln(N_c/N_d) + q\phi_c \qquad (68)$$

Furthermore, an incremental change dT in the temperature results in an incremental change $qN_i(E_{Fi})dE_{Fi}$ in the interface charge (Fig. 20), for a constant surface potential. This incremental charge gives rise to an incremental voltage displacement $d\Delta V_g$ in adjacent (the T and T + ΔT characteristics) C_T-V_g characteristics, given by

$$dV_g = qN_i(E_{Fi})dE_{Fi}$$

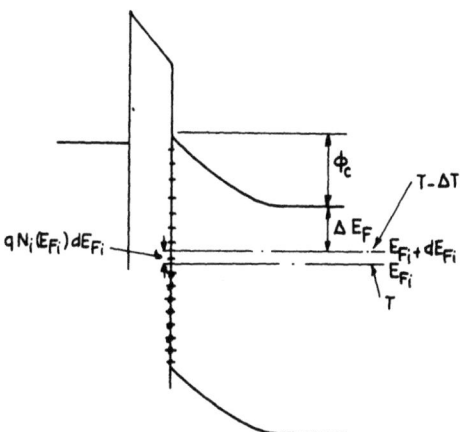

Fig. 20. Energy diagram for MOS capacitor at different temperatures for constant surface potential.

or

$$\frac{dV_g}{dE_{Fi}} = qN_i(E_{Fi}) \tag{69}$$

Differentiating Eq. (68) and using the result in Eq. (69) yields

$$\frac{dV_g}{dT} = qkN(E_{Fi}) \ln(N_c/N_d)$$

Hence, plotting V_g vs. T for <u>constant surface potential</u>, and then differentiating the curve yields the trap distribution. This is the basis of the Gray and Brown technique, which in fact uses the <u>flat-band</u> capacitance as reference capacitance, that is, $\phi_c = 0$. Of course, this method is subject to the same limitations as the isothermal C_T vs V_g method discussed above. As it happens, the flat-band capacitance is a function of temperature is given by

$$\frac{1}{C_{T,FB}} = \frac{1}{C_{ox}} + \frac{1}{C_{s,FB}}$$

so the reference capacitance changes with temperature. The flat-band capacitance as a function of temperature is shown dotted on the curve in Fig. 19(a). Following the above procedure, the trap distribution shown in Fig. 19(c) was obtained (dotted line). It is seen that this distribution is in good agreement with that obtained from the isothermal method (Fig. 19(b)).

C. Low Frequency C_T-V_g Characteristic

In contrast to the definition of high-frequencies, here low frequencies are defined as frequencies sufficiently low that both the interface charge and the inversion charge can follow the small-signal; in other words, the system is in quasi-equilibrium with the small-signal (as well, as with the large biasing signal).

1. <u>Ideal Low Frequency C-V Curve</u>. Differentiating Eq. (39) yields the capacitance of the semiconductor space charge region:

$$C_s(\phi) = \frac{dQ_s}{d\phi} = \left(\frac{N_d q^2 \epsilon_s}{2kT}\right)^{1/2} \frac{[1 - e^{-\lambda\phi} + (n_i/N_d)^2(e^{\lambda\phi}-1)]}{[(e^{-\lambda\phi}+\lambda\phi-1) + (n_i/N_d)^2(e^{\lambda\phi}-\lambda\phi-1)]^{1/2}} \quad (70)$$

Using Eq. (70) and (47), the ideal low-frequency C_T-V_g curves shown in Fig. 21 were generated. Also, shown is the corresponding ideal high-frequency curve. It is noted that the two curves are essentially identical in accumulation, depletion and weak inversion since the inversion charge is negligible. Only when the device enter the strong inversion mode do the two curves deviate.

When deeply in the strong inversion mode, it is seen that $C_T \simeq C_{ox}$. This is because of the fact that inversion charge, with its strong dependence on changes in surface potential, supplies practically the entire semiconductor charge required by the a.c. signal. Consequently, nearly all the a.c. voltage appears across the oxide, with the consequence that $C_T \simeq C_{ox}$, as demonstrated.

2. <u>Non-Ideal Characteristics</u>. Since the system is assumed to be in quasi-equilibrium with the a.c. signal. then the occupancy of a discrete trap at the interface is determined from Fermi-Dirac statistics:

$$n_i = \frac{N_i}{1 + e^{(E_{ti}-E_{Fi})/kT}} \quad (71)$$

Hence, assuming acceptor-type traps,

$$Q_i = q(N_i - n_i) = \frac{qN_i}{1 + e^{(E_{Fi}-E_{ti})/kT}} \quad (72)$$

or since $E_{Fi} = q\phi + \Delta E_F$

CHARACTERIZATION OF MOS DEVICES

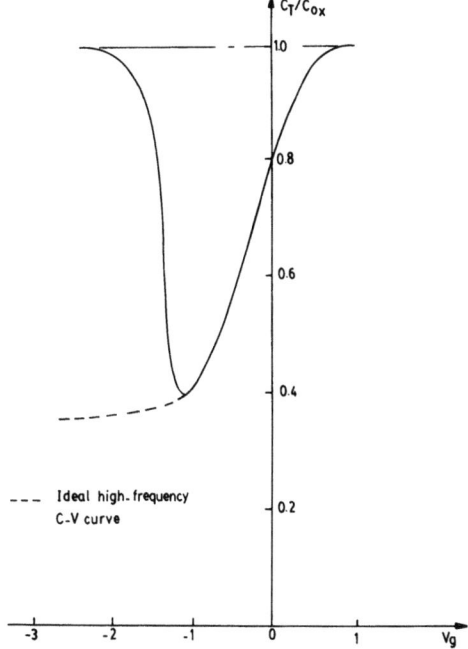

Fig. 21. Ideal low-frequency C_T-V_g curve.

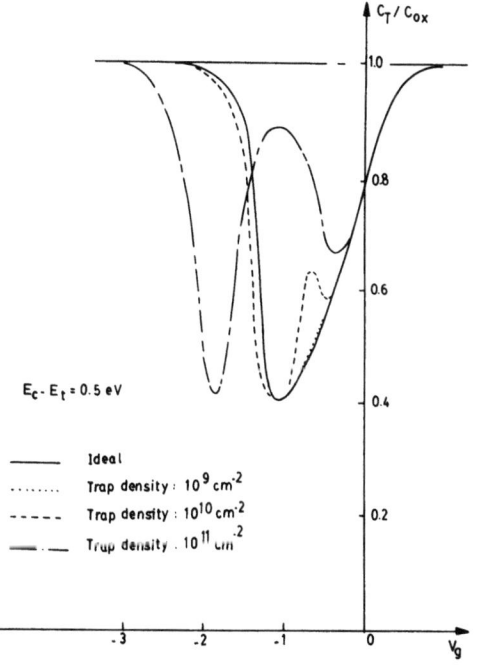

Fig. 22. Low frequency C_T-V_g curve for MOS capacitor, for various discrete trap densities.

$$Q_i = \frac{qN_i}{1 + e^{(q\phi+\Delta E_F - E_{ti})/kT}} \tag{73}$$

from which we obtain

$$C_i = \frac{dQ_i}{d\phi} = \frac{q^2 N_i}{kT} \frac{e^{(\Delta E_F + q\phi - E_{ti})/kT}}{(1 + e^{(\Delta E_F + q\phi - E_{ti})/kT})} \tag{74}$$

Using Eq. (74) with (70) in (47) yields the low-frequency C_T vs. V_g curve for a MOS system with a discrete trap.

The above procedure can be generalized to include many discrete traps. For such a system the system capacitance is given by

$$\frac{1}{C_T} = \frac{1}{C_{ox}} + \frac{1}{C_s + \sum_n C_i(N_{in}, E_{in})} \tag{75}$$

where

$$C_i(N_{i,n}, E_{i,n}) = \frac{q^2 N_{in}}{kT} \frac{e^{(\Delta E_F + q\phi - E_{in})/kT}}{(1 + e^{(\Delta E_F + q\phi - E_{in})/kT})} \tag{76}$$

and $N_{i,n}$ and $E_{i,n}$ are the density and energy of the nth discrete trap.

Figure 22 illustrates the effect of discrete traps of various densities at an energy $E_c - E_{ti} = 0.5$ eV below the bottom of the conduction band. It is seen that the trap level results in a pronounced peak in the curve, which increases and broadens with increasing trap density. This is a consequence of the trap capacitance exhibiting a maximum as a function of the surface potential. The position of the maximum shifts towards more negative voltage levels the deeper the trap level.

The maximum value of C_i may be obtained by differentiating Eq. (76) with respect to ϕ and solving for ϕ to obtain

$$\phi = E_{ti} - E_F \tag{77}$$

which corresponds to the Fermi level lying in the trap level. As we noted in Section II B from high-frequency C_T-V_g measurements we can always determine the depletion capacitance and, hence, the for a given voltage bias. Consequently, from the voltage at which the maximum in the C_T-V_g curve occurs, the trap energy may be determined. Furthermore, substituting Eq. (76) into (74) yields for maximum trap capacitance, C_{imax}

$$C_{imax} = \frac{q^2 N_i}{4kT} \tag{78}$$

Thus, the capacitance of the maximum of the peak provides the trap density.

In the case of distributed traps we note that an incremental change dE_{Fi} in the Fermi level at the interface causes a change $dQ_i(E_{Fi})$ in the charge at the interface given by

$$dQ_i(E_{Fi}) = qN_i(E_{Fi})dE_{Fi} \tag{79}$$

where $N_i(E_{Fi})$ now is in units of traps per unit area per unit energy. Since $E_{Fi} = q\phi + \Delta E_{Fi}$ then we have

$$dQ_i = q^2 N_i d\phi \tag{80}$$

so Eq. (79) may be written as

$$C_i = \frac{dQ_i}{d\phi} = q^2 N_i(E_{Fi}) \tag{81}$$

Hence, we see the trap capacitance is directly proportional to the interface trap density at E_{Fi}.

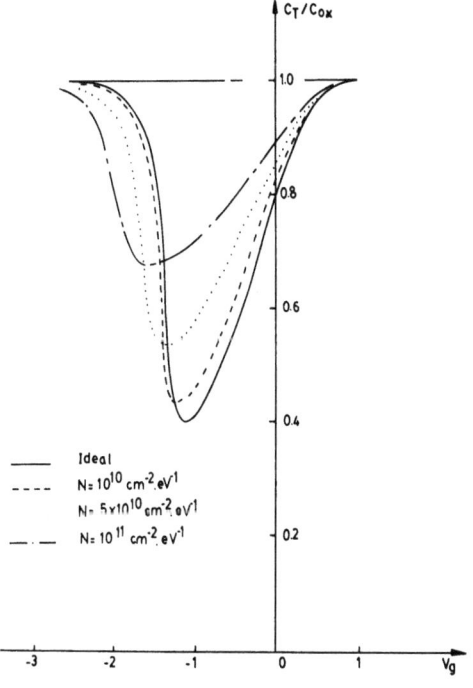

Fig. 23. Low frequency C_T-V_g curve for MOS capacitor, for various constant trap densities distributions.

Figure 23 illustrates the effect of distributed traps on the $C_T\text{-}V_g$ characteristics for various constant trap densities, i.e.,

$$N_i = \text{constant}$$

It is seen that the general effect of the traps is to broaden and make shallower the whole curve, without distorting its shape.

As a practical example of this technique we refer to Fig. 24 (a), which illustrates the low-frequency $C_T\text{-}V_g$ curve obtained from an MNOS device in which the interface density is known to be high. Figure 24(b) shows the ϕ vs. V_g curve obtained from the high-frequency $C_T\text{-}V_g$ measurements as described in Section II B. Figure 24(c) shows the C_T vs. ϕ derived Figs. 24(a) and (b). The chain-dotted line in Fig. 24(c) is the ideal C_T vs. ϕ curve for the sample, the doping density and insulator thickness having been determined from high frequency measurements as described in Section II B. Using Eq. (47) in the form, when the device is in depletion or inversion ($\phi > 0$)

$$C_i = \frac{C_T, C_{ox}}{C_{ox} - C_T} - C_d$$

where C_d, the depletion capacitance, is obtained from the high-frequency $C_T\text{-}V_g$ curve in the manner described in Section II B, we obtain N_i as a function of C_d and hence of ϕ. The result of this procedure is shown in Fig. 24(b).

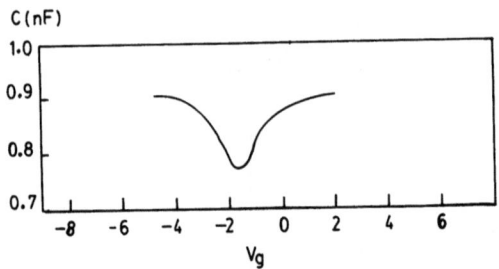

Fig. 24(a)

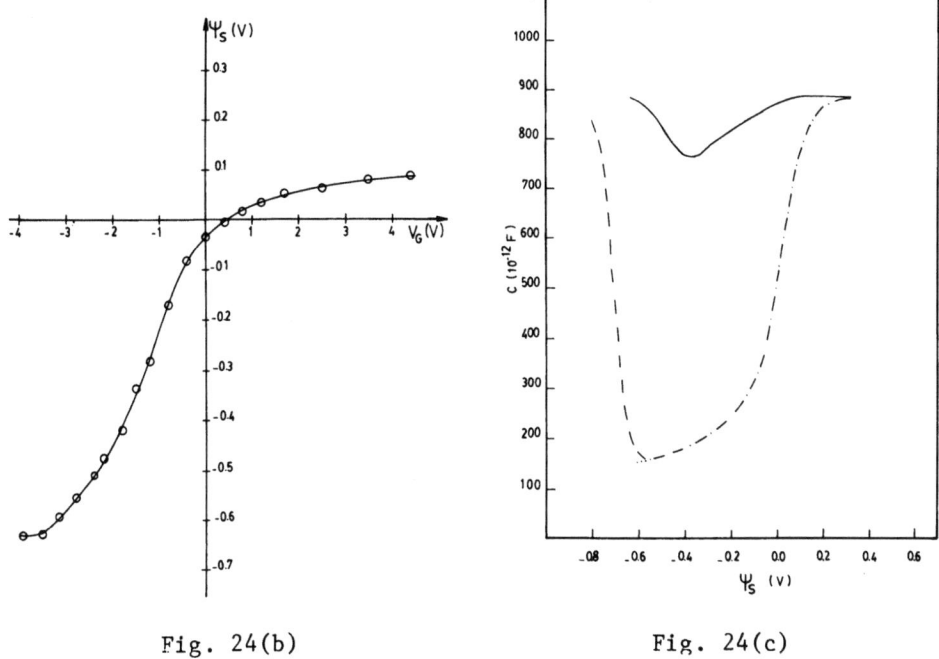

Fig. 24(b) Fig. 24(c)

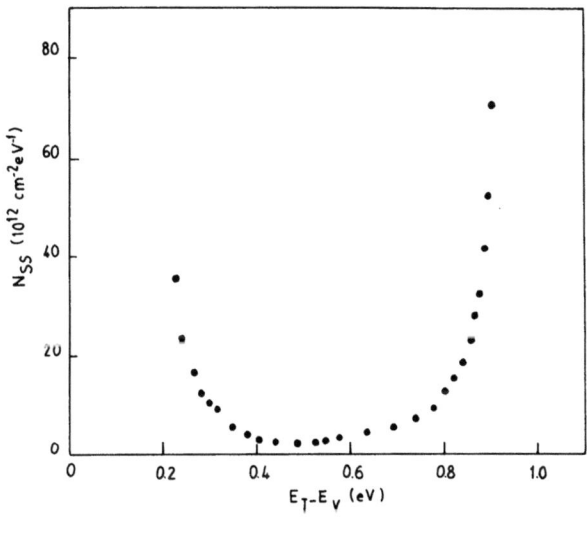

Fig. 24(d)

Fig. 24. Characteristics of an MNOS Device: (a) low frequency C_T-V_g characteristics; (b) ϕ_S vs. V_g characteristics; (c) C_T vs. ϕ (full line) and ideal (dotted line) C_T-ϕ curves; (d) trap distribution in the device.

III. CURRENT VOLTAGE MEASUREMENTS

As we showed in the previous chapter, low-frequency measurements yield a considerable amount of information on the properties of interface states in MOS systems. However, small-signal frequencies of less than one or two cycles per second are required to be used, in order to ascertain the system is in quasi-equilibrium. Hence, relatively elaborate and sensitive detection techniques are required.

A more elegant technique which achieves the same results and requiring a much simpler experimental technique is the method suggested by Kuhn [14], often referred to as the quasi-static technique, and which we will describe in the next section.

A. Quasi-Static Method

With this technique a linear varying voltage ramp is applied to the device (see Fig. 25):

$$V_g = V_o \pm \alpha t \tag{82}$$

where V_o is the initial applied voltage and α the (constant) voltage rate. The displacement current is measured as a function of the applied voltage. There is thus a simple relationship between current and capacitance as follows:

$$I = \frac{dQ_g}{dt} = \frac{dQ_g}{dV_g}\frac{dV_g}{dt} = \pm\alpha\frac{dQ_g}{dV_g} = \pm\alpha C_T \tag{83}$$

since by definition dQ_g/dV_g is the small-signal capacitance of the system. Consequently, provided that the voltage ramp is applied

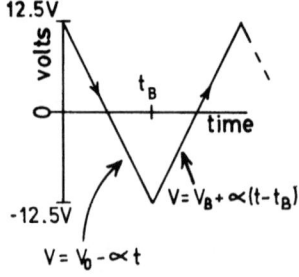

Fig. 25. Voltage waveform for quasi-static I-V curves.

CHARACTERIZATION OF MOS DEVICES

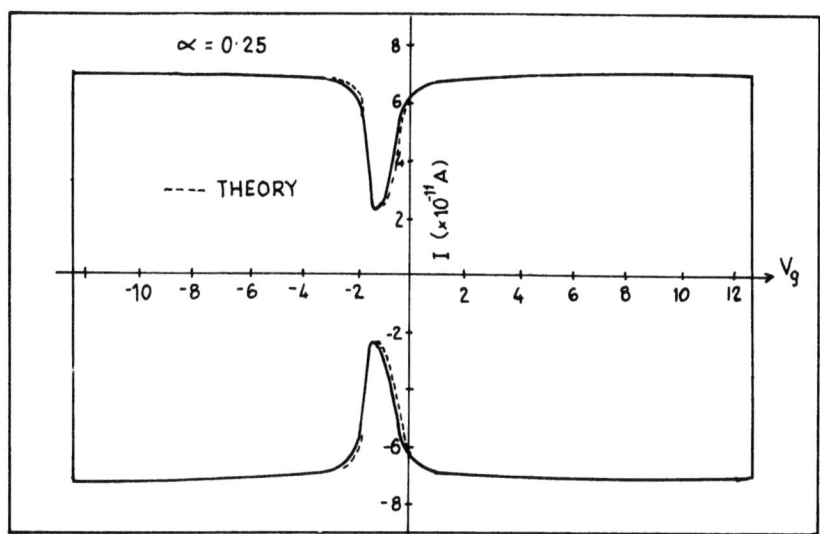

Fig. 26. Quasi-equilibrium current-voltage curve of the MOS sample at 330°K.

slowly enough so that the system stays in quasi-equilibrium, then the current is proportional to the low-frequency small-signal capacitance of the system. That this is the case is demonstrated in Fig. 26, which shows the quasi-static I-V curve obtained from a relatively good device, that is one with a low density of interface states. Note that the curve is symmetric about the voltage which is due to the fact that a triangular ramp (Fig. 25) was applied (see section III B). Superimposed on the curve is the I-V_g characteristic obtained using the relationship

$$I = \pm \alpha C_s$$

where C_s is given by Eq. (70). It is seen that the two curves correlate quite closely, the slight deviation being due to the low-density interface states.

B. Non-Steady-State I-V Curves

In generating the quasi-static curves described above, as mentioned the rate of change of the voltage ramp was kept sufficiently slow in order that the system remains in quasi-equilibrium. Under this condition, holes are being generated in the depletion region of the silicon at a sufficiently fast rate to satisfy the charge

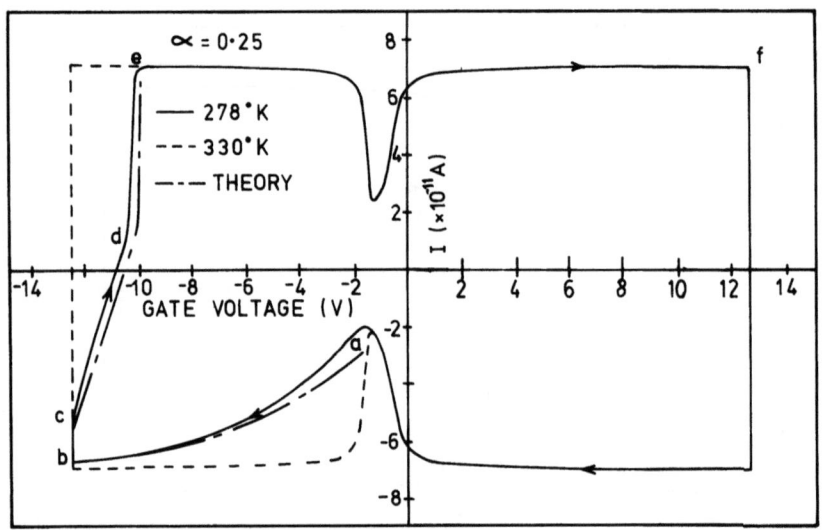

Fig. 27. Comparison between current-voltage curve at 278°K and quasi-equilibrium curve.

requirements at the silicon/silicon-oxide interface. It follows, however, that a particular value of α that satisfies quasi-equilibrium requirements at a particular temperature will not necessarily do so at a lower temperature, since the rate of generation of electron-hole pairs in the depletion region decreases as an exponential function of temperature.

This effect is demonstrated in the curves shown in Fig. 27. The dotted curve taken at 300°K is the quasi-equilibrium curve taken from Fig. 26. The curve taken at 278°K is strongly non-symmetric and thus a non-equilibrium curve. It is seen that on the negative-going voltage ramp, while the silicon is being depleted (0 to -12.5V, portion a-b of the curve), the current scales to levels below that of the quasi-equilibrium curve on account of the lower generation rate of electron hole pairs in the depletion region of the silicon. It should be noted that while the negative-going voltage ramp is applied, the current I flowing in the device comprises (see Fig. 28), both the current I_d discharging the donor centers at the edge of the depletion region, and the generation current I_g:

$$I = -(I_d + I_g)$$

On reversal of the voltage sweep, at -12.5V, the mirror-image of the I-V characteristic is not generated as in the quasi-equilibrium case.

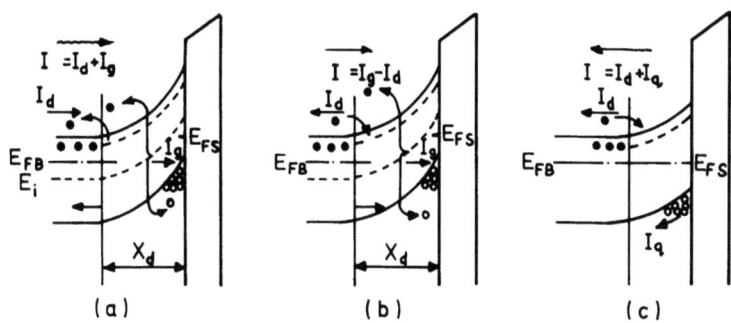

Fig. 28. Band diagram of MOS device with linear voltage sweep.
(a) end of forward sweep; (b) Start of reverse sweep
(c) Return to equilibrium.

This is because the generation current continues to flow in the same direction as before reversal, but I_d reverses since the donor centers at the edge of the depletion region now have to be filled: hence, the current I is now given by

$$I = I_g - I_d$$

Consequently, immediately on reversal of the voltage sweep, the current drops abruptly, as illustrated by the section b-c of the curves, and then increases positively as indicated by section c-d of the curves.

Clearly, in the initial stages of the positive-going sweep (-12.5-0V), $I_g > I_d$ since I is still negative. Only at the instance when the curve crosses the voltage axis does $I_g = I_d$, that is, I=0. Thereafter, I continues to increase positively, and when the Fermi level E_{FS} in the bulk of the silicon lines up with the Fermi level E_{FS} at the interface, the current rises almost vertically (section d-e of the I-V curve), until it reaches the quasi-equilibrium curve. The system is now in the quasi-equilibrium state, so now the I-V characteristic follows that of the quasi-equilibrium curve (portion e-f).

As the temperature is lowered, the I-V characteristics show distinctive traits, as illustrated by the curves in Fig. 29. These curves have been separated in the figure simply for reasons of clarity.

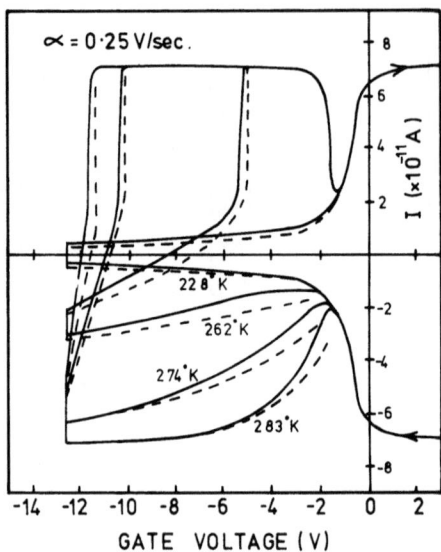

Fig. 29. Comparison of experimental and theoretical current-voltage curves at various temperatures. (full line - experimental; dotted line - theoretical). Parameters for theoretical curves $N_D = 7 \times 10^{14} \text{cm}^{-3}$, $v_{th} \sigma N_t = 2.5 \times 10^5 \text{sec}^{-1}$, $E_i - E_t = 0$.

It is seen that the portions of the curves corresponding to a-b of Fig. 29 scale to lower current levels with decreasing temperature because of the decrease in the electron-hole pair generation rate. The precipitous change in the current corresponding to portion b-c of Fig. 27 decreases with decreasing temperature, while the point at which the curves cut the voltage axis and the fast rise of the current (see portion d-c of Fig. 27) towards equilibrium occur at lower voltages the lower the temperature.

The fundamental equations governing the non-steady-state mode are obtained as follows:

$$I = \dot{Q}_g = -\dot{Q}_{inv} - \dot{Q}_d = -\dot{Q}_{inv} - qN_d\dot{X}_d$$

The rate of increase of inversion charge $\dot{Q}_i$ is due to generation of holes within the depletion region

$$\dot{Q}_{inv} = qU_g(X_d - X_0)$$

where $(X_d - X_o)$ is the effective width of the generation region and U_g is the bulk generation rate give by

$$U_g = \frac{\sigma N_b v_{th} n_i}{2 \cosh(\frac{E_b - E_{mid}}{kT})}$$

where E_{mid} is the energy at mid gap. Hence, we have

$$I = -qU_g(X_d - X_o) - qN_d \dot{X}_d \tag{84}$$

The gate voltage is obtained from Eq. (41) and (49c):

$$V_g = \frac{V_g}{C_{ox}} - \frac{qN_d X_d^2}{2\varepsilon_s} \tag{85}$$

which after differentiation with respect to time and transposing yields

$$I = C_{ox} - \frac{qC_{ox} N_d X_d \dot{X}_d}{\varepsilon_s} \tag{86}$$

Solution of Eq. (84), (85) and (86) yields the I-V_g characteristic.

Using the parameters, as determined above, and $N_d = 7 \times 10^{14} \text{cm}^{-3}$ $v_{th}\sigma N_b = 2.5 \times 10^5 \text{sec}^{-1}$, $E_{mid} - E_t = 0$, the theoretical plots illustrated by the dotted lines in Fig. 29 were generated. It is seen that the correlation between the theoretical and experimental curves are very good. To obtain a fit over the range of voltages and temperatures indicated requires a selection of the necessary parameters to within a few percent. In other words, the method is a very sensitive way of determining the bulk trap parameters. While it may be possible to fit the data using different parameters to those given above over the section say a-b of the I-V_g characteristic at a particular temperature, in all probability the other sections of the curve will not yield to the theoretical curves. In the propitious case where such a fit should occur over all sections of the I-V_g characteristic, the data taken at other temperatures will certainly not correlate with the theory.

The theoretical curves shown in Fig. 30 (a-d) demonstrate the sensitivity of the curves to variations in the trap parameters. A study of the trend of these curves with variation in trap parameters assists in eliminating the tedium that might appear to be associated with selection of the trap parameters. For example, an inspection of Figs. 30(a) and (b) clearly indicates that moving the trap away from mid-gap causes the section a-b of the curves to scale to lower current levels, a decrease in the drop b-c in current as the voltage ramp changes direction, and a reduction in the slope of the

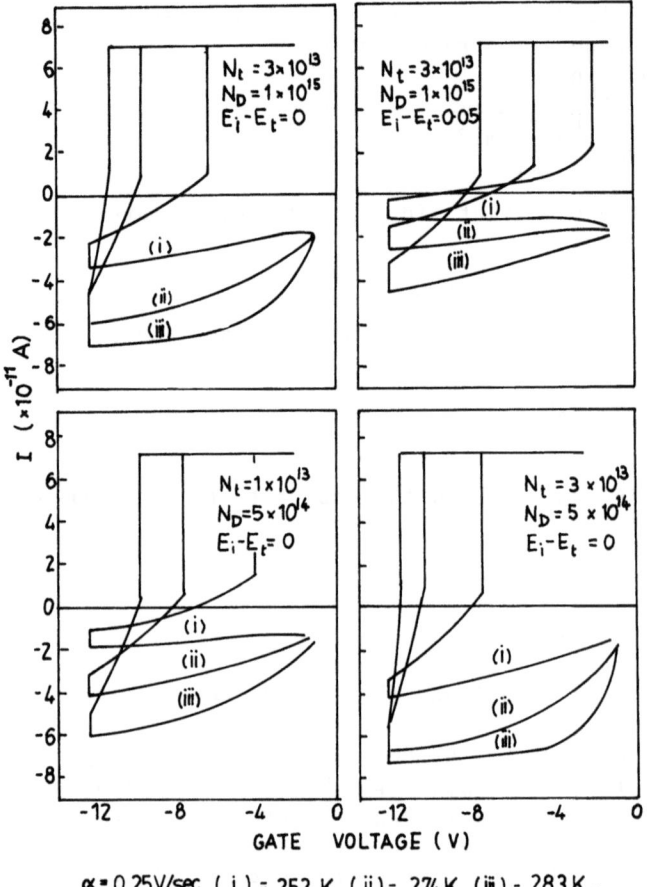

Fig. 30. Theoretical curves showing the effect of variations in N_t, N_D and $E_i - E_t$ at various temperatures.

section c-d of the curves. From the parameters used in generating the theoretical curves, we find that, since $E_{mid} - E_b = 0$ gives the best fit, that the traps involved in the generation process are at mid-gap. This does not, of course, necessarily imply that these are the only traps in the band-gap since distributed traps would also yield the same result. This is because, in such cases, the most efficacious traps involved in the generation process are those located a few kT in energy about the mid-gap, the generating efficiency of the traps $|E_{mid} - E_b|$ in energy away from mid-gap, falling off as $\exp(-|E_{mid} - E_t|/kT)$. Consequently, the free-carrier

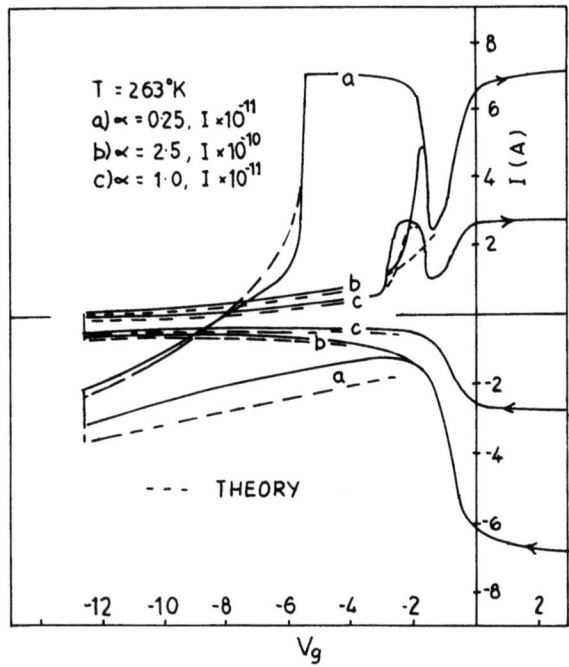

Fig. 31. Comparison of experimental and theoretical current-voltage curves at 263°K for various sweep rates.

generation lifetime may now be written as $\tau = (v_{th}\sigma N_t)^{-1} = 4 \times 10^{-6}$ sec. Furthermore, if we assume $v_{th} = 10^7$ cm sec^{-1} and $\sigma = 10^{-15}$ cm^2, then $N_b = 2.5 \times 10^{13}$ cm^{-3}.

C. Effect of Sweep Rate

A further test of the reliability of the theory and, hence, the trap parameters obtained from the temperature data discussed above, is how well the theory fits the data with sweep rate used as variable rather than temperature.

The full lines in Fig. 31 illustrate the experimental plots for various sweep rates, taken at T = 263°K. The dotted curves illustrate the theoretical curves for the trap parameters given above. Again we note the acceptable agreement between the experimental and theoretical plots.

D. Effect of Gold Doping

The devices, described earlier, hereafter referred to as the control devices, were doped with gold to illustrate the efficiency of the I-V method in studying both interfacial and bulk effects.

1. <u>Quasi-Static Measurements</u>. In Figures 32(a) and (b) we show the salient portions of the quasi-static curves for the control and doped devices, respectively. It is seen from Fig. 32(b) that the gold has introduced a relatively strong sharply peaking trap level at the interface, as indicated by the bump appearing the characteristic. Following the methods described in Section II in connection with the low-frequency method, the trap distribution corresponding to this bump was determined and the result is shown in Fig. 33(c). It is seen the trap distribution peaks sharply below the bottom of the conduction band, which is very close to the value reported for the location of the acceptor level introduced by gold dopant in the bulk silicon.

2. <u>Non-Steady-State Measurements</u>. Figure 33(b) illustrates the non-steady-state I-V characteristics for the gold-doped samples. An inspection of these curves shows that they are similar to those for the control devices (Fig. 27) but for a given temperature the current levels are higher (compare for example experimental control and gold curves at 262°K from Figs. 27 and 33(a)), reflecting the

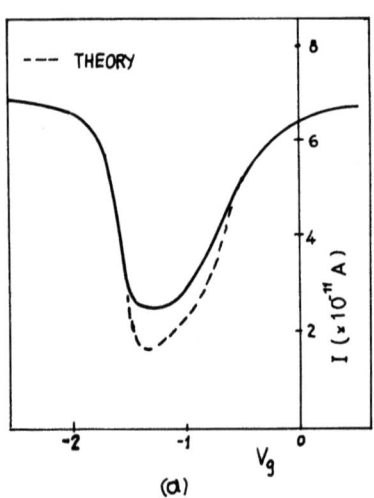
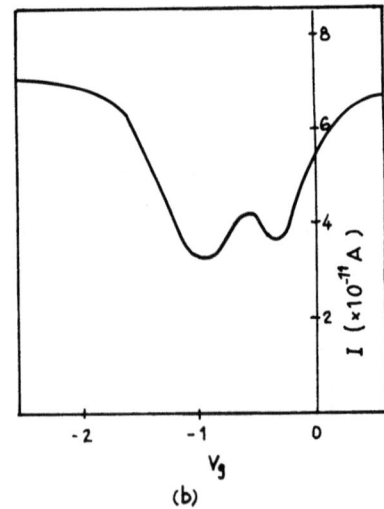

Fig. 32. Quasi-static I-C curves: (a) control sample; (b) gold-doped sample.

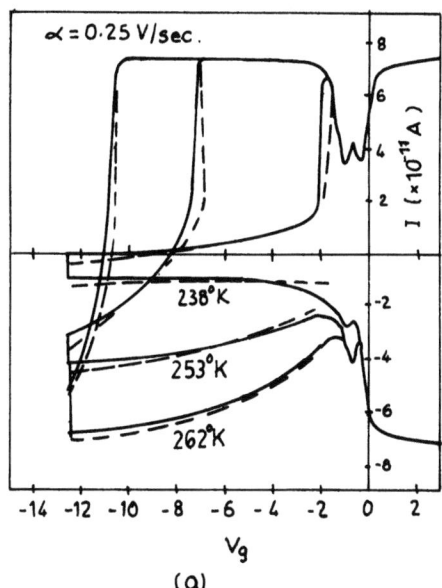 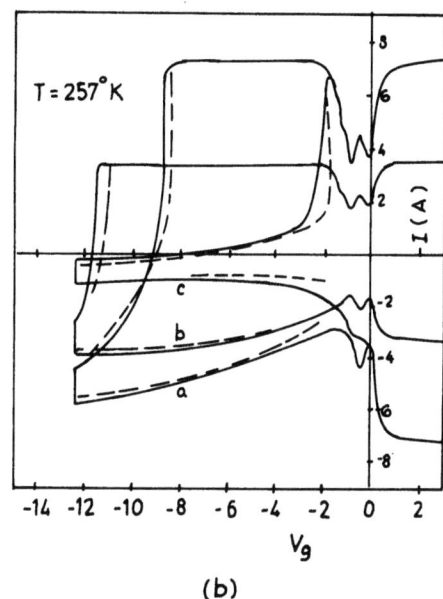

Fig. 33. Non-steady-state I-V curves for gold-doped devices: (a) at various constant temperatures; (b) at various constant sweep rates. (full line - experimental, dotted line - theoretical). Parameters for theoretical curves: $N_d = 5 \times 10^{-14} \text{cm}^{-3}$, $\sigma_p v_{th} N_{Au} = 1.1 \times 10^6 \text{sec}^{-1}$, $E_t - E_i = 0.02$ eV.

fact that the gold centers, and the bump due to gold centers at the interface is still apparent. Again these curves correlate well with the theory, as indicated by the dotted lines. The best fit to the experimental curves was obtained using $E_t - E_i = 0.02$ eV and $\sigma_p v_{th} N_{Au} = 1.1 \times 10^6 \text{sec}^{-1}$ where N_{Au} is the concentration of gold centers in the silicon, yielding a free-carrier generation lifetime of $\tau = (\sigma_p v_{th} N_{Au})^{-1} = 9.1 \times 10^{-7}$ sec. Hence, gold doping the silicon has reduced the lifetime by a factor of 4. If we use our estimate of $N_{Au} = 1 \times 10^{14} \text{cm}^{-3}$, we find the value for

$$\sigma_p = (\tau v_{th} N_{Au})^{-1} = 1.1 \times 10^{-15} \text{cm}^2$$

which compares favorably with previously published values [16-19].

A further test of the reliability of the method, is how well the theory fits the data with sweep rate used as variable rather than temperature. The full lines in Fig. 33(b) illustrate the experimental plots for various sweep rates, taken at T = 257°K. The dotted curves illustrate the theoretical plots for the trap parameters given above. Again we note the acceptable agreement between experimental and theoretical plots.

REFERENCES

1. C.G.B Garrett and W.H. Brattain, Phys. Rev. 99, 376 (1955).
2. R. Lindner, Bell Syst. Tech. J. 41, 830 (1962).
3. L.M. Terman, Solid-St. Electron. 5, 285 (1962).
4. K. Lehovec, A. Slobodskoy and J.L. Sprague, Phys. Status Solidi, 3, 447 (1963).
5. F. P. Heiman and G. Warfield, IEEE Trans. Electron Devices ED-12, 165 (1965).
6. K.H. Zaininger and G. Warfield, IEEE Trans. Electron Devices, ED-12, 179 (1965).
7. A.S. Grove, B.E. Deal, E.H. Snow and C.T. Sah, Solid-St. Electron. 8, 145 (1965).
8. M.V.Whelan, Philips Res. Rep. 20, 562 (1965).
9. C.N.Berglund, IEEE Trans. Electron Devices ED-13, 701 (1966).
10. P.V. Gray and D.M. Brown, Appl. Phys. Lett. 8, 31 (1966).
11. E.H. Nicollian and A. Goetzberger, Appl. Phys. Lett. 10, 60 (1967).
12. E.H. Nicollian and A. Goetzberger, Bell Syst. Tech. J. 46, 1055 (1967).
13. E. Arnold, IEEE Trans. Electron Devices ED-15, 1003 (1968).
14. M. Kuhn, Solid-St. Electron. 13, 873 (1970).
15. S. Luby, R.N. Lovjagin, N. Doshdikova, L.N. Alexandrov and J. Cervenak, Solid-St. Electron. 14, 571 (1971).
16. J. Koomen, Solid-St. Electron. 14, 571 (1971).
17. D.M. Brown and P.V. Gray, J. Electrochem. Socl. 115, 760 (1968).
18. A. Goetzberger and J.C. Irvin, IEEE Trans. Electron. Devices ED-15, 1009 (1968).
19. B.E. Deal, E.L. Mackenna and P.L. Castro, J. Electrochem. Soc. 116, 997 (1969).
20. J.G. Simmons and L.S. Wei, Solid-St. Electron. 16, 43 (1973).
21. D.K. Schroder and J. Guldberg, Solid-St. Electron. 14, 1285 (1971).
22. H. Preier, IEEE Trans. Electron Devices ED-15, 990 (1968).
23. F. P. Heiman, IEEE Trans. Electron Devices ED-14, 781 (1967).
24. C. Jund and R. Poirier, Solid-St. Electron. 9, 315 (1966).
25. J. Grossvalet, C. Jund, C. Motach and R. Poirier, Surface Sci. 5, 49 (1966).
26. A. Goetzberger and E.H. Nicollian, Bell Syst. Tech. J. 46, 513 (1967).
27. M. Zerbst, Z. Agnew, Phys. 22, 30 (1966).
28. J.G. Simmons and L.S. Wei, Solid-St. Electron. 17, 117 (1974).
29. J.G. Simmons and G.W. TAylor, Solid-St. Electron. 17, 125 (1974).
30. H. Mar and J.G. Simmons, Solid-St. Electron. 17, 131 (1974).

31. J.G. Simmons and L.S. Wei, Solid-St. Electron, 16, 53 (1973).
32. H.A. Mar and J.G. Simmons, Solid-St. Electron, 17 1181 (1974).
33. H.A. Mar and J.G. Simmons, Solid-St. Electron, 17, 1181 (1974).
34. S.M. Sze, Physics of Semiconductor Devices (J. Wiley 1969).

Chapter 5

SEMICONDUCTOR MATERIAL EVALUATION BY MEANS OF SCHOTTKY CONTACTS

Klaus Heime

Chair of Solid State Electronics

University of Duisburg

D 4100 Duisburg, Federal Republic of Germany

I. SCOPE OF THE PAPER

The large variety of possible application of semiconductor devices -- from high current, high voltage to microwave devices, from integrated optoelectronics -- is to a large extent due to the fact that semiconductors can be doped to a desired type and magnitude of conductivity. This is done by incorporating impurities whose energy levels in the forbidden gap are close to valence or conduction band and which are therefore completely ionized at normal temperatures of device operation. These levels are called shallow levels. The first part of this paper will discuss the principal methods for the evaluation of the concentration and local distribution of these shallow levels by means of Schottky contacts.

Simultaneously or independently during material or device processing another type of impurities can be incorporated with energy levels deep in the forbidden gap. These levels are called deep levels. Generally, deep levels are not fully ionized but their charge state can be influenced by electrical or magnetic fields, optical illumination and temperature variations during device operation. It is this change in charge state that may have deleterious effects on mobility, conductivity, device behaviour under dc bias, efficiency and life-time of optoelectronic devices. The second part of this paper will concentrate on a discussion of the methods by which the concentrations and the energy levels of "deep impurities" can be evaluated by means of Schottky contacts.

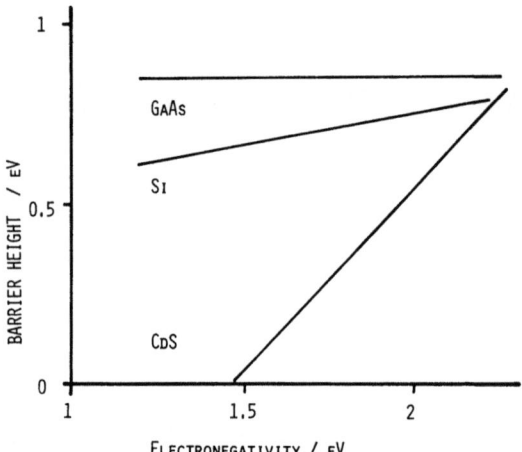

Fig. 1. Barrier height of metal contacts on semiconductors (simplified).

This diagnosis of deep levels is <u>one</u> step towards a thorough therapy of the semiconducting patient. It is a necessary step in our aim at completely healing the semiconductor of the "deep level malady".

II. INTRODUCTION

At semiconductor surfaces or at interfaces between a semiconductor and another solid, deviations from the bulk equilibrium values of charge and potential distributions are expected [1].

Energetic levels different from those in the semiconductor bulk are generated by a number of reasons: the interruption of the periodicity of the crystal lattice at the surface or interface; the adsorption of atoms and molecules from the ambient; the chemical reaction between the semiconductor and the ambient (including solids); work function differences between the semiconductor and a solid. If the surface or interface levels are charged (surface or interface charges) their charge has to be compensated by a charge of opposite polarity but of the same magnitude. This is <u>the principle of charge neutrality</u> which is a very effective tool for the calculation of the surface properties. Under zero bias conditions the surface or interface charges are compensated by charges in the semiconductor bulk close to the surface. These compensating charges can be:

a) majority carriers accumulated near the surface;

b) ionized donors or acceptors unneutralized by mobile charge carriers;

c) minority carriers accumulated near the surface.

If we have an N-type semiconductor doped with N_D [cm^{-3}] shallow donors and with N_S [cm^{-2}] surface (interface) states which are partially negatively ionized ($N_S^- \leq N_S$), then this surface charge has to be compensated. The compensation is possible by removing electrons from a region in the semiconductor close to the surface, thus producing a space-charge region of unneutralized donors of thickness ℓ. Charge-neutrality demands that

$$N_S^- = N_D^+ \cdot \ell \tag{1}$$

The space-charge region depleted of mobile charge carriers is called the depletion region. With high surface charge densities N_S^- compensation can be accomplished not only by donors but also by mobile defect electrons (holes). Then an <u>inversion</u> layer (P-type instead of N-type in the bulk) is generated near the surface. If the surface states are positively charged the charge-

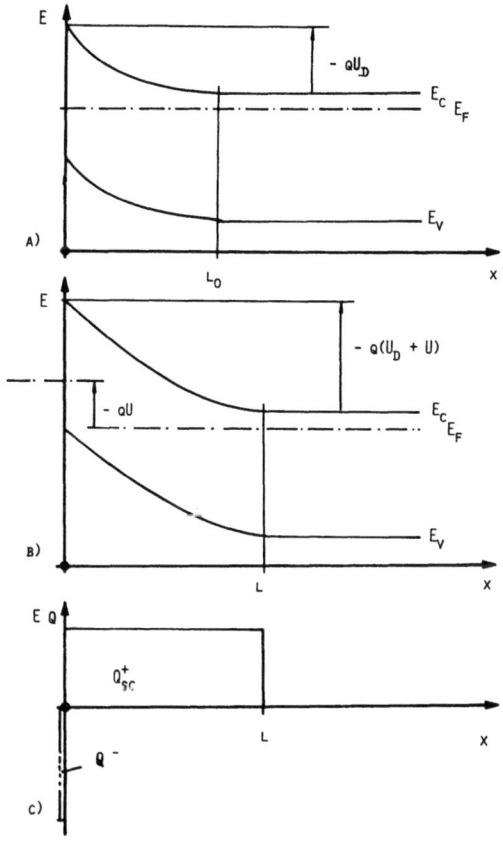

Fig. 2. (a,b) Band diagram of a Schottky contact for $U = 0$; $U < 0$.
(c) Charge distribution for $U < 0$

neutrality is effected by an <u>accumulation</u> of electrons. The potential distribution $\phi(x)$ [V] and the energy band bending $E(x)$ [V], with $E \sim -q\phi$, are related to the carrier concentration n by

$$n = N_c \exp(-\zeta/kT) \tag{2}$$

$$\zeta = E_c - E_f \tag{3}$$

$$N_c = 2(2\pi m_n^+ kT/h^2)^{3/2}$$

where N_c equals the effective density of states of the conduction band, m_n^+ equals the electron effective mass, k equals the Boltzmann constant, T is the absolute temperature, h is the Planck constant, E_c is the conduction band edge and E_f is the Fermi level.

Hence a decrease of carrier concentration towards the surface is followed by an "upward bending" of the conduction band edge E_c. If we assume that the band gap $E_g = E_c - E_v$ is not influenced by the surface, the bending of the valence band edge E_v is parallel to E_c. As a result, a potential difference

$$\Delta\phi = [E_c(x \to \infty) - E_c(x=0)] q^{-1} \tag{4}$$

$$= [E_v(x \to \infty) - E_v(x=0)] q^{-1}$$

exists between the surface and the semiconductor bulk. If the work functions of the metal and semiconductor are ϕ_{MeVac} and ϕ_{HLVac}, respectively, then the difference in work functions

$$\phi_{MeHL} = \phi_{Me\ Vac} - \phi_{HL\ Vac} \tag{5}$$

must be added to the potential difference $\Delta\phi$ (equation 4).

In 1929, Schottky suggested [2] that this work function difference explained the nonohmic current-voltage characteristics of metal-semiconductor contacts (Schottky contacts). If this were true, $\phi_{Me\ HL}$ for a given semiconductor ($\phi_{HL\ Vac}$ = const) should be proportional to $\phi_{Me\ Vac}$.

Figure 1 shows that this is obeyed for some II-VI compounds but is not true for the III-V semiconductors under consideration [3]. In III-V semiconductors, surface charges play an important role in the formation of surface band bending. In this paper we are interested in analyzing the semiconductor bulk close to the surface. We simply assume that a certain band bending is given. Its origin will not be investigated. The system under investigation is defined as shown in Figure 2.

Near the surface of the semiconductor to be evaluated, a depletion region and band bending of magnitude ($-qU_D$) exists. The band bending is due to one or more of the facts discussed above. If we apply an external voltage across the Schottky contact additional charge is induced on the metal plate. Due to the principle of charge neutrality, an equal charge of opposite polarity is induced in the semiconductor "plate", thus increasing or decreasing the zero-voltage space-charge and the thickness of the space-charge region. According to the definition in the insert of Fig. 2(a), $U > 0$ means the forward direction for the diode current and $U < 0$ the reverse direction.

Varying the space-charge Q_{SC} by an ac-voltage u of frequency $f = \omega/2\pi$ gives rise to a capacitance

$$C_{SC} = \frac{dQ_{SC}}{dU} \tag{6}$$

$$Q_{SC} = qN_D^+ \ell \tag{7}$$

Hence a measurement of the space-charge capacitance yields information on N_D. If deep levels are present and their charge state can be influenced externally, an additional change in the space-charge capacitance C_{SC} is observed. Therefore the measurement of the capacitance C_{SC} of a Schottky contact is one of the most effective tools for material evaluation.

III. EVALUATION OF SHALLOW LEVELS

A space-charge region is a three-dimensional structure. In general the dopant concentration may vary in all three directions. The total dopant concentration can be composed of several different dopant species: both shallow and deep donors and acceptors. The calculation of the capacitance-voltage characteristics of Schottky contacts will start with the simplest case: one kind of shallow levels with locally constant concentration. The more complicated cases will be introduced step by step.

A. One-Dimensional Theory

1. <u>One kind of shallow levels with locally constant dopant concentration</u>. We assume a n-type semiconductor uniformally doped with N_D completely ionized donors

$$N_D = N_D^+ \tag{8}$$

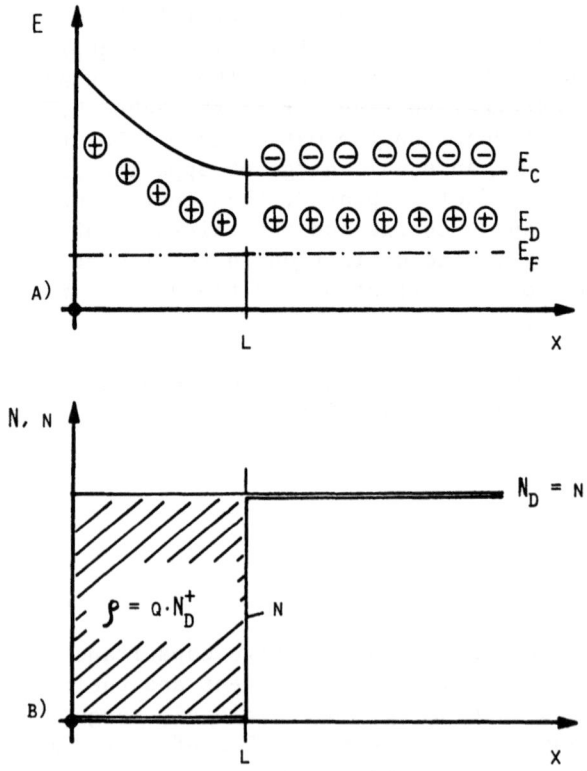

Fig. 3. (a) Band diagram and charge distribution in a Schottky contact. (b) Space charge in the depletion region.

If N_D is much larger than the intrinsic carrier concentration n_i, we can write without serious error

$$n = N_D^+ = N_D; \quad p = n_i /N_D^+ \tag{9}$$

Surface charges and/or work function differences will have generated a depletion region close to the surface. Figure 3 shows the charge distribution and the band bending in the system under consideration. The energy E of the electrons are related to the potentials in the semiconductor by

$$E = -q\phi \tag{10}$$

The curvature of the potential due to space charges is described by Poisson's equation:

$$\frac{d^2\phi}{dx^2} = \frac{-\rho}{\varepsilon_r\varepsilon_0} \tag{11}$$

where ρ is the space charge density in cm^{-3} ε_0 is the permittivity of free space ε_r is the dielectric constant of the semiconductor.

In the Schottky diode under consideration, a space-charge exists in those regions where $N_D^+ \neq n$. For simplicity we assume that the carrier concentration n increases very rapidly from $n = 0$ to $n = N_D^+$ at the end of the space-charge region. The justification of this simplification will be discussed in this chapter as well as possible consequences in section III-2b. The simplification yields

$$\rho = \begin{cases} qN_D^+ & \text{for } x \leq \ell \\ 0 & \text{for } x \geq \ell \end{cases} \tag{12}$$

Hence equation (11) reads

$$\frac{d^2\phi}{dx^2} = -\frac{qN_D^+}{\varepsilon_r\varepsilon_0} \quad \text{for } x \leq \ell \tag{13}$$

Integrating equation (13) twice with respect to x yields

$$\phi(x) = -\frac{qN_D^+}{2\varepsilon_r\varepsilon_0} x^2 \tag{14}$$

i.e., a parabolic band bending occurs in a locally constant space-charge layer. Including the boundary conditions (see Fig. 2(a)).

$$\phi(x = \ell_0) = 0; \quad \phi(x = 0) = -U_D; \quad U = 0 \tag{15}$$

yields the total potential drop (called diffusion voltage) across the depletion region:

$$U_D = -\frac{qN_D^+}{2\varepsilon_r\varepsilon_0} \rho_0^2 \tag{16}$$

Introducing the Debye length L_D

$$L_D = \sqrt{\frac{2\varepsilon_r\varepsilon_0 kT}{qN_D^+ q}} = \sqrt{\frac{2\varepsilon_r\varepsilon_0 U_T}{qN_D^+}}; \quad U_T = \frac{kT}{q} \tag{17}$$

and inserting this into equation (16), results in

$$U_D = -U_T \left(\frac{\ell_0}{L_D}\right)^2 \tag{18}$$

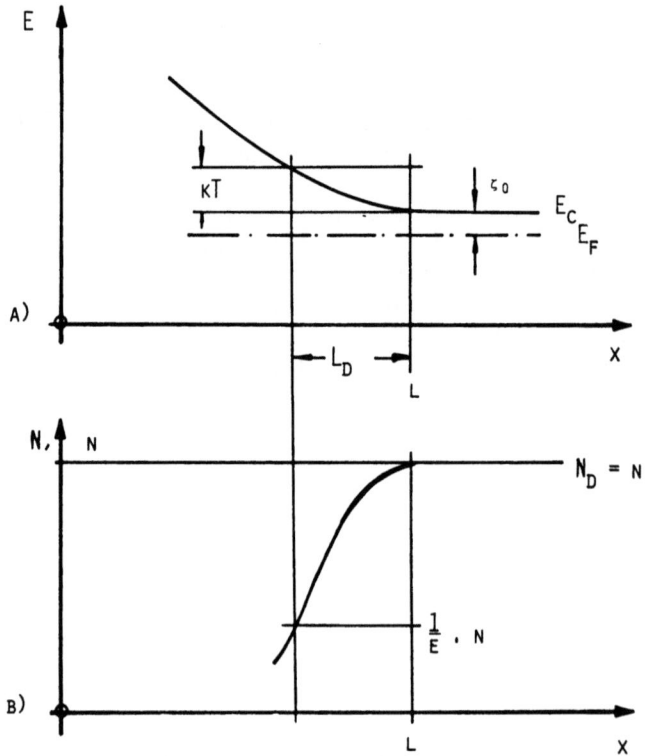

Fig. 4. Definition of the Debye length L_D.

L_D is the distance in which the carrier concentration decreases by a factor of e^{-1} at the edge of the depletion region. This can be shown as follows (See Fig. 4): according to equation (2) the carrier concentration is uniquely related to the difference between the conduction band edge and the Fermi level E_F.

In the semiconductor bulk

$$n = N_c e^{-\zeta_0/kT} = N_c e^{-\zeta_0/qU_T}$$

To what value does n decrease if ζ_0 increases to $\zeta_0 + qV_T$

$$\frac{n(\zeta_0 + qU_T)}{n(\zeta_0)} = \frac{\exp(+\frac{\zeta_0}{qU_T})}{\exp(+\frac{\zeta_0}{qU_T} + 1)} = e^{-1}$$

Using equation (18) with $U_D = U_T$ shows that this decrease in carrier concentration occurs within a distance $\ell_0 = L_D$. Consequently, if the barrier height qU_D is much larger than qU_T, then the thickness

ℓ_0 of the space-charge layer is much larger than L_D and the assumptions in equation (12) are justified. If an external voltage is applied, U_D has to be replaced by $U_D + u$ and ℓ is a function of u:

$$U_D + u = -\frac{U_T}{L_D^2} \cdot \ell^2(u) \tag{19}$$

The capacitance C_{sc} of the space-charge region is obtained from equation (6):

$$C_{sc} = \left|\frac{dQ_{sc}}{du}\right| = qN_D^+ \left|\frac{d\ell}{du}\right|$$

The magnitude $|d\ell/du|$ is used here, because C_{sc} is defined as a positive value, only. From equation (19) it follows

$$\ell = L_D \sqrt{\frac{-U_D - u}{U_T}} \; ; \quad \ell \geq 0 \text{ for } u \leq -U_D \tag{20}$$

and

$$\left|\frac{d\ell}{du}\right| = \frac{L_D}{2U_T(\frac{-U_D-u}{U_T})^{1/2}} \tag{21}$$

Inserting equation (21) into equation (6) and using equation (17) for L_D we arrive at

$$C_{sc} = \left(\frac{\varepsilon_r \varepsilon_0 qN_D^+}{2(-U_D-u)}\right)^{1/2} \; ; \quad u \leq -U_D \tag{22}$$

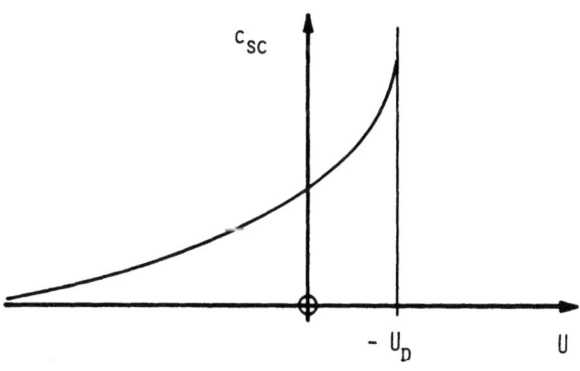

Fig. 5. Capacitance-voltage relation of a Schottky contact.

This equation is plotted in Figure 5. If equations (17) and (19) are used to eliminate N_D^+ and $U_D + u$ from equation (22) one obtains

$$C_{sc} = \frac{\varepsilon_r \varepsilon_0}{\ell} \qquad (23)$$

The space-charge capacitance behaves like a parallel plate condensor with a separation ℓ between the plates. If a "forward" voltage $u = -U_D > 0$ is applied no band-bending exists. This is called the "flat-band" case. Simultaneously $\ell = 0$ and $C_{sc} \to \infty$. For voltages $u > U_D$, an accumulation layer of electrons instead of a depletion layer is generated. Equation (12) and consequently equation (22) are no longer valid.

The dopant concentration $N_D = N_D^+$ can easily be determined by plotting C_{sc}^{-2} as a function of u. From equation (22) it follows (see Fig. 6):

$$\frac{1}{C_{sc}^2} = K(-U_D - u); \quad K = \frac{2}{\varepsilon_r \varepsilon_0 q} \frac{1}{N_D^+} \qquad (24)$$

The slope of the straight line is determined by the dopant concentration:

$$N_D = N_D^+ = \frac{2}{\varepsilon_r \varepsilon_0 q} \frac{1}{K} = \frac{2}{\varepsilon_r \varepsilon_0 q} \left(\frac{d(\frac{1}{C_{sc}^2})}{du}\right)^{-1} \qquad (25)$$

From equation (24) and Fig. 6 the potential difference U_D is easily determined from the intersection of the straight line with the voltage axis:

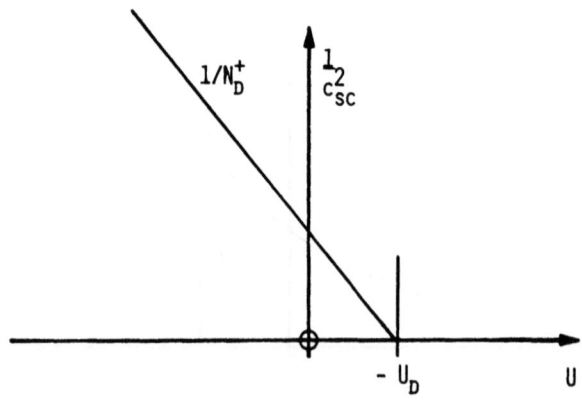

Fig. 6. Same as Fig. 5, but $1/C_{sc}^2$ vs. U.

$$u(C_{sc}^{-2} = 0) = -U_D \tag{26}$$

Conclusion: From a measurement of the capacitance-voltage characteristic of a Schottky contract the dopant concentration N_D of the semiconductor and the diffusion voltage U_D can be determined.

Hint: A P^+N-junction (with $N_A^- \gg N_D^+$) can be used for the evaluation of the dopant concentration N_D, too, because for this type of junction the space-charge layer extends almost completely into the low doped N-region as it does for metal-semiconductor junction. The formulas deduced above hold for the P^+N-junction as well [4].

2. One Kind of Shallow Levels with Locally Varying Dopant Concentration.

2 a. Smoothly varying dopant concentration. If the dopant concentration N_D varies smoothly in x-direction but remains constant in planes parallel to the metal contact, such that

$$N_D^+(x) = N_D^-(x) \quad \text{for all } x \tag{27}$$

then equation (23) remains valid:

$$C_{sc} = \frac{\varepsilon_r \varepsilon_0}{\ell}. \tag{23}$$

From equation (25) it follows that

$$N_D(\ell) = N_D^+(\ell) = \frac{2}{\varepsilon_r \varepsilon_0 q} \cdot \frac{d(1/C_{sc}^2)^{-1}}{du} \tag{28}$$

The plot $C_{sc}^{-2}(u)$ is no longer a straight line (see Figure 7), but the slope of the curve is still determined by the dopant

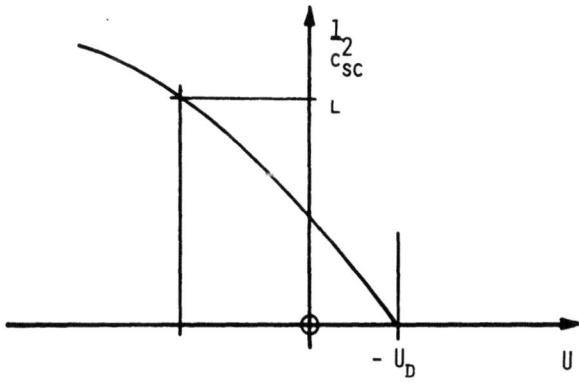

Fig. 7. Same as Figure 6, but N_D locally varying.

concentration at the edge of the depletion region, while the thickness of the depletion region is given by equation (23). Hence the local variation $N_D(x)$ is obtained.

2 b. __Abruptly Varying Dopant Concentration__. The situation becomes more complicated if the dopant concentration N_D varies abruptly, i.e., if N_D increases by orders of magnitude within a few Debye lengths. Then the assumption given by equation (12) is no longer valid, and a more detailed analysis is necessary [5].

The situation is illustrated in Figure 8(a). The donor concentration increases abruptly from a low value N_{Dlow} to a high value N_{Dhigh}. It is again assumed that all donors are completely ionized. The electron concentration, $n_o(x)$, increases from $n_o = N_{Dlow}^+$ to $n_o = N_{Dhigh}^+$ over a distance of several Debye lengths L_{Dlow} and L_{Dhigh} respectively. Therefore, a space-charge region

$$\rho = q[N_D^+(x) - n_o(x)] \tag{29}$$

is established. If a Schottky contact is applied to the low concentration side of the semiconductor, the electron and hole densities in the depletion region of the contact are zero. At some reverse voltage u_1 the electron distribution $n_1(x)$ is slightly different from $n_o(x)$ under equilibrium conditions. At $u_2 = u_1 + du$, there is a new electron distribution $n_2(x)$. The difference in space charge is

$$dQ = -q\,dn(x) = -q[n_2(x) - n_1(x)] \tag{30}$$

and is shown schematically in Figure 8(a). The distributions n_1, n_2 reflect the true dopant profile $N_D(x)$ very poorly. Detailed information on $N_D(x)$ cannot be obtained to better than several Debye lengths. The problem becomes even clearer if the Schottky contact is applied to the highly doped side of the semiconductor (see Fig. 8(b)). Then n_1, n_2 are different not only from N_D but also from n_o and therefore neither N_D nor n_o can be evaluated.

The problem of abruptly varying dopant concentrations can be treated in the same manner as constant concentrations, however, the electron concentration has to be included. The space-charge under applied voltage U is now (with $p(x) \equiv 0$):

$$\rho = q[N_D^+(x) - n(x)] \tag{31}$$

For the electron concentration we may write (see equation (2)):

$$n(x \to \infty) = N_c \exp(-\zeta_0/kT) \tag{32}$$

$$= N_D^+(x \to \infty) = N_D(x \to \infty)$$

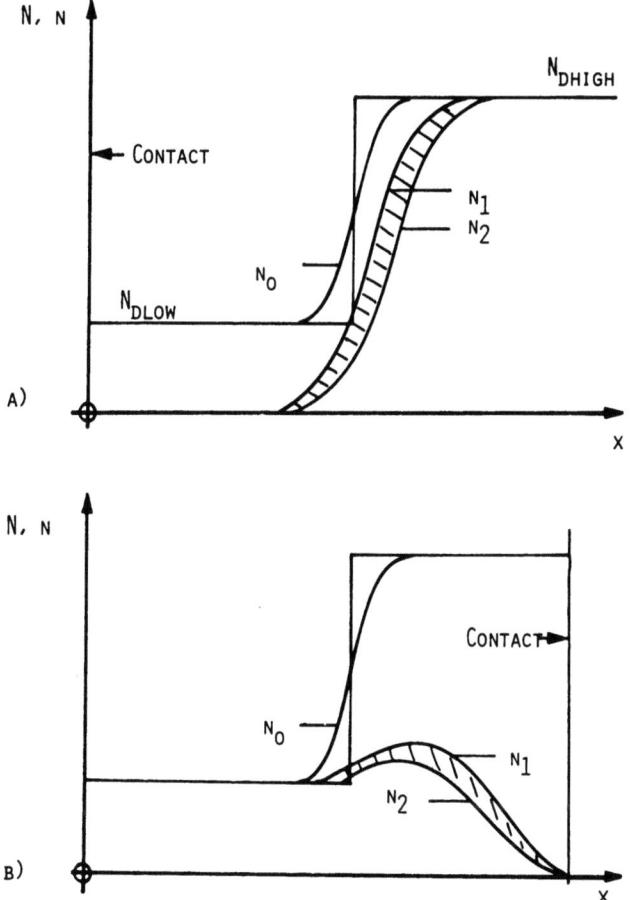

Fig 8. Donor and electron distribution with the Schottky contact on:
(a) the low concentration side; (b) the high concentration side [5].

$$n(x) = N_C \exp[-(\zeta_0 + q\phi)/kT] \qquad (33)$$

$$= N_D(\infty) \exp(-q\phi/kT)$$

Thus the Poisson equation is

$$\frac{d^2\phi}{dx^2} = -\frac{q}{\varepsilon_r \varepsilon_0} [N_D^+(x) - N_D(\infty) \exp(-q\phi/kT)]$$

The boundaries for the solution of equation (34) are:

$$\phi = 0 \quad \text{for } x \to \infty \tag{34a}$$

$$\phi = + U_D + U \quad \text{for } x = 0 \tag{34b}$$

Numerical solutions of equation (34) were obtained for several dopant profiles. All quantities are normalized:

$$N_{Dn} = \frac{N_D}{N_{Dhigh}} \qquad x_n = \frac{x}{L_{Dhigh}}$$

$$n_n = \frac{n}{N_{Dhigh}} \qquad u_n = \frac{U_D + u}{U_T}$$

$$n_{no} = \frac{n_o}{N_{Dhigh}}$$

N_{Dn}^* is the normalized dopant profile deduced from capacitance-voltage plots via equations (23), (28). The following examples were calculated numerically:

1. The concentration N_D varies abruptly by a factor of 100, the Schottky contact is on the low concentration side (see Figure 9),

2. Same as 1. However, the Schottky contact is on the highly doped side (Figure 10).

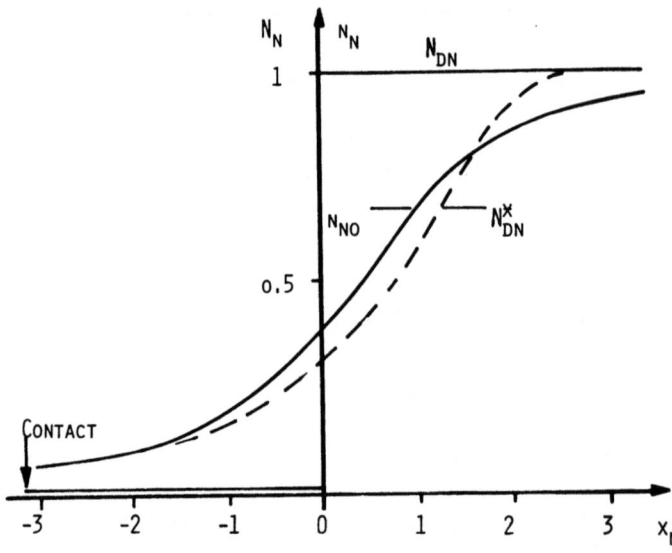

Fig. 9. Description see text.

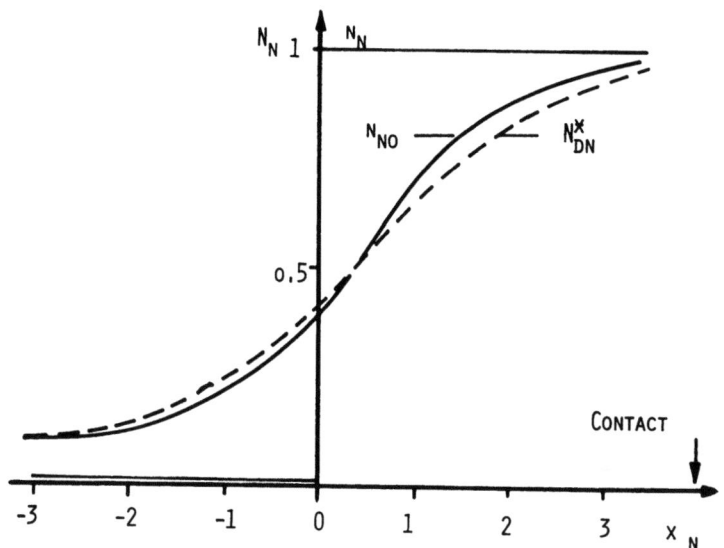

Fig. 10. Description see text.

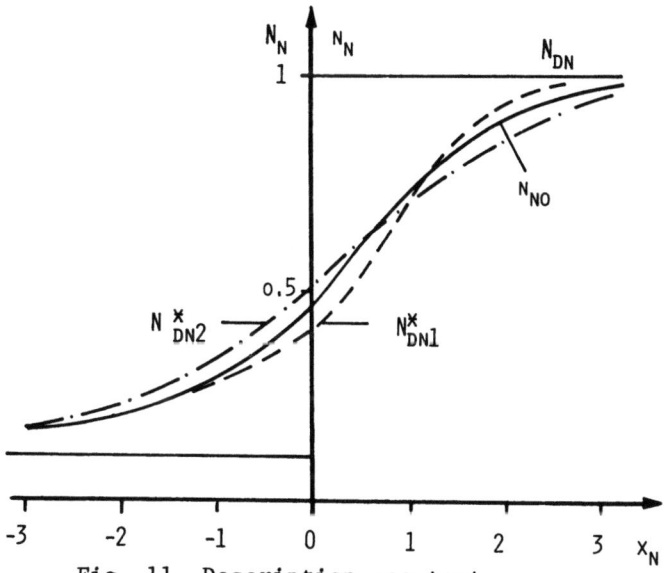

Fig. 11. Description see text
N_{Dn1}^*: Depletion from the low doped side
N_{Dn2}^*: Depletion from the high doped side.

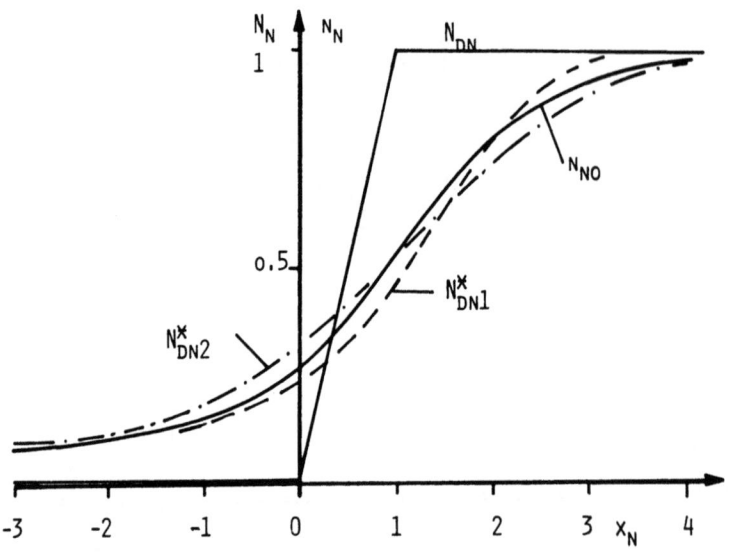

Fig. 12. Description see text.

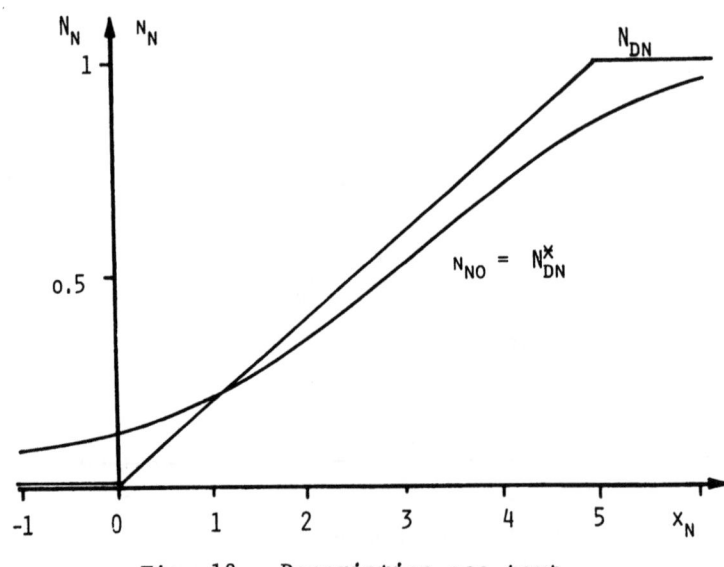

Fig. 13. Description see text.

3. Same as 1 and 2. However, $N_{Dhigh}/N_{Dlow} = 10$ (Figure 11).

4. The concentration varies by a factor of 100 within one Debye length L_{Dhigh} (Fig. 12).

5. The concentration varies by a factor of 100 within five Debye lengths L_{Dhigh} (Fig. 13). $N_{Dn1}{}^*$; $N_{Dn2}{}^*$; n_{no} coincide within a few percent.

The results very clearly show that the apparent profile is almost independent of details of the real profile within 5 - 10 L_D.

Conclusion: If dopant profiles vary considerably within a few Debye-lengths, capacitance-voltage measurement do not resolve the true donor profile but rather an "average" profile.

3. More Than One Kind of Shallow Levels. If several shallow levels of both donor or acceptor type are present, then equation (12) for the space-charge in the depletion region has to be replaced by

$$\rho = q(\sum_i N_{Di}{}^+ - \sum_k N_{Ak}{}^-) \qquad (35)$$

i.e., the net space-charge. Consequently, capacitance-voltage measurements yield the net donor or acceptor concentration (whatever is larger). Most frequently one donor and one acceptor is present:

$$\rho = q(N_D{}^+ - N_A{}^-) \qquad (35a)$$

The net donor concentration $N_D{}^+ - N_A{}^-$ is equal to the electron concentration n (with p << n) in equilibrium outside the depletion region: $n = N_D{}^+ - N_A{}^-$.

B. Corrections To The One-Dimensional Theory

1. Edge Effect. So far, the calculations presented have been one-dimensional, including variations of concentrations, fields and potentials in x-direction only. However, the Schottky contact is a two-dimensional structure of finite dimensions. The depletion region extends not only in x-direction but also in the y-z plane (Fig. 14).

Thus an error is introduced if the area A is used in the calculation of the capacitance by equation (23):

$$C_{sc} = \frac{C_{sc}}{A} \qquad (36)$$

Where C_{sc} is the measured capacitance and A is πr^2, the area of the metal contact. This error depends on the applied voltage u, the diameter of the metal contact and the dopant concentration.

For the following calculation, a locally constant dopant concentration is postulated. Then the depletion length is the same in all directions (Fig. 14). C_{sc} is composed of two parts:

$$C_{sc} = C_{sc_0} + C_{sc_1} = C_{sc_0}(1 + \frac{C_{sc_1}}{C_{sc_0}}) \tag{37}$$

with

$$C_{sc_0} = \frac{\varepsilon_r \varepsilon_0 A}{\ell} = \left(\frac{\varepsilon_r \varepsilon_0 q N_D^+(\ell)}{2(-U_D - u)}\right)^{\frac{1}{2}} \cdot A \tag{37a}$$

the capacitance from which N_D^+ is to be determined, and C_{sc_1} an additional capacitance due to the edge effect. For a very approximate solution [6] we define

$$C_{sc_1} = \frac{\varepsilon_r \varepsilon_0 A'}{\ell} \tag{38}$$

where A' is the area of the interface formed by the curved part of the depletion region edge.

$$A' = 2(r + \ell) \cdot \frac{2\pi\ell}{4} = \pi^2 \ell(r + \ell) \tag{39}$$

Inserting A' into equation (33) yields

$$C_{sc_1} = \varepsilon_r \varepsilon_0 \pi^2 (r + \ell) \tag{40}$$

Inserting into equation (37) leads to

$$C_{sc} = C_{sc_0}\left(1 + \pi \frac{(r + \ell)\ell}{r^2}\right) \tag{41}$$

$$C_{sc} = C_{sc_0}\left(1 + \frac{\pi\ell}{r}\right) \qquad \ell \ll r \tag{42}$$

Thus C_{sc} has to be divided by a factor $(1 + \pi\ell/r)$ in order to obtain C_{sc_0} from which N_D^+ is determined. Since C_{sc_1} was calculated as if it were a parallel plate capacitor with plate area A', the correction factor is too large. More accurate calculations [7] have shown that the true correction factor is

$$(1 + b\frac{\ell}{r}) \quad ; \quad b = 1.5 \tag{43}$$

EVALUATION BY SCHOTTKY CONTACTS

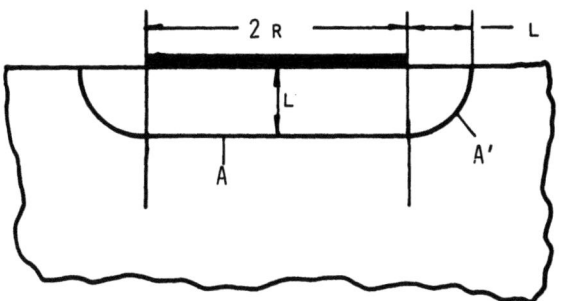

Fig. 14. Lateral extension of the depletion region of a Schottky contact.

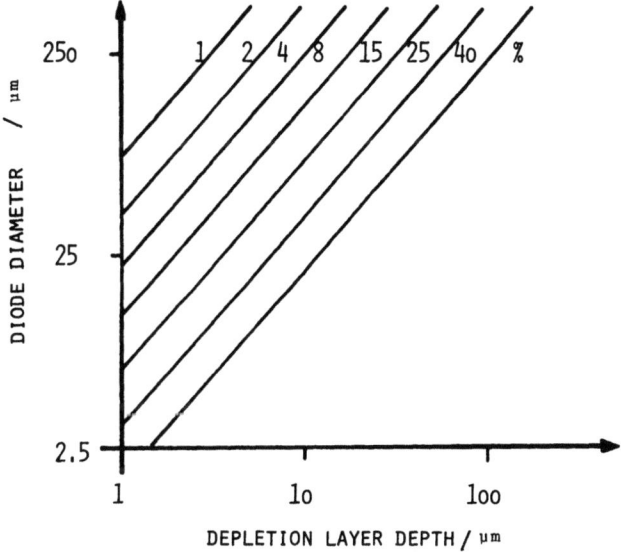

Fig. 15. Error (%) due to the edge effect [7].

Since the apparent dopant concentration N_{Dapp}^+ is

$$N_{Dapp}^+ = \frac{2}{\varepsilon_r \varepsilon_0 q} \left(\frac{d(1/C_{sc}^2)}{du} \right)^{-1} = \frac{C_{sc}^3}{\varepsilon_r \varepsilon_0 q \; dC_{sc}/du}$$

the correction factor to N_D^+ is

$$(1 + b\frac{\ell}{r})^3 \approx 1 =+ 3b\frac{\ell}{r} \qquad \text{for} \quad \ell \ll r$$

This correction factor is shown in Figure 15 [7]. N_D^+ can be obtained from N_{Dapp}^+ by

$$N_D^+ = \frac{N_{Dapp}^+}{1 + 3b\frac{\ell}{r}} \tag{44}$$

If N_D^+ decreases by several orders of magnitude from the contact surface into the semiconductor, severe erros occur (See Figure 16), because C_{sc0}, determined by N_{Dlow}^+ and ℓ_{low} may be smaller than C_{sc1}, determined by N_{Dhigh}^+ and ℓ_{high}. In this case N_{Dapp}^+ may increase even if N_D^+ decreases further.

Conclusion: Due to the finite area of Schottky contacts, an edge capacitance has to be included in the calculation of dopant concentration. The apparent dopant concentration is larger than the true one.

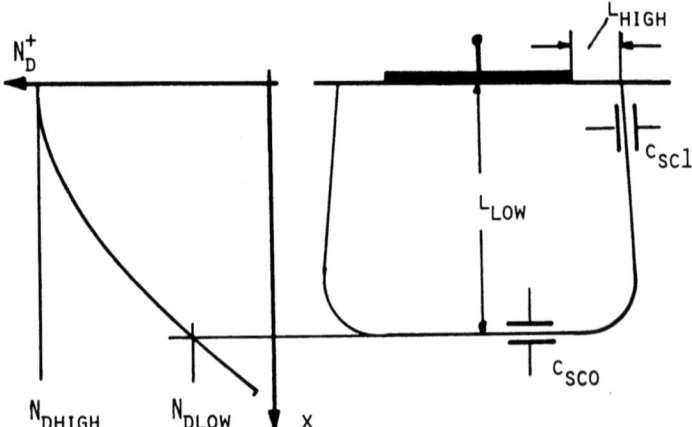

Fig. 16. Edge capacitance C_{sc1} and doping profile [8].

C. Influence of Parallel and Series Resistances

The Schottky diode used for the C-V measurements is a structure that consists not only of the space-charge capacitance but also of several parallel series resistances (Fig. 17 and Fig. 18(a,b). The external voltage applied to the diode is therefore divided into

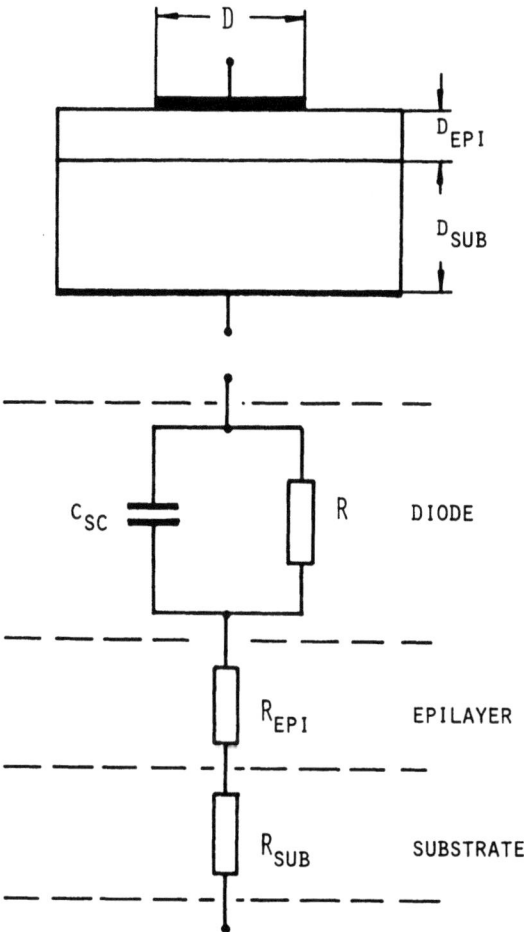

Fig. 17. Equivalent circuit of a Schottky diode on a n-epilayer - n^{++}-substrate [8].

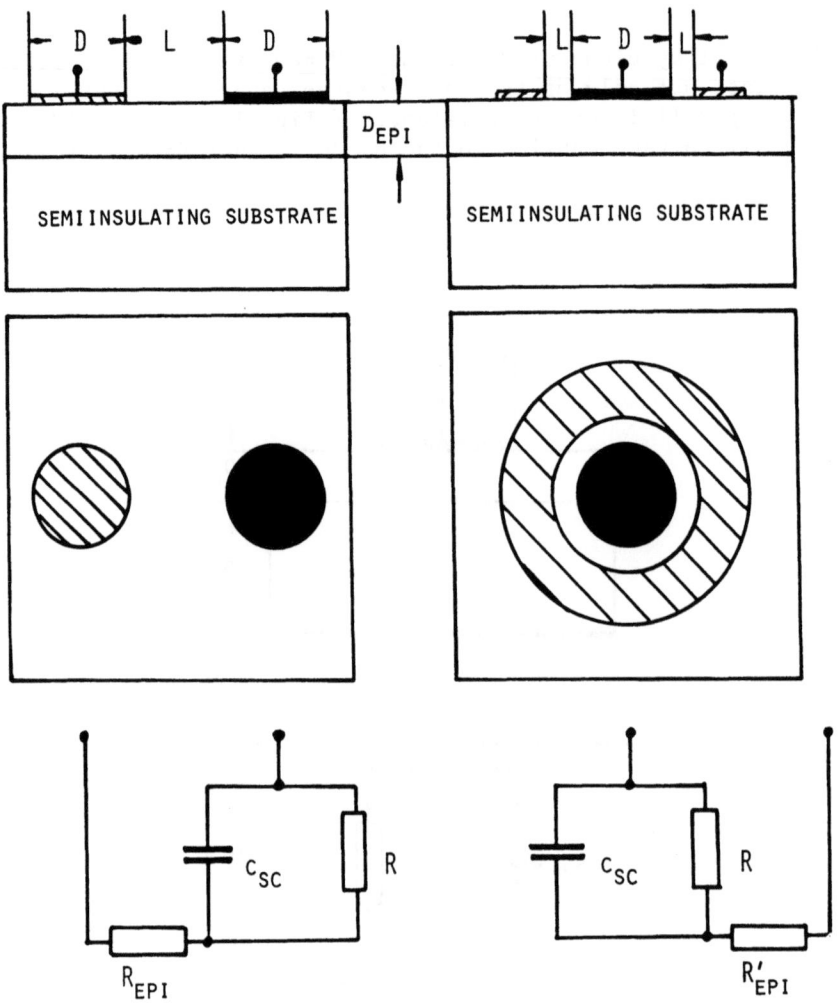

Fig. 18. Equivalent circuit of a Schottky contact on an n-epilayer -- Si-substrate [8].

several parts, but only one of them is used to influence the diode capacitance. The possible error due to the resistances will be discussed in the following section [8].

1. <u>Parallel Resistance R</u>. The parallel resistance R describes the conduction current due to electrons being emitted over the potential barrier of the contact or tunneling through the barrier. Since both emission and tunneling are field-dependent [9], R is field (and voltage) dependent. In general, R can be neglected

(i.e., $R \gg 1/\omega C_{sc}$) as long as the diode is not driven into breakdown.

2. Series Resistances R_{epi}, $R_{substrate}$.

2 a. Conducting substrate. The C-V method is used mainly to evaluate the properties of epitaxial layers deposited onto a substrate. This substrate can be either conducting (Fig. 17; e.g., rectifiers, LEDs) or semi-insulating (Fig. 18; e.g., GaAs field-effect transistors) In practical diodes, the substrate resistance is usually negligible

$$R_{sub} \ll R_{epi}$$

The resistance R_{epi} is that part of the epilayer not depleted of mobile carriers and is to be compared to the impedance of the space-charge layer $1/\omega C_{sc}$. R_{epi} is negligible if

$$R_{epi} \ll 1/\omega C_{sc} \qquad (45)$$

A general solution of equation (45) is not possible; an easy experimental test of the validity of equation (45) is the variation with frequency f: if the same N_D^+ (1) is obtained for all frequencies, then R_{epi} can be neglected.

2 b. Semi-insulating substrates. With semi-insulating or insulating substrates both the Schottky and the ohmic contact have to be deposited onto the conducting layer (Fig. 18). Now R_{epi} between the two contacts may be larger than with conducting substrates since the distance L between the contacts is usually larger than the thickness d_{epi} of the epitaxial layer. For an estimate of typical values we assume for the case of Fig. 18(a) a contact diameter D = 0.6 mm, a distance L = 1 mm, f = 20 MHz, μ_n = 7000 cm/V$_s$ (good quality GaAs). From

$$R_{epi} = \frac{1}{nq\mu} \cdot \frac{L}{Dd_{min}} \leq (\omega C_{sc})^{-1} \qquad (46)$$

it follows

$$d_{min} \geq \frac{\omega C_{sc} L}{nq\mu D} \qquad (47)$$

For the case of Fig. 18(b), we assume L = 0.1 mm with the other quantities unchanged:

$$R'_{epi} = \frac{1}{nq\mu} \cdot \frac{L}{2\rho D d'_{min}} \leq (\omega C_{sc})^{-1} \qquad (48)$$

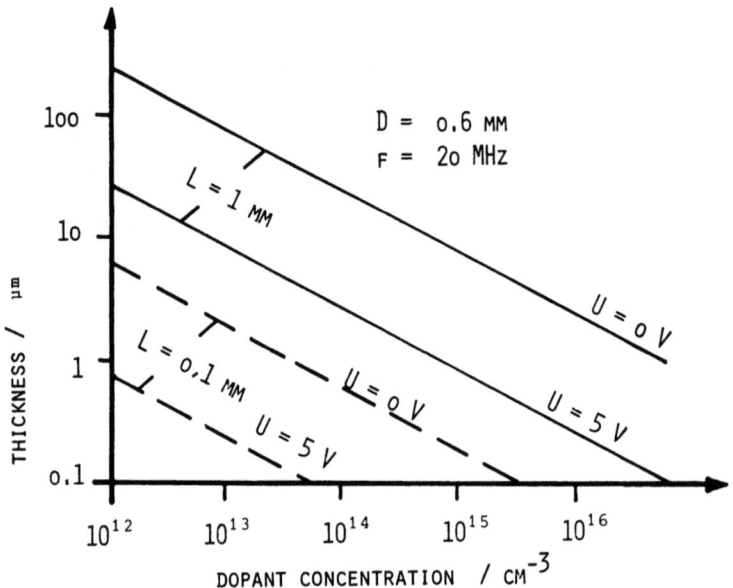

Fig. 19. Minimum thicknesses d_{min}, d'_{min} [8].

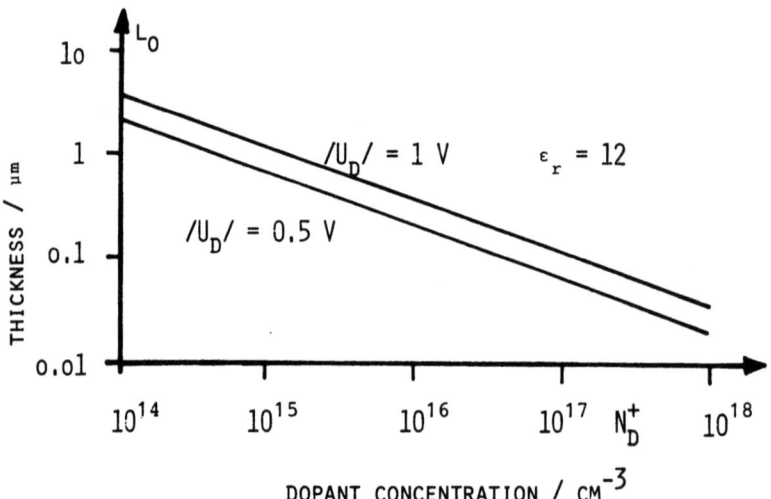

Fig. 20. Depletion layer thickness ℓ_0 at zero bias for various diffusion voltages U_D and donor concentrations.

$$d'_{min} \geq \frac{\omega C_{sc} L}{nq\mu 2\pi D} \tag{49}$$

In Figure 19, the two minimum thicknesses d_{min}, d'_{min} are plotted as functions of $n = N_D^+$. GaAs devices such as field effect transistors or planar GUNN devices have $N_D^+ \sim 10^{16} - 10^{17}$ cm^{-3} and $d_{epi} \sim 0.1 - 1$ μm. Thus R_{epi} can again be neglected.

The estimate of R_{epi} does not include the resistance of the epilayer under the Schottky contact. This is possible as long as the depletion layer does not approach the semi-insulating substrate. Once more, a check of the C-V characteristic at various frequencies or at contacts with different distances L gives information about the influence of the series resistance R_{epi}.

D. Limitations To The Method

There are several limitations to the use of the method for practical purposes, which will be summarized here.

1. Lower and Upper Limit of the Depletion Layer Width. Due to the built-in diffusion voltage U_D, a depletion layer of thickness ℓ_0 already exists under zero bias. The variation of ℓ_0 with N_D^+ for two values of U_D is given in Fig. 20 (for the calculation, the equation

$$\phi_{MeH1} = U_D + \zeta/q$$
$$\zeta = \frac{kT}{q} \ln \frac{N_c}{n} \tag{50}$$

was used).

Dopant concentrations in the range $0 \leq x \leq \ell_0$ can, in principle, be measured if the diode is forward biased. But now the parallel resistance (Fig. 18) decreases exponentially with increasing forward voltage and the capacitance measurement, either by a bridge or by the detection of the displacement current through the diode, is severely hindered. In practice reverse bias is applied only, and Fig. 20 yields the lower limits of ℓ_0.

The upper limit of ℓ is given by the fact that above a certain voltage U_{Br} (break-down voltage), the electric field inside the depletion layer increases to values at which either impact ionization or tunneling through the forbidden gap occurs. Both effects lead to a strong increase in reverse current and hence a decrease of the parallel resistance R similar to the decrease under forward bias. The break-down voltage is related to N_D^+ and, with a fixed value of

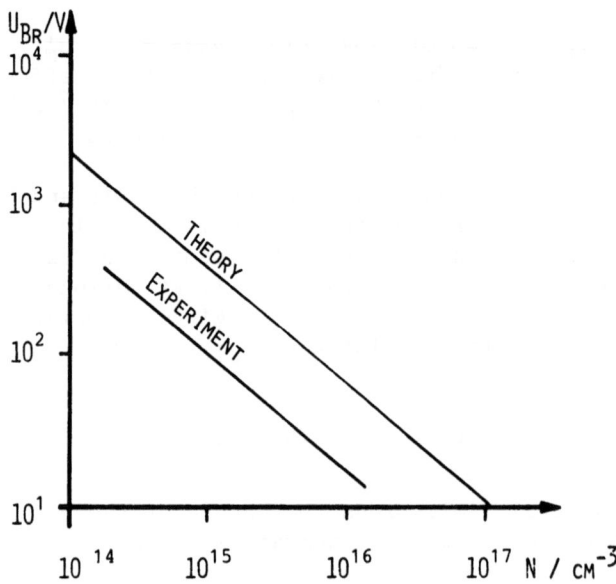

Fig. 21. Breakdown voltage U_{Br} versus donor concentration [8].

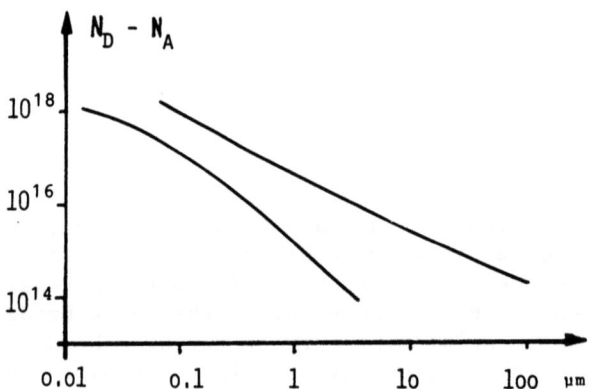

Fig. 22. Minimum and maximum depletion layer depths obtainable with Au - GaAs Schottky contacts [12].

U_D to the thickness of the depletion layer ℓ_{Br} at breakdown voltage U_{Br}. This relation is plotted in Figure 21. In Fig. 22, the lower and upper limits of ℓ obtained with a metal - n-GaAs contact is shown [12].

2. Possible Range of Diffusion Voltages U_D. In the deduction of of the C-V relation it is assumed that ℓ is larger than L_D. This assumption is equivalent to the assumption that the diffusion voltage U_D is much larger than the "thermal voltage" U_T:

$$\ell \gg L_D \rightarrow U_D \gg U_T = \frac{kT}{q}$$

For metal contacts on Si and GaAs these conditions usually hold. It was further assumed that n is determined by N_D ($n = N_D^+$) and that holes can be neglected. This assumption is true only if:

(a) the intrinsic level E_i (i.e., the Fermi level under intrinsic conditions, $n = p - n$;) given by

$$E_i = E_V + \tfrac{1}{2} E_g + 3/4 \, kT \ln \frac{m_p^*}{m_n^*} \tag{51}$$

lies below the Fermi level E_F in the depletion layer (see Fig. 23). Otherwise an inversion layer with mobile holes close to the surface is formed. In this case the space-charge is given by

$$\rho = q(N_D^+ + p(x)) \tag{52}$$

and the C-V relation does not reflect the N_D^+. With Si and GaAs-Schottky-diodes at room temperature and under zero bias $E_F \gg E_i$. Care must be taken with large reverse voltage u.

(b) the intrinsic carrier concentration given by

$$n_i = (N_C N_V)^{1/2} e^{-E_g/2kT}$$

(N_C, N_V are the effective density of states in the conduction or valence band, respectively) is smaller than the carrier concentration due to the dopants:

$$n_i \ll n = N_D^+$$

$n_i(T)$ is shown in Figure 24 [31].

For Si and GaAs the intrinsic concentration n_i at room temperature and below is much smaller than the lowest doping levels used for practical purposes. InSb, e.g., is in the intrinsic region at room temperature. Thus C-V measurements can be made only at lower temperatures (77 K).

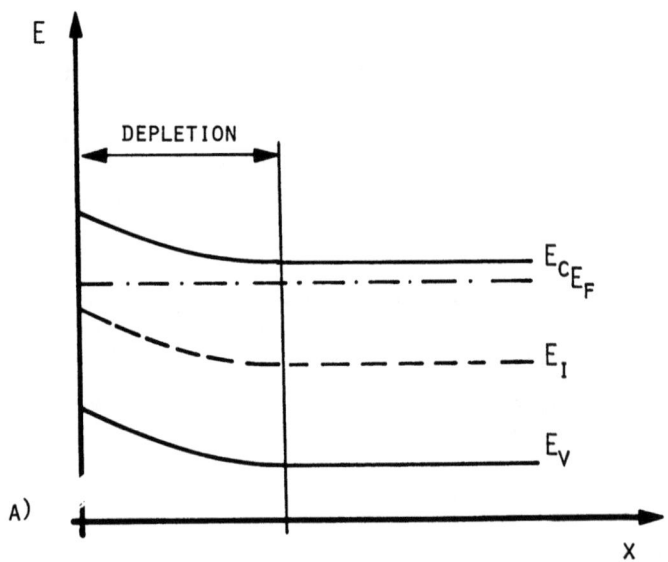
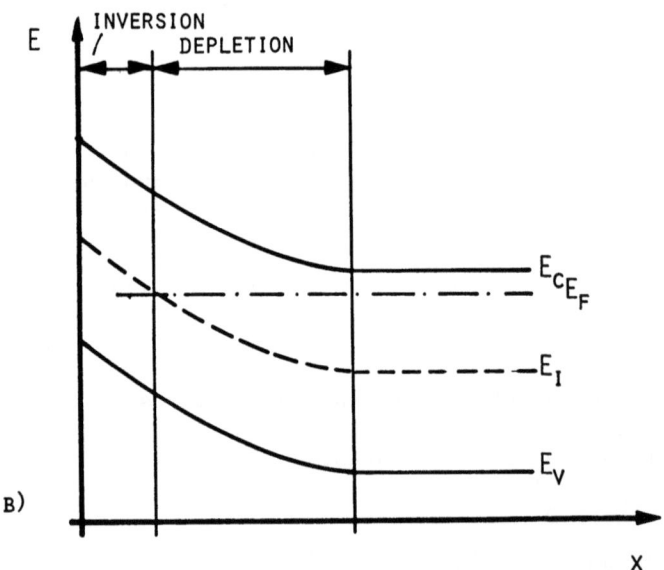

Fig. 23. (a) Barrier with a depletion region only.
(b) Formation of an inversion layer.

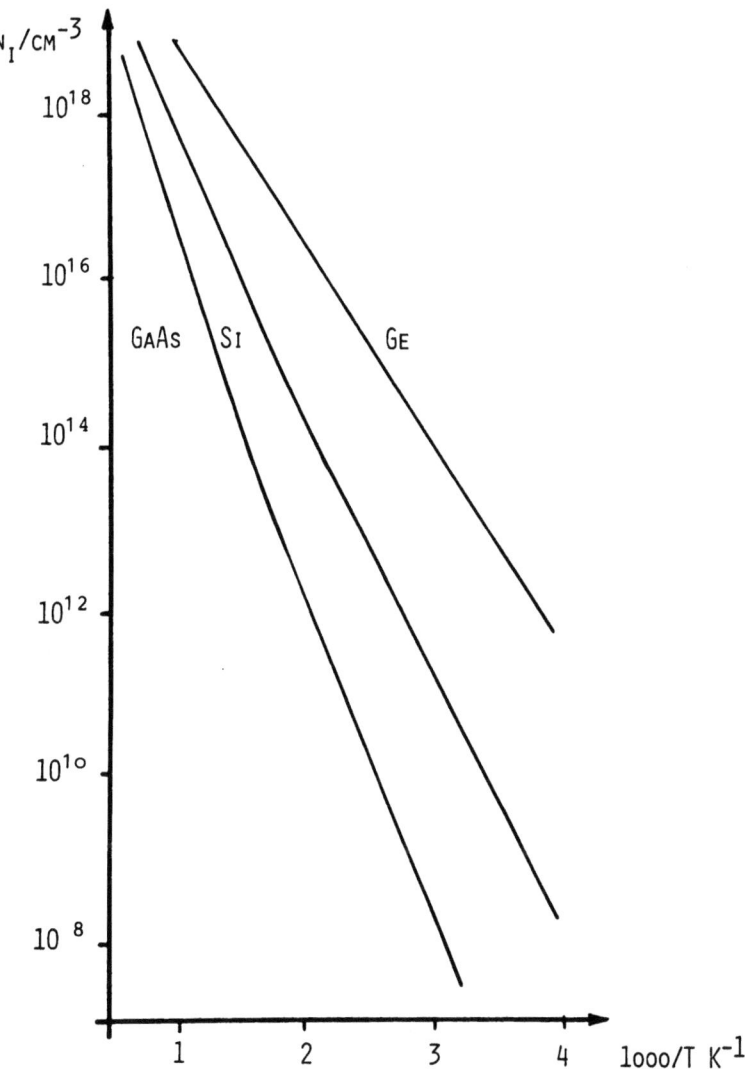

Fig. 24. Intrinsic concentration in GaAs, Si and Ge.

E. Experimental Methods

1. <u>Preparation of the Contacts</u>. A Schottky diode needs two contacts: an ohmic one and the Schottky contact itself. For accurate measurements, especially at different temperatures, the ohmic contact is an alloyed contact, with the alloy containing a dopant in order to minimize contact resistances. The Schottky contact is deposited in vacuum by evaporation or sputtering. For purposes of

good adhesion to the semiconductor, high temperature stability and formation of good bond contacts, multi-layer Schottky-contacts are used [10]. A typical Schottky-contact to GaAs is composed of a 10 nm Cr layer and a 200 nm Au layer. Cr is used to improve the adhesion, Au is suitable for bonding Au wires to the contact. Contacts with a refractory metal (Mo, W) instead of Cr greatly improve the temperature stability (up to 450 °C).

2. <u>Measurement of the Capacitance-Voltage Relation and Evaluation of the Dopant Concentration</u>. For the C-V measurements, a dc voltage U is applied to the diode, which defines ℓ (equation 20). A small ac voltage is superimposed on the dc voltage to measure the capacitance C_{sc} (equation 6). The frequencies used are 1-10 MHz. At lower frequencies, charging or discharging of traps may introduce errors (Section IV).

In the pioneering days of "Schottky profiling", capacitance bridges with manual balancing were used. While the accuracy of the bridges could be very high, the evaluation of the dopant profile from the measured data was time consuming. A graphical analysis of the C-V plot was more rapid yet less accurate. Nowadays, self-balancing bridges with digital data outputs are available. The data can be fed into a computer which directly plots the doping profile Self-balancing bridges have the disadvantage that some time is necessary for sampling the capacitance value. If the capacitance changes rapidly no balance is possible and error is introduced into the measurement. A typical example is shown in Fig. 25. If, however, the rate of the voltage increase is controlled by the bridge itself, these difficulties can be overcome [11].

These difficulties can be prevented if direct indicating capacitance meters are used. They are based on the fact (Fig. 26) that the current i through the diode is a pure displacement current if the parallel resistance R is neglected. Then the voltage u_B across a load resistor R_L is a measure of C [8].

The current i through the diode is (see Fig. 26)

$$i = \omega C_{sc} (u_A - u_B) \tag{53}$$

The voltage drop across R_L is

$$u_B = i R_L \tag{54}$$

or with the use of equation (53)

$$u_b = \omega C_{sc} R_L (u_A - u_B) \tag{55}$$

If $R_L \ll 1/\omega C_{sc}$, then $u_B \ll u_A$

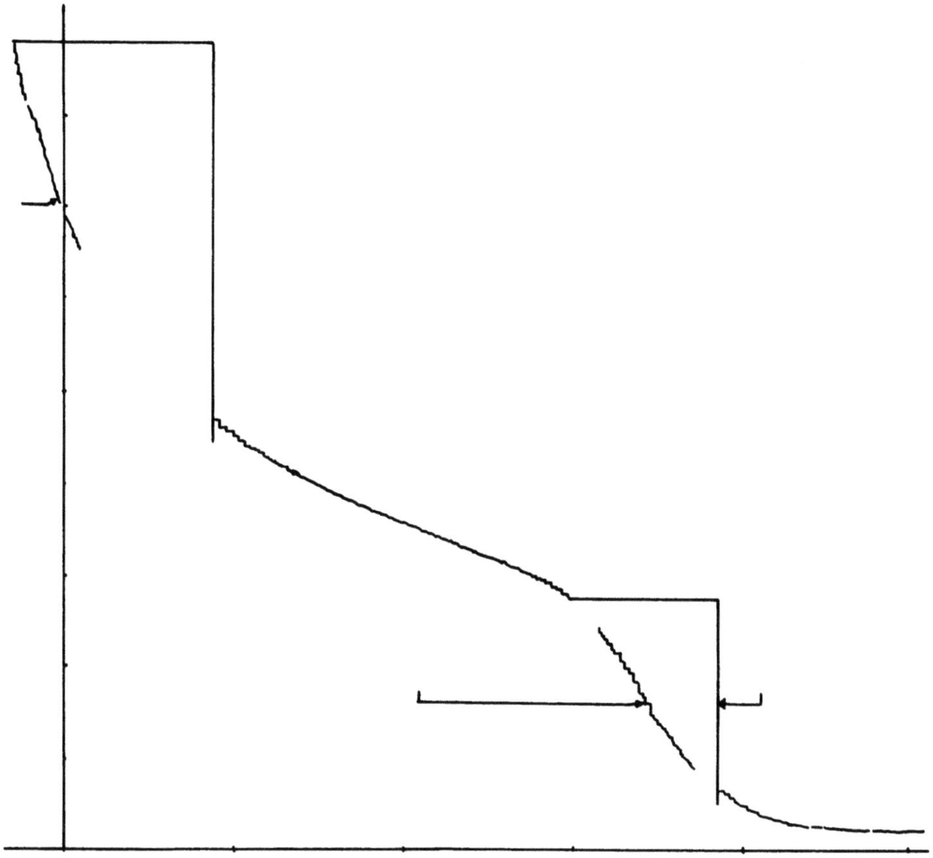

Fig. 25. $1/C_{sc}^2$ vs. U plot obtained by using two different rates of voltage increase

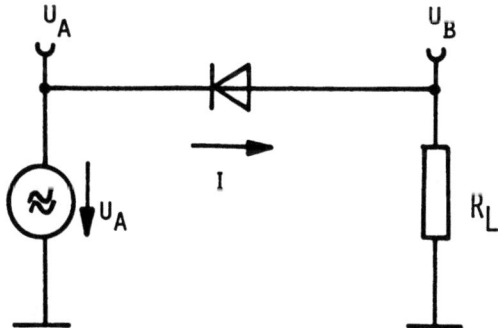

Fig. 26. Set-up for a direct measurement of the capacitance [8].

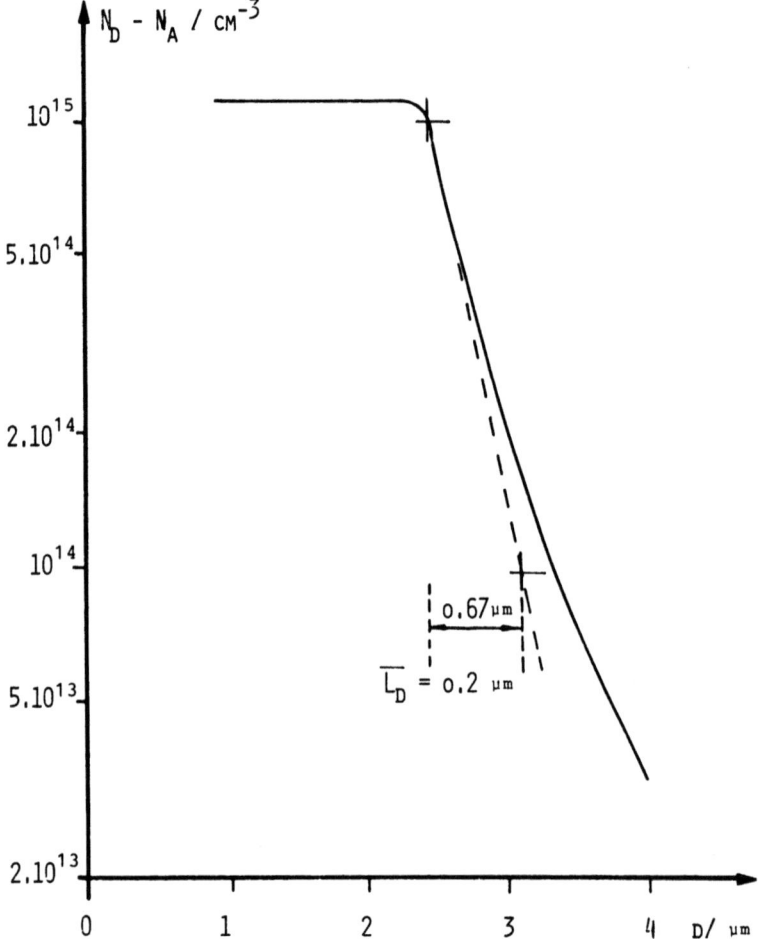

Fig. 27. Dopant concentration at an epilayer - Si-substrate [13].

$$u_B = \omega C_{sc} R_L u_A \tag{56}$$

if ω and u_A are held constant, one obtains

$$u_B = \text{const } C_{sc} \tag{57}$$

The voltage can be measured and indicated as

$$C_{sc} = u_B/\text{const}$$

or the data are fed into a digital computer for the calculation of $N_D^+(1)$. An example of a result obtained in this way on an epitaxial GaAs layer on a semi-insulating substrate is shown in Fig. 27 [13].

The figure also shows the local resolution at an abrupt dopant transition. The deviations from the exponential decrease expected from the influence of the Debye length L_D are due to edge and series resistance effects described in Section IIIB-C.

IV. EVALUATION OF DEEP LEVELS

In the early days of transistor technology it was already concluded from experiments that the lifetime of carriers injected into a semiconductor was a "structure sensitive property of material" [14]. The reason for this conclusion was the fact that the lifetime was much shorter than was expected from the band-band transition probability in indirect semiconductors such as Ge and Si. Therefore Shockley and Read [14] introduced the concept of recombination through the mechanism of trapping.

The question how a fairly large amount of energy can be dissipated by a non-radiative transition through traps will not be answered here. For details of the mechanism, i.e., cascade capture [15], Auger recombination [16] or multiphonon emission [17,18] the reader is referred to the literature.

A. The Shockley-Read Model of Recombination and Trapping at Deep Levels

According to Schockley and Read [14] a deep level can experience the following four processes (Fig. 28):

(1) electron capture
(2) electron emission
(3) hole capture
(4) hole emission

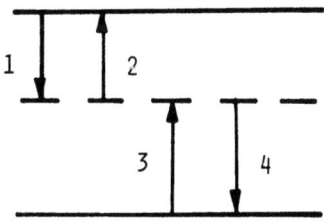

Fig. 28. Electron/hole emission/capture at a trap.

If an electron capture (1) is followed by a hole capture (3), the electron is recombined with the hole via the deep level which is called a <u>recombination center</u>. (Example: Au levels in the collector region of bipolar transistors, which enhance the recombination and diminish the storage time of the switching process).

For III-V devices another kind of deep level is important, i.e., levels for which an electron capture is not followed by a hole capture, or vice versa. With such levels the reemission of the electron is more likely than the hole capture. This type of deep level is called a <u>trap</u>. If electron capture and emission is predominant, the level is called a <u>donor-like trap</u> T:

(1) electron capture: $\quad T_D^+ + e^- \quad T_D^0$

(2) electron emission: $\quad T_D^0 - e^- \quad T_D^+$ (58)

Similar equations hold for acceptor-like traps. The relative importance of capture and emission can be influenced by temperature variations, electric fields or illumination. In general, variations in carrier concentration and mobility are a consequence which may cause deleterious effects on device behaviour.

For the processes given in equation (58), the rate equations are [14]:

(1) electron capture $\quad r_1 = C_n n N_T^+ \; [cm^{-3} s^{-1}]$ (59)

where r_1 is the rate by which the electron concentration changes due to the capture by traps, C_n is the capture coefficient, σv_{th}, σ is the capture cross section, v_{th} is the thermal velocity of the electrons and N_T^+ is the concentration of ionized donor-like traps.

(2) electron emission $\quad r_2 = e_n n_T$ (60)

Where e_n is the emission rate and n_T is the concentration of electrons capture by traps. N_T^0 is the concentration of neutral traps. Since the trap has two charge states, only (positively charged and neutral), the sum of the two concentrations N_T^0 and N_T^+ equals the total trap concentration N_T:

$$N_T = N_T^0 + N_T^+ \qquad (61)$$

The change of electron population in the trap n_T is equal to the change of the free carrier concentration n:

$$\frac{dn_T}{dt} = r_1 - r_2 = C_n n N_T^+ - e_n n_T \qquad (63)$$

EVALUATION BY SCHOTTKY CONTACTS

Since it is our intention to investigate the behaviour of traps in the space-charge region of a Schottky contact we may assume n = 0 in the depletion layer. Then equation (63) reduces to

$$\frac{dn_T}{dt} = -e_n\, n_T = -n_t/\tau \tag{64}$$

As the simplest case, it is assumed that the emission rate e_n and the corresponding lifetime τ of an electron in a trap are constant at constant temperature for a given trap. Integration of equation (64) yields

$$n_T(t) = n_T(t=0)\, e^{-t/\tau}$$

$$n_T(t) = N_T(t) = N_T^0 - N_T^+(t) \tag{65}$$

$$n_T(t=0) = N_T$$

we arrive at

$$N_T^+(t) = N_T(1 - e^{-t/\tau}) \tag{66}$$

The initial condition $n_T(t = 0) = N_T$ means that at the beginning of the experiment (t = 0) all traps are filled with electrons. The time constant τ depends on the energy difference $\Delta E_T = E_C - E_T$ and the temperature T. This is easily shown by reformulating equation (63) in terms of Fermi-Dirac statistics.

The probability f(E) that a quantum state at energy E is occupied is a function of E and E_F [19].

$$f(E) = \frac{1}{1 + \exp[(E - E_F)/kT]} \tag{67}$$

Consequently the probability that a quantum state is empty is

$$f^*(E) = 1 - f(E) = f(E)\, \exp[(E - E_F)/kT] \tag{68}$$

For a trap at energy E_T we obtain

$$N_T^+ = N_T\, f^*(E) \tag{69}$$

$$N_T^0 = n_T = N_T\, f(E) \tag{70}$$

Inserting this into equation (63) results in

$$\frac{dn_T}{dt} = r_1 - r_2 = C_n n N_T f^*(E_T) - e_n N_T f(E_T) \tag{71}$$

After the establishment of equilibrium $dn_T/dt = 0$ and from equation (71):

$$e_n = C_n n \frac{f^*(E_T)}{f(E_T)} + C_n n \exp[(E_T - E_F)/kT] \tag{72}$$

with $n = N_C \exp[(E_C - E_F)/kT]$

we finally arrive at

$$e_n = C_n N_C \exp[-(E_C - E_F)/kT] = C_n N_C e^{-\Delta E_T/kT} \tag{73}$$

or

$$\tau = \frac{1}{C_n N_C} \exp\frac{\Delta E_T}{kT} \tag{74}$$

From equation (74), $C_n = \sigma v_{th}$ can be determined, too.

<u>Conclusion</u>: The lifetime τ of an electron in a trap depends exponentially on the activation energy ΔE_T of the trap and the reciprocal temperature. Consequently, the observation of the emission process at different temperatures results in the determination of the activation energy ΔE_T of the trap.

B. Deep Level Analysis by Schottky Contact Techniques

1. <u>One Donor-Like Trap of Locally Constant Concentration</u>. The simplest case is the case of one kind of trap (here: donor-like) with locally constant concentration N_T. The shallow level concentration is assumed to be constant, too. All shallow levels are fully ionized. According to the Fermi distribution function, levels several kT above E_F are empty (ionized), those several kT below E_F are full (neutral). The distribution of neutral and ionized traps in the space-charge region of a Schottky contact is shown in Figure 29 for two different reverse voltage U_1, U_2. It is evident that by increasing the reverse voltage from U_1 to U_2 more traps emerge above the Fermi level. The emission of electrons from these levels above E_F obeys equation (74). The charged traps contribute

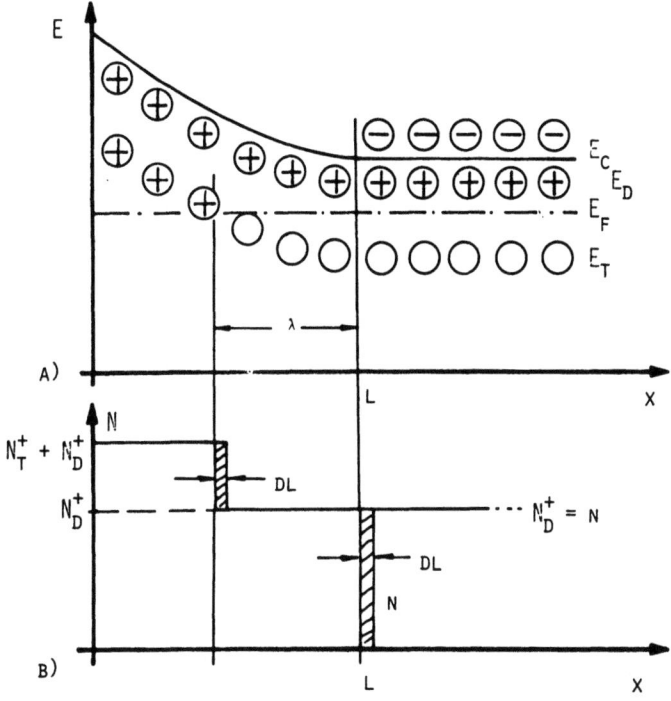

Fig. 29. Band diagram and charge distribution in a depletion layer containing traps.

to the space-charge and to the capacitance. Hence the time behaviour of the capacitance is governed by the time-constant τ of electron emission.

1 a. <u>Stationary Equilibrium</u>. At first it is assumed that the capacitance is measured a long time after the application of the reverse voltage. The occupation of the traps has come to an equilibrium value. It is further assumed that the frequency and amplitude of the ac voltage necessary for the capacitance measurement do not influence the trap occupation. Under these conditions the contact region can be divided into three different parts (see Fig. 29) [20]:

TABLE 1

	1. $0 < x < \ell-\lambda$	2. $\ell-\lambda < x < \ell$	3. $\ell < x$
ρ	$q(N_D^+ + N_T^+)$	qN_D^+	0

In equilibrium all traps in region (1) are ionized, i.e., $N_T^+ = N_T$. If shallow acceptors are present, N_D^+ is to be replaced by $N_D^+ - N_A^-$ for $0 < x < \ell$.

The Poisson equation now reads (with $d\phi \equiv du$)

$$\frac{d^2 u}{dx^2} = -\frac{\rho}{\varepsilon_r \varepsilon_0}$$

$$du = -\frac{\rho}{\varepsilon_r \varepsilon_0} x \, dx \tag{75}$$

$$= -\frac{q}{\varepsilon_r \varepsilon_0} [N_D^+ \ell d\ell + N_T^+ (\ell - \lambda) d\ell]$$

$$\ell d\ell = \tfrac{1}{2} d(\ell^2)$$

$$du = -\frac{2}{2\varepsilon_r \varepsilon_0} [(N_D^+ + N_T^+) d(\ell^2) - N_T^+ \frac{\lambda}{\ell} d(\ell^2)] \tag{76}$$

with

$$C_{sc} = \frac{\varepsilon_r \varepsilon_0}{\ell} \; ; \; \ell^2 = \left(\frac{\varepsilon_r \varepsilon_0}{C_{sc}}\right)^2 \; ; \; d(\ell^2) = (\varepsilon_r \varepsilon_0)^2 \, d\!\left(\frac{1}{C_{sc}^2}\right)$$

equation (76) finally reads

$$-\frac{du}{d\left(\frac{1}{C_{sc}^2}\right)} = \frac{q \varepsilon_r \varepsilon_0}{2} [N_D^+ + N_T^+] + N_T^+ \frac{q\ell}{2} C_{sc} \tag{77}$$

A plot of $y = du/d(1/C_{sc}^2)$ versus C_{sc} for different reverse voltages is a straight line (Figure 30):

$$y = a + b C_{sc} \tag{78}$$

$$a = \frac{q \varepsilon_r \varepsilon_0}{2} [N_D^+ + N_T^+]$$

$$b = N_T^+ \frac{q\lambda}{2}$$

$N_D^+ + N_D$ can be determined seperately at times $t \ll \tau$, i.e., before an appreciable number of traps have emitted the captured electrons.

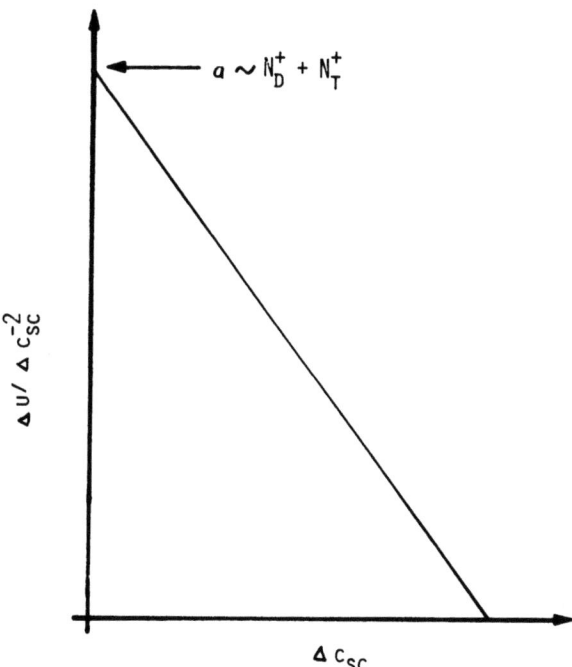

Fig. 30. Graphical representation of equation (77) [20].

It then follows that

$$N_T^+ = N_T = \frac{2}{q\varepsilon_r\varepsilon_0} \cdot a - N_D^+ \qquad (79)$$

and

$$\lambda = \frac{2}{qN_T} \cdot b \qquad (80)$$

Conclusion: Capacitance-voltage measurements in stationary equilibrium allow the determination of the trap density N_T.

1 b. <u>Relaxation into equilibrium (STEPCAP)</u>. Next it is assumed that the reverse voltage across the contact is changed abruptly (step-like). The capacitance relaxes according to the time constant into the new equilibrium value (STEPCAP). Then $N_T^+ = N_T$ has to be replaced by equation (66):

$$N_T^+ = N_T(1 - e^{-t/\tau})$$

As a consequence of the increasing degree of ionization of the traps with time t the width λ becomes time-dependent:

$$\lambda \to \lambda(t) \tag{81}$$

Equation (77) now reads

$$\frac{-du}{d(\frac{1}{C_{sc}^2})} = \frac{q\varepsilon_r\varepsilon_0}{2}[N_D^+ + N_T(1 - e^{-t/\tau})] + \frac{q}{2} N_T(1 - e^{-t/\tau})\lambda(t)C_{sc}(t) \tag{82}$$

Once again a plot of the left hand side of equation (82) versus $C_{sc}(t)$ at fixed times yields straight lines similar to those in Figure 30:

$$\frac{-du}{d(\frac{1}{C_{sc}^2})} = a(t) + b(t) C_{sc}(t) \tag{83}$$

$$a(t) = \frac{q\varepsilon_r\varepsilon_0}{2} [N_D^+ + N_T(1 - e^{-t/\tau})]$$

Combining the results for $a(t \to \infty)$ and $a(t)$ results in

$$\frac{2}{q\varepsilon_r\varepsilon_0} [a(t \to \infty) - a(t)] = N_T e^{-t/\tau} \tag{84}$$

A plot of the left-hand side of equation (84) in a logarithmic scale versus t yields a straight line, the slope of which is proportional to τ^{-1} and the value of which at $t \to 0$ corresponds to N_T. If these measurements are done at different temperatures T, the activation energy ΔE_T is deduced from $\tau(T)$ through equation (74).

In many cases the second term on the right hand-side of equation (82) is neglected, $\lambda(t) \to 0$ [22]. This is possible under large reverse voltages ([U] > 1V), because the voltage drop across the region $\lambda(t)$ is $(E_T - E_i)/q$ which is of the order 0.5 - 1V in Si, GaAs. Hence $\lambda(t) \ll 1$. In this case it follows from equation (82)

$$C_{sc}^2(t) = \frac{q\varepsilon_r\varepsilon_0}{2(-U_D-U)} [N_D + N_T(1 - e^{-t/\tau})] \tag{85}$$

with

$$C_{sc}^2(t=0) = \frac{q\varepsilon_r\varepsilon_0}{2(-U_D-U)} N_D$$

$$C_{sc}^2(t\to\infty) = \frac{q\varepsilon_r\varepsilon_0}{2(-U_D-U)} (N_D + N_T)$$

Using these relations one obtains from equation (85)

$$C_{sc}^2(t\to\infty) - C_{sc}^2(t) = \frac{q\varepsilon_r\varepsilon_0}{2(U_D-U)} N_T e^{-t/\tau} \qquad (86)$$

Through this equation τ is determined more easily.

Conclusion: Observation of the time evolution of the capacitance under constant dc voltage allows the determination of both the trap concentration N_T and the time constant τ. From the temperature dependence of τ the activation energy ΔE_T of the trap is determined.

1 c. <u>Enhanced Carrier Emission or Capture by Temperature Variations (TSCAP)</u>. The method of trap evaluation discussed in section IVB-2b uses the relaxation of the capacitance after a change of the applied voltage.

Another method is based on the shift of the Fermi level $E_F(T)$ due to changes in temperature. Again donor-like traps are assumed. If $E_F(T_1) > E_T$ (Fig. 31 a) all traps are full; if however $E_F(T_2) < E_T$ all traps are ionized. The cross-over of Fermi and trap level occurs at a distinct temperature T_m which in turn indicates the position E_T of the trap. It is possible to collect the electrons emitted from the trap (thermally stimulated current), if the concentration of free carriers is smaller than the emitted one. For "high conductivity" material like GaAs it is more appropriate to measure the contribution of the charged traps (after the emission of the electron) to the depletion layer capacitance (<u>T</u>hermally <u>S</u>timulated <u>Ca</u>pacitance → TSCAP) (Fig. 31 b).

The Poisson equation now includes a space-charge (T) which is temperature dependent. Similarly to Table 1, the space charge region is divided into three parts:

	1. $0 \leq x < \ell(T) - \lambda(T)$	2. $\ell(T) - \lambda(T) < x < \ell(T)$	3. $\ell(T) < x$
ρ	$q(N_D^+ + N_T^+(T))$	qN_D^+	0
	$N_T^+ = N_T(1 - f(E_FT))$		

TABLE 2

In region 1, the degree of ionization depends on the temperature and the position of the Fermi level.

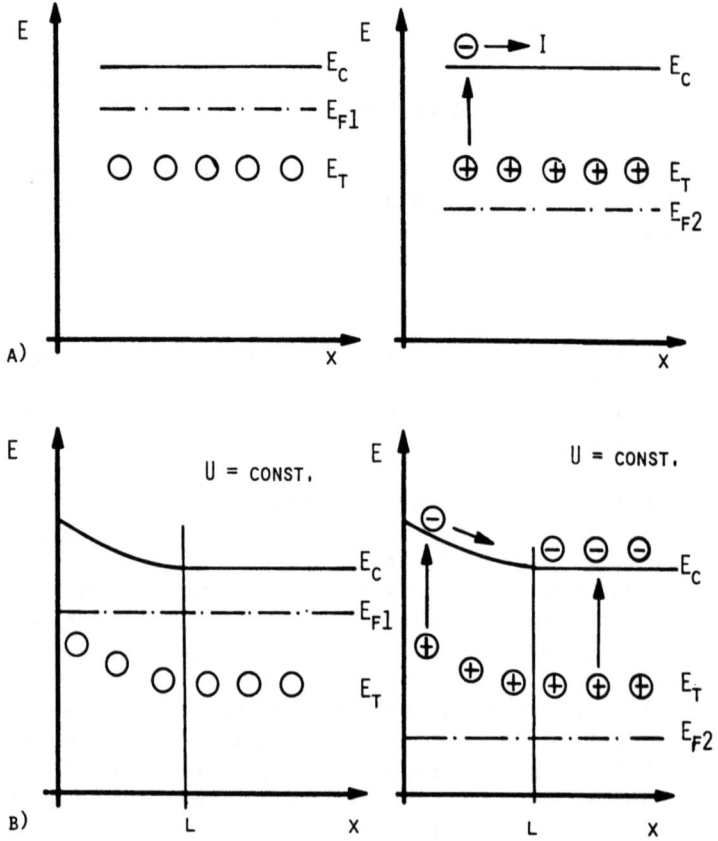

Fig. 31. Principle of thermally stimulated current (a) and thermally stimulated capacitance (b).

Equation (82) can be rewritten for this case [20]:

$$\frac{-du}{d(\frac{1}{C_{sc}^2})} = \frac{q\varepsilon_r\varepsilon_0}{2} \{N_D^+ + N_T [1 - f(E_T,T)] [1 - \frac{\lambda(T)}{\ell(T)}]\} \quad (87)$$

In this formula, C_{sc} in the second term of equation (82) is replaced by $C_{sc}(T) = \varepsilon_r\varepsilon_0/l(T)$. Instead of equation (85) we now obtain

$$C_{sc}(T) = \frac{q\varepsilon_r\varepsilon_0}{2(-U_D(T)-U)} \{N_D^+ + N_T [1 - f(E_T,T)] [1 - \frac{\lambda(T)}{\ell(T)}]\} \quad (88)$$

In the evaluation of N_T from equation (88) two difficulties arise:

(1) The diffusion voltage U_D is temperature dependent
(2) The ratio $\lambda(T)/l(T)$ is unknown.

ad (1): The temperature dependence of U_D can be determined separately. For Ni contacts on n-GaAs it was found [12] that

$$\frac{d(-U_D)}{dT} = -7.4 \times 10^{-4} \text{V/k}$$

similar values apply for other metals.

ad (2): Since it is the purpose of TSCAP experiments to ionize the deep levels completely the factor $\lambda(T)/l(T)$ tends to zero. For the temperature region of maximum electron emission from the traps it is therefore assumed

$$\frac{\lambda(T)}{l(T)} = \sigma$$

Hence the capacitance in TSCAP experiments finally reads:

$$C_{sc}^2(T) = \frac{q\varepsilon_r\varepsilon_0}{2(-U_D(T)-U)} \{N_D^+ + N_T [1 - f(E_T, T)]\} \tag{89}$$

The trap parameters can be determined from equation (89) in the following way (Fig. 32):

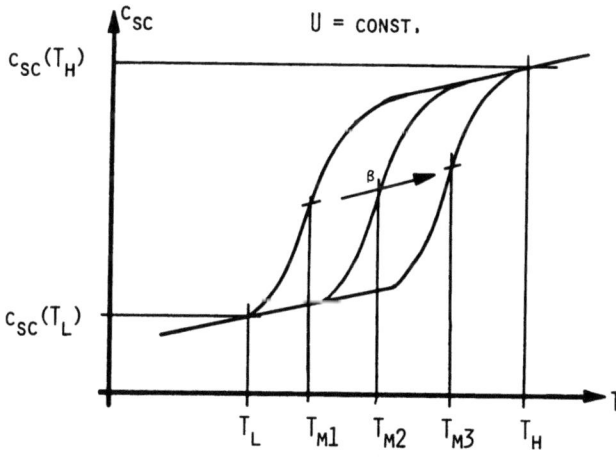

Fig. 32. Capacitance-temperature variation for different heating rates β in normal TSCAP experiments.

(a) **trap concentration N_T**

At low temperatures $\quad (T \leq T_\ell) : f(T_\ell) \to 1$

$$C_{sc}^2(T_\ell) = \frac{q\varepsilon_r\varepsilon_0}{2(-U_D(T_\ell)-U)} \cdot N_D^+ \qquad (90)$$

At high temperatures $\quad (T \geq T_h) : f(T_h) \to 0; \; N_T - N_T^+$

$$C_{sc}^2(T_h) = \frac{q\varepsilon_r\varepsilon_0}{2(-U_D(T_h)-U)} (N_D^+ + N_T^+) \qquad (91)$$

and finally

$$N_T^+ = N_T = \frac{2(-U)}{q\varepsilon_r\varepsilon_0} [U_D(T_\ell) - U_D(T_h)] [C_{sc}^2(T_h) - C_{sc}^2(T_\ell)] \qquad (92)$$

$C_{sc}(h)$, $C_{sc}(T_\ell)$ are determined from Fig. 32), $U_D(T)$ is known, U is a fixed dc voltage.

(b) **activation energy ΔE_T**

From equations (89) and (90) one obtains 23

$$\frac{C_{sc}(T)^2}{C_{sc}(T_\ell)} = \frac{N_D^+ + N_T \; 1 - f(E_T,T)}{N_D^+} \qquad (93)$$

Differentiation with respect to the temperature yields

$$\frac{d}{dT} \frac{C_{sc}(T)}{C_{sc}(T_\ell)}^2 = -\frac{N_T}{N_D^+} \frac{df(E_T,T)}{dT} \qquad (94)$$

since $n_T = N_T \, f(E_T,T)$: equation (70)

$$\frac{dn_T}{dT} = N_T \frac{df(E_T,T)}{dT} \qquad (95)$$

and equation (94) now reads

$$\frac{d}{dT} \frac{C_{sc}(T)}{C_{sc}(T_\ell)}^2 + = -\frac{1}{N_D^+} \frac{dn_T}{dT} \qquad (96)$$

At a certain temperature T_m, the Fermi level crosses the trap level. At this temperature the trap discharging and the capacitance

EVALUATION BY SCHOTTKY CONTACTS

changes will be maximum. The maximum is determined by setting the second derivative equal to zero:

$$\frac{d^2}{dT^2}\left[\frac{C_{sc}(T)}{C_{sc}(T)}\right]^2 = -\frac{1}{N_D^+}\left.\frac{d^2 n_T}{dT^2}\right|_{T=T_m} = 0 \rightarrow \left.\frac{d^2 n_T}{dT}\right|_{T=T_m} = 0 \quad (97)$$

It is now assumed that the temperature varies linearly with t with a constant heating rate β

$$T(t) = T_\ell + \beta t \; ; \quad dT = \beta dt \quad (98)$$

With the help of equation (98) the differentiation of equation (97) with respect to temperature is transformed into a differentiation with respect to time. The time behaviour of n_T is known from equation (64):

$$\frac{dn_T}{dt} = -e_n n_T$$

Also the temperature behaviour of τ is known from equation (74):

$$e_n(T) = C_n N_C e^{-\Delta E_T/kT} = \sigma_n(T) v_{th}(T) N_C(T) e^{-\Delta E_T/kT} \quad (74)$$

It follows

$$\left.\frac{d^2 n_T}{dT^2}\right|_{T=T_m} = n_T \frac{e_n^2}{\beta} - \left.\frac{de_n}{dT}\right|_{T=T_m} = 0 \quad \text{or} \quad (99)$$

$$\left.\frac{de_n}{dT} - \frac{e_n^2}{\beta}\right|_{T=T_m} = 0 \quad (100)$$

For the solution of the differential equation (100) it is assumed that
   (a) The capture cross section is temperature independent
   (b) $v_{th} \sim T^{1/2}$
   (c) $N_C \sim T^{3/2}$

Hence equation (74) reads

$$e_n(T) = BT^2 e^{-\Delta E_T/kT} \quad (101)$$

Solving equation (100) leads to

$$\ln\left(\frac{T_m^4}{\beta}\right) = \frac{\Delta E_T}{kT_m} = \ln\left\{\frac{\Delta E_T}{kB}\left(1 + \frac{2kT_m}{\Delta E_T}\right)\right\} \quad (102)$$

Measurements of the capacitance-temperature variations for different heating rates β yield different values of T_m. A plot of $\ln(T_m^4/\beta)$ over $1/T_m$ for different β yields straight lines the slope of which is proportional to the activation energy ΔE_T.

(c) <u>capture cross-section σ_n</u>

Inspection of equation (102) shows that information about the capture cross-section σ_n is available, too. One should bear in mind, however, that equation (102) was deduced under the assumption of a temperature-independent cross-section.

Conclusion: Measurement of the thermally-stimulated capacitance variations (TSCAP) allows the determination of the trap concentration N_T, the activation energy ΔE_T and the capture cross-section σ_n.

As important conditions it was assumed

(a) that the shallow levels are completely ionized in the whole range of temperatures investigated,

(b) that all traps are ionized at the highest temperature used in the experiment,

(c) that the emission rate does not depend on the electric field in the depletion region (in strong fields the Schottky effect increases the emission rate).

2. <u>Several donor- and acceptor-like traps</u>. If more than one trap is present the TSCAP method can be used, too, if the traps are separated by

$$\Delta E_{Ti} - \Delta E_{T(i+1)} \gg kT \quad (103)$$

Now each time the Fermi level crosses a trap level ΔE_{Ti} a capacitance change similar to the one shown in Figure 32 occurs. If condition (103) holds, each level can be analyzed separately using the methods derived in section IVB-1(c).

If acceptor-like traps are present in a n-type semiconductor which emits holes into the valence band, the ionization ($N_T \rightarrow N_T^- + e^+$) generates negative space-charge which partly compensates the positive one. This in turn leads to a decrease of the capacitance instead of an increase in the case of donor-like traps. The evaluation of the levels follows the principles described in the preceding sections.

3. **Traps With Locally Varying Concentration.** Locally varying trap concentrations can be evaluated both by STEPCAP and TSCAP methods. We describe here a modified TSCAP method [12,24,25] which is the most elegant and most successful method for trap profiling.

For the normal TSCAP method it was assumed that the reverse voltage U is held constant during a heating cycle. The corresponding capacitance variations are measured. According to the relation

$$C_{sc}(t) = \frac{\varepsilon_r \varepsilon_0}{\ell(T)} \tag{104}$$

the depletion layer width $\ell(T)$ varies if $C_{sc}(T)$ varies. This is of no importance as long as N_T is constant in the semiconductor sample. If, however, the trap concentration varies locally it is no longer possible to relate $C_{sc}(T)$ and N_T. A simple modification of the normal TSCAP easily allows the determination of trap profiles: charge, capacitance and applied voltage are connected through

$$U(T) = \frac{Q(T)}{C_{sc}} \tag{105}$$

The measurement is done in a way which holds the capacitance constant by properly controlling the applied voltage. Then $U(T) \sim Q(T)$ (Fig. 33). If two measurement cycles are done with slightly different capacitance values and hence slightly different depletion layer widths ℓ_1, ℓ_2 all differences between the two results are due to traps within the region $(\ell_2 - \ell_1)$. The equations relating the observed voltage variations to the trap concentrations $N_T(x)$ will not be deduced here. The derivation follows the principles explained in the preceding sections.

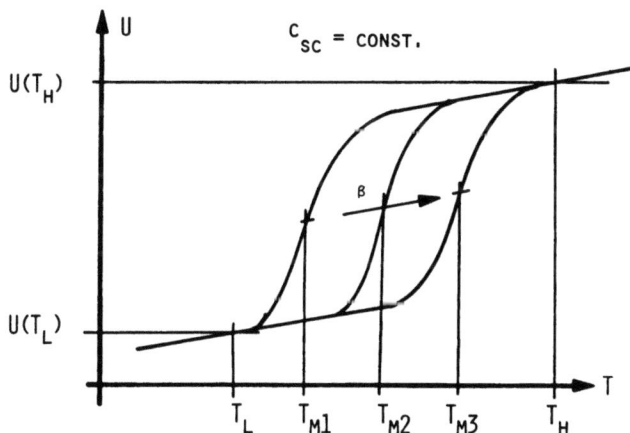

Fig. 33. Control voltage-temperature variation for different heating rates β in modified TSCAP experiments.

For a fixed capacitance C_{sc1} (ℓ_1) the mean trap concentration within the region ℓ_1 is given by

$$\frac{\overline{N_T(\ell_1)}}{N_D^+} = \frac{[U_D(T_h) - U(T_h)] - [-U_D(T_\ell) - U(T_\ell)]}{-U_D(T_\ell) - U(T_\ell)} \quad (106)$$

From the mean concentrations $\overline{N_T(\ell)}$ the true trap profile $N_T(\ell)$ can be obtained:

$$N_T(\ell) = \overline{N_T(\ell)} + \tfrac{1}{2} \frac{d\overline{N_T(\ell)}}{d\ell} \ell \quad (107)$$

Typical experimental results are shown in Figures 34 through 36 [26]. Figure 34 shows the control voltage U necessary to keep the capacitance C_{sc} constant. The parameter is ℓ. All curves were measured with a constant heating rate. For the evaluation of the activation energy ΔE_T according to equation (102), similar curves with different heating rates have to be measured. The trap profiles deduced from the data of Fig. 34 are given in Fig. 35 together with the shallow level profile which was determined seperately. Finally the activation energy ΔE_T is determined in Fig. 36. In order to achieve sufficient accuracy it is necessary to collect all data by a computer and to evaluate the trap parameters numerically. The curves in Figures 34 -36 are replicas of computer plots.

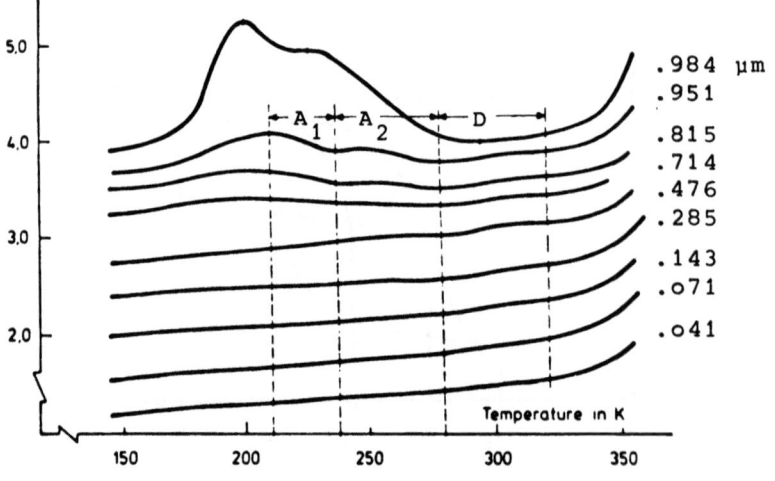

Fig. 34. Control-voltage versus temperature, parameter: depth D.

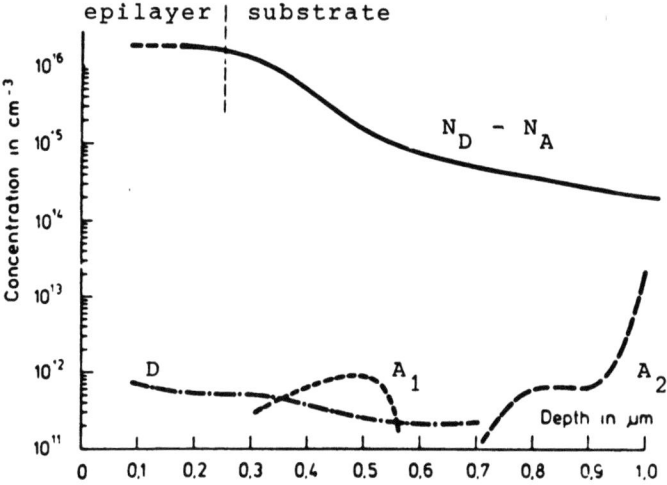

Fig. 35. Shallow and deep level profiles deduced from Fig. 34.

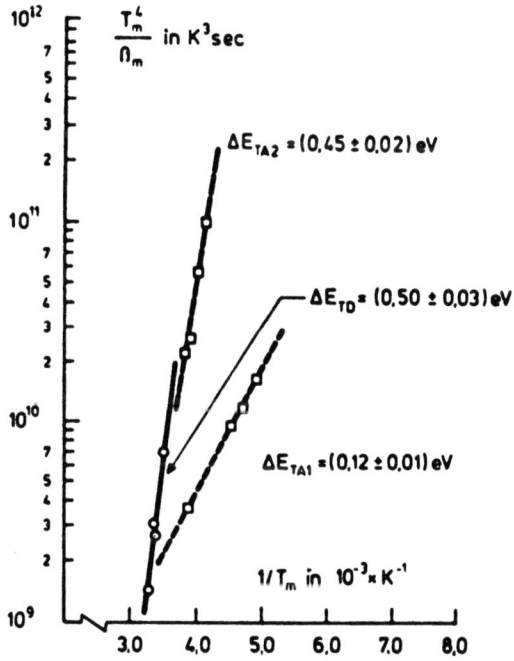

Fig. 36. Activation energy deduced from Fig. 34. (34-36 from Reference [26].)

The accuracy of the procedure is determined by the accuracy with which the capacitance (TSCAP) or the voltage (modified TSCAP) can be measured. For the modified TSCAP it follows that with a 100 μV resolution the minimum detectable ratio $\overline{N_T/N_D^+}$

$$10^{-5} \leq \frac{N_T}{N_{Dnet}^+} \leq 10^{-3} \tag{108}$$

In very pure GaAs material, $N_D^+ - N_A^- = N_{Dnet}^+ \geq 10^{13} \text{cm}^{-3}$. Thus trap densities down to 10^{10}cm^{-3} or less should be detectable. All other methods for trap evaluation have detection limits several orders of magnitude above the limit of the TSCAP method.

C. Other Methods Using Schottky Contacts

Lang [27] has described a very refined method which is based on the STEPCAP method but which evaluates the time constants electronically by comparing the experimental time constant with a predetermined one. If at a certain temperature the two time constants coincide the corresponding electronic signal has a maximum. T_m and τ are easily determined. This method is known as DLTS (deep level transient spectroscopy). It has been further refined by Lefevre and Schulz [28]. They use the differences in the capacitance relaxation due to two consecutive pulses of different amplitude to exclude contact effects and field-dependence of capture cross section.

Another method principally different from those discussed above is the photocapacitance method [29,35]. Here the discharging of traps is not generated by shifting the Fermi level due to temperature variations (TSCAP) or external voltages (STEPCAP), but by irradiating monochromatic light of frequency ν onto the Schottky contact. If the energy of the photons equals the activation energy ΔE_T an electron or hole emission occurs.

The two processes

(1) $N_{TD} + h\nu \rightarrow N_{TD}^+ + e^-$

(2) $N_{TA} + h\nu \rightarrow N_{TA}^- + e^+$

can be distinguished by their influence on the total space charge and the sign of the corresponding capacitance change ΔC_{sc}. However, the method cannot distinguish between the activation of an electron from the valence band into a charged donor-like trap:

(a) $N_{TD}^+ + e^- + h\nu \rightarrow N_{TD}$: $\Delta Q = -e$

and the emission of a hole:

(b) $N_{TA} + h \quad N_{TA}^- + e^+$: $Q = -e$

By using two beams of different energies 30 process (a) may be converted into

(c) $N_{TD}^+ + h\nu + e^- \rightarrow N_{TD}$

$$\Delta Q = \sigma$$

$N_{TD} + h\nu \rightarrow N_{TD}^+ + e^-$

$h\nu_1 + h\nu_2 = E_g$ (in the simplest case)

while process (b) remains unchanged. Thus donor- and acceptor-like traps can be distinguished. A rather new technique for detecting minority-carrier traps was described by Mitonneau, Martin and Mircea [34].

D. Experimental Results

In this final section some of the recent results on GaAs and GaP evaluation are presented. They were collected by Ikoma et. al. [18]. For comparison results on Si [33] obtained by implanting known species are given. Whereas a number of deep levels are known in the III-V compounds almost none of them are clearly identified as being due to a single impurity or complex.

The capacitance method itself is fully exploited. Other methods and combinations of those methods to gether with capacitance measurements are necessary for a better understanding of deep levels in III-V compounds.

IV. CONCLUSION

Measurements on Schottky contact capacitances are a very powerful, very sensitive tool for the detection of shallow and deep levels in semiconductors. For shallow levels, the concentration can be evaluated which controls the free carrier concentration and the mobility of the sample.

TABLE 3

Electron and hole traps in various sorts of n-GaAs crystals
(0: detected, -: not detected) Ref. [18].

CZ is the bulk material grown by the Czochralsky method
HB is the horizontal boat grown bulk material
VPE is the vapor phase epitaxial layer
LPE is the liquid phase epitaxial layer

Samples	Electron Traps (eV)						Hole Traps (eV)				
	0.30	0.53	0.62	0.65	0.75	0.82	0.39	0.45	0.64	0.75	0.77
CZ	-	0	-	0	0	-	-	0	-	0	-
HB	0	0	0	-	-	0	-	0	-	-	0
VPE	-	-	-	-	0	-	-	0	-	0	-
LPE	-	-	-	-	-	-	0	0	0	-	-

TABLE 4

Activation energies of various electron traps in GaP [18].

Electron Traps in GaP

E_c 0	18	36	37	38
		0.27	0.27	
		0.36		
			$\underline{0.42}_N$	
				Presence $\underline{0.45}$ 0 --
0.5 --------	-----	------	------	------
			0.56	
	$\underline{0.65}$	$\underline{0.64}$	$\underline{0.63}$ Lattice Relaxation	
		$\underline{0.72}$	$\underline{0.71}$	
		$\underline{0.80}$ 0 Presence		
	$\underline{0.82}$		$\underline{0.89}$	$\underline{0.90}$ 0 -
		$\underline{0.90}$		
1.0 --------	-----	------	------	------
				2.0 0 --

TABLE 5

Activation energies of various hole traps in GaP [18].

Hole Traps in GaP

	18	36	37	39	40	41
1.0	---	---	---	1.0	---	---
					0.97	
						0.82
	0.76	degrada- tion	0.71	0.75 Lifetime killer	0.72	
	0.65					0.62
	0.55	0.55 Cu Diffusion			degrada- tion	0.50
0.5	---	---	---	---	---	---
		0.39	0.39 0.33		0.41	Cu-doped
		0.29				
	0.23	0.22				
eV 0	---	---	---	---	---	---

TABLE 6

Electron (D) and hole traps (A) in silicon due to the implantation of various impurities. D, A without numbers indicate levels close to band edges [33].

Element	E_C					E (Midgap)				E_V
Au						A.50		D.35		
Ba			D.28				D.45			
Be							A.45			
Bi	D									
C			D.25							
Cr					D.38					
Co						A.50		A.34		
Cs			D.28			D.50				
Cu			D.28			A.50			A.21	
Fe	D				D.50					
Ge			D.27			D.50				
Ir					D.50				A.24	
K			D.25					D.35		
Mn						D.50				
Mo				D.30				D.33		
Na	D							D.35		D
Nb			D.26				D.40			
Ni			D.28						A.21	
Pd			F.29							
S	D		D.27							
Se			D.27							
Si				A.34		D.49			D.19	
Sn	D						A.27			
Sr			D.28			D.50				
Ta			D.29				D.40			
Te	D									
Ti	D									
Tl			D.28					A.30		
Zn						A.55		A.28		
	Conduction Band					Midgap				Valence Band

For deep levels, the concentration, the activation energy and the capture cross-section can be determined. The sensitivity is high; i.e., trap concentrations several orders of magnitude smaller than the shallow level concentration are detectable. Capacitance measurements are much more sensitive than other methods such as SIMS, Auger-spectroscopy, luminescence spectroscopy and others. However, information about the physical and chemical nature of the level is not available directly. Here a comparison with standard samples containing known deep levels is necessary. A comparison with theoretically obtained trap parameters is impossible at present because a self-consisting theory of deep levels is not yet available [31].

REFERENCES

1. A. Many, Y. Goldstein, N. B. Grover, Semiconductor Surfaces, Amsterdam, London, New York (1971).
2. W. Schottky, W. Deutschmann, Phys. Z. 30, 839 (1929).
   W. Schottky, Z. Phys. 118, 539 (1972).
3. K. Heime, Solid-State Electronics 13, 1505 (1970) (with further references to the problem).
4. W. Schockly, Electrons and Holes in Semiconductors, New York (1950).
5. W. C. Johnson, P. T. Panousis, IEEE Trans. ED-18, 965 (1971).
6. A. M. Goodman, J. Appl. Phys. 34, 329 (1963).
7. J. A. Copeland, IEEE Trans. ED-17, 404 (1970).
8. K. Heime, Z. Angew, Physik 32, 374 (1972).
9. C. R. Crowell, S. M. Sze, Solid-State Electronics 9, 1035 (1966).
10. K. H. Bachem, J. Engemann, K. Heime, Proc. 5th Int. Conf. Solid-State Devices, Tokyo, 1973; Suppl. Jap. J. Appl. Phys. 43, 222 (1974).
11. J. Baston, K. Heime, private communication
12. J. Engemann, Dissertation, RWTH Aachen (1975).
13. L. M. F. Kaufmann, K. Heime, J. Crystal Growth 42, 321 (1977).
14. W. Shockley, W. T. Read, Jr., Phys. Rev. 87, 835 (1972).
15. M. Lax, Phys. Rev. 119, 1502 (1960).
16. T.P. Landsberg, A. R. Beattie, J. Phys. Chem. Solids 8, 73 (1959).
17. C. H. Henry, D. V. Lang, Proc. 12th Int. Conf. Phys. Semiconductors, Stuttgart, 411 (1974).
18. T. Ikoma, M. Takikawa, T. Okumura, Proc. 8th Int. Conf. Solid-State Devices, Tokyo, 1976, Suppl. Jap. J. Appl. Phys. 16, 223 (1977).
19. Ch. Kittel, Thermal Physics, New York (1969).
20. R. R. Senechal, J. Basinski, J. Appl. Phys. 39, 458 (1968).
21. C. T. Huang, S. S. Li, Solid-State Electr. 16, 1481 (1973).
22. C. T. Sah. L. Forbes, L. L. Rosier, A. F. Tasch, Jr., Solid-State Electr. 13, 759 (1970).
23. M. G. Buehler, Solid-State Electr. 15, 69 (1972).

24. G. Goto, S. Yanagisawa, O. Wada, H. Takanashi, Appl. Phys. Letters 23, 150 (1973).
25. G. Goto, S. Yanagisawa, Jap. J. Appl. Phys. 13, 1127 (1974).
26. J. Engemann, K. Heime, CRC Crit. Rev. Solid-State Sciences 5, 485 (1975) (II. Annual Conf. on the Phys. of Compound Semic. Surf., San Francisco, (1975)).
27. D. V. Lang, J. Appl. Phys. 45, 3023 (1974).
28. H. Lefevre, M. Schulz, Applied Phys. 12, 45 (1977).
29. cf [23] or A. G. Milnes, Deep Impurities in Semiconductors, New York (1973).
30. A. M. White, P. J. Dean, P. Porteous, J. Appl. Phys. 47, 3230 (1975).
31. A. B. Roitsin, Sov. Phys. Semicond. 8, 1, (1974).
32. S. M. Sze, Physics of Semiconductor Devices, New York, (1969)
33. M. Schulz, Inst. Phys. Conf. Series No. 22, 226 (1974).
34. A. Mitonneau, G. M. Martin, A. Mircea, Inst. Phys. Conf. Series 33a 73, (1977) (GaAs and related compounds, Edinburgh (1976)).
35. H. G. Grimmeiss, C. Ovren, J. W. Aller, J. Appl. Phys. 47, 1103, (1976).
36. E. Fabre, R. N. Bhargava, W. K Zwicker, J. Electron. Mater. 3, 409 (1974).
37. B. L. Smith, T. J. Hayes, A. R. Peaker, D. R. Wight, Appl. Phys. Lett. 26, 122 (1975).
38. H. Kukimoto, M. Mizuta, Proc. 5th Conf. Solid-State Dev., J. Jap. Soc. Appl. Phys. 43, 95 (1974), Suppl.
39. B. Hamilton, A. R. Peaker, S. Bramwell, W. Harding, D. R. Wight, Appl.-Phys. Lett. 26, 702 (1975).
40. C. H. Henry, P. D. Dapkus, J. Appl. Phys. 47, 4067 (1976).
41. B. W. Wessels, J. Appl. Phys. 47, 1131 (1976).

After the completion of the manuscript, the following valuable papers were brought to the author's attention:

42. H. G. Grimmeis, Ann. Rev. Mater. Sci 7, 341-76 (1977). (A review paper: Deep Level Impurities in Semiconductors).
43. F. Lau, H. Poth, P. Balk, to be published (Isothermal and non-isothermal C-V trap measurements -- A critical comparison).

Chapter 6

NOISE

A. D'Amico

Laboratorio di Elettronica dello Stato Solido

Consiglio Nazionale delle Ricerche

Via Cineto Romano, 42

00156-ROMA Italy

I. INTRODUCTION

During the past 25 years, developments in the invention and production of solid state devices such as diodes, transistors, lasers, integrated circuits, etc., have been extremely rapid and there seems little reason to believe that there will be an appreciable slackening in the near future. Strong competition at the market level has pushed both researchers and producers into providing better and better devices tending towards what might be considered a theoretical "limit of optimum performance". Consequently the problem of evaluation has become of major importance and there is growing difficulty in determining the long term reliability of the devices. One of the most interesting and stimulating aspects of active and passive semiconductor electronic devices is that of noise. The understanding of this particular topic is most useful from a physical point of view but, above all, it is important in characterizing devices. It is in fact, well known that noise imposes a lower limit for detecting and amplifying weak signals and, as a result, noise measurements are of fundamental importance in determining the performances of semiconductor devices. The utilization of noise measurements to characterize semiconductor devices and materials has been considered in exploring diagnostic methods for nondestructive evaluation. Noise analysis in semiconductors, more precisely the determination of the spectral density, can point out the presence of anomalies connected with the fabrication processes of devices and materials and can also indicate the appearance of defects due to continued usage. Furthermore, noise measurements can enable us to understand phenomena related to generation-recombination processes and help clarify the interface state behaviour in MOS or MIS e.g. structures where the semiconductor can be either mono- or polycrystalline. Finally noise measurements can evaluate the limits of performance of radiation detectors, linear amplifiers, integrated circuits, parametric amplifiers, etc.

We shall begin by looking at some statistics in order to define the subject we shall be dealing with as clearly as possible. We will then describe those techniques normally employed for evaluating noise quantitatively. The most common forms of noise frequency distribution (thermal noise, shot noise, $1/f^a$ noise, burst noise, g-r noise) will be examined after which we will look at noise behaviour in several electronic deivces with the view toward establishing a simplified model for their noise properties. Finally we introduce a new nondestructive method (third harmonic index method) and the corresponding measurement chain. We will see how, in weakly non-linear devices, it is possible to detect and measure the first component of an odd harmonic spectrum. A knowledge of the amplitude of this component (which itself shows a low frequency fluctuation during the stabilization time) amy prove useful for evaluating long time stability.

We should like to point out that this paper is intended as an introductory survey on low and intermediate frequency noise behaviour. A bibliography is provided for those who wish to take the subject further.

II. STOCHASTIC PROCESSES, STATIONARITY, ERGODICITY [1]

Let us consider an experiment described by its results η which form the S space, a set of events which for a subsystem into the S space, and the probability of these events. To each result η let us assign the time function

$X(t,\eta)$

which can be, real or complex but t itself comes from the real numbers ensemble. The function $X(t,\eta)$ can be [1]:

1. a family of time dependent functions (t, η, variables),

2. a single time dependent-function (t variable, η constnat)

3. a random variable (t constant, η variable) or

4. a number (t constant, η constant).

NOISE

In the following we shall assume X(t) represents a real process. The first order distribution function of the random variable X(t) will be:

$$F(x,t) = P\{x(t) \le x\} \quad (1)$$

and the corresponding density function will be:

$$f(x,t) = \frac{\partial}{\partial x} F(x;t) \quad (2)$$

Analogously, if we consider two random variables $X_1(t_1)$, $X_2(t_2)$, we shall deal with the second order distribution function which depends on t_1, t_2 and the corresponding density function will be given by:

$$f(X_1, X_2; t_1, t_2) = \frac{\partial^2}{\partial X_1 \partial X_2} F(X_1, X_2; t_1, t_2) \quad (3)$$

In the case of two variables $X_1(t_1)$, $Y_1(t_2)$ the density function will be given by:

$$f(X_1, Y_1; t_1, t_2) = \frac{\partial^2}{\partial X_1 \partial Y_1} F(X_1, Y_1; t_1, t_2) \quad (4)$$

The following functions are fundamental to stochastic processes:

(a) the mean value $\quad \overline{X}(t) = E\{x(t)\} = \int_{-\infty}^{\infty} X(t) f(X;t) dX \quad (5)$

which is in general a time function

(b) the autocorreleation function of a process X(t)

$$R(t_1, t_2) = E\{X_1(t_1) X_2(t_2)\} = \int_{-\infty}^{\infty} X_1(t_1) X_2(t_2); t_1, t_2) dX, dX_2 \quad (6)$$

(c) the autocovariance of X(t):

$$C(t_1, t_2) = E\{[X_1(t_1) - \overline{X}_1(t_1)][X_2(t_2) - \overline{X}_2(t_2)]\} \quad (7)$$

which can be written as:

$$C(t_1, t_2) = R(t_1, t_2) - \overline{X}_1(t_1) \overline{X}_2(t_2) \quad (8)$$

(d) the variance fof a process X(t) which is given by:

$$\sigma_{X(t)}^2 = C(t,t) = R(t,t) - \overline{X}^2(t) \quad (9)$$

(e) the crosscorrelation function of a process

$$R_{XY}(t_1, t_2) = E\{X_1(t_1), Y_1(t_2)\} = \int_{-\infty}^{\infty} X_1(t_1) Y_1(t_2)$$

$$f_{XY}(X_1, Y_1; t_1, t_2) dX \; dY \quad (10)$$

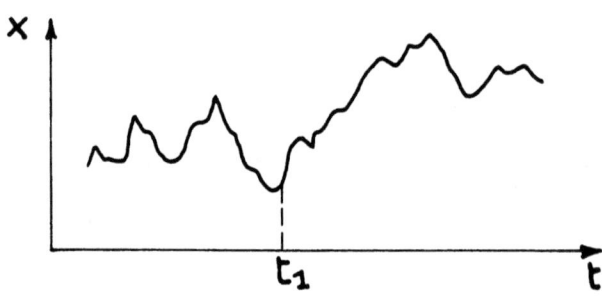

Fig. 1

Now if the probability of the following events

$$\{X = + \infty\} \quad \text{and} \quad \{X = - \infty\}$$

is equal to zero for each value of t, then the family of functions $X(t,\eta)$ is a STOCHASTIC PROCESS. Two stochastic processes $X(t)$, $Y(t)$ are identical for each event if

$$X(t,\eta_i) = Y(t,\eta_i)$$

Generally speaking, functions of a stochastic process are complicated.

Let us consider, for instance, Brownian motion. In this case the motion of one particle is represented by the curve in Fig. 1 and this motion is not regular. As a consequence, it is impossible to give a representative analytical expression for it because if we know a value of $X(t)$ for $t < t_1$, it is not possible to predict future values. But not all processes have this property and some of them are very simple.

Let us consider as an example a family of sinusoidal current generators, then $i(t) = A\ I_m\ \exp[j(\omega t+\phi)]$ where A, ω, ϕ are random variables. This is a "regular" stochastic process. As a matter of fact, if $X(t)$ is known at $t \le t_1$; then it also is known at $t > t_1$.

If $R_{xy}(t_1,t_2)$ is equal to zero, then the two processes are orthogonal. For an uncorrelated process

$$C_{XY}(t_1,t_2) = R_{XY}(t_1,t_2) - \bar{X}_1(t_1)\bar{Y}_1(t_2) = 0 \tag{11}$$

A stationary random process is one whose statistical properties do not vary by a time translation. As a consequence, both the distribution function and the density functions are time independent and only depend upon the time differences. In the case of a two variable, stationary random process we have

$$f(X_1,X_2;t_1,t_2) = f(X_1,X_2;t_1+\tau,t_2+\tau) = f(X_1,X_2;\tau) \tag{12}$$

where $\tau = t_1 - t_2$. Reduced to the case of only one variable, we have

$$f(X;t) = f(X;t+\tau) \tag{13}$$

which tells us that $f(X;t)$ is time independent. An important consequence is:

$$E\{X(t)\} = \bar{X}(t) = \text{constant} \tag{14}$$

Generally speaking erogodic systems, are concerned with the time evolution of Gibbs ensembles and that means roughly that if the system is left to itself for a long enough period, it will pass as close as we choose to all the dynamical states compatible with conservation of energy. In other words ergodocity is concerned with the problem of determining the statistical properties of a process by observing only one function of the process. We can say that a stochastic process is ergodic if we can utilize the mean time average over a singel function of the process instead of ensemble averages. Let us consider now a stochastic process $X(t)$ and its time average over a period $2T$:

$$h_T = \frac{1}{2T} \int_{-T}^{T} S(t)dt \tag{15}$$

then we have

$$\lim_{T\to\infty} \frac{1}{2T} \int_{-T}^{T} X(t)dt = E\{X(t)\} = \bar{X} \tag{16}$$

if and only if

$$\lim_{T\to\infty} \frac{1}{T} \int_{0}^{2T} \{1 - \frac{\tau}{2T} [R(\tau) - \bar{X}^2]\}d\tau = 0 \tag{17}$$

This is the ergodic theorem for the mean value. To verify the ergodicity of the mean value it is sufficient to know $\bar{X}$ and R. To be ergodic, a stochastic process must be stationary but the contrary is not true. Let us now introduce ensemble averages denoted by <> and time averages denoted by bars. Thus if we are dealing with a voltage v(t); the time average assumes the following form:

$$\bar{V}(t) = \lim_{T \to \infty} \frac{1}{T} \int_0^T V(t)dt \qquad (18)$$

and if we are interested only in a finite period we can write:

$$\bar{V}_{\Delta T} = \frac{1}{\Delta T} \int_t^{t+\Delta t} V(t)dt \qquad (19)$$

If now $V_i(t)$ is the voltage in the ith system of the ensemble at t, we can write the ensemble average as follows:

$$<V(t)> = \frac{1}{n} \sum_{i=1}^{n} V_i(t) \qquad (20)$$

and if we need mixed averages, we can for instance write

$$\overline{<V(t)>} = \frac{1}{\Delta t} \int_t^{t+\Delta t} \frac{1}{n} \sum_{i=1}^{n} V_i(t)dt \qquad (21)$$

To better understand the difference between time and ensemble averages, we can consider the following simple signal:

$$V(t) = \text{Re } V_0 e^{j\omega t} \qquad (22)$$

for which the ensemble average is

$$<V(t)> = \text{Re } V_0 e^{j\omega t} \qquad (23)$$

The time average over an interval Δt given by equation 21 is

$$V_{\Delta t} = V_0 \text{ Im}(e^{j\omega(t+\Delta t)} - e^{j\omega t})/\omega \Delta t \qquad (24)$$

If V(t) represents a stationary process, all times are equivalent so we can form an ensemble by a set of observation at different time $t_1, t_2, \ldots t_n$ provided that $t_i - t_j > T_c$ where T_c is the characteristic time of the process. As far as the time average $\bar{V}$ is concerned, we have

$$\bar{V} = \frac{1}{\tau} \int_0^\tau <V>dt = <V> \qquad (25)$$

So for stationary processes, time and ensemble average are equivalent. It is interesting to point out at this time that the

NOISE 263

three fundamental sources of noise in semiconductor devices, i.e., Johnson, shot, and $1/f^a$ noise, are stationary processes.

In statistics, the most important averages are $\bar{X}$ and $\bar{X^2}$. If $\bar{X}$ is not equal to zero it is convenient to introduce $(X - \bar{X})$ as a new new variable and in this case the significant average is the variance of X(t) defined as:

$$\sigma^2 = \overline{(X - \bar{X})^2} \qquad (26)$$

If we consider two variables X(t), Y(t) the averages are defined from equation 6 as follows:

$$\overline{X^n Y^m} = \int\!\!\int_{-\infty}^{\infty} X^n Y^m f(X,Y) dX dY \qquad (27)$$

Now if $\bar{X} = \bar{Y} = 0$ the most important averages are: $\bar{X^2}$, $\bar{Y^2}$, $\overline{XY}$. In case $\overline{XY} = 0$, the random variables X(t), Y(t) are uncorrelated but, if $\overline{XY} \neq 0$, we have correlation and the following coefficient determines the degree of correlation

$$C = \frac{\overline{XY}}{\sqrt{\overline{X^2}\overline{Y^2}}} \qquad (28)$$

Another very useful average for stationary random processes is the autocorrelation function (from eq. 6 also)

$$\overline{X(t)X(t+\tau)}$$

which has the following interesting properties:

$$\overline{X(t) \ X(t+\tau)} = \bar{X^2} \text{ if } \tau = 0 \text{ and if } \overline{X(t) \ X(t+\tau)} \neq A\,d(\tau) \qquad (29)$$

$\overline{X(t) \ X(t+\tau)}$ is symmetrical in τ which means that

$$\overline{X(t) \ X(t+\tau)} = \overline{X(t)X(t-\tau)} \qquad (30)$$

$\overline{X(t) \ X(t+\tau)}$ is a continuous function or a delta function in τ.

The autocorrelation function $R_{XX}(\tau)$ can be written as

$$\overline{X(t) \ X(t+\tau)} = R_{XX} = \lim_{T \to \infty} \frac{1}{T} \int_0^T X(t) X(t+\tau) dt \qquad (31)$$

where T is the averaging time, τ is the delay or time shift and X(t) is the waveform of the time function. The difference between

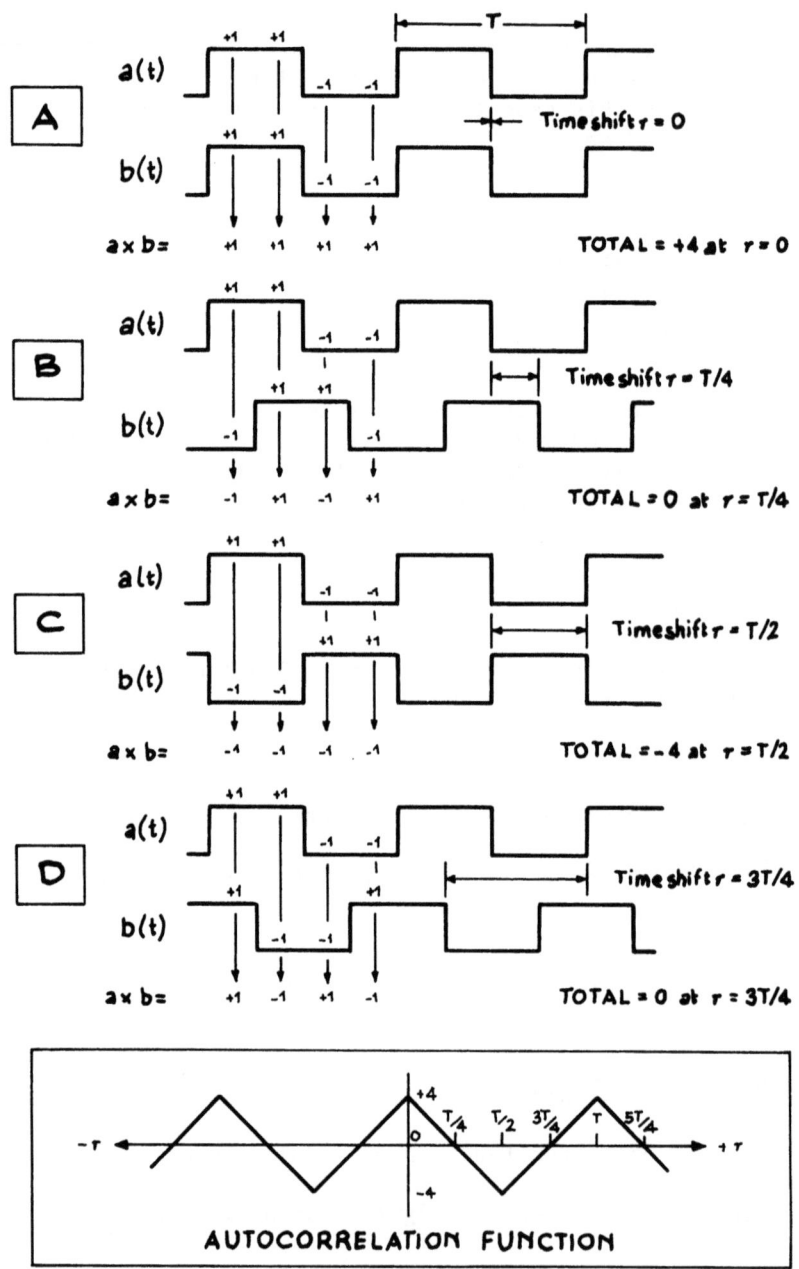

Fig. 2

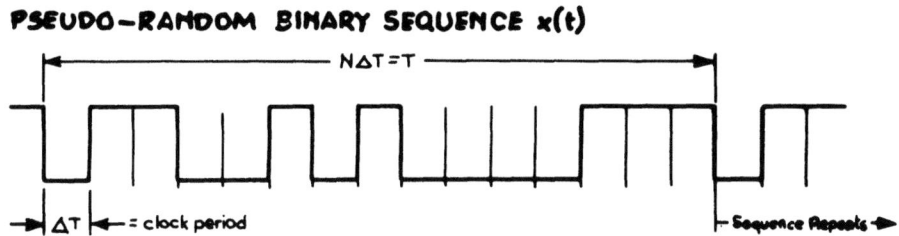

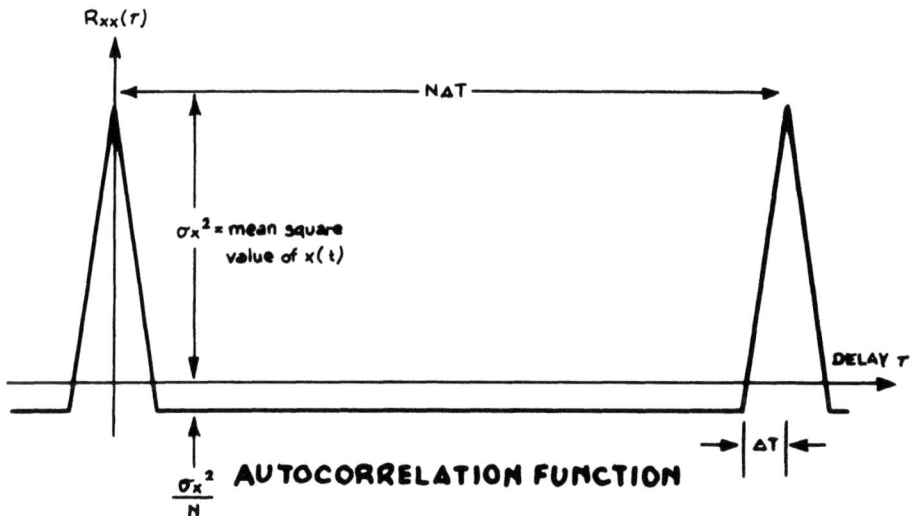

Fig. 3

equation 31 and 6 is the constancy of the distribution function in 6. $R_{XX}(\tau)$ represents a graph of the similarity between the waveform and a time-shifted version of itself as a function of this time shift. The auto-correlation function of a periodic waveform is periodic, but its shape is not necessarily the same as that of the original waveform. For instance the autocorrelation of a square wave yields a triangular autocorrelation function and the autocorrelation of a pseudo-random binary sequence gives a "spiky" function. (Figures 2 and 3). The autocorrelation function has a positive maximum at $\tau = 0$ where it is equal to the mean square value of the signal from which it is derived.

$$R(0) = \sigma^2 \geq R(\tau) \tag{32}$$

In general, we can obtain the Fourier transform of a random variable over a suitably large frequency range, as in the case of ideal white noise. However, in dealing with actual electronic amplifiers and quadratic detectors which have a finite bandwidth, any Fourier analysis of $X(t)$ must be carried out in the finite time interval $0 \leq t \leq T$.

So we have

$$X(t) = \int_{-\infty}^{\infty} x_n \exp(j\omega_n t) \tag{33}$$

where

$$x_n = \frac{1}{T} \int_0^T X(t) \exp(-j\omega_n t) dt \quad \begin{array}{l} \omega = 2\pi n/T \\ n = 0, \pm 1, \pm 2 \ldots \end{array} \tag{34}$$

The knowledge of x_n will enable us to determine the spectral density $S_{X(t)}(f)$ which can be defined by:

$$S_{X(t)}(f) = \lim_{T \to \infty} 2T x_n x_n^* \tag{35}$$

We can obtain another expression for $S_{X(t)}(f)$ (the Wiener-Khinchin theorem)

$$S_{X(t)}(f) = 4 \int_0^\infty \overline{X(t) X(t+\tau)} \cos\omega\tau d\tau \tag{36}$$

which by inversion gives:

$$\overline{X(t) X(t+\tau)} = \int_0^\infty S_X(f) \cos\omega\tau df \tag{37}$$

In measurements, one determines $S_X(f)$ and calculates the correlation function.

If in Eq. 37 we put $\tau = 0$ we get the result that the mean square value can be obtained by a simple integration of $S_X(f)$:

$$\overline{X(t)^2} = \int_0^\infty S_X(f) df \tag{38}$$

In an actual measurement, $X(t)$ is applied to the input of an arbitrary linear system (an amplifier) with a transfer function $h(f)$. Let $Y(t)$ denote the signal coming out of the system. If $S_X(f)$, $S_Y(f)$, X_n, Y_n are the spectral densities and the Fourier coefficients, then

$$Y_n = X_n h(f) \tag{39}$$

$$S_Y(f) = S_X(f) |h(f)|^2 \tag{40}$$

By using eq. 38 we have:

$$\overline{Y^2}(t) = \int_0^\infty S_X(f) |h(f)|^2 \, df \tag{41}$$

Now $\overline{Y^2}(t)$ can be measured with a quadratic detector. If the amplifier is sharply tuned at the frequency f_0, $S_X(f) \simeq S_X(f_0)$ so that

$$\overline{Y^2(t)} = S_X(f_0) \int_0^\infty |h(f)|^2 df = S_X(f_0) \, h_0^2 B_{eff} \tag{42}$$

where $h_0 = h(f_0)$ is the midband response and B_{eff} is the effective bandwidth of the system, defined as:

$$B_{eff} = \frac{1}{h_0^2} \int_0^\infty |h(f)|^2 \, df \tag{43}$$

The value of h_0 and B_{eff} can easily be determined with the help of a signal generator. Combining the measurement of these quantities with $\overline{Y^2}$, we can get $S_X(f_0)$ from equation 42. In the case of noise measurements it can be useful to determine the fraction of time spent by the signal at all possible amplitudes during a finite period of time. In practice, this means totalizing the time spent by the signal in a selection of narrow ΔV amplitude windows, and then dividing the totals for each window by the measurement time. The curve obtained by plotting the window totals against amplitude is known as the probability density function of the signal, - PDF. The most frequently encountered PDF is the Gaussian Distribution (Fig. 4, Fig. 5). The PDF can be used to yield information about non-linearities in a system. The integral of the PDF of a signal is called the cumulative probability distribution function CDF. The CDF represents the probability that a signal will be at or below a certain amplitude.

A. Amplifier Noise Model

There are numerous noise models for any two-port network. The network is usually considered as a noise free black box and the internal sources of noise are represented by two noise generator, normally located at the input port. (Fig. 6)

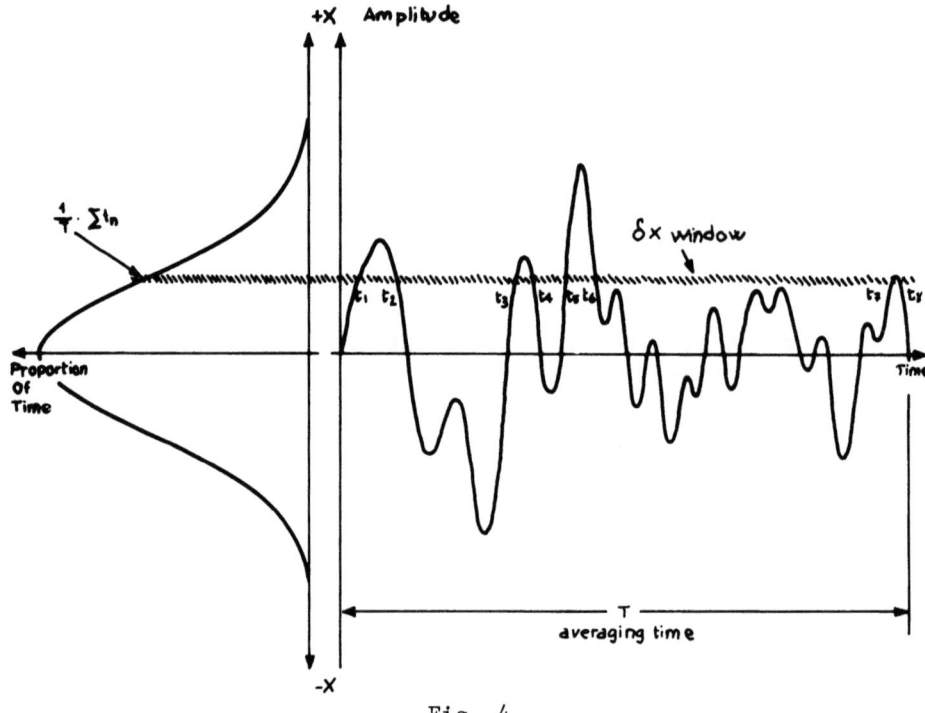

Fig. 4

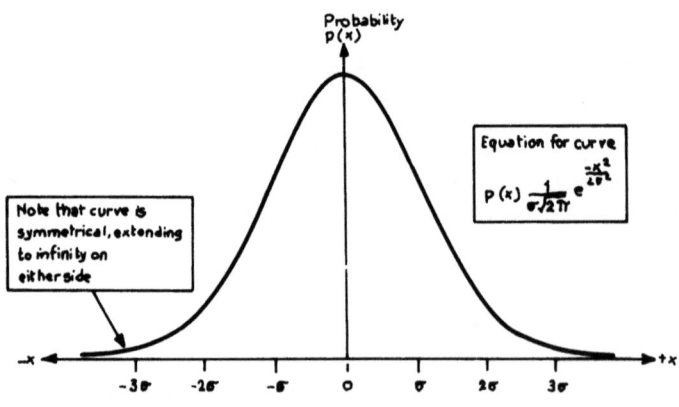

Fig. 5

NOISE

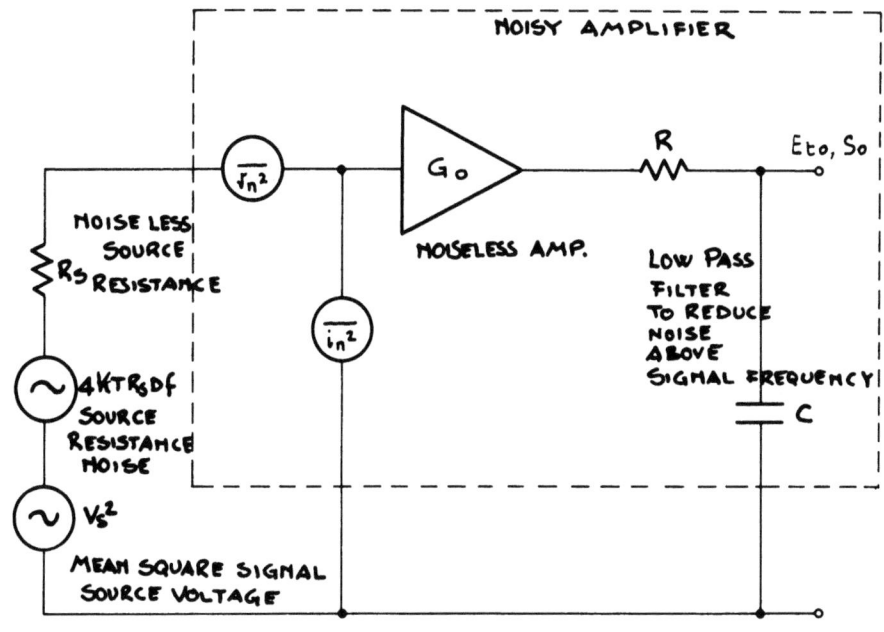

Fig. 6

The noise density generator is represented in terms of which are expressed in units of rms Volts^2/ hertz and rms Ampere^2/ hertz, respectively. This form of characterization simplifies the expression for the SNR (signal to noise ratio) and the NF (noise figure). We assume that v_n, i_n are derived from white noise sources whose crosscorrelation is zero.

Consider a conventional noisy amplifier represented as shown in Figure 6. It is easy to see from Fig. 6 that the total noise-output voltage, V_{no} is given by:

$$V_{no} = [4KTR_S + v_n^2 + i_n^2 R_S^2]^{1/2} G_0 \Delta f^{1/2} \qquad (44)$$

where G_0 is the midband gain of the ideal amplifier and Δf is the noise bandwidth, that is $1/4RC$. It can be shown that

$$\Delta f = \frac{1}{G_0^2} \int_0^\infty |G(f)|^2 \, df$$

Using a standard form for the frequency dependence of the amplifier gain, $G(f)$, and assuming $G_0 = 1$ then

$$\Delta f = \int_0^\infty \frac{df}{1 + (\omega RC)^2} = \frac{1}{4RC} \qquad (45)$$

At the output, the signal is:

$$S_0 = v_S G_0$$

Consequently, the output signal-noise voltage ratio $(SNR)_0$ is

$$(SNR)_0 = \frac{v_S}{[4KTR_S + v_n^2 + i_n^2 R_S^2]^{1/2} \Delta f^{1/2}} \tag{46}$$

The equivalent noise voltage referred to the amplifier input terminals V_{ni} is then:

$$V_{ni} = V_{no}/G_0 \tag{47}$$

It is clear from equation 44 that the total noise output depends upon the R_S value (source resistance), so we can reduce the noise by reducing R_S. In particular the SNR is a maximum when $R_S = 0$.

B. Noise Figure

A useful figure of merit to describe the noise performance of an amplifier is its noise figure [2,3]. While many definitions can be given, the two followings are quite common: the voltage noise figure is

$$NF = 20 \log_{10} \frac{\text{input-voltage SNR (without amplifier)}}{\text{voltage SNR (at amplifier output terminals)}} \tag{48}$$

and the power noise figure is:

$$NF = 10 \log_{10} \left(\frac{\text{input power SNR (without amplifier)}}{\text{voltage SNR (at amplifier output terminals)}}\right) \tag{49}$$

by using eqs. 46 and 49 we have

$$NF = 10 \log_{10} \frac{v_S^2/4KTR_S \Delta f}{v_S^2/[4KTR_S + v_n^2 + i_n^2 R_S^2] \Delta f} \tag{50}$$

We can also define the noise power factor F_p as:

$$F_p = 10^{NF/10} = 1 + \frac{v_n^2 + i_n^2 R_S^2}{4KTR_S} \tag{50a}$$

or in terms of voltage

$$F_V = 10^{NF/20}$$

The total equivalent noise referred to the amplifier input terminals becomes:

$$V_{ni} = (4KTR_S \Delta f)^{\frac{1}{2}} 10^{NF/20} \tag{51}$$

which tells us that, with a given source resistance, the least noise amplifier is the one with the smallest NF.

C. Noise Figure Contours

Equations 48-50a indicate that $NF = NF(R_S, f)$. The frequency dependence derives from the frequency dependence of the noise generators. For fixed NF, there will be a relation $R_S = R_S NF = $ constant. This is the noise figure contour, the loci of points of constant noise figure as a function of source resistance and frequency. By using these contours, it is possible, for example, to determine suitable points of operation, such as source resistance and operaintg frequency.

D. Optimum Source Resistance and Matching Problems

The optimum source resistance can be found be differentiating eq. 50 and setting the derivative to zero. As a result we get $R_{opt} = v_n/i_n$ and the minimum noise factor (power) will be given by:

$$F_{min} = 1 + \frac{v_n i_n}{2KT} = 1 + \frac{2v_n^2}{4KTR_{opt}} \tag{52}$$

If we need to improve the SNR, we must use a matched impedance transformer which allows us to use R_{opt}.

E. Noise Measurements

There are two fundamental methods for measuring the noise in an amplifier. The first one makes use of a sinusoidal voltage generator, the second utilizes a noise generator. In the first method, shown in Fig. 7, we measure first the total gain of the amplifier and then the total output noise V_{no}. Then we calculate the equivalent input noise V_{ni} which is given by

$$V_{ni}^2 = v_{th}^2 + v_n^2 + i_n^2 R_S^2 + 2C \, v_n i_n R_S \tag{53}$$

where $v_{th}^2 = 4KTR_S \Delta f$, the Johnson noise.

If $C = 0$ (cross correlation coefficient), we measure V_n by shorting the input, (i.e., $R_S = 0$) and only V_n remains. We open circuit

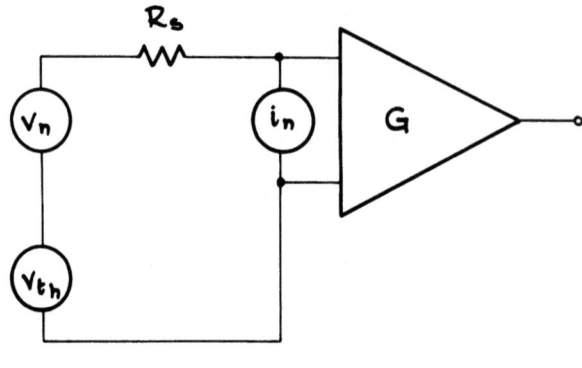

Fig. 7

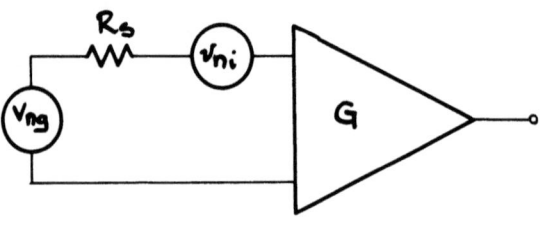

Fig. 8

the input, i.e. set $R_S \to \infty$, which, if $C = 0$, allows us to measure I_n directly. The second method of measurement is shown in Fig. 8. By using a noise voltage generator (V_{ng}) we can determine V_{n01}^2 (the output voltage due to V_{ng} and V_{ni}).

$$V_{n01}^2 = G^2(V_{ni}^2 + V_{ng}^2) \tag{54}$$

By putting $V_{ng} = 0$, we determine V_{n02}^2 (the output voltage due to V_{ni})

NOISE   273

$$V_{no2}^2 = G^2 V_{ni}^2 \tag{55}$$

From equations 54 and 55, we get

$$G^2 = \frac{V_{no1}^2 - V_{no2}^2}{V_{ng}^2} \tag{56}$$

then

$$V_{ni}^2 = \frac{V_{no2}^2}{G^2} = \frac{V_{no2}^2 V_{ng}^2}{V_{no1}^2 - V_{no2}^2} \tag{57}$$

If we make $V_{no1}^2 = 2V_{no2}^2$ then we get

$$V_{ni}^2 = V_{ng}^2 \tag{58}$$

In other words, if we measure the output noise of the amplifier with the noise generator voltage equal to zero and then introduce a white noise voltage which increases the output voltage by 3dB, then the additional noise voltage produced by the signal generator is equal to the equivalent input noise of the amplifier.

Once having characterized the preamplifier, in other words after the determination of the equivalent input noise current and the equivalent noise voltage, it is possible to determine the noise of other components.

Three kinds of noise sources can be considered: i.e., resistive capacitance and inductive sources. Depending on the kind of noise source we are working with, a suitable preamplifier circuit must be used. Fig. 9 shows a typical resistive source J FET amplifier with noise sources. The current generator (i_n) gives current to both R_g and R_i which can be considered in parallel (Fig. 9).

The total equivalent input noise voltage source is given by:

$$V_{nt} = \left[V_{nig}^2 + v_n^2 + V_{ng}^2 + \frac{V_{nD}^2}{A_V^2} + i_n^2 (R_i \| R_g)^2 \right]^{1/2} \tag{59}$$

where V_{nig}^2 is the noise of the parallel connection of V_{ni} and V_{ng}, v_n^2 is the noise voltage of the J FET, V_{ng}^2 is the noise of the source resistor R_S, V_{no}^2/A_V^2 is the noise at the drain (due to the thermal noise of R and that due to the second stage) related back to the input and $i_n^2(R_i \| R_g)^2$ is the noise current contribution of the two resistance R_i and R_g in parallel.

Fig. 10 shows a typical preamplifier equivalent circuit for capacitive sources. The preamplifier must be a current preamplifier

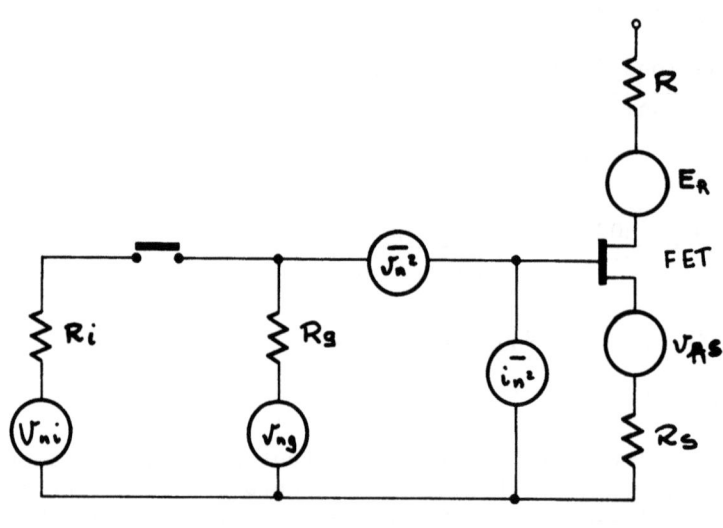

Fig. 9

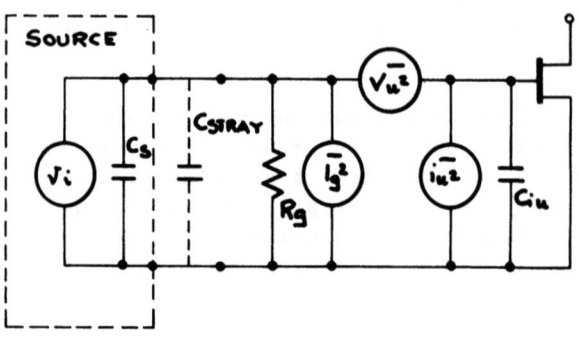

Fig. 10

with very high input resistance R_g and controlled input capacitance in C_{in} order to match the source.

The source capacitance C_S should be equal to the sum of the stray capacitance C_{stray} and C_{in} in order to achieve the maximum power transfer from signal to source. If we assume that R_g is large enough not to load the capacitance source, then the input noise voltage is

$$V_{nt} \cong [v_n^2 + (i_n^2 + i_g^2) \frac{R_g^2}{1 + \omega^2 R_g^2 C^2}]^{1/2} \qquad (60)$$

where

$$C = C_S + C_{in}$$

and where the input signal is given by:

$$V_i \cong I_i (\frac{R_g^2}{1 + \omega^2 R_g^2 C^2})^{1/2} \qquad (61)$$

A similar amplifier circuit can be considered in the case of the inductive source. (Fig. 11)

In case we are dealing with very low source impedances, transformers may be used with J FET amplifiers to minimize noise. The noise voltage is transformed by the turns ration N and the resistance is transformed by N^2. These can be used to match very low values of source resistance to a low noise amplifier and still maintain a good signal to noise ratio, i.e., the total noise at the source by taking into account an ideal transformer is

$$V_{nt}^2 = v_n^2 (R_S) + \frac{V_{n\,amp}^2}{N^2} \qquad (62)$$

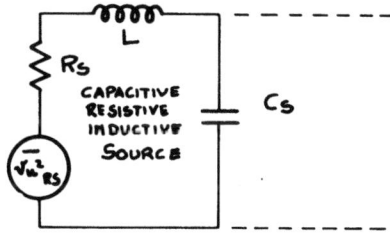

Fig. 11

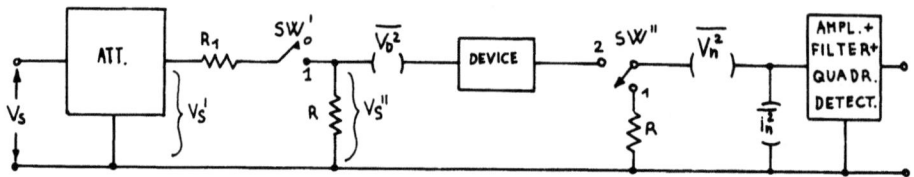

Fig. 12.

System used to measure noise.

F. Noise Measurements (Systems)

There are a variety of methods for measuring noise. In this Section, we will review a number of experimental configurations that are commonly employed.

Fig. 12 describes a circuit which is very suitable for noise measurements. The preamplifier noise is represented by a total noise voltage input generator $\bar{V}_n^2$. The noise of the devices is represented by a voltage $\bar{V}_D^2$. R is a wound resistor (assumed to be noiseless in this calculation). Let us suppose the system be tuned at frequency f with bandwidth B_{eff} and gain $G(f)$.

Using equation 41 and the appropriate transfer function for a noise measurement band on a quadratic detector, the noise power will be given by:

$$P_o \simeq \int S_V(f) \, |G(f)|^2 \, df \tag{63}$$

By assuming a small bandwidth, we have:

$$P_o \simeq S_V(f_0) \, G_0^2 B_{eff} \simeq K G_0^2 \bar{V}^2 \tag{64}$$

By means of three measurements, it is possible to calculate $S_V(f)$ i.e., the spectral density of the sample

(1) $P_1 = KG_0^2 (\overline{v_n^2} + \overline{i_n^2} R^2)$ SW' → 0   SW" → 1

(2) $P_2 = K G_0^2 (\overline{v_n^2} + \overline{V_d^2} + \overline{i_n^2} R^2)$ SW' → 0   SW" → 2

(3) $P_3 = K G_0^2 (\overline{v_n^2} + \overline{V_d^2} + \overline{V_S^2} + \overline{i_n^2} R^2)$ SW' → 1   SW" → 2

(4) $V_S" = V_S' \dfrac{R}{R_1+R}$

(5) $\overline{V_d^2} = S_V(f_0) B_{eff}$

By manipulating 1, 2, 3, 4, 5, we finally get:

$$S_V(f_0) = f\left(\dfrac{1}{B_{eff}}, V_S R, R_1, P_1, P_2, P_3\right)$$

which shows the dependence of $S_V(f_0)$ on various constants and the measurements of P_1, P_2, P_3.

A method employing two thermocouples is shown in Fig. 13. A voltage is applied to a preamplifier chain which puts power into R_1. The output voltage of the operational amplifier drives resistors R_2 by a voltage $-A(V_1-V_2)$. At equilibrium the output voltage is proportional to the input noise voltage. This method is accurate to approximately 5-6% and has 2% linearity.

Integrated circuits are available which can perform the same measurement and a typical example is shown in Fig. 14. The current produced by the noise voltage heats the base of transistor T_1. Consequently the collector current I_c increases and produces a voltage drop across R_2. The output voltage of the amplifier A warms up R_4 and consequently we have a voltage drop across R_3. When steady state is reached, $\overline{V_o^2}$ is proportional to $\overline{v_n^2}$.

Figure 15 shows a quadratic detector made with field effect transistors whose output power is proportional to the applied noise voltage. With this kind of circuit, the accuracy is given by $2B^{-\frac{1}{2}}$ where B is the bandwidth and τ is the time constant.

Figure 16 shows a simple chain where the noise voltage is first tape recorded and then processed by a spectrum analyzer and an integrator. Figure 17 shows another simple chain where an analog to digital converter is set between the noise source and a computer. By analyzing the digital signal, we can readily extract the power spectrum and the root mean square of the noise. This system can be very accurate (3%). Another system employs two identical preamplifier-amplifier chain (Fig. 18), a crosscorrelator and a

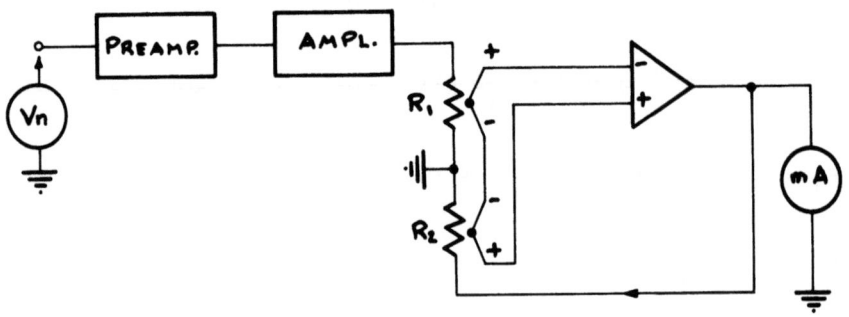

Fig. 13

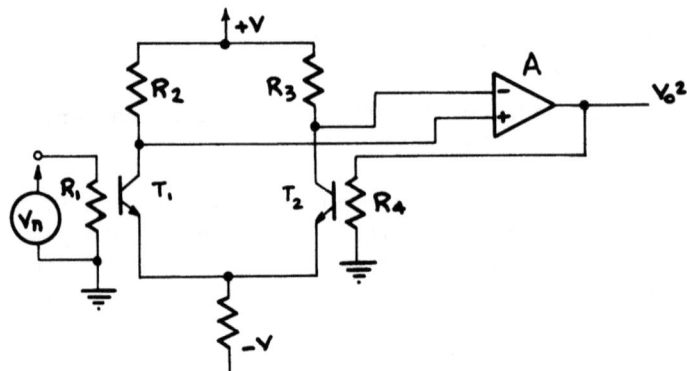

Fig. 14

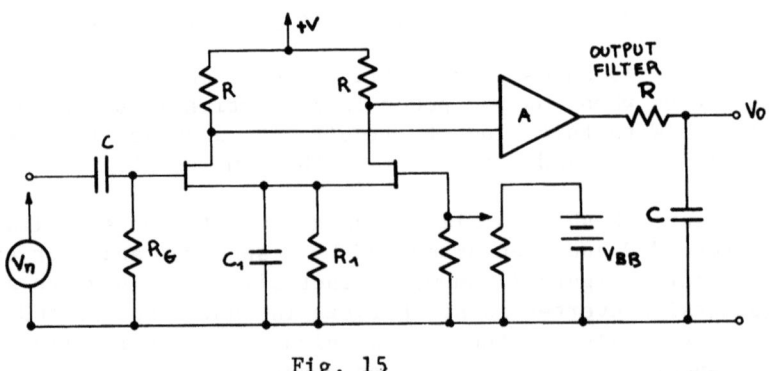

Fig. 15

NOISE

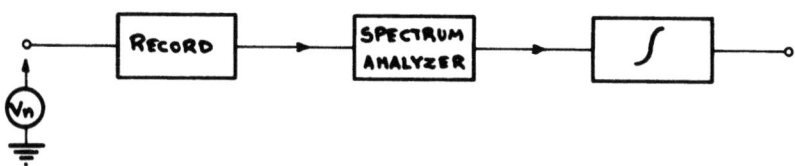

Fig. 16

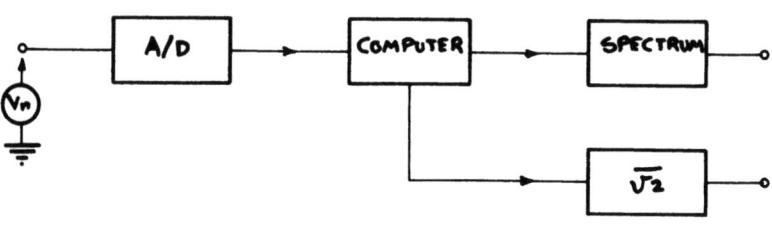

Fig. 17

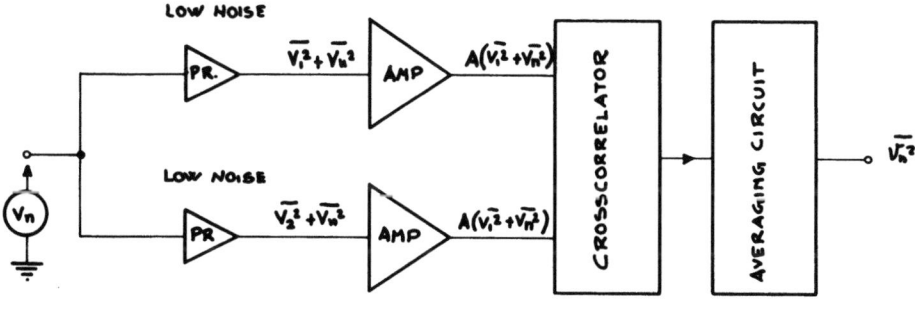

Fig. 18

spectrum analyzer. Looking at the Fig. 18, it is easy to see that each chain amplifies its own uncorrelated noise and the applied noise. Since the cross-correlator erases the uncorrelated noise, we have only the contribution of the applied noise voltage to be measured at the output. This system can give an accuracy of better than 3%.

In these cases where thermal noise is dominant with respect to other kinds of noise, it is convenient to use A.C. techniques which can discriminate between the different noise sources by a suitable choice of mid-band frequency.

Both crosscorrelation and lock-in techniques can be successfully used in this type of measurement. Fig. 19(a) shows a block scheme which utilize an internal reference driving lock-in system.

At microwave frequency, concepts related to noise measurements remain unchanged but the interpretation of the results require a field theoretical approach instead of circuit theoretical methods.

G Thermal Noise

Thermal noise (sometimes known as Johnson noise) is caused by the random movement of electrons as a result of their thermal agiation energies. This kind of noise [4,5,6,7] is common to all conducting media and is found to be a function of the temperature and the resistance of the element only. Let us see how it is possible to derive the magnitude of Johnson noise in a resistor at a temperature T. I would like to point out that the method used here is closely related to that of Nyquist [4].

Consider two resistances with the same value R connected by means of a lossless transmission line (Fig. 20). Assume that the resistor R_A will be the source of a certain amount of average noise power $<P_A>$ which goes to the resistor $R_B (T_A > T_B)$ where it will be completely dissipated.

When equilibrium ($T_A = T_B$) has been reached, we may use a statistical mechanical argument that every mode of a system at equilibrium at temperature T will have an average energy equal to:

$$\varepsilon = \frac{hf}{\exp(\frac{hf}{KT}) - 1} \tag{65}$$

where K is the Boltzman constant and h is Planck's constant. If we short the transmission line at both ends, the energies associated with the power $\overline{P_A}$ and $\overline{P_B}$ are trapped as standing waves. In this case, we can say that the total energy trapped is given by:

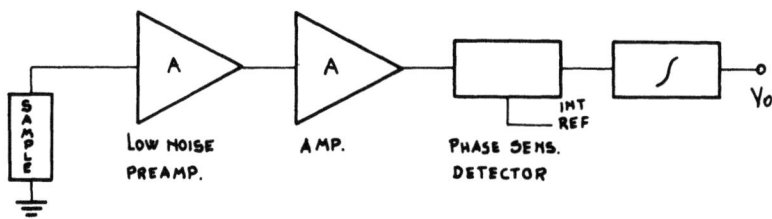

Fig. 19

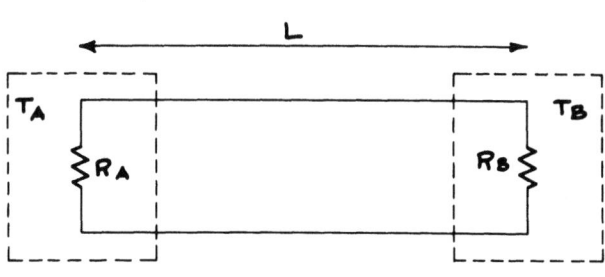

Fig. 20

$$(\overline{P_A} + \overline{P_B}) \Delta t = 2\overline{P_A} \frac{L}{C} \tag{66}$$

where Δt is the time required for the energies to transfer from R_1 to R_2 and vice versa. Considering the interval of frequencies between f and $f + \Delta f$ where Δf is the bandwidth. The number of standing waves Δn on the transmission line is given by

$$\Delta n = \frac{2L}{C} \Delta f \tag{67}$$

Between f and $f + \Delta f$, the total average energy trapped on the transmission line will then be:

$$\epsilon \Delta n = \frac{hf \Delta n}{\exp(\frac{hf}{KT} - 1)} = \frac{2L}{C} \frac{hf \Delta f}{\exp(\frac{hf}{KT}) - 1} = 2\overline{P_A} \frac{L}{C} \tag{68}$$

from which

$$\overline{P_A} = \frac{hf\Delta f}{\exp(\frac{hf}{KT}) - 1}$$

Now $\overline{P_A}$ results from a voltage source $\overline{v_n^2}$, associated with R and since the transmission line is a load matched to this voltage source, we can write

$$\overline{P_A} = \frac{\overline{v^2}}{4R} \tag{70}$$

then from equation 69

$$\overline{v^2} = 4R \frac{hf\Delta f}{\exp(\frac{hf}{KT})-1} \tag{71}$$

If hf << KT, it is possible to write:

$$\overline{v^2} = 4RKT\Delta f \tag{72}$$

In terms of the current, we have:

$$\overline{i^2} = 4KT\Delta f/R \tag{73}$$

where $\overline{i^2}$ represents the noise current that will flow across the shorted terminals of a resistor R. The two equivalent ways of representing a noisy resistor are shown in Fig. 21.

In order to make any meaningful measurements of noise, we must restrict our measurements of noise power to some limited band of frequency. Using this approach, we can derive Equations 72 and 73 from a circuit theory consideration. If we measure the available power in a certain resistor R at a temperature T(°K) across a band of frequency B centered at f, we shall find that the average power P is given by:

$$P = KT\Delta f \tag{74}$$

It must be noted that the average power available at a temperature T does not depend on the value of the resistor. Moreover it is proportional to the bandwidth over which measurements are being made and it does not depend on the center frequency. This means that in a band of frequency from 100 KHz to 1 MHz or from 1000 kHz to 10 MHz there is exactly the same thermal noise power.

It is also important to observe that the power which we are considering is the power available from the resistor. The available power is defined as the maximum power which may be drawn from the source by variation of a resistor R_1 connected across its terminals (Fig. 22). The power dissipated in the resistor R_L is given by:

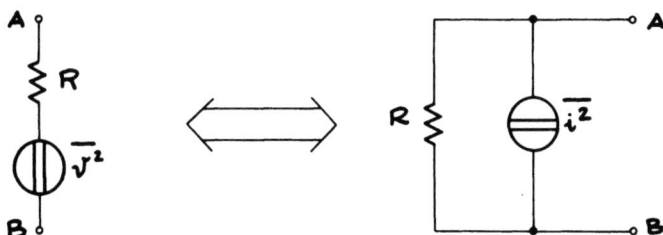

Fig. 21

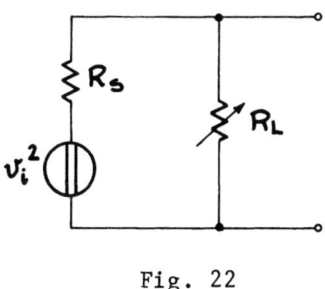

Fig. 22

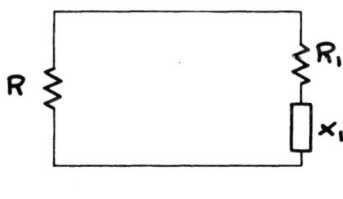

Fig. 23

$$P = \frac{v^2 R_L}{(R_S + R_L)^2} \tag{75}$$

and the maximum value is obtained when $R_S = R_L$. Thus the available power from this source will be

$$P_{max} = \frac{v^2}{4R_S} \tag{76}$$

Now if we consider a noisy resistor R, which gives an open circuit mean square voltage $\overline{v^2}$, the available power, according to what has been said before will be:

$$\overline{P}_{noise} = \frac{\overline{v^2}}{4R}$$

Combining this with Equation 74 we have Equation 72. If our noise source is directly connected to an impedance $Z = R_1 + jX_1$ (Fig. 23), then the power dissipated in Z will be given by:

$$P = \frac{\overline{v^2}}{|R+Z|^2} R_1 = \frac{\overline{v^2}}{(R+R_1)^2 + X_1^2} R_1$$

where $\overline{v^2} = 4KTR\Delta f$. The power transfered from the impedance Z, to R under watched load conditions is

$$\frac{\overline{v^2}}{R} = \frac{\overline{v_1^2}}{R_1} = 4KT\Delta f$$

from which we have that the equivalent circuit of an impednace $Z = R_1 + jX_1$ is given by a noise voltage generator

$$\overline{v_1^2} = 4RKT\Delta f \ ReZ \tag{77}$$

I would like to observe here that the expression cannot give us, in practice, an infinite voltage response as R goes to infinity (open circuit) because of the capacitance which limits this voltage. As a matter of fact an actual noisy resistance must be represented by the circuit shown in Fig. 24, where v_{th} is the thermal noise voltage generated by R and V_{out}^2 is given by:

$$\overline{V_{out}^2} = \frac{4KTR\Delta f}{(1-\omega^2 LC)^2 + \omega^2 R^2 C^2} \tag{78}$$

If $\omega^2 LC \ll 1$, by allowing Δf to become a differential and integrating over the entire frequency spectrum, we get the total noise energy from Equation 78

$$\int_0^\infty \frac{4KTRdf}{1+\omega^2 R^2 C^2} = \frac{KT}{C} \tag{79}$$

which is limited by both R and C, and does not depend on the value of R where the available noise power is generated.

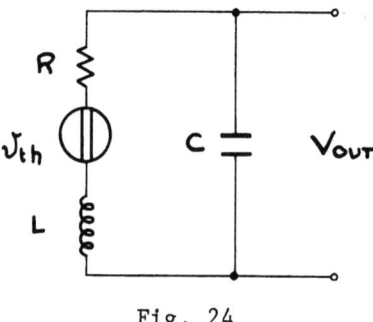

Fig. 24

Fig. 25

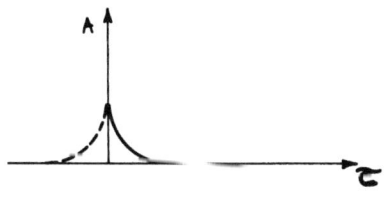

Fig. 26

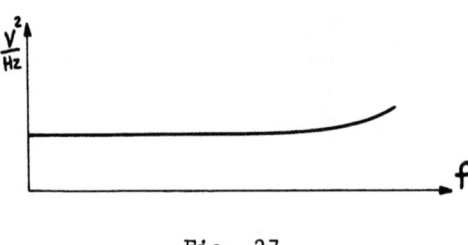

Fig. 27

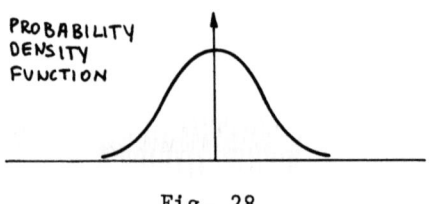

Fig. 28

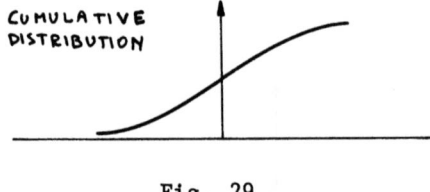

Fig. 29

NOISE

Let us consider now some characteristics of the thermal noise. Its time dependence is represented in Fig. 25, its autocorrelation function in Fig. 26, its frequency spectrum in Fig. 27, its probability density function in Fig. 28 and finally its cumulative distribution (Fourier Transform of the autocorrelation function) in Fig. 29.

The Johnson noise voltage spectral density in a semiconductor is

$$S_V(f) = 4\,KT\,R_0 = 4\,KT\,\frac{L^2}{e(n_0\mu_n + p_0\mu_b)} \tag{80}$$

where R_0 is the sample dark resistance, n_0 and p_0 are the total number of electrons and holes in the sample, respectively, μ_n and μ_p are the electron and hole mobilities, respectively and L is the semiconductor length.

It may be interesting to derive the Johnson noise behaviour by using a simple physical model [2]. Let us consider a resistor of volume $V = AX$ which contains N free electrons per unity volume.

Each electron has an average kinetic energy

$$\overline{E} = \frac{3}{2}KT = \frac{1}{2}m \sum_{j=xyz} \overline{v_j^2} \tag{81}$$

The resistance arises from one or more of the following scattering mechanisms: electron-electron, electron-ion, electron-phonon interactions which are characterized by the mean scattering time τ. As a consequence, the conductivity will be given by

$$\sigma = \frac{ne^2E}{m}\,\overline{\tau} \tag{82}$$

Correspondingly, the resistance is

$$R = \frac{L}{\sigma A} = \frac{mL}{ne^2\overline{\tau}A} \tag{83}$$

where A is the area of the semiconductor.

Let us assume that the time between two successive scatterings is equal to τ, then it is possible to show that the pulse current in an external circuit is

$$i(t) = \begin{cases} ev_i/L & 0 \le t \le \tau \\ 0 & \text{for all other values} \end{cases} \tag{84}$$

Taking the Fourier transform we have:

$$I(\tau, v_i, \omega) = \int_0^\tau i(t) e^{-j\omega t} dt \tag{85}$$

By introducing Equation (84) and squaring we get:

$$I^2(\tau, v_i, \omega) = -\frac{2e^2 v_i^2}{\omega^2 L^2} (1 - \cos\omega\tau) \tag{86}$$

Now we need to average $I^2(\tau, v_i, \omega)$ over both the scattering times and the velocities.

Assuming that τ and v_i are independent variables, i.e., the probability function

$$P = P(\tau, v_i) = H(\tau) \, g(v_i)$$

is the product of the individual probabilities and assuming that $H(\tau) = \exp(-\tau/\tau_0)/\tau_0$, by averaging over τ we obtain:

$$|I(v_i, \omega)|^2 = \int_0^\infty H(t) \, |I(\tau, v_i, \omega)|^2 \, d\tau = \frac{2e^2 v_i^2 \tau_0^2}{L (1 + \omega^2 \tau_0^2)}$$

and by averaging over v_i^2 where $v_i^2 = KT/m$ we get:

$$|I(\omega)|^2 = \frac{2e^2 \tau_0^2 KT}{mL^2 (1 + \omega^2 \tau_0^2)} \tag{88}$$

Now the power in the frequency interval from f to $f + \Delta f$ associated with the current is $S(f) \, df$ where:

$$S(f) = 2\overline{N} \, \overline{|I(\omega)|^2} \tag{89}$$

and $\overline{N}$ is the average number of scattering events per second given by

$$\overline{N} = \frac{NV}{\tau_0}$$

By considering the approximation $\omega\tau < 1$ we finally have from eq. 88 and 89:

$$\overline{I^2} = 4KT\Delta f/R \tag{91}$$

III. $1/f^a$ NOISE (FLICKER NOISE)

For many years, much experimental and theoretical work has been done on $1/f^a$ noise which arises in all electronic devices such as photoconductors, junction diodes, transistors J FET, MOSFET, Gunn diodes, Zener diodes, hot electron diodes. Many attempts to establish a comprehensive theory have been made but there is insufficient knowledge to get a real picture of its physical origin and its intrinsic mechanism [8-26].

If we look at the work of the past ten years, we find that there are four approaches used to explain 1/f noise. They are:

(a) mathematical models based on signal theoretical aspects [27,28,29] or on non stationary stochastic processes [30,31]

(b) models based on the purely empirical relation $(\Delta R/R)^2 \simeq (1/N)(\Delta f/f)$ where N is the total number of mobile charge carriers in the sample (Hooge) [32].

(c) models which are based on the fluctuation of the temperature of the sample. In this case those electrical properties depending on temperature fluctuations which produce the temperature dependent resistance [33] (Clark and Woss)

(d) models which introduce, as the main cause, a particular distribution of surface states at the semiconductor-oxide interface or in semiconductor-insulator structures (McWhorter) [34,35].

The spectrum of the $1/f^a$ noise with "a" sometimes close to 1 holds down to extremely low frequencies (10^{-4}Hz) and as far as statistical properties are concerned, it seems to be a stationary random process which gives a Gaussian distribution. Another characteristic behaviour attributed to $1/f^a$ noise is that the power spectral density of the current fluctuation $Si(f)$ varies as I^γ where I is the dc constant current and $1 \leq \gamma \leq 2$. The problem of whether the noise generation mechanism is linear or not is not yet completely solved. Experimental results show that in FET's, a linear mechanism is possible while in PN junction devices a non linear mechanism seems to be involved [36].

In order to start to examine 1/f noise processes, we will employ the 4th approach mentioned above. In line with this model, we shall consider a semiconductor-oxide system. In this system there are surface states associated with the interface region between semiconductor and oxide. The important states for our purposes are those states that can communicate with the bulk rapidly. It is believed that the physical origin of these states must lie within a few angstroms of the semiconductor surface. It is also hypothesized that states occurring in the oxide layer are slow states which can only communicate with the semiconductor bulk by tunneling through the insulator. The dependence of $1/f^a$ noise on the current is generally found to be proportional to the square of the steady state d.c. current. The observed $1/f^a$ noise could be due to conductivity fluctuations (fluctuations in number of carriers) or mobility fluctuations in the semiconductor due to variations in the surface state density.

Let us suppose that the number n of carriers in a semiconductor fluctuates by an amount Δn due to the bulk-surface states interaction and suppose a single time constant is involved in the process.

By using the autocorrelation function (Eq. 31), we have:

$$\overline{\Delta n(t) \Delta n(t+s)} = \overline{\Delta n^2} e^{s/\tau} \tag{93}$$

and by using the Wiener-Khintchine theorem (Eq. 36) we obtain for the carrier spectral density:

$$S_n(f) = 4 \overline{\Delta n^2} \frac{\tau}{1+\omega^2\tau^2} \tag{94}$$

Let the current through the semiconductor be I_o, (the subscript o refers to the equilibrium value). The voltage across the sample will be proportional to $1/N$ and its fluctuation is given by:

$$\Delta V = \frac{V_o}{N_o} \Delta n$$

Then the voltage spectral density will be

$$S_V(f) = \left(\frac{V_o}{N_o}\right)^2 S_n(f) = \left(\frac{V_o}{N_o}\right)^2 4\overline{\Delta n^2} \frac{\tau}{1+\omega^2\tau^2} \tag{95}$$

Now we assume

$$\overline{\Delta n^2} = \Gamma(N_o)N_o \tag{96}$$

where Γ is the efficiency of the oxide states-free carrier interaction.

McWhorter used a model for $1/f^a$ noise in semiconductors based on slow surface states. Fluctuations in the capture and release of electrons of the bulk region by these slow states produce fluctuations in the concentrations of the electrons in the bulk and consequently in the conductivity.

We assume a tunneling process as the mechanism by which bulk electrons interact with the oxide states. Since the tunnel probability depends exponentially on the distance x over which tunneling occur, the following dependence of τ on x may be written as:

$$\tau = \tau_0 e^{a'x} \tag{97}$$

where a' is of the order of 10^8 cm^{-1}.

By assuming a uniform trap distribution in the region $x_1 \leq x \leq x_2$, the normalized distribution function is

$$g(x) dx = \frac{dx}{x_2-x_1} \quad x_1 < x < x_2$$
$$= 0 \quad \text{otherwise}$$

From equation 97

$$x_2 - x_1 = \frac{1}{a'} \ln \frac{\tau_2}{\tau_1} \tag{98}$$

$$dx = \frac{1}{a'} d\tau/\tau$$

Then

$$g(\tau) d\tau = \frac{d\tau}{\tau \ln \tau_2/\tau_1} \tag{99}$$

which is the normalized distribution in τ. By averaging $S_V(f)$ in Equation 95 over all τ's yields:

$$\{S_V(f)\} = \frac{4\Gamma(N_o)V_o^2}{N_o \ln \tau_2/\tau_1} \int_{\tau_1}^{\tau_2} \frac{\tau}{1+\omega^2\tau^2} \frac{d\tau}{\tau} = \frac{4\Gamma(N_o)V_o}{N_o \ln \frac{\tau_2}{\tau_1} 2\pi f} \tag{100}$$

$$[\tan^{-1}\omega\tau_2 - \tan^{-1}\omega\tau_1]$$

which reduces in the approximation $1/\tau_2 < \omega < 1/\tau_1$ to

$$\{S_V(f)\} = \frac{K}{N_o} \frac{V_o^2}{f} = \frac{KR_o^2}{N_o} \frac{I_o^2}{f} \tag{101}$$

where

$$K = \frac{\Gamma(N_o)}{\ln \tau_2/\tau_1}$$

It is important to point out that Eq. 101 contains the factor $\Gamma(N_o)$ which, according to equation 96, is related to the trap density and is itself proportional to the surface state density. The derivation of the formula assumes that the surface is uniformly active, but if that is not the case, the factor K must be reconsidered.

It is interesting now to consider the well known expression proposed by Hooge for $1/f^a$ noise. Hooge established that $1/f^a$ noise arises in the bulk and, where homogeneous material is concerned, he gave the following empirical expression for the $1/f^a$ noise power spectral density:

$$S_V(f) = \frac{KV_o^b}{f^a} = \frac{K}{f^a} (I_o R_o)^b \tag{102}$$

where $b \simeq 2$, $a \simeq 1$ and K is an empirical coefficient given by

$$K = \frac{C}{N_{tot}} \simeq \frac{2 \times 10^{-3}}{N_{tot}} \tag{103}$$

and $N_{tot} = N_o L \omega \tau$.

Hooge concluded that $1/f^a$ noise is associated with electrical conduction process. In particular he argues that the fluctuation producing the 1/f noise arises in the carrier mobility rather than in the number of carriers. The same conclusion was reached by G. M. Kleinpenning [36] in examining both intrinsic germanium and extrinsic germanium and silicon.

A. Generation-Recombination Noise (g-r)

g-r noise is basically a thermal effect [37,38,39]. In this process electrons and holes appear and disappear in pairs due to spontaneous fluctuations in the generation, recombination and trapping rates of carriers. If trapping is neglected we find that $\Delta n = \Delta p$ where Δn and Δp are fluctuations around the equilibrium values n_0, p_0. The fluctuation ΔR in resistance is:

$$\Delta R = \frac{e(\mu_n + \mu_p)R_0^2}{L^2} \Delta n \tag{104}$$

and the corresponding fluctuation in voltage ΔV is

$$\Delta V = I_0 \Delta R \tag{105}$$

As a result, the g-r noise spectral density will be given by:

$$S_V(f) = \frac{e(\mu_n + \mu_p)R_0^2}{L^2} I_0^2 S_n(f) \tag{106}$$

where $S_n(f)$ is the spectral density of Δn.

From the theory of g-r noise we have that

$$\Delta n = 4 \, g(N_0) \frac{\tau_0^2}{1 + \omega^2\tau_0^2} \tag{107}$$

where $g(N_0)$ is the thermal generation rate and τ_0 is the lifetime of added carriers. Combining equations 104, 106 and 107 we have:

$$S_V(f) = \left(\frac{e(\mu_n + \mu_p)R_0^2}{L^2}\right)^2 4g(N_0) \frac{\tau_0^2 I_0^2}{1 + \omega^2\tau_0^2} \tag{108}$$

Noise measurements performed on photoconductive PbSnTe (single crystal) films at 77° show an I_0 dependence as predicted by the 108 (Fig. 30).

In "good" transistors this kind of noise is generally negligible; it appears most often in transistors having very low current gain and falls off rapidly at high frequency.

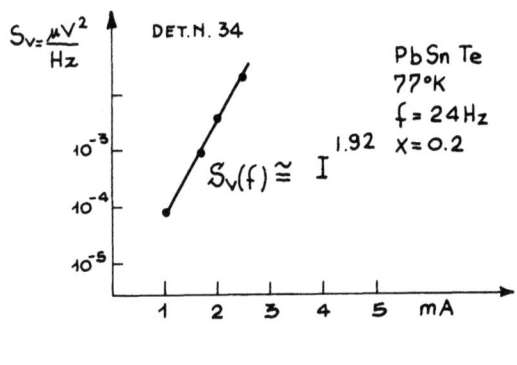

Fig. 30

B. Shot Noise

Shot noise [6] in semiconductor is mainly associated with current flow across a potential barrier. Such a barrier is common in p-n junction devices (junction diodes, transistors, polycrystalline photoconductors). In discussing noise in p-n junctions, $p \ll N_d$, (the donor density), in the n-region and $n \ll N_a$, (the acceptor density), in the p-region. High injection means $p \gg N_d$ in part of the n-region of a p^+n diode and $n \gg N_a$ in part of the p-region of a pn^+ diode.

At low injection, full shot noise will be associated with the current which flows through the devices. The passage of carriers across the barriers can be considered as independent random events (corpuscular approach). This approach is no longer valid at high injection but it could be a sufficient first approximation in some cases.

Another approach (collective model) employs the random processes of diffusion and recombination. It is possible to show that the two approaches give rise to the same results only when low injection is concerned.

C. Burst Noise

This kind of noise also called popcorn noise was first observed by Pay [40] in germanium point contact diodes. Since then it has been found in tunnel diodes, in forward biased junction silicon diodes, in junction transistors and in linear integrated circuits. Many silicon transistors, particularly those of the planar diffused type, polarized by means of large value base resistance show this kind of noise which can be characterized as a low frequency noise.

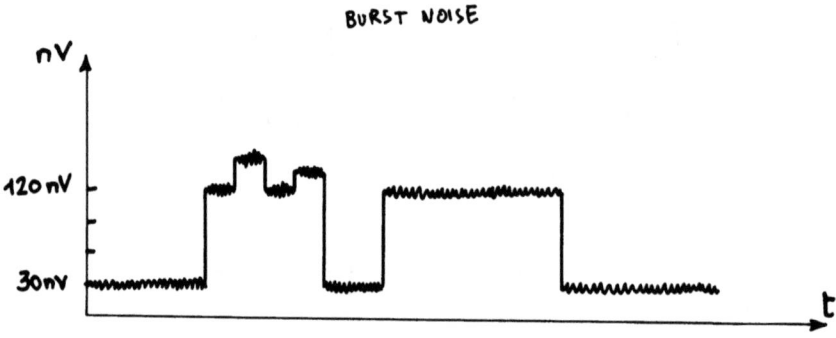

Fig. 31

It consists of random pulses of variable duration and, very frequently, equal amplitude which are in many cases, ten times more than the pedestal noise. For instance, in pn junction diodes under forward bias, it has been found that burst current amplitudes are no higher than a few tenths of microamperes.

The length of burst pulses, which can vary from a few µs to minutes, and their frequency are random events. The power spectral density, in the low frequency range, has a $1/f^a$ behaviour, where a can vary from 1 to 2. Fig. 31 shows a typical burst noise with multilevel events.

A simplified circuit which takes into account burst noise in transistors is shown in Fig. 32. This scheme considers the possibility of multi-level burst noise which is not always present in practice. As a matter of fact, each current generator represents its own contribution for each multilevel burst. Data obtained from experimental results made on a silicon planar transistor show the following behaviours [41,42]:

(a) the burst noise amplitude I_{BB} (VBB) depends on the temperature as shown in Fig. 33 (b) and the results suggest that the dependence follows is $I_{BB} = A_o \exp(-a^*T)$ where a* does not seem to be dependent on the different transsistors used.

(b) the burst noise amplitude does not depend on the emitter-collector voltage.

(c) the burst noise amplitude is related (at a constant temperature) to collector current by the following empirical relation $I_{BB} = K\, I_c^P$, where $p = 0.6$;

(d) the average burst rate is linearly dependent upon collector current (at a constant temperature) and Fig. 34 shows a typical experimental result;

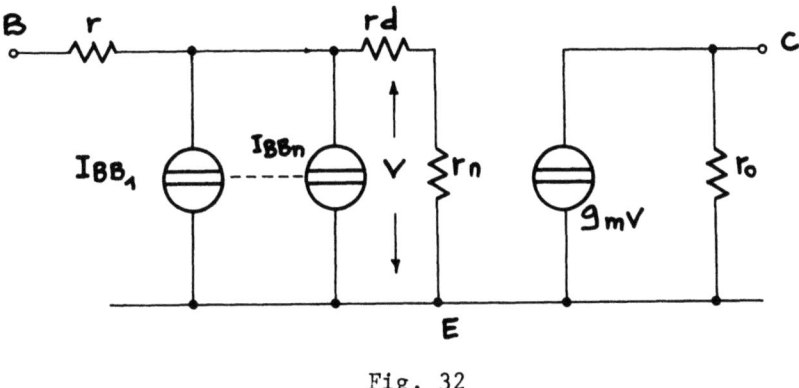

Fig. 32

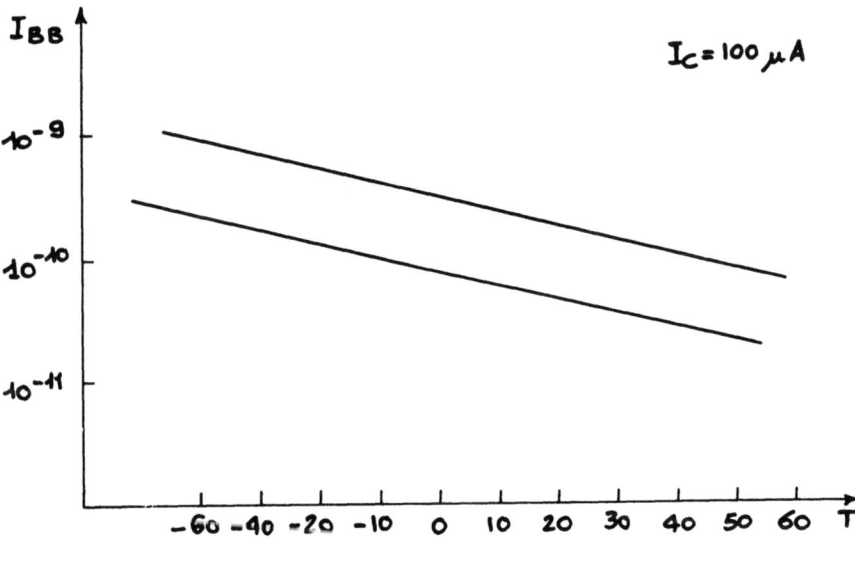

Fig. 33

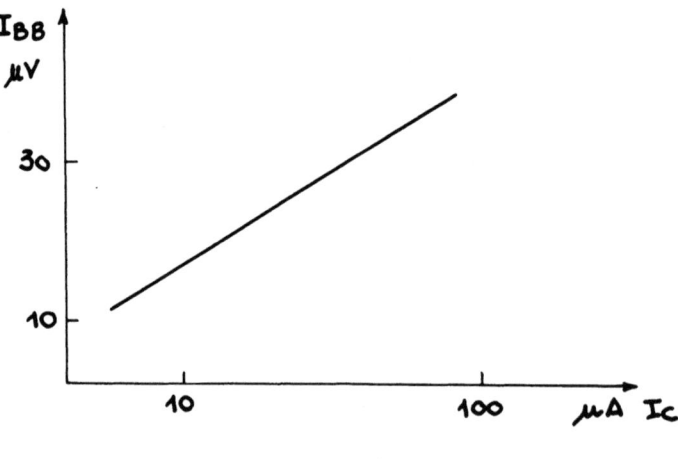

Fig. 34

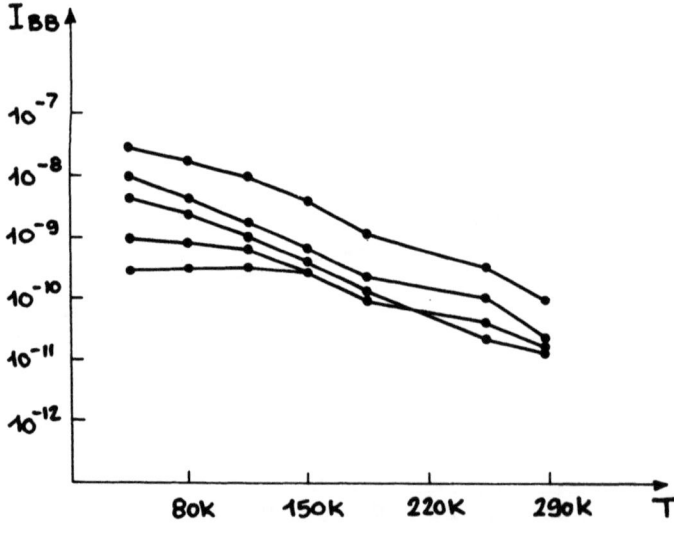

Fig. 35

(e) burst noise has also been recently observed here in Pb/PbSnTe Schottky diode polycrystalline film. Fig. 35 shows the temperature dependence of the burst noise amplitude from 300°K down to 77°K;

(f) the distribution of the numbers of burst against their pulse duration seems to be close to a stationary Poisson distribution

$$E\{N_b\} = N_\ell \, e^{-m\ell} \tag{109}$$

where N_ℓ is the number of bursts whose length is greater than ℓ; ℓ is the pulse width, N_b is the total number of bursts observed, m is the density factor of the Poisson process.

A burst noise model has been proposed by G. Doblinger who postulated a particular region with reduced band gap in conjunction with a nearby trap. This region would shunt the base emitter junction, drawing in this way more current. This current is modulated by the changes in the charge state of the trap. The band gap reduction seems to be related to localized compressive stress due to crystallographic imperfections.

By considering these comments on the observation of burst noise, se can obtain information on the choise of components, especially if these componenets are to be used as a low level preamplifier in an amplifier chain. Moreover, burst noise can also be correlated to failure mechanism.

Other authors have attributed the phenomenon to defects located in the neighbourhood of emitter-base junction; but opinions on the nature of these defects vary. For example, Brodersen, Cook, Chenette [56] and others have suggested metallic precipitates as the cause, while Martin, Blasquez et. al. found a correlation between the cryotallography and defect density and the number of transistors which exhibited burst noise. Crystallographic defects come from the substrate (grown - in defects) and from manufacturing processes of the type described in other papers of this Institute.

If n is the proportion of devices (transistors) having burst noise and D_d the crystallographic defect density, the data indicates that the correlation between n and D_d may be related to the distribution of defects within transistors with evident burst noise [43]. While metallic impurities can effect the burst noise of transistor characteristics, it is not a strong relationship. Metallic impurities are known to act as generation-recombination centers if their concentration is low but if their concentration is sufficiently high, precipitates can occur which strongly influence the junctions locally [44,45,46].

Other investigations are needed in order to better understand the different contributions to the burst noise, although it does appear that crystallographic defects are more significant. Sources of data on burst noise intermittency and its statistics in NPN and PNP planar transistors are available in a recent work by K. F. Knott [47].

D. Noise in MS Diodes (Schottky Barriers)

In a metal-semiconductor diode, the following relation holds

$$I = I_0(V) \exp\left(\frac{qV}{mKT}\right) - 1 \; ; \; h = f(I) \tag{110}$$

It is easy to see two contributions:
(a) one due to carriers going from the semiconductor to the metal

$$I + I_0 = I_0(V) \exp\left(\frac{qV}{mKT}\right) \tag{111}$$

(b) another contribution due to carriers going in the opposite sense

$$I_0(V)$$

If we suppose that each current can fluctuate independently and that carriers give rise to a shot phenomena, it is possible to write:

$$S_i(f) = 2q \Sigma I = 2q(I + 2I_0) \tag{112}$$

Remembering that the conductance g (in the low frequency range) can be written as

$$g = \frac{dI}{dV} = \frac{dI_0}{dV} \{\exp\left(\frac{qV}{mKT}\right) - 1\} + \frac{qI_0}{mKT} \exp\left(\frac{qV}{mKT}\right) \approx \frac{q}{mKT}(I+I_0)$$

we have

$$S_i(f) = 2mKT \, (g + g_0) \; ; \; g_0 = \left.\frac{dI}{dV}\right|_{V=0}$$

which gives the thermal noise when I = 0.

The diode may have a series resistance r_i which shows thermal noise

$$S_V = 4KTr_i \tag{113}$$

In the cases where r_i is significant, the total spectral intensity $S_t(f)$ of the short-circuit noise current is

$$S_t = \frac{2mKT\ R_o + 4\ KT\ r_i}{(R_o + r_i)^2} \tag{114}$$

if

$$R_o \gg r_i \qquad S_t(f) = 2mKT/R_o$$

if

$$r_i \gg R_o \qquad S_t(f) = 4KT/r_i$$

E. Noise in p-n Junction Diodes at Low Injection

In this case we can apply the same results used in the last section. For simplification we will consider a pn$^+$ diode, so the current is due to electrons. Again the current can be written as

$$I = I_o \{\exp(\frac{qV}{mKT}) - 1\} \tag{115}$$

consists of two parts:

(a) a part ($I_o \exp(qV/mKT)$) due to electrons injected into the p region and recombining there,

(b) a part ($-I_o$) due to electrons generated in the p-region and collected in the n-region.

These two currents show shot noise, so we can use with m = 1

$$S_i(f) = 2KT\ g_o\ \frac{I + 2I_o}{I + I_o} \tag{116}$$

The series resistance also generates thermal noise. At high frequency an additional source of noise due to electrons injected into the p-region and returning to the n-region due to the drift field reversal before they recombine. One can take into account this phenomena by writing the current spectral density

$$S_i = \frac{2KT\ R_o + 4\ KT\ r_i + 4\ KT(g - g_o)}{(R + r_i)^2} \tag{117}$$

Van der Ziel has examined this in greater detail using a corpuscular approach.

F. Noise in (p-i-n) Diodes

In this case we have the following expression of the spectral current density:

$$S_i(f) = 2q(I + 2I_o) + 4m'KT(g - g_o) \tag{118}$$

where m' takes into account the admittance H.F. behaviour.

G. Noise In Tunnel Diodes

In this device the main characteristic is that the two opposite currents depend on bias in a different way. The spectral current density has the following expression [55]:

$$S_i(f) \simeq 2qI \, \text{Coth} \, \frac{qV}{2mKT} \qquad (119)$$

In the approximation $qV \gg KT$ it reduces to the full shot noise expression.

$$S_i(f) \simeq 2qI$$

In modern tunnel diode [48] it has been found that the noise contribution depends on the quiscent point. At the University of Cosenza [51] the noise contribution of the valley region has been measured which seems to give the following spectral current density (Fig. 36):

$$S_i(f) = 2qI(p) \qquad 1.16 < p < 1.64 \qquad (120)$$

As we can see from Fig. 36 the tunnel behaves more noisy as we go close to both the valley and the peak point.

H. Noise In Transistors At Low Injection

We will consider a p-n-p transistor where all the current is carried by holes. The emitter current and the collector current have the following expressions:

$$I_F = E_{ES} \exp\left(\frac{qV_{EB}}{KT}\right) - I_{BE} \qquad (121)$$

$$I_C = \alpha_F I_{ES} \exp\left(\frac{qV_{EB}}{KT}\right) + I_{BC} = \alpha_F I_E + I_{CO} \qquad (122)$$

where $I_{CO} = \alpha_F I_{BE} + I_{BC}$ is the saturation current at the collector. The emitter current is due to holes injected into the base and holes generated in the base region and collected by the emitter.

The first term in Equation 122 for collector current I_C is due to the collection of holes injected by the emitter where α_F is the forward current amplification factor while the second term is due to holes which are generated in the base region and collected by the collector.

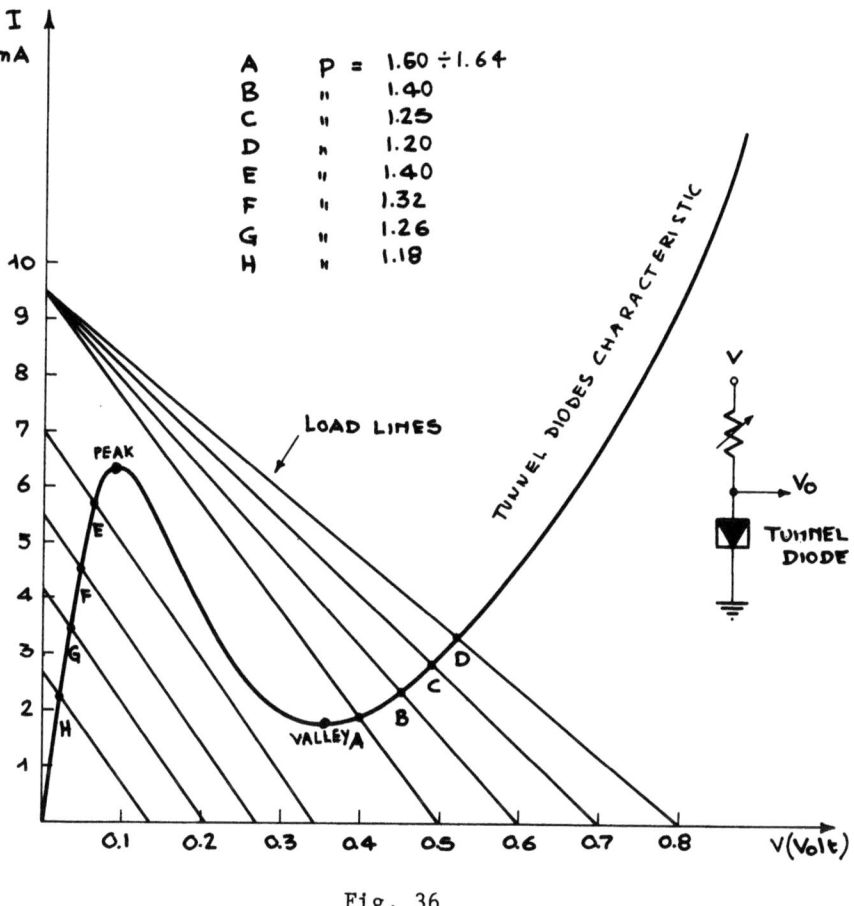

Fig. 36

The low frequency emitter conductance is

$$g_{EBO} = \frac{\partial I_E}{\partial V_{EB}} = \frac{q}{KT}(I_E + I_{EB}) \qquad (123)$$

and the low frequency transconductance g_{mo} is given by:

$$g_{mo} = g_{ECO} = \frac{\partial I_C}{\partial V_{EB}} = \frac{q}{KT} I_{ES} \exp(\frac{qV_{EB}}{KT}) = \alpha_F g_{EBO} \qquad (124)$$

the low frequency current amplification factor is:

$$\alpha_F = \frac{\partial I_C}{\partial I_E}$$

Since I_C, I_E show full shot noise at low frequency, we have the following current spectral density:

$$S_{I_C}(f) = 2qI_{ES} \exp(\frac{qV_{EB}}{KT}) + 2qI_{BE} = 2q(I_E + 2I_{BE}) \quad (125)$$

$$S_{F_C}(f) = 2q\alpha_F I_{ES} \exp(\frac{qV_{EB}}{KT}) + 2qI_{BC} = 2qI_C \quad (126)$$

By performing the cross correlation operation we get:

$$S_{I_C,I_E} = 2q\alpha_F I_{ES} \exp(\frac{qV_{EB}}{KT}) = 2KTg_m = 2KT \frac{\partial I_C}{\partial V_{EB}} \text{ (L.F.)} \quad (127)$$

At high frequency, the emitter base admittance Y_{EB} becomes complex and its real part is proportional to the frequency. This behaviour is due to holes injected by the emitter and returning by diffusion before their collection at the collector. Since diffusion is a thermal process, we can finally write for the current spectral density

$$S_{I_E}(f) = 2q(I_E + 2I_{BE}) + 4KT(g_{EB} - g_{EBO}) \quad (128)$$

At high frequency the cross correlation term must be modified by taking into account the transconductance variation

$$g_{mo} \rightarrow Y_{EC} \rightarrow g_{mHF} + j\beta$$

so we have

$$S_{I_C,I_E}(f) = 2KT \, Y_{EC} = 2KT\alpha_F Y_{EB} \quad (129)$$

At high frequency there is also a variation in the current amplification factor which follows the law:

$$\alpha_{HF} \simeq \frac{\alpha_f}{1 + \alpha \, f/f_\alpha} \quad (130)$$

where f_α is the cut-off frequency of the devices.

Now it is possible to consider an equivalent circuit of the transistor as shown in Fig. 37 where the current generators have the following meaning:

$$\overline{I_E^2} = S_{I_E}(f)\Delta f; \quad \overline{I_C^2} = S_{I_C}(f)\Delta f; \quad \overline{I_E^* I_C} = S_{I_C,I_E}(f)\Delta f$$

and where the thermal noise generator is due to the base resistance R_b:

$$\nu = \sqrt{4KTR_b \Delta f}$$

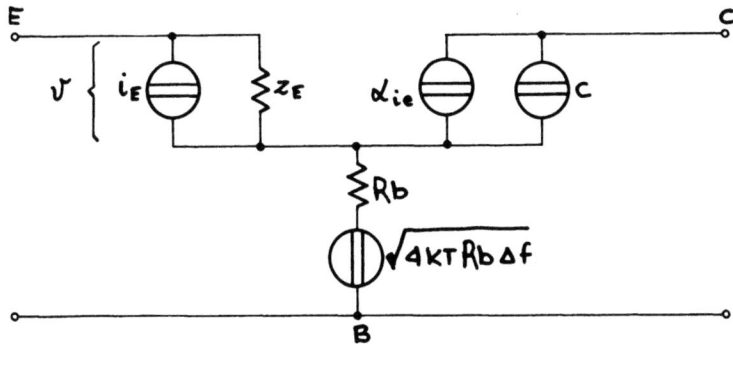

Fig. 37 (a)

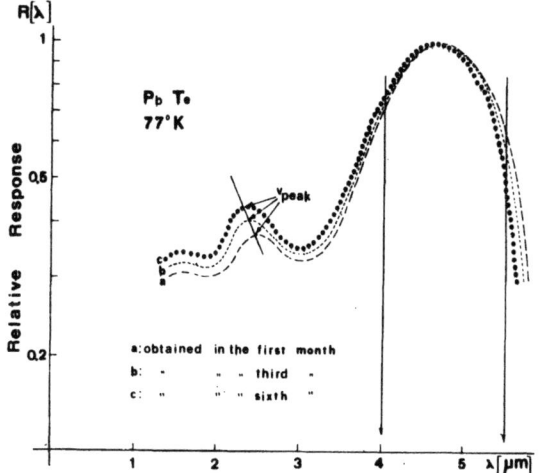

Fig. 37(b). Photoconductive relative response of a PbTe polycrystalline film vs. wavelength after three different time periods.

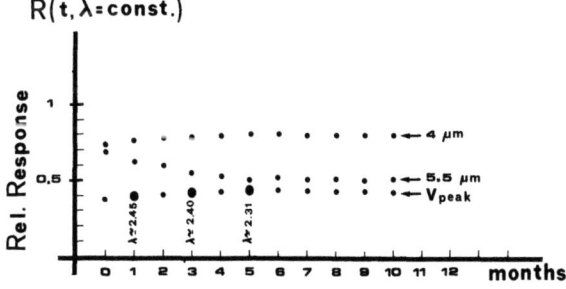

Fig. 37(c). Drift of the relative response for two different wavelengths and V_{peak} vs. time (months).

At low frequencies

$$\overline{\nu^2} = \overline{I_E^2}|Z_E|^2 = 2KT\, g_{CBO} \left| \frac{I_E + 2I_{BE}}{I_E + I_{BE}} \right| \Delta f \tag{131}$$

$$\overline{I^2} = 2q\, \{I_E \alpha_F (1 - \alpha_F) + I_{CO}\}\, \Delta f \tag{132}$$

experimental results (Montgomery and Clark, Van der Ziel) [48] have confirmed the above relation.

I. Noise In FETs and MOSFETs

If we look at the conduction mechanism in junction FETs where the width of conducting channel is modulated by the applied gate voltage, we expect the limiting noise of junction FETs to be thermal noise.

But if we look at a MOSFET, since the conducting channel is present when the gate voltage is properly chosen one would expect the noise to be diffusion noise. In MOSFET's, the noise associated with the inversion layer is also thermal in character.

In junction FETs, thermal generation of carriers produces a leakage current which goes from gate to channel in such a way as to produce a gate noise and, as a consequence, a drain noise. In MOSFET thermal generation of carriers produces a leakage current which goes from substrate to the channel and, as a consequence it produces drain noise. In present MOSFETs and MOS-V-FETs these effects are important only at high temperature. The most important source of noise is the thermal noise of the channel which may be written as

$$\overline{\nu^2} = \frac{8}{3}\, g_{max}\, KT \tag{133}$$

Derivation of 1/f noise in FETs is based on McWhorter's theory modified in such a way that free carriers are randomly trapped at the SiO$_2$-Si interface by means of a tunneling process into the oxide.

By using the Van der Ziel theory outlined above, it is possible to obtain a general form for the current spectral density in MOSFETs.

$$\overline{i^2} = 4\left(\frac{qI_D}{L}\right)^2 d \int_{F=-\infty}^{\infty} \int_{x=0}^{\infty} \int_{y=0}^{t_{ox}} N \frac{f_t(1-f_t)\tau}{tg(V_x)^2(1-\omega^2\tau^2)}\, dxdydE \tag{134}$$

where q is the electron charge; E is the energy; L and d is the length and width on the channel; f_t is the fractional occupation of trapping centers and V9x) the potential in the channel.

An evaluation of the integrals requires a knowledge of the energy and geometric dependencies for each set of bias conditions, i.e.

(a) $0 \sim V_{DS} < V_{GT}$ (heavy uniform inversion)

(b) $V_{DS} = V_{GT}$ (onset of current saturation)

(c) $0 < V_{GT} < V_{DS}$ (deep saturation)

where V_{DS} is the drain source voltage and V_{GT} is the gate bias. For (a), N_t is constant in $0 < y < d$ and τ is exponentially dependent on y and E, we have in the approximation based on equation 97

$$e^{-a'd/\tau} << \omega << 1/\tau$$

that

$$\overline{i^2} \simeq (\frac{g_m q}{C_{ox}})^2 \frac{KTA}{\alpha' f} N_t(E_f) \qquad (135)$$

where g_m is the transconductance, C_{ox} is the oxide capacitance, A is the area of the gate, $N_t(E_t)$ is the actual trap density near the Fermi level.

The noise voltage spectral density referred to input is then

$$\overline{e^2} = \frac{\overline{i^2}}{g_m^2} = (\frac{q}{C_{ox}})^2 \frac{A}{\alpha' f} N_t \qquad (136)$$

We conclude this short review by making some general comments. The expression for the noise voltage spectral density referred to input can assume the general form:

$$\overline{e^2} = K \, F_1(d,L,t_{ox}) \, F_2(V_{GS}, V_{DS}, N_t, f) \qquad (137)$$

where k is a constant, F_1 is a geometry factor and F_2 is a function of bias, top density, and frequency. It is clear that $\overline{e^2}$ is proportional to the fast surface state density. The signal source resistance has no effect on the inherent 1/f noise generation of the devices unless feedback becomes important. 1/f noise in MOSFETs is dominant from below 1 Hz to 1000 Hz, the lower frequency limit has not yet been determined. As far as the temperature dependence of 1/f noise is concerned, we do not yet have a reasonable theoretical basis for interpreting this phenomena. Measurements have been reported in the literature and attempts have been made to explain the changes in both p and n-MOSFETs with temperature. In order to minimize $\overline{e^2}$, N_{ss} and the oxide thickness t_{ox} must be minimized and the area maximized. But there is a limitation in reducing A because of the requirements related to gain-band width product and load and source impedance. As a result, t_{ox} and N_t are the main intrinsic characteristics to adjust. We conclude by noting that the product $N_{ss}t_{ox}$ is sometimes considered as a reasonable noise figure of merit for MOSFETs.

J. Non-Linearity Measurements by Third Harmonic Index Method

With requirements for increasingly reliable electronic components, there is a need for new methods of measurements which can be more accurate in estimating the quality of components. Current noise measurements are used a present as a measure of the reliability of resistances and contacts between metal and semiconductors.

Non-linearities are sometimes very small and cannot be evaluated from the D.C. V-I characteristics. Therefore alternating currents can be used for such measurements [49,50]. As a matter of fact, when they are applied to non-linear devices, odd harmonic frequencies are generated which are more easily measured. In order to show the presence of the third harmonic, J. S. Anderson and V. Rysanek proposed the following very simple model [52]. They supposed the existence of grain boundaries in a resistor represented by a potential barrier ϕ_i. By applying voltage V, we have

$$I_i = I_i^+ - I_i^- = 2A \exp\{-\frac{e\phi_i}{KT}\} \sinh \frac{eV}{KT} \qquad (138)$$

Now if there are N_t barriers, and if $\phi_i = \phi$, $I_o = \sum_0^{N_t} I_i$, we can write:

$$I_o = 2N_t A \exp\{-\frac{e\phi}{KT}\} \sinh \frac{eV}{KT} \qquad (139)$$

Assume that N_t can be expressed in the form

$$N_t = N_o \exp(-bt/\tau)$$

where τ = average time of reaction and

$$b = K \exp\{-E_a/KT\} \qquad \text{(Ahrrenius Law)} \qquad (140)$$

If we suppose that the resistance R_B due to barriers goes as the number of barriers we can write:

$$R_B = R_o \exp\{-bt/\tau\} \qquad (141)$$

where R_o is the initial value (t=0). The total resistance of the sample R_{tot}, can be written by Matthiessen law as:

$$R_{tot} \approx R_B + R_T$$

and, if T is constant:

$$\frac{dR_T}{dt} = \frac{dR_B}{dt} = -\frac{b}{\tau} R_B \tag{142}$$

So we can see the relation between the resistance of the sample and the activation energy E_a. The current in the sample, I_B, can be written as

$$I_D = 2A' N_o \exp\{-(\frac{bt}{\tau} + \frac{e\phi}{KT})\} [\frac{eV}{KT} + \frac{e^3V^3}{3!(KT)^3} + \ldots] \tag{143}$$

If the applied voltage is sinusoidal $V(t) = V\sin\omega t$, from the above equation we get the third harmonic component

$$I_D = \frac{A' N_o e^3 V^3}{2 \cdot 3! (KT)^3} \exp\{-(\frac{bt}{\tau} + \frac{e\phi}{KT})\} \sin 3\omega t \tag{144}$$

The coefficient of the third harmonic component depends on the initial defect concentration, N_o and the activation energy ϕ. The third harmonic index, THI, is defines as

$$THI = \frac{V_{eff,3\omega}(\mu N)}{V_{eff,\omega}^n V}$$

where $V_{eff,3\omega}$ is the voltage of the third harmonic component and $V_{eff,1\omega}$ is the potential of the fundamental. In a resistor, $n \simeq 3$ which assures us that the THI is dimensionless. The quantity $\exp(-bt/\tau)$ in Equation (144) suggests that the THI will give an indication of the long term stability of the resistor.

We have conducted an experiment which confirms the usefulness of the third harmonic method in determining the reliability of resistors. A similar theory has been developed in studying thin film gold-semiconductor-gold systems where the semiconductor is polycrystalline. The above theory gives the following expression:

$$V_{3\omega}(eff) = \frac{1}{3} \frac{kT}{q} \frac{V_{eff}^3}{r^3} (\frac{1}{i_{SD}} + \frac{N_o}{i_{SG}} \exp\{-\frac{bt}{\tau}\}) \tag{145}$$

where i_{SD} is the saturation current due to contact diodes and i_{SG} is the saturation current due to the contact representing the grain boundary.

Measurements done on many samples show that the third harmonic amplitude does vary with time. The time stability in such components have also been observed in photoconductive spectral response measurements. (Fig. 37a)

K. Experimental Apparatus

The measurement assembly consists of an alternating generator connected to a resistance bridge and a differential selective amplifier.

The resistance bridge consists of four resistors, one of which is the component R_x to be analyzed. Resistance R_1, R_2 have both the "same" resistive value and nonlinearities. The resistance R_p is used to balance the bridge. (Fig. 38).

L. The Measurement Technique

The G voltage is adjusted to the required value $V_1(t)$ measured on resistor R_x using channel B which is tuned to the required frequency.

This voltage is set in branch A by changing the ohmic value of R_p. In order to reach the minimum difference between the voltage in branch A and B, the differential voltage between both branches is determined at a higher amplification and is adjusted to a minimum by fine tuning of R_p so that the residual voltage $V_{IR} = 0$ (valid if $R_p = R_x$). In order to balance the resistors of the bridge, the odd harmonic voltages are measured across R_x.

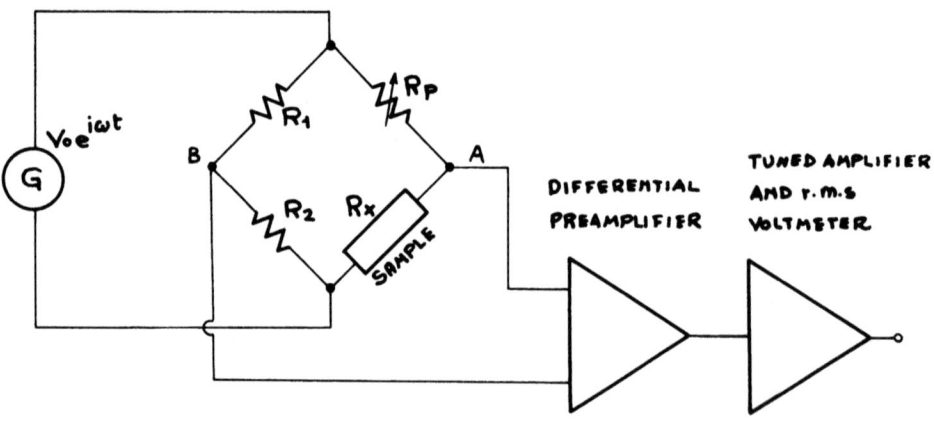

Fig. 38

NOISE

Thus maximum accuracy is achieved, depending on the magnitude of the noise level. The tuned amplifier must have a Q-factor as high as possible. Then the frequency of the selective amplifier is changed to the third harmonic of G in order to measure the third harmonic voltage for resistor R_x.

According to Figure 39, the value of the third harmonic amplitude for R_x is given by:

$$V_{3h} = V_{3hm} - K_3 V_{IR}$$

where K is a coefficient depending upon the value of Q. If Q = 100, K_3 = 0.003, K_2 = 0.002 (K, K can be experimentally determined or taken from P.A.R. lock-in Amplifier applications notes.

The method is valid if both R_1 and R_2 have the same linearity and if the nonlinearity of R_x is greater than that of $R_{1,2}$. Now by plotting V_3 against V_1, one gets results like those which appear in Fig. 40.

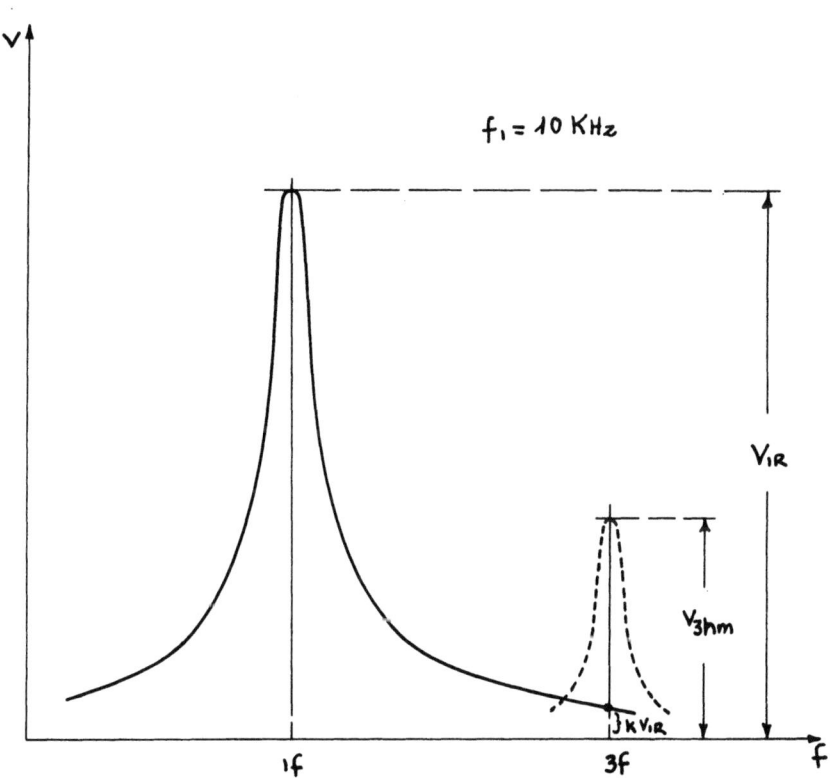

Fig. 39

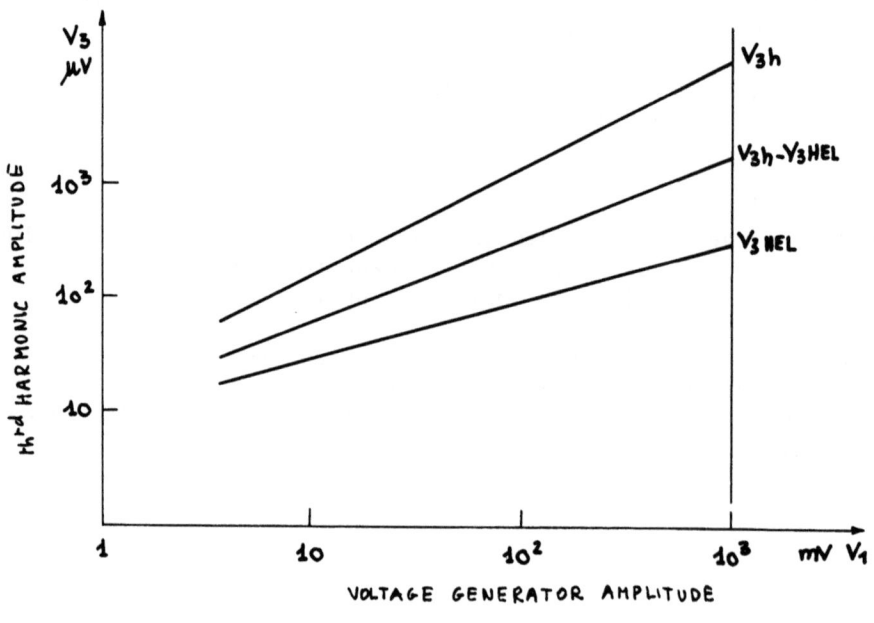

Fig. 40

From Fig. 40, it is possible to obtain the following information

$$V_3 = V_{3h} - V_{3HEL}$$

where V_{3HEL} is the nonlinearity of the helipot. Then it is possible to write:

$$V_3 = A V_{if}^n$$

Where A is a constant characterizing the nonlinearity. By setting V_{if} equal to 1V we have immediately

$$A = V_3 \quad \text{for any n}$$

and the value of nonlinearity can be expressed by the voltage of the third harmonic at V_{if} = 1 Volt.

NOISE

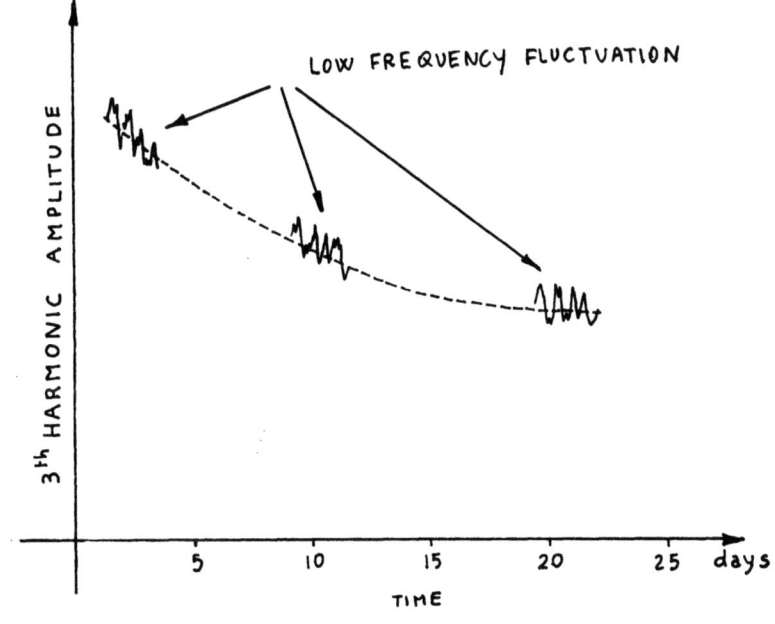

Fig. 41

REFERENCES

1. A. Papoulis, Probability Random Variables and Stochastic Processes - M.G. Hill (1965).
2. F.N.H. Robinson, Noise and Fluctuations in Electronic Devices and Circuits, Oxford (1974).
3. A. T. Starr, "Radio and Radar Techniques", Pitman & Sons, Ltd., London (1953), Appendix 9, "Noise".
4. H. Nyquist, Phs. Rev. 32, 110 (1928).
5. J. B. Johnson, Phys. Rev. 32, 97 (1928).
6. A. Van Der Ziel, "Noise," Prentice-Hall, Inc., (1953).
7. S. O. Rice, "Mathematical Analysis of Random Noise," Bell System Technical Journal, 23-4 (July 1944).
8. G. Abowitz, E. Arnold, and E. A. Leventhal, "Surface States and 1/f Noise in MOS Transistors", IEEE Trans. Electron Devices, Vol. ED-14, pp. 775-777, (November 1967).
9. R. D. Baertsch, "Low-frequency Noise Measurements in Silicon Avolanche Photodiodes, " IEEE Trans. Electron Devices, Vol. ED-13, pp. 383-384 (March 1966).
10. E. R. Chenette, "Measurement of the Correlation between Flicker Noise Sources in Transistors," Proc. IRE, Vol. 46, p. 1304 (June 1958).
11. W. H. Fonger, "A Determination of 1/f Noise Sources in Semiconductor Diodes and Transistors" in Transistor I. Princeton, NJ RCA Labs pp. 239-297 (1956).
12. I. D. Guttoyv, "Investigation of the 1/f Low Frequency Noise of Back Biased Germanium p-n Junctions," Radioteku Electron", Vol. 12, pp. 946=948 (May 1967) Radio Eng. Electron Phys. (Engl. Transl.) Vol. 12, pp. 880-882 (May 1967).
13. H. E. Hallsday and A. Van Der Ziel, "Field-deparmtne Mobility Effects in the Excess Noise of Junction-gate Field-Effect Transistors," IEEE Trans. Electron Devices, Vol. ED-14, pp. 110-111, (February 1967).
14. S. T. Hsu, D. J. Fitzgerald, and A. S. Grove, "Surface State Related 1/f Noise in p-n Junctions and MOS Transistors," Appl Phys. Lett., Vol. 12, pp. 287-289, (May 1968).
15. E. A. Leventhal, "Derivation of 1/f Noise in Silicon Inversion Layers from Carrier Motion in a Surface Band," Solid State Electron, Vol. 11, pp. 621-627 (June 1968).
16. E. M. Nicollian and H. Melchior, "A Quantitative Theory of 1/f Type Noise Due to Interface States in Thermally Oxidized Silicon," Bell System Tech. Journal, Vol. 46, pp. 2019-2033 (November 1967).
17. C. T. Sah and F. H. Hilscher, "Evidence of the Surface Origin of the 1/f Noise," Phys. Rev. Lett., Vol. 17, pp. 956-958 (October 1966).
18. A. Van Der Ziel, "Carrier Density Fluctuation Noise in Field Effect Transistors," Proc. IEEE, Vol. 51, pp. 1670-1671 (November 1963).

19. C. T. Sah, "Theory of Low Frequency Generation Noise in Junction-gate Field-effect Transistor," Proc. IEEE Vol. 52, pp. 795-814 (July 1964).
20. S. Christenson, I. Lundstrom, et. al., "Low Frequency Noise in MOS Transistors," Solid State Electron, Vol. 11, pp. 797-812, (1968).
21. C. T. Sah and F. H. Hielscher, "Evidence of the Surface Origin of the 1/f Noise," Phys. Rev. Lett., Vol. 17, pp. 956-958 (1966).
22. S. T. Hsu, "Surface State Related 1/f Noise in MOS Transistor," Solid State Electron, Vol. 13, p. 1451 (1970).
23. F. N. Hooge, "Amplitude Distribution of 1/f Noise," Physica 42, (1969).
24. L. Bess, Phys. Rev. 103, 72 (1956).
25. R. F. Voss, "Linearity of 1/f Noise Mechanism," Fifth Int'l Conf. on Noise in Physical System. Germany (1978).
26. R. F. Voss and J. Clarke, Phys. Rev. 1313, 556 (1976).
27. B. Mandelbrot, IEEE Trans. Inform. Theory IT 13, 289-298 (1967).
28. T. H. Bell, Jr., J. Appl. Phys. $\underline{45}$, 1902-1904, (1974).
29. A. Ambroxy, Electronic Letters $\underline{13}$, 137-138 (1977).
30. S. T. Hsu, IEEE Trans. Elec. Dev. ED-18, 882-887 (1971).
31. M. Agu and T. Kinoshita, Proc. Symposium on 1/f Fluctuations, Tokyo pp. 41-48 (1977).
32. F. N. Hooge, Phys. Lett. $\underline{29A}$, 139-140 (1969).
33. J. Clark and R. F. Voss, Phys.Rev. Letters $\underline{33}$, 24-27 (1974).
34. A. L. McWhorter, "1/f Noie and Related Surface in Germanium," MIT Lincoln Lab., Lexington, MA Rep. 80 (May 1955).
35. A. L. McWhorter, "Semiconductor Surface Physics", Philadelphia, PA, Univ. of Pennsylvania Press,(1957).
36. G. M. Kleinpenning, "1/f Noise...," Physica 77, 78-98 (1974).
37. K. M. Van Vliet, "Noise and Admittance of the Generation-Recombination Current Involving SRH Centers in the Space-Charge Region of Junction Devices," IEEE Trans. Elec. Dev., Vol. ED-23, pp. 1236-1246 (No.v 1976).
38. P. O. Lauritzen, "Noise Due to Generation and Recombination of Carriers in p-n Junction Transistion Regions," IEEE Trans.Elec. Dev., Vol. ED-15, pp. 770-776 (Oct. 1968).
39. K. Van Vliet and van der Ziel, IEEE Trans. Elec. Dev. Vol. ED-24, n. 8.
40. R. G. Pay, MSC Thesis, University of Birmingham (1956).
41. J. C. Martin, D. Esteve and G. Basquez, "Burst Noise in Silicon Planar Transistors," Conf. On Physical Aspects of Noise in Electronic Devices, University of Nottingham, Eng.and, (1968).
42. J. F. Schenck, "Burst Noise and Walkout in Degraded Silicon Devices," Sixth Annual Reliability of Physics Symp. Proc., pp. 31-35 (1967).
43. S. T. Hsu, and R. J. Whittier, "Characterization of Burst Noise in Silicon Devices," Solid State Elec. 12 November (1969).
44. S. R. Hsu, R. J. Whittier and C. A. Mead, "Physical Model for Burst Noise in Semiconductor Devices," Solid State Elec., (13 July 1970).

45. A. S. Growe, Physics and Technology of Semiconductor Devices, Chap. 10, Wiley (1967).
46. A. Lvque, et. al., Electronics Letter 6, 176 (1970).
47. K. F. Knott, "Characteristics of Burst Noise Intermittancy," Solid State Elec. Vol. 21, No. 8 (August 1978).
48. A. Van Der Ziel, "Noise in Solid-State Devices and Lasers," IEEE Proceedings, Vol. 58 (1970).
49. Vladimir Rysanek, Carlo Corsi and Arnald D'Amico, Electrocomponent Science and Technology, Vol. 5 (1978).
50. A. D'Amico, et. al., "Experimental Evidence of Third Harmonic Variations at 770K and 300K in -PbTe- and structures.
51. A. D'Amico, A. Bvonomo, "Peak and Valley Noise in Tunnel Diodes," To be published.
52. J. C. Anderson and V. Rysanek, "Prediction of the Stability of Thin Film Resistor," Radio Electron. Eng. 39,6 (June 1970).

USEFUL BOOKS

53. A. Van der Ziel, Noise Sources Characterization and Measurements, Prentice Hall (1970).
54. W. Beck, Statistical Mechanis, Fluctuations and Noise, Halsted Press (1976).
55. A. VanDer Ziel, "Fluctuation Phenomena in Semiconductors, (Lonston (1959).
56. Dietrich Wolf, Noise in Physical Systems, Proc. of the Fifth Intern'. Conf. on Noise, Baol Nauheim. Fed. Rep. of Germany (1978).

Chapter 7

OPTICAL CHARACTERIZATION OF SEMICONDUCTORS

E. D. Palik and R. T. Holm

Naval Research Laboratory

Washington, DC 20375

I. REVIEW OF BASIC OPTICAL PROPERTIES

A. General Discussion of Reflection and Transmission

Most samples are in the form of a slab or a thin film on a slab. We first review the optical properties of such samples described only by a complex index of refraction $\tilde{n} = (n + ik)$. Later, we will determine the origin of n and k in a semiconductor so that we may characterize it in terms of free-carrier density, mobility, donor and acceptor densities, alloy homogeneity, etc.

1. <u>A slab</u>. We present results [1-7] for reflectance (R) and transmittance (T) of a slab when a plane electromagnetic wave (p and s polarization) described by $\exp i(qr - \omega t)$ is incident at an angle θ on a slab of thickness d_2. In Figure 1, the amplitude coefficients for the single interface between two semi-infinite media are

$$r_s = \frac{E_{rs}}{E_{is}} \qquad t_s = \frac{E_{ts}}{E_{is}}$$

$$r_p = \frac{H_{rp}}{H_{ip}} \qquad t_p = \frac{H_{tp}}{H_{ip}} \tag{1}$$

These become for light incident from medium j to k

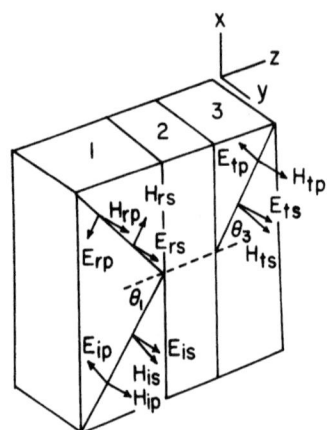

Fig. 1. Schematic representation of a plane wave incident from medium 1 on medium 2 and exiting into medium 3. Electric (E) and magnetic (H) fields for s and p polarization are indicated.

$$r_{jk} = \frac{g_j - g_k}{g_j + g_k} \qquad t_{jk} = \frac{2g_j}{g_j + g_k} \qquad (2)$$

where $g_j = \hat{n}_j \cos\theta_j$ (s polarization)

$$g_j = \frac{\cos\theta_j}{\hat{n}_j} \qquad \text{(p polarization)}. \qquad (3)$$

Snell's law is $n_j \sin\theta_j = n_k \sin\theta_k$. The intensity coefficients are

$$R = |r|^2 = r^*r$$

$$T = \frac{\text{Re}(g_k)}{g_j} |t|^2 = \frac{\text{Re}(g_k)}{g_j} t^*t \qquad (4)$$

For a thick absorbing or wedged sample where the back interface can be ignored, and no multiple reflections take place, these results give the single-surface reflectivity.

When two interfaces are present and now medium 2 has a thickness d_2, the amplitudes are given by

$$r = \frac{r_{12} + r_{23}e^{2i\phi_2}}{1 + r_{12}r_{23}e^{2i\phi_2}} \quad (5)$$

$$t = \frac{t_{12}t_{23}e^{i\phi_2}}{1 + r_{12}r_{23}e^{2i\phi_2}}$$

where $\phi_2 = (\omega/c)d_2\tilde{n}_2\cos\theta_2$. The forms of the r_{jk} are given for each interface by Eq. (2). Multiple-reflection effects are treated coherently, i.e., amplitudes are added. For a slab media 1 and 3 are usually air. The forms of R and T at normal incidence are familiar [1-7].

$$R = R\frac{(1 - e^{-2\omega k d/c})^2 + 4e^{-2\omega k d/c}\sin^2(\omega n d/c)}{[1 - R^2 e^{-2\omega k d/c}]^2 + 4R^2 e^{-2\omega k d/c}\sin^2[(\omega n d/c) + \phi_r]} \quad (6)$$

$$T = e^{-2\omega k d/c}\frac{(1 - R)^2 + 4R\sin^2\phi_r}{[1 - R^2 e^{-2\omega k d/c}]^2 + 4R^2 e^{-2\omega k d/c}\sin^2[(\omega n d/c) + \phi_r]}$$

where $\tan\phi_r = 2k/(1 - n^2 - k^2)$ and $R = [(n-1)^2 + k^2]/[(n+)^2 + k^2]$. For cases when spectral bandpass is larger than the interference-fringe spacing Equation (6) can be integrated over one cycle of nd/c to obtain [6,7]

$$R = \frac{R\{1 + [1 - 2(\cos^2\phi_r - \sin^2\phi_r)R]e^{-4\omega k d/c}\}}{1 - R^2 e^{-4\omega k d/c}}$$

$$T = \frac{[(1-R)^2 + 4R\sin^2\phi_r]e^{-2\omega k d/c}}{1 - R^2 e^{-4\omega k d/c}} \quad (7)$$

An interesting problem arises here when the medium is lossy. Poynting's vector contains a cross-product term proportional to k which cannot be associated with energy propagating to the left or to the right in Figure 1 and the proper disposition of this term is ambiguous when one tries to calculate Equations (7) by the addition of intensities [2]. This is not the case when amplitudes are considered as in Equations (6)

2. A Film on a Slab. For three interfaces the reflection amplitude is

$$r = \frac{a(g_1-g_2)e^{-i\phi_2} + b(g_1+g_2)e^{i\phi_2}}{a(g_1+g_2)e^{-i\phi_2} + b(g_1-g_2)e^{i\phi_2}} \tag{8}$$

where

$$a = (g_2+g_3)(g_3+g_4)e^{-i\phi_3} + (g_2-g_3)(g_3-g_4)e^{i\phi_3}$$

and

$$b = (g_2-g_3)(g_3+g_4)e^{-i\phi_3} + (g_2+g_3)(g_3-g_4)e^{i\phi_3},$$

which can be rewritten as

$$r = \frac{r_{12} + r_{24}\,e^{2i\phi_2}}{1 + r_{12}r_{24}e^{2i\phi_2}} \tag{9}$$

where

$$r_{24} = \frac{r_{23} + r_{34}e^{2i\phi_3}}{1 + r_{23}r_{34}e^{2i\phi_3}}, \quad r_{12} = \frac{g_1-g_2}{g_1+g_2}, \quad r_{23} = \frac{g_2-g_3}{g_2+g_3}, \quad r = \frac{g_3-g_4}{g_3+g_4}.$$

This is the same form as equation (5). Note that medium 2 can be the film and medium 3 can be the substrate, or turning the sample around, medium 2 can be the substrate and medium 3 the film.

The transmittance amplitude is

$$t = \frac{8g_1g_2g_3}{a(g_1+g_2)e^{-i\phi_2} + b(g_1-g_2)e^{i\phi_2}}$$

where a and b are given in equation (8). This can be rewritten as

$$t = \frac{t_{12}t_{23}t_{34}\,e^{i\phi_2}e^{i\sigma_3}}{\{1+r_{23}r_{34}e^{2i\phi_3}\} + r_{12}\{r_{23}+r_{34}e^{2i\phi_3}\}e^{2i\phi_2}} \tag{11}$$

where

$$t_{12} = \frac{2g_1}{g_1+g_2}, \quad t_{23} = \frac{2g_2}{g_2+g_3}, \quad \text{and } t_{34} = \frac{2g_3}{g_3+g_4}.$$

Note that s and p polarization are still contained in the choice of the g_j in Equation (3).

For normal incidence the Eqs. (9) and (11) simplify considerably but we do not write them down.

The substrate multiple-reflection effects can be treated incoherently (intensities added) by integration over the appropriate one cycle of substrate fringe [2,8]. The results have been obtained by adding the intensities in the substrate and the amplitudes in the films [8,9]; then

$$R' = \rho_{is} + (\tau_{is} \tau_{si} R_{si} \eta^2/D)$$

$$R'' = R_{is} + (T_{is} T_{si} \rho_{si} \eta^2/D) \quad (12)$$

$$T' = \eta \tau_{is} T_{si}/D$$

$$T'' = \eta \tau_{si} T_{is}/D$$

but $T' = T''$. Here $\eta = e^{-\alpha L}$, $\alpha = 4\pi k_s/\lambda$ is the substrate absorption coefficient, L is the substrate thickness, k_s is the substrate extinction coefficient, λ is the wavelength of light in vacuo, s refers to substrate, f to film and 1 to air on both sides, $D = 1 - \rho_{si} \sin^2$. Also, $\tau_{is} = |\varepsilon/\Delta|^2 \text{Re}(t_{1f} t_{fs} t_{f1}^* t_{sf}^*)$ and $\rho_{is} = |r_{1f} + (\xi^2 r_{fs} t_{f1} t_{kf}/\Delta)|^2$ with appropriate interchanges of subscripts to obtain ρ_{s1} and τ_{s1}. Also,

$$\Delta = 1 - r_{f1} r_{fs} \xi^2, \quad \xi^2 = e^{-2\pi i \tilde{n}_f \ell/\lambda},$$

where ℓ is the film thickness. Also,

$$T_{1s} = \frac{|\tilde{n}_1 - \tilde{n}_s|^2}{|\tilde{n}_1 + \tilde{n}_s|^2} \qquad R_{1s} = 4\text{Re} \frac{\tilde{n}_1 \tilde{n}_s^*}{|\tilde{n}_1 + \tilde{n}_s|^2}$$

$$t_{1f} = \frac{2\tilde{n}_1}{(\tilde{n}_1 + \tilde{n}_f)} \qquad r_{1f} = \frac{(\tilde{n}_1 - \tilde{n}_f)}{(\tilde{n}_1 + \tilde{n}_f)}$$

with appropriate interchanges to obtain T_{s1}, R_{s1}, t_{f1}, t_{fs}, t_{sf}, e_{f1}, r_{fs}. R' is the reflectance when the light is incident on the film and R'' is the reflectance when the light is incident on the substrate. This asymmetry in R occurs only when there is a finite extinction coefficient in the film and/or substrate. Although not obvious, Eq. (9) is also different for the two cases of film first and substrate first. Note that the transmittance is symmetric and not dependent on the direction of the beam. Such asymmetric effects have been used to determine n and k for metal films on glass substrates [3,5,10].

In Fig. 2, we illustrate the dependence of R' and R'' on n, k and ℓ of a film which is on a silicon substrate with $n_s = 3.418$, $k_s = 0.000017$ and $L = 381$ μm. The calculation is made at 1416 cm^{-1} in the infrared. The curve for a given ℓ is the locus for $R' = R''$.

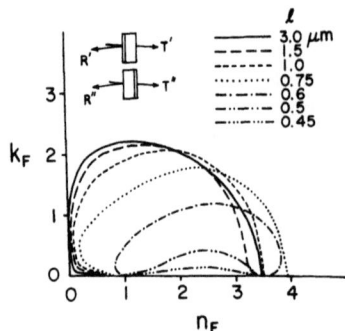

Fig. 2. The various curves are the loci for R' = R" for a series of films of various thicknesses ℓ, index of refraction n_F and extinction coefficient k_F on a Si substrate. Above a given curve, R' is greater than R", while below the curve, R' is less than R".

For a given ℓ curve in the region above the curve R' is greater than R" and in the region below the curve R' is less than R". Note that for a given n, k set, it is often possible to vary ℓ to produce R' > R" or R' < R". Since T' always equals T", this leads to the absorptances A' and A" of the film being asymmetric. It is interesting to note that workers who have measured thin films of insulators and semiconductors on substrates of semiconductors and insulators have typically measured R' and T' and extracted n and k without noting the asymmetric effects.

B. Absorption Mechanisms in Semiconductors

The values of n and k in a semiconductor are determined by the properties of the energy bands (in the UV, visible, and near IR), optical phonons (in the middle and far IR), free carriers (in the near, middle and far IR) and impurities (in the far IR).

1. <u>Band Edge</u>. Fundamental-band-edge absorption for a direct allowed transition between simple, parabolic valence and conduction bands is typically described by [11,13]

$$n\alpha = C(\hbar\omega - E_g)^{\frac{1}{2}}/\hbar\omega \qquad (13)$$

where C is a constant factor characteristic of the particular band structure, E_g is the band gap, n is the index of refraction and $\alpha = 4\pi k/\lambda$ is the absorption coefficient. Both n and α are frequency dependent, but often times n is assumed to be a constant, so the α can be calculated. A typical absorption-edge α calculated

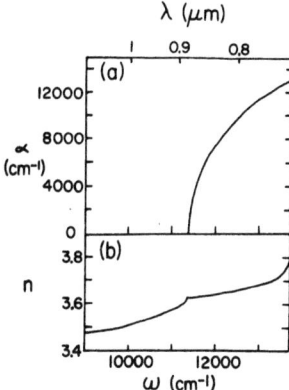
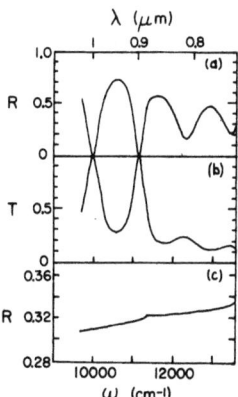

Fig. 3. (a) Schematic form of band-edge absorption coefficient for GaAs. (b) Index of refraction in the band-edge region.

Fig. 4. (a) Calculated reflectance of a 1 μm thick GaAs-like film in the band-edge region. (b) Calculated transmittance of this film. (c) Calculated reflectance of a semi-infinite sample.

for GaAs parameters is shown in Fig. 3(a). (For GaAs we use $n = 3.63$, $E_g = 1.41$ eV and $C/n = 41425$ cm$^-$(eV)$^{\frac{1}{2}}$.) Note that ω has units of rad/s, but we list cm^{-1}. Although we have assumed n to be a constant to do this, we can Kramers-Kronig analyze k to obtain n from the equation

$$n(\omega) - 1 = \frac{2}{\pi} \int_0^\infty \omega' k(\omega') (\omega'^2 - \omega^2)^{-1} d\omega \qquad (14)$$

which is shown in Fig. 3(b). The detailed assumptions about handling the limits in Eq. (14) are discussed by Stern [14]. For the calculation of Fig. 3, we have included the absorption to higher energy due to the higher band gaps of GaAs. One would actually measure k and Kramers-Kronig analyze it to obtain n. Or, given the analytical expression (13) for nα (imaginary part of the dielectric function), this can be K-K analyzed to obtain the real part of the dielectric function), this can be K-K analyzed to obtain the real part of the dielectric function [11,14]. Then both n and k can be extracted from these quantities. The model for Equation (13) omits phonon-assisted absorption, indirect band-gap transitions, and excitonic structure near the gap. The reflectance and transmittance of a semiconductor such as GaAs is schematically shown in Fig. 4 for the cases described by Equation (13). We will ultimately (Section VA) be interested in the effects that free carriers in the conduction band have on the interband absorption.

2. **Optical Phonons.** The Lorentz model is adequate to account for much of the optical-phonon structure in polar semiconductors. The complex dielectric function is given by [15]

$$\varepsilon = (n + ik)^2 = \varepsilon_\infty \left[1 + \frac{\Omega^2}{\omega_{TO}^2 - \omega^2 - i\Gamma\omega} \right] \tag{15}$$

where ε_∞ is the high frequency dielectric constant determined by the interband transitions; ω_{TO} is the transverse optical-phonon frequency; Γ is the phonon damping constant (assumed to be a constant); $\Omega^2 = \omega_{LO}^2 - \omega_{TO}^2$ is an oscillator-strength term defined in terms of both ω_{TO} and ω_{LO}, the longitudinal optical-phonon frequency. The real (ε') and imaginary (ε'') parts of ε are shown in Figure 5(a) and (b) for GaAs. Here we use $\omega_{TO} = 268.2$ cm^{-1}, $\omega_{LO} = 291.5$ cm^{-1}, $\Gamma = 2.3$ cm^{-1} and $\varepsilon_\infty = 11.1$. The corresponding values of n and k are shown in Fig. 6(a) and (b). R and T for a sample 5 μm thick are shown in Fig. 7 (a) and (b). Note the interference fringes, since Eqs. (6) were used for this calculation. Also indicated in Fig. 7(c) is the bulk reflectance when the sample is very thick (no multiple reflections).

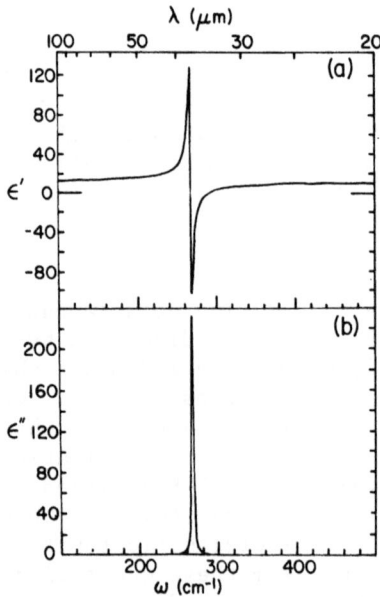

Fig. 5. (a) The real part of the dielectric function ε' for GaAs with no free carriers; (b) The imaginary part of the dielectric function ε''.

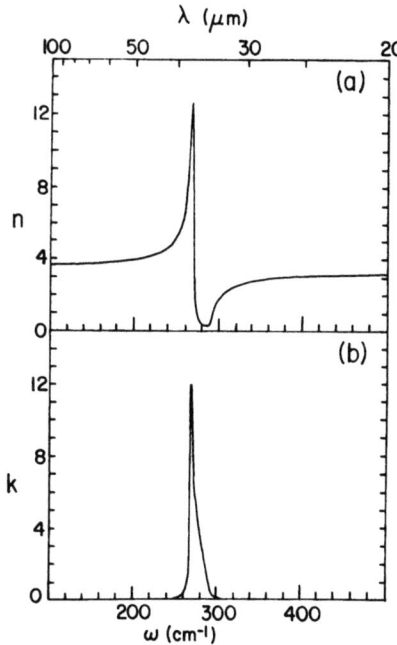

Fig. 6. (a) The index of refraction n for semi-insulating GaAs.
(b) The extinction coefficient k.

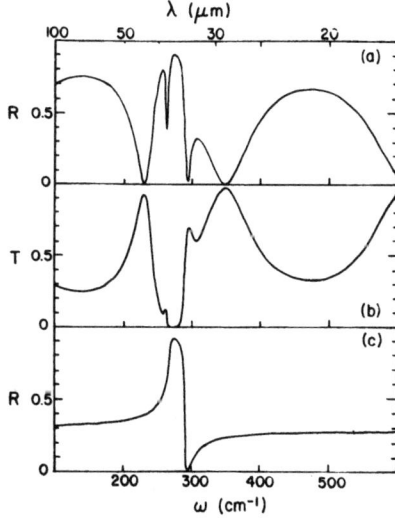

Fig. 7. (a) The calculated reflectance of a 5 μm thick GaAs sample.
(b) The transmittance of the same semi-insulating sample.
(c) The reflectance of bulk GaAs.

Multiphonon absorption occurs due to anharmonicity in the force constant or in the effective charge [16, 17]. In this case overtones of ω_{TO} can occur as well as various combinations of ω_{TO}, ω_{LO}, and ω_{TA} and ω_{LA}, the corresponding acoustic phonons. These can be located at the zone center as well as at the zone edge of the Brillouin zone. Multiphonon absorption is reviewed in detail by Spitzer [17].

An impurity atom in a substitutional position in the lattice can vibrate locally producing an absorption line at a frequency characteristic of its mass and force constant [18]. While not strictly an optical phonon, as the impurity density increases, this vibration mode becomes characteristic of the optical phonon of the alloy system giving one-or-two-mode behavior for such alloys [19].

3. <u>Free Carriers</u>. In its simplest form, free-carrier absorption is described by the Drude model [5]. The dielectric function is

$$\varepsilon = (n + ik)^2 = \varepsilon_\infty \left(1 - \frac{\omega_p^2}{\omega(\omega+i\gamma)}\right) \qquad (16)$$

where $\omega_p^2 = 4\pi Ne^2/m^*\varepsilon_\infty$ defines the plasma frequency, N is the free-carrier density, m^* is the effective mass, and γ is the free-carrier damping constant. This model has worked well for many metals

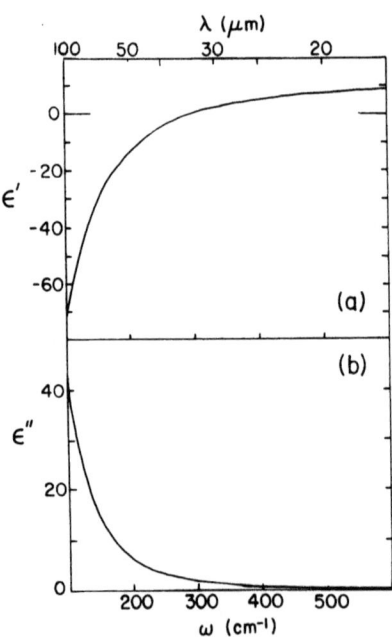

Fig. 8. (a) The real part of the dielectric function ε' for a GaAs-like semiconductor with free carriers but no polar optical phonons. (b) The imaginary part of the dielectric function ε.

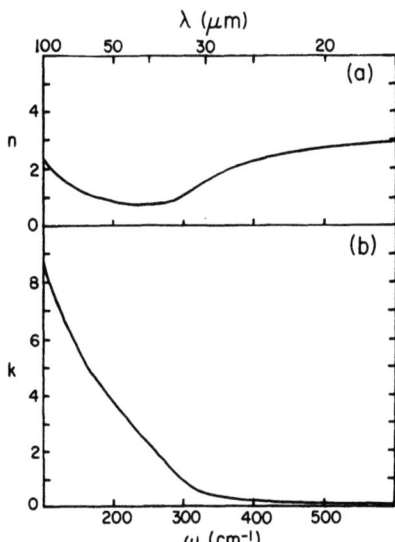

Fig. 9. (a) The index of refraction n for a GaAs-like semiconductor with same parameters as Fig. 8. (b) The extinction coefficient k.

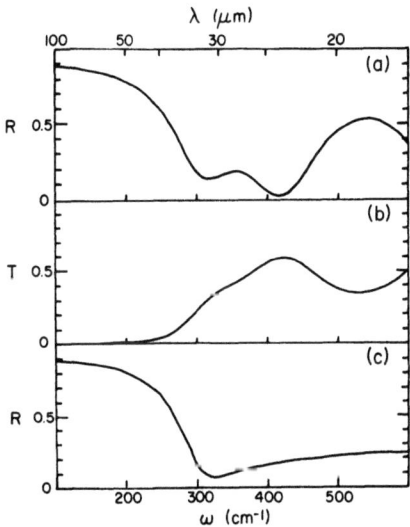

Fig. 10. (a) The calculated reflectance of a 5 μm thick GaAs-like semiconductor sample. (b) The transmittance of the same sample. (c) The reflectance of bulk GaAs.

and semiconductors. The form of ε is shown in Fig. 8 (a) and (b) for a typical non-polar semiconductor. The plasma frequency occurs at the frequency where ε' = 0. It is obvious that as N increases, the frequency where ε' = 0 increases. The corresponding n and k are given in Fig. 9 (a) and (b). The reflectance and transmittance described by Eq. (6) for a sample with ℓ = 5 μm are given in Figure 10 (a) and (b). The interference fringes are well defined when the film becomes transparent. The reflectance of semi-infinite sample is given in Fig. 10 (c). The characteristic plasma reflection edge and minimum occur near 325 cm^{-1} and can be used to determine N and γ. The analysis of a complete spectrum will also give N and γ. For Figs. 8-10 ω_p = 300 cm^{-1} and γ = 50 cm^{-1}.

The application of a magnetic field parallel to the direction of the light propagation (Faraday configuration) alters the dielectric function to the form [20]

$$\varepsilon_\pm = (n_\pm + ik_\pm)^2 = \varepsilon_\infty \left(1 - \frac{\omega_p^2}{\omega(\omega \pm \omega_c + i\gamma)}\right) \quad (17)$$

where ± indicate the left- or right-circular sense (CRA or CRI) appropriate for electrons. Here ω_c = eB/m*c is the cyclotron frequency. We note that a resonance occurs at ω = ω_c in ε_+ for electrons and ε_- for holes. A number of magnetooptical effects including cyclotron resonance, plasma reflection, helicon propagation, Faraday effect and Voigt effect have been used to characterize new materials over the past twenty years as to N, m* and γ [20]. With a well-understood material such as GaAs, epitaxial films or ion-implanted layers can also be studied by these techniques.

The Faraday rotation is obtained using Eq. (17) in the formula for the rotation of the plane of polarization of an electromagnetic wave incident in a gyro-electric medium. Then the angle of rotation

$$\phi = \frac{\omega L}{2c} (n_- - n_+) \quad (18)$$

in the region of approximation ω << ω_c, ω_p becomes

$$\phi = \frac{e^3 N L \lambda^2 B}{2\pi c^4 n \, m^{*2}} \quad (19)$$

This experiment thus yields NL. For GaAs with N = 1 x 10^{17}cm^{-3}, m*/m = 0.066, L = 1 mm, B = 10 kG at 15 μm in the IR, we get φ = 21.4°. Of more interest is the possibility of characterizing N for a 1 x 10^{17}cm^{-3} film with ℓ = 0.5 μm on a semi-insulating GaAs substrate which shows negligible free carriers. Then φ = 0.001° indicating that one must still increase B an order of magnitude and go to longer wavelength (perhaps 60 μm) to take advantage of the λ^2 dependence. Under these circumstances it would be possible to determine N, since it is possible to measure a rotation of a degree.

When ω_c approaches the observation frequency ω, ϕ increases more rapidly than linearly with B and goes through a dispersive oscillation as ω_c passes through ω. Here rotation is not given by Eq. (19), and the full Eq. (17) must be used in Eq. (18).

The helicon region can also be used to characterize a slab. In this case dimensional resonances (interferences effects) complicate the issue but, in fact, allow one to determine N. In the region γ, $\omega << \omega_c << \omega_p^2/\omega$

$$\varepsilon_+ = \frac{4\pi Nec}{\omega B} \tag{20}$$

Since at a Farbry-Perot resonance

$$m\lambda = 2n_+ L \tag{21}$$

it follows that adjacent fringe maxima occur occur at magnetic fields B_{m+1} and B_m such that

$$B_{m+1}^{-\frac{1}{2}} - B_m^{-\frac{1}{2}} = \frac{1}{2L} \left(\frac{e\omega N}{\pi c}\right)^{-\frac{1}{2}} \tag{22}$$

Thus, the fringe pattern yields the carrier density directly. At 8 mm wavelengths, for example, this technique works for thicknesses of the order of millimeters.

4. <u>Hydrogenic Impurities</u>. An impurity atom (acting as a donor or acceptor of one electronic charge) in the lattice of a host crystal behaves much like a hydrogen atom with energy levels of the form

$$E = \frac{2\pi^2 e^4 m^*}{h^2 \varepsilon^2 n^2} - \frac{Rm^*}{m_0 \varepsilon^2 n^2} \tag{23}$$

where R is the Rydberg (equal to 13.6 eV), m_0 is the free-electron mass, n is a principal quantum number, and ε is the background dielectric constant. Because of the large dielectric constant and small effective mass, the energies are quite small ($\sim$50 meV) for the usual transition corresponding to the first line of the Lyman series (n = 1 $\rightarrow$ 2), and the orbits are correspondingly large extending over many lattice spacings. The orbit radius is

$$r = \frac{h^2 \varepsilon^2 n^2}{4\pi^2 m^* e^2} = \frac{a\varepsilon^2 n^2}{m^*} \tag{24}$$

where $a = 0.528$ Å is the Bohr radius. Thus, for a typical semiconductor with $\varepsilon \sim 10$ and $m^* = 0.1\ m_0$, we find $4 \sim 500$ Å.

Impurity spectra for many different impurities in Si, Ge, InSb and GaAs have been measured at low temperature when the electron (hole) has become localized on the impurity atom [21]. Numerous magnetooptical Zeeman-type experiments have been used to identify the transitions [22].

II. INFRARED REFLECTION AND TRANSMISSION IN BULK MATERIALS

Infrared (IR) techniques have been used for over twenty years to determine free-carrier densities, mobilities and effective masses [23,24]. Since semiconductors are highly absorbing in the vicinity of ω_{TO} and ω_p and films less than 10 μm thick are hard to prepare by grinding and polishing, much of the work concentrated on the measurement of reflectivity, which could be analyzed by oscillator models which gave dielectric functions of the form of Eqs. (15) and (16). Alternatively, the reflectivity could be Kramers-Kronig analyzed to provide the optical constants n and k [25].

A. Infrared Reflection of GaAs

1. <u>Drude Model</u>. Most work utilized the Drude model to account for the reflection and transmission properties due to free carriers [26]. From Eq. (15) we obtain the real ε' and imaginary ε'' parts of the dielectric function

$$\varepsilon' = n^2 - k^2 = \varepsilon_\infty - \frac{\varepsilon_\infty \omega_p^2}{(\omega^2 + \gamma^2)} \quad (25)$$

$$\varepsilon'' = 2nk = \frac{\varepsilon_\infty \omega_p^2 \gamma}{\omega(\omega^2 + \gamma^2)}$$

The exact expression for n and k are complicated:

$$2n^2 = \varepsilon' + (\varepsilon'^2 + \varepsilon''^2)^{\frac{1}{2}}$$
$$2k^2 = -\varepsilon' + (\varepsilon'^2 + \varepsilon''^2)^{\frac{1}{2}} \quad (26)$$

In the approximation $\omega > \omega_p \gg \gamma$ and $k^2 \ll 1$, it follows that

$$n^2 = \varepsilon_\infty \left(1 - \frac{\omega_p^2}{\omega^2}\right) \quad (27)$$

and the reflectivity (bulk reflectance) can be used to determine ω_p and ε_∞ from the formula [26]

$$\frac{1+R^{\frac{1}{2}}}{1-R^{\frac{1}{2}}}{}^2 = n^2 = \varepsilon_\infty - \frac{\varepsilon_\infty \omega_p^2 \lambda^2}{4\pi^2 c^2} \qquad (28)$$

Note the λ^2 dependence of this function.

2. <u>Extraction of n, k, N and μ</u>. The reflectance of three bulk samples of N-type GaAs are shown in Fig. 11 [27]. The sample surfaces were chemically polished with a bromine-methanol solution. A least-squares fit of the reflectance is indicated by the solid curves as obtained using Eqs. (15) and (16). The fits are good over a wide spectral range. Two reflection minima are invariably present and can themselves be used to characterize new material [23,24] since their frequency positions are dependent on the plasma frequency ω_p (carrier density N) and their reflectance magnitudes are dependent on the damping constant γ (mobility μ). This is shown in Fig. 12. To use this curve one must measure both the

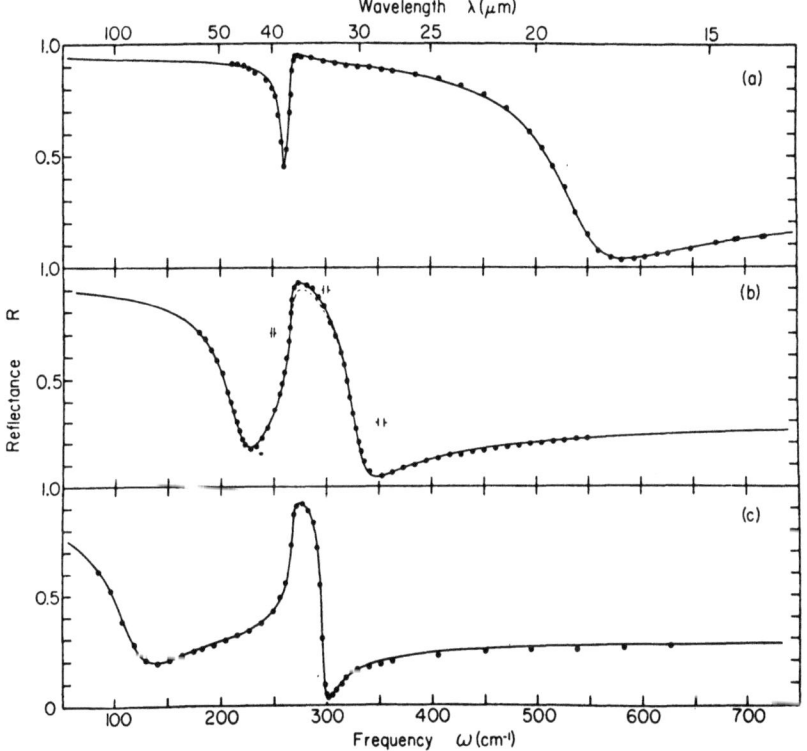

Fig. 11. (a) The reflectance of GaAs sample (1). (b) The reflectance of sample (3). (c) The reflectance of sample (4). Parameters are listed in Table 1.

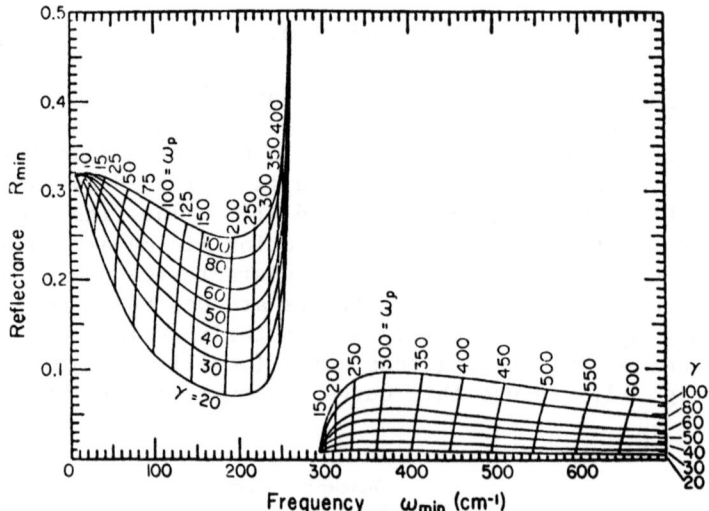

Fig. 12. Graph for determining ω_p and γ from the reflectance minima positions and magnitudes.

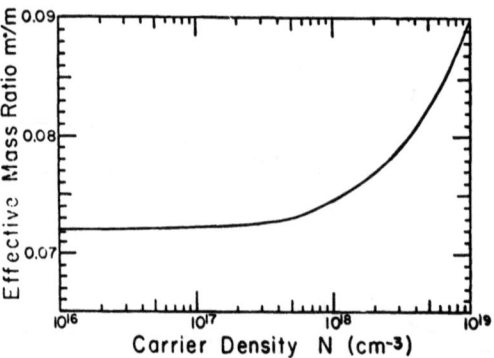

Fig. 13. Relation of effective-mass ratio m^*/m_0 to carrier density N.

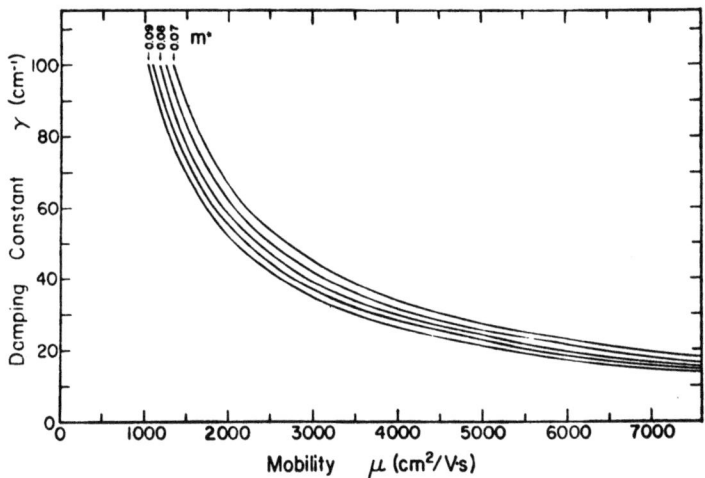

Fig. 14. Relation of damping constant to mobility for various effective-mass ratios m*/m.

position and magnitude of the reflectance minimum. Locating this point on the graph gives ω_p and γ. These need to be translated to N and μ with the aid of Figs. 13 and 14 which relate effective mass m* to N and μ. These determinations can be made from an examination of only one reflectance minimum. The entire analysis of Fig. 11 will also give ω_p and γ. The value of ω_p can be translated to carrier density with Fig. 15. In this treatment the damping constant has been assumed to be independent of frequency.

A comparison of values of N and μ obtained from Hall and conductivity measurements and optical measurements is given in Table 1. The N_{el} and N_{op} are in good agreement. For samples 1 through 4, N_{op} is less than N_{el} but probably within the uncertainties of both measurements. Also, for samples 1 through 4, μ_{op} is less than μ_{el} by roughly 15%. We cannot account for this latter discrepancy if it is real. This implies that the high-frequency damping constant is somewhat larger than the dc damping constant. The plasma reflection edge for sample 5 was below 100 cm^{-1}, and the uncertainty in the data do not warrant making comparison with trends in the other four samples. For low carrier densities (<5 x 10^{16}cm^{-3}) a Fourier-transform-spectrometer measurement is necessary to obtain reliable data [28]. The phonon parameters ω_{TO} = 268.2 cm^{-1}, ω_{LO} = 291.5 cm^{-1} and Γ = 2.3 cm^{-1} were obtained from fitting the reflectance of a semi-insulating sample.

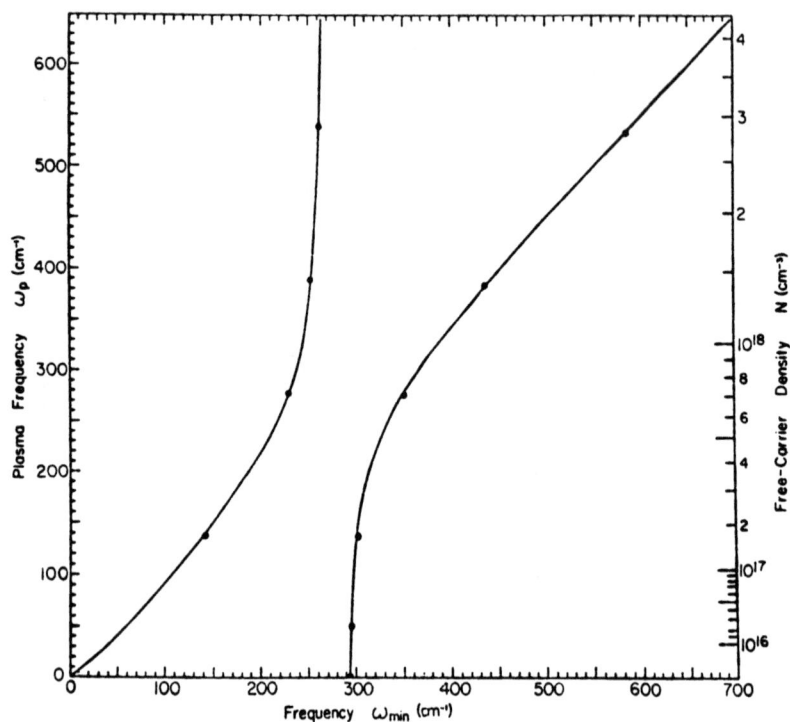

Fig. 15. Frequency position of reflection minima as a function of plasma frequency ω_p and free-carrier density N.

3. <u>Effects of Mechanical Polishing</u>. In many optical studies of semiconductor reflectance, early workers mechanically polished samples with polishing grits such as Linde A with 0.3 m particle size. It was recognized that such treatment changed the reflection properties slightly [29,30]. We have repeated such measurements on several samples of different carrier densities [31]. We assumed the polishing produced a damaged surface region which had different optical constants from the bulk. This introduced three more adjustable parameters ω_{pd}, γ_d and ω_d for the damaged layer in addition to the two parameters ω_{ps} and γ_s already known for the chemically polished surface. For sample 3 of Figure 16, we see significant changes in R upon polishing with Linde A ($\sim$ 0.3 μm grit size) and Linde C ($\sim$ 1 μm grit size). The most interesting feature is the decrease in frequency and magnitude of the reflection

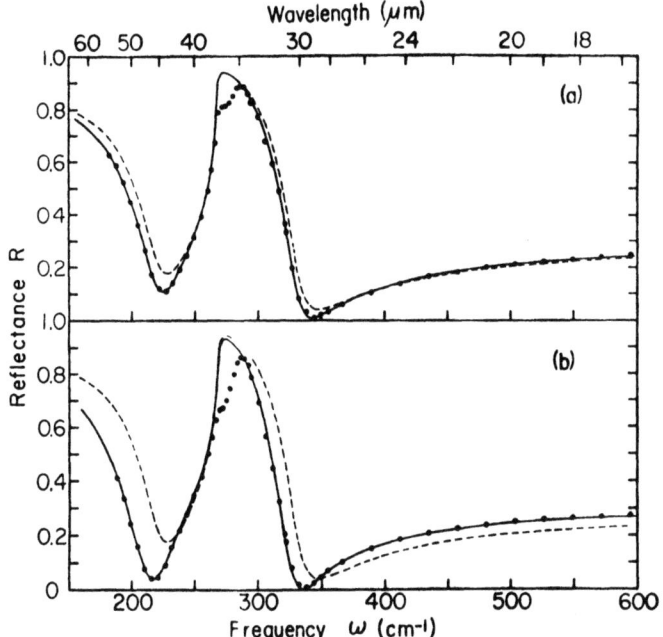

Fig. 16. (a) The reflectance (dashed curve) of chemically polished GaAs sample (3). The solid curve is after polishing with Linde A. (b) The reflectance (dashed curve) of the same chemically polished sample. The solid curve is after polishing with Linde C.

minima. In fact, the reflection minimum going to zero at 340 cm^{-1} suggests a very high mobility (small γ) which is not reasonable. When analyzing the reflectance of a mechanically polished sample, if the damage is assumed negligible or to extend very deep, then the values of N and μ are not too meaningful compared to transport-measurement values [30]. However, the damage-layer model gives the parameters of Table 2 and the results indicated by the solid curves which are good fits except near the ω_{TO} frequency. The results suggest a damage layer for free carriers comparable to the grit size and a large decrease in N_d and μ_d. The same type of results were obtained for sample 1 illustrated in Fig. 17. Note again that the reflectance is zero at the plasma reflection minimum near 560 cm^{-1}.

For a semi-insulating sample the only change occurred at the reststrahlen maximum as shown in Figure 18 for sample 6. This decrease in R chould be fitted with the dashed curve assuming that

TABLE 1

Plasma frequency ω_p, damping constant γ, carrier density N_{opt}, mobility μ_{opt} for bulk n-type GaAs samples along with N_{el} and μ_{el} determined from Hall and conductivity measurements.

Sample	ω_p cm^{-1}	γ cm^{-1}	N_{opt} 10^{-3} cm^{-3}	N_{el} 10^{-3} cm^{-3}	μ_{opt} cm^2/Vs	μ_{el} cm^2/Vs
1	533	61	2.7	2.8	1950	2360
2	383	48	1.4	1.44	2600	3020
3	274	51	0.70	0.83	2500	2750
4	135	51	0.16	0.2	2520	2900
5	49	32	0.021	0.018	4000	3800

TABLE 2

Bulk plasma frequency ω_p and damping constant γ for CM polish damaged-layer plasma frequency ω_{pd}; damping constant γ_d and thickness ℓ for LA and LC polish for samples 3 and 1 (as obtained from least squares fit to infrared reflectance).

Polish		Sample 3	Sample 1
CM	ω_p	274 cm^{-1}	534 cm^{-1}
	γ	51 cm^{-1}	62 cm^{-1}
LA	ω_{pd}	127 cm^{-1}	383 cm^{-1}
	γ_d	492 cm^{-1}	184 cm^{-1}
	ℓ	2420 Å	2120 Å
LC	ω_{pd}	137 cm^{-1}	452 cm^{-1}
	γ_d	330 cm^{-1}	125 cm^{-1}
	ℓ	6050 Å	5890 Å

OPTICAL CHARACTERIZATION OF SEMICONDUCTORS 335

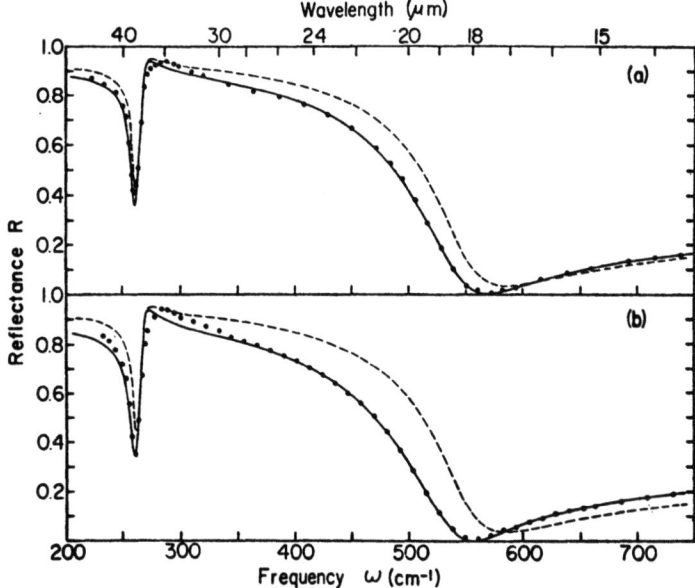

Fig. 17. (a) The reflectance (dashed curve) of chemically polished GaAs sample (1). The solid curve is after polishing with Linde A. (b) The reflectance (dashed curve) of the same chemically polished sample. The solid curve is after polishing with Linde C.

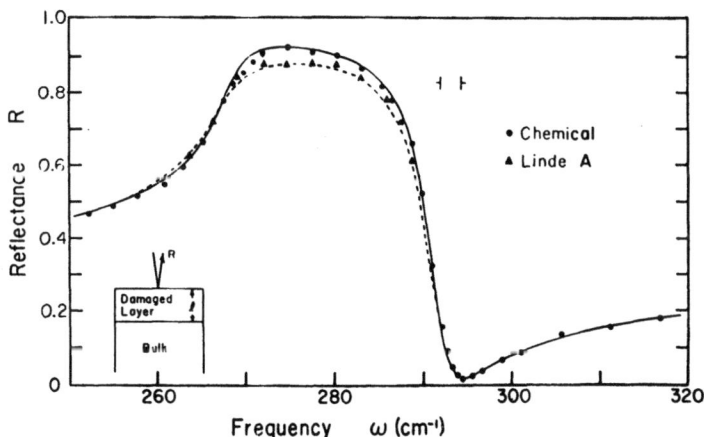

Fig. 18. The reststrahlen region of semi-insulating GaAs chemically polished (solid curve) and mechanically polished with Linde A (dashed curve).

Γ increases to 5.0 cm^{-1} in a layer with ℓ_d = 3000 Å. A puzzle in this work is the behavior of the reflectance maximum near ω_{TO} in all three samples. While the damage to the reststrahlen is accounted for by a simple increase in Γ in sample 6, Fig. 18, the damage in the other two samples seems to be different. The fit includes the Γ = 5 cm^{-1} for samples 3 and 1 in Figures 16 and 17, respectively. Shifts in ω_{TO} in the layer due to damage can produce a notch as in Figs. 16 and 17 but such a shift is not required for sample 6 in Fig. 18.

4. **Depletion-layer Effects.** Mechanical polishing essentially produces a slight depletion layer. We noticed that even in fitting the chemically polished samples, the calculated reflectance minima (dashed curve) were slightly higher than the experimental points [27] as indicated in Fig. 19. Usually, in measuring low reflectance, filtering problems lead to data points which are too high. We attempted to fit the reflection minima assuming that a completely depleted layer existed at the surface (solid curve). Such a layer is suggested by backscattering Raman studies of optical phonons in semiconductors [32,33]. The result of reflectance fits to four samples is shown in Fig. 20 where the depletion layer thickness is plotted against carrier density N. For one sample we actually made CV measurements on a Schottky barrier to get the triangle datum point. Raman results are given by the square data points. We find that the zero-bias depletion depth $\ell = \sqrt{\phi \varepsilon_0 / 2\pi e^2 N}$ is given by the solid line in Figure 20 for a barrier height ϕ = 0.8 eV and dc dielectric constant ε_0 = 13.1. A slightly smaller ϕ would fit a little better. The origin of a free-surface depletion layer is unknown, but might be due to adsorbed impurities or the natural oxide on the surface.

B. Infrared Transmission of GaAs

1. **Free-carrier Absorption.** While the Drude model has been shown to fit reflectance data very well, the reflectance above the higher reflectance minimum is not very sensitive to k as was noted in Section IIA-1. On the other hand, transmittance of a slab is quite sensitive to k and has often been used to measure the wavelength dependence of free-carrier absorption.

From Eq. 26 it follows that in the approximation $\omega \gg \omega_p$, $\gamma = 1/\tau$ and $\varepsilon' \gg \varepsilon''$ that the absorption coefficient is

$$\alpha = \frac{\varepsilon_\infty^{1/2} \omega_p^2 \tau \lambda^2}{4\pi^2 c^3} \qquad (29)$$

Rather than being λ^2, the wavelength dependence is often found experimentally to be λ^n where $1 < n < 3$. One way to amend the model has been to assume various energy dependences for the scattering time $\tau(\xi)$ [34,35]. For example, for ionized impurity scattering $\tau \sim \xi^{3/2}$; for polar optical phonon scattering $\tau \sim \xi^{1/2}$; for acoustic phonon scattering $\tau \sim \xi^{-1/2}$. With such alterations of the model, it is possible to average various functions of τ which appear in ε' and ε'' (or σ' and σ'') to obtain the observed wavelength dependence. For example, for acoustic phonons this $\alpha \sim \lambda^n$ dependence can vary from $1.5 < n < 3$ depending on whether $\hbar\omega \gg kT$ or $\hbar\omega \ll kT$ holds.

An extension of this has been done for the case of p-Te where reflection and transmission experiments in the far infrared have been carried out [36] to determine the real and imaginary parts of ε. These are related to the dynamical conductivity σ and its reciprocal, the dynamical resistivity ρ. It is found that τ is a constant below a frequency ω_0 which locates the zero in the total dielectric function and varies above this as $\omega^{3/2}$ (ionized impurity scattering) and $\omega^{1/2}$ (polar optical phonon scattering) depending on the temperature and polarization. This gives wavelength dependences for α at 5 K of λ^4 and λ^5, respectively for light polarized $\perp$ and $\parallel$ to the c axis.

The absorption cross section for free carriers has been calculated quantum mechanically for various kinds of scattering mechanisms [37,38]. The index of refraction is not calculated, but in principle, a Kramers-Kronig analysis of k would yield n. Specific theories have been applied to a number of semiconductors [39].

For GaAs it is found [40,42] that below the reststrahlen region, n varies as λ^2 indicating good Drude-model behavior, while above the reststrahlen region α varies as λ^3 [43-46]. This λ^3 dependence indicates that polar-optical-phonon and ionized impurity scattering are important at room temperature. It has been fitted with a quantum model [37,38]. In Fig. 21, we have collected examples of free-carrier absorption (References 40-43) to illustrate the change in slope in the vicinity of ω_{LO}. Such wavelength dependencies vary with temperature: for example, as the temperature is lowered to 4.2 K, optical-phonon scattering becomes less important.

2. *Laser Scanning of Free-Carrier Absorption.* Reflection scanning at the plasma edge is a sensitive method of determining free-carrier density inhomogeneity [27,47]. This is illustrated in Fig. 22 where the surface was scanned at a frequency where $R \approx 0.5$. Since R is much more sensitive to N than to μ, we can translate the variation in R into the variation in N. However, the wavelengths are long (> 15 μm) for carrier densities less than $\sim 3 \times 10^{18} cm^{-3}$, and the lack of spectral energy and inherent diffraction limits do not allow scanning with spot sizes of the order of micrometers which are typical for visible-laser scanning. Thus, it is possible to scan a sample with a millimeter spot size at 25 μm wavelengths, but even with a better detector and diffraction-limited optics, it

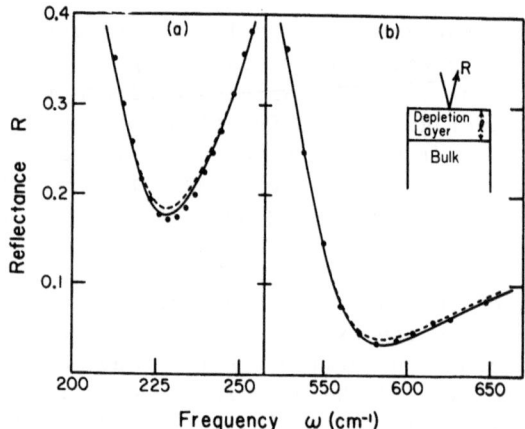

Fig. 19. Reflectance minima of chemically polished GaAs samples (a) (3) and (b) (1) indicating calculated fits (dashed curves) similar to those of Figs. 16 and 17. The data points are slightly below the fit. When a depletion layer is assumed, the calculated minima are deeper, in better agreement with observation (solid curve).

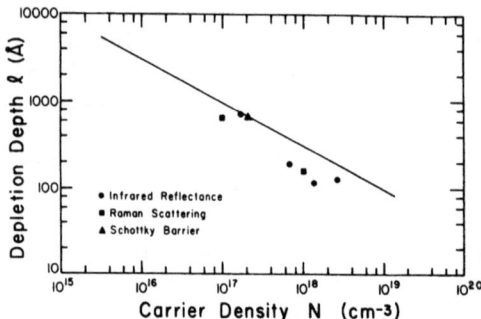

Fig. 20. Depletion layer obtained from fitting the relectance minima vs. carrier density ●; Raman data ■; Schottky-barrier data ▲.

is likely that spot sizes of less than 100 μm would be difficult. If low-carrier-density material is used ($N < 10^{17} cm^{-3}$), the wavelength must be longer than 100 μm, so the spot size is proportionally larger. To measure free-carrier effects at shorter wavelengths to achieve smaller spot sizes, one must go to the transmission technique with CO_2 (10.6 μm), CO (5 μm) or YAG (1.06 μm) lasers.

The transmission of a slab is sensitive to the free-carrier density. This effect has been used to measure free-carrier inhomogeneity with a focussed CO_2 laser in n-type CdTe 48. The experimental, room-temperature free-carrier absorption at 10.6 μm is described by the absorption coefficient $\alpha = 13.5 \ (N/10^{17})^{1.3} cm^{-3}$. A 1 mm thick wafer with $2 \times 10^{17} < N < 5 \times 10^{17} cm^{-3}$ can be examined readily, whereas a thinner wafer is needed for larger N. Variations in transmittance of a wafer 3.5 cm in diameter were observed which corresponded to variations in free-carrier density as indicated in Fig. 23. One wafer, cut at an angle of 30° with respect to the growth axis, showed large variations in transmission, while another, cut in the growth plane, showed little variation. The uniformity in N between 1.5 and 2.5 cm in Fig. 23(b) is better than ±0.5%. Note how small the transmission is, thus decreasing multiple-reflection effects Spot-size resolution was about 300 μm. In principle, resolution an order of magnitude better than this should be achievable.

It is of interest to note that it would be useful to be able to scan an epitaxial film of GaAs on a semi-insulating substrate to determine inhomogeneity. The film thickness ℓ is now of the order of 1 μm or less and N is typically less than 10^{17} cm^{-3} for FET applications. Thus, we are down three orders of magnitude in ℓ compared to the CdTe sample discussed above. A calculation in the 10.6 μm region assuming $\alpha = 12$ cm^{-1} yields the results shown in Fig. 24. The calculated transmittance (solid curve) of a reference substrate is shown with L = 500 μm. The slab is assumed to be plane parallel. When the epitaxial film ($\ell = 1$ μm) is added to the substrate, the interference fringes are shifted significantly, since this is about the 300th order of interference. Thus, it is possible to determine the film thickness (and presumably film-thickness inhomogeneity) from such a measurement. However, the film absorption effects are very small, the peak transmittance at 942.2 cm^{-1} changing from 0.9274 to 0.9262. Thus, such a film-substrate sample could not be scanned to determine inhomogeneity in free-carrier absorption, especially since in an actual sample the substrate thickness variation might be significant.

An IR microscope has been used at the 3.39 μm line of the He-Ne laser to examine free-carrier inhomogeneities in heavily doped Si and GaAs (N $\sim 10^{18}$cm^{-3}) samples 0.5 - 2.0 mm thick [49]. Resolution of about 12 μm was achieved in this case. A raster-scan signal was presented on a CR screen. Qualitative information on such inhomogeneities was obtained in this way.

The same ideas [50] were used to study carrier densities in InAs samples 0.5 mm thick with densities as low as 4.6 x 10^{18}cm^{-3} at 3.39 μm wavelengths. In this case the Burstein-Moss effect shifts the band edge causing increased transmission as the carrier density is increased. This type of band-edge effect is discussed in Section V-A.

C. Local Modes in GaAs

Newman, et al [51-54] have measured the local-mode vibration lines of Si and B in GaAs. Heavily doped Si (10^{18}cm^{-3}) samples were electron irradiatiated to compensate the high carrier density and render the samples transparent in the 10-30 μm region. Numerous

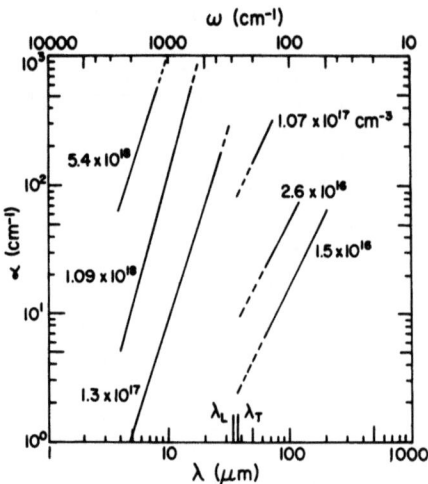

Fig. 21. Schematic representation of GaAs free-carrier absorption coefficient at room temperature for various carrier densities. The solid lines are extracted from data in references 40-43 and are extrapolated as dashed lines to emphasize the change in slope from 3 to 2 at ω_{LO}.

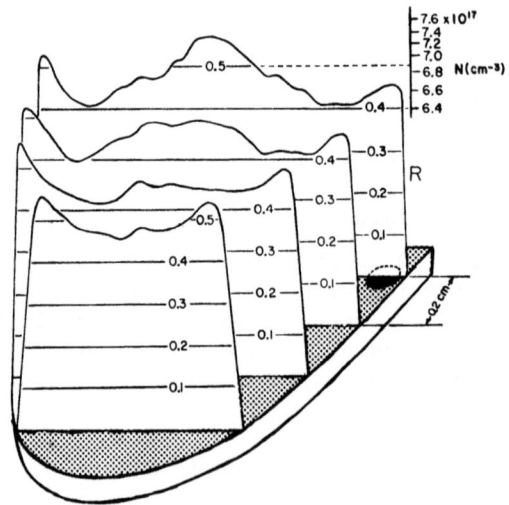

Fig. 22. The variation of carrier density measured in a GaAs wafer at the R = 0.5 level of the plasma reflection edge at 320 cm^{-1}.

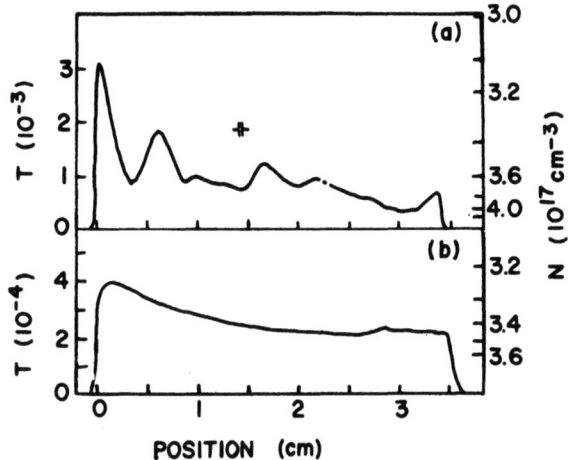

Fig. 23. (a) Transmittance of Br-doped CdTe translated into carrier-density variation for a wafer cut at an angle of about 30° with respect to the growth plane. (b) Wafer cut in the growth plane, so the scan is across the boule perpendicular to the growth direction.

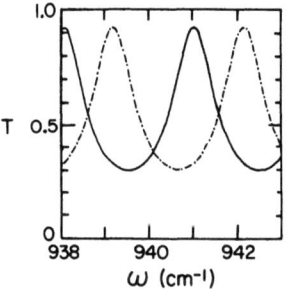

Fig. 24. The calculated transmittance of a reference substrate of a GaAs (solid curve), the transmittance with an epitaxial film added to the substrate with $N_f = 0$ (dashed curve), and the transmittance of the film and substrate with $N_f = 1 \times 10^{17} cm^{-3}$ (dotted curve). Dotted curve is nearly coincident with dashed curve.

lines due to Si on Ga and As sites, B on Ga sites, and pairs (complexes) of Si-B were assigned. The main lines of ^{28}Si(Ga) and ^{28}Si(As) are found at 383.7 and 398.2 cm^{-1}. The main lines of ^{11}B(Ga) and ^{10}B(Ga) are at 517.0 and 540.2 cm^{-1}. Samples were typically < 0.5 mm thick and cooled to near liquid nitrogen temperatures. By controlled doping it is possible to calibrate peak absorption coefficient or integrated absorption to determine the Si content of uncharacterized samples.

More recently, it has become of interest to determine that Si and B impurities in semi-insulating GaAs. In this case the material is already compensated, so that the net carrier density is less than $\sim 10^{14}$ cm^{-3}. However, the acceptor and donor concentrations may be as large as 10^{17} cm^{-3}. Using results of Newman, et. al. [51-54] it is possible to find Si [55] and B [56] in semi-insulating GaAs and to estimate densities in the low 10^{16} cm^{-3} range.

D. Far-Infrared Impurity Spectra

The principal interest in impurity spectra is for the identification of specific impurities at various sites in the host lattice. The energy levels and consequently the spectral lines can be chemically shifted (central cell correction) because of deviations of the potential from a simple Coulombic relation close to the impurity site. These effects are most important for the ground state because of the appreciable amplitudes of the 1S wave functions close to the origin. The chemical shifts of individual donors are about 1 cm^{-1}. The sharpness of the lines due to lack of strain effects allows shifts as small as 0.1 cm^{-1} to be determined in the submillimeter wave spectral region. The chemical shifts can be determined by back-doping experiments [57-60] and these results then serve as a calibration for identifying the chemical impurity and its concentration. Dopants such as Sn, Pb, Se and Ge have been measured in GaAs in the concentration range $< 10^{15}$cm^{-3}. Frequently, a magnetic field has been used to unravel the lines and aid in their identification [22].

III. INFRARED REFLECTION AND TRANSMISSION OF A THIN FILM ON A SUBSTRATE

The characterization of an epitaxial film of GaAs on a GaAs substrate can now be attempted since we know that the bulk substrate can be analyzed with the appropriate model. Although the lattice-vibration properties of both film and substrate are the same, the differences in free-carrier densities will produce differences in the optical constants which can be measured in reflection and transmission experiments.

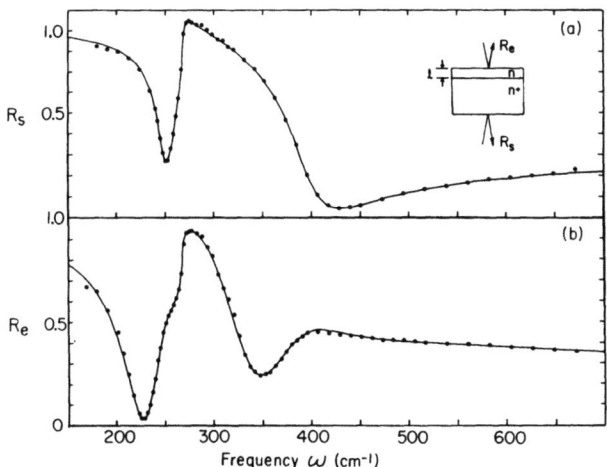

Fig. 25. (a) Reflectance from substrate side of an N-GaAs film on an N+-GaAs substrate. (b) Reflectance from the film side. Analysis yields the solid curves and the parameters for sample A in Table 3.

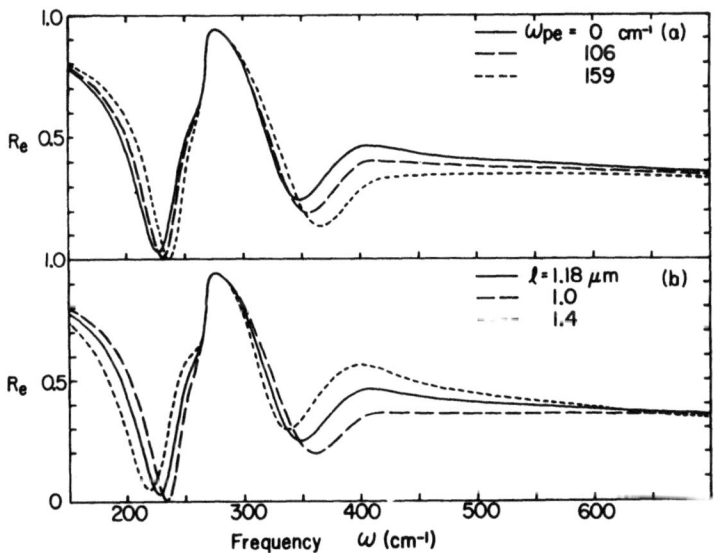

Fig. 26. Calculated dependence of the film reflectance on ω_{pf} (a) and ℓ (b).

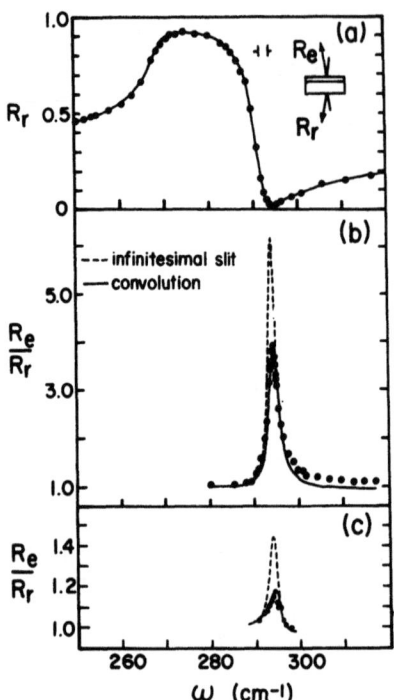

Fig. 27. (a) Reflectance of a semi-insulating sample of GaAs.
(b) Ratio reflectance of N^+/N sample (c) and the reference sample of part a. (c) Ration reflectance of sample (a) of N^+/N GaAs and the reference sample of part a. Table 3 lists the parameters obtained.

A. Epitaxial Film of GaAs on a GaAs Substrate

1. <u>IR Reflectance of N/N^+ and N^+/N Samples</u>. A low-carrier-density N film on a high-carrier-density N^+ substrate is quite straightforward [27,61]. One example is given in Fig. 25. The reflectances R_s and R_f on each side of the sample are measured and found to be quite different. Fitting the reflectances with a two-layer model gives ω_{ps} and γ_s and ω_{pf}, γ_f and ℓ. The results for a series of N/N^+ samples are shown in Table 3. In Figure 26, the sensitivity of the spectrum to changes in ω_{pf} and ℓ is seen to be significant. The measurement typically gives a good value of ℓ, N and μ, but as the film becomes thinner ($\sim 1\mu m$), the fitting program begins to give unreasonable values for μ. We then fix μ

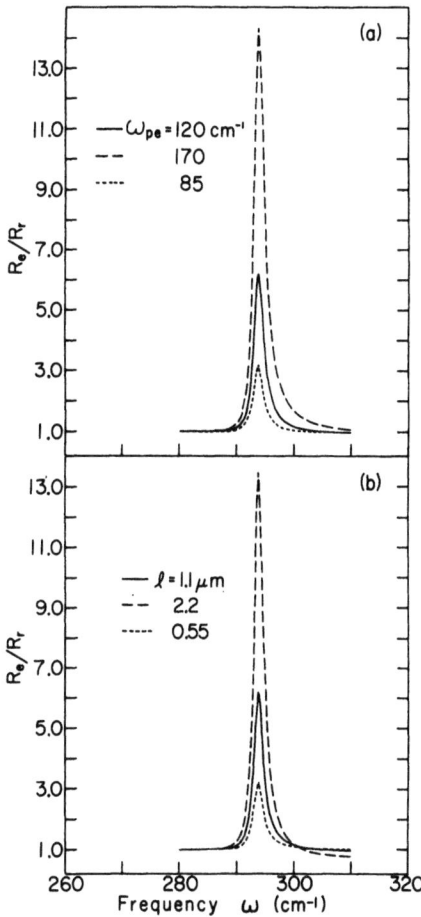

Fig. 28. (a) Calculated dependence of the ratio reflectance on ω_{pf} for an N^+/N GaAs sample. (b) Calculated dependence of the ratio reflectance on ℓ.

to a reasonable value to continue the fitting. The spectrum is especially sensitive to film thickness as the interference effects suggest. In fact, as we note shortly, it is possible to scan the sample at a relatively high spectral frequency above 600 cm^{-1} and see small variations in film thickness.

The spectrum for an N^+/N sample is entirely different as indicated in Fig. 27 [61]. Fig. 27(a) shows the reflectance of a semi-insulating substrate while Fig. 27 (b) and (c) show the reflectances of two different filmed samples. The substrate and film spectra differ significantly only near the reststrahlen minimum and at

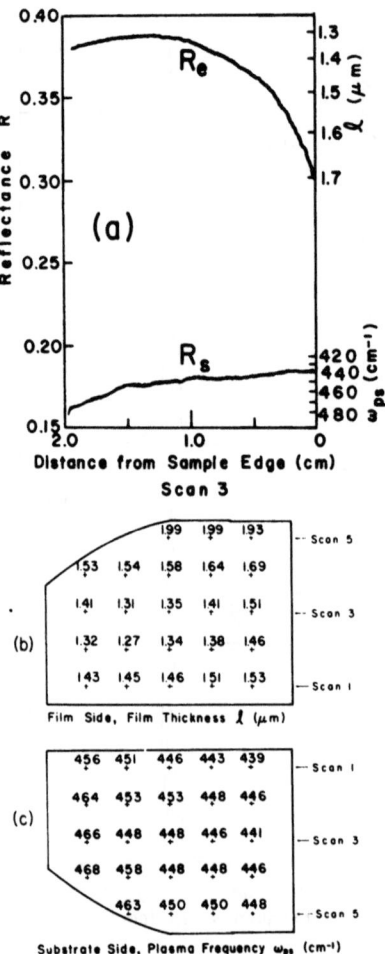

Fig. 29. (a) A reflectance scan of the substrate and film sides of sample (B) at a frequency of 862 cm^{-1} revealing the inhomogeneity in ω_{ps} and ℓ. (b) The variation in ℓ. (c) The variation in ω_{ps}.

low frequencies below 100 cm^{-1}. In this case, however, the substrate is transparent and multiple-reflection effects in the substrate must be considered. The prominent feature in the spectrum is a spike at 293 cm^{-1} which is a sensitve function of both ω_{pf} and ℓ. It is much less sensitive to γ_f. In fact, we could not extract a good value of γ_f in the fitting process, so we usually fixed γ_f at some reasonable value. The behavior of the peak for variations in ω_{pf} and ℓ was similar, as shown in Fig. 28 and made the extraction

TABLE 3

Parameters of epitaxial films and substrates of n-type GaAs.
The growth column indicates how the film was grown: VPE, vapor phase epitaxy; LPE, liquid phase epitaxy; MBE, molecular beam epitaxy. The electrical column indicates how the film carrier density and/or mobility were measured: C-V, capacitance-voltage; vdP, van der Pauw.

Sample	ω_p (cm^{-1})	γ (cm^{-1})	ℓ_{op} (μm)	N_{op} (cm^{-3})	N_{el} (cm^{-3})	μ_{op} cm^{2}V^{-1}s^{-1}	μ_{el} cm^{2}V^{-1}s^{-1}	ℓ_{mech} (μm)	Growth	Electrical
N/N$^+$										
A Film	127	46	2.9	1.4×10^{17}	1.5×10^{17}	2800	—	2.8	VPE	C-V
Substrate	401	71	—	1.5×10^{18}	2×10^{18}	1700	—	—	—	
B Film	90	(50)	1.2	7.0×10^{16}	1.8×10^{17}	(2550)	—	1.4	VPE	C-V
Substrate	452	70	—	1.9×10^{18}	—	1750	—	—	—	
C Film	47	(60)	1.8	$<2\times10^{16}$	1.0×10^{17}	(2150)	—	0.5	VPE	C-V
Substrate	381	65	—	1.4×10^{18}	2×10^{18}	1850	—	—	—	
D Film	214	900	0.32	$\sim4\times10^{17}$	—	~150	—	0.30	MBE	
Substrate	535	57	—	2.7×10^{18}	—	2000	—	—	—	
N$^+$/N										
a Film	30	(30)	[2.8]	8.0×10^{15}	1.2×10^{16}	(4300)	4930	2.8	LPE	vdP
b Film	130	(50)	[1.7]	1.5×10^{17}	2.8×10^{17}	(2550)	2760	1.7	LPE	vdP
c Film	120	(50)	[1.1]	1.3×10^{17}	1.2×10^{17}	(2550)	2190	1.1	VPE	vdP
d Film	320	(50)	[1.4]	9.4×10^{17}	4.8×10^{17}	(2550)	—	1.4	VPE	C-V

of each one somewhat ambiguous. We therefore were forced to fix
ℓ (known from a cleave and stain measurement). However, subsequent refining of the fitting process suggests that both ω_{pf} and ℓ can be extracted. Because the spike feature is sharp and spectral resolution was limited to ~ 2.3 cm^{-1}, a convolution had to be included in the analysis. This would not be necessary if resolution could be improved to ~ 1 cm^{-1}. Results are listed in Table 3.

While in Section IIB-2 we discussed scanning of a sample near the plasma reflectin edge where $R_s = 0.5$ to reveal carrier-density inhomogeneity, this can be done at any higher frequency where R_s is dependent on ω_{ps}. In Fig. 29, the reflectance of the substrate at 682 cm^{-1} reveals the inhomogeneity in ω_{ps}. When the film is scanned across the same region, a more rapid variation in R_f is observed, due primarily to non-uniform thickness, since the free-carrier effects in the film are negligible at this frequency. Assuming that the carrier density of the substrate is constant through the thickness of the substrate, an analysis can give the variation in ℓ and in ω_{ps} as indicated in Fig. 29 (b) and (c).

2. <u>Microwave Magnetoplasma Reflection and Transmission</u>. The reflectance and transmittance of an N$^+$/N sample of GaAs is calculated in Fig. 30 with $N_f = 1 \times 10^{17}$cm^{-3} ($\omega_p = 106$ cm^{-1}), $\mu = 3300$cm^2/Vs ($\gamma = 40$ cm^{-1}) $\ell = 1.0$ µm and L = 300 µm. These are indicated by the solid curves in Figs. 30 (a) and (b) obtained with Eqs. (6), (9) and (11), which are only calculated to 20 cm^{-1}. For comparison, the dashed curves are the R and T for the substrate itself. The differences should be measurable. Note that the interference-fringe spacing is about 5 cm^{-1}. Microwave experiments discussed below were done at 0.33 and 1.25 cm^{-1} in the vicinity of the 1st order fringe. In Fig. 30 (c) and (d) we have averaged out the fringes with poor resolution (> 5 cm^{-1}) by using Eqs. (7) and (12) and have extended the frequency range to 200 cm^{-1}. Again, there are differences between the substrate spectrum and the film-on-a-substrate spectrum which would be measurable and therefore, characterization would be possible based on the Drude model.

It has been demonstrated that the mobility of an epitaxial film on a semi-insulating substrate can be obtained from a microwave reflection experiment [62]. This measurement is based on the magnetic-field dependence of the dielectric function given by Eq. (16). We remember that

$$\varepsilon = \varepsilon_\infty - \frac{i4\pi\sigma}{\omega} \qquad (30)$$

where σ is the free-carrier conductivity. Then

$$\sigma = \frac{i\varepsilon_\infty \omega_p^2}{4\pi(\omega \pm \omega_c + i\gamma)} \qquad (31)$$

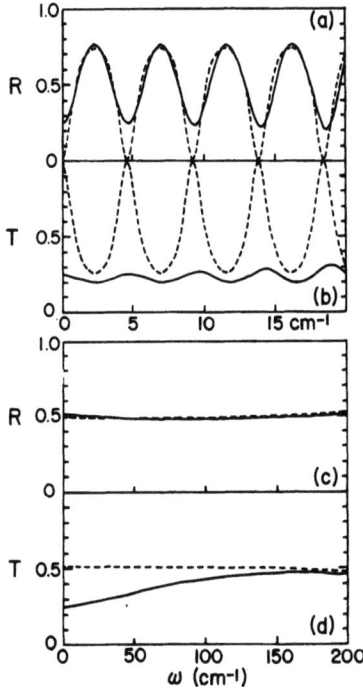

Fig. 30. Calculated reflectance R and transmittance T of an N^+/N GaAs sample in the submillimeter region. Solid curves are for sample and dashed curves are for substrate alone. Fields treated coherently in a and b and incoherently in c and d.

and the real part of σ is

$$\sigma' = \frac{\varepsilon_\infty \omega_p^2 \gamma}{4\pi[(\omega \pm \omega_c)^2 + \gamma^2]} \tag{32}$$

Replacing γ by $1/\tau$ and assuming $\omega \ll \omega_c$ and $\omega\tau \ll 1$ for the microwave region gives

$$\sigma' = \frac{Ne^2\tau}{m^*} \frac{1}{1 + (\omega\tau)^2} = \frac{\sigma_0}{(\omega\tau)^2} \tag{33}$$

where σ_0 is the dc conductivity. Thus, since the change in reflectance is proportional to the power absorption, which is proportional to σ', the magnetic field at which the signal decreases to 0.5 of its initial value is a direct measure of $\omega_c\tau = \mu B/10^8$ and thus of the mobility. This is shown in Fig. 31 for four samples

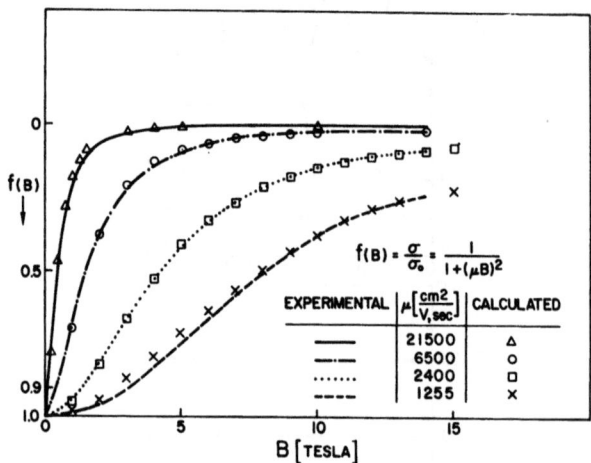

Fig. 31. The calculated and experimental variations of power absorption for four samples of epitaxial films on semi-insulating substrates as a function of applied magnetic field at a microwave frequency of 11 GHz.

of epitaxial films of GaAs on GaAs semi-insulating substrates. Measurements were made at 11 GHz (3.3 cm wavelength). Note that the first sample listed is at low temperature while the other three are at room temperature. These results are in good agreement with electrical transport measurements.

The corresponding magnetotransmission experiment has been performed at 8 mm wavelength (35 GHz) at 77 K in an attempt to characterize similar samples [63]. The analysis utilizes the dielectric function of Eq. (17) and a two-layer model described by Eq. (11) but only $N\ell$ and μ can be obtained. Information on N or ℓ must be obtained independently. At 35 GHz, the substrate with $L = 300$ μm and $\varepsilon(0) = 14.0$ does contribute some multiple-reflection effects. In Fig. 32 are shown the left circular polarized (CRA) and right circular polarized (CRI) transmittances as measured and calculated assuming $N_f = 1.6 \times 10^{15} cm^{-3}$ as determined from van der Pauw measurements. Then $\mu_f = 8 \times 10^4$ cm^2/Vs and $\ell = 30$ m are obtained. This experiment gives only intensity information. It was also possible to measure the Rayleigh interference patterns for CRA and CRI configurations, i.e., the beat pattern of the transmitted signal with a reference signal. Some results are given in Fig. 33 where four transmittance curves for CRA are obtained for four settings of the relative phase of the two signals when $B = 0$. In this case, the signal has phase as well as amplitude information.

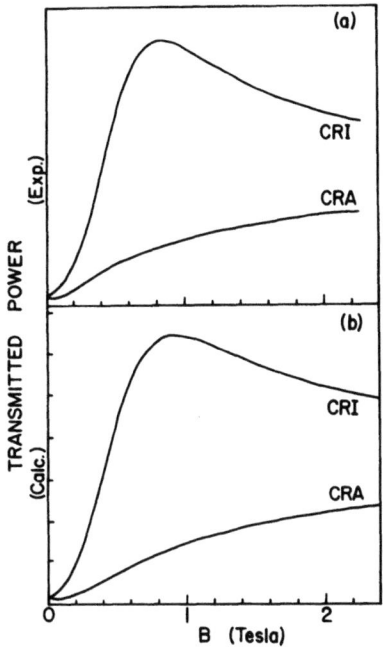

Fig. 32. (a) The experimental CRA and CRI transmittances of an epitaxial film on a semi-insulating substrate at 35 GHz.
(b) The calcualted CRA and CRI transmittances.

3. <u>Submillimeter-Wave Cyclotron Resonance</u>. From Eq. (6) and Eq. (17), it is possible to compute the magnetotransmission of a GaAs slab as shown in Fig. 34 [64]. In this case the frequency is fixed at 84 cm^{-1} and the magnetic field B is swept. The sample has N = 5 x 10^{14} electrons/cm^3, μ(78K) = 4.5 x 10^{4}cm^2/Vs, ℓ = 10^{-3}cm, m*/m -= 0.069, ε_∞ = 13.1. The resonant field is B$_c$ = 62 kG. The absorption indicated by the dotted curve is obtained. The half width at half maximum ΔB$_{1/2}$ is a direct measure of mobility μ through the equations

$$B_c/\Delta B_{1/2} = \frac{\nu c}{\nu} = \omega_c \tau = \mu B/10^8 \tag{34}$$

where ν is the wave number.

The relative intensity is given by

$$\frac{I(B)}{I_o} \approx \exp \frac{-2e^2 N \ell}{\varepsilon_\infty^{1/2} c^2 m^* \gamma} \frac{(B_{1/2})}{(B - B_c)^2 + (\Delta B_{1/2})^2} \tag{35}$$

and yields the product Nℓ. Since such a thin film cannot be

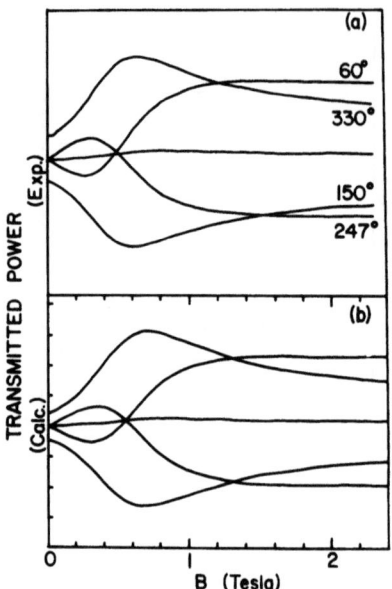

Fig. 33. (a) The experimental CRA transmittance Rayleigh interference patterns. (b) The calculated CRA transmittance Rayleigh interference patterns.

self-supporting easily, it is usually grown on a semi-insulating substrate of GaAs. To analyze this sample requires use of a multilayer model given by Eq. (11) with Eq. (17) used to describe the dielectric function of the film. A calculation for a substrate with L = 0.0385 cm yields the solid curve of Fig. 34. The substrate actually produces multiple-reflection effects which are minimized in the calculation by choosing its thicknes to give an integral number of wavelengths inside the sample.

Measurements were performed on two samples, one consisting of a Cr-doped semi-insulating substrate implanted with 1×10^{13} S^+ ions cm^{-2} at 100 keV and annealed at 800°C (sample 1). The implanted layer was buried $\sim$1000 Å below the surface. The results of a fit, which did not include the actual Gaussian distribution of activated free carriers, yielded values of $N\ell$ and μ give in Table 4. The results are in good agreement with transport measurements. A second sample consisted of a Cr-doped substrate with an n-type epitaxial film approximately 10 μm thick which was implanted with 2.5×10^{12} S^+ ions cm^{-2} at 100 keV and annealed at 800°C (sample 2).

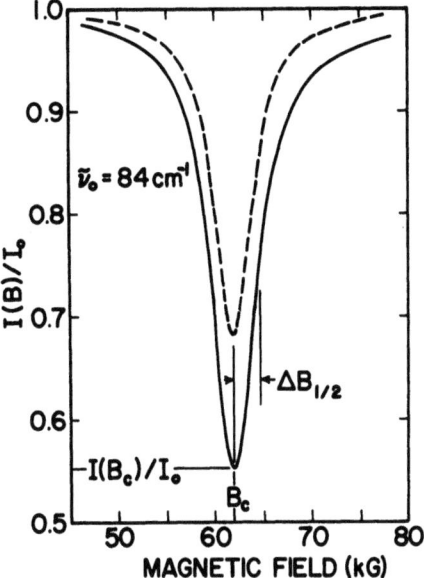

Fig. 34. Calculated cyclotron resonance absorption line for a free GaAs film (dashed curve) and for a film on a semi-insulating GaAs substrate (solid curve).

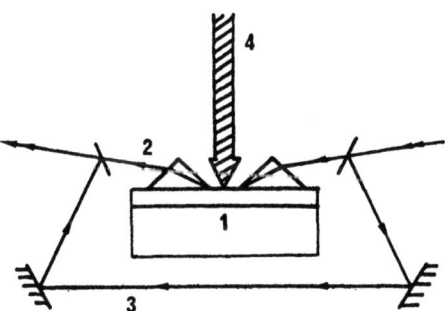

Fig. 35. Schematic diagram of the measurement of wave-guide propagation in a GaAs film on a GaAs substrate. The ruby-laser beam produces free carriers in the top of the film, changing the transmittance properties of the film to IR radiation launched into and coupled out of Ge prisms.

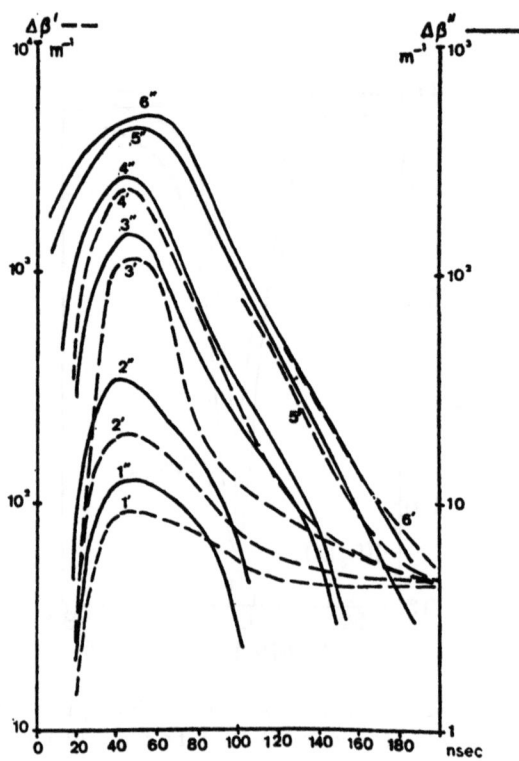

Fig. 36. The change in complex propagation constant $\beta = \beta' + i\beta''$ as a function of time for different excitation intensities: 1 – 64 W cm^{-2}; 2 – 200 cm^{-2}; 3 – 810 cm^{-2}; 4 – 2.0 kW cm^{-2}; 5 – 7.9 kW cm^{-2}; 6 – 14 kW cm^{-2}.

The sheet carrier density $N_e \ell_e$ of the epitaxial layer and $N_i \ell_i$ of the implanted layer did not differ appreciably, so the observed resonance was a superposition of the two absorption lines. The implanted layer was removed by anodic oxidation of the outer 2000 Å of the epitaxial film and stripping it. The remaining sample acted as a standard for the epitaxial-film cyclotron resonance. The results are shown in Table 4 and are compared with transport measurements. A further step for improvement would be to model the Gaussian free-carrier density instead of assuming it to be uniform.

4. <u>Guided Waves in Films</u>. The characteristic parameters of free carriers produced by a pulsed ruby laser in an epitaxial film of N-GaAs (N = 1.5 x 10^{15}cm^{-3}, ℓ = 16 μm) on an N$^+$-GaAs (N = 2 x 10^{18} cm^{-3}) substrate have been determined by measuring the wave-guide

TABLE 4

Parameters for GaAs as determined from transport and cyclotron-resonance measurements.

Sample	Method	$n\ell$ (cm^{-2})	μ cm^2/Vs
1	Transport	3.0×10^{12}	3400
(57K)	Cyclotron resonance	2.3×10^{12}	3400
2 (epi-film)	Transport	0.76×10^{12}	14000
(70K)	Cyclotron resonance	1.1×10^{12}	45000
2 (implants)	Transport	7×10^{11}	17000
(70K)	Cyclotron resonance	2×10^{11}	17000

properties of the film with a second beam of IR light 65,66.
A schematic diagram of the experiment is shown in Fig. 35. A Q-switched ruby laser beam is incident on the film. This beam is absorbed within the optical skin depth of the film which is about 0.6 μm at 6900 wavelength. The hole-electron pairs formed diffuse into the film and cause absorption in the IR beam which is launched through a Ge coupling prism and coupled out through a second Ge prism some 5 mm away. From an analysis of the waveguide propagation properties of the film it was possible to extract the change in complex propagation constant $\beta = (n + ik)\, \omega/c$ in $\exp i(\omega t - \beta x)$ as a function of time as shown in Fig. 36 and as a function of laser pulse intensity (not shown). The changes in the real $\Delta\beta'$ and imaginary $\Delta\beta''$ parts of β are seen to increase rapidly at first, peak and then decay more slowly. This slow decay is more pronounced in $\Delta\beta'$ than $\Delta\beta''$. With the assumption $\Delta p = \Delta n$ and $\Delta p/\tau_p = \Delta n/\tau_n$, where $\Delta p, \Delta n$ are the local excess hole and free-electron densities and τ_p, τ_e are lifetimes of these carriers, the ambipolar diffusion coefficient D and the ambipolar mobility μ become $D = (n + p)D_p D_n (nD_n + pD_p)$ and $\mu = (n-p)\mu_p\mu_n/(n\mu_n + p\mu_p)$ where D_p and D_n are the hole and free-electron diffusion coefficients, μ_p and μ_n are the hole and free-electron mobilities, and n and p are the local concentration of holes and electrons. The dielectric function for an N-GaAs sample becomes

$$\varepsilon = \varepsilon_\infty [1 - \frac{4\pi e^2 (n_o + \Delta n)}{m_e \varepsilon_\infty \omega (\omega - i\gamma_e)} \quad \frac{4\pi e^2 \Delta p}{m_p \varepsilon_\infty \omega (\omega - i\gamma_p)}] \quad (36)$$

where γ_p, γ_e are the damping constants of the holes and free electrons, n_o is the equilibrium free-carrier density, and m_e and m_p are the effective masses for electrons and holes. For γ_p, $\gamma_e \gg \omega$, ε becomes

$$\varepsilon = \varepsilon_\infty [1 - \frac{4\pi e^2 n_o}{m_e \varepsilon_\infty \omega (\omega - i\gamma_e)} \quad \frac{4\pi e^2 \Delta n}{m\omega (\omega - i\gamma_c)}] \quad (37)$$

where $1/m = 1/m_e + 1/m_p$ and $\gamma_c = \gamma_e m/m_e + \gamma_p m/m_p$. Thus, the photoexcited system can be described in terms of one type of carrier out of equilibrium

Photocarriers generated near the surface diffuse into the film a penetration depth δ before recombining. We consider the carrier density as exponential into the surface so that $\Delta n'/\delta = N_s$, where N_s is an average volume density and $\Delta n'$ is the initial surface density excited by the laser light in the optical skin depth. A calculation of $\Delta\beta'$ and $\Delta\beta''$ as a function of N_s is shown in Fig. 37 assuming four penetration depths of A, 2.0 μm; B, 2.5 μm; C, 3.0 μm and D, 3.5 μm. Curves 1-3 are for $\Delta n' = N_s \delta$ = constant for laser output powers of 1, 800 kW; 2, 1.0 MW and 3, 1.2 MW.

It is possible to extract several pieces of information from the measurements. (1) The ratio $\Delta\beta'/\Delta\beta''$ depends primarily on the collision frequency γ_c, being almost independent of δ, so that $\gamma_c = 2 \times 10^{13}$ rad/s can be obtained. (2) The dependence of $\Delta\beta'$ and $\Delta\beta''$ on laser pulse intensity (not shown) and on time after the pulse (Fig. 36) and the calculation of Fig. 37 for $\Delta\beta'$ and $\Delta\beta''$ can be used to determine the number $\Delta n'$ of electron-hole photopairs generated per cm^2 and the penetration depth δ = 2.2-3 μm. Note that this gives $N_s \approx 10^{16}$cm^{-3}. (3) The time of build-up of $\Delta\beta'$ and $\Delta\beta''$ in Fig. 36 is described by a time constant τ_1 = 25 ns, while the decay from maximum values is described by a much longer time constant $\tau_2 = \sim 10$ μs. The slow return of β to its equilibrium value is presumably due to the holes being trapped out which slows down the process of recombination. From the residual value of δ, the concentration of hole traps $N_p = 7.5 \times 10^{15}$cm^{-3} in the layer can be determined.

The advantage of this method is that the waveguide technique gives a long path length for a thin-film sample allowing one to measure free-carrier effects which would be orders of magnitude smaller by direct transmission and reflection. It would be difficult, however, to measure n_o itself because of the problems in coupling IR light into and out of the film and determining quantitative numbers.

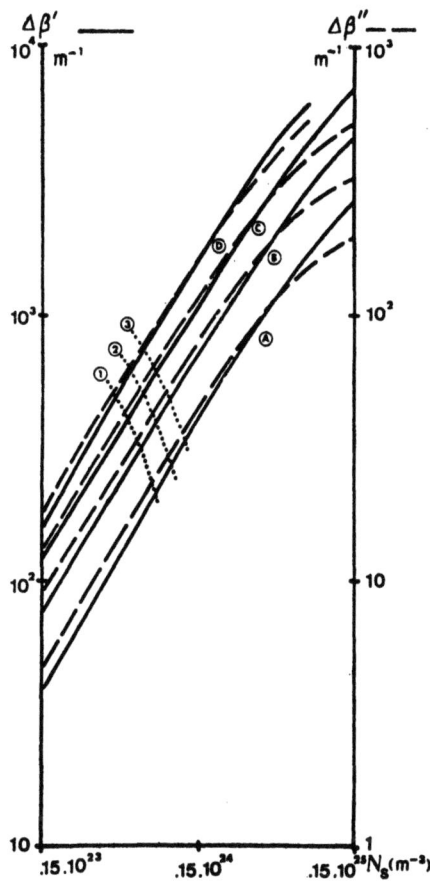

Fig. 37. Calculated changes in $\Delta\beta'$ and $\Delta\beta''$ as a function of free-carrier density at the surface of the film for different depths of penetration of the photo-induced hole-electron pairs: A, $\delta = 2.0$ μm; B, $\delta = 2.5$ μm; C, $\delta = 3.0$ μm; D, $\delta = 3.5$ μm. Curves 1, 2, 3 are for $\Delta n' = N_s \delta =$ constant. Laser output: 1, 0.8 MW; 2, 1.0 MW; 3, 1.2 MW.

B. Multiple Films of GaAs on a GaAs Substrate

The electrical and optical results for a GaAs film on a GaAs substrate are mostly in reasonable agreement, although the N_{op} often appears to be a factor of two or more less than N_{el}. We have tried to see if more complicated multilayers might be amenable to optical characterization [67]. In this case three layers consisting of active, buffer and substrate materials were studied both electrically and optically. Detailed electrical measurements and spherical drilling followed by chemical delineation yielded the

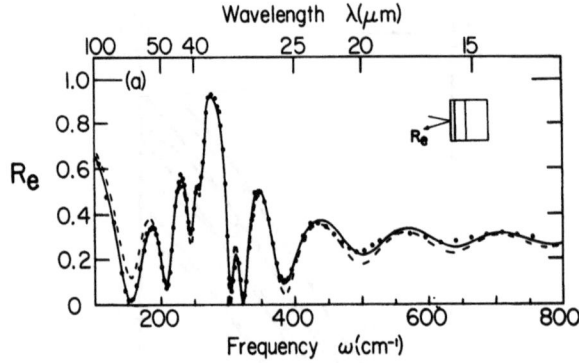

Fig. 38. The reflectance of an active N-GaAs film on an N+-GaAs buffer layer on an N+-GaAs substrate. Parameters obtained from the fitting process are listed in Table 5.

TABLE 5

Sample 1 parameters. ω_p and γ are in cm^{-1}, N is in $10^{17} cm^{-3}$ and ℓ is in μm. Under the Delineation and C vs. V column the ℓ's are from delineation measurements and N_a is from C vs. V. The values listed under Specified were supplied by the manufacturer. The parameters obtained from the three- and two-media fits are given by the first and second Infrared columns, respectively.

Parameter	Infrared (3-layer fit)	Infrared (2-layer fit)	Delineation & CV	Specified by vendor
ω_{ps} (cm^{-1})	315 ± 5	326 ± 14		
N_s $(10^{17} cm^{-3})$	9.1 ± .3	9.8 ± 1.1		10 - 22
γ_s (cm^{-1})	68 ± 11	81 + 39		
ω_{pb}	180 ± 2	--		
N_b	3.25 ± .05	--		15
γ_b	40 ± 4	--		
ℓ_b (μm)	8.71 ± 0.28	0	7.8	7.65
ω_{pa}	105 ± 20	164 ± 3		
N_a	1.0 ± 0.4	2.4 ± 0.1	1.0	1.0
γ_a	75 ± 34	47 ± 7		
ℓ_a	1.31 ± 0.27	10.06 + 0.16	1.4	1.45

active layer N_a and the layer thickness ℓ_a. The analysis of IR reflectance provided the carrier densities and thicknesses of the three layers. Some reasonable values of mobility were also obtained although mobilities are hard to determine for thin layers. A typical result is shown in Fig. 38 for sample 1. The solid curve is a fit with three layers, while the dashed curve is a fit with only two layers. It is not possible to fit the data with two layers (including substrate). Good results are obtained in most cases between those parameters specified by the supplier, those determined by delineation and CV measurements and those determined by IR reflectance measurements. Results for sample 1 are given in Table 5 for two-layer and three-layer fits.

C. Graded Free-Carrier Plasmas

The graded free-carrier plasma has been widely discussed in the context of radiowave propagation through various layers of the ionosphere [68], and gaseous plasmas [69]. Some techniques have been developed to determine analytical forms for R and T. This has also been done for semiconductor materials [70-72] in which it is often possible to grade a free-carrier density by diffusion doping and ion implantation. Basically, one requires a solution for a spatially varying index of refraction. Computationally, it is easier to model a graded index of refraction with a multilayer model such as developed by Wolter [1]. Then one can calculate R and T by dividing the sample up into a number of thin, uniform layers, each described by a slightly different n and k, which approximate the desired grading of the complex index of refraction. Then the number of layers can be increased until R and T do not change significantly. We illustrate how the reflectance of a sample of GaAs might change for different forms of grading shown in Fig. 39. For a thin film of high carrier density on a semi-insulating substrate, we have assumed uniform density, and two linearly graded densities keeping the graded thickness $\ell = 4$ μm and keeping the sheet density constant at $4 \times 10^{13} cm^{-2}$. The peak densities for the three cases shown in Fig 39 are $N = 1 \times 10^{17} cm^{-2}$ and $N = 2 \times 10^{17} cm^{-3}$. Differences occur in the reflectances which are uncharacteristic of a uniform density. Even for a sheet density a couple of factors-of-two smaller, the differences would be noticed by a trained eye. Gradations which occur in depletion layers associated with Schottky barriers are usually too narrow to give easily measurable reflectance effects. Tennant, et. al. [73] were able to analyze a film of p - $Pb_{0.82} Sn_{0.18}$ Te on a p^+ - $Pb_{0.8} Sn_{0.2}$ Te substrate and infer some gradation of carrier density at the interface from the behavior of interference fringes in the far IR.

In a similar vein, Amirtharaj, et. al. [74] studied the case of an InAs film on a semi-insulating GaAs substrate. In the optical-

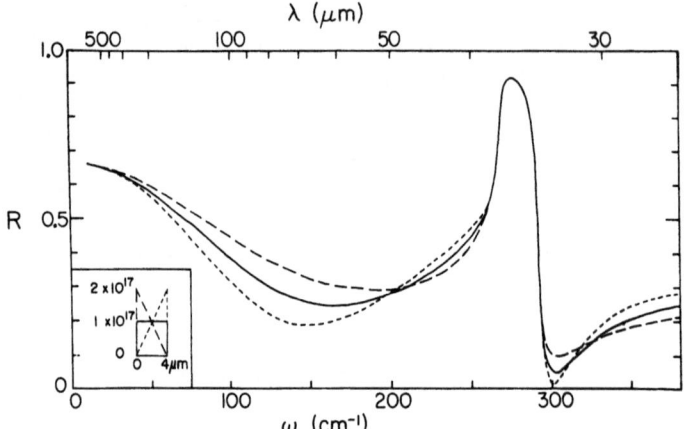

Fig. 39. The calculated reflectance of a semi-infinite, semi-insulating GaAs sample with a 4 μm thick uniform carrier density $N = 1 \times 10^{17} \text{cm}^{-3}$ (solid curve). The carrier density is graded two different ways (dashed curve, dotted curve) as indicated in the inset.

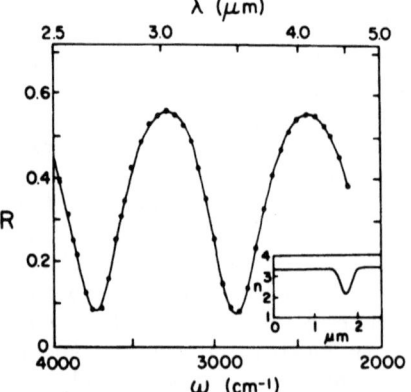

Fig. 40. Interference fringes due to a buried Si_3N_4 layer in a Si substrate. The index of refraction profile obtained from the fit is indicated in the inset.

phonon region they could not fit the reststrahlen spectrum assuming two uniform layers, suggesting that some alloying had occurred as the InAs film grew forming a transition layer of $In_xGa_{1-x}As$. Free carriers were not pertinent in this experiment.

D. Ion-Implanted Layers

1. *Interference-Fringe Effects*. An ion-implanted layer provides a graded plasma of a Gaussian shape. A detailed analysis of R would require a multilayer model to approximate the distribution in N. Horowitz [75] has shown that an implanted layer can be treated as a uniform layer to a good approximation and has obtained layer thickness, free-carrier density and mobility for N/N^+ samples. Results are very similar to the case of an N/N^+ epitaxial film as discussed in Section IIIA-1. The samples consisted of a substrate with $N = 2.7 \times 10^{18} cm^{-3}$ and $\mu = 1400$ cm^2/Vs as determined from an analysis of R before implantation. The sample was then implanted with 2 MeV oxygen ions (fluence of $8 \times 10^{12} cm^{-2}$) but was not annealed. The effect was to reduce the carrier density in the implanted region. Analysis of R in this case gave $\ell = 2.6$ µm, $N = 5 \times 10^{16} cm^{-3}$ and $\mu = 150$ cm^2/Vs. Another sample was implanted with oxygen ions (fluence of $2 \times 10^{14} cm^{-2}$) and annealed in an rf-sputtered, encapsulating Si_3N_4 film at 750°C for 30 minutes. An analysis of R gave $\ell = 2.5$ µm, $N = 1.15 \times 10^{18} cm^{-3}$ and $\mu = 1700$ cm^2/Vs.

The optical effects of a heavily implanted nitrogen layer in Si have been investigated by Hubler, et. al. [76]. In this case the Si substrates were held at 700°C during implantation with 0.67-3.17 MeV nitrogen ions. Fluences of 2.5×10^{17}- 1.55×10^{18} ions cm^{-2} were used. The nitrogen produced Si_3N_4, both crystalline and amorphous forms, but no significant numbers of free carriers. The buried layer of Si_3N_4 produced a Gaussian distribution in the index of refraction which, in turn, produced interference fringes in the near IR. The fringes could be analyzed to yield the distribution of the index of refraction, its width, and its depth into the surface. Results are shown in Fig. 40 where the fringes are fit with both a Gaussian-layer model as indicated in the inset. A uniform-layer model does not fit the data nearly as well (not shown). The Gaussian-layer fit is verified by Rutherford-backscattering profiling measurements.

It is also possible to implant Si with Si and P ions to obtain buried damage layers [77-79]. In this case the variation in index of refraction is due to an amorphous Si layer rather than a new chemical compound. In this example [79] the implantation was carried out so that a uniform disordered region was produced whose index was graded as shown in the inset of Fig. 41. The sloping region was assumed to be described by a half Gaussian of the form $n = n_S - (n_S - n_D) \exp[-(z-z(1))^2/2\sigma^2]$. The interference-fringe

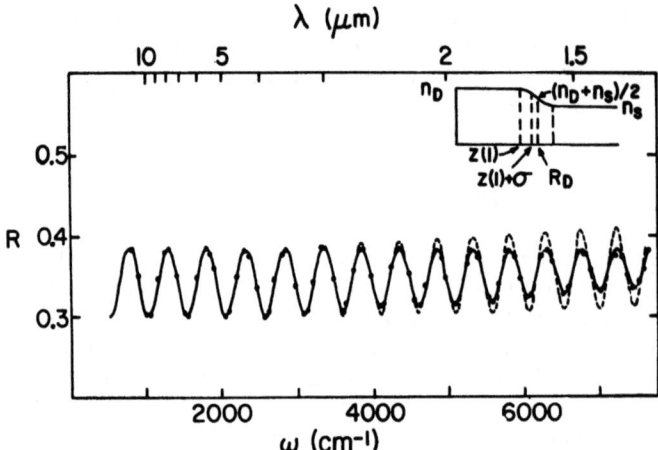

Fig. 41. Interference fringes due to an implanted region in Si indicated in the inset. The decreasing fringe amplitude to high frequency (solid curve) indicates a Gaussian gradation of carrier density. The dashed curve is for an abrupt junction.

system is shown in Fig. 41 where the experimental points are fitted with $\sigma = 0$ and $\sigma = 0.053$. The distance $z(1)$ was $= 2.46$ µm, $n_D = 3.87$ at 4000 cm^{-1}, $n_S = 3.418$, $R_D = 2.52$ µm. Thus, the width of the transition region is 0.12 µm. The damping of the fringes as the frequency increases (wavelength decreases) is the hint that a Gaussian gradation is present rather than an abrupt stepped gradation. The long wavelengths are not too sensitive to the detailed shape of the transition, but the shorter wavelengths are. Another implanted sample is indicated in Fig. 42 before (dashed curve) and after (solid curve) annealing. The anneal produced a high carrier density in which the free-carrier plasma frequency is about 2000 cm^{-1}. The interference effects now make it possible for R to approach unity [78].

2. <u>Lattice-damage Effects in the IR</u>. The lattice damage due to ion implantation of GaAs has been studied by Kachare, et. al. [80,81]. In this case a number of semi-insulating samples were ion implanted with nitrogen ions at 1-3 MeV with fluences ranging from 3.3×10^{13} to 2.0×10^{17} cm^{-2}. Reststrahlen spectra were measured as illustrated in Fig. 43 showing drastic changes in reflectance. A damage-layer model was used which comprised one or two damage layers and the optical parameters n and k of each layer were adjusted to fit the experimental spectra. From such an analysis

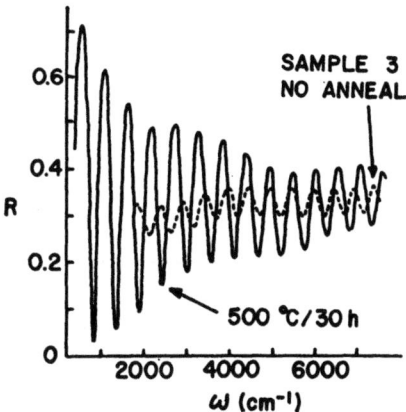

Fig. 42. Interference fringes due to an unannealed implanted layer (dashed curve) and annealed layer (solid curve) in Si producing free carriers in the implanted layer. The plasma frequency is 2000 cm^{-1}.

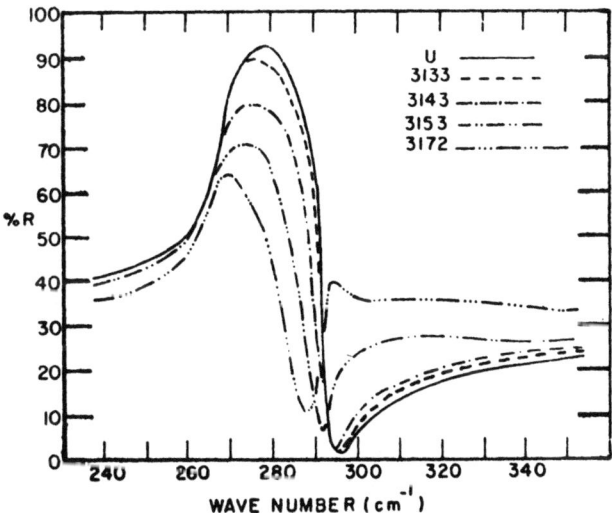

Fig. 43. Reststrahlen spectra of nitrogen-implanted GaAs for a series of fluences. Lattice damage is apparent and can be fitted with one or two damage layers.

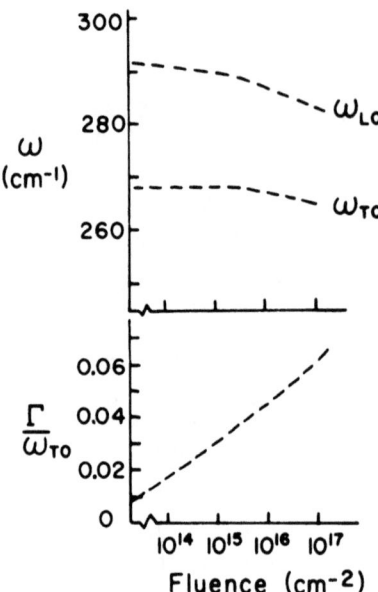

Fig. 44. The variation of ω_{TO}, ω_{LO} and Γ with fluence damage.

it was found that ω_{TO} and ω_{LO} decreased in frequency and Γ increased as shown in Fig. 44. Upon annealing, the lattice damage was removed.

For low fluences, the lattice damage is presumably smaller. To determine the changes in the reststrahlen region, it is necessary to measure the ratioed reflectance of an implanted and a reference sample [82]. This is shown in Fig. 45 for GaAs implanted with S at 200 keV with a fluence of $1 \times 10^{13} \text{cm}^{-2}$. Examining the data of Fig. 44 we can expect ω_{LO} and Γ to be changed slightly. Therefore, we have decreased ω_{LO} by 1 cm^{-1} and increased Γ to 16 cm^{-1} to produce the solid calculated curve in Fig. 45. The dotted curve is obtained if only ω_{LO} is reduced by 1 cm^{-1} and Γ is kept fixed at 2.3 cm^{-1}. Thus, the observed effects seem to require a significant increase in Γ, much larger than suggested by Fig: 44. Here, uniform implantation was assumed from the surface to 2500 Å. Upon low-temperature annealing, this damage was removed and the ratioed reflectance was a straight line at unity. High-temperature annealing activated the free carriers.

3. <u>Lattice-Damage Effects in the Visible</u>. When GaAs is implanted with berylium, a lattice-damage distribution is produced which is somewhat shallower than the Be distribution. This is

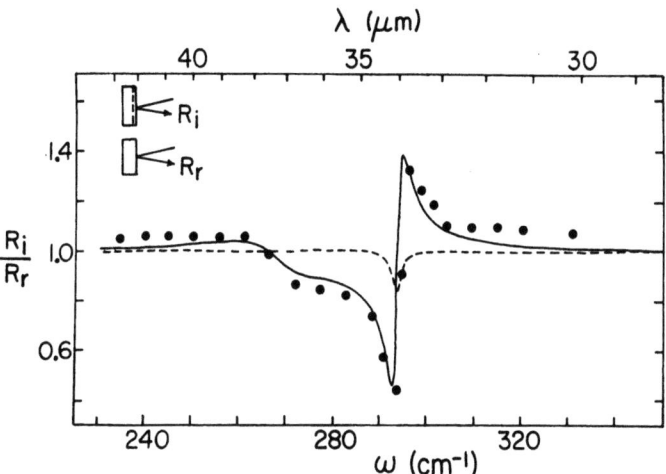

Fig. 45. The ratio spectrum of a low-fluence implanted GaAs sample showing changes in the reststrahlen spectrum. The dashed curve is obtained if only ω_{LO} is decreased to 290.5 cm^{-1} and Γ is kept at 2.3 cm^{-1}. The solid curve is obtained assuming Γ also increases to 16 cm^{-1} due to damage.

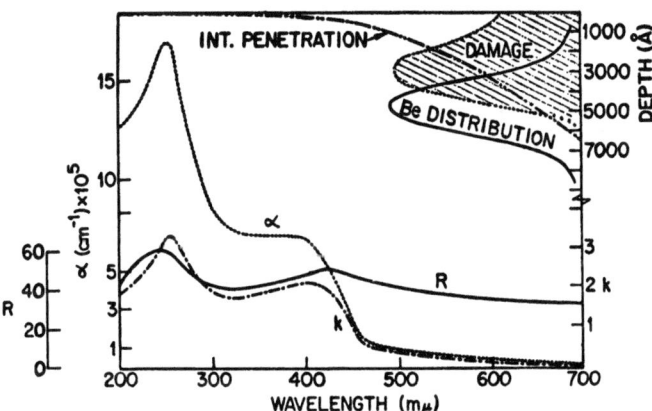

Fig. 46. The optical constants of GaAs in the visible region along with intensity penetration depth and relectance. The regions of lattice damage and Be density are also indicated for implanted Be.

shown schematically in Fig. 46. The optical constants n and k and
the reflectance R of GaAs are also shown here, along with the intensity
penetration depth of visible light. Lattice damage decreases the
reflectance peak near 4200 Å due to the E_1 and $E_1 + \Delta$ interband
transitions. Since the light penetration depth here is about
300 Å, the reflected light can act as a probe of the damaged region.
Molnar [83] has measured this change in reflectance for various
implantation depths by a series of etch-back experiments and was
able to locate the peak in the damage distribution and determine
its shape. While this experiment was destructive, it might be
possible to model the variation in n and k in a Gaussian distribution, so as to be able to account for the reflectance changes for
much shallower implants. Also, ellipsometry measurements would
yield both n and k as discussed in Section IVA.

IV. MORE OPTICAL CHARACTERIZATION TECHNIQUES

A. Ellipsometry

Ellipsometry is widely used to characterize a thin film on a
substrate (i.e., SiO_2 on Si and Si_3N on Si) in the visible spectral
region [84,85]. In such cases the substrate optical constants are
assumed to be known, the film is transparent, and only n and ℓ need
to be determined. If the film is absorbing, the difficulties are
compounded and two-angle measurements are needed. If the film
thickness is determined by other means or from a measurement at a
wavelength at which the film is transparent, then wavelength scanning
into the band-edge region can be done to determine n and k. By
such a wavelength-scanning method m and k and ℓ have been obtained
for anodic oxide on GaAs in the spectral region 2000 to 8000 Å [86].
Typically, the dispersion in n has a Sellmeier-type dependence
below the band gap of many insulator films.

Ellipsometry is not widely used in the infrared region, no
doubt because of the inferior quality of IR polarizers, phase
shifters and detectors [87]. Such measurements could be done at
10.6 μm to determine the free-carrier contribution to n and k.

The precise measurement of n and k for bulk semiconductors must
include a consideration of the natural-oxide film present in non
ultra-high vacuum (UHV) measurements [88].

Standard ellipsometry has been used to study the behavior of
the optical constants of GaP upon ion implantation with Be, Mg,
Cd and Te [89]. Substrate material was implanted with Be or Mg
to a depth of ~ 1460 Å and with Cd and Te to a depth of ~ 150 Å.
Fluences ranged from 10^{12} to 10^{15} cm^{-2}. The implanted layer had
a Gaussian distribution centered at these depths. Measurements

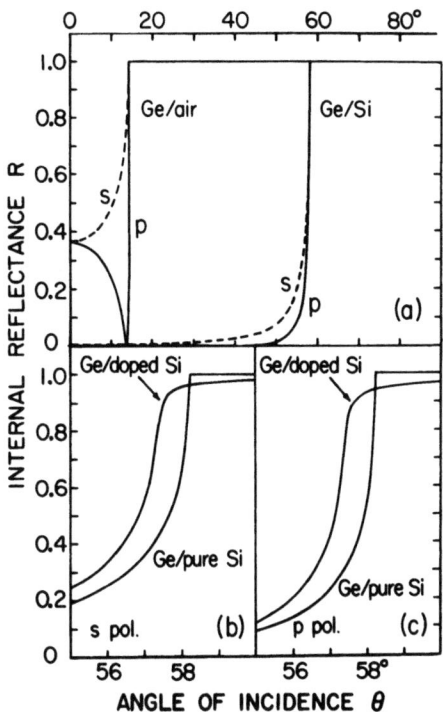

Fig. 47. (a) Calculated internal reflectance of a Ge prism facing air for s and p polarization and also facing pure Si for a wavelength of 3.39 µm. (b) Calculated internal reflectance for s polarization for a Ge prism facing pure Si and heavily doped Si. (c) Calculated internal reflectance for p polarization for a Ge prism facing pure Si and heavily doped Si.

were made at 6328 Å where the penetration depth is $\sim$500 Å. It was necessary to assume that the implanted region was uniform into the surface to a distance greater than the penetration depth of light in order to extract values of n and k. No modeling was attempted by using the known Gaussian distribution of the damage. The initial values of n and k were 3.2 and 0.06, respectively. The results indicated that n increased steadily with increasing fluence from 10^{12} to 10^{15} cm^{-2} for all four ions; for example, from 3.2 to 3.5 for Mg. For Be and Mg, k increased steadily; e.g., from 0.15 to 1.35 for Mg. However, for Cd and Te, k peaked near 10^{13} cm^{-2} fluence, decreased somewhat and then continued to increase; e.g., from 0.2 to 1.0 for Cd. Since ion implantation eventually produces amorphous material, it is likely that these changes in n and k reflect the trends in amorphous GaP itself. Upon annealing the samples, the optical constants return to their pre-implanted values. Evidently it is not possible to detect the impurity layer itself by its perturbation of the optical constants of GaP.

B. Internal Reflection

Internal-reflection spectroscopy has been used to determine optical constants since its developement in 1961 [90-92]. An example is provided by the work of Gupta [93] who measured the angular dependence of the s- and p-polarized internal reflection in a transparent Ge hemicylinder facing air in the first case and facing a Si sample in the second case. In Fig. 47 the s and p internal reflection in the near IR for Ge-air behave in typical ways illustrated in many textbooks with p polarization showing Brewsters angle near 14°. When a flat Si sample is pressed against the prism, the reflectance is changed with the critical angle ($\sin \theta_c = n_{Si}/n_{Ge}$) becoming 59°. The spectra of pure Si and doped Si near the critical angle are significantly different as illustrated in Fig. 47 for a wavelength of 3.39 μm and a free-carrier density of 5×10^{18} electrons per cm^3. The shift of the leading edge in Fig. 47 (b) and (c) is due primarily to the index of refraction while the rounding off is due to the extinction coefficient. Thus, the optical constants of an uncharacterized Si sample can be obtained by analyzing this reflectance change. Then, using a reliable model for free-carrier effects [94], one can determine the carrier density and mobility in a surface region probed by the evanescent wave. Above the critical angle at 70° the penetration depth of the evanescent wave from the Ge is about 0.9 μm. Of course, below the critical angle the light propagates across the interface through the Si. Critical-angle ellipsometry [95] seems to be another way to determine optical constants.

C. Laser Scanner for Semiconductor Devices

A laser scanner has been developed to examine semiconductor devices with a small spot size [96-98]. The system, illustrated in Fig. 48 incorporated two cw He-Ne lasers, one operating at 0.633 μm and the other at 1.15 μm. The laser beam was deflected sequentially from two mirrors oscillating in orthogonal directions to generate a raster scan, and the raster was focussed on the device. A display of the photoresponse of the sample was obtained by amplifying current variations in the power supply to the device being scanned and using this signal to moderate the intensity of the CRT display screen. The system also had a reflected-light circuit consisting of a lens and photocell which allowed one to pinpoint the regions of photoresponse in relation to topographical device features such as metal electrodes. Resolution was 1.5 μm for the 0.633 μm light and 3 μm for the 1.15 μm light.

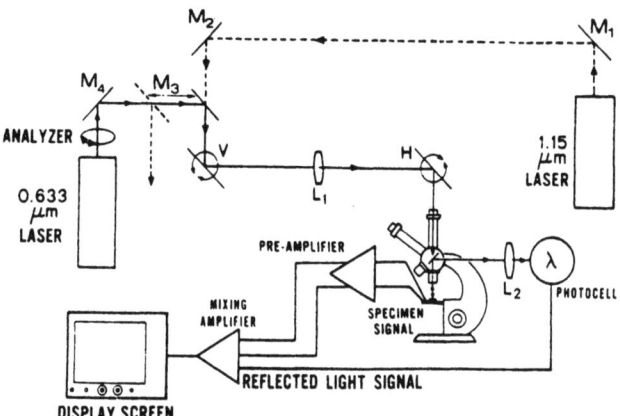

Fig. 48. Schematic diagram of laser scanner for measurement of photoconductivity of microcircuits.

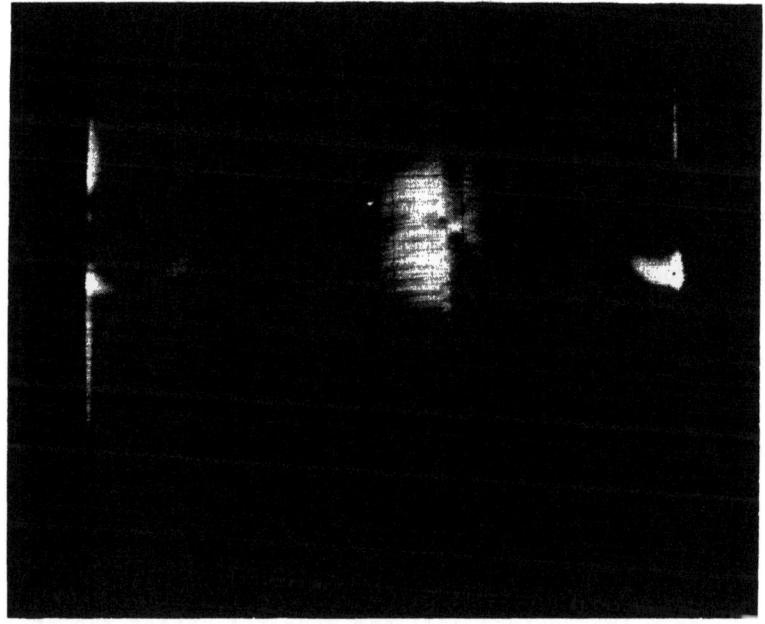

Fig. 49. Hot spot in a power transistor as revealed by bright area due to increased photoconductivity.

The 1.15 μm laser light is weakly absorbed by Si at room temperature. However, as the temperature rises, the bandgap decreases which produces an increase in hole-electron pairs. This increases the photoresponse in regions which are warmer. Fig. 49 shows a portion of a UHF power transistor designed to furnish 5 W at 1.1 GHz [97]. There are four inline cells (two shown) connected in parallel with a total active area 1.2 mm long and 0.15 mm wide. The device is an interdigitated structure with emitter fingers coming in from the right and base fingers from the left. The change in voltage across a 60 Ω collector load as the laser scans the device provides the signal to modulate the CRT display screen. For Fig. 49 the device was operated at V_{CE} = 26 V and I_C = 250 mA and showed the enhanced 1.15 μm photoresponse due to a hot spot caused by current crowding as a brighter area.

The laser scanner was also used to monitor the flow of logic in a MOS shift register [98]. The shift register was a static dual 128 bit p-MOS ion-implanted device. In order to observe the logic flow in the device, the package leads were connected as appropriate for normal circuit operation. The variations in the power supply current to the device provided the information for the display. Fig. 50 shows the display obtained when scanning about one third of one of the registers and some associated input/output circuitry. A random pattern of logical 1's and 0's can be seen which appear as displaced dark and light areas. The logic flows along the snake-like path which begins in the left-hand column at the bottom and flows up the column, then down the second column, up the third and down the fourth, where it switches the output circuitry at the bottom of the fourth column. The contrast in photoconductivity of the transistors forming the logical 1 and the logical 0 is due to the different states of saturation, so that in one case, the laser light produces a measurable change in current and in the other case does not. After a single clock pulse was injected through the appropriate inputs, all the patterns were seen to shift one position.

It was discovered that internal logic states can be selectively changed in a nondestructive manner. Logical 1's can be changed to 0's and vice versa be decreasing the laser scan raster to a sensitive region on the logic cell which one wishes to change. The laser intensity is then increased to change the data in the cell, and then returned to previous intensity. The laser raster can then be restored to its original size. This operation changes the data in only the desired cell and does not affect the state of any other cells in the register. The intense laser beam produces enough current to cause the set of transistors in the flip-flop circuit to switch to the other state, so that the subsequent photoconductivity is different.

Fig. 50. A random pattern of logical 1's and 0's appear as displaced dark and light areas in a shift register scanned by the laser beam.

D. Scanning of Wafer Inhomogeneities

1. <u>Scanning for Resistivity and Carrier Lifetime</u>. To scan wafers for inhomogeneities due to variations in resistivity, a flying-spot scanner has been developed [99]. The principle is illustrated in Fig. 51. A clean Si wafer 3 inches in diameter is mounted between a metal plate M and a fine wire mesh on a glass plate. The contacts on the mesh form a multiplicity of Schottky barriers about 1 mm apart. Visible-near IR light is modestly focussed and scanned across the sample. Photons produce hole-electron pairs which move to opposite sides of the sample under bias. If there is a variation in resistivity, the depletion depths will vary producing a varying photovoltage as the scanning spot moves. Alternately, if there are lifetime variations spatially, the Schottky barriers are all considered of equal depth and signal differences are due to the greater equilibrium electron-hole populations within a region of longer lifetime. The voltage variations across the wafer package are then amplified and displayed on a CRT screen. To evaluate the signal on the scope screen one has to measure the resistance across the sample under bias but without light injection. This resistance corresponds to the internal resistance of the Schottky barriers under reverse bias. A normal 4 Ωcm wafer yields a 10^4 Ω resistance with 10 V applied. The variation in voltage at the sample contacts can be translated into a resistance change as indicated in Fig. 52 where isoresistance lines are plotted. The contour interval is 1%. The dark line represents a surface resistance of 20.29 Ω/sq.

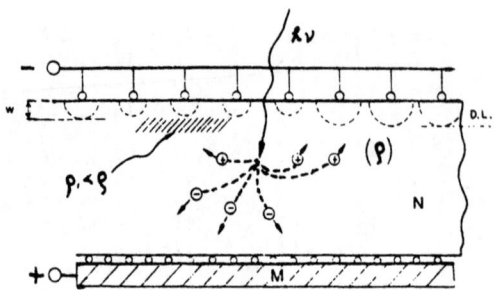

Fig. 51. Schematic diagram of a method of measuring resistance variations in a Si wafer by pressing a metal grid against the Si surface to produce an array of Schottky barriers. Variations in resistance produce variations in depletion depths, detectable in photovoltage experiments.

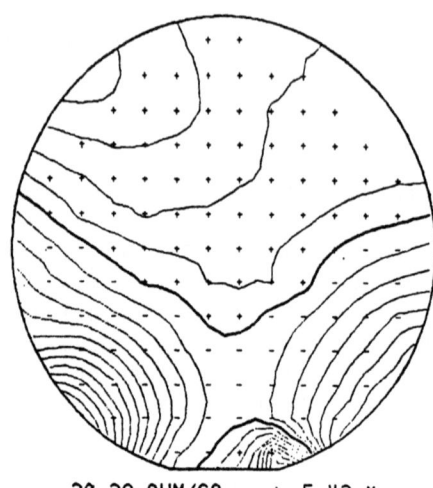

20.29 OHM/SQ.   ± 5.43 %

Fig. 52. Inhomogeneity of resistance in a Si wafer. The contour intervals represent 1% changes in resistance.

The spatial variation of carrier lifetime has been measured using a photovoltage approach [100] similar to the one discussed above. Light from a quartz-halogen lamp was focussed to ∼10 µm diameter on a sample placed in an electrolytic cell. The electrolyte allowed a high voltage to be established across the Schottky depletion region. The photovoltage signal ϕ_{pv} in first approximation is given by

$$\phi_{pv} = \frac{KI\lambda\alpha L}{1 + \alpha L} \qquad (38)$$

where I is the light intensity, λ is the wavelength, α is the

the absorption coefficient, L is the ambipolar diffusion length and K is a constant depending on the material. Also,

$$L = \frac{KI\mu\tau_b}{e}^{1/2} \quad (39)$$

so that it is possible to determine the bulk recombination time τ_b of the generated carriers if the mobility μ is known. Equation (38) can be rewritten in the form

$$\frac{1}{\alpha} = \frac{KI\lambda L}{\phi_{pv}} - L \quad (40)$$

A measurement of ϕ_{pv} as a function of λ and thus α yields a straight line with the $1/\alpha$ intercept given by L. For a Si wafer immersed in a Na_2SO_4 electrolyte, it was found that L = 364 µm at a point where the surface was smooth and L = 174 µm under a scratch put on the surface. This demonstrated how mechanical imperfections shortened the diffusion length. A commercial Si MOSFET was also proved to map out the carrier diffusion length between the gate and the source-drain electrodes.

Another variation of probing the surface photovoltage has been given by Philbrick and DiStefano [101]. In this case the Si wafer is coated with a thin SiO_2 film, and the capacitance of a small capacitor constructed near one edge of the wafer is monitored as a laser beam is scanned across the sample.

E. Thermal Imaging of Microcircuits

The location of hot spots in a microcircuit sample has been accomplished with the use of an infrared microscope with spatial resolution of ~ 10 µm under best conditions [103]. Such instruments are commercially available [104,105]. Typically, light emitted in the spectral range 1.8 - 5.5 µm is collected from a small spot and measured by an InSb photovoltaic detector. For sample temperatures up to about 200 C, $Hg_{1-x}Cd_xTe$ detectors are somewhat better in sensitivity [106]. The light is chopped, so that a signal is obtained alternately from the sample and from a standard black body at room temperature or 77 K. A visible microscope is built into the instrument, so that the spot of emission on the microcicuit can be identified. Samples can be scanned with oscillating mirrors or by moving the sample stage. Since such a microscope collects all wavelengths in the near IR region, it is vital that the emissivity of the material being examined be known since the emissivity of Si, SiO_2, Al, etc. are different [106]. Indeed, some of the layers are somewhat transparent and more detailed corrections involving multiple-reflection emittance must be considered in order to assign an absolute temperature [107]. This has sometimes been averted by coating the entire device with a blackening, so as to make the emissivity close to unity [103]. Alternately, the device can be electrically disconnected and then heated up uniformly to the range of operating temperatures [104]. Then a scan produces an emittance calibration which can be compared with the scan when the device is operated electrically. Operating temperatures of devices can be in the range of 40-90°C and temperature differences of $\sim 1°C$ are resolved. Thus, in scanning numerous identical elements, it is easy to see that one of them is operating a few degrees hotter than the others. Typically, a Ge filter is used to block recombination radiation from Si during electrical operation. Often, hot spots are traceable to mechanical failures such as poor heat sinks or bond failure.

F. Free-Carrier Raman Scattering

It has been shown [108] that the position and shape of the TO Raman line of Si is dependent on the free-carrier density. This effect occurs at high doping levels of the order of $10^{20} cm^{-3}$. The effect has been explained in terms of an interaction of the discrete phonon states and the continuum states due to intervalence and interconduction-band electronic states. Thus the effect occurs in both n- and p-type Si. The theory involves two asymmetry parameters q and Γ, where q is the ratio of the transition amplitude of the discrete state to that of the continuum states and Γ is a measure of the strength of the coupling between the two kinds of states. Γ is proportional to the free-carrier density. The phonon resonance shift to ω_T' from ω_T is given by $\omega_T' - \omega_T = \Delta\omega + \Gamma/q$ where $\Delta\omega = \Gamma\Delta'/\Delta''$ and Δ' and Δ'' are the real and imaginary parts of the electronic susceptibility. Both Δ' and Δ'' also depend on free-carrier density.

Boron was implanted into Si with 5×10^{15} ions cm^{-2} and annealed to 1000°C [109]. Raman backscattering experiments were performed with the 4880 Å line of an argon laser. The light penetration depth was ~ 0.5 μm, so the average free-carrier density of this region was obtained. While no detailed calculations were made, the measurement of the shift and width of the line demonstrated that $\omega_T' - \omega_T$ was a roughly linear function of N. For example, for p-type Si, $\omega_T' - \omega_T$ was 2 cm^{-1} at $10^{20} cm^{-3}$ and 11 cm^{-1} at $4 \times 10^{20} cm^{-3}$.

In the same measurement, it was possible to see the local mode of substitutional B after annealing but not before. In fact, as the annealing temperature was raised toward 1000°C, the free-hole density as determined from the frequency shift was observed to rise in the same manner as the intensity of the B local mode (fraction of B atoms in substitutional positions), indicating that the free-hole density corresponded to the boron atom density. This is illustrated in Fig. 53.

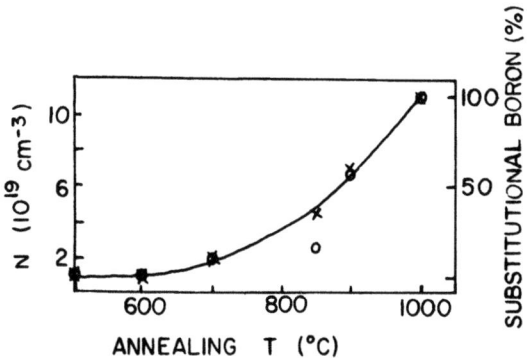

Fig. 53. The free-hole density due to B impurities in Si and the intensity of the B local mode as a function of annealing temperature. There is a one to one correspondence of the increase of free-hole density and local-mode intensity.

G. Magnetooptical Studies of a Si MOSFET

In a p-type Si surface covered by a thin insulator (SiO_2) and a metal film (source, gate and drain), it is possible to apply large electric fields which can bend the valence and conduction bands downward so that a thin (tens of Angstroms) surface region of the Si inverts to electrons of high mobility. This two-dimensional plasma has been of great theoretical and experimental interest. Two main optical experiments have been the study of (1) the transitions of electrons between the electric subband levels formed in the potential well of the inversion layer (polarization of light perpendicular to the surface) [110,111] and (2) cyclotron resonance of these electrons confined to the plane of the plasma layer (polarization of light parallel to the surface) [112-114]. Such experiments are confined to the far infrared-submillimeter spectral region.

V. CHARACTERIZATION AT THE BAND EDGE

The band-edge region can also be exploited to characterize a semiconductor. Free carriers produce the Burstein-Moss effect; the band gap is sensitive to the inhomogeneities in alloy composition; photo and cathodoluminescence yield impurity-level information.

A. Burstein-Moss Effect

Band-edge absorption is described by Eq. (13) where the parameter

$$C = \frac{2e^2}{m_o^2 c} \left(\frac{2m^*_{cv}}{h^2}\right)^{3/2} |P_{cv}|^2 \qquad (41)$$

is fixed for a given band structure [12]. Here, m_o is the free electron mass, $(m^*_{cv})^{-1} = (m^*_c)^{-1} + (m^*_v)^{-1}$ is a reduced effective mass given in terms of the conduction and valence band effective masses, respectively, and $|P_{cv}|^2$ is a momentum matrix element describing the transition probability. In Fig. 3 (a), α is shown for GaAs.

When free carriers are added to the conduction band, some interband transitions are blocked and the interband absorption coefficient α_N can be written as

$$n_N \alpha_N = n_o \alpha_o \{1 + \exp[(E_F - E)/kT]\}^{-1} \qquad (42)$$

where at absolute zero $E_f = (h^2/2m^*_c)(3N/8\pi)^{2/3}$ is the Fermi energy for a total carrier density N, E is the energy in the conduction band and $n_o \alpha_o$ is given by Eq. (13). Both E_F and E are measured from the band edge. Since $h\omega - E_g = (1 + m^*_c/m^*_v)E$, Eq. (42) may be rewritten as

$$n_N \alpha_N = n_o \alpha_o \left(1 + \exp \frac{E_F' - (h\omega - E_g)}{\{1 + (m^*_c/m^*_v)\} kT}\right)^{-1} \qquad (43)$$

where $E_F' = [1 - (m^*_c/m^*_v)E_F]$ is the photon energy relative to the band gap, at which the absorption is just half the value of the pure crystal.

The above formulas apply for the case of degeneracy [12,13]. For GaAs we can estimate whether the Fermi level is in the conduction band by calculating a carrier density vs. temperature plot of $\eta = E_F/kT$ (Fig. 54), so that we can determine E_F to see if the sample is degenerate enough [115]. For $\eta = 0$, E_F is at the bottom of the conduction band; for positive η, E_F is in the conduction band; for negative η (not shown), E_F is in the band gap. We can see in Fig. 54 that for 10^{17} N-GaAs, η is less than zero at 300 K; whereas at 77 K, η is about unity and at 4 K, η is much greater than 10. Therefore, optical experiments to utilize the Burstein-Moss effect as described by Eq. (43) would need to be done at low temperatures in low-carrier-density GaAs.

The effects of carrier density and temperature on α_N are illustrated in Fig. 55. In Fig. 55(a) α_N is given for various carrier

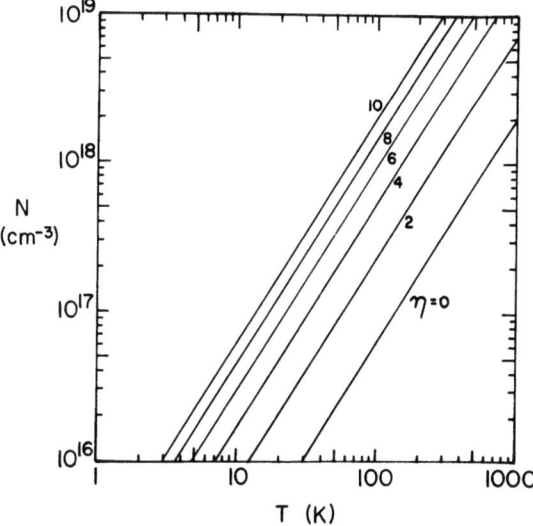

Fig. 54. The carrier density vs. temperature for various values of η for GaAs parameters. The upper left corner is the region of high degeneracy. When η = 0, the Fermi level is just at the conduction-band minimum.

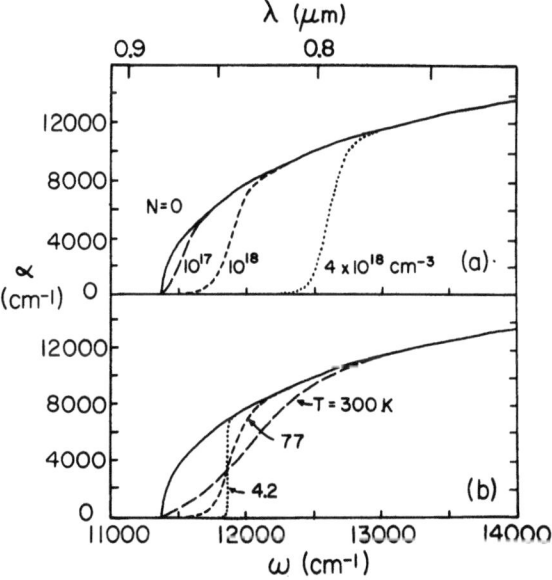

Fig. 55. (a) The calculated room-temperature Burstein-Moss effect in a GaAs-like semiconductor for various carrier densities. (b) The calculated band-edge absorption coefficient for a given carrier density of $1 \times 10^{18} cm^{-3}$ as a function of temperature.

densities at 77 K. Parabolic bands with $m^*_c = 0.066\ m_0$ and $m^*_v = 0.7\ m_0$ are assumed. While the free-carrier absorption does increase slightly in the vicinity of the band edge (this is ignored), the band-edge absorption decreases greatly. The band edge for a carrier density of $1 \times 10^{18} cm^{-3}$ as a function of temperature is shown in Fig. 55 (b). There would be corresponding variations in n also (not shown).

If the substrate is a larger band-gap material such as GaAs, then a film of $GaAs_xSb_{1-x}$ could be analyzed in transmission and compared with a pure film to determine the free-carrier density. For $N < 10^{17} cm^{-3}$ this difference becomes very small and low temperatures would be needed to sharpen the effect. An experiment along these lines has been done by Leheny and Shah [116]. A ~ 0.5 μm GaAs film was supported by two layers of $Al_{0.12}Ga_{0.88}As$ each ~ 2 μm thick. The GaAs substrate was etched away to provide a window. They used one tunable, pulsed dye laser to pump the GaAs film to produce a high electron density in the conduction band and a second tunable dye laser to measure the band-edge absorption coefficient. In this manner they could see the Burstein-Moss effect as a function of pump intensity and also observed the subsequent luminescence when the electrons returned to the valence band. A peak power of $\sim 5 \times 10^5 Wcm^{-2}$ produced an electron density somewhat greater than $\sim 10^{17} cm^{-3}$. At 2 K the exciton structure was seen and the actual shape of the absorption edge did not have a simple square-root dependence. A more correct form of α_0 [117] was used in the analysis by Eq. (43).

While a thin film of GaAs on a GaAs substrate cannot be studied near the band edge in transmission, in principle the reflectance would be measured to determine the peak in n and thus the carrier density. Ideally, ellipsometry measurements with a tunable laser could give by both n and k.

B. Homogeneity in Alloy Semiconductors

1. A Film of $GaAs_{1-x}Sb_x$. In heterojunction devices where bandgap variation is achieved by control of the alloy composition, the grading of material properties may extend a micrometer or more from the metallurgical interface. If a film is graded, so that the band gap varies with distance, the band-gap absorption would look smeared out compared to a film of one composition of the same thickness, as indicated in Fig. 56. Here, the form of the band-gap absorption is assumed the same for all alloys of $GaAs_{1-x}Sb_x$ with only the band gap varying as $E_g = 1.41 - 1.72\ x + 1.02\ x^2$. Then, an effective absorption coefficient α_{eff} can be calculated which describes the absorption properties. Thus, if the bandgap is linearly graded from $E_g(GaAs)$ to $E_g'(GaAs_{0.9}Sb_{0.1})$ over a film thickness of 2 μm, the absorption coefficient has a weaker frequency

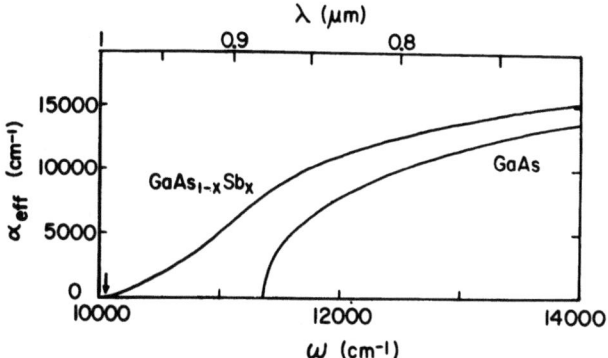

Fig. 56. The absorption coefficient for a linear gradation of band gap in a $GaAs_{1-x}Sb_x$-like semiconductor. A film 2 μm thick linearly graded from $x = 0$ to $x = 0.1$ would have the effective absorption coefficient shown. E_g for $x = 0.1$ is indicated by the arrow.

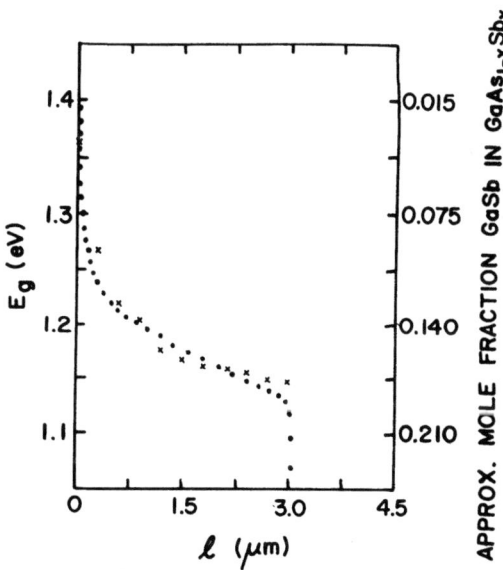

Fig. 57. Analysis of the gradation in a $GaAs_{1-x}Sb_x$ film on a GaAs substrate from transmission experiments. The crosses indicate the profile from x-ray emission analysis across the cleaved edge.

dependence than any individual alloy. Hall, et. al. [118] have measured the transmission of a 3 µm thick GaAs$_{1-x}$Sb$_x$ film with x = 0.15 on a GaAs substrate. With the assumption that the form of the band-edge absorption is the same for GaAs and the film and that only the gap shifts to lower energy for the alloy, they were able to analyze the absorption edge to obtain the composition profile shown in Fig. 57. The points represent the band-gap profile computed from the optical data. The crosses represent the profile obtained using x-ray emission across the cleaved edge of the sample excited by the electron beam in a scanning electron microscope.

2. <u>AlAs-GaAs Monolayers</u>. By molecular-beam epitaxy it is possible to grow alternate monolayers of GaAs and AlAs on a GaAs substrate, so as to construct an artificial, periodic crystal [119]. The average monolayer thickness was 2.83 Å, and up to 6000 periods were laid down. The band edge of AlAs is 2.43 eV, while that of GaAs is 1.56 eV. The band-edge region of the film was measured by etching away the substrate over a region of about 4.5 mm^2 leaving only the film. The direct band gap of the AlAs-GaAs monolayers is consistently higher in energy than the gap of the random alloy Al$_x$Ga$_{1-x}$As having comparable average Al content. This agrees with the premise that the band gaps of the ordered monolayer structure should not exhibit the downward bowing of the band gap which is present in the alloy due to disorder.

The index of refraction was obtained from an analysis of the Fabry-Perot fringes seen in transmission below the band gap. A birefringence was measured when the fringes were measured when the fringes were measured with light plane polarized parallel and perpendicular to the plane of the monolayers, indicating strain effects due to contraction of the GaAs substrate with respect to the monolayers during cooling.

C. Electroreflectance in Alloy Semiconductor Films

The electroreflectance effect has been used to determine band-gap information for about 13 years [25,120]. A voltage is applied between the back of a sample and a transparent conducting electrode on the front surface. A depletion region is formed in which a high electric field can be sustained to produce the Franz-Keldysh effect, a slight shift in the band edge. This results in a change in n and k producing a reflection feature which locates the band gap.

1. <u>Stoichiometry in Hg$_{1-x}$Cd$_x$Te</u>. The variation of the E_0, E_1 and E_0' structures of Hg$_{1-x}$Cd$_x$Te are shown in Fig. 58. Vanier, et. al. [121], have measured electroreflectance to determine the variation of the positions of the E_1 gap as a function of spatial position of a probing light beam focussed to a spot size of about

150 μm. The samples were insulated with paraffin wax except for the front surface and mounted on a movable stage in an optical cell with an electrolyte and a platinum electrode. The electrolyte was 1 part concentrated HNO_3 to 5000 parts of methanol by volume.

Spectra were taken in the region of the E_1 gap and the shifts in energy of the spectral features were correlated with the known composition dependence of the E_1 peak to determine the composition change Δx. For samples in the range $x = 0.2-0.3$, a 7.5 meV shift of the peak corresponded to $\Delta x = 0.01$. A contour map is shown in Fig. 59, where the nominal value of x is 0.31 and the intervals are $\Delta x = 0.005$.

It is likely that with a tunable laser, the spatial resolution could be greatly improved making this technique very useful for determining alloy homogeneity.

2. $GaAs_{1-x}Sb_x$ Multilayers. A variation of these technique has been applied by Bottka and Hills [122] to study epitaxial films of $GaAs_{1-x}Sb_x$ on GaAs substrates. It is illustrated in Fig. 60. Three

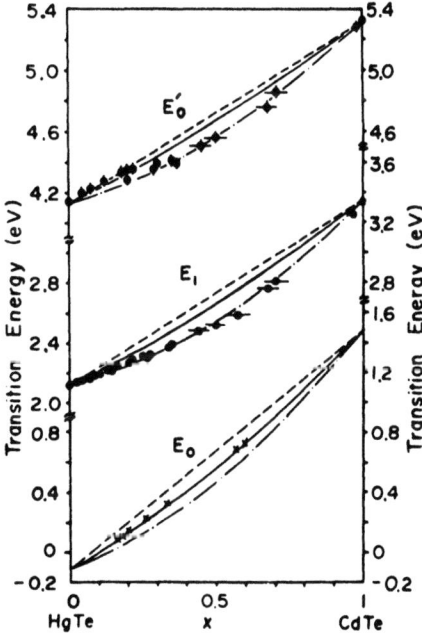

Fig. 58. The variation of the E_0, E_1 and E_0' band gaps of $Hg_{1-x}Cd_xTe$ as a function of x.

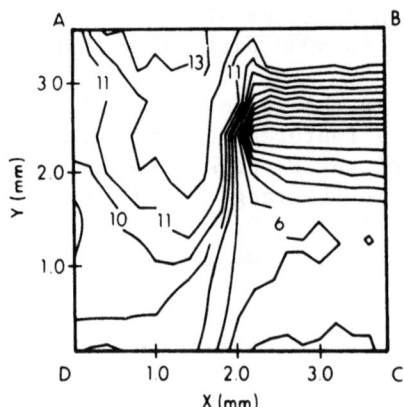

Fig. 59. A contour map of the carrier density inhomogeneity in $Hg_{0.69}Cd_{0.31}Te$. The intervals are $\Delta x = 0.005$.

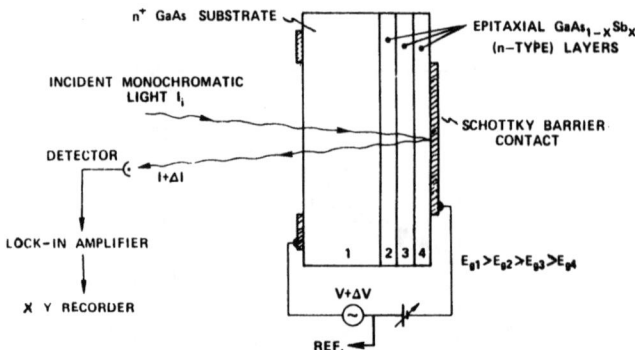

Fig. 60. Schematic diagram of electro-absorption experiment for $GaAs_{1-x}Sb_x$ multilayers.

layers of n-type $GaAs_{1-x}Sb_x$ of a different alloy composition are grown on an N^+ GaAs substrate. One metal electrode forms a Schottky-barrier contact, so that a dc and a square-wave electric field can be applied across the package. For transmission we choose $E_{g1} > E_{g2} > E_{g3} > E_{g4}$ so that the band edge of each can be probed. Layer 4 will be depleted first as the bias voltage is increased. As the electric field is modulated, the band-edge absorption is

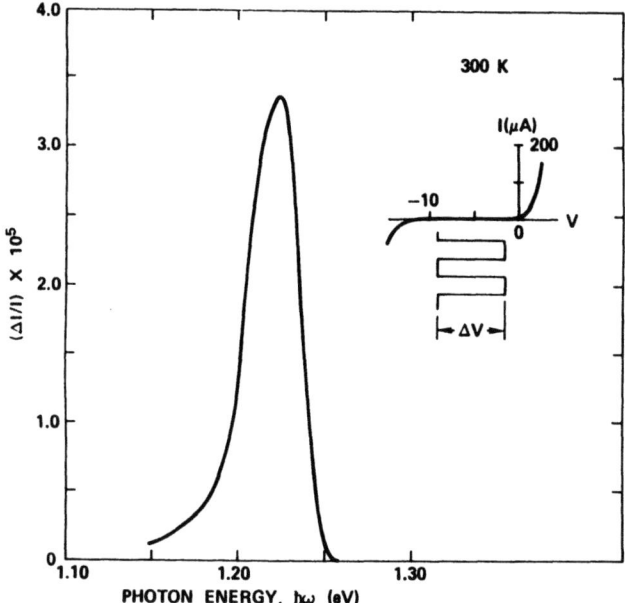

Fig. 61. Electroabsorption signal locating the band gap of layer 4 at $E_{g4} = 1.22$ eV at room temperature.

modulated, so that a change in intensity $\Delta I/I$ is measured. This change appears as a peak in transmission and locates the band gap of layer 4 at ∼1.22 eV, as shown in Fig. 61 for a 300 K sample. For N_D a constant and $\Delta I/I \ll 1$ we get

$$\frac{\Delta I}{I} \frac{1}{\Delta V} = \frac{2}{F_S} \Delta\alpha \tag{44}$$

where ΔV is the amplitude of the applied square-wave voltage on top of the applied dc bias voltage V_A. F_S is the maximum electric field across the Schottky-depletion layer, and $\Delta\alpha$ is the change in absorption coefficient. The maximum electric field is

$$F_S = [2\ eN_D(V_B + V_A)/\varepsilon_\infty]^{\frac{1}{2}} \tag{45}$$

Here, V_B is the built-in potential

$$V_B = e\phi - (E_c - E_F) \tag{46}$$

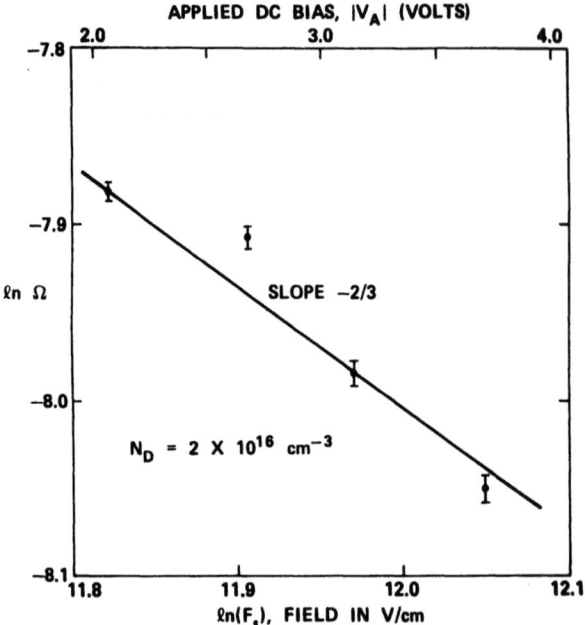

Fig. 62. Determination of carrier density in layer 4 from a measurement of the intensity of the peak in Fig. 61 as a function of applied voltage.

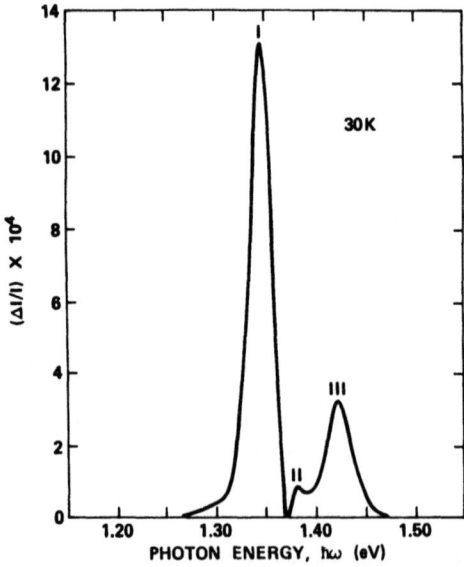

Fig. 63. Electroabsorption signal locating the band gap of layer 4 (peak I) and layer 3 (peak III) at low temperature.

For the Franz-Keldysh effect at the band gap

$$\Omega = A \frac{E_g}{\Delta V} \frac{\Delta I}{I}\bigg|_{max} = F_S^{-2/3} \qquad (47)$$

where A is a material parameter. Eqs. (44-47) indicate that one can obtain N_D as shown in Fig. 62.

If the layer 4 is thin enough, the depletion region can be driven across layer 4 into layer 3. Then the results of Fig. 63 are obtained. Peak I is due to layer 4, but E_{g4} is shifted to 1.34 eV at low temperature. Peak III is due to layer 3 and locates E_{g3}. Peak II is not explained.

It is quite feasible to scan a small spot across such a sample to determine its alloy homogeneity, as has been done in electroreflectance [121].

D. Photo and Cathodoluminescence

Photoluminescence is a powerful tool for studying impurities in GaAs [123,124]. It has been possible through selective doping and implanting [125] to identify a number of emission lines. We can only discuss two aspects, the measurement of luminescence properties in implanted layers [126] and the determination of the densities of acceptors and donors.

1. <u>Ion-implanted Layers</u>. Bishop, et. al. [126] have measured the luminescence due to 100 keV implanted Be in semi-insulating Cr-doped bulk GaAs substrates. The fluences ranged from 3×10^{12} to $1 \times 10^{15} cm^{-2}$. After encapsulation in pyrolytic SiO_2 and annealing for 30 minutes at 800°C, the photoluminescence spectra were obtained at 4.2 K. Exciting radiation was provided by 6471 or 4762 Å lines of a Kr^+ laser with maximum power of about 150 mW. The distribution of impurities was Gaussian and centered around 3200 Å for fluences $<10^{11} cm^{-2}$ but for fluences $>10^{14}$ had a plateau-like tail deeper into the substrate due to the annealing process. Fig. 64 shows spectra for an unimplanted sample and several fluences of implanted samples. The 6471 Å line was used. A shallow Si acceptor gives rise to the line at 1.487 eV, while a weaker peak at 1.515 eV is associated with excitons bound to acceptors. With increasing Be fluences, the shallow-acceptor line increases in intensity and shifts to higher energy approaching 1.493 eV which has been reported for implanted Be [123].

Fig. 65 shows spectra for the shallow acceptor for a fluence of $1 \times 10^{15} cm^{-2}$ obtained for a range of excitation intensities. The shift in energy of the line can be interpreted in terms of a

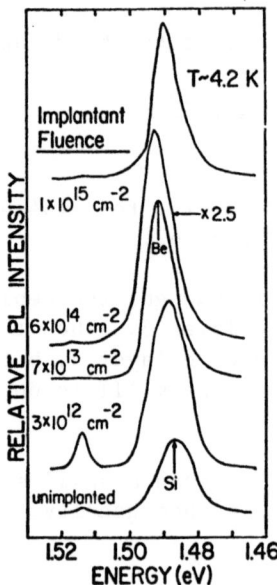

Fig. 64. Low-temperature photo-luminescence of semi-insulating GaAs substrate for several fluences of implanted Be.

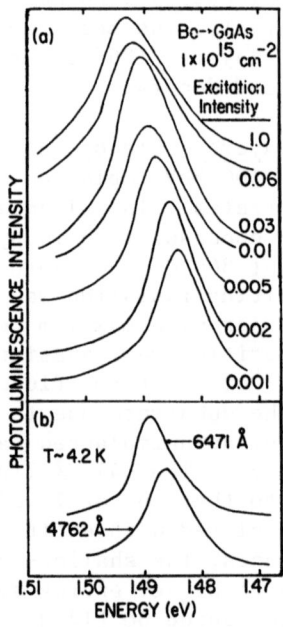

Fig. 65. (a) Photoluminescence spectra from shallow acceptors in GaAs as a function of intensity of the 6471 Å exciting light. (b) Photoluminescence spectra for the same sample excited by 6471 and 4762 Å light of the same power ($\sim$0.75 W/cm^2).

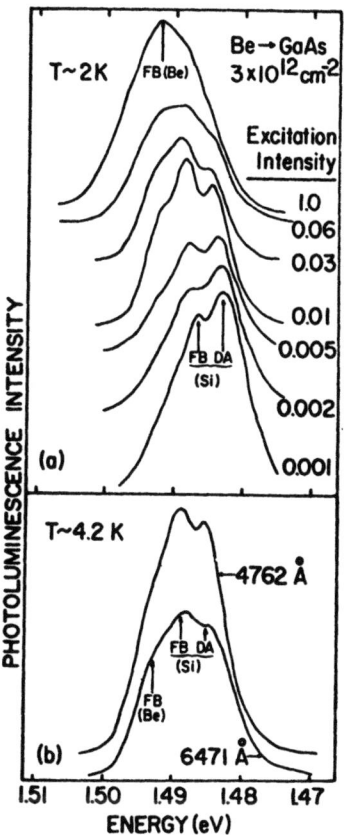

Fig. 66. (a) Photoluminescence spectra from shallow acceptors in GaAs as a function of intensity of the 6471 Å light. Low-intensity spectra are dominated by Si acceptors near the surface. High-intensity spectra originate primarily from the implanted Be acceptors. (b) Photoluminescence spectra for the same sample excited by 6471 and 4762 Å light of the same power ($\sim$0.75 W/cm^2).

power-dependent profiling effect. If the diffusion length of photoexcited carriers is small in comparison to the layer thickness the emitted light should depend on the distribution of emitting centers and the 1/e penetration depth of the exciting light, which is about 3000 Å (see Fig. 46).

The spectra of Fig. 65 were obtained with two wavelengths of light but at the same intensity and again show a shift in line position. Thus it appears that the shallow-penetrating light (4762 Å) excites centers closer to the surface near the peak of

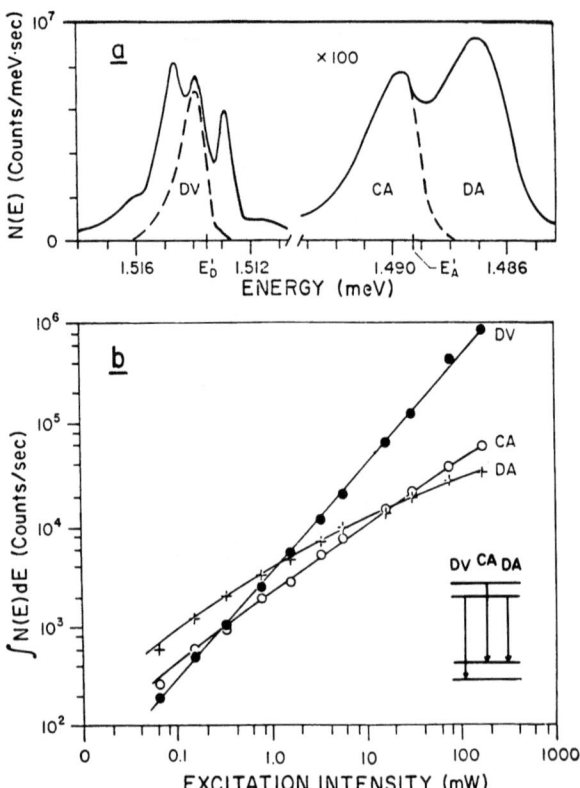

Fig. 67. (a) Low-temperature photoluminescence of a GaAs epitaxial film. (b) Integrated emission intensities due to the conduction-band-to-neutral-acceptor (CA) transition, neutral-donor-to-valence-band (DV) transition. These levels are indicated schematically in the inset.

the distribution while the deeper penetrating light (6471 Å) excites centers throughout the implant distribution including the long plateau of centers behind the peak. The peak energy and width of the shallow acceptor bands are known to be strongly dependent upon acceptor concentration [123] at high concentrations. Hence, the bands shift in energy as the Be acceptors in spatial regions of different concentrations dominate the spectrum.

For low-fluence samples the results of Fig. 66 (a) were obtained. The low-intensity spectra are dominated by the Si acceptors (both donor-acceptor (DA) transitions and free electron-bound

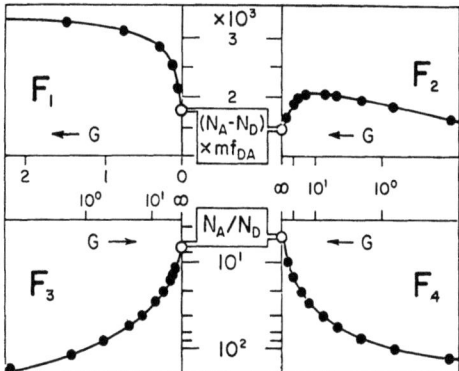

Fig. 68. The functions F_1, F_2, and F_3 and F_4 vs. the excitation intensity G. In the limits of G→0, ∞, the values of $N_A - N_D$ as determined from the limiting values of F_1 and F_2 should approach each other. Also, the values of N_A/N_D as determined from the limiting values of F_3 and F_4 should approach each other.

hole (FB) transitions) near the surface of the substrate while high-intensity spectra originate primarily from the implanted Be acceptors (FB) deeper in.

The spectra of Fig. 66 were obtained with two different wavelengths and tend to confirm the spatial variation of the Be centers. It is obvious that step-wise etching of the surface region would also tend to confirm such distributions. In this case it would be useful to use an even shorter wavelength to confine the excitation region closer to the surface to improve spatial resolution.

2. <u>Donor and Acceptor Densities</u>. Photoluminescence has also been used to determine the concentration of donors and acceptors in GaAs [127]. In Fig. 67 (a) the photoluminescence spectrum of an epitaxial film of GaAs as obtained with an Ar laser at 5145 Å is shown. Three transitions are noted: conduction band to neutral acceptor (CA), netural donor to valence band (DV), and neutral donor to neutral acceptor (DA). In Fig. 67 the integrated intensity of these lines as a function of excitation intensity is given. The three transitions are shown in the inset. It is possible to calculate such integrated emission intensities which are given by

$$I_{CA} = \int dE\, N_{CA}(E) = m^2\, nn_A\, f_A^2$$
$$I_{DV} = \int dE\, N_{DV}(E) = m^2\, pn_D\, f_D\, f_{DA}^2$$
(48)

$$I_{DA} = \int dE\, N_{DA}(E) = m^2\, n_A n_D\, f_{DA}^2$$

where the N's are the emission intensities per spectral band width, m^2 is the mean value of the square of the transition matrix element times a constant related to the effective masses, the band gap and the dielectric constant. The f_D and f_A are the mean values of the squares of the amplitudes of the impurity wave functions in k space for donors and acceptors, and f_{DA} is the summation of the square of the overlapping integral of the donor and acceptor wave functions. These quantities can be calculated and are functions of the donor (E_D) and acceptor (E_A) binding energies.

During the excitation-emission process, charge neutrality is preserved so that

$$n + N_A - n_A = p + N_D - n_D$$

where n_D, n_A, n and p are the concentration of the neutral donors, neutral acceptors, electrons in the conduction band and holes in the valence band, respectively, and N_D and N_A are the donor and acceptor densities.

Since the experimental values of I depend upon the excitation intensity, the determination of N_D and N_A can be done with Eqs. (48) as follows: In the high-excitation limit we expect $n_D \to N_D$ and $n_A \to N_A$ and the charge neutrality condition yields $p \to n$. In the low-excitation limit, $n_D \to N_D - N_A$ and $n_A \to 0$ for n-type material or $n_D \to 0$ and $n_A \to N_A - N_D$ for p-type material. It is then possible to define F functions in terms of f's and I's which are measured (I's) or calculated (f's) so that

$$\lim_{G \to 0} F_1 = (N_A - N_D)\, mf_{DA} = \lim_{G \to \infty} F_2$$

$$\lim_{G \to \infty} F_3 = N_A/N_D = \lim_{G \to \infty} F_4$$

(49)

Thus, in Fig. 68 the values of F's so obtained are indicated as a function of excitation intensity, and at the two limits the values of F_1 and F_2 are quite close as are the values of F_3 and F_4. The self-consistent values of N_D and N_A are $N_D = 1.1 \times 10^{14} cm^{-3}$ and $N_A = 6.5 \times 10^{14} cm^{-3}$. These are the concentrations of radiative donors and acceptors only.

REFERENCES

1. H. Wolter, in Handbuch der Physik, edited by S. Flugge (Springer, Berlin, 1956), p. 461.
2. P. H. Berning in Physics of Thin Films, edited by G. Hass, Vol. 1, p. 69, Academic (1963).
3. O. S. Heavens, Optical Properties of Thin Films, Dover (1965).
4. A. Vasicek, Optics of Thin Films, North Holland (1960).
5. F. Abeles, in Progress in Optics, Vol. II edited by E. Wolf, p. 251 North Holland (1963).
6. H.Y. Fan, Repts. Prog. Phys. 19, 107 (1956).
7. T. S. Moss, Optical Properties of Semiconductors, Academic (1959).
8. L. Harris, J. K. Beasley and A. L. Loeb, J. Opt. Soc. Am. 41, 604 (1951).
9. H. B. Rosenstock (Private Communication).
10. C. Hilsum, J. Opt. Soc. Am. 44, 181 (1954).
11. F. Bassani and G. Pastori Parravicini, Electronic States and Optical Transitions in Solids, Pergamon (1975).
12. E. Johnson, in Semiconductors and Semimetals, edited by R.K. Willardson and A.C. Beer, Vol. 3, p. 153, Academic (1967).
13. E.D. Palik and D. L. Mitchel, in Physics of Solids in Intense Magnetic Fields, edited by E.D. Haidemenakis, p. 90, Plenum (1969).
14. F. Stern, Phys. Rev. 133, A1653 (1964).
15. R. W. Christy, Am. J. Phys. 40, 1403 (1972).
16. S. S. Mitra, in Optical Properties of Solids, p. 333 Plenum (1969).
17. W. G. Spitzer, in Semiconductors and Semimetals, Vol. 3, edited by R. K. Willardson and A.C. Beer, p. 17, Academic (1967).
18. W. G. Spitzer, Festkorperprobleme XI, 1 (1971).
19. S. S. Mitra, Advances in Physics, 20 359 (1971).
20. E. D. Palik and J. K. Furdyna, Repts. Prog. Phys. 33, 1193 (1970).
21. E. Burstein, G. Picus, B. W. Henvis and R. F. Wallis, J. Phys. Chem. Solids 1, 65 (1956): G. Picus, E. Burstein and B. W. Henvis 1, 75 (1956).
22. M. S. Skolnick, A. C. Carter, Y. Couder and R. A. Stradling, J. Opt. Soc. Am. 67. 947 (1977).
23. J. F. Black, E. Lanning, and S. Perkowitz, Infrared Phys. 10, 125 (1970).
24. S. Perkowitz and J. Breecher, Infrared Phys. 13 321 (1971).
25. M. Cardona, in Solid State Physics, Suppl. II, Modulation Spectroscopy, Academic (1969).
26. J. R. Dixon, in Optical Properties of Solids, edited by S. Nudelman and S. S. Mitra, p. 61, Plenum (1969).
27. R. T. Holm, J. W. Gibson, and E. D. Palik, J. Appl. Phys. 48, 212 (1977).
28. S. Perkowitz, Phys. Rev. B12, 3210 (1975).
29. C. E. Jones and A. R. Hilton, J. Electrochem. Soc. 112, 908 (1965).
30. A. Klotinsh, V. Petrov, and I. Feltinsk, Latv. PSR Zinot. Akad. Vestis, Fiz. Tek. Zinat. Ser. (USSR) No. 3, 40 (1974).
31. R. T. Holm and E. D. Palik, J. Vac. Sci. Technol. 13, 889 (1976).
32. K. Murase, S. Katsyama, Y. Ando and H. Kawamura, Phys. Rev. Lett. 33, 1481 (1974).

33. R. Tsu, H. Kawamura and L. Esaki, Solid State Comm. **15**, 321 (1974).
34. R. A. Smith, Semiconductors, Cambridge University Press, (1959).
35. B. Donovan and N. H. March, Proc. Phys. Soc. **76**, 528 (1956).
36. E. Gerlach and P. Grosse, Festkorperproblem XVII, 157 (1977).
37. E. Haga and H. Kimura, J. Phys. Soc. Japan, **18**, 777 (1963); **19**, 471, 658, 1596 (1964).
38. B. Jensen, Ann. Phys. **80**, 284 (1973); **95**, 229 (1975); Phys. Stat. Sol. (b) **86** 291(1978).
39. R. M. Culpepper and J. R. Dixon, J. Opt. Soc. Am. **58**, 96 (1968).
40. S. Perkowitz, J. Phys. Chem. Solids, **32** 2267 (1971).
41. H. Sobotta, Phys. Lett. **32A** 4(1970).
42. R. N. Zitter and K. As'Saadi, J. Phys. Chem. Solids **35**, 1593 (1974).
43. W. G. Spitzer and J. W. Whelan, Phys. Rev. **114**, 59, (1959).
44. M. G. Mil'vidskii, F. B. Osvenskii, E. P. Rashevskaya and T. G. Yogova, Fiz. Tverd, Tela **7**, 3448 (1965); (Sov. Phys. - Solid State **7**, 2784 (1966).
45. O. V. Vakulenko and M. P. Lisitsa, Fiz. Tverd, Tela **9**, 979 (1967); (Sov. Phys. - Solid State **9**, 769 (1967).
46. E. P. Rashevskaya and V. I. Fistul', Fiz. Tverd. Tela **9**, 3618 (1967); (Sov. Phys.-Solid State **9**, 2847 (1968).
47. D. F. Edwards and R. D. Maker, J. Appl. Phys. **33**, 2466(1962).
48. D. L. Spears and A. J. Strauss, Solid State Research Report, Lincoln Laboratory, MIT (1974:3)
49. B. Sherman and J. F. Black, Appl. Opt. **9**, 802 (1970).
50. E. D. Jungbluth and J. F. Black, Solid State Comm. **13**, 1099 (1973)
51. R. C. Newman, F. Thompson, M. Hyliands and R. F. Peart, Solid State Comm. **10**, 505 (1972).
52. F. Thompson and R. C. Newman, J. Phys. C: Solid State Phys. **5**, 1999 (1972).
53. S. R. Morrison, R. C. Newman and F. Thompson, J. Phys. C; Solid State Phys. **7**, 633 (1974).
54. K. Laithwaite, R. C. Newmann, J. F. Angress and G. A. Gledhill, in <u>Gallium Aresnide and Related Compounds</u> (Edinburgh) 1976, edited by C. Hilsom, Conference Series No. 33a (The Institute of Physics, Bristol 1977), p. 133.
55. M. M. Kreitman, K. K. Bajaj and C. W. Litton, Bull. Am. Phys. Soc. **23**, 225 (1978).
56. B. D. McCombe, <u>Characterization of III-V Materials</u>, NRL Memorandum Report 3701 (Feb. 1978).
57. G. E. Stillman, C. M. Wolfe and D. M. Korn, Proc. 13th Conf. Phys. Semicond., Rome (Tipografia, Rome, 1976), p. 623.
58. R. A. Cooke, R. A. Hoult, R. F. Kirkman and R. A. Stradling, J. Phys. D: Appl Phys. **11**, 945 (1978).
59. J. H. M. Stoelinga, D. M. Larsen, W. Walukiewicz and R. L. Aggarwal, J. Phys. Chem. Solids (in press).
60. R. A. Stradling, L. Eaves, R. A. Hoult, N. Miura, P. E. Simmonds and C. C. Bradley, in <u>Gallium Arsenide and Related Compounds</u>, Conference Series No. 17, (The Institute of Physics, London, 1973).

61. E. D. Palik, R. T. Holm and J. W. Gibson, Thin Solid Films 47, 167 (1977).
62. B. Molnar and T. A. Kennedy, J. Electrochem. Soc., (in press).
63. J.K. Furdyna, private communcation.
64. R. Kaplan and R. J. Wagner, J. Vac. Sci. Technol. 13, 899 (1976).
65. J. Botineau, F. Gires and C. Vanneste, C. R. Acad. Sc. Paris 278B, 171 (1974).
66. A. Azema, J. Botineau, F. Gires, A. Saissy and C. Vanneste, Appl. Phys. 9, 47 (1976).
67. R. T. Holm and J. A. Calviello, J. Appl. Phys. (in press).
68. V. L. Ginzburg, Propagation of Electromagnetic Waves, Gordon and Breach (1960).
69. C. D. Taylor and C. W. Harrison, J. Appl. Phys. 42, 2676 (1971).
70. F. Flores, F. Garcia-Moliner and G. Navascues, Surf. Sci. 24, 61 (1971).
71. G. Navascues and F. Flores, Solid State Comm. 9, 1267 (1971).
72. N. S. Kochneva and V. M. Kochetkov, Fiz. Tekh. Poluprovodn. 9, 1821 (1975).
73. W. E. TEnnant and J. A. Cape, Appl. Phys. Letters 26, 694 (1975).
74. P. M. Amirtharaj, B. J. Bean and S. Perkowitz, J. Opt. Soc. Am. 67, 939 (1977).
75. G. Horowitz, Phys. Stat. Sol. 39, 533 (1977).
76. G. K. Hubler and P. R. Malmberg, private communication.
77. V. M. Gusev, L. N. Strel'tsov and I. B. Khaibullin, Fiz. Tekh. Poluprovodn. 5, 832 (1971); Sov. Phys - Semicond. 5, 737 (1971).
78. W. G. Spitzer, C. N. Waddell, G. H. Narayanan, J. E. Fredrickson and S. Prussin, Appl. Phys. Lett. 30, 623 (1977).
79. G. K. Hubler, C. N. Waddell, W. G. Spitzer, R. G. Wilson, S. Prussin and J. E. Fredrickson, (private communication).
80. A. H. Kachare, W. G. Spitzer, F. K. Euler and A. Kahan, J. Appl. Phys. 45, 2938 (1974).
81. A. H. Kachare, W. G. Spitzer, J. E. Fredrickson and F. K. Euler, J. Appl. Phys. 47, 5374 (1976).
82. R. T. Holm, J. W. Gibson and E. D. Palik, Bull. Am. Phys. Soc. 20, 811 (1975).
83. B. Molnar, Report of NRL Progress, March 1975, p. 21.
84. D. E. Aspnes, in Optical Properties of Solids: New Developments, edited by B. O. Seraphin, North Holland (1976) p.799.
85. W. A. Pliskin, in Physical Measurement and Analysis of Thin Films, edited by E. M. Must and W. G. Guldner, p. 1, Plenum (1969).
86. D. E. Aspnes, B. Schwartz, A. A. Stdna, L. Derick and L. A. Koozi, J. Appl. Phys. 48 3510 (1977).
87. M. E. Pedinoff, M. Braunstein, and O. M. Stafsudd, Appl. Opt. 16, 2849 (1977).
88. H. R. Philipp, J. Appl. Phys. 43, 2835 (1972).
89. B. C. Dobbs, W. J. Anderson and Y. S. Park, J. Appl. Phys. 48, 5052 (1977).
90. J. Fahrenfort, Spectrochem. Acta. 17, 698 (1961).

91. J. Fahrenfort and W. M. Visser, Spectrochm. Acta 18, 1103 (1962); 21, 1433 (1965).
92. N. J. Harrick Internal Reflection Spectroscopy, Wiley (1967).
93. D. C. Gupta, Solid State Electr. 13, 543 (1970).
94. W. G. Spitzer and H. Y. Fan, Phys. Rev. 106, 882(1957).
95. G. T. Ayoub and N. M. Bashara, J. Opt. Soc. Am. 67, 1430 (1977). Abstract.
96. D. E. Sawyer and D. W. Berning, NBS Special Publication 400-24 (Feb. 1977).
97. D.E. Sawyer, D. W. Berning and D. C. Lewis, Solid State Technol. Vol. 20, No. 6, p. 37 (1977).
98. D. W. Sawyer and D. W. Berning, Proc. IEEE 64 1634 (1976).
99. H. F. Matare, Solid State Technology, p. 56 (Sept. 1977).
100. D.L. Lile and N. M. Davis, in Proceedings of the Society of Photo-Optical Instrumentation Engineers, Vol. 62, Modern Utilization of Infrared Technology, p. 117 (1975).
101. J. W. Philbrick and T. H. Di Stefano, in Proceedings of the 13th Annual Reliability Physics Conference, Las Vegas 1975,p.159; T. H. DiStefano, NBS Special Publication 400-23, p. 197 (March 1976).
102. R. F. Greene, J. N. Zemel, A. D'Amico and N. Ginsburg, (private communication).
103. D. Peterman and W. Workman, Microelectronics and Reliability 6, 307 (1967).
104. S. V. Bearse, Microwaves, Jan. 1976, P. 14.
105. H. Kaplan, in Proceedings of the Society of Photo-Optical Instrumentation Engineers, Vol. 62, Modern Utilization of Infrared Technology, p. 238 (1975).
106. E. E. Anderson and P. S. Castro, in Proceedings of the Society of Photo-Optical Instrumentation Engineers, Vol 62, Modern Utilization of Infrared Technology, p. 231 (1975).
107. A. Marek, A. A. Jaecklin and J. Cornu, IEEE Transactions on Electron Devices ED-21, 54 (1974).
108. M. Balkanski, K. P. Jain, R. Beserman and M. Jouanne, Phys. Rev. B12, 4328 (1975).
109. R. Beserman and T. Bernstein, J. Appl. Phys. 48 1548 (1977).
110. J. Kotthaus, in Second International Conference of "Electronic Properties of Two-Dimensional Systems", Berchtesgaden, (1977); Surface Sci. (in press).
111. D. C. Tsui, S. J. Allen, R. A. Logan, A. Kangar and S. N. Coppersmith, in Second International Conference of "Electronic Properties of Two-Dimensional Systems", Berchtesgaden, (1977); Surf. Sci. (in press).
112. P. Kneschaurek, in Second International Conference of "Electronic Properties of Two-Dimensional Systems", Berchtesgaden, (1977); Surf. Sci. (in press).
113. C. C. Hu, J. Pearse, K. M. Cham and R. G. Wheeler, in Second International Conference of "Electronic Properties of Two-Dimensional Systems", Bechtesgaden, (1977); Surf. Sci. (in press).

114. E. Gornik and D. C. Tsui, in Second International Conference of "Electronic Properties of Two-Dimensional Systems", Berchtesgaden, (1977); Surf. Sci. (in press).
115. J. S. Blakemore, Semiconductor Statistics, p. 75, Pergamon (1962).
116. R. F. Leheny and J. Shah, Solid State Electr. 21, 167 (1978).
117. J. O. Dimmock, in Semiconductors and Semimetals, Vol. 3, edited by R. K. Willardson and A. C. Beer, p. 290, Academic (1967).
118. W. F. Hall, W. E. Tennant, J. A. Cape and J. S. Harris, J. Vac. Sci. Technol. 13, 914 (1976).
119. J. P. van der Ziel and A. C. Gossard, J. Appl. Phys. 48, 3018 (1977).
120. B. O. Seraphin, in Semiconductors and Semimetals, edited by R. K. Willardson and A. C. Beer, Vol. 9, p. 1, Academic (1972).
121. P. E. Vanier, F. H. Pollak and P. M. Raccah, Appl. Opt. 16. 2858 (1977).
122. N. Bottka and M. Hills, Bull. Am. Phys. Soc. 23, 291 (1978).
123. D. J. Ashen, P. J. Dean, D. T. J. Hurle, J. B. Mullin and A. M. White, J. Phys. Chem. Solids 36, 1041 (1975).
124. P. W. Yu and Y. S. Park, J. Appl. Phys. 48, 2434 (1977).
125. P. W. Yu, J. Appl. Phys. 48, 5043 (1977).
126. S. G. Bishop, J. Comas, S. Sundaram and B. D. McCombe, Appl. Phys. Lett. 31, 845 (1977).
127. S. B. Nam, D. W. Langer, D. L. Kingston and M. J. Luciano, Appl. Phys. Lett. 31, 652 (1977).

Chapter 8

USE OF PHOTOEMISSION AND RELATED TECHNIQUES TO STUDY

DEVICE FABRICATION

W. E. Spicer

Stanford W. Ascherman Professor of Engineering

Stanford University

Stanford, California 94305

I. INTRODUCTION

This paper will be different from many of the others in this book. Although we share the interest of all the authors in the performance and testing of semiconductor devices, our approach is somewhat different. Rather than concentrating on examining completed devices, our approach is to examine, on a fundamental and atomic basis, the formation of a device structure or even a subcomponent of a device structure. The work was motivated by the realization of two factors: (1) that so many steps in a semiconductor device fabrication had become highly empirical, and (2) new tools have appeared which could be developed to investigate the device components and their fabrication on a detailed (atomic) level which had not been previously available. Rather than writing in the abstract, we will concentrate in this chapter on two major device components--the semiconductor-oxide interface and the Schottky barrier.

The performance of MOS devices, so essential to integrated circuits, can be dominated by states at the interface between the semiconductor and insulator. Electrical measurements can provide information on the position, energy, and density of the states, but can give us little direct information on the physical nature of the defects or impurities which give rise to these levels. In fact, we know precious little about the detailed chemical and physical makeup of the interfaces between the semiconductor and the passivating oxide. If nature is generous, as is the case with the Si-SiO$_2$ interface, a strictly empirical approach may be the

most reasonable as technology is being built up. However, as
the demands of that technology become more and more exacting (in
terms of performance, size, reliability, production yield, etc.),
the cost of using a purely empirical approach becomes more and more
important and the need for more basic knowledge (if only to help
guide empirical approaches) becomes increasingly valuable. For
example, if basic knowledge can help choose the one or two most
promising of a number of empirical approaches, the savings can be
very large

An interestingly different case is that of Schottky barriers
in particular and metal-semiconductor contacts in general. Schottky
barriers have been in use for about four decades and there is an
extensive literature concerning the physics (as well as other
aspects) of these devices. However, it has become clear recently
that the conventional ideas concerning the mechanism by which the
Fermi level of these devices is pinned is definitely incorrect for
the 3-5 compound semiconductors and is highly suspect for both
compound semiconductors and for silicon. This is a result of
studies reported within the past year. As will be discussed in
detail later, these studies have lead to new models of Schottky
barrier pinning.

In this chapter we will emphasize the work which, in recent
years, has begun to give us fundamental insight into both the
oxide-semiconductor interface and the Schottky barriers. The
theme of this chapter is the use of photoemission spectroscopy, and
thus we will place particular emphasis on this technique. However,
to give overall perspective we will also mention other critical
techniques such as Auger spectroscopy which can be combined with
ion sputtering techniques to mill away material to give a chemical
analysis as a function of depth into the material (depth profiling).
In the last two years these sputtering techniques have reached the
stage where, with extreme care, depth resolutions of a very few
atomic or molecular layers has been achieved even after milling
through as much as 1000 Å of material. For brevity these are
usually described as sputter-Auger techniques.

In accord with the outline given above, this chapter will
be organized as follows: In Section II we will discuss the physics
of photoemission. This chapter will also include a short description
of conventional radiation sources. This will be followed by a
section (III) in which the new synchrotron radiation sources will
be discussed and the unique capabilities they bring to these studies
discussed. Next, a short section giving a description of sputter-
Auger techniques.

After building this necessary foundation, we will next discuss
in Section V the use of synchrotron radiation and allied experimental
techniques to study oxide formation on the 3-5 compounds. In that

section we will include a comparison of the results from electrical studies of actual device oxides with the fundamental studies reported here. This will be followed by a section in which a shorter account of studies of oxygen on Si is given. Here sputter-Auger techniques have played a major role.

Section VII will cover the study of Schottky barrier formation on 3-5 compounds. This will include the presentation of a new model for the mechanism of Schottky barrier pinning. Again, this will be followed by a short discussion of Schottky barrier formation on Si and short reference to work on 2-6 compounds. Section IX will give a short summary.

The author will draw very heavily on the work done by his group at Stanford University since it is this with which he is most familiar. He will attempt to mention and/or reference other work; however, the article is intended to be more tutorial in nature than a comprehensive survey of the field. In any case, it would not be proper to attempt a comprehensive review since the field is moving so fast that much of the Stanford work drawn on here is not published at the time of this writing.

II. THE PHYSICS OF PHOTOEMISSION

A. Introduction

In the last 15 years, ultraviolet photoemission spectroscopy (UPS) and x-ray photoemission spectroscopy (XPS) have emerged as powerful tools for the investigation of the electronic structure of solids. Until about five years ago, the emphasis in photoemission spectroscopy (PS) has been on determining the bulk electronic structure of solids. More recently, increased interest has developed in the application of UPS to understanding the clean surfaces of solids, the development of an understanding of gas adsorption and the formation of oxides or other compounds. In this section we will emphasize UPS since, in this way, we can make most the natural contact with other sections in this book; however, the basic physics is the same for UPS or XPS.

B. The Physical Mechanisms of Photoemission

The ultimate objective of any spectroscopic technique is to determine the quantum states of the system under study. One cannot consider the surface without understanding the bulk excitation. Thus, we will start at that point. The valence and conduction states of solids present special difficulties for the interpretation of adsorption or reflection spectra because the quantum states form continuous bands of significant width rather than the discrete

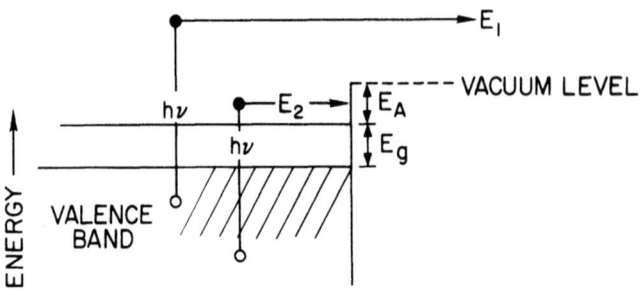

$\eta(h\nu, E) dE$ PROBABILITY $h\nu$ EXCITING ELECTRON TO FINAL STATE AT ENERGY E TO E+dE

$\varepsilon_2 \longleftarrow \int \eta(h\nu, E) dE$

$N(E, h\nu) dE \longleftrightarrow \eta(h\nu, E) dE$

$N(E, h\nu)$ = ENERGY DISTRIBUTION OF PHOTOEMISSION

Fig. 1. The essence of the use of photoemission to study the electronic structure of solids is determining $\eta(h\nu, E)$ the probability of a photon of energy $h\nu$ exciting an electron to a final state at energy E. In contrast, it is the integral of $\eta(h\nu, E)$ over all final states which is related of the optical parameters such as ε_2.

states which characterize atoms (see Fig. 1). Because one has transitions from one band to another, it is much more difficult, in general to interpret "conventional" optical spectral data from solids than that from atoms. By "conventional" spectral data we mean optical adsorption and reflection spectra and the optical constants (ε_1, ε_2, α, η, etc.) which one obtains from such spectra. This difficulty is, of course, just a reflection of the fact that, for valence to conduction band transition, one has usually a large range of possible initial and final state energies for a given photon energy. Modulation optical studies provide another approach to this difficulty [1].

Figure 1 illustrates this point. Consider, the probability, $\eta(h\nu, E)$ of a photon $h\nu$ exciting an electron of final state energy E [2]. The optical constants are determined by the integral of $\eta(h\nu, E)$ over all final states. For example

$$\varepsilon_2 \propto \int \eta(h\nu, E) dE \qquad (1)$$

where ε_2 is the imaginary part of the dielectric constant. It is because of this integration over final states that it is often so difficult to obtain unambiguous information on the quantum states

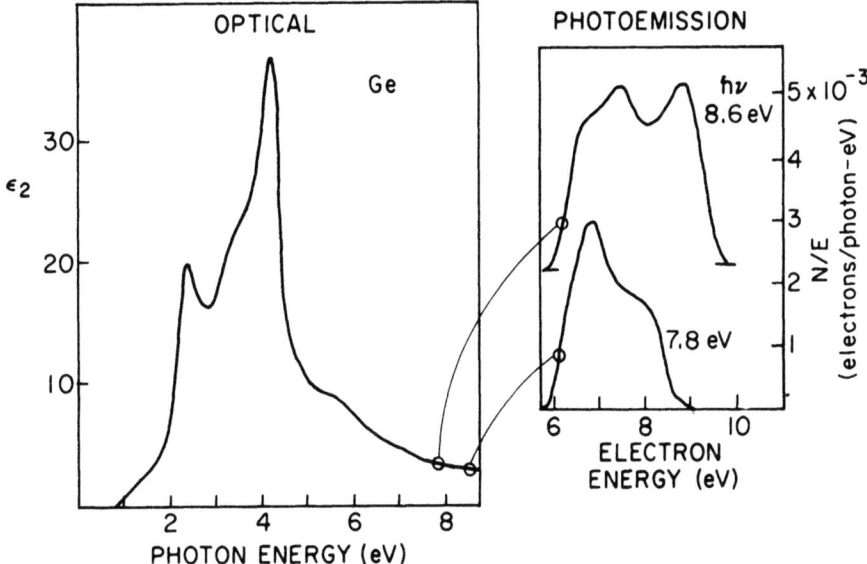

Fig. 2. A comparison of optical and photoemission data. The left panel gives ε_2 for Ge versus $h\nu$. In the right hand panel, photoemission energy distribution curves (EDCs) for photon energies of 7.8 and 8.6 eV are given. Note that there is no structure in ε_2 between 7 and 9 eV but that the EDCs in this $h\nu$ range show strong structure.

from reflection or adsorption measurements or from the optical constants themselves. In photoemission, the excited electrons are allowed to escape into vacuum (see Fig. 1) where their energy can be measured. From this measurement, one obtains the energy distribution of the emitted electrons $N(h\nu, E)$. In a properly calibrated experiment, $N(h\nu, E)$ gives the number of electrons emitted per photon adsorbed in a given interval of energy. If all the electrons excited escaped without energy loss,

$$N(h\nu, E)dE = \eta(h\nu, E)dE \qquad (2)$$

Thus, photoemission "removes the integral sign in Eq. (1) and in this way removes much of the ambiguity inherent in attempting to determine the quantum states from optical constants alone. In fact, there is inelastic scattering as the electron escapes from the solid and this modifies the relation between $N(h\nu, E)$ and $\eta(h\nu, E)$.

This is illustrated by Fig. 2 where ε_2, for Ge [3] is plotted as well as photoemission energy distribution curves (EDCs) [4], for $h\nu$ = 7.8 and 8.6 eV. Note that for $h\nu$ > 6 eV, there is no structure in ε_2; however, the EDCs for 7.8 and 8.6 eV show strong structure. This structure changes with $h\nu$. The EDCs give the distribution of emitted electrons as a function of their energy and thus are closely related to $\eta(h\nu, E)$ whereas ε_2 is related to the

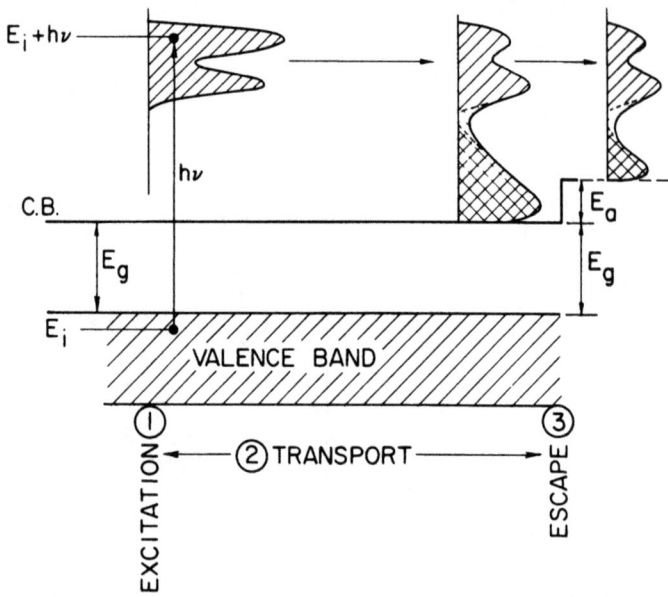

Fig. 3. A schematic diagram illustrating the three step photoemission process. An initial optically excited distribution is shown on the left. Changes in this distribution as it approaches the surface and after it has escaped into vacuum are indicated.

integral of the transition probability per unit time over all possible final states.

Previously we mentioned the inelastic scattering of the electron between the moment of excitation and escape into vacuum. Such events must be taken into account since the optical excitation extends about 100 Å or more into the solid and the escaping electrons must pass through the solid before escaping into vacuum. When considering such excitation, the "three step" model of photoemission gives an extremely useful approximation which is capable of giving an adequate treatment of most existing photoemission as well as good overall perspective [5,6,7]. This model, which is described in more detail elsewhere, approximates the photoemission as three successive events. The first event is that of optical excitation. The second is the transport of the electrons to the surface. In this step, the electrons can suffer inelastic scattering. The last step is the escape of the electrons across the potential barrier at the surface into vacuum.

Table 1

Type of Scattering	Threshold	Typical Energy Loss (per event)	Typical Scattering Lengths
Electron-Electron	Metal-none nonmetal bandgap	Large fraction of excitation energy electron volts	See Fig. 4 can be few Å
Electron-phonon	None	Small (tens of meV)	10-100 Å
Electron-plasmon	$> h\omega_p$	$h\omega_p$	From a few Å to 100 or more Å

$h\omega_p$ = plasmon energy

In Figure 3, we indicate schematically what happens to an energy distribution excited from the valence band in UPS between the time it is excited and when it reaches the surface and escapes into vacuum. On the left of the figure, a hypothetical $\eta(h\nu, E)$ has been sketched. The details of $\eta(h\nu, E)$ depend on the details of the electronic structure. As they move towards the surface, the excited electrons lose energy due to inelastic scattering [6,8,9]. Electron-electron (el-el) scattering between the excited electron and the valence electrons, and electron-phonon (el-ph) scattering between the crystal lattice and the electron are the two principle events. Scattering between the electron and lattice defects or impurities is also possible, as are plasma or other collective scattering events. The latter scattering becomes extremely important for XPS.

In Table 1, we list the characteristics of the electron-electron, electron-plasmon, and electron-phonon events. One important difference between the two former and the latter scattering events is the magnitude of the energy loss. As shown in Table 1, the loss in lattice scattering is small (order of 0.01-0.05 eV) compared to the excitation energy ($h\nu$); whereas, in the electron-electron or electron-plasma scattering, the loss may be a good fraction of the excitation energy; i.e., many electron volts or more. Thus for the higher values of $h\nu$, which we will consider here, electron-phonon scattering will not be of prime importance.

To a first approximation, the probability S(E, x) of scattering after traveling a distance x without scattering can be defined in terms of a scattering length L(E) [5,6,10],

$$S(E, x) = \exp\{-x/L(E)\} \quad (3)$$

As can be clearly seen from Fig. 4, where the available data on L(E) is presented [11], L(E) depends strongly on hν. Figure 4 also emphasizes the small (a few Å) minimum values of L(E) which occur.

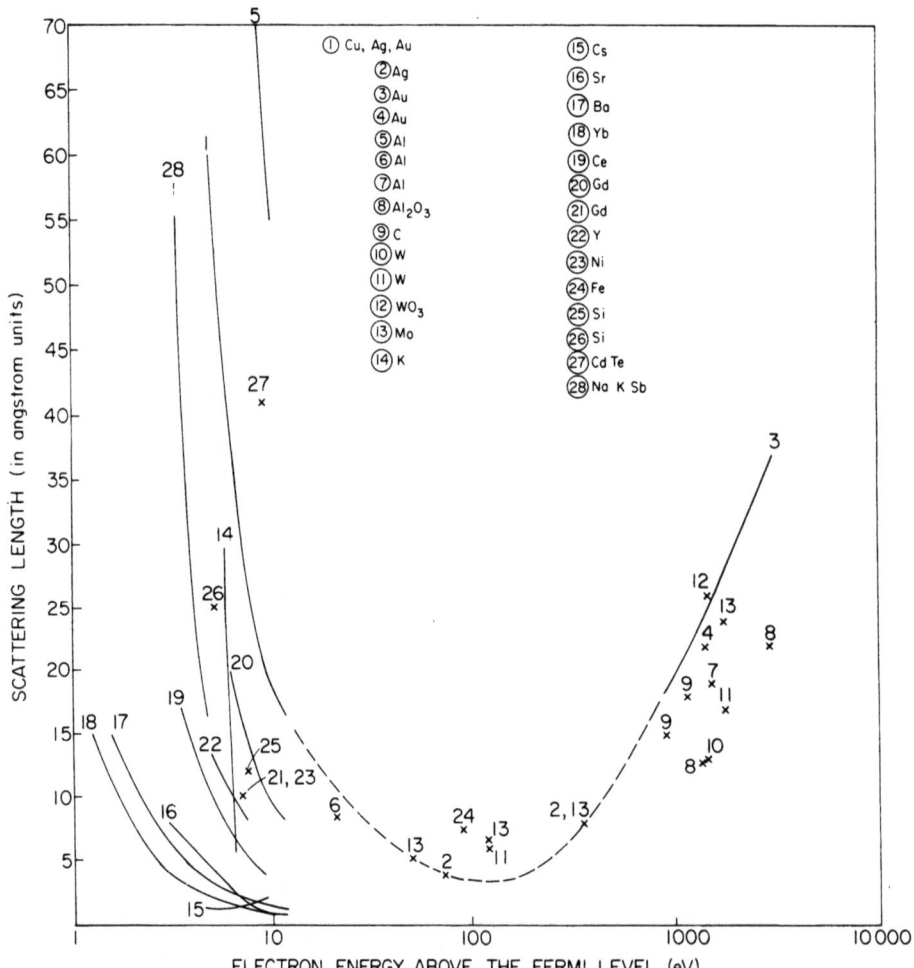

Fig. 4. This gives a summary of the available information on electron escape depth versus photon energy. The solid curve on the right had side of the figure is curve number 3. The curve is taken from I. Landau and W. E. Spicer (1974).

Note in Fig. 4 that only for relatively low energies, i.e., $E \gtrsim 6$ eV, is the el-el $L(E)$ large compared with the typical el-ph scattering lengths. Only under these conditions does loss of energy to phonons become of major importance. As soon as the el-el length becomes comparable to the el-ph length, el-el scattering becomes dominant due to the large energy loss per event. In this paper, we will concentrate on this case since it is most common. The reader is referred to the literature for cases where el-ph scattering dominates [5,6,12], e.g. DiStefano and Spicer (1973). CsI is a good example of such a case. Monte Carlo calculations give insight into escape with el-ph and el-el scattering [10].

In Fig. 3, the low energy peak which appears as the excited electrons move to the surface is due principally to el-el scattering. Because of the large energy loss associated with the el-el event, the scattered electrons are often effectively removed from the distribution or else simply provide a monotonic background. This is important because the electrons are normally removed from the EDCs without producing new structure which would confuse the interpretation of the data. As shown in Fig. 3, structure due to the excitation probability, $\eta(h\nu, E)$, remains in the EDCs after scattering and a new peak of low energy electrons is produced near the low energy cutoff of the EDC. In cases where there are large peaks in the final density of states into which the electrons scatter, this can produce corresponding peaks in the distributions of scattered electrons [7,12,13]. At higher energy, discrete loss due to plasmons appear as will structure to photon excited Auger transitions.

To summarize, scattering will remove electrons from the original distribution. This can cause broadening in structure and changes in the relative intensities of peaks in that distribution. Except for these changes, there is a close one-to-one relationship between structure in the measured EDCs and that in $\eta(h\nu, E)$.

Before leaving Fig. 4, we should mention that the excitation function itself will, in general, change as one approaches the surface. Because of the termination of the lattice at the surface, new electronic states may appear and/or the electronic structure associated with the bulk will be gradually changed. This will take place in the last few atomic layers. As one can see from Fig. 4, the minimum values of $L(E)$ correspond to the last or last few atomic layers. Thus by "tuning" $L(E)$ through selection of E, one has the possibility of alternately looking at bulk or surface electronic structure. The shorter the escape depth, the less meaningful it is to separate the photoemission process into excitation, transport, and escape; thus, the three step model becomes a progressively poorer approximation as the escape length becomes shorter and approaches atomic or lattice dimensions. However, the three step model provides an invaluable tools for determining when the photoemission is dominated by the last few atomic layers.

The last step in photoemission is the escape of the electrons across the potential barrier at the surface. The principal effect here is to cut off the low energy portion of the EDCs in such a way that the magnitude of the EDCs increases gradually from their threshold value. This is indicated in Fig. 3. Detailed models have been used to calculate the escape function [7,14].

It must be realized that the separation of the photoemission event into three processes which are then treated independently is an approximation. As Schaick and Ashcroft [17], Ashcroft and C. Caroli [10], D. Lederer-Rozenblatt, B. Roulet and D. Saint-James [19] and Feibelman [20] and others have indicated, one would want to treat these as coupled events ideally, and, take such processes as interference and/or multiple scattering effects due to the surface into account; however, this is a very difficult theoretical problem.

C. Radiation Sources and the Work Reported Here

In the work reported here, the emphasis will be placed on working at the minimum of the escape depth (see Fig. 4) so as to emphasize the surface. Much of the work reported will depend on the use of photoexcitation from core rather than valence levels (see Chapters 3, 5, etc.). In order to minimize the escape depth for both valence and core electrons, it is necessary to work at different photon energies. One must also consider the photon energy dependence of the optical excitation matrix elements from the valence various core levels [21]. For example, the valence band excitation matrix elements of the covalent semiconductors falls off very rapidly as $h\nu$ increases above about 5 eV. Thus it is usually optimum to study the valence surface electronic structure in the range $20 \lesssim h\nu \lesssim 30$ eV. This gives the best compromise between short escape depth and practical optical excitation probability. This will become more apparent when we present an escape depth curve for GaAs in the next chapter. In contrast, the 3 d core levels of greatest interest here have matrix elements which are fairly constant ranging up to several hundred eV of their appearance in the photoemission spectra. The 4 and 5 d's are sharply peaked about 50 eV above their threshold [21]. Since the As and Ga 3 d levels have binding energies of about 40 and 20 eV respectively, we obtain optimum results studying them with $h\nu \approx 100$ eV. Thus, the optimum energy for studying these states is about $h\nu = 130$ eV.

In order to obtain the optimum energies for a wide range of experiments, conventional sources of radiation are inadequate. These are line sources produced by either electrical breakdown in gases or discrete X-rays source produced by electronic excitation of deep core levels. A few practical emission lines from gas

discharges (He) lie between 20 and 50 eV [22]. These can prove very
useful in certain cases but do not give the wavelength flexibility
nor range needed here. The only nearly practical X-ray sources
give a few discrete lines between 1 and 2 KeV. Therefore, the
conventional radiation sources are inadequate as the sole source of
radiation. Rather, a source with a continuous, relatively flat
emission continuum for the range $6 \lesssim h\nu \lesssim 500$ eV or higher would be
optimum. As will be discussed in the next chapter, synchrotron
radiation provides just such a source. In our own laboratory, we
have found an arc equipped with a monochromator to be extremely
valuable as an auxiliary to the synchrotron source since the latter
is not nearly as easy to use as the arc.

D. Auger Emission and Photoemission Constant Final State (CFS) or Partial Yield Spectroscopy

In the Auger process [23] an electron is excited from a core
state leaving a hole in that level. Recombination in the Auger
process occurs when an electron drops from a higher lying state
into the hole and the excess energy is given directly to an electron
in a higher lying state. This "Auger" electron normally has sufficient energy to escape from the solid into vacuum.

The first use of Auger for surface studies was to determine
the chemical composition of the material sample. The 3-step model
can be applied to the Auger as well as the photoemission process so
long as the Auger rather than the direct optical excitation process
dominates. (The original hole in the Auger process can be produced
by an electron or photon.) Most importantly, it is the escape
depth L(E) (see Fig. 4) which determines the depth sampled. Thus,
Auger electron spectroscopy (AES) is only a surface tool if the
energy of the Auger electron is such that the escape depth of the
outgoing electron is small. As can be seen from Fig. 4, this L(E)
can vary between ten to several hundred $\overset{\circ}{A}$ depending on the solid
under study and the energy. Thus, AES is only a surface tool when
the energy of the outgoing electron is properly chosen.

The Auger transition can be used in a different way to investigate the lowest available empty states at the surface. This
process is illustrated in Fig. 5. Here one uses the Auger process
to determine the threshold for the optical transition from the core
state into the lowest available state at the surface. The onset
and details of the transition are detected by examining the number
of electrons emitted at a fixed and sufficiently low energy final
state as $h\nu$ is increased. Once an optical transition is available,
producing a core hole, the Auger process "turns on" and the yield
of photoelectric electrons at a properly chosen final state
increases quite strikingly. By measuring this increase of

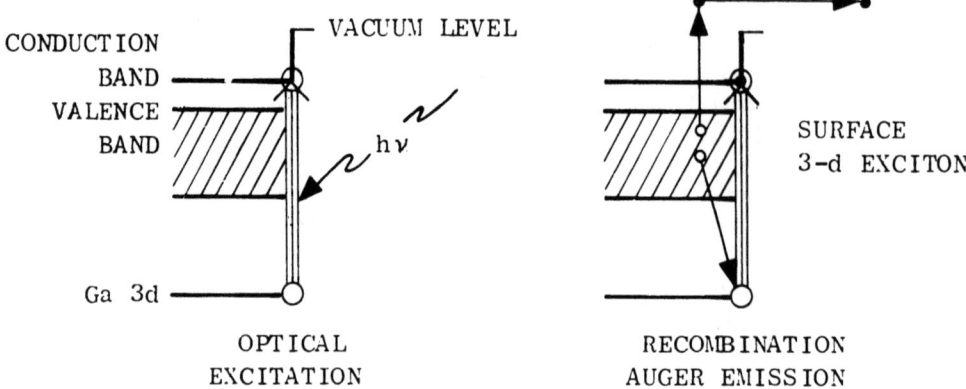

Fig. 5. Illustration of use of electron emission to measure surface optical transition. The threshold of excitation from the Ga 3d state involves a surface exciton. The probability of formation of a hole at the surface is proportional to the surface optical transition probability. Formation of a hole may lead to Auger recombination (second panel) resulting in production of an electron which escapes into vacuum and is measured. Thus the number of such electrons is proportional to the surface optical adsorption. In this spectroscopy the electron emission is measured at a constant (final) kinetic energy as the photon energy is continuously varied; thus the name -- Constant Final State Spectroscopy.

photoemission yield as a function of energy, the probability of an Auger process and thus the optical transition probability is traced out as a function of hν. This technique is a special application [26] of the Constant Final State spectroscopy (CFS) developed by Lapeyre and his group.

Near the threshold of this core transition, an exciton is created which is characteristic of the surface. The reader is reminded that the binding energy of an exciton goes inversely as the square of the dielectric constant of the material in which it is excited. Thus, a bulk exciton in GaAs has only tens of millivolts binding energy. However, at the surface, the exciton sees, for example, only a fraction of the bulk binding energy because the dielectric constant of vacuum is unity. The electron effective mass may well be increased also. Thus, binding energy of the exciton is increased to 0.5 eV or more. Theoretical work has just begun on theses excitons [28] which are also seen in electron energy loss experiments [29].

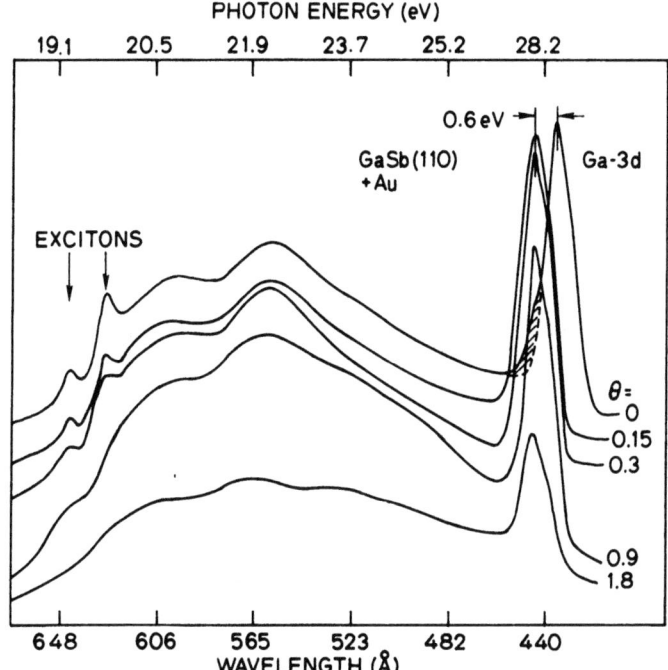

Fig. 6. Constant final state spectra of GaSb as a function of Au coverage.

The surface exciton will be made up from core states and the lowest available empty states. It is usually assumed that these are the empty surface states; but, as will be discussed in more detail later, it now appears that the surface states lie well above the conduction band minimum (CBM); thus, the lowest lying conduction band states rather than surface states may be involved in the formation of the exciton.

In Fig. 6, we give an example of CFS data from a 3-5 semiconductor. The two pieces of sharp structure are two excitonic peaks resulting from the spin-orbit splitting of the core d-level [30]. Figure 6 illustrates the utility of such data. One curve was taken from a clean surface, the others are for metal overlayers of increasing thickness. We will draw upon this data later in this chapter. Its importance lies in determining whether or not the intrinsic surface states (these will be defined and discussed in more detail in Section V) are moved from the conduction band into the bandgap region by the overlayer. If this occured, the lowest energy states from which the surface excitons could be formed would be the empty surface states. As a result of the lowered energy of the final states, the excitonic peaks would move to lower energy. As can be seen from Fig. 6, this does not happen. The energy of the excitonic peaks remains constant as the metal coverage is increased and begin to disappear when the coverage approaches a monolayer. As will be mentioned in discussing Schottky barriers, this is important because it suggests that massive movement of the intrinsic surface states into the bandgap cannot explain the Schottky barrier pinning. This conclusion will be strongly reinforced when we see that the pinning stabilizes for only about 0.1 of a monolayer metal coverage [31,32].

E. Angle Resolved and Angled Integrated Energy Distribution Curves (EDC's)

All of the results here will be from measurement of the EDC's which have been integrated over a wide range of angles. Additional information on the propagation vector or reciprocal lattice vector on $\bar{K}$ of the excited electron inside the solid may be obtained by studying the EDC's as a function of the emission angle [33]. Such techniqes have proved useful for studying, e.g. the details of surface states [34], and will clearly become increasingly important in the future; however, for the studies we report here, such sophisticated and time consuming techniques were not necessary.

III. SYNCHROTRON RADIATION SOURCES

A. Introduction

In the last 15 years, the use of synchrotron radiation to study the electronic structure of solids has become increasingly important. Originally, the radiation was obtained from synchrotrons designed for high energy studies (such as that at the National Bureau of Standards in the U.S. and the much higher energy source, DESY, in Germany). More recently, storage rings (usually built for high energy physics) have proven much more satisfactory as radiation sources [35]. Such sources are now operating or are under construction in at least eight countries making them available to larger parts of the scientific community.

As mentioned earlier, the advantage of such sources is the continuum of radiation which they provide. Of critical importance is the high energy cut off of this radiation. This is set by the energy of the stored electrons and the binding radius of electrons in the machine. If the effective high energy cutoff comes before 200 eV, studies such as those reported here are very difficult to carry out. As far as the author knows, the Stanford Synchrotron Radiation Laboratory (SSRL) and Doris at Hamburg were the first high energy storage rings to be utilized as synchrotron sources (1974)

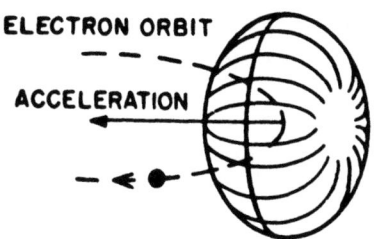

CASE I : $\frac{v}{c} \ll 1$

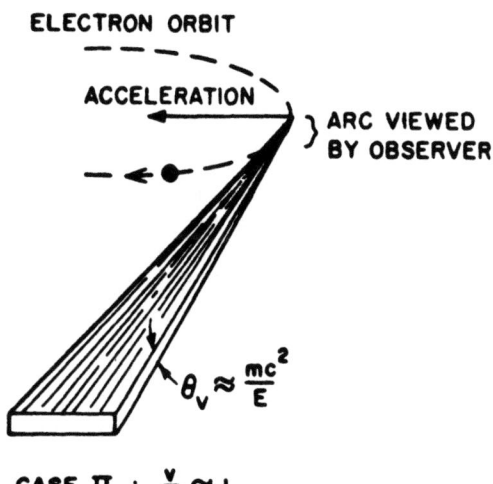

$\theta_v \approx \frac{mc^2}{E}$

CASE II : $\frac{v}{c} \approx 1$

Fig. 7. Radiation emission pattern by electrons in circular motion. Case I is for the classical case (v = speed of electron, c = speed of light); Case II is for the relativistic case. θ_v is the vertical divergence in radians.

Other storage rings came into operation. Since all of the work reported here was done on SSRL, we will limit ourselves to the description of this storage ring, since the general principles of all storage rings are the same.

B. Introduction: The Use of Synchrotron Radiation to Study 3-5 Surfaces and Interfaces

The key to the studies to be described here is the use of synchrotron radiation to investigate the surface electronic structure and chemistry of 3-5 surfaces by examining the last few atomic layers preferentially through the use of a large range in photon energies. In this way, we can look in a microscopic (atomic) way at materials processes which have previously been examined on only a macroscopic basis [36]. Our object in these studies has been to examine phenomena which are very difficult or impossible to examine using more conventional tools. For example, other techniques can explore the oxides or other insulating layers built up on a 3-5 surface. However, how is such a layer bound to the 3-5 host crystal? Our first experiments were designed to examine this question after gaining an understanding of the clean free surface. In order to evaluate the work on oxygen uptake, it was absolutely necessary to start with a detailed knowledge of the clean surface.

Before describing the experiments, their interpretation, and implications for passivation, we will describe, briefly, SSRL [37] with emphasis on the part of the facility utilized in these experiments. SSRL is located at Stanford University and is a national facility supported by the National Science Foundation for the utilization of synchrotron radiation. Synchrotron radiation is produced by electrons in the SPEAR storage ring [38] as they are accelerated to keep them in circular orbit. Since the electrons are highly relativistic (energies between 1.5 and 4.0 GeV), they radiate in a very narrow (on the order of 2×10^{-4} radians) cone in the forward direction as indicated in Fig. 7. This characteristic is of enormous benefit in designing experimental stations to make use of the radiation, e.g., monochromators. However, the most critical characteristic of the synchrotron radiation is that it provided the first high intensity continuum of radiation extending from the infrared to the hard X-ray spectral region. This is illustrated in Fig. 8. We and others have made use of that part of the spectrum essential to our experiments (approximately 6 to 300 eV). A large number of other groups have made use of the higher energy X-rays.

On Fig. 9, we give a schematic drawing of the beam line which we have used at SSRL. We will report results obtained on two of the ports coming off this beam line. The first is labeled UV radiation and the second UV + soft-x-ray. The first is labeled UV for $8 \text{ eV} \lesssim h\nu \lesssim 35 \text{ eV}$ and the latter for $40 \lesssim h\nu \lesssim 800 \text{ eV}$. Each beam line is equipped with a monochromator [39] so that monochro-

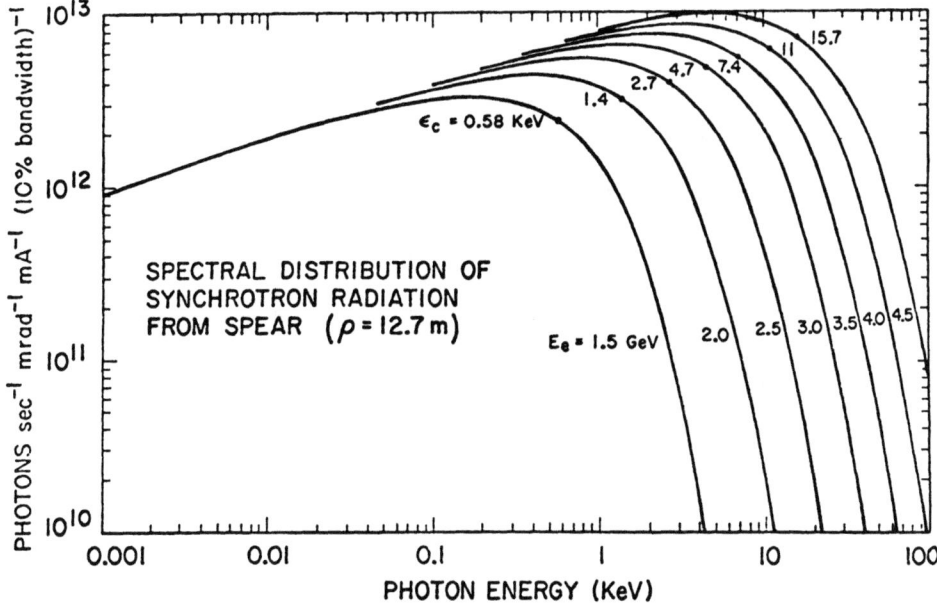

Fig. 8. Spectral distribution of radiation from SPEAR. E_e is the energy of the electrons stored in the ring. On the vertical axis, mA^{-1} is the current in the ring (typically 20 ma, at present, but expected to increase by an order of magnitude within a few years) and mrad is the angle of radiation intercepted in the horizontal plane. This is typically a few milliradians.

matic radiation can be obtained in a continuous manner over the photon energy range indicated.

C. Essential Capabilities Provided by SSRL

The experiments that we will describe involve photoemission spectroscopy in which the valence and core levels of the last few atomic layers are examined. As mentioned earlier, the reason why continuous tunability over a wide spectral range was necessary for these measurements was due to the difference in binding energy of the various levels of interest (see Fig. 10), the fact that the minimum L(E) occurs for kinetic energies of about 60 eV (see Fig. 11) [40] and to the energy dependence of the matrix elements. Because of basic similarity in band structure, the escape depth curves for the other two materials studied in this work, InP and GaSb, should be similar.

In section II, we defined the escape depth. Since only the electrons which escape without inelastic scattering give direct information on the energy structure of the solid, the escape depth

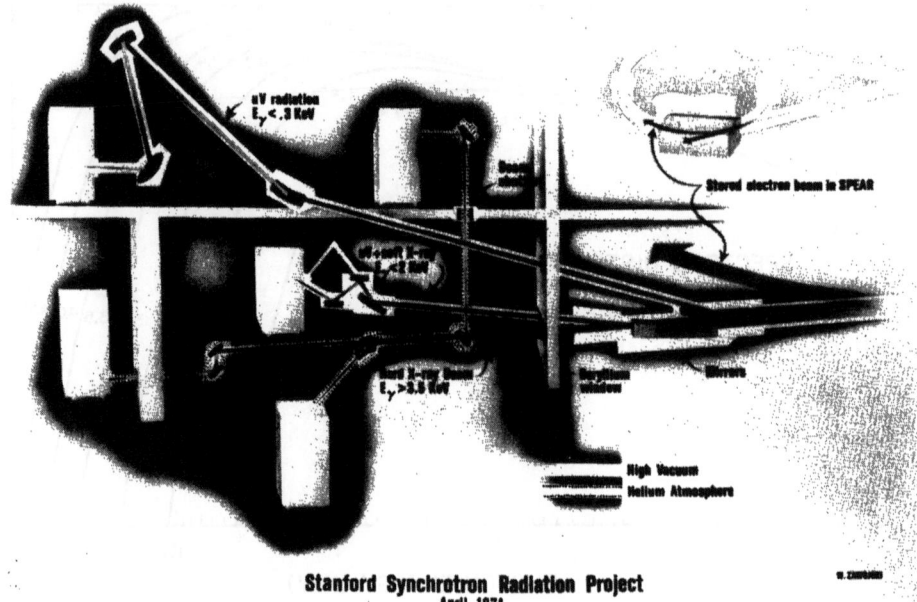

Fig. 9. The beam line at SSRL used in this work. The insert at the upper right shows the entire SPEAR storage ring; the building housing shows the beam line. The work described in this article was done on the UV and UV + soft x-ray lines used in this work. Both lines are equipped with monochromators.

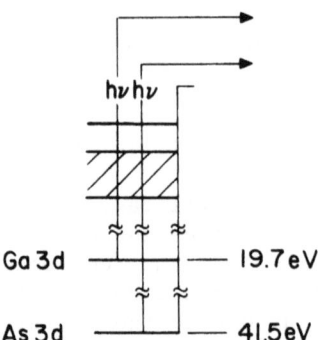

Fig. 10. Schematic of valence and core level energy diagram for GaAs. The cross-hatched area is the valence band, the top of which lies about 5 eV below the vacuum level and is about 12 eV wide. The energy (with respect to the valence band maximum) of the highest lying core (3d) levels are indicated.

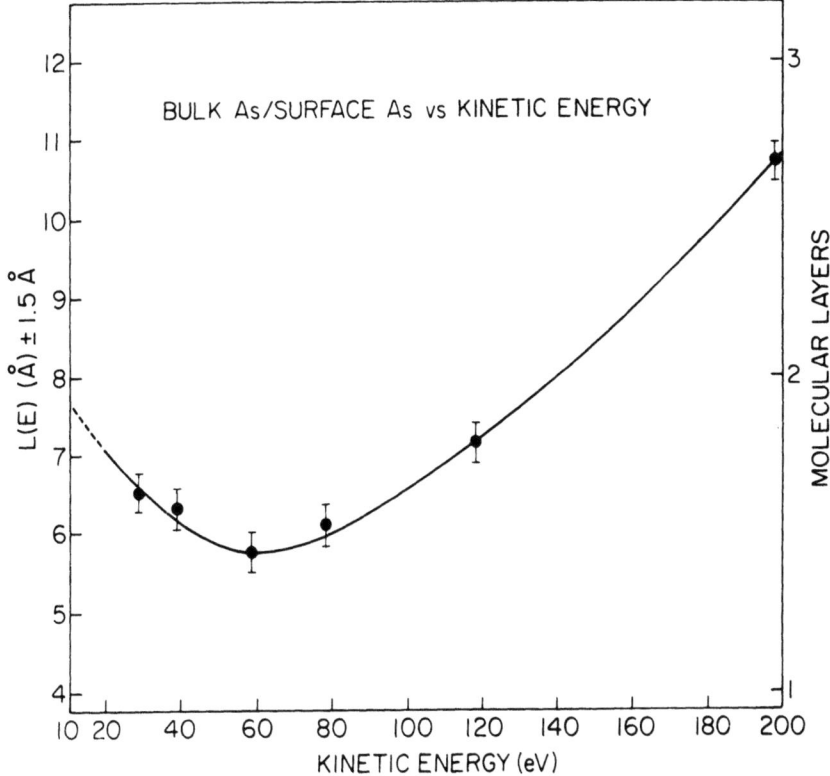

Fig. 11. The escape depth as a function of kinetic energy for GaAs.

determined the depth "probed" in our experiments. As can be seen from Fig. 11, the minimum escape depth is about 5 to 6 Å, i.e., about one-and-a-half molecular layers. Thus one can examine preferentially the last couple of molecular layers of the semiconductor.

We will be interested in both the valence band electronic structure at the surface (which provides, for example, information on the perfection of the crystal lattice near the surface) and the core levels (which gives us information on the surface chemistry). Note (see Fig. 10) that these levels lie at different energies so that for GaAs one would, for example, use photon energies of 65 and 108 eV to obtain the minimum escape depth for emission from the top of the valence band and As d levels, respectively. However, other considerations must also be taken into account such as the strength of the matrix elements for a given level as a function of hν and the desirability of examining the core level of both Ga and As at a single photon energy. Fortunately, the minimum in the

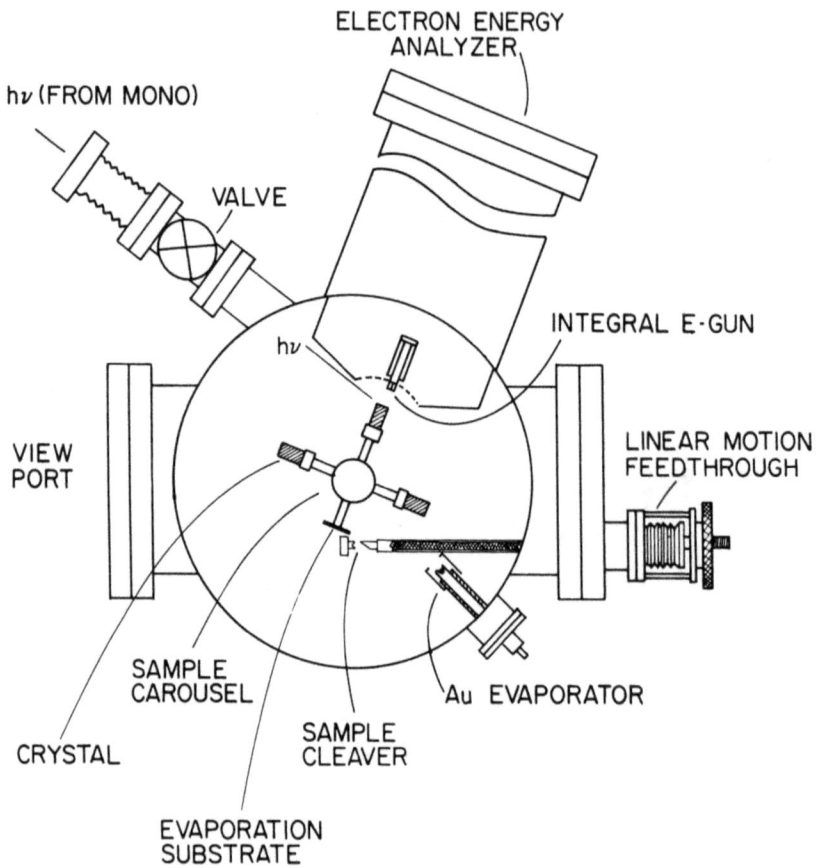

Fig. 12. Schematic of the experimental chamber used in this work. The integral electron gun allows Auger measurements to be made. Not shown is the LEED arrangement in the chamber. The chamber is shown in a configuration for cleaving single crystals. Five to eight cleaves could typically be made on each of the three crystals. The fourth position on the carousel is for a reference sample formed by Au evaporation.

escape depth curve is relatively broad. Thus, the valence band is often examined with $h\nu$ = 21.2 eV since the valence band matrix elements fall off very rapidly for higher values of $h\nu$. Also, $h\nu$ = 100 eV is used to examine simultaneously the 3d levels of Ga and As since both levels have strong matrix elements for this value of $h\nu$ and the escape depths are reasonably close to their minima. For InP, the highest lying phosphorous core level used in our study is the 2p which is located 129 eV below the valence band maximum (VBM) whereas the In 4d is about 18 eV below the VBM. Quite different values of hv must be used to examine each core level for optimal surface sensitivity. A further advantage is the spectral resolution achievable for core spectroscopy (0.1 to 0.3 eV) which is much better than that usually available in X-ray photoemission spectroscopy (CPS) 0.5 to 1.0 eV.

D. The Experimental Chamber

A schematic of the experimental chamber [41] used in this work is shown and described in the caption of Fig. 12. An important aspect of synchrotron radiation is that the source is inherently ultra-high vacuum since the storage ring is at a high vacuum in order for the electrons to circulate in the ring for hours without being knocked out of orbit by collision with a gas molecule. The experimental chamber was specifically designed to have ultra high vacuum capability for use at SSRL. When the vacuum valve shown is open, radiation from the monochromator is directed on to the sample. There is no window between the storage ring and the crystal being studied, and the chamber can be attached or removed from SSRL without breaking vacuum.

IV. SPUTTER-AUGER TECHNIQUES

We have already discussed the basics of the Auger process in Section II D. Here we will discuss two aspects of the use of AES to study surfaces and interfaces.

The first of these is the chemical shift which occurs in Auger structure due to the chemical state of the element involved in the Auger transition. For example, a Si atom in silicon has characteristic Auger electron lines which have about 20 eV more energy than the Auger electrons from Si in SiO_2 (the Si LVV transition). Making use of this characteristic of Auger emission (when it is sufficiently apparent) allows one to obtain not only a measurement of the amount of an element in a given sample, but to determine its chemical state. This has proven important in probing the Si-SiO_2 interface in device oxides [42]. As will be seen in Section V, the chemical shifts seen in photoemission also play a key role in understanding oxygen chemisorption and oxide formation.

The second aspect of sputter-Auger techniques which will be described here is the ion sputtering (ion etching) technique which can be used to remove the 1000 Å of oxide on a practical structure so that the Si-SiO_2 interface can be studied using Auger [42] (or other spectroscopies). In recent years it has been established that properly done ion milling can provide very uniform craters [42,42].

With optimum sputtering voltage and minimum Auger excitation currents (so that the perturbations caused are minimized), raw experimental data shows heterojunction widths as small as 13 Å for molecular beam epitaxially grown GaAlAs-GaAs heterojunctions [44]. Similar reproducibility of a few Å has been obtained on Si-SiO interfaces after sputtering through 500 to 1000 Å of SiO_2 [45]. It is beyond the scope of this report to describe the detailed techniques used. However, they are available in the literature now or will be shortly [46].

V. OXIDES AND OXYGEN ADSORPTION ON GaAs, InP, AND GaSb

A. Introduction

The GaAs (110) free clean surface has become, perhaps, the best understood of all semiconductor surfaces. This is the result of work from many research groups using LEED, photoemission, theory and other tools. We will first give an overview of the principle results of these studies. Then, having laid the proper foundation by examining the clean (110) surface, we will examine in more detail the recent experimental efforts by the Stanford group to follow the early steps in oxygen sorption in order to make use of our understanding of the free surface and, through the use of synchrotron radiation, to explore oxidation and other phenomena in a manner which has not been previously possible. Finally, correlations will be made between the surface pinning formed by small fractions of a monolayer of oxygen and the interface states found at the semiconductor-insulator interface in real device structures. Models for the interface states will emerge.

Our approach in this article, is to start from an extremely well-founded knowledge of the free surface and proceed via experiments and their interpretation toward the more complex real configurations. Of necessity, the explanation of phenomena will become more speculative as we proceed.

The experimental techniques used in this work will not be described here as they are well described elsewhere [30,37,38-52]. However, it should be emphasized again that certain aspects of this work were only made possble by use of synchrotron radiation providing 6 ≳ hν ≳ 300 eV from the Stanford Synchrotron Radiation Laboratory (SSRL).

B. The (110) Surface of GaAs

Figure 13 gives a schematic indication of our present understanding of the GaAs (110) free (clean) surface. Although careful

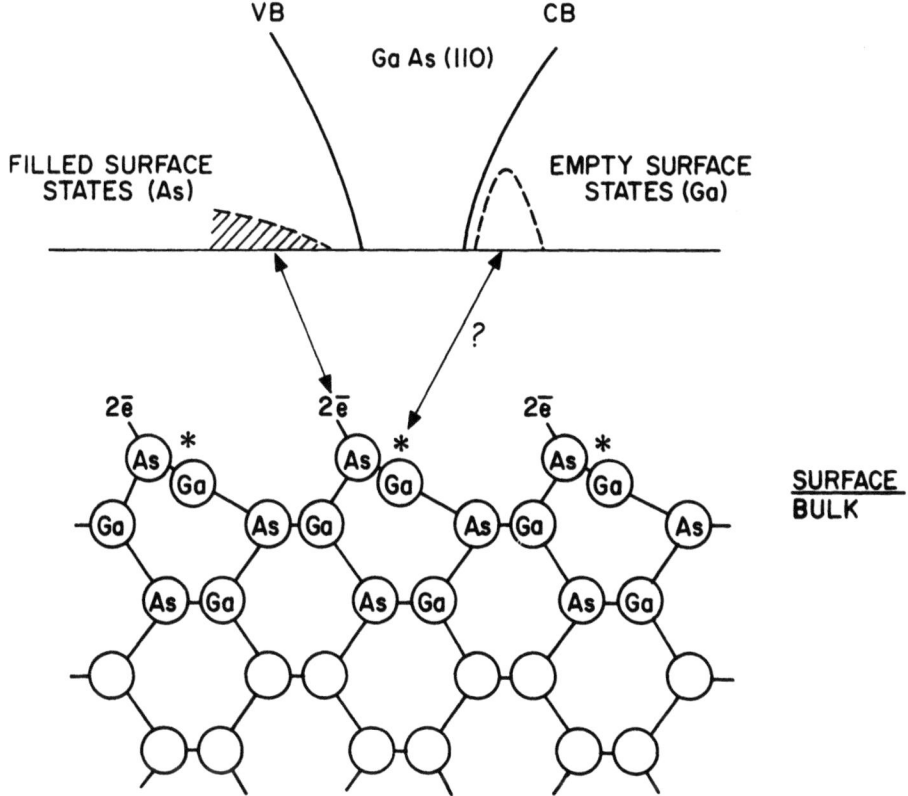

Fig. 13. Schematic of the surface electronic and lattice structure of (110) GaAs. Other 3-5 compounds are thought to be similar. Note the strong rearrangement of the surface atoms and that there are no intrinsic surface states in the band gap. Thus, there can be no intrinsic surface state pinning of the Fermi level. Note also that the dangling bond electrons are associated principally with the As surface atoms, wheareas the empty surface states are predominantly assoicated with the Ga surface atoms.

measurements and calculations have been made systematically only on GaAs (110), it is likely that the surface rearrangement is similar for other 3-5 compounds and, consequently is the surface effects on the electronic structure for 3-5 materials (at least those with band gaps no greater than GaAs). Of critical importance is the rearrangment of the surface atoms. This is intimately connected with and in a certain sense driven by the electronic rearrangement [10-13]. When one covalent bond of each of the surface atoms is broken to form the surface, the surface atoms tend toward their atomic configuration consistent with retention of the covalent bonds with each of their three remaining nearest neighbors. Thus, five electrons will be associated with the surface As and three with the surface Ga. It is this surface electronic rearrangement which is the surface rearrangment. The final positions and bond configurations of the surface atoms are determined by achieving the lowest energy state consistent with minimizing the bond and strain energies. The strain, which is rather large, is due to the lattice mismatch between the rearranged surface and the bulk lattice atoms.

As a result of the surface electronic rearrangement, the surface As and Ga atoms, respectively, move toward p^3s^2 and sp^2 configurations. In these configurations, there are five and three valence electrons associated with the surface As and Ga atoms, just as in the case of the free atoms. Thus, the Ga surface atom is constrained to move toward a sp^2 planar configuration because it must use all of its available electrons to form covalent bonds with its three remaining nearest neighbors. In contrast, the As surface atom needs only three of its five valence electrons to form covalent bonds with its three nearest neighbors. As Harrison [53] and others have shown, the lowest As bond energy is obtained by this electron taking up a s^2p^3 arrangement with the three p electrons forming the bonds and s electrons forming filled s^2 orbitals. As a result of the As bonding electrons moving toward a p^3 configuration and the Ga toward sp^2, the As moves outward by about 0.5 Å and the Ga inward by about 0.3 Å. Low energy electron diffraction (LEED) work [54-57] has played the essential role in establishing the rearrangement of the surface atoms and, although there may still be discussions among the LEED and other workers as to the fine details of the rearrangement, there is no argument over the essential features of the rearrangement. Mark et. al., concluded from their recent LEED studies that the lattice distortions extend three molecule levels below the surface [58]. The filled surface states are associated with the As atoms and the empty states largely with the Ga atoms as indicated in Figure 13 [59] however, Bauer et. al. [60], have shown that the empty surface states may have considerable amplitude on the As as well as Ga atoms.

Even more important, independent confirmation for this model is found through the results of photoemission [61-63] and contact potential [64-66] measurements followed by theoretical calculations [67] of the electronic structure. As can be seen, a striking feature of Fig. 13 is the lack of surface states in the band gap region. After much confusion due to the extrinsic surface states (i.e., states due to defects or impurities) and intrinsic surface states (i.e., states characteristic of the perfect rearranged surface), it was established through careful photoemission [61-63] measurements that there were no surface states in the band gap in agreement with the original work of van Laar and Scheer [64] and van Laar and Huijser [66,67].

One striking characteristic of the results summarized in Fig. 13 is the reversal of the chemical nature of the As and Ga atoms at the surface [61]. Note that the As has electrons not involved in covalent bonds, whereas Ga has none; thus, the surface As atoms appear more metallic than the Ga atoms. As will be reported later, this can be tested by chemisorbing oxygen on the GaAs (110).

In order to appreciate the progress made, it is interesting to compare Fig. 13 to the early GSCH (Gregory, Spicer, Ciraci and Harrison) model proposed in 1974. That early model did not include the lattice reconstruction and had the empty intrinsic surface states in the upper half of the band gap. However, it conforms to Fig. 13 in all other key ways. In fact, based on the GSCH model [59], the first suggestions were made that the oxygen would chemisorb on the As rather than Ga surface atoms.

The fact that Fermi level pinning was sometimes found on GaAs despite the lack of intrinsic surface states focused attention on the important concept of extrinsic as well as intrinsic surface states, i.e., it became apparent that surface states due to impurities and defects as well as intrinsic surface states must be considered. As we will see, these concepts became very important in considering passivation and Schottky barriers.

After the surface rearrangement and lack of intrinsic surface states in the gap was established, new calculations of the surface electronic structure were undertaken and these showed the movement of surface states out of the band gap region due to the surface rearrangement [67]. Thus, one emerges with strong foundations for the results summarized in Fig. 13 in which the results from various experiments and theory all converge to give a strong self-consistent picture.

There is a final feature of Fig. 13 which needs emphasis; this is the large lattice mismatch between the rearranged surface atoms and the bulk lattice. This leads to considerable strain in the surface and can be a driving force in producing extrinsic surface states. For example, the strain may be relieved by formation of local surface defects which, in turn, may pin the Fermi level. As will be seen later, these effects [55,66] can also be important when oxygen is chemisorbed on the surface.

C. Sorption of Oxygen and GaAs, GaSb, and InSb (110) Surfaces

1. <u>Introduction</u>. From a fundamental point of view, the sorption of oxygen on the (110) face of GaAs gives a further test of the results for the free surface outlined in Section B above. It also provides a first step in making connection between knowledge of the free surface and the interface states produced by growing native oxides on the 3-5 surfaces to make practical devices.

The very special capabilities of SSRL were used in this work. As will be seen, both the highest lying core states and the valence states are of interest. The chemical shifts on the core states allowed us to follow the chemical reactions in great detail; whereas the changes in the valence structure with oxygen exposure as well as the CIF studies give some indication of the effect of the oxyygen adsorption on long range order.

In most cases, the electrons of interest were emitted with kinetic energies near that which gives minimum escape depth (about 5 Å at 60 eV); however, since this minimum is rather shallow, it was often advantageous to move somewhat away from it in order to optimize the matrix element for the transition of interest. In all cases, most of the electrons came from within the first one to three molecular layers of the semiconductor [36,38,55].

2. <u>Core Levels and Surface Chemistry</u>.

2a. <u>Oxygen Chemisorption on As</u>. In the course of these investigations, we found that the state of excitation of the oxygen could play a key role in determining the type of sorption which took place [68]. Figure 14 shows the core levels after exposure to unexcited oxygen [21,40,42,48,68]. In order to insure that the oxygen was unexcited, the Alpert ionization gauge was turned off, the vac-ion pump sealed off, and all other possible sources of excitation removed. Details are given elsewhere [40,41,52,68].

The data of Fig. 14 are important in that they give strong evidence that the oxygen chemisorbs on the As and not Ga sites. Note that, as the oxygen exposure increases, an As peak shifted to higher binding energy by 2.9 eV appears and grows with an accompanying decrease in intensity of the unshifted As peak. After an exposure of 10^{12} L (L = Langmuir = 10^{-6} torr-sec) of oxygen, the shifted and unshifted peaks are approximately equal in amplitude. At this point, the surface As atoms are approximately saturated with oxygen and the near equality of the shifted and unshifted peaks is as would be expected for an escape depth of approximately two molecular layers [21,40,41].

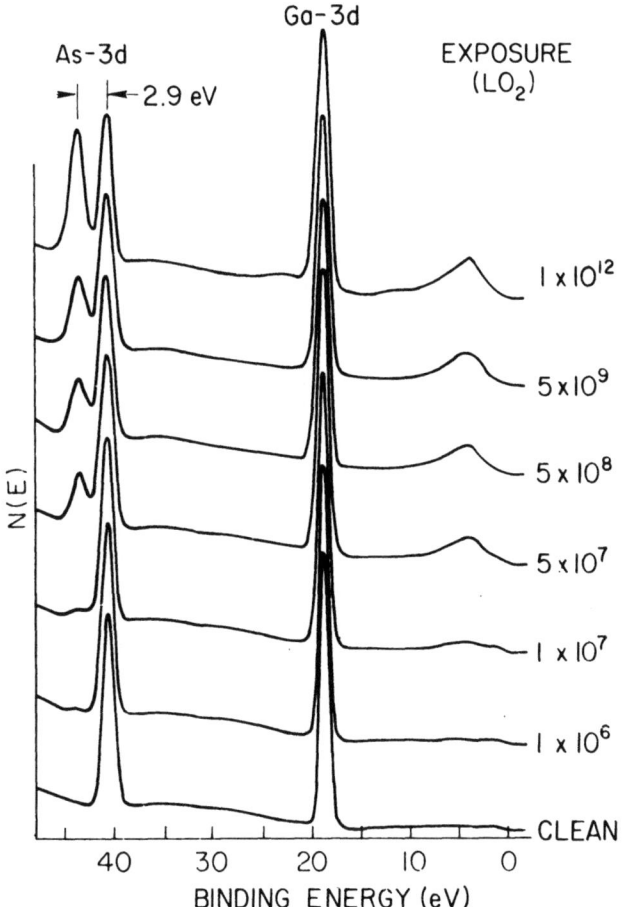

Fig. 14. Effect of O_2 exposure on cleaved GaAs (110). The exposures are given in terms of Langmuirs (L), i.e., 10^{-6} torr-sec. This is approximately the coverage necessary to form a monolayer if every oxygen striking the surface stuck to it. Note the well-defined As shift and lack of Ga shift at any exposure. This indicates chemisorption on the As atoms with little breaking of the bonds binding the surface As's to the surrounding crystal. The matrix elements for valence band excitation are so small that the valence band of GaAs can hardly be seen at this value of $h\nu$ (100 eV); however, the excitation from the oxygen 2p near 5 eV can be seen to grow with oxygen exposure.

Whereas a chemical shift of 2.9 eV is observed on the As peak, only a broadening is observed on the Ga peak at higher coverages. This is indicative of second order effects taking place on the Ga site due to chemisorption on the As site, as suggested by Chye et. al. [29]. This effect on Ga is also detected in energy loss spectroscopy [69].

Recent theoretical work by Mele and Joannopoulos [71] has confirmed that the oxygen will chemisorb on As rather than Ga sites and has also shown that "second-order" effects will take place on the Ga site. This will affect the excitonic transition from the Ga 3d states to the empty states above the conduction band minimum as seen in either energy loss [70] or CFS spectroscopy [26,30,60]. In particular, it appears that the disappearance of this feature from the respective spectra is due to these "second-order" effects on the Ga surface atoms. Goddard et. al. [71] have also shown theoretically that the oxygen will chemisorb on the As. There still appears to be disagreement between the two theoretical groups [71,72] as to whether the chemisorption involves oxygen atoms or molecules.

It is appropriate that we conclude this section by noting that an oxide or other passivating and protective layer could be added to the GaAs via bonds with the As surface atoms. In this way, strong bonds can be made without (in principle) breaking bonds in the GaAs [73]. As will be seen later, such bonds may be broken and/or other defects produced by secondary effects such as strain.

2b. <u>Kinetics of Oxygen Sorption and State of Oxygen Excitation</u>. The fact that only chemisorption takes place with oxygen in the ground state helps focus attention on a critical aspect of surface chemistry. This is the fact that one cannot simply apply equilibrium thermodynamics but must consider activation energy barriers which have to be surmounted before thermodynamic equilibrium can be approached. Consider the barriers for the formation of bulk Ga and As oxides. These barriers involve breaking of the bonds holding the O_2 molecule together as well as those bonding the Ga and As atoms to each other. The occurrence of only chemisorption indicates that the energy is not available to surmount all of the activation barriers necessary to obtain bulk oxides if only unexcited oxygen is used. However, as Figures 15 and 16 show, [40,68], the use of excited oxygen provides the energy necessary to break down these barriers.

Details of what is meant by "excited" oxygen are given elsewhere [40,41,68,69]. Suffice to say here that the apparatus was built so that the sample was well out of the line of sight from the source of excitation and that gas scattering from many metallic surfaces had to occur before the excited oxygen could reach the sample. Because of this, as well as other factors, we believe that by far the dominant excited species is the singlet state of the molecule (the ground state is a triplet) which has a lifetime of approximately 15 minutes and which has an energy of about 1 eV above that of the ground state [30].

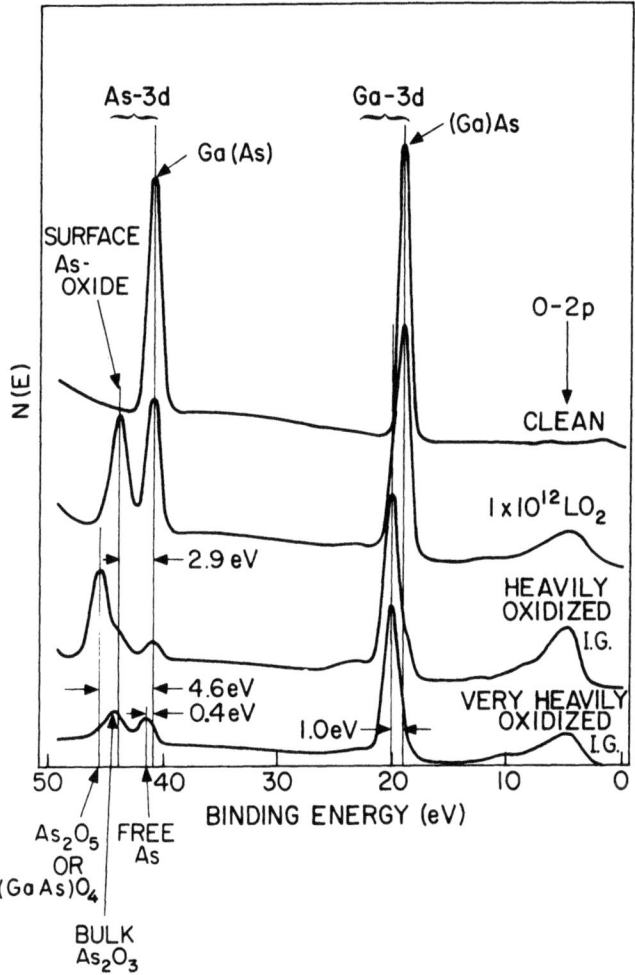

Fig. 15. Difference in the core levels for exposure to unexcited O_2 and to excited oxygen. The curve a is for clean GaAs. Curve b is for exposure of 10^{12} L of unexcited oxygen. Curve c is for an exposure of only 5×10^5 L of oxygen with the ion gauge on after the 10^{12} L exposure and an ionization current of 0.4 ma. The curve d is for a freshly cleaved exposed to 1×10^5 L with 4 ma of ionization gauge on, whereas As and Ga oxides as well as elemental As are formed with the ionization gauge on. The effect of the excited oxygen is much greater when the surface is already saturated with chemisorbed oxygen as seen in Fig. 4.

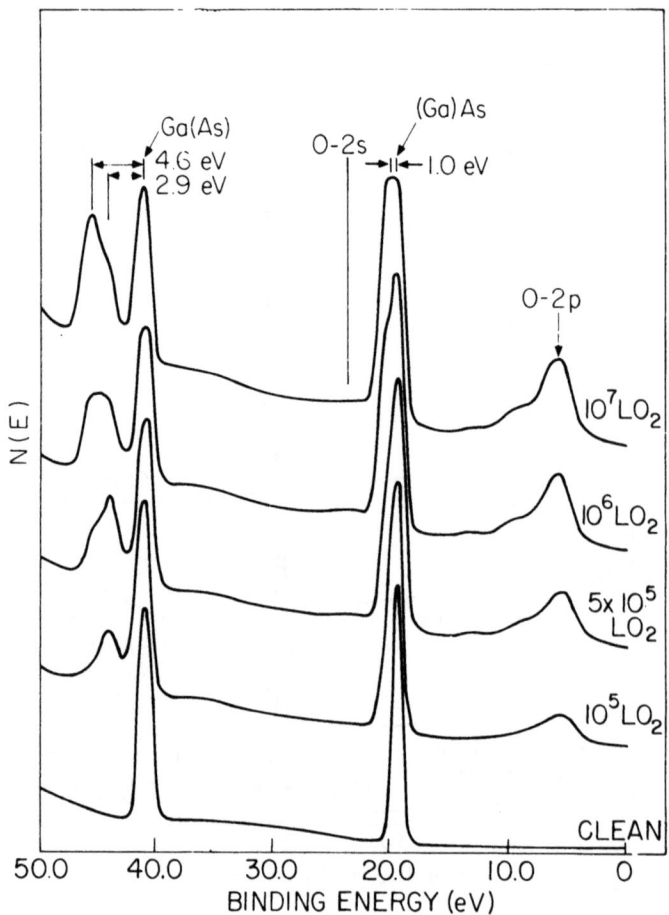

Fig. 16. EDCs of clean p-type GaAs (110) and the clean surface exposed to excited oxygen at $h\nu = 100$ eV. Note that the chemisorbed state appears first followed by oxide formation. However, the effect of the excited oxygen is not nearly as large as for Fig. 3.

In Fig. 15 we show the effect of excited oxygen. As can be seen, Ga_2O_3 and a number of As oxides as well as elementary As is formed [3,24]. It is likely that the species which appears early in the exposure to excited oxygen (curve c of Fig. 15) may be (GaAs) O_4--the equivalent of SiO_2 where each Ga is bonded to each As through a single O, although other possibilities exist [40,73,74]. This species appears to be unstable and to break down into the separate As_2O_3 and Ga_2O_3 oxides (see Fig. 15). Since As_2O_3 is volatile, it probably evaporates from the surface, leaving a deficit of As. Since the heat of formation of Ga_2O_3 is much larger than that of As_2O_3, the Ga_2O_3 probably "steals" oxygen from the As oxides, leaving metallic As. Interestingly, the effects demonstrated in Figure 15 seem to characterize many of the phenomena observed in native oxide growth for practical devices.

Figure 16 shows [68] the effect of exposure of a clean surface to smaller amounts of excited oxygen. Unfortunately, the instrumental resolution in this figure was not as good as that for Figures 14 and 15 as it was taken earlier in our experimental program. It is useful to compare Fig. 16 to Figs. 14 and 15. Note that, just as in Figure 14, the first strong structure produced in Figure 16 by the excited oxygen is the 2.9 eV shifted As peak, indicating chemisorbed oxygen. However, comparison with Fig. 14 shows that this occurs at exposures approximately 500 times less than that necessary with unexcited oxygen. Also, note that the 2.9 eV peak in Fig. 16 is asymmetric toward lower energy and that the Ga peak has broadened toward lower energy. Both effects indicate the formation of relatively small amounts of bulk Ga and As oxides. This is even more pronounced with increasing exposure where 4.6 and 1.0 eV shifted As and Ga peaks become dominant, indicating that the oxides are becoming the dominant species.

At this point, it is important to make note of possible connections between the data described here and recent developments in techniques of growing native oxides on GaAs by use of plasma discharges. It has been recently found that high grade and perhaps quite useful oxides can be produced by having a reasonable pressure ($\sim$1 torr) plasma separated from the GaAs by a reasonable distance (for example, 5 to 10 cm) containing neutral gas [75]. It seems likely that the main function of this neutral gas is to reduce greatly the number of ions or atoms striking the GaAs so that the dominant excited species will be the oxygen singlet states. This probably allows for a "gentle" oxide formation, thus minimizing the amount of defect formation.

Oxygen sorption on InP and GaSb has also been studied with respect to the formation of bulk oxides and the use of excited oxygen -- the results are summarized in Table 2. As can be seen, there is a strong relationship between the kinetics of oxidation and the heat of formation. More details are given elsewhere [74].

3. Fermi Level Pinning Due to Oxygen Chemisorption.
3a. Pinning Due to Sorption of Less Than a Monolayer of Oxygen.
For most practical devices, it is the traps at the semiconductor-insulator interface which is critical. As Figure 13 indicates, there are no intrinsic surface states in the GaAs band gap. As was mentioned earlier, the empty extrinsic surface states (due to defects and impurities) on some cleaved surfaces are low enough ($\gtrsim 10^{12}/cm^2$) so that the Fermi level is not pinned at the surface for n-type samples. For p-type GaAs samples, it has generally been found

TABLE 2

Correlation between the heat of formation, H, of semiconductors and the sorption of oxygen. GaSb with H = 10 Kcal/mol will form bulk oxides with unexcited oxygen. GaAs needs excited oxides to form bulk oxides even with excited oxygen.

Compound	Heat of Formation (Kcal/mol)	Formation of Bulk Oxides	
		With Unexcited Oxygen	With Excited Oxygen
GaSb	10 (0.5 eV)	Yes	Yes
GaAs	17 (0.8 eV)	No	Yes
InP	22 (1.0 eV)	No	No

[21,59,76] also that the Fermi level is not pinned at the cleaved surface. Thus, it is interesting to: (1) see how the Fermi level pinning is affected by oxygen chemisorption, (2) to attempt to understand the observed pinning, and (3) to see if this pinning can be related to that observed on practical device interfaces. In this section, we will review work aimed at these three objectives.

Figure 17 shows typical data on the change of the Fermi level pinning position with exposure to oxygen. The n-type samples were taken with unexcited oxygen [40,68]. The p-type data [76] was taken before the importance of oxygen excitation was understood; therefore, the oxygen excitation was not so systematiclaly controlled; however, it is thought to have been absent because of the use of a Redhead rather than an Alpert pressure gauge. From studies of the shifted versus nonshifted As peak on an n-type sample in which unexcited oxygen was used (Fig. 15), an estimate of the coverage was obtained [40,68]. The coverage at 10^7 L exposure was small, about 3% of a monolayer. This is a very striking result in that the pinning takes place after very little oxygen adsorption. Note also that the final pinning on the n-type sample is independent of whether or not the Fermi level was pinned on the free surface suggesting the final extrinsic pinning states do not depend on any initial extrinsic states.

As Table 2 suggests, the kinetics of oxygen uptake is very different on GaAs, InP, and GaSb. GaSb forms [40,61] bulk oxides

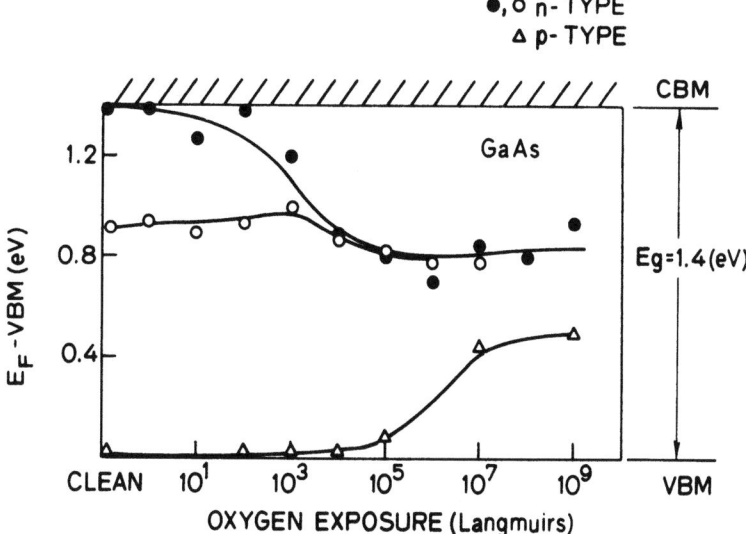

Fig. 17. Illustration of the Fermi level pinning produced on n- and p-type GaAs by oxygen adsorption. The pinning saturates for oxygen coverages which are a small fraction of a monolayer and is independent of whether or not the Fermi level was pinned on the "free" surface.

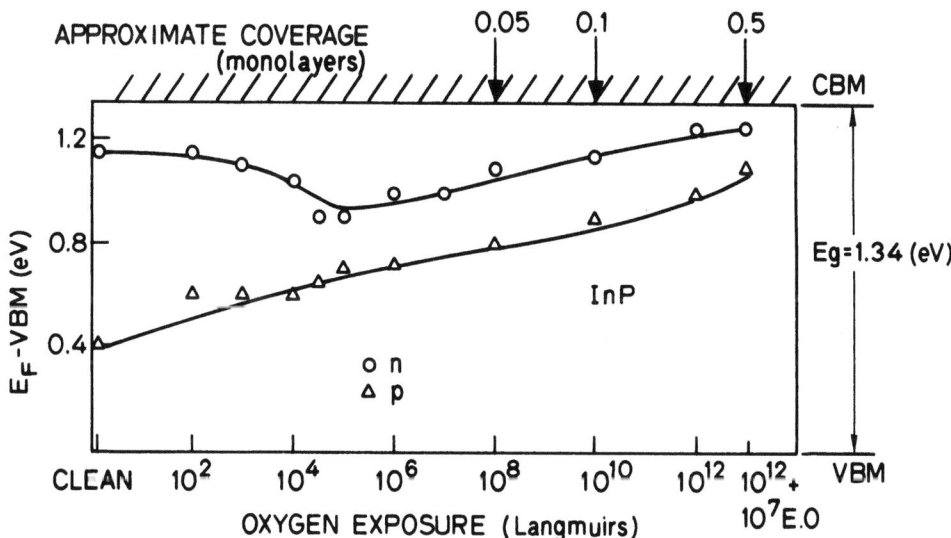

Fig. 18. The movement of the Fermi level on InP as a function of oxygen exposure. 10^8 L corresponds to approximately a a tenth of a monolayer.

even with unexcited oxygen. As might be expected, it also takes up oxygen much more readily. For example, 10^5 L exposure results in 0.5 monolayer coverage. More details are given elsewhere. In contrast, InP [74,77] was not found to form bulk oxides even when excited oxygen was used, as would be expected since the rate of oxygen uptake was slower. For example, 10^{10} L of unexcited oxygen gave only about 0.1 of a monolayer coverage.

The Fermi level behavior with oxygen exposure was also characteristic of each material. For GaSb, the Fermi level pins near the valence band maximum (VBM) for both n- and p-type samples at relatively low (about 10^4 L) exposures, corresponding to about 0.05 monolayer of sorbed oxygen. GaSb is different from GaAs and InP in that the pinning position rises for exposures over 10^5 L. This may be due to bulk oxide formation [77].

For InP, Figure 18 shows the change of Fermi level pinning with exposure [74,77]. The striking thing here is that the pinning position rises to near the conduction band minimum. For n-type InP, the final position is within 0.1 eV of the conduction band minimum.

3b. <u>Comparison With Pinning of Practical Device Oxides</u>. Since an ultimate object of this work is to understand the Fermi level pinning of practical device oxides it is useful to compare the pinning positions found with less than a monolayer of oxygen and that found on practical thick device-type oxides. Most of the pinning positions for the thick oxides were obtained from the recent review article of Weider [78]. The fact that a pinning position can be obtained in these samples with interface state densities which are reasonably well controlled. The criteria for pinning on the thick device oxides is that of Weider [78]. Only GaAs and InP are included in Table 3 since "thick device oxide" data were only available for these oxides.

The relatively good correlation between the pinning position obtained for the "thick device oxides" and that for surfaces with less than a monolayer are rather striking. If the correlation is not accidental, it suggests that the levels leading to the Fermi level pinning may be similar for the two sets of samples. Thus, in the next section, we will present further data concerning the changes in valence band structure at the surface with oxygen coverage and on this basis, make a general suggestion as to the source and nature of the levels which pin the Fermi level for both submonolayer and thick oxides. This will be followed by a section summarizing the data and suggesting possible levels responsible for the pinning.

4. <u>The Effect of Oxygen Chemisorption on the Valence Band at the Surface: Development of a Model for Interfacial States</u>.
Again making use of the tunability of synchrotron radiation, we can examine the valence electronic structure at the surface as a func-

TABLE 3

Fermi Level Pinning Energy* After Indicated Surface Treatment

Note the similarity between the Fermi level pinning position for thick oxides formed by various techniques and that obtained by a fraction of a monolayer of oxygen. Note also that the pinning position for Schottky barriers (the example here is Cs) is close to that for Oxygen (see Ref. 28).

Material	Thick Oxide	Submonolayer of Chemisorbed Oxygen	Submonolayer to Many Monolyaers of Cesium
GaAs (n)	0.83 (Ref. 32) 0.8 (Ref. 33) 0.5 (Ref. 34)	0.65 to 0.8 (Refs. 7,9,35) and Fig. 13	0.6 to 0.8 (Ref. 36) 0.7 (Ref. 37)
InP	0.14 (Ref. 30) 0.087 (Ref. 30) 0.075 (Ref. 30) 0.1 (Ref. 29) 0.1 (Ref. 31)	0.05 to 0.2 (Ref. 26) and Fig. 14	0.15 to 0.45 (Ref. 26)

tion of oxygen chemisorbed. Since the matrix elements for excitation from the valence band maximum fall off swiftly with increasing photon energy, as low a photon energy as possible was used, consistent with a short escape depth. Thus, most of the studies were made with radiation at $h\nu = 21$ eV (where the escape depth is about 7 Å, i.e., about 1.5 molecular layers) [40].

The valence band structure of many cleaved surfaces has been studied at 21 eV as a function of the oxygen exposure. The complete results are given elsewhere [21,40,41,73]. In Figure 19, we present a typical result. The characteristic of Fig. 19 which

---

* Below conduction band minima.

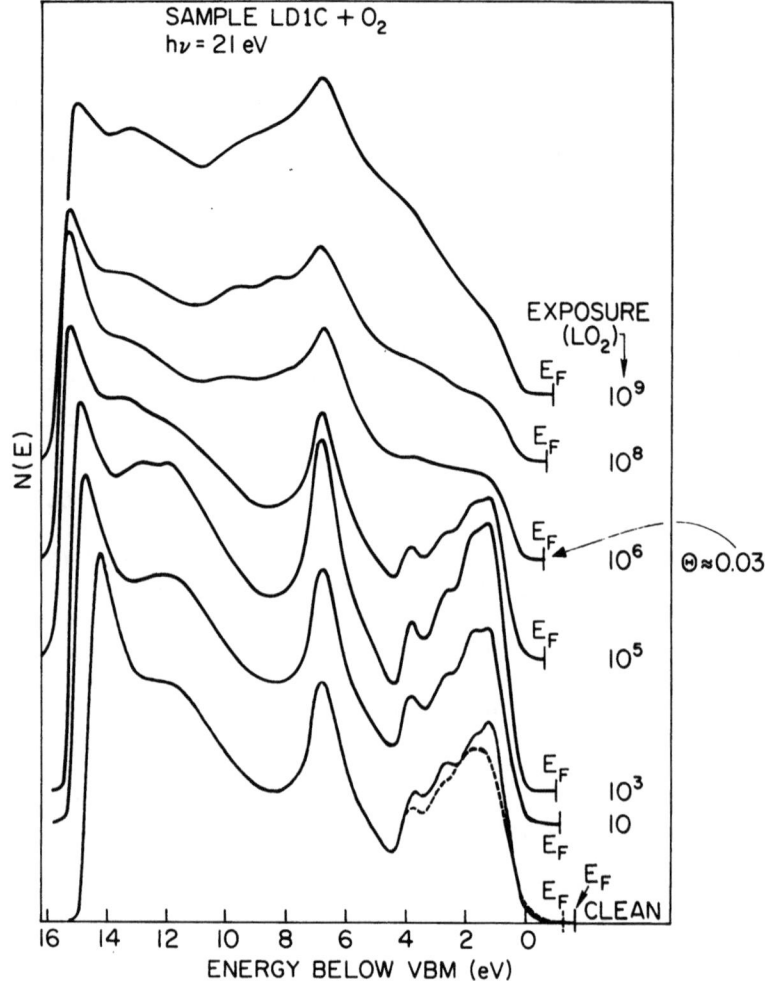

Fig. 19. The change in the surface valence band structure in GaAs due to oxygen adsorption. The structure most affected is near the uppermost part of the valence band (within 5 eV of the VBM). Note the sharp change in structure for exposures between 10^6 and 10^7 L (coverages of approximately 5 percent of a monolayer or less). This is attributed to a local disordering of the surface. A similar transformation was seen for all samples studied.

we want to emphasize is the sudden and extreme change in the top part of the valence band (i.e., within 4 eV of the VBM) which occurs at an exposure of 10^6 to 10^7 L (coverage of about 0.03 of a monolayer). Although there are second-order changes in the valence band for exposures < 10^6 L, the change between exposures of 10^6 and 10^7 L is first order, suggesting a very strong change in the valence band structure; e.g., something approaching a phase change near the surface [73]. Most important, it is just in the range of this coverage that the Fermi level becomes pinned.

We suggest that the surface which is already [21,73] highly strained (see Fig. 13 and Section VI B) is brought under additional strain by the chemisorption of oxygen and that, with a few percentages of a monolayer, defects are produced by the strain which causes the surface to "break up" -- losing much of its long-range order. These defects are formed in sufficient density to stabilize the Fermi level pinning. We further suggest, based on the correlations with the "thick device" oxides (see Table 3), that these defects are the same as those producing the pinning on the "thick device" oxides.

In order to support this line of reasoning, let us examine in more detail the atomic rearrangement on the free surface and the effect that small amounts of oxygen chemisorption will have on this. In Figure 20, we show symbolically (due to the need to project three-dimensional phenomena on two-dimensional paper) three arrangements of a GaAs six atom ring at the surface. Our purpose is to give better insight into the mechanism which produces the rearrangement on the free surface and, most importantly, how the chemisorption of oxygen may change that rearrangement. As indicated by Fig. 16, As on the free surface moves outward because of p^3 bond orbitals. However, when the chemisorbed oxygen effectively removes electrons from the As (as shown by the large, 2.9 eV, As chemical shifts), the As has less than five valence electrons, and $p^3(s^2)$ bond orbits are no longer possible. As a result, the bonding scheme must move back toward (sp^3). The importance of this is that it must lead to a new rearrangement at the site of the oxygen chemisorption. We again symbolically indicate this in the bottom part of Fig. 20. Although it is not apparent from the two-dimensional drawings of Fig. 20 (one should make a three-dimensional model to see the effects) additional strain will be produced by the second rearrangement at the chemisorption site in an otherwise perfectly rearranged free surface. As more and more chemisorption events take place, the lattice will be placed under increasing strain until the "disordering phase change" identified in the experimental data of Fig. 19 takes place. Defects may be produced in this process and extrinsic surface states in the band gap will be associated with these defects. With the "phase change," the defects will be produced in sufficient number to pin the Fermi level, and this pinning mechanism is postulated to maintain (or reestablish) itself for "thick device" oxides.

We have outlined a model above and made plausibility arguments for it based on available experimental results. However, it must

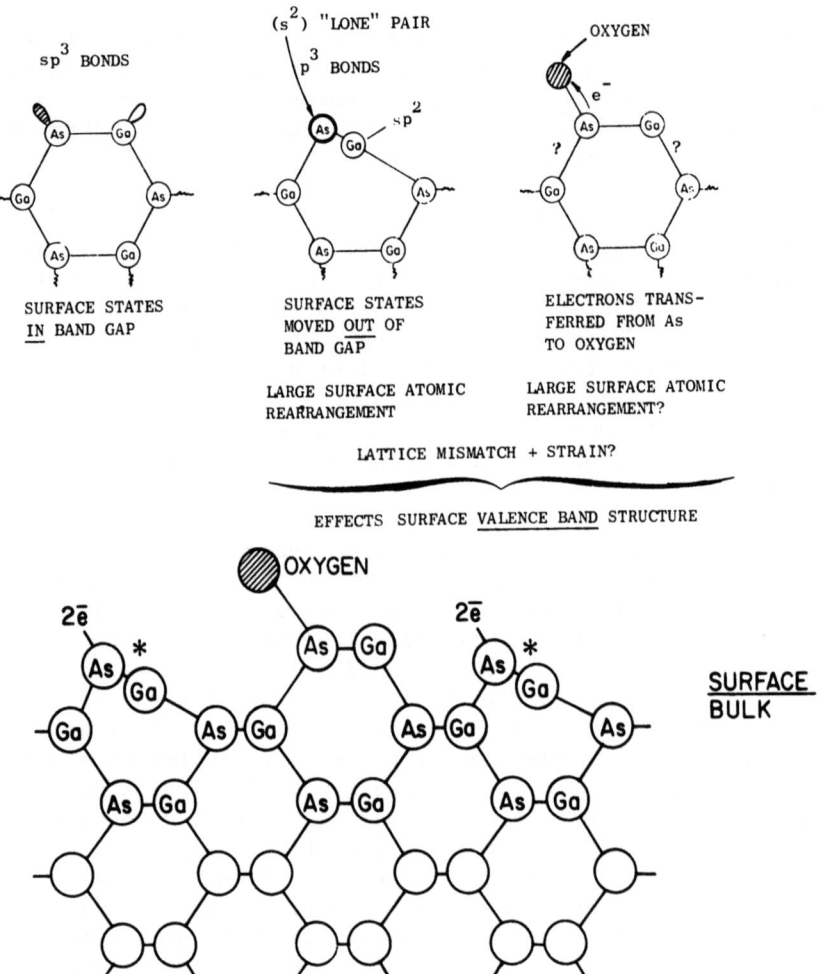

Fig. 20. A schematic representation of the various atomic arrangements which might occur within the unit cell at the surface. In the top row, the first sketch indicates an unrearranged surface for which theory indicates that the surface states lie within the band gap. The middle sketch indicates the electronic and atomic rearrangement which takes place on the (110) GaAs surface. This rearrangement moves the surface states out of the band gap. The third sketch indicates the fact that a strong change in the rearrangement probably takes place when oxygen is chemisorbed. The bottom row attempts to give an impression of the oxygen adsorption on the surrounding lattice.

be considered as a beginning model to be modified and finally proven or disproven by subsequent results. Its importance lies in making a somewhat systematic connection between fundamental and applied results and, hopefully, providing a stimulus for further work.

VI. SPUTTER-AUGER STUDIES OF THE $Si-SiO_2$ INTERFACE ON REAL DEVICE OXIDES

A. Introduction

For perspective, we will examine briefly some of the studies of the $Si-SiO_2$ interface for real devices. Here the history contrasts sharply with that of the passivation of the 3-5's. Whereas the more fundamental studies of the 3-5 started before practical 3-5 MOS devices were in production (and, in fact, they still are not in production), such studies on the $Si-SiO_2$ interfaces did not become common until integrated circuits (IC's) using MOS structure had been successfully put into production. However, research of a more basic nature has been growing very noticeably in recent years.

Very sophisticated methods have been developed to obtain electrical characteristics of the interface in terms of such things as interface state densities, slow and fast traps, and electrical transport properties near the interface. These question are dealt with elsewhere in this volume (DeClerck, Chapter 3). Rather we will concentrate on attempts to define the materials parameters which characterize the interface and determine the electric characteristics.

To make it tractable, we will put another restriction on this discussion. We will restrict ourselves to materials studies of real device oxides. For a more complete review and references to the literature, the reader is referred to the Proceedings of the Topical Conference on the Physics of SiO_2 and Its Interfaces, sponsored by the American Physical Society at Yorktown Heights, N. Y., March, 1978 [79]. More recent work is also underway, for example, in studying the first steps in the oxidation of the cleaved (111) [80] surface similar to that reported in Section V for the 3-5 compounds. Other Auger studies are underway on very thin oxides [81, 82]. However, real oxides have thickness of the order of 1000 Å, and only He backscattering experiments [83] have the ability to probe the interface nondestructively through such thicknesses. A discussion of this technique is given in Chapter 12 (Nicolet). Suffice it to say that it is difficult to obtain the depth resolution necessary to define the transition region well from such backscattering experiments. Further, one cannot obtain information on the chemical state of the atoms by this technique.

Returning to photoemission and Auger techniques, one can see from the escape depth data of Fig. 4 why one cannot use these techniques for studies of an interface separated from the outer surface by 1000 Å of oxide. The energies available for photoemission or Auger normally lie below 2 KeV; the escape depth is many time smaller than the oxide thickness. Thus, even if the exciting radiation could penetrate to the interface, the excited electrons could not escape. The only solution is to remove the oxide until one can study the interface. This is normally done by very careful sputtering techniques; although, chemical etching has also been used [79]. Here we will concentrate on the sputtering technique [43,46] and the use of AES to study the interface. More information could be obtained if UPS and XPS were also used. However, to the author's knowledge, these techniques have not yet been used in conjunction with the very precise sputtering techniques necessary to obtain good depth resolution [43,46]. Once again, we will draw on work done at Stanford and Varian to illustrate the sputter-etching studies of the Si-SiO$_2$ interface. This is both because the author is most familiar with these results and because, as far as he knows, the results represent the forefront of the existing art [42-46].

Until a few years ago, it was generally thought that the depth resolution possible using sputter etching was 5 to 10% of the depth through which the sputtering had been done. This meant a depth resolution of 50-100 Å for practical oxides. As we will see the depth resolution has been increased over an order of magnitude for SiO$_2$ in the last few years. The secret has been obtaining a very uniform sputter crater [43,46]. This is achieved by using an ion beam with a very constant current density across it and to raster the ion beam over several millimeters in order to form a very uniform crater. This is shown schematically in Fig. 21 [84]. Here the transition region between the SiO$_2$ and Si is called the "interface". However, through the use of the Auger chemical shift [85],

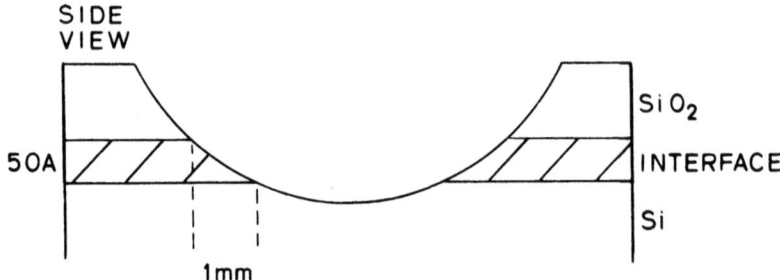

Fig. 21. Schematic illustration of the crater formed by optimum sputter-etching through SiO$_2$ to the Si interface of an integrated circuit wafer.

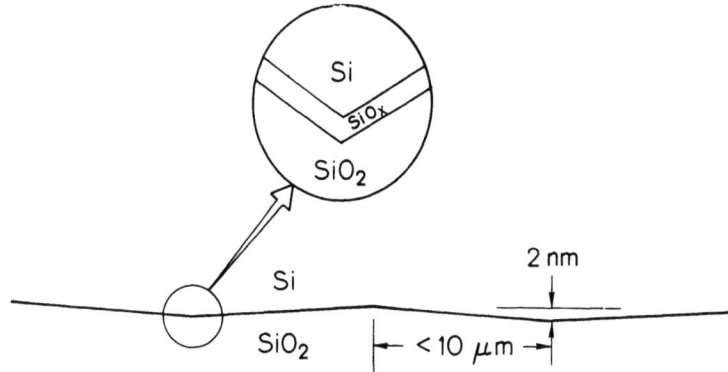

Fig. 22. Model of the Si-SiO$_2$ interface. The total transition region is about 2.0 nm wide due to long wavelength undulations. The SiO region may be as narrow as one monolayer.

it was possible to show that the width of this was caused by undulations in the interface. Figure 22 gives a detailed schematic of this interface region for the case where it is 20 Å [86] rather than 40 Å wide. Returning to Fig. 21, one can see that if one scans [43] the Auger beam over a small region near the center of the crater, the variation in thickness will be very small (of order of a few Angstrom).

In Section V, we illustrated how the core shifts could be used to determine the chemical state of atoms. Similar techniques can be used in AES. Figure 23 shows a sputter profile through a ~1000 Å device oxide. The Auger system was programmed so that it only detected signals in certain discreet energy ranges. For example, it picked up the Si$_{LVV}$ line only when this came from Si in Si. (The horizontal Si$_{LVV}$ signal from zero to about 950 Å was for zero signal.) In contrast, for the Si$_{KLL}$, the energy range was set so that it swept across the signal from Si in SiO$_2$ and in Si (see the upper part of Fig. 23). The O$_{KLL}$ was for oxygen bonded to two silicon atoms.

The drop in the O$_{KLL}$ and rise in Si$_{LVV}$ gives the width of the transition between SiO$_2$ and Si. Correcting for escape depths, this is found to be 23 Å [45]. However, is this width due to SiO or SiO$_x$ ($0 < x \leq 2$)? Or is it due to a slightly undulating interface as is indicated in Figure 20? We can distinguish by examining the Si$_{KLL}$ spectra as one moves through the transition region. Spectra are given in the upper part of Fig. 23 for the positions indicated on the Si$_{KLL}$ curve in the bottom of the figure. The spectra (this

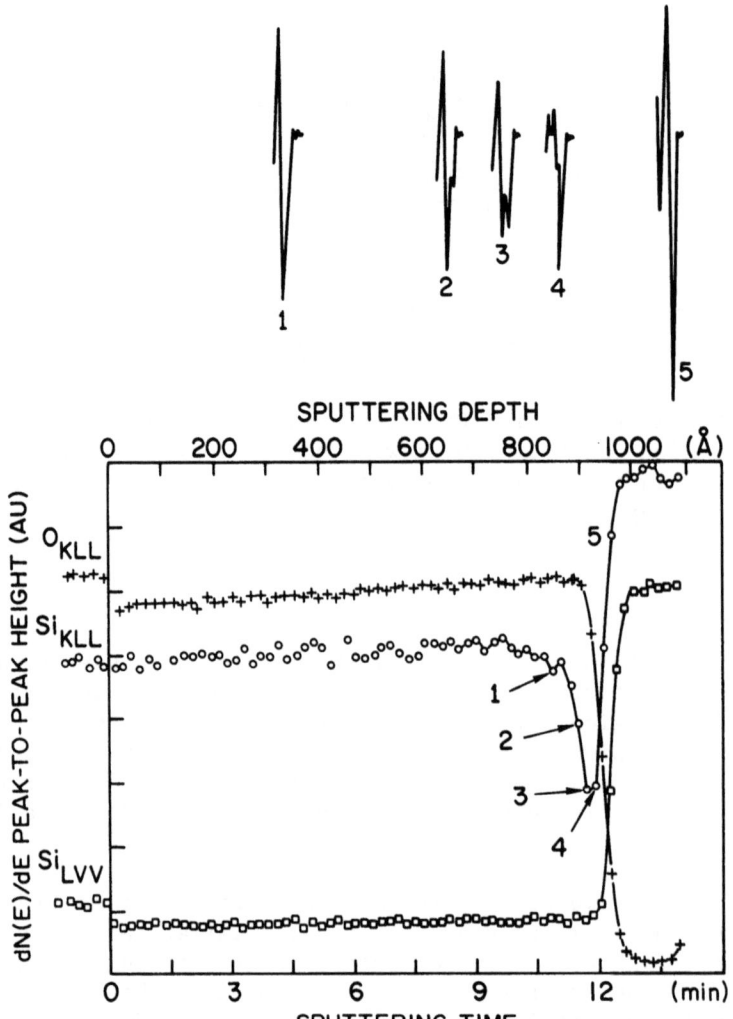

Fig. 23. Chemical depth profile through 1000 Å of thermally grown SiO_2 on Si (100). Sputtering rate is 77 Å/min. Above depth profile is shown a series of Si $KL_{2,3}L_2$ Auger spectra from the SiO_2-Si interface region. Spectra are numbered, and positions in depth from which spectra were taken are indicated on depth profile.

is the first derivative of the energy distribution curve, N(E)) for position 1 is characteristic of Si in SiO_2 and that of position 5 of Si in silicon. The fact that for positions 2, 3, and 4, the dominant peaks are those for Si in silicon or SiO_2 shows that the transition region was predominantly Si or SiO_2. In fact, the interface region in which Si is bonded in intermediate oxides can be limited to 8 Å [45,46]. This is the maximum width for the region labeled SiO_x in Fig. 20.

Using the data handling capabilities of the Varian Automated Auger Microprobe computer, we have recently been able to determine for the first time, rather exactly the chemical state of the Si in this < 8 Å transition region and show that the connective region between the SiO_2 and Si may be as narrow as one monolayer [86]. In order to do this, it was necessary to study energy distribution N(E) rather than the normal derivative dN(E)/dE spectra using the new equipment.

VII. A NEW MECHANISM FOR FERMI LEVEL PINNING OF SCHOTTKY BARRIERS

A. Introduction

The most widely accepted prior model of Schottky barrier pinning has been the "Bardeen" model which assumes that intrinsic surface states characteristic of the free surface pin the Fermi level in Schottky barriers. As emphasized in Section V B, there are no intrinsic surface states in the band gap of the 3-5's understudy here. Therefore, we can reject the "Bardeen" model for the 3-5's.

One might argue that the addition of the metal relaxes the surface and moves the intrinsic surface states into the band gap where they pin the Fermi level. However, as we will show later in this section, we can rule this out by means of the CFS data (Fig. 6, Section II).

In Section V, we have produced a rather detailed model suggesting that the pinning in "thick device" oxides of high practical quality is due to defects produced by strains produced at the interface between the oxide and the 3-5 compound. We were led to this model by new experimental data made possible by the use of synchrotron radiation. We have likewise been led to propose a new model for Schottky level pinning by recent data. Again, this has been obtained chiefly from studies using synchrotron radiation.

As with the oxide phenomena, the new model involves pinning by defects. In this case, the defects are produced by the energy released by the heat of condensation of the metal placed on the surface to form the Schottky barrier [31]. Let us start by examining the experimental data which led us to this conclusion. We will start with GaSb [87] since the results and their explanation are particularly clear cut for this material.

B. An Experimental Examination of Schottky Barrier Formation on GaSb on an Atomistic Level

Figure 24 show the surface Fermi level pinning versus metal coverage for two metals (Au [30,31,87] and Cs [88]) on GaSb. As can be seen, the Fermi level drops very fast as the metal is added, and the final pinning position is reached with about 0.1 of a monolayer coverage. Since the Cs atoms wet the semiconductor surfaces well and are repelled from each other due to the Cs dipole induced by the strong attraction of the semiconductor for the Cs outer most electron, the Cs does not clump but gives a uniformly dispersed coverage until the first monolayer is formed. By studying the d-band structure of the Au atoms for coverages of about 0.1 monolayer and higher, it can also be shown that [30,31], near the coverages at which pinning takes place, the Au atoms have the characteristics of well separated atoms [89] with little Au-Au overlap. This was also found to be true for Au on GaAs, and InP, details of which are given elsewhere [30]. Thus, we conclude that the pinning takes place with close to atomic-like dispersion of the metal and, that any model of Fermi level pinning based on a metal with band-like properties in contact with a semiconductor is inappropriate.

As mentioned above, the fact that there are no surface states in the band gap of the free surface argues against the "Bardeen" model. However, it might be argued that the addition of the metal relaxes the surface rearrangement and moves the intrinsic states into the band gap where they can pin the Fermi level. However, the CFS data of Fig. 6 argues against this since for coverages where the pinning has been completed, there is no movement of the surface excitons. If the intrinsic surface states moved into the band gap, they would provide lower energy states for the exciton so that it would move to lower energies.

The result reported above showing pinning taking place with highly dispersed metal is also important since, in the mid sixties, Heine [90] proposed a model for the pinning which depended not on intrinsic surface states of the semiconductor, as did the "Bardeen" model, but rather on the wavefunctions of the metal inducing new interface states in the semiconductor. In recent years, a large number of detailed calculations have appeared based on this general approach but with large variations in detailed method [47,91-94]. From the results of Fig. 24 and very similar results found for GaAs and InP, [30,31] we concluded that such calculations are not realistic insofar as they start from the assumption of an interface between the semiconductor and a continuous band-like metal film. Calculations based on isolated metallic atoms on the semiconductor surface would be of great interest but, to our knowledge, no such calculations have been performed.

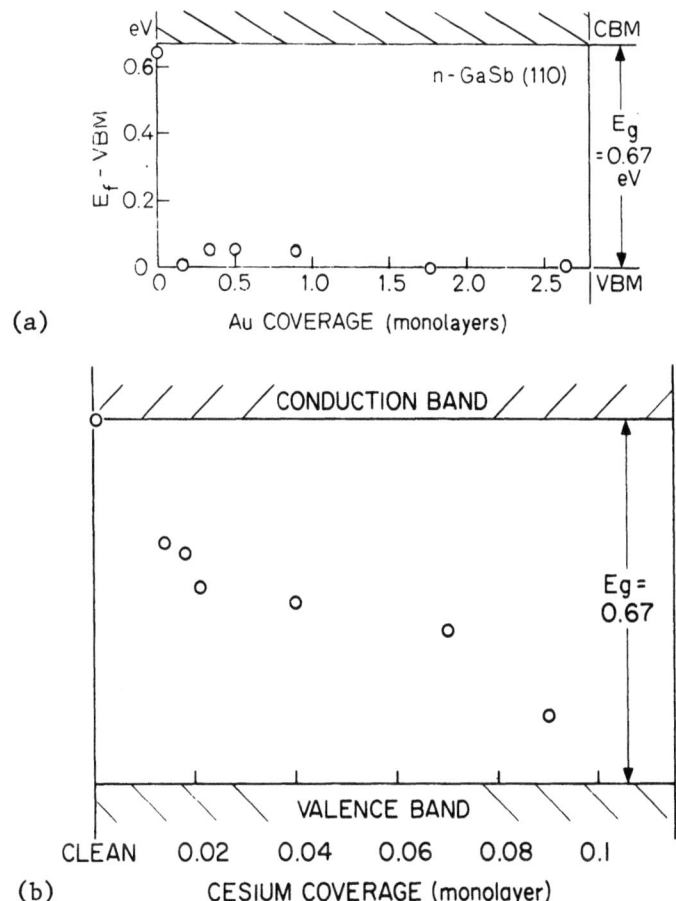

Fig. 24. (a) The Fermi level position for Au metal overlayers on the n-GaSb(110) surface is shown as a function of coverage. (b) The Fermi level position for Cs metal overlayers on the n-GaSb(110) surface is shown as a function of coverage (notice the difference for the coverage scales in a and b).

Another assumption implicit in the model of Schottky barriers is that of an abrupt planar interface between the metal and semiconductor. In Figures 25 and 26, we show experimental data which indicate that the interface is also nonideal. Therefore, calculations based on an ideal metal-semiconductor interface also appear to be inappropriate. In Fig. 25, we present spectra taken with synchrotron radiation for $h\nu$ = 120 eV [30,87] showing core levels and the Au 5d states as a function of Au coverage (indicated

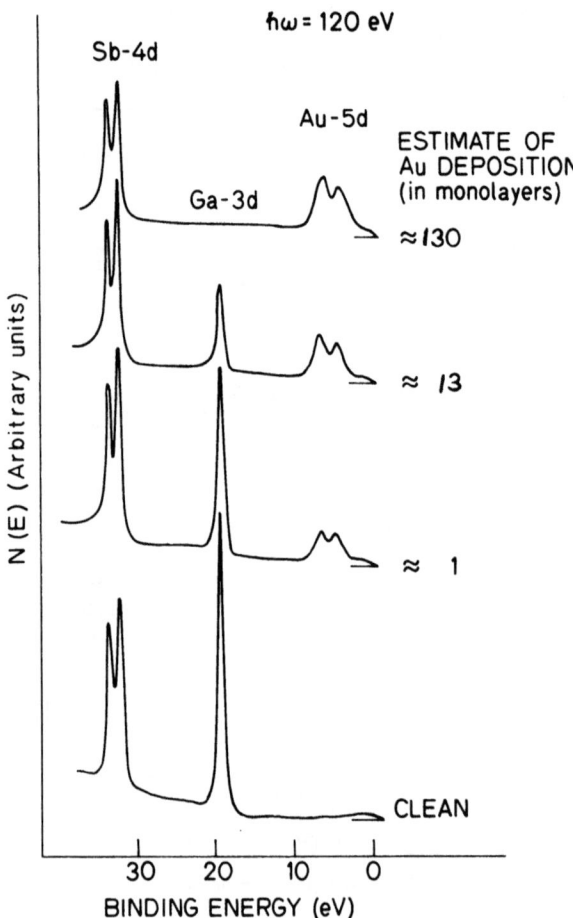

Fig. 25. Photoemission spectra taken at a photon energy of 123 eV for GaSb(110) exposed to different amounts of Au. Note the preferential movement of Sb to the surface.

on the right-hand side of the spectra in terms of monolayers). We have mentioned that, at these energies, the escape depth of the electrons through the 3-5's is less than about two molecular layers (i.e., about 6 Å); for Au, it is comparable or a little less, i.e., about 5 Å [11,40]. Thus, we are sampling only about two or three atomic layers into the Au.

The striking characteristic of Fig. 24 is that the Ga 3d peak drops relatively rapidly with Au coverage, whereas there is only a slight drop in the Sb peak even after over a 100 monolayers of Au have been deposited. The conclusion is inescapable--Sb is being

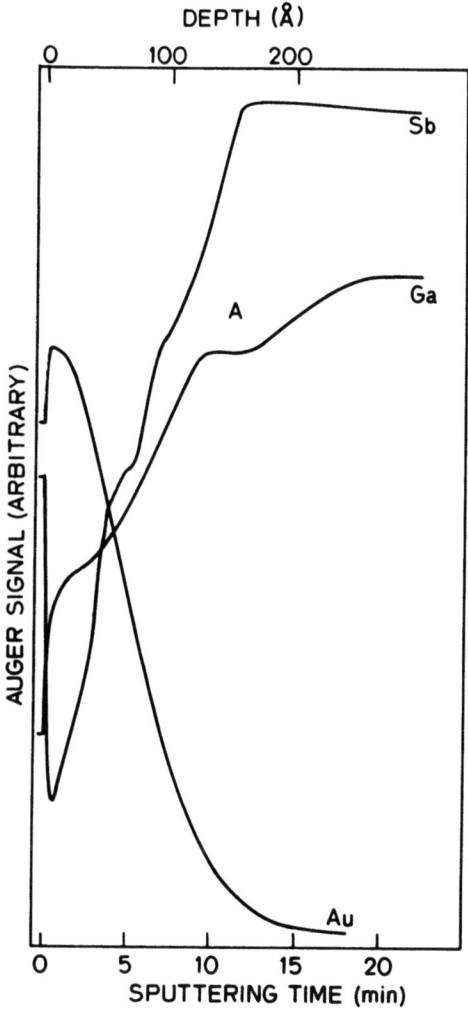

Fig. 26. Compositional depth profile of Au covered GaAs obtained in situ with Auger electron spectroscopy in combination with argon ion etching. Note the sharp SB build-up at the vacuum surface.

removed from the interface and placed on the surface of the Au film. (We will give a detailed discussion of this later.) The fact that the Ga peak height is reduced by only a factor of 1/4 after 12 monolayers have been deposited, suggests that some Ga is included in the Au although there is not a heavy concentration near the surface as is the case for Sb. In fact, with an Au film of thickness greater than about 100 layers, no Ga can be detected within the escape depth; whereas, the Sb intensity is diminished by less than a factor of two.

Several conclusions can be derived from Fig. 25. Most important, any model of a Schottky barrier based on an ideal abrupt-planar interface between the GaSb and the Au must be suspect since semiconductor material is being removed from the interface. (We will presently report the use of sputter-Auger techniques to examine the interface.) Further, the semiconductor material is not necessarily being removed in stoichiometric quantites. Even when the Au thickness is only about three monolayers (so that almost half of the measured electrons are from the GaSb), the Sb signal is hardly reduced whereas the Ga signal is reduced by almost a factor of two. This shows clearly a preferential removal of Sb in the early stages of Au deposition. This is not surprising if we recall that, in the early sixties, when the growth of semiconductor grade GaSb was being perfected, great difficulty was experienced in producing anything but p-type GaSb and that this was due to a natural tendency of the material to grow with an Sb deficit. Clearly, the present experiments show that, when a GaSb (110) surface is perturbed by metal deposition (the nature of this perturbation will be discussed in more detail presently), Sb is preferentially removed. Further, the pinning position shown in Fig. 22 and also found in Schottky barriers is, within experimental error, identical to that found in GaSb with a Sb deficit [30,87].

Thus, one is led to a new model for Schottky barrier pinning [30-32]. This is that defect centers are produced by the deposition of the metal on the semiconductor and these lead to the Schottky barrier pinning. The fact that the pinning is essentially completed at much less than a monolayer coverage is indicative of the fact that thick metallic film is not necessary and that the pinning is basically due to "atomic" like interaction between metal and semiconductor. Remember that it takes only about $10^{12}/cm^2$ defects to produce the pinning. Details as to the depth of penetration of the Sb deficit from the surface and the importance of interaction between the "deficit" center and the metal must await more detailed studies. It is generally thought that a Sb deficit produces a Ga atom on a Sb site [30-32,87].

C. Generalization to GaAs, InP, and Other 3-5 Compounds

We suggest that similar Schottky barrier pinning mechanisms are general for the 3-5 materials. Studies of GaAs and InP [30-32] showed both semiconductor constituents coming out of the 3-5 compound and that the pinning took place with a small fraction of a monolayer of atomically dispersed metal. Details are given elsewhere [30].

USE OF PHOTOEMISSION AND RELATED TECHNIQUES   445

In the case of GaAs and InP, it was not as easy to identify the nature of the deficit center formed. However, studies of Bachrach [96] on p-type GaAs and Woodall et. al. [97] on n-type GaAs as well as more recent studies by ourselves on both n- and p-type GaAs show that using Ga as a Schottky barrier material tends to supress Schottky barrier pinning. This suggests that a Ga deficit is reponsible for the GaAs pinning [30-32]. However, this can be treated as a very preliminary result, requiring much more study.

Likewise, the position of the pinning level near the CBM for InP and the fact that P stabilized molecular beam epitaxially grown samples show ohmic contacts with Ag [98] suggest that a P deficit is involved here. However, again this is the most speculative of our suggestions on the nature of surface defects.

Figure 26 shows the depth profile obtained for GaSb which was cleaved and had approximately 100 Å of Au evaporated on it in vacuo [30,32,99,100]. In studying this curve, two facts should be kept in mind:

(1) The Schottky barrier pinning was completed with only about 0.1 of a monolayer of Au on the surface. Thus, the pinning phenomena had taken place long before the thick Au layer of Fig. 26 was formed.

(2) The actual sputtering and Auger profiles may have been distorted by inherent to the measurements. Future studies should clarify this point.

However, Fig. 26 shows at least two essential points:

(1) The Sb buildup at the surface

(2) the very diffuse nature of the Au-GaSb interface.

It is unlikely that this last point is completely an experimental artifact of the sputtering. For example, GaAs-GaAlAs interfaces of less than 13 Å width have been measured with the same apparatus [101].

D. Mechanism of Defect Formation

There is a very important question which we still must answer; this is the source of the energy necessary to break 3-5 bonds and disassociate this material from the bulk 3-5 crystal. We have examined this question and, in the process, tied our observations together with the classical results of Mead and his coworkers. In Fig. 27, we present a modified version of the classical curve of Mead et. al. [102]. Here, a parameter, S, which specifies the tendency of the Schottky barrier to be pinned is plotted versus the heat of formation of the semiconductors and insulators (in the

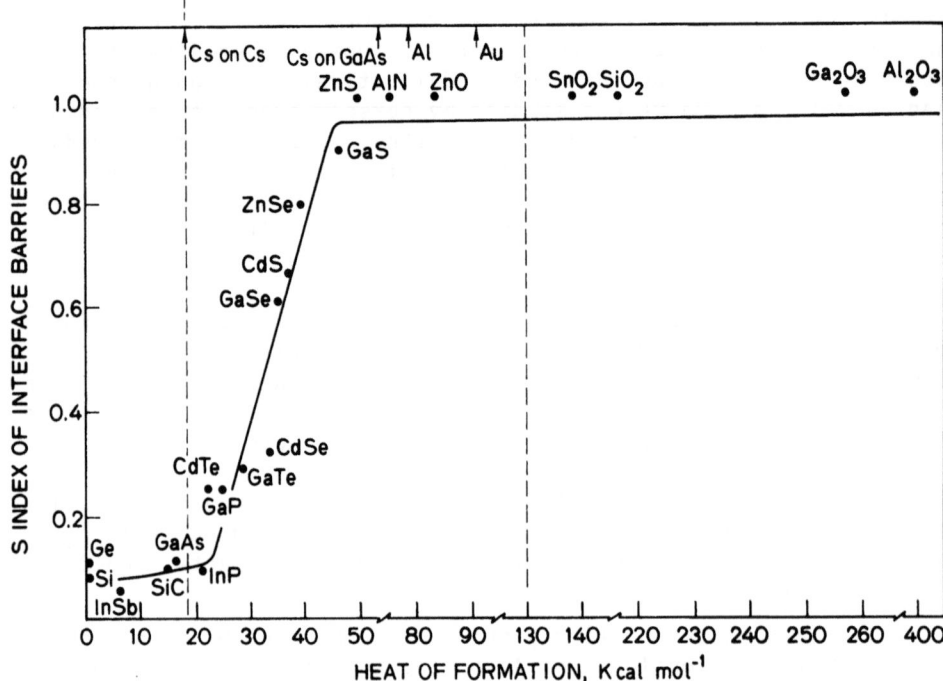

Fig. 27. The index of interface behavior, S, plotted against the heat of formation. If S is small, strong pinning takes place. Also shown is the range of heat of condensation for various metals. These are for the metal on themselves except for Cs as noted. Note that the heats of condensation span the pinned to unpinned transition.

original work of Mead et. al. [102], S was plotted versus electronegativity). As can be seen, the experimental data fall on the S shaped curve as well as in the original work of Mead et. al. (A similar plot has been independently published by Brillson [103].)

The key point, however, is the heats of condensation of the metals. With one exception, the only data available was for the metal on itself. However, for Cs, careful [104] studies have been made by Derrien and d'Avitaya for Cs on GaAs, and this showed the heat of condensation to be 60 Kcal/mol for small coverages, as compared to 18 Kcal for Cs on Cs metal. The important point is that the heats of condensation straddle the "break" region between the materials which pin and those which do not pin. (The metals with low heat of condensation on themselves will be expected to have higher heats of condensation on the semiconductor, as is exemplified by Cs.)

Thus, it is suggested that the energy necessary to break the
3-5 bonds to form the defect states is provided by the heat of
condensation of the metal on the semiconductor or insulator. Further, it is suggested that the difference in behavior between the
semiconductors with low heats of formation and the insulators with
high heats of formation is the magnitude of the heat of condensation of the metal relative to the heat of formation [31,32].

E. Dynamics of Defect and Schottky Barrier Formation

One has to think about the dynamics of defect and Schottky
barrier formation. The metal atom condenses on the surface and produces a "thermal spike". If the energy available from this is sufficiently large compared to the heat of formation of the semiconductor, there is the possibility of bonding being disturbed and defects being formed which will pin the Fermi level. The data can be
explained if only about one defect is formed for 100 metal atoms
condensing on the surface.

On the basis of the data and analysis given above, we propose
that Schottky Fermi level pinning for most of the 3-5's is basically due to defect formation at the metal-semiconductor interface
and suggest that the activation energy necessary for formation of
the defects comes from heat of condensation of the metal. Further,
the metal-semiconductor interfaces appear to be far from ideal with
much intermixing of metal and semiconductor. One manifestation of
this non-ideality is the selective appearance of approximately a
monolayer of semiconductor material at the vacuum-metal surface;
this can be explained by the lowering of surface energy by placing
the less strongly bonded material on the surface (e.g., Sb on Au),
as has been well established for metallic alloys such as NiAu and
NiCu [105].

Brillson [103] has also recently studied Schottky barrier formation with most attention give to the II-VI compounds. He has
emphasized the heat of formation of compounds and/or alloys formed
between the semiconductor compounds and the metal in explaining the
Schottky barrier pinning. While such considerations may be important in the 3-5's in determining the detailed distribution of composition (see Fig. 26 and Ref. 9) through the metal and the metal-semiconductor interface, we do not believe that they provide the
prime mechanism of pinning in the 3-5's.

Much more work must be done on the defect model put forth
above. If it is proven basically correct, much work is still needed
to understand all of the details; for example, what is the detailed nature of "defect" centers leading to Fermi level pinning?
However, if proven correct, it could lead the way to scientifically

engineered Schottky barriers so that the barrier height can be adjusted to give optimum performance for a given device.

It should also be recognized that practical Schottky diodes are usually fabricated on 3-5 surfaces containing at least thin oxide layers. We are beginning to study such diodes [106]; however as will be indicated in the Conclusions (Section IX), the pinning position produced by oxides or metals are surprisingly similar.

VIII. SCHOTTKY BARRIERS ON Si AND II-VI COMPOUNDS

In this section we will briefly review the work on fundamental studies recently done on Si and the II-VI semiconductor compounds. There are definite similarities between these results and those reported here on 3-5 compounds. Brillson [103] who has done the 2-6 work, finds intermixing between the semiconductor and the metal to be important. However, he emphasizes compound formations between the metal and the components of the semiconductor as the driving force. Thus, there is general agreement between ourselves and Brillson that the conventional models for Schottky barrier pinning are not correct, in general, and that the "atomic" interactions between the semiconductor components and the metal provide the key to understanding the pinning. However, Brillson emphasizes chemical reactions (even for GaAs), whereas we emphasize defect formation. In fact, compound formation may be more likely for the 2-6 than for the 3-5 compounds.

The situation for Si is considerably different from that for the compound semiconductor in that it has surface states in the gap [18,107,108]. Therefore, one cannot at once rule out the "Bardeen" model. However, three studies [109,110,111] have all found evidence that deposition of the metal tends to remove surface states from the band gap region. Rowe [109] has studied Al, Ga, and In on Si and observed a decrease of emission from the surface states. Garner et. al. [110] has studied Au and Al on Si and observed a very sharp decrease in the filled surface states. They also found evidence for intermixing between the semiconductor and metal; the intermixing was decreased by the addition of a half monolayer of oxygen to the Si surface before deposition of the metal. Wagner and Spicer [111] observed a decrease in emission from the filled surface states when cesium was placed on the surface.

It is clear that much remains to be done in order to understand the Schottky barrier pinning for Si. However, it appears that Heine's suggestion that the intrinsic surface states will be greatly perturbed by the metal is well founded [90]. There is also a question whether the conventional picture of an abrupt planar Si-metal interface is justified. The possibility of Si-metal intermixing must be investigated further, but evidence is available which suggests this [110].

IX. CONCLUSIONS

In this chapter, we have concentrated on two phenomena which seem to be dissimilar at first sight -- interface states formed by oxygen and the mechanism of Schottky barrier formation with emphasis on the 3-5 compounds. However, one common denominator has appeared in the 3-5 work--the conclusion that defect states play a key role in both processes. There are also suggestions that defects are or may be important in other cases. Figure 28 indicates the pinning positions [30-32] found for four different surface materials, oxygen, Cs, Au and Al on GaAs, GaSb, and InP, respectively. The striking thing is the similarity of the pinning position for the oxygen and the metals. (Remember that this pinning position is obtained for a small fraction of a monolayer and maintained) and is also found in thick "practical" overlayers. Considerable spread is shown in the pinning position of Figure 28. This is due to a combination of factors such as experimental uncertainty as well as "second order" variations with composition of the adlayer.

Based on the data of Fig. 28, we make the tentative suggestion that the same basic defects in the 3-5's are responsible for pinning for both oxygen and the Schottky barriers. We say "the same basic defects" since we recognize that the energy levels may be varied, plus or minus a few tenths of an eV by detailed interactions between the defect and its surroundings. For example, it is well established that there is a small change in the Schottky barrier pinning position based on the electronegativity of the metal [102]. This would also explain why the Schottky barrier pinning position is only affected to a small degree by a thin native oxide, as is usually present on the 3-5 surface before the metal is deposited [106].

The difference between oxygen on Si and the 3-5 compounds is interesting. On a clean cleaved surface of n-type Si, oxygen removes the surface states and the Fermi level pinning disappears [107]. By contrast, on n-type GaAs for example, the Fermi level may be unpinned on the clean cleaved surface; but it always becomes pinned near mid-gap with oxygen addition (see Fig. 5). Further more studies of interface state densities at a Si-SiO$_2$ device interface usually show the interface density decreasing monotonically as one moves away from either the valence or conduction band edge. A similar phenomena appears to occur in the 3-5 compounds; however, in addition, there appears to be superimposed on this a rather sharp peak in the density of interface states. It is this

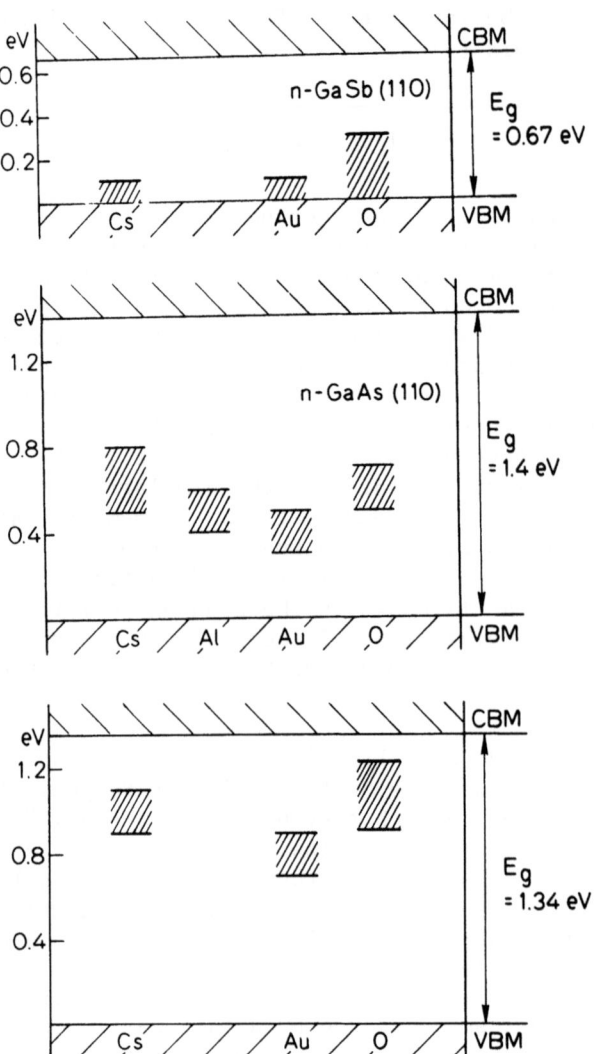

Fig. 28. The Fermi level position is shown for different atoms: Cs, Al, Au, and O, on the (110) surfaces of GaAs (a) GaSb (b) and InP (c). Note that the pinning position appears to depend on the semiconductor much more than the metal.

peak which we associate directly with the atomic defects. Its position for GaAs, GaSb, and InP is given roughly by the data in Figure 28.

Thus, tentative models connecting MIS interface and Schottky barriers have been developed for the 3-5's. Much more work must be done to test them and use the knowledge to practical advantage. For Si, the situation is considerably different; here much more must be done before as much knowledge has been acquired about Schottky barrier pinning as is now the case for the 3-5's. However, it appears that it is important to apply the new tools, such as those discussed here, in order to obtain a fundamental understanding (on an atomic level) of the materials phenomena and characteristics which determine the electrical characteristics of device structures.

We hope we have given the reader insight on how research proceeds. It is desired that this presentation will stimulate others to critically examine the interpretations as well as to develop new approaches of their own aimed at increased understanding. In this way, one may proceed to a good fundamental understanding as rapidly as possible and advance towards our ultimate objective of increased scientific engineering of device structures. In this process, the inputs and results of the workers more directly concerned with practical devices must play an essential role.

ACKNOWLEDGEMENT

Support from the U. S. Defense Advance Projects Agency, Office of Naval Research, and Army Office of Research is gratefully acknowledged. The facilities at SSRP were made possible by support of the National Science Foundation. The help of all my colleagues at Stanford is also gratefully acknowledged.

REFERENCES

1. W.E. Spicer, in *Optical Properties of Solids - New Developments*, (B.O. Seraphin, ed.), North-Holland (1975).
2. W.E. Spicer, J. de Physique 34, C6-19 (1973). E.E. Spicer, K.Y. Yu, I. Lindau, P. Pianetta, and D.M. Collins, in Vol. V of "Surface and Defect Properties of Solids," (J.M. Thomas and M.W. Roberts, Eds.), W.E. Spicer, CRC Reviews in Solid State Sci. 6, 317 (1976).
3. H. R. Philipp and H. Ehrenreich, Phys. Rev. 129, 1550 (1963).
4. T. M. Donovan, J. Matsuzaki, and W.E. Spicer (unpublished), see also Ref. 2.
5. W.E. Spicer, Phys. Rev. 112, 114 (1958); J. Appl. Phys. 31, 2077 (1960).
6. W.E. Spicer and F.E. Wooten, Proc. IEEE 51, 1119 (1963).
7. C.N. Berglund and W.E. Spicer, Phys. Rev. 136, A1030 and A1044 (1964).
8. W.E. Spicer and R.C. Eden, Proc. of 9th Intl. Conf. on Physics of Semiconductors, Moscow, 1, 65.
9. W.E. Spicer, 139, in Electronic Density of States, (L.H. Bennett, Ed.) (U.S. Govt. Printing Office, SD Catalog No. C 13: 323) (1971).
10. R. Stuart, F. Wooten, and W.E. Spicer, Phys. Rev. B 4, 4390 (1964).
11. I. Lindau and W.E. Spicer, J. Electron Spectroscopy 3, 409 (1973).
12. T.H. Di Stefano and W.E. Spicer, Phys. Rev. B 7, 1554 (1973).
13. R.A. Powell, W.E. Spicer, G.B. Fischer, and P. Gregory, Phys. Rev. B 8, 3987 (1972).
14. W. Krolikowski and W.E. Spicer, Phys. Rev. 185, 882 (1969).
15. A.R. Williams, J.F. Janak, and V.L. Moruzzi, Phys. Rev. Lett. 28, 671 (1972).
16. W.D. Grobman, D.E. Eastman, and J.L. Freeouf, B 12, 4405 (1975).
17. W.L. Schaich and N.W. Aschroft, Phys. Rev. B 3, 2452 (1971): N.W. Aschroft in *Vacuum Ultraviolet Radiation*, (E.E. Koch, Haensel, and C. Kunz, Eds.), Pergamon Press (1974).
18. *Photoemission and the Electronic Properties of Surfaces*, (B. Feuenbacker, B. Fitton, and R.F. Willis, Eds.), Wiley and Co. (1978); G.D. Mahan, *Theory of Photoemission*, 1-51 in Electron and Ion Spec. of Solids, NATO Adv. Stud. Inst. Series B, 32, (L. Freimans, J. Vennik, and W. Dekeyser, Eds.), Plenum (1978).
19. C. Caroli, D. Lederer-Rozenblatt, B. Roulet, and D. Saint James, Phys. Rev. 138 , 4552 (1973).
20. P.J. Feibleman, Phys. Rev. Lett. 34, 1092 (1975); Phys. Rev. B 14, 762 (1976).
21. W.E. Spicer, I. Lindau, J.N. Miller, D.T. Ling, P. Pianetta, P.W. Chye and C.M. Garner, Phsica Scripta 16, 388 (1977).
22. J.N. Miller and I. Lindau, (to be published).
23. T.E. Gallon, in *Electron and Ion Spectroscopy of Solids*, NATO Adv. Stud. Inst. Series B, Phys. 32, (L. Freimans, J. Vennik, and W. Debeyser, Eds.), Plenum (1978).
24. *Electron Spectroscopy*, (D. Shirley, Ed.)North-Holland, American Wlsevier (1972).

25. C.C. Chang in *Characterization of Solid Surfaces*, (P.F. Kane and G.B. Larrabee, Eds.), Plenum (1974).
26. D.E. Eastman and J.L. Freeouf, Phys. Rev. Lett. **34**, 1624 (1975).
27. G.J. Lapeyre, J. Anderson, P.L. Govvy, and J.A. Knopp, Phys. Rev. Lett. **33**, 1290 (1974).
28. E. Tosatti et al, J. Vac. Sci. and Technol. **15**, (1978).
29. P.W. Chye, P. Piannetta, I. Lindau, and W.E. Spicer, J. Vac. Sci. and Technol. **14**, 917 (1977).
30. P.W. Chye, I. Lindau, P. Pianetta, C.M. Garner, C.Y. Su, and W.E. Spicer, (in press, Phys. Rev. B).
31. I. Lindau, P.W. Chye, C.M. Garner, P. Pianetta, C.Y. Su, and W.E. Spicer, J. Vac. Sci. and Technol. **15** (1978).
32. W.E. Spicer, P.W. Chy, C.M. Garner, I. Lindau, and P. Pianetta, submitted to Surface Sci.
33. N.V. Smith and P.K. Larsen, in *Photoemission and the Electronic Properties of Surfaces*, (B. Feverbacher, B. Fitton, R.F. Willis, Eds.), John Wiley and Sons (1978).
34. J.E. Rowe, M.M. Traum, and N. Smith, Phsy. Rev. Lett. **33**, 1333 (1973); M.M. Traum, J.E. Rowe, and N.V. Smith, J. Vac. Sci. Technol. **12**, 298 (1975).
35. See. for example, *Vacuum Ultraviolet Radiation* (E.E. Koch, R. Haensel, and C. Kunz, Eds.), Pergamon Press (1974); and *Synchrotron Radiation Research*, (A.N. Mancini and I.F. Quercia, Eds.), Alghero, Italy, (1976).
36. See for example, C.W. Wilsmen, Thin Solid Films **39**, 105 (1976); C.C. Chang, P.H. Citrin and B. Schwartz, J. Vac. Sci. Technol. **14**, 943 (1977).
37. S. Doniach, I. Lindau, W.E. Spicer, and H. Winick, J. Vac. Sci. Technol. **12**, 1123, (1975); I. Lindau and H. Winick, J. Vac. Sci. Technol. **15**, 1 (1978); W.E. Spicer, I. Lindau, and C.R. Helms, Research and Developments **28**, 12, 20-31 (1977).
38. The reader is possibly familiar with SPEAR, since it was for work utilizing it that Dr. B. Richter of Stanford received a recent Nobel Prize. Dr. Richter also led the group which designed and constructed SPEAR.
39. F.C. Brown, R.Z. Backrach, S.B.M. Hagstrom, N. Lien, and C.H. Pruett, *Vacuum Ultraviolet Physics*, (E. E. Koch, R. Haensel, and C. Kunz, Eds.), Pergamon Press (1974); V. Rehn, A.D. Baer, J. L. Stanford, D.S. Kyser, and V.O. Jones, *ibid*, 780.
40. P. Pianetta, I. Lindau, C.M. Garner, and W.E. Spicer, Phys. Rev. B, (in press).
41. P. Pianetta, Ph.D. Dissertation, Stanford University (unpublished 1976); also, SSRL Report No. 77/17 (1977).
42. C. R. Helms, S. Swartz, W.E. Spicer, Appl. Phys. Lett., (in press).
43. N.J. Taylor, J.S. Johannessen, and W.E. Spicer, Appl. Phys. Lett. **29**, 497 (1976).
44. C.M. Garner, C.Y. Su, Y.D. Shen, C.S. Lu, G.L. Pearson, W.E. Spicer, D.D. Edwall, D. Miller and J.S. Harris, submitted to J. Appl. Phys.

45. C.H. Helms, W.E. Spicer, and N.M. Johnson, Solid State Comm. 25, 673 (1978).
46. National Bureau of Standards Report on Sputter-Auger Techniques and Their Applications to Semiconductor Structures, (C.R. Helms and W.E. Spicer, (in preparation).
47. See, for example, the July-August Volumes of the J. Vac. Sci. and Technol. for 1976,1977,1978; each volume consists solely of the Proc. of the Annual Conference on the Physics of Compound Semiconductor Interfaces for the indicated year.
48. I. Lindau, P. Pianetta, C.M. Garner, P.W. Chye, P.E. Gregory, W.E. Spicer, Suf. Sci. 63, 45 (1977).
49. P. Pianetta, I. Lindau, and W.E. Spicer, ASTM Special Technical Publication 643, 105-123, (N.S. McIntyre, Ed.) (march 1977).
50. W.E. Spicer, 54-89 in Electron and Ion Spectroscopy of Solids, (L. Freimans, J. Vennik, and W. Debeyser, Plenum (1978).
52. C.M. Garner, Ph.D. Dissertation, Stanford Univ. (unpublished, 1978(; also, SSRL Report NO. 78/05 (1978).
53. W.A. Harrison, Surf. Sci. 55, 1 (1976); Proc. of 13th Intn. Conf. on Phys. of Semicond. 111 (1976); J. Vac. Sci. and Technol. 14, 883 (1977).
54. C.B. Duke, J. Vac. Sci. and Technol. 14, 870 (1977); C.B Duke, A. Lubinsky, B.W. Lee and P. Mark, J. Vac. Sci. and Technol. 13 761 (1976); P. Mark, G. Cisneros, M. Bonn, A. Kahn, C.B. Duke, G. Patton, and A.R. Lubinsky, J. Vac. Sci. and Technol. 14, 910 (1977).
55. W.E. Spicer, I. Lindau, J.N. Miller, D.R. Ling, P. Pianetta, P.W. Chye, and C.M. Garner, Physica Scripta 16, 388 (1978).
56. P. Mark, P. Pianetta, I. Lindau, and W.E. Spicer, Sur. Sci. 69, 735 (1977); P. Skeath, W.A. Saperstein, P. Pianetta, I. Lindau, and W.E. Spicer, J. Vac. Sci. and Technol. 15, 4 (in press).
57. S.Y. Tong, private communication; A. Kahn, E. So, P. Mark, C.B. Duke, and R. Meyer, J. Vac. Sci. and Technol. 15, 4 (in press).
58. P. Mark et al, CRC Reviews in Solid State Physics, (in press).
59. P.E. Gregory, W.E. Spicer, S. Ciraci, and W.A. Harrison, Appl. Phys. Lett. 25, 511 (1974).
60. A.S. Bauer, R.Z. Bachrach, S.A. Flodstrom, and J.C. McMenamin J. Vac. Sci and Technol. 14, 378 (1977); R.S. Bauer, J. Vac. Sci. and Technol. 14, 899 (1977).
61. W.E. Spicer, I. Lindau, P.E. Gregory, C.M. Garner, P. Pianetta, and P.W. Chye, J. Vac. Sci. and Technol. 13, 780 (1976).
62. W. Gudat and D.E.Eastman, J. Vac. Sci. and Technol. 13, 831 (1976).
63. See references 19 and 20.
64. J. van Laar and J.J. Scheer, Sur. Sci. 8, 342 (1967).
65. J. van Laar and A. Huijser, J. Vac. Sci. and Technol. 13, 769 (1976).
66. A. Huijser and J. van Laar, Sur. Sci. 52, 202 (1976); J. van Laar, A. Huijser, and T.L. von Rooy, J. Vac. Sci. and Technol. 14, 894 (1977); W.E. Spicer, P. Pianetta, I. Lindau, and P.W. Chye, J. Vac. Sci. and Technol. 14, 885 (1977).

67. J.R. Chelikowsky, S.G. Louie, and M.L. Cohen, Phys. Rev. B 14, 894 (1977); D.J. Chadi, J. Vac. Sci. and Technol., (to be published); E.J. Mele and J.D. Joannopoulos, Phys. Rev. B 17, 1816 (1978).
68. P. Pianetta, I. Lindau, C.M. Garner, and W.E. Spicer, Phys. Rev. Lett. 35, 1356 (1975); 37, 1166 (1976).
69. R.M. Badger, A.C. Wright, and R.F. Whillock, J. Chem. Phys. 43, 4345 (1965).
70. R. Ledeke, Solid State Comm. 21, 815 (1971).
71. E.F. Mele and J.D. Joannopoulos, Phys. Rev. Lett. 40, 361 (1978); J.D. Joannopoulos and E.J. Mele, J. Vac. Sci. and Technol. 15, 4, (1978).
72. W.A. Goddard and T. McGill, J. Vac. Sci. and Technol. 15, (1978).
73. W.E. Spicer, P. Pianetta, I. Lindau, and P.W. Chye, J. Vac. Sci. Technol. 14, 885 (1977); I. Lindau, P. Pianetta, W.E. Spicer, P.E. Gregory, C.M. Garner, and P.W. Chye, . Elect. Spect. and Related Phenomena 13, 155 (1978).
74. W.E. Spicer, I. Lindau, P. Pianetta, P.W. Chye and C.M. Garner, Thin Solid Films (in press).
75. T. Sugano, private communication.
76. P.E. Gregory and W.E. Spicer, Phys. Rev. 13, 725 (1976).
77. P.W. Chye, et al, to be published.
78. H. Wieder, J. Vac. Sci. and Technol. 15 (1978).
79. Proc. of Topical Conference on the Physics of SiO_2 and Its Interfaces, (S.T. Pantolides, Ed.), to be published by Amer. Inst. of Phys.
80. C.M. Garner, I. Lindau, C.Y. Su, J.N. Miller, P. Pianetta, and W.E. Spicer, Phys. Rev. Lett. 40, 403 (1978); C.M. Garner, I. Lindau, C.Y. Su, P. Pianetta, and W.E. Spicer, submitted Phys. Rev.
81. See articles and references in Ref. 79.
82. J.F. Wagner and C.W. Wilmsen, J. App. Phys., (Dec. 1978), to be published.
83. T.W. Signmon, W.K. Chu, E. Lugujjo, and W. Mayer, Appl. Phys. Lett. 24, 105 (1974); F. Offerman, J. Appl. Phys. 48, 1890 (1977); see also Ref. 79.
84. W.E. Spicer, I. Lindau, and C.R. Helms, Research/Development 28. 20 (1978).
85. J.S. Johannessen, W.E. Spicer and Y. Strausser, J. Appl. Phys. 47, 3028 (1976).
86. C.R. Helms, Y.E. Strausser, and W.E. Spicer, Appl. Phys. Lett. (in press).
87. P.W. Chye, I. Lindau, P. Pianetta, C.M. Garner, and W.E. Spicer, Phys. Rev. B 17, 2682 (1978).
88. P.W. Chye, T. Sukegawa, I.A. Babalola, H. Sunami, P.E. Gregory, and W.E. Spicer, Phys. Rev. B 15, 2118 (1977).
89. P.W. Chye, I. Lindau, P. Pianetta, C.M. Garner, and W.E. Spicer, Phys. Lett. 63A, 387 (1977).
90. V. Heine, Phys. Rev. 138, A1639 (1965).

91. J.C. Inkson, J. Phys. C 5, 2599 (1972); C 6, 1350 (1973); J. Vac. Sci. Technol. 11, 943 (1974).
92. E. Louis, F. Yndurain, and F. Flores, Phys. Rev. B 13, 4408 (1976).
93. J.M. Anderson and J.C. Phillips, Phys. Rev. Lett. 35, 56 (1975.
94. S.G. Louis, J.R. Chelikowsky, andM.L. Cohen, Phys. Rev. B 15, 2154 (1977); J.R. Chelikowsky, Phys. Rev. B 16, 3618 (1977).
95. P.W. Chye, Ph.D. Dissertation, Stanford Univ. (1978), unpublished.
96. R.Z. Backrach, J. Vac. Sci. and Technol. 15 (1978).
97. J.M. Woodall, C. Lanza andJ.L. Freeouf, J. Vac. Sci. and Technol. 15 (1978).
98. R.H. Williams, R.R. Varma and A. McKinley, J. Phys. C 10, 4545 (1977).
99. Hiraki et al (Ref. 100) recently reported Sputter Auger studies showing 3-5 material distributed in the metal of Schottky barriers. However, their sample preparation and measurement was not done in situ as was ours. We tentatively attribute differences between our two studies to this.
100. A. Hiraki, K. Shuto, S. Kim. W. Kammura, and M. Iwani, Appl. Phys. Lett. 31, 611 (1977).
101. C.M. Garner, C.Y. Su, Y.D. Shen, T. Lee, G.L. Pearson, W.E. Spicer, D.D. Edwall, D. Miller, and J.S. Harris, Jr., submitted for publication.
102. C.A. Mead, Solid State Elec. 9, 1023 (1966); S. Kurtin, T.C. McGill, and C.A. Mead, Phys. Rev. Lett. 22, 1433 (1969).
103. L.J. Brillson, Phys. Rev. Lett. 40, 260 (1978); Phys. Rev. B, (in press); J. Vac. Sci. Technol. 15 (1978).
104. J. Derrien and F. Armand D'Avitaya, Sur. Sci. 65, 668 (1977).
105. W.M.H. Sachtler, G.J.H. Dorgelo, and R.J. Jongepier, J. Catal. 4, 100 (1965); F.L. Williams and M.J. Boudart, J. Catal. 30, 438 (1973); C.R. Helms, J. Catal. 36, 114 (1975); K.Y Yu, D.T. Ling, and W.E. spicer, J. Catal. 44, 373 (1974).
106. C.M. Garner, C.Y. Su, W.A. Saperstein, K.G. Jew, C.S. Lee, G. L. Pearson and W.E. Spicer, submitted to J. Appl. Phys.
107. L.F. Wagner and W.E. Spicer, Phys. Rev. Lett. 28, 1381 (1972); Phys. Rev. B 9, 1512 (1975).
108. D.E. Eastman and W.E. Grobman, Phys. Rev. Lett. 28, 1378 (1972).
109. J.E. Rowe, J. Vac. Sci. and Technol. 13, 248 (1976).
110. C.M. Garner, I. Lindau, P.W. Chye, P. Pianetta, and W.E. Spicer, submitted to Phys. Rev. B.
111. L.F. Wagner and W.E. Spicer, unpublished.

Chapter 9

SCANNED PHOTOVOLTAGE AND PHOTOEMISSION

T. H. DiStefano

IBM Thomas J. Watson Research Center

Yorktown Heights, New York 10598

I. INTRODUCTION

Focused beams of light or electrons [1] are useful in probing semiconductor materials and devices in order to obtain information on crystalline defects, surface irregularities, minority carrier lifetime, and resistivity variations as well as gross device defects. Generally, in this type of measurement, the beam is scanned across the surface to generate hole-electron pairs in the material. For the standard configuration, a junction near the surface is used to collect a current which is displayed as a function of position, producing an image of the defects or inhomogeneities in the material. The structure of the image is due to fluctuations of the current produced by material inhomogeneities in the vicinity of the scanned beam. Of the two modes of scanning, the electron beam offers a higher resolution; on the other hand, the optical beam is a simpler measurement which can be used for probing semiconductor material under various insulating materials found in actual devices. Together, the techniques provide complimentary information on the microscopic structure and the electronic propoerties of defects and inhomogeneities in semiconductor materials.

Of the optical techniques, scanned internal photoemission primarily determines the presence of small inhomogeneites in the contact barrier between two materials, one of which is a semiconductor or an insulator. The photoemission technique is useful in imaging local variations of a barrier due to contamination, impurity segregation, and structural or crystallographic defects. Photoemission images can be obtained from Schottky barrier of MOS structures in which the metal is semitransparent.

Beam induced current in a p-n junction or a Schottky barrier has been widely used to obtain images of crystallographic defects in a semiconducting material. The electron beam induced current (EBIC) technique is able to image structure in a depletion region with a resolution of about 0.1 μm. Unfortunately, the technique is not easily applied to the study of semiconductor surfaces on which there is no existing junction.

Recently, scanned surface photovoltage has been used to probe free surfaces of semiconductors, without the necessity of using a junction to detect the induced signal. Since the photovoltage is measured capacitively, no direct contact to the sample is required and the technique is completely non-destructive. Typical defects seen in photovoltage images include dislocations, stacking faults, precipitates, surface scratches, and grain boundaries. Scanning surface photovoltage can be used to examine and screen silicon wafers in the early states of processing.

II. INTERVAL PHOTOEMISSION PROBE OF INTERFACES

A. The Photoemission Process

Photoemission [2] and internal photoemission [3] yield spectroscopy have long been useful in examining the properties of surfaces and interfaces. More recently, photoemission images have been used to study defects and nonuniformities on surfaces and interfaces. In order to understand photoemission imaging and the effects which determine the contrast forming the image, we will first examine the photoemission process and the parameters which influence it. Generally, photoemission images are formed by measuring and displaying the local microscopic photocurrent produced by monochromatic light. Any factor or contrast mechanism which influences the yield is mirrored in the image. From such an image, information about the structure and defects on the surfaces can be inferred with the aid of an understanding of the contrast mechanisms which produce details in the image.

The process of internal photoemission [4,5] can best be understood in terms of a three step process [6]: photoexcitation of electrons, transport to the surface, and transmission over the barrier. The process is represented schematically in Figure 1, showing the absorption of a photon of energy $\hbar\omega$ at x, with the electron being excited in the direction θ. Following the development by Williams [3] for internal photoemission from a metal, we assume an isotropic direction and a constant transition probability. For this simple model, the distribution of excited electrons is uniformly constant in energy E from the Fermi level up to an energy $\hbar\omega$. The yield Y is then the fraction of excited electrons which arrive at the surface

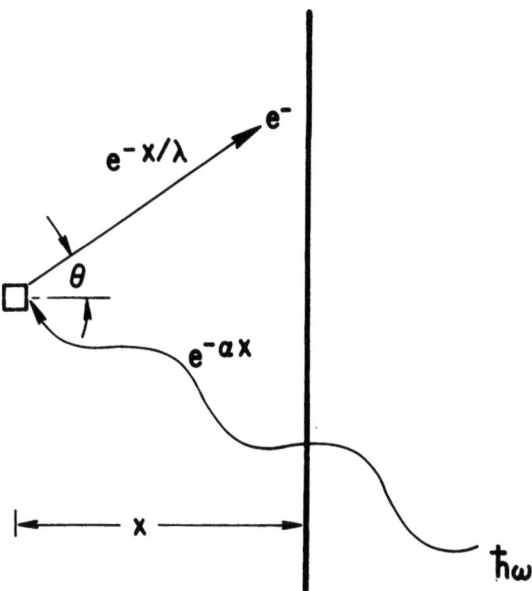

Fig. 1. Fundamentals of the photoemission process, showing photoexcitation at position x and transport to the surface at x = 0.

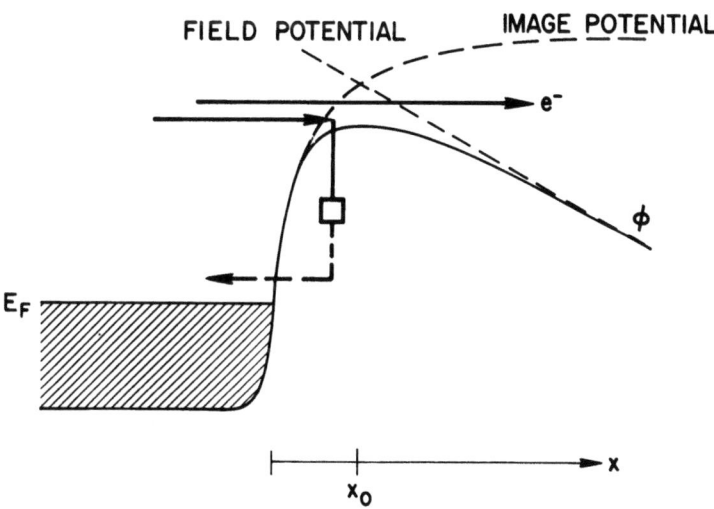

Fig. 2. Schematic representation of the process of transmission over the barrier in internal photoemission. Electrons with sufficient energy to escape over the barrier may be captured by a trap in the insulator, and subsequently decay back to the cathode.

with sufficient energy normal to the surface to surmount the barrier Φ. In transport to the surface, excited electrons are assumed to scatter from phonons and electrons over a total mean free path λ; the emission of scattered electrons are not considered. The yield is then,

$$Y(\hbar\omega) = \alpha \int_{\Phi}^{\hbar\omega} \int_{\Phi/E}^{1} \int_{0}^{\infty} e^{-\alpha x} e^{-x/(\lambda \cos\theta)} dx \left(\frac{d\cos\theta}{2}\right)\left(\frac{dE}{\hbar\omega}\right) \qquad (1)$$

where α is the photon attenuation constant. The analysis is made at 0°K because thermal broadening effects are generally insignificant in yield spectra. A coordinate transformation is introduced to define θ in terms of energy;

$$E_\perp = E \cos^2\theta \qquad (2)$$

where $E_\perp$ is the component of electron energy normal to the surface. After integration over x, and the use of the transformation, the yield is,

$$Y(\hbar\omega) = \frac{1}{4\hbar\omega} \int_{\Phi}^{\hbar\omega} \int_{E_\perp}^{\hbar\omega} \frac{\alpha\lambda}{\alpha\lambda\sqrt{E_\perp/E}+1} \frac{dE}{E} \, dE_\perp \qquad (3)$$

Integration over E leads to,

$$Y(\hbar\omega) = \frac{\alpha\lambda}{2\hbar\omega} \int_{\Phi}^{\hbar\omega} \ln\left\{1 + \frac{1}{\alpha\lambda+1}\left(\frac{\hbar\omega}{E_\perp} - 1\right)\right\} dE_\perp \qquad (4)$$

which can be expanded near threshold;

$$Y(\hbar\omega) \cong \frac{1}{4(\hbar\omega)^2} \frac{\alpha\lambda}{1+\alpha\lambda} \int_{\Phi}^{\hbar\omega} \left\{\left(\frac{\hbar\omega-E_\perp}{\hbar\omega}\right) + \frac{3\alpha\lambda+2}{4\alpha\lambda+4}\left(\frac{\hbar\omega-E_\perp}{\hbar\omega}\right)\right\} dE_\perp \qquad (5)$$

This integral can be understood in terms of an electron distribution in normal energy $E_\perp$ which is proportional to the bracketed expression. That portion of the distribution with $E_\perp > \Phi$ is collected. The quantum yield is then,

$$Y(\hbar\omega) \cong \frac{1}{8(\hbar\omega)^2}\left(\frac{\alpha\lambda}{1+\alpha\lambda}\right)(\hbar\omega-\Phi)^2 + \frac{1}{6\hbar\omega}\left(\frac{3\alpha\lambda+2}{\alpha\lambda+1}\right)(\hbar\omega-\Phi)^3 \qquad (6)$$

The first term in Eq. (6) is ($\Phi/\hbar\omega$) times that obtained if the electrons were assumed to be attenuated by a one dimensional scattering length λ as they are transported to the surface. Near threshold, the second term is negligible.

Photoemission from a semiconductor can be analyzed in a similar manner except that the initial distribution of excited electrons is not simply isotropic and constant in energy but has a more complex shape and directional dependence, depending upon the band structure of the material and the nature of the optical excitation [7]. Kane finds that, for indirect optical excitation [7],

$$Y(\hbar\omega) \propto (\hbar\omega - \Phi)^{5/2} \qquad (7)$$

where Φ is the photoelectric threshold. The additional power of ½ is due to the density of states near the valence band edge. For direct transitions [7],

$$Y(\hbar\omega) \propto (\hbar\omega - \Phi) \qquad \text{(no scattering)} \qquad (8)$$

$$(\hbar\omega - \Phi)^2 \qquad \text{(elastic scattering)}$$

where the barrier surface is normal to a unit vector of the crystal.

The third step in the emission process, transmission over the barrier, is influenced by several factors which are represented in Fig. 2. The total electronic potential Φ includes the field potential $x\xi$, the image potential $e/(16\pi\varepsilon_0\varepsilon_\infty x)$, and any interface polarization dipole layer $\Delta\Phi_0$; ε_0 is the permittivity of free space and ε_0 is the high frequency dielectric constant of the insulator. Charge neutrality in the insulator is assumed; this condition is not always rigorously valid [8]. The barrier potential,

$$\Phi(x) = (\Phi_0 - \Delta\Phi_0) - \frac{e}{16\lambda e_\infty e_0 x} - \xi x \qquad (9)$$

and the total effective barrier is,

$$\Phi = \Phi_0 - \Delta\Phi_0 - \left(\frac{e}{4\lambda e_\infty e_0}\right)^{1/2} \sqrt{\xi} \qquad (10)$$

The Schottky barrier reduction, given by the last term in Equation (10) has been confirmed experimentally [9,10] for SiO_2, where the high frequency dielectric constant $\varepsilon_\infty = 2.15$ is determined [9] by a fit to the field dependence of the photoemission threshold.

In moving to the insulator barrier, the escaping electron may be captured by a trap from which it tunnels as shown in Fig. 2. It is assumed that the trap is relatively shallow, and the electron is lost if it is captured before reaching the potential maximum.

The transmission probability T for the density of traps N_t with cross section b reduces the yield in equation 6 so that the actual yield is a product of

$$T = e^{-bN_t x_0} \cong (1 - bN_t \frac{e}{16\pi e_\infty e_0 \xi}) \quad (11)$$

and equation 6. Typically, the Coulomb capture cross section [11] in an insulator is about 10^{-12} to 10^{-14} cm for injected electrons [5,12,13]. Thus trapping would be significant in photoyield for concentrations of N_t above about $10^{18}/cm^3$.

Experimentally, internal photoemission yield spectra from clean interfaces are found to have a parabolic or cubic dependence on energy above threshold. For a given system, a power law fit (where the power is between 2 and 3) is used to determine the threshold [14,15,3]. Typical yield spectra are shown in Fig. 3, for Al-SiO$_2$, at several values of applied field. A third power fit to the spectra determines a threshold of 3.19 eV at zero field; the field dependence of the threshold follows the Schottky barrier reduction given by the dashed line.

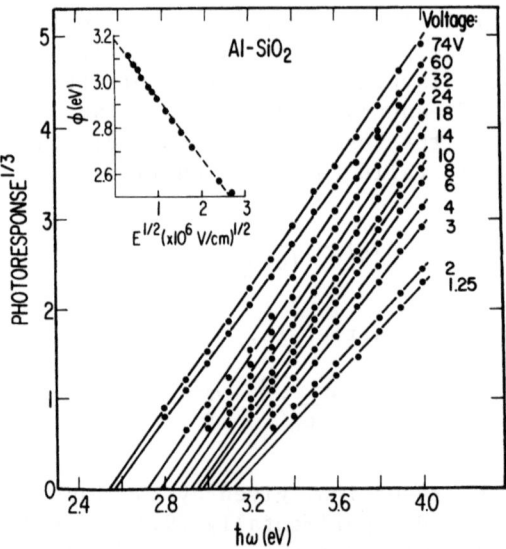

Fig. 3. The yield for internal photoemission from Al into SiO$_2$. The yield is approximately a cubic function of energy above threshold, which is lower than the energy barrier by an amount equal to the Schottky barrier reduction, shown as the dashed line. (from Ref. 16)

B. Contrast Mechanisms in Photoemission

In examining the photoemission process, we find several factors which influence the spectral yield. It is the lateral inhomogeneity of these factors which are mirrored in photoemission images. The various mechanisms can be examined by studying a combination of equations 6, 10 and 11.

$$Y(\hbar\omega) \cong \frac{1}{8(\hbar\omega)^2} \left(\frac{\alpha\lambda}{1+\alpha\lambda}\right) \left[\hbar\omega - \Phi_0 + \Delta\Phi_0 + \left(\frac{e}{4\pi e_\infty e_0}\right)^{1/2} \xi^{1/2}\right]^2 \quad (12)$$

$$\times \left[1 - bN_t \left(\frac{e}{16\pi e_\infty e_0 \xi}\right)^{1/2}\right]$$

Where the second term in the square bracket of equation 6 is negligibly small. For small local deviations of ΔY of the yield from Y_0, due to inhomogeneities in the material

$$\frac{\Delta Y}{Y_0} \cong \left(\frac{\lambda'-\lambda}{\lambda}\right) - bN_t \left(\frac{e}{16\pi e_\infty e_0 \xi}\right)^{1/2} + \frac{2\Delta\Phi_0}{\hbar\omega-\Phi_0} + \left(\frac{e}{4\pi e_\infty e_0 \xi}\right)^{1/2} \frac{\Delta\xi}{\hbar\omega-\Phi_0} \quad (13)$$

where λ' is the local mean free path, $\Delta\Phi_0$ is the local dipole layer and $\Delta\xi$ is the local enhancement of the field in the insulator.

Probably, the most significant of the several factors is the dipole layer, $\Delta\Phi_0$. Near threshold, a small local fluctuation in Φ leads to a large change in photocurrent, as is illustrated in Fig. 4. The distribution with respect to the normal component of energy is shown before and after the barrier, for both the full and reduced barrier. Several phenomena can produce an interface dipole, including impurity segregation on the surface, decomposition of a compound material at the interface, and interfacial reactions between the two materials.

The traps near the interface would appear in a photoemission image as a reduction in current due to the second term in Eq. 13. These traps may be introduced due to contamination of the insulator, interdiffusion across the interface and vacancies or nonstoichiometry of the insulator. Such traps have been a problem in silicon integrated circuits, and they are an obstacle to fabrication of MOS devices on III-V materials. They appear as slow surface states in MOS measurements.

Interface field enhancement leads to an increased photocurrent. The fractional increase is large for photon energies near threshold, as seen in the last term in Eq. 13. A local field enhancement can be caused by a surface structural irregularity in the cathode. For

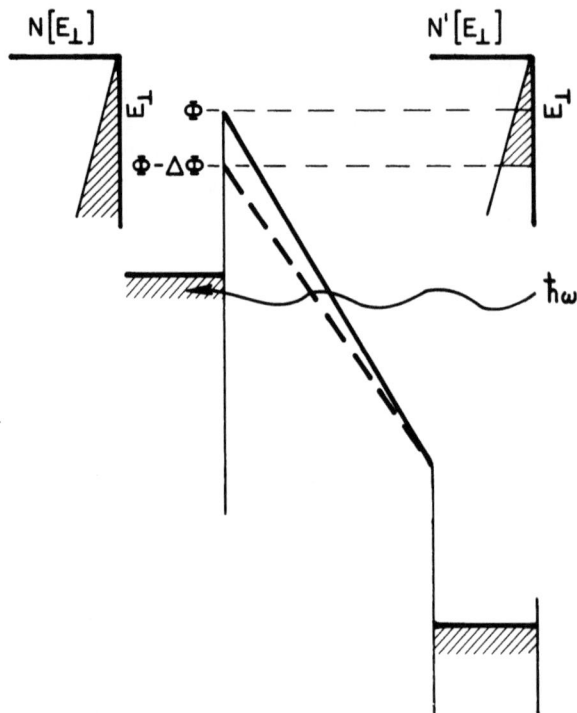

Fig. 4. A representation of the increased photoyield for internal photoemission produced by a reduction of the barrier by $\Delta\Phi$. N and N´ show the distribution of electrons in the component of energy normal to the interface, for excited and collected electrons respectively.

example, under conditions of high temperature and applied voltage, Schottky barrier diodes on silicon can develop spikes or needles of electrode material projecting into the silicon [17]. A field enhancement at the tip of such a needle leads to a local increase in the photoyield, and a feature in the photoemission image. Photoemission imaging allows the non-destructive examination of Schottky barrier rectifiers for such spikes as they develop.

Another contrast mechanism, although it would be difficult to observe, is due to local modulation of λ in the cathode as represented by the first term in Equation 13. A local change in λ may be caused by grain boundary segregation in a polycrystalline material, or by phase separation of the material. Unfortunately, these features are too small to be resolved by the scanned photoemission technique. However, they maybe resolved by vacuum photoemission microscopy.

Factors other than those considered above may effect the quantum yield for specific types of interfaces. For a semiconductor contact, photoemission may be influenced by areas of band bending [18-20] or of emission from the conduction band [21]. Surface band bending in p-type material increase the photocurrent because electrons excited in the bulk gain kinetic energy from the internal field ξ_s in transit to the surface. Following Gobeli and Allen [20], the yield from an indirect transition semiconductor is

$$Y(\hbar\omega) \propto \int_0^\infty (\hbar\omega - \Phi_0 + x\xi_s)^{5/2} e^{-x(\alpha+1/\lambda)} dx \qquad (14)$$

For the case in which λ is much less than the extent of the band bending region, the yield is approximately,

$$Y(\hbar\omega) \propto (\frac{\alpha\lambda}{1+\alpha\lambda}) (\hbar\omega - \Phi_0 + \frac{\xi_s \lambda}{1+\alpha\lambda})^{5/2} \qquad (15)$$

or, the band bending reduces the apparent threshold by about $\xi_s \lambda$. Local band bending caused, for example, by Fermi level pinning at crystalline defects would appear (for p-type material) in a photoemission image.

C. The Influence of Interface Layers

1. **Impurity Induced Interface Dipole Layer.** Mobile charged impurities in insulators generally will segregate onto the interface contacts because of an image field driving force and, as a result, they induce a change in Φ on the contact. As an example, sodium, a common impurity in SiO_2 films in MOS devices, is found to be concentrated at the two contacts [23,24]. This sodium can be drifted from one contact to the other by the application of a voltage across the film [23,25]. A fractional monolayer of sodium on the cathode will significantly reduce the barrier by producing a dipole layer similar to that produced by cesium on the surfaces of various metals. The dipole is due to the polarization of the valence electrons on the ion by the lower electron potential in the metal or cathode material. The dipole layer is seen in a photoemission image as an increased current due to the reduced barrier.

As an example, consider the change in threshold observed by Viswanathan and Ogura for the Au-SiO_2 interface [26]. In drifting ions from one electrode to the other by applying a field at 250 °C, the threshold is reduced by 0.15 eV. The polarity of the applied field is consistent with the drift of Na to the Au-SiO_2 interface, although the density of the ions is unknown. A similar barrier reduction is seen in the influence of Na on the Si-SiO_2 interface

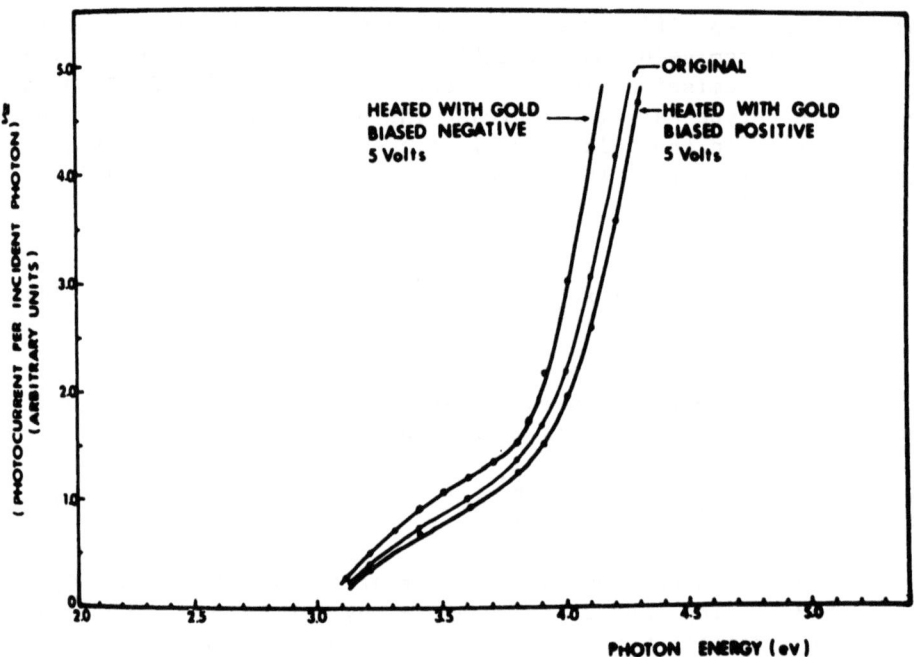

Fig. 5. Internal photoemission yield spectrum for Au-SiO$_2$ with several conditions of pretreatment. The effective barrier is reduced by annealing at a bias and temperature which would drift cations to the interface. (After Viswanathan and Ogura, Ref. 26.)

[27,28,24], as shown in Figure 6. The effective threshold is reduced monotonically with increasing sodium to an energy of 2.6 eV at approximately one monolayer coverage. Correspondingly, thermionic emission measurements determine a barrier of 2.7 eV for the same coverage (about 1.3×10^{15}Na/cm^2). Even a relatively small amount of sodium ($\sim$0.001 monolayer) is able to significantly reduce the threshold.

2. <u>Segregation at Interfaces</u>. Impurity ions in concentrations sufficient to influence internal photoemission may be present in the material near interfaces due to any of several processes. Doping or impurity atoms may segregate onto an interface during annealing and cooling of the system. Impurities may build up at an interface due to the movement of the interface, such as is found in the oxidation of doped silicon [29]. Also, interdiffusion of material across the interface introduces ions which may be detected by internal photoemission [16]; such interdiffusion commonly occurs between various metals and SiO$_2$.

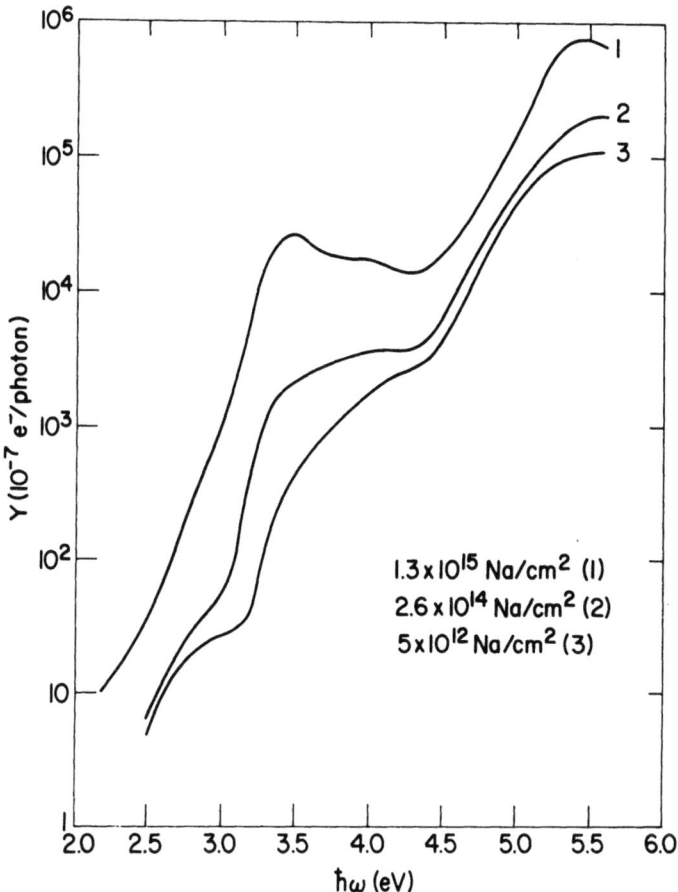

Fig. 6. Photoemission yield from Si-SiO interfaces which were uniformly covered by three different concentrations of Na. The measurements determine threshold energies of 2.6, 2.9, and 3.3 eV when extrapolated to zero electric field. (From Ref. 24)

As an example of impurity build-up at a technologically important interface, consider excess phosphorus at an Si-SiO_2 interface grown on P-doped Si [30,31]. Auger profiling measurements have shown that the phosphorus lies in the SiO_2 within ~40 Å of the interface [30,32], and that the peak concentration is higher than one would expect based on a diffusion-redistribution theory [32]. The formation of a separate phase at the interface may account for the results [32]. Calculations show that a build-up of impurity in the SiO_2 also occurs for oxide growth on boron doped silicon [33].

III. SCANNED ELECTRO-OPTICAL TECHNIQUES

A. Optical Scanning

Many scanned optical techniques have been used to probe semiconductor surfaces, but because of the low signal levels involved, a laser source is advantageous in achieving a high resolution with a reasonable signal-to-noise ratio. An early technique used for forming photovoltage images of junctions in relatively large devices is a CRT flying spot scanner in which a rastered CRT screen is demagnified and imaged onto the surface of the device [34,35]. The flying spot method is simple, but limited because of the radiation density at the spot and large wavelength spread. Other techniques such as a rotating mirror [36], a Nipkow disc [37], and mechanical sample translation [38] have been used to scan a focused spot of light over a semiconductor surface, again with somewhat limited resolution. Narrow band light from a monochromator has been used, at the expense of a low sensitivity/resolution, to probe surfaces. For each of the scanning methods, except laser scanning, the resolution at a reasonable S/N is limited to about 10 µm.

A laser source is obviously superior to noncoherent sources for electro-optical scanning because it allows a significant photon flux to be focused into a defraction limited spot. Potter and Sawyer have used rotating mirrors to deflect a laser beam entering the eyepiece of the microscope in order to obtain a raster pattern on the device under test [42]. To optimize the resolution of the system, the rotating mirrors should be located at the entrance pupil of the eyepiece, with the beam uniformly filling the pupil aperture [43]. The system used by Phelan and DeMeo, outlined in Figure 7, is capable of rastering a surface at 30 frames per second [43] to provide a real time image. For high resolution, a large numerical aperture lens is required, leading to limited field of view. A metallograph microscope objective with a numerical aperture N.A.=0.36 will resolve about 1000x1000 spots. To scan a larger field, a mechanical arrangement may be used for moving the sample in a raster pattern; a scan speed of 0.4 mm/sec has been obtained with such a mechanical scan system [44].

For a spot diameter significantly greater than the photon wavelength, a larger field can be scanned because of the lower N.A. required for the objective lens. The diffraction limit of the lens is $\lambda/((2N.A.)$. Also, for rasters larger than 1000x1000 spots, galvanometer driven rotating mirrors are at present inadequate. Several techniques, including rotating mirrors [18] and holographic scanning [45] are capable of deflecting light onto arrays of $10^4 \times 10^4$ points or larger, but such large fields have yet to be used in probing semiconductor surfaces. However, these techniques may be necessary for future applications; for example, an array of $10^4 \times 10^4$

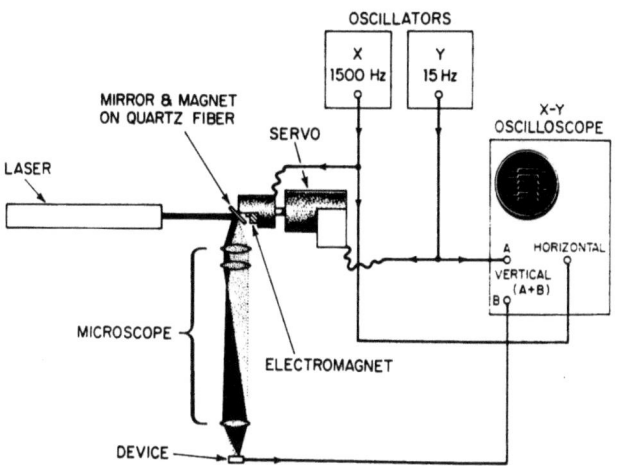

Fig. 7. Schematic representation of the optical scanning system used by Phelan and DeMeo. (Ref. 43)

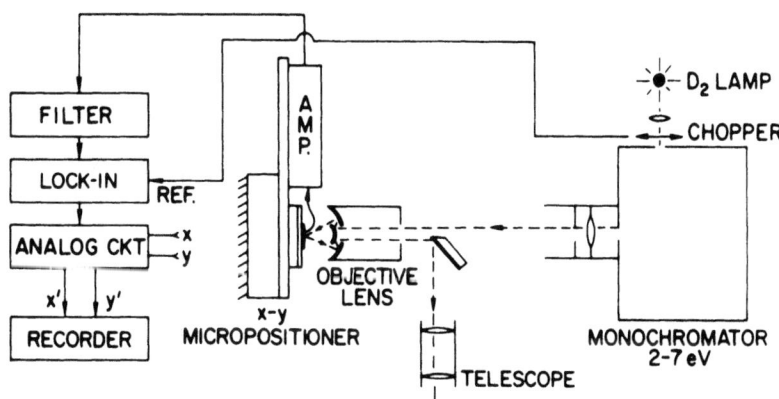

Fig. 8. Apparatus for the measurement of scanned internal photoemission. A monochromator is used to select an appropriate photon energy. The light spot is scanned by mechanically moving the sample. (From Ref. 40.)

spots is required to raster a 100 mm diameter silicon wafer at a
resolution of 10 μm.

B. Scanned Internal Photoemission

Scanned internal photoemission measurements [40,31,46] are obtained by scanning a spot of monochramatic light over an MIM sandwich structure or a Schottky barrier. One of the electrodes is semitransparent to allow light to penetrate to the cathode interface; typically, for metals the semitransparent electrode is between 100 and 200 Å thick. The electrons emitted from the illuminated spot on the cathode are detected as a current which is influenced by the various factors determining yield (as discussed in Section II-B). In order to increase the sensitivity of the measurement to local inhomogeneities, the photon energy is set slightly above threshold.

A monochromator source can be used, at a sacrifice of resolution and S/N, to form photoemission images. The system represented in Figure 8 has been used [40] to obtain both images and local spectral yield measurements. Chopped light from a monochromator fills the aperture of a reflecting objective lens, which was chosen so that the system remains in focus over a scan of wavelength. Light reflected from the sample is deflected by a moveable mirror, and viewed through a telescope in order to focus and position the spot. The sample is rastered by moving it with a motor driven x-y micropositioner. Because of low signal levels compared to the 1/f noise, phase detection is often required. When one of the electrodes is a semiconductor, care must be exercised to minimize the extraneous photovoltage; a high doping level, a low chopping frequency, and a reference phase set to null the photovoltage are usually sufficient. The signal is displayed as a function of position on a recorder or a CRT. The x and y drive for the display are derived from position sensors on the mechanical scanner.

For probing interfaces having a threshold near a laser line, laser scanning [46] offers greatly improved S/N and resolution in the photoemission image. Of course local spectral yield measurements, even with a tuneable laser, are difficult. However, in many cases, a high resolution image obtained at a single wavelength is sufficient to provide significant information on an interface. Laser scanning measurements are described in section III-C.

C. Photovoltage Scanning Techniques

1. <u>Measurement Systems</u>. Photovoltage scanning systems have successfully used various types of light sources, including CRT's

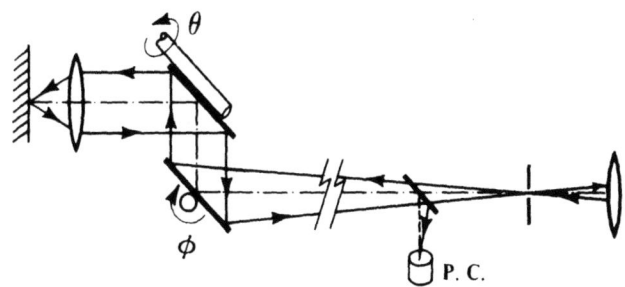

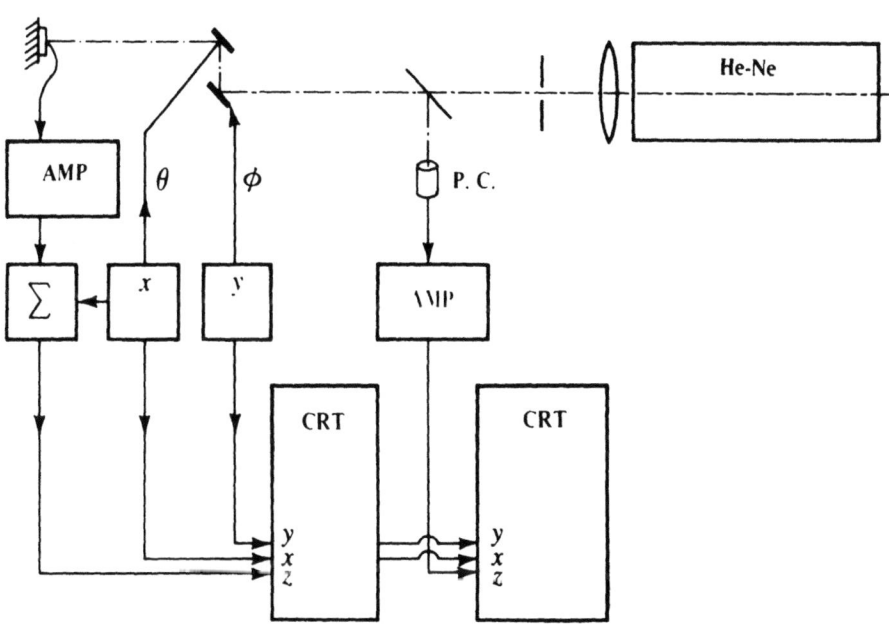

Fig. 9. Laser scanning apparatus used to obtain scanned photoemission and photovoltage images. (a) The rotating mirror arrangement for scanning the light spot. (b) The electronic arrangement for obtaining photoemission, photovoltage, and reflectivity images. (From Ref. 31)

arc lamps, monochromators, and lasers. Typically, high resolution photovoltage scans are made with a laser source at a wavelength such that the light is absorbed near the surface. The wavelength dependence of the local photovoltage has also been used [41] to obtain carrier lifetime information on a scale of the diffusion length of minority carriers. Since high resolution photovoltage scanning systems generally use a laser source, we will examine one such system in detail. With simple modifications, this system has also been used for scanned photoemission measurements.

Schematic representations of a typical laser scanning system and associated electronics [31] are shown in Figures 9(a) and 9(b). For the optical path in Fig. 9(a), the laser beam is focused through a 25 μm pinhole to eliminate the widely divergent light. The beam may or may not be chopped, depending upon the mode of measurement. The light is then reflected from two galvonometer driven scanning mirrors, rotating about orthogonal axes, placed immediately behind the objective lens. The light uniformly fills the lens aperture over the full range of deflection angles. The sample is brought into focus by moving the sample holder on a translation stage. The focus condition is difficult to discern, particularly for a UV laser. To detect the focus, the time-reversal property of the optical system was used, as represented in Fig. 10. The object pinhole is deflected and brought into focus as a demagnified image on the sample surface. Light reflected from the surface retraces the ray paths through the system and is refocused on the pinhole, independent of the mirror positions. Then, in adjusting the focus, the sample is moved to minimize the size of the light spot which falls on a fluorescent screen at the pinhole. A portion of the reflected light is deflected by a beam splitter, detected, and used to form a reflectivity image of the sample. The capability of obtaining both reflectivity and electro-optical images on the same coordinate system is an advantage since comparisons can easily be made between the two.

The deflection apparatus and sample holder displayed in Fig. 11 show the galvanometer motors, deflection mirrors, reflecting objective lens, and sample holder. The photovoltage signal is detected by a modified Kiethley 18000 high speed picoammeter mounted behind the sample holder. After differentiation, filtering and addition of a blanking pulse the signal is displayed as z modulation on a CRT which is rastered by x and y voltages obtained from the galvanometer drivers. Photographs are taken of the image as it is scanned at 20 lines per second.

2. <u>Sample Preparation and Electrode Structure</u>. The measurement of photovoltage images from samples, such as power devices or solar cells, which have large area p-n junctions, is relatively simple - the junction voltage or current is detected directly by contacting the junction. In samples not containing a junction, a probe contact in the vicinity of the scanned area is often sufficient to detect

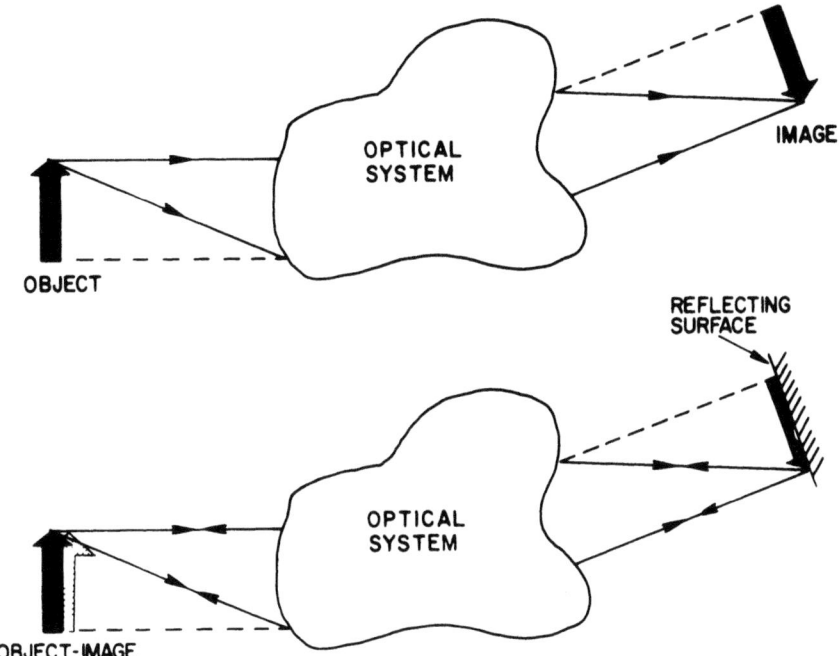

Fig. 10. General representation of the method used for focusing the laser onto the sample: (a) when the system is in focus, the object pinhole is focused onto the sample, and (b) the light is then reflected back through the system to re-focus on the pinhole.

Fig. 11. View of the laser scanner, lens sample holder, and positioning stages used for obtaining photovoltage images. (From Ref. 52)

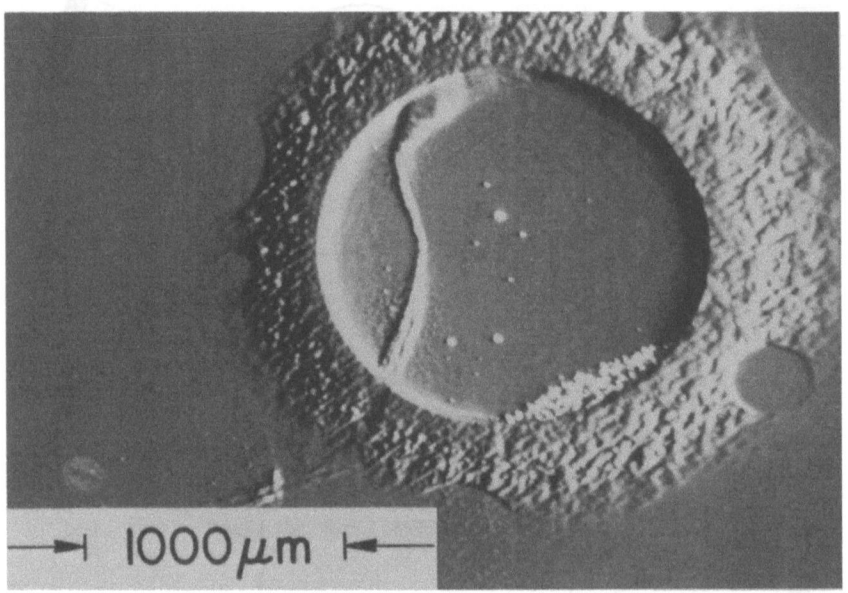

Fig. 12. Surface photovoltage image of saw cut silicon wafer. The electrode dot in the center is sufficient to obtain an image of the material within about 1000 μm of the electrode.

a signal. For example, the photovoltage image shown in Figure 12 shows detail up to about 1000 μm from the electrode. Direct contact probes can also be used in detecting bulk photovoltage due to resistivity variations [47]. The situation is somewhat more difficult in the optical scanning of LSI circuits, where the junctions are interconnected in a complex pattern; in such cases, a photoinduced signal is measured through an external contact such as a power supply line, and the resulting image interpreted in terms of the circuit operation [44,48,49].

For insulator or oxide coated semiconductor surfaces, photovoltage images may be obtained by scanning the surface through counter-electrodes of a semitransparent metal film [50] or an electrolyte solution [41]. The bias on the counter-electrode is adjusted in order to slightly deplete the semiconductor surface. As the light spot is scanned, the photovoltage produced across the depletion region is detected as a current induced in the counter electrode. In this MOS arrangement, care must be exercised that the charge in the surface channel is not contacted through pinholes or other defects, because the conductive channel formed by the light would then be shunted.

SCANNED PHOTOVOLTAGE AND PHOTOEMISSION 475

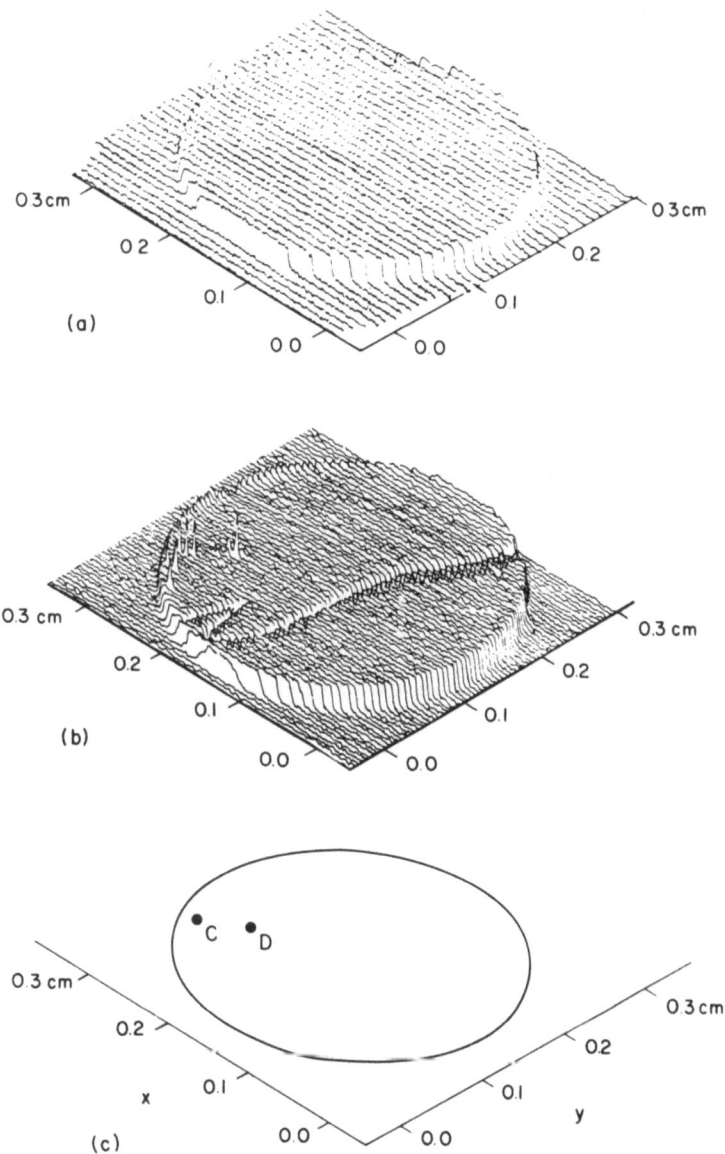

Fig. 13. Scanned internal photoemission maps of the Si-SiO$_2$ interface, obtained (a) in the initial condition and (b) after drifting sodium to the interface for 40 h at 150°C. The dots C and D mark the points at which the sample failed in dielectric breakdown. (from Ref. 54)

Perhaps the most interesting configuration for nondestructive evaluation is a scanned photovoltage probe of the semiconductor surface in which the surface photovoltage is detected as a displacement current in noncontacting capacitor electrodes 51,52 The photovoltage, which develops across a small depletion region on the surface, is somewhat sensitive to surface conditions. To obtain a uniform image, the entire surface must be depleted to the same depth. Nonuniformities in the surface depletion are easily eliminated by illuminating the surface with a low noise dc light source to create a conductive, equipotential surface channel. However, it is still essential that the entire surface area to be probed is depleted. This depletion can be accomplished in several ways including surface treatment, thin conductive surface films, or electric fields. Dry oxidation of p-type silicon normally leaves the surface depleted, so that the surface of wafers being processed for n-channel MOS LSI circuits already have a surface suitable for noncontacting photovoltage scanning. The surface may also be depleted by a field induced by a voltage on the electrode; ionic charge introduced into the air dielectric greatly enhances the effect of the field in charging the surface. Although it may introduce contamination, a conductive surface film such as a detergent residue [53], can be used to establish an equipotential which will uniformly deplete the surface. Any of these techniques can be successfully used to deplete the surface of a silicon wafer. Ideally, a wafer in process can be passed between two capacitor plates and examined for defects by the scanned photovoltage technique, without making physical contact to the wafer.

IV. PHOTOEMISSION IMAGES

A. Interface Images

1. <u>The Si-SiO$_2$ Interface</u>. Scanned internal photoemission has been used to produce images of a variety of interfaces, of which the most technologically important is Si-SiO$_2$. Sodium, a common contaminant in MOS structures, is known to reduce the interface barrier. The influence of residual sodium contamination was probed by by scanned photoemission maps in Figures 13 (a) and (b) [54]. The map in 12(a) was obtained initially by scanning the sample with a 20 μm diameter spot of 3.5 eV light, using the system shown in Fig. 8. The sample is an Al-SiO$_2$-Si capacitor, which was scanned through the semitransparent 160 Å thick Al electrode. After 40h at 150°C under an applied field of 10^6V/cm, the map in Fig. 13(b) shows areas of enhanced photocurrent, apparently due to a reduction of the Si-SiO$_2$ barrier caused by residual sodium ions which were drifted to the interface under the applied field. Spectral photoyield measured at a peak in a similar map matches that [24] from an Si-SiO$_2$ interface uniformly coated with 1.3 x 10^{15}Na/cm^2, adding

further evidence that the peaks are due to sodium. After the final map in Fig. 13(b) was obtained, the sample was driven to breakdown by a linearly increasing voltage; breakdown occurred at points "C" and "D" at an average field of about 3×10^6V/cm. Williams and Woods [46] have used a scanned He-Cd laser at 3.81 eV to obtain similar results, indicating a local, sodium-induced barrier reduction. The various types of barrier nonuniformities seen in the Si-SiO$_2$ system appear as "surface states" in MOS electronic measurements [55]. The capacitance-voltage curve is shifted and broadened for samples with nonuniform interfaces because it is simply a composite of elements from each small area for which the flat-band voltage is shifted by an amount dependent upon the local barrier [46]. Defects on the Si-SiO$_2$ interface imaged by internal photoemission, using the 3.81 eV line of a He-Cd laser cannot be directly observed because the threshold is 4.3 eV[3]. Some means such as a high electric field (about 3.50×10^6V/cm) or a sodium dipole layer must be used to lower this threshold energy. The photoemission image of Fig. 14 shows a line defect on a (100) silicon interface which has been "stained" with a uniform coverage of about 5×10^{12}Na/cm^2 in order to lower the threshold below the 3.81 eV laser photon energy. In the image, the light areas correspond to areas of increased photocurrent; the background has been removed electronically. The line, lying along the <110> direction of the surface, is identified as a stacking fault or a microsplit [56], either of which would be preferentially oriented in the <110> direction. The contrast mechanism responsible here is due to either an excess of sodium accumulated at the defect or a band bending caused by Fermi level pinning at the defect.

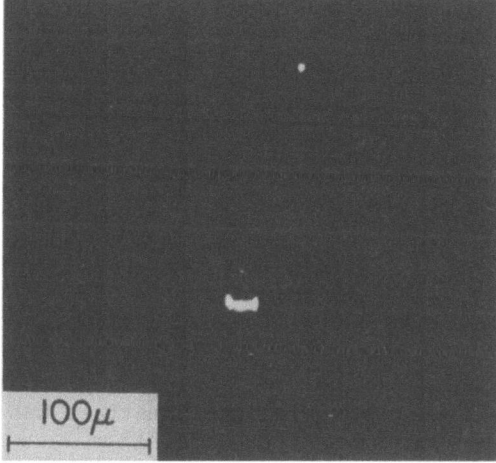

Fig. 14. Photoemission images of Si-SiO$_2$ interface showing a line of enhanced emission in the <100> direction. The scan was obtained with a He-Cd laser at 3.81 eV.

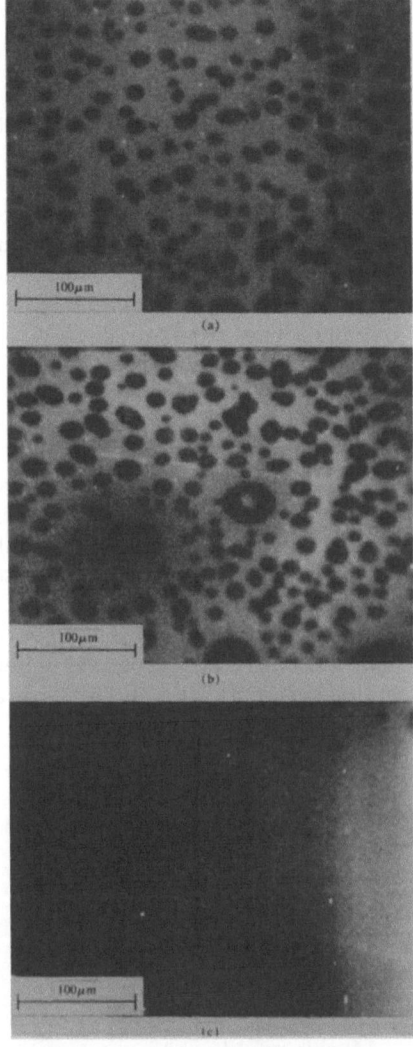

Fig. 15. Photoemission images of Si-SiO$_2$ interfaces formed by oxidation of silicon doped with approximately 10^{20}cm^3 phosphorus. The images (a)-(c) were obtained at a photon energy of 3.81 eV. In order to make the structure visible, the interface was "stained" with 4 x 10^{12}Na/cm^2. (From Ref. 31)

Some rather peculiar patterns have been observed in photoemission images of Si-SiO$_2$ interfaces grown by thermal oxidation of heavily phosphorus doped silicon [31]. Typical images, shown in Figs. 15 (a)-(c), were measured at 3.81 eV from several interfaces grown on $\sim 10^{20}$P/cm^2 doped (100) silicon, and "stained" with $\sim 4 \times 10^{12}$Na/cm^2. The dominant dark spots covering the surface are thought to be accumulations of a phosphorus-rich phase near the interface. Further evidence that the spots are due to phosphorus is that the fractional coverage of the surface increases with phosphorus doping concentration up to complete occlusion [57] at about 5×10^{20}P/cm^3. Phosphorus segregation seen by Auger profiling measurements [30,31] of the Si-SiO$_2$ interface has been ascribed [31] to the segregation of a phosphorus-vacancy complex at the interface. The size of the islands in the images indicates that the phosphorus has an appreciable mobility at the oxide formation temperature of 1100°C. The contrast mechanism which forms the dark spots is either a local neutralization of the sodium ion stain by the phosphorus or an increased scattering of the photoelectrons by the phosphorus impurity.

2. <u>Metal-Insulator Interfaces</u>. Photoemission images of metal-insulator interfaces typically have insufficient resolution to show details associated with the individual metal grains. Gross inhomogeneities such as contamination or nonstoichiometry is occasionally seen. For example, an image of a Bi-Nb$_2$O$_5$ interface is shown in Figure 16. This interface is of interest because Bi-Nb$_2$O$_5$-Nb capacitors display a bistable conductivity. The image was obtained by scanning a 1.96 eV He-Ne laser over a 200 Å thick

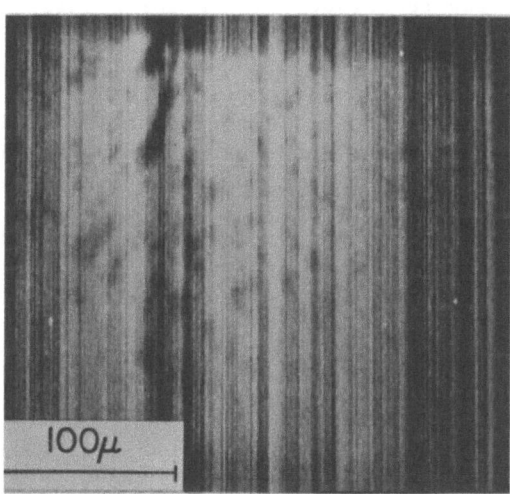

Fig. 16. Internal photoemission map of an Nb$_2$O$_5$-Bi interface, measured at a photon energy of 1.96 eV. (From Ref. 31)

bismuth electrode. The spotty pattern is due to areas of lower photoemission yield, possibly because of an enhanced trapping of the photoelectrons in these areas. The dark spots in the photoemission image correspond directly to areas of lower reflectivity seen in a high contrast reflectivity image of the same interface. Physically, the spots in both images may be caused by a reaction at the interface, leading to a high trap density in the Nb_2O_5. Interestingly, the capacitor "forms" bistable filaments in the dark areas of the image, where the $Bi-Nb_2O_5$ reaction has occurred.

B. Vacuum Photoemission Images

Scanned photoemission techniques have been used to obtain images of the free surfaces of materials in vacuum. Although scanned optical techniques are relatively simple, other nonscanned techniques described below are more attractive in providing high resolution images of surfaces in vacuum. In a reasonably direct application of the technique, a scanned (He-Ne) laser beam has been used to map the response of an $InAsP-Cs_2O$ photo-cathode in vacuum [58]. The image provides information on the uniformity of the Cs-O surface treatment. Another application of scanned photoemission is in imaging the slip lines in the surface of a metal which has undergone strain [59,60]. The photostimulated exoelectron emission is enhanced along newly formed slip lines because the passivating oxide over them is ruptured, exposing a clean metal surface [61,62]. Emission from the line first rises as oxidation reduces the work function, and then falls as the oxide becomes thicker.

A more promising technique for examining surfaces in vacuum involves focusing the photoemitted electrons to obtain an image, rather than scanning a light spot and measuring total current. The resolution is then limited by electron optics rather than by optical diffraction, making possible the formation of images on almost an atomic scale. Baxter and Rouze have used a Phillips thermionic emission microscope, with an electron multiplier array image intensifier, to produce photoelectron images of uniformly illuminated samples [63]. Images of photostimulated electron emission were obtained [64,65] from various metals after the samples were deformed. The images in Figs. 16(a)-(d) were measured on an aluminum surface after four successive strains of the sample. Photoemission microscopy is seen to provide information about metal fatigue and failure on a microscopic level.

A photoemission microscope proposed by Letokhov, and perhaps others, uses spherical fields to obtain a high resolution [66]. The proposed arrangement, represented schematically in Fig. 17, is similar to other point-emission techniques. It may be possible with high resolution microscopy, to display the effect on work function of local stress around dislocation and lattice imperfections.

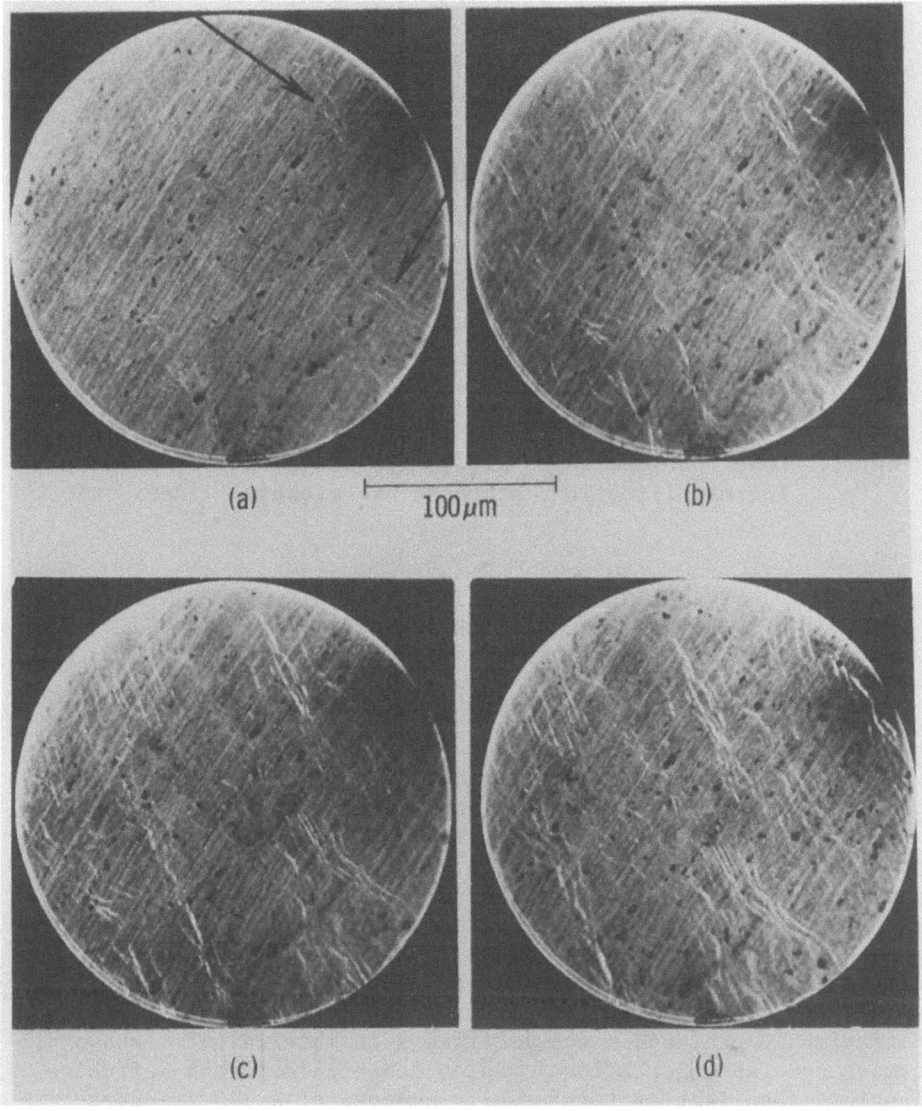

Fig. 17. Photostimulated exoelectron emission images of a surface of aluminum after four successive stages of plastic deformation. The image diameter is 225 μm. (After Baxter and Rouze, Ref. 62)

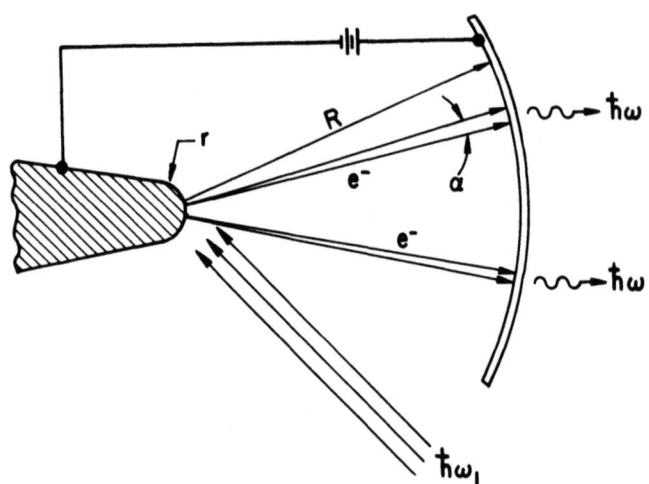

Fig. 18. Laser photoemission microscope using radial fields for magnification, proposed by Letokhov. (Ref. 64)

V. SCANNED PHOTOVOLTAGE ANALYSIS OF SEMICONDUCTOR SURFACES

A. Probe of Minority Carrier Parameters

1. <u>Low Frequency Optical Probe</u>. An early classic technique used to characterize semiconducting material involves a measurement of the decay length of minority carriers generated by a spot of light. Following a suggestion by Shockley, Goucher used a point contact to collect holes generated in n-type germanium by a spot of light at a distance x from the contact [67]. From the exponential decay of the collected voltage or current with x, the diffusion length L_p was determined. This quasistatic scanned spot experiment is useful in determining local minority carrier diffusion length, and hence lifetime τ in material of a known diffusion constant, for a local region comparable to the diffusion length. The light spot is usually chopped at a low frequency $f \ll 1/\tau$, to facilitate the measurement. In order to make an accurate measurement, it is necessary that the scanned spot does not come closer than a diffusion length from the crystal boundaries, and that the probed surface is passivated so that the recombination velocity is negligible. Also, since actual measurements are often not rigorously one dimensional, some corrections are necessary, particularly near the collecting contact. For a stripe of light absorbed near the surface, the collected signal has a Hankel function

dependence on the stripe-collector separation [68]. At large distances, this signal asymptotically approaches an exponential function of the separation between the light spot and the collector; the diffusion length L is the characteristic length of the exponential decay.

As the light spot scan velocity v is increased, the quasi-static approximation is no longer strictly valid, and the collected current is not symmetrical for ±v. By using this asymmetry, Adam was able to determine both the lifetime τ and the diffusion constant D for minority carriers in germanium [36]. A rotating mirror was used to scan a stripe of light across the surface on both sides of a rectifying contact, and the exponential decay lengths l_1 and l_2 were determined for the stripe moving toward and away from the contact, respectively. In the moving frame of the stripe, the diffusion equation is,

$$D \frac{\partial^2 n}{\partial x^2} - v \frac{\partial n}{\partial x} - \frac{n}{\tau} = 0 \qquad (16)$$

yielding solutions for the excess minority carrier density n,

$$n \propto e^{-(\pm \frac{v}{2D} + \frac{v^2}{4D^2} + \frac{1}{D\tau}) x} = e^{-x/l} \qquad (17)$$

From eq. 17, the lifetime and diffusion constant are determined;

$$\tau = \frac{1}{v} (l_1 + l_2) \qquad (18)$$

and

$$D = v \left(\frac{l_1 l_2}{l_1 + l_2}\right) \qquad (19)$$

This scanned photovoltage technique can be used to characterize material on a scale of several diffusion lengths, but it cannot easily yield truly microscopic information.

2. High Frequency Optical Probe. From results of a scanned photovoltage measurement using light chopped at a relatively high rate, Avery and Gunn have been able to obtain both the diffusion constant and lifetime of minority carriers in a local region [69]. Carriers excited in a small spot on the surface diffuse in a spherically symmetric pattern as illustrated in Fig. 19. Recombination at the surface is assumed to be small. Following Avery and Gunn [69], the excess minority carrier density n diffuses according to

$$D \frac{1}{r} \frac{\partial}{\partial r} \left(r \frac{\partial n}{\partial r} \right) - \frac{n}{\tau} = j2\pi f n \qquad (20)$$

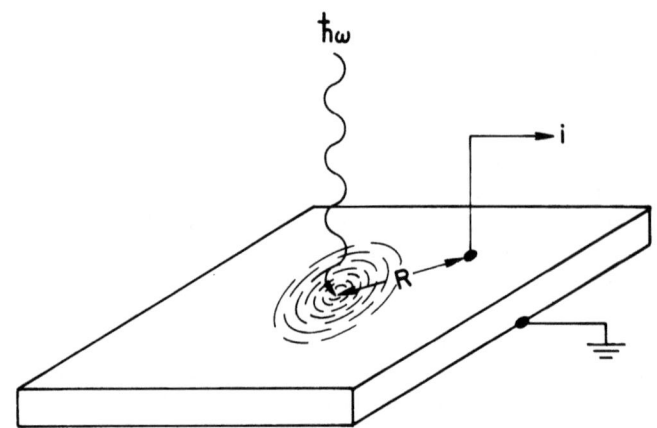

Fig. 19. Scanned spot photovoltage measurement in which a point probe at junction detects current induced by a light spot at a distance r from the probe.

where r is the radial distance from the light spot. The carrier density is then,

$$n \propto \frac{1}{r} e^{-\frac{1}{\sqrt{D\tau}}(1+j2xf\tau)^{\frac{1}{2}}r} \qquad (21)$$

and the collected current $i = |i| \exp(j\xi)$ is proportional to the excess minority carrier density near the contact at r_0. After solving Eq. 21, we find from the real and imaginary components of the exponent,

$$-\frac{r_0}{\ell} = -\frac{1}{\sqrt{D\tau}} (1+\pi f\tau)^{\frac{1}{2}} r_0 \qquad (22)$$

$$\xi = \frac{1}{\sqrt{D\tau}} (xf\tau)^{\frac{1}{2}} r_0 \qquad (23)$$

The phase shift ξ is seen to increase linearly with r_0, and the weighted signal $r_0 i$ decays with an exponential characteristic length ℓ. From an experimental determination of the phase ξ and decay length 1, the lifetime and diffusion constant can be determined;

$$D = \pi f \left(\frac{r_0}{\xi}\right)^2 \qquad (24)$$

and

$$\tau = \frac{1}{\pi f \left[\left(\frac{r_0}{\ell \xi}\right)^2 - 1\right]} \tag{25}$$

Thus, the phase and amplitude dependence of photovoltage near a rectifying contact is sufficient to characterize the material in that region.

Local fluctuations in the minority carrier lifetime caused by bulk or surface defects will return the photovoltage measured near the collecting probe. By comparing the spatial dependence of either the measured amplitude or phase with that expected from homogeneous material, it is possible [70] to detect the presence of local defects or inhomogeneities. Similar techniques can also be used with electron beam scanning [71,72,73]. In application, it is somewhat difficult to obtain images directly from these measurements because the signal analysis depends upon spot position. The derivative of the phase angle ξ with respect to position r_0 is a constant in homogeneous material; deviations from this constant derivative have been used to form a map of local variations in lifetime [70]. Also, uncorrected photovoltage amplitude measurements have been used [74] to map defects in $Pb_{1-x}Sn_xTe/PbTe$ heterojunctions, showing recombination at slip dislocations with a resolution of about 20-40 μm.

3. <u>Material Analysis by Surface Photovoltage</u>. Both the lifetime and diffusion length of minority carriers can be determined by a measurement of surface [75-78] photovoltage under a planar electrode on a semiconductor surface. In contrast to the scanned spot methods discussed above in sections V-A-2,3, the planar photovoltage method can be made less sensitive to surface conditions and somewhat easier to analyze. Instead of using the dependence of photovoltage on the light spot-to-collector separation to determine material parameters, each of several planar methods uses, in effect, the dependence of the carrier collection efficiency on the diffusion length. Generally, a photoresponse is measured as a function of photon energy or the frequency at which the light is chopped. For the case of measurements on MOS structures, the photovoltage is somewhat sensitive to the condition of the semiconductor surface [79,80]. We will not consider here the use of photovoltage spectroscopy to investigate homogeneous surface state density [81,82] or the spectral density at occupied surface traps in wide band gap materials [83]. By an extension of the planar photovoltage techniques, a small scanned light spot can be used to characterize semiconductor material and map the diffusion length or lifetime over the surface [80,84] with a spatial resolution approximately equal to the diffusion length.

Consider the photocurrent collected by a rectifying barrier on a surface uniformly illuminated by light of energy $\hbar\omega$. The light of intensity $I(\hbar\omega)$ is chopped at a rate f to produce a periodic fluctuation in excess minority carrier density n. Diffusion of the periodic component near the surface in extrinsic material is described by,

$$D \frac{\partial^2 n}{\partial x^2} - \frac{n}{\tau} - j2\pi f n = \frac{2\alpha(1-R) \, I(\hbar\omega)}{\pi \hbar \omega} e^{-\alpha\pi} \tag{26}$$

where R is the surface reflectivity and where n is subject to the boundary condition that n = 0 at the collecting junction at x = 0. The solution is easily obtained by Laplace transform methods;

$$n = \frac{2\alpha\tau(1-R) I(\hbar\omega)}{\pi\hbar\omega(1+j2\pi f \tau - \alpha^2 L^2)} \left[e^{-\alpha x} - e^{-(1+j2\pi f \tau)^{\frac{1}{2}} \frac{x}{L}} \right] \tag{27}$$

where $L^2 = D\tau$. The collected current is i,

$$i = \frac{2(1-R) I(\hbar\omega)}{\pi\hbar\omega} \frac{\alpha L}{L + (1+j2\pi f \tau)^{\frac{1}{2}}} \tag{28}$$

The expression for the current in Eq. 28 is valid for a Schottky barrier illuminated at a photon energy less than twice the bandgap so that there is no pair multiplication. For a small ac signal measured on a deeply depleted surface, the current i_s is,

$$i_s = \left(\frac{C_0}{C_0 + C_s}\right) \left(\frac{j2\pi f \tau_s}{1 + j2\pi f \tau_s}\right) i \tag{29}$$

where C_0 and C_s are the insulator and the depletion layer capacitances, respectively, and τ_s is the minority carrier lifetime in the channel. At relatively high measuring frequencies, where $f \gg \tau_s$, the effects of surface recombination are minimized so long as the surface capacitance or surface potential are unaffected. For presently available silicon MOS structures, a frequency above 1000 Hz is usually sufficient to avoid extraneous effects of surface states.

Measurements of the surface photoresponse at several wavelengths can be used to determine L. At measurement frequencies $f \ll (1/\tau)$,

$$i(\hbar\omega) \propto \frac{(1-R) I(\hbar\omega)}{\hbar\omega} \frac{\alpha(\hbar\omega) L}{1 + \alpha(\hbar\omega) L} \tag{30}$$

By measuring the photoresponse at two photon energies for which the absorption coefficents are significantly different, the minority carrier diffusion length L is determined [77]. In order to minimize the influence of surface states on the MOS measurements, the intensity I is usually adjusted to equalize the photoresponse at each photon energy. The diffusion length is then determined as the inverse slope of a plot of the generation rate $(1-R)I/(\hbar\omega)$ (normalized to the saturated value at $\alpha\to\infty$) as a function of $(1/\alpha)$. The technique has been extended to the characterization of epitaxial layers of silicon [85].

The frequency dependence of photovoltage has been used to determine both the lifetime and the diffusion constant for minority carriers [86,87]. For the case of weak absorption, where $\alpha L \ll 1$, the phase shift of the signal is $\tan^{-1}\pi f\tau$; the lifetime τ is determined from the slope of a plot of the phase shift or from the frequncy at which the amplitude falls to $1/\sqrt{2}$ of saturation w869. The diffusion constant can be found similarly for the case in which α is adjusted to be relatively large, $\alpha L \gg 0$. Then the frequency at which the amplitude falls by a factor of $\sqrt{2}$ determines $D = 2\pi f/\alpha^2$.

B. Scanned Surface Photovoltage

1. <u>Local Excitation of Excess Minority Carriers</u>. Recently, several scanned surface photovoltage techniques have been used to obtain images of recombination at surface defects on semiconductors [41,88,50]. The surface is scanned with a spot of either chopped or dc light to produce a local excess density of minority carrier in the surface depletion region. The carriers produce a photovoltage which is detected capacitively and displayed as a function of position of the light spot. The measurement is represented schematically in Fig. 20. As the scanned spot traverses a defect, the resulting drop in photovoltage is detected and displayed in the photovoltage image. An analogous electron beam technique has also been used to image electrically active defects in free surfaces of silicon [89].

The excess real carrier density $\eta(x,y)$ on a homogeneous surface is obtained from the diffusion equation in the moving frame of the light spot;

$$D\nabla^2\eta - v\frac{dn}{dx} - \frac{\delta_0}{\delta}\eta = j2\pi f\eta \qquad (31)$$

where D is the diffusion constant in the channel, V is the light spot velocity, f is the chopping rate, and δ is the surface channel width. A transformation of $\eta = \chi\exp(-vx/D)$ is convenient. Then

$$\nabla^2\chi - \frac{1}{\Lambda^2}\chi = 0 \qquad (32)$$

where the effective diffusion length Λ is given by

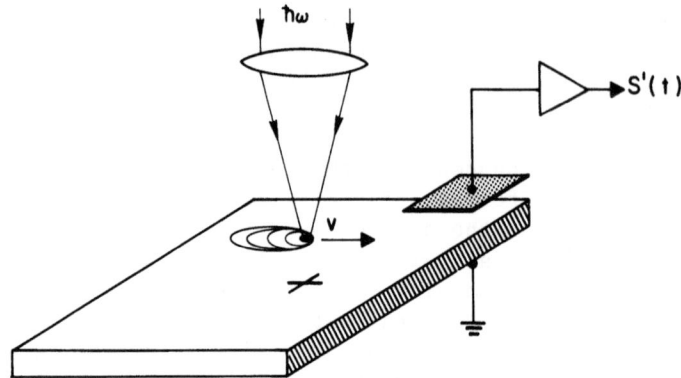

Fig. 20. The scanned surface photovoltage measurement. The light spot, which may be chopped, is scanned at a linear velocity v. The resulting surface charge is detected by measuring the current induced in an electrode capacitively coupled to the surface.

$$\Lambda = \left(\frac{D\tau}{1+j\frac{2\pi f \delta}{S_0}}\right) \; 1 + \frac{V^2 \delta}{4DS_0} \left(\frac{1}{1+j\frac{2\pi f \delta}{S_0}}\right)$$

In cylindrical coordinates, the solution is a modified Bessel function $K_0(r/\Lambda)$ of the zero order and the second kind. The excess carrier density is then,

$$\eta = B \, e^{\frac{Vx}{2D}} \, K_0(r/\Lambda) \tag{33}$$

where

$$B = \frac{(1-R)I(\hbar\omega)}{\pi^2 \hbar\omega \Lambda^2} \; \frac{\frac{\delta}{S_0}}{1+\left(\frac{j2\pi f \delta}{S_0}\right)}$$

and where B is multiplied by $\pi/2$ for unchopped light. The excess charge is a comet shaped cloud of extent $|\Lambda|$ that follows the scanned spot. The asymptotic solutions for η are

$$\eta \cong B \left(\frac{\pi\Lambda}{2r}\right)^{\frac{1}{2}} e^{-r/\Lambda} \, e^{Vx/2D} \qquad r \to \infty$$

$$\eta \cong B \ln\left(\frac{r}{2\Lambda}\right) + .5772B \qquad r \to 0 \tag{34}$$

It is interesting to note that the extent of the charge cloud is reduced, except for the tail, by an increased scan velocity.

2. _Chopped Light_. The signal in the scanned surface photovoltage technique is typically detected as an induced current on an electrode of an MOS capacitor. The surface potential is developed across a depletion region in the surface by the excess surface charge. The Dember potential is negligible in almost all cases. The signal from a deeply depleted surface is,

$$i = \left(\frac{C_0}{C_0+C_s}\right) \frac{dQ_s}{d\tau} \tag{35}$$

where Q_s is the surface charge and C_0 and C_s are the insulator and depletion layer capacitances, respectively. The change of surface charge Q_s is simply the difference between generation and recombination,

$$\frac{dQ}{dt} = \frac{2(1-R)I(\hbar\omega)}{\pi\hbar\omega} - \frac{S_0}{\delta} Q_s - \iint \frac{1}{\delta} S'(x,y) \, \eta\,(x-x_0, y-y_0) dx dy \tag{36}$$

where the last term represents the excess recombination in the stationary frame. From Eqs. (35) and (36), the detected current i is,

$$i = \left(\frac{C_0}{C_0+C_s}\right) B \; 2\pi\Lambda^2 - \frac{1}{(1+\frac{j2\pi f\delta}{S_0})} \iint e^{\frac{V(\lambda-x_0)}{2D}} K\left(\frac{r-r_0}{\Lambda}\right) \frac{S'(x,y)}{S_0} dx dy \tag{37}$$

For the chopped light scanned photovoltage, the excess local recombination velocity $S'(x,y)$ causes a dip Δi in the signal as the spot scans over the site of the local recombination. For a point defect at (x,y) scanned at a low velocity,

$$\frac{\Delta i}{i} \propto e^{\frac{V(x-x_0)}{2D}} K_0\left(\frac{r-r_0}{\Lambda}\right) \tag{38}$$

and asymptotically,

$$\frac{\Delta i}{i} \propto r^{-\frac{1}{2}} e^{\frac{V(x-x_0)}{2D}} e^{-\left(\frac{r-r_0}{\Lambda}\right)} \qquad r \to \infty \tag{39}$$

The resolution of the chopped light technique is on the order of the corrected diffusion length Λ.

3. _Scanned Constant Intensity Light Spot_. The photovoltage induced in a surface by a scanned beam of light has been used to

produce high resolution images of defects on silicon surfaces [50,52]. The analysis is similar to the case for chopped light, except that the generation rate for the chopped case is reduced by a factor of $2/\pi$, the first Fourier coefficient of the square optical pulse. Again, the change in excess surface charge is the balance of generation over recombination;

$$\frac{dQ_s}{dt} = \frac{(1-R)I}{\hbar\omega}\frac{S_0Q_s}{\delta} - \frac{1}{\delta}\iint n(x-x_0, y-y_0)S'(x,y)\,dxdy \tag{40}$$

There are two approximate solutions to Eq. (40); a slow scan regime where recombination dominates and a fast scan regime where charge acummulation is more important.

Consider the slow scan limit, where the first term in Eq. (36) is negligible. Then the signal current is

$$i = -\left(\frac{C_0}{C_0+C_s}\right)\frac{(1-R)I(\hbar\omega)}{2\pi\hbar\omega(D+\frac{V^2\delta}{4S})}\frac{d}{dt}\iint e^{\frac{V(x-x_0)}{2D}}\,K\left(\frac{r-r_0}{\Lambda}\right)$$

$$\left(\frac{S'(x,y)}{S_0}\right)\,dxdy \tag{41}$$

where now, the effective surface diffusion constant is,

$$\Lambda = \left(\frac{DS}{S}\right)^{\frac{1}{2}}\left(1 + \frac{V}{4D}\right)^{\frac{1}{2}} \tag{42}$$

For a linear scan in the x direction,

$$i \propto V \iint \left(\frac{S'(x,y)}{S_0}\right)\left\{\frac{\partial}{\partial x_0}\left(e^{\frac{-V(x-x_0)}{2D}}\,K_0\left(\frac{r-r_0}{\Lambda}\right)\right)\right\}\,dxdy \tag{43}$$

From Eq. (43) it is seen that the signal displayed in the scanned photovoltage image is a convolution of the local recombination $S'(x,y)$ with the sampling function in brackets. The magnitude of this sampling function is shown in Fig. 21 for several scan velocities. Because the sampling function falls rapidly at distances greater than the light spot radius, the broadening of the image is on the order of the dimension of the light spot rather than the diffusion length Λ. Also, the tail of the charge cloud is negligible in the image. The resolution of the technique can be estimated by considering the asymptotic limit $r \to 0$;

$$\left.\frac{\partial n}{\partial x}\right|_{y=0} \propto \frac{1}{x-x_0} \tag{44}$$

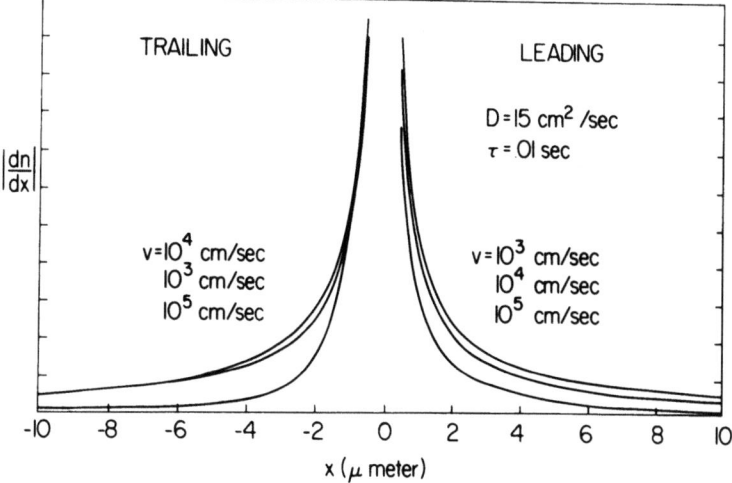

Fig. 21. The absolute value of the sampling function $d\eta/dt$ used to determine the response of the dc scanned surface photovoltage. The function is plotted for three scan velocities. (From Ref. 50)

so that the sampling function broadens the image by about 2 to 3 times the light spot radius. Examples of high resolution images measured with this technique are shown in Section VI.

The signal obtained in the fast scan regime is analyzed in a similar manner, except that the image is formed from the first derivative of the measured current. In this case, the homogeneous recombination $S_0 Q_s$ in Eq. (40) is insignificant. The image is a display of (di/dt);

$$\frac{di}{dt} = \left(\frac{C_0}{C_0 + C_s}\right) \frac{1}{\delta} \frac{d}{dt} \iint \eta(x-x_0, y-y_0) S'(x,y) \, dx \, dy \tag{45}$$

As before,

$$\frac{di}{dt} \propto V \iint \left(\frac{S'(x,y)}{S_0}\right) \left\{ \frac{\partial}{\partial x_0} \left(e^{\frac{V(x-x_0)}{2D}} K\left(\frac{r-r_0}{\Lambda}\right) \right) \right\}$$

and the image is broadened by an amount only 2-3 times the light spot radius.

Either the scanned surface photovoltage or its first derivative is capable of resolving electrically active local defects on a silicon surface, an MOS structure or a free surface. The resolution is sufficient to show structure of crystallographic defects which extend more than several micrometers.

C. Bulk Photovoltaic Effect

Doping inhomogeneities in semiconducting material produce local gradients in potential which will cause a charge separation current to flow if excess minority carriers are created in the area. As recognized by Tauc [90], the photovoltage produced across a sample by a scanned spot of light can be used to determine quantitatively the resistivity gradient and from it, the resistivity of the bulk material. A simple scanned optical measurement is able to map doping or resistivity variations in semiconductor wafers [38,91]. Following Tauc [92], the technique can be analyzed in one dimension, with light falling on a small length of material. Since no current is drawn, $i_p + i_n = 0$ or

$$-\sigma \frac{d\phi}{dx} + eD_n \frac{dn}{dx} - eD_p \frac{dp}{dx} = 0 \tag{46}$$

where μ_n and μ_p are the hole and electron mobilities, respectively. The potential gradient is expressed as a perturbation on the system before being exposed to light;

$$\frac{d\Phi}{dx} = \frac{\sigma_0 \frac{d\Phi_0}{dx} + kT(\mu_n \frac{d\Delta n}{dx} - \mu_p \frac{d\Delta p}{dx})}{\sigma_0 + \Delta\sigma} \qquad (47)$$

where the relationship,

$$\frac{dn_0}{dx} = \frac{e}{kT} n_0 \frac{d\Phi_0}{dx} \qquad (48)$$

and the corresponding relationship for holes were used. For low levels of excitation, we expand Eq. (47),

$$\frac{d\Phi}{dx} \cong \frac{d\Phi_0}{dx} - \frac{\Delta\sigma}{\sigma_0} \frac{d\Phi_0}{dx} + \frac{kT}{\sigma_0}(\mu_n \frac{d\Delta n}{dx} - \mu_p \frac{d\Delta p}{dx}) \qquad (49)$$

The gradient is integrated over x to find the total photovoltage. The first right hand term integrates to zero since it is the unperturbed potential. In extrinsic n-type material, the measured photovoltage is V,

$$V \cong \int \frac{\Delta\sigma}{\sigma_0}(\frac{kT}{en_0}\frac{dn_0}{dx}) + \frac{kT}{e\sigma_0}(\frac{\mu_n - \mu_p}{\mu_n + \mu_p})\frac{d\Delta\sigma}{dx} \, dx \qquad (50)$$

where approximate charge neutrally has been assumed. After integration by parts of the last term, we have

$$V = \frac{kT}{e} \int \frac{\Delta\sigma}{\sigma_0^2}\frac{d\sigma_0}{dx} - (\frac{\mu_n - \mu_p}{\mu_n + \mu_p})\frac{\Delta\sigma}{\sigma_0^2}\frac{d\sigma_0}{dx} \, dx \qquad (51)$$

or simply,

$$V = \frac{kT}{e} \frac{2}{(1+\frac{\mu_n}{\mu_p})} \int \frac{\Delta\sigma}{\sigma_0^2}(\frac{d\sigma}{dx}) \, dx \qquad (52)$$

By measuring both the photovoltage and the increased conductivity produced by a spot of light, the conductivity gradient is easily obtained.

Mapping the doping profile of a two dimensional wafer is a bit more difficult. Blackburn, Schaftt, and Swartzendruber have used transform techniques to determine the local photovoltage, produced by a scanned spot, from measurements of the voltage induced at the edge of a wafer [47]. The local resistivity obtained by integrating the gradient along a diametric scan of a silicon is shown in Fig. 22, along with a resistivity profile obtained by a four point probe. Similar bulk photovolate measurements are used to map inhomogeneities on semiconductor wafers.

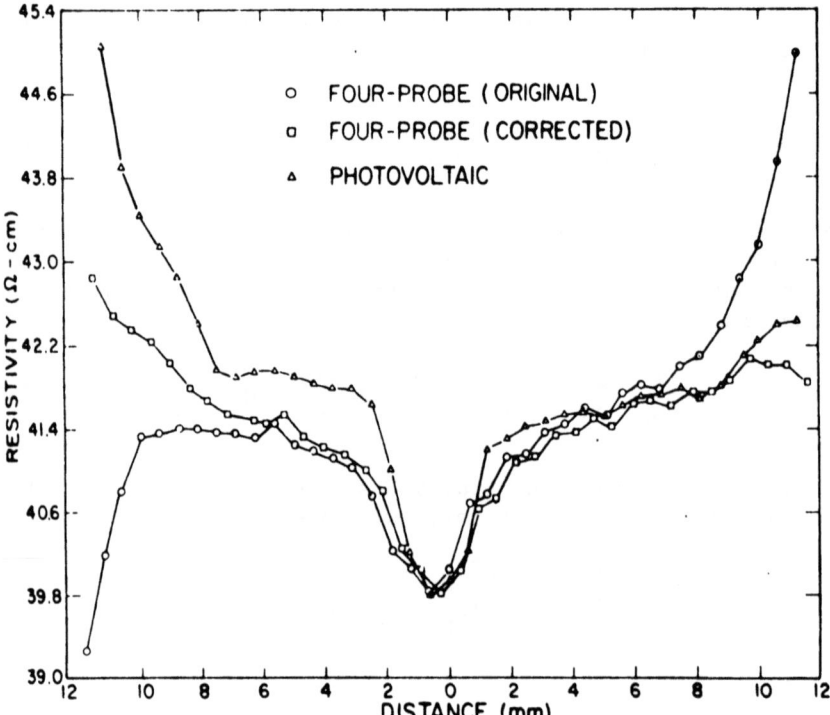

Fig. 22. A resistivity profile through the diameter of a silicon wafer obtained by integrating the bulk photovoltage across the wafer produced by a scanned light spot. The resistivity is in overall agreement with four point probe measurements. (After Blackburn, Schafft, and Schwartzendruber, Ref. 47)

Since the doping inhomogeneities typically found in a semiconductor wafer grown by the Czochralski method are predominantly cylinderically symmetric, two orthogonal diametric scans will provide a significant amount of information. As a practical consideration, the sensitivity and resolution of the bulk photovoltage technique are limited by the relatively low S/N ratio of the typical measurement.

Bulk photovoltage measurements have been used to probe inhomogeneities. However, with this sort of bulk photovoltage imaging method, it is difficult to arrange the sensing electrodes in a configuration such that they do not introduce an apparent inhomogeneity. Additionally, the bulk photovoltage measurement includes detail which is introduced by recombination on crystallographic defects in the material. A measurement of the photovoltage produced by a spot of light scanned across a uniform, partially transparent Schottky barrier is capable of providing some measure of doping uniformity, but has the disadvantage that the Schottky barrier must be removed without damaging the surface if devices are to be fabricated on the material.

VI. SCANNED PHOTOVOLTAGE IMAGES

A. Crystallographic Defects

The scanned surface photovoltage technique is capable of imaging electrically active defects in a semiconductor surface with a sufficiently high resolution to allow the identification of many common crystallographic defects. Several examples of photovoltage images of defects, imperfections and inhomogeneities on silicon surfaces demonstrate the resoltuion and capability of this technique. The images to be described were obtained by the method of Section V B 3, using a 5 mW HeNe laser source. At the laser wavelength of $\lambda = 6328$ Å, the attenuation length in silicon is approximately 1.3 μm. The silicon, except for the polycrystalline sample, is device grade p-type (100) material.

The simplest observed defects are dislocations such as the ones seen as recombination sites in the photovoltage image shown in Fig. 23. The calibration marker is orientated in the <110> direction. The cluster of emergent dislocations penetrating the surface was produced by first sand blasting a small spot on the reverse side of an oxidized wafer, and then annealing it in dry He for 30 min. at 800°C. During the annealing step, dislocation loops produced by the mechanical damage grow by gliding along the (111) planes. Eventually, the loop penetrates the front surface [94], forming a pair of emergent dislocations. The emerging end

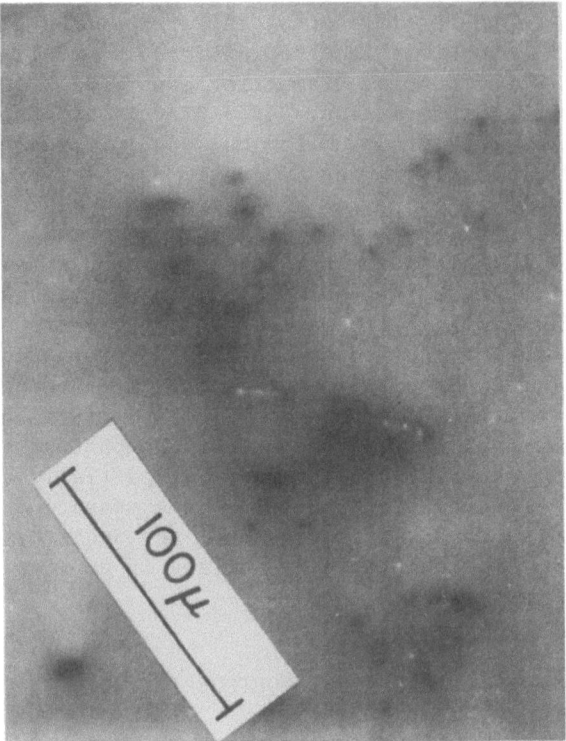

Fig. 23. Scanned surface photovoltage image of several clusters of emergent dislocations through the (100) surface of silicon. The dislocations were produced by sand blasting a small spot on the reverse side of the wafer so that the loops generated by the damage grow and penetrate through the wafer surface. (From Ref. 52).

of the dislocation appears as a point in the photovoltage image. The resolution of the image and the attenuation length of the light are not sufficient to resolve the direction of the dislocation below the surface. This limitation on the depth of sensitivity is an advantage in analyzing silicon material for use in integrated circuits, where the circuits are sensitive to defects only within several microns of the surface. The dislocations were not intentionally decorated; the uniformity of the dislocation images indicates that the recombination producing the images is likely an intrinsic property of the dislocation structure, and due to impurity atom decoration.

Stacking faults of various types are a serious problem in the fabrication of silicon devices, both MOS and bipolar. Stacking faults can be caused by precipitates in the silicon material as well

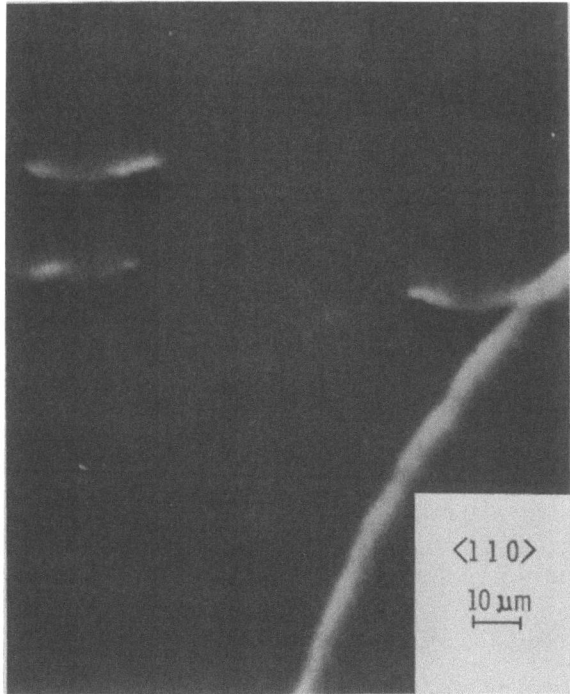

Fig. 24. Photovoltage image of oxidation induced stacking faults in silicon (100). The pair of stacking faults in the upper left grew from an oxide inclusion below the surface. The curved line in the lower right is the edge of the surface electrode. (From Ref. 50.)

as by the high temperature processing steps of epitaxial layer growth, oxidation, and diffusion [95,96]. For a photovoltage study, large stacking faults were produced by oxidation of (100) Czochralski grown silicon. The sample was oxidized in dry oxygen for 20h at 1050 C, quenched to room temperature, and then oxidized again for 20h at 1050 C. Several of the large stacking faults formed by the oxidation are visible in the photovoltage image of Fig. 24. The stacking faults shown in Fig. 24 were grown to a very large size in order to show the structure of the defect, but in processed silicon, typical stacking faults are considerably smaller. Recombination is seen on the Frank loops bounding the stacking faults, which lie in the (111) plane. The Frank loop is only faintly visible over the midsection where it extends more deeply in to the silicon; the ends of the Frank loop are clearly visible where they intersect the surface.

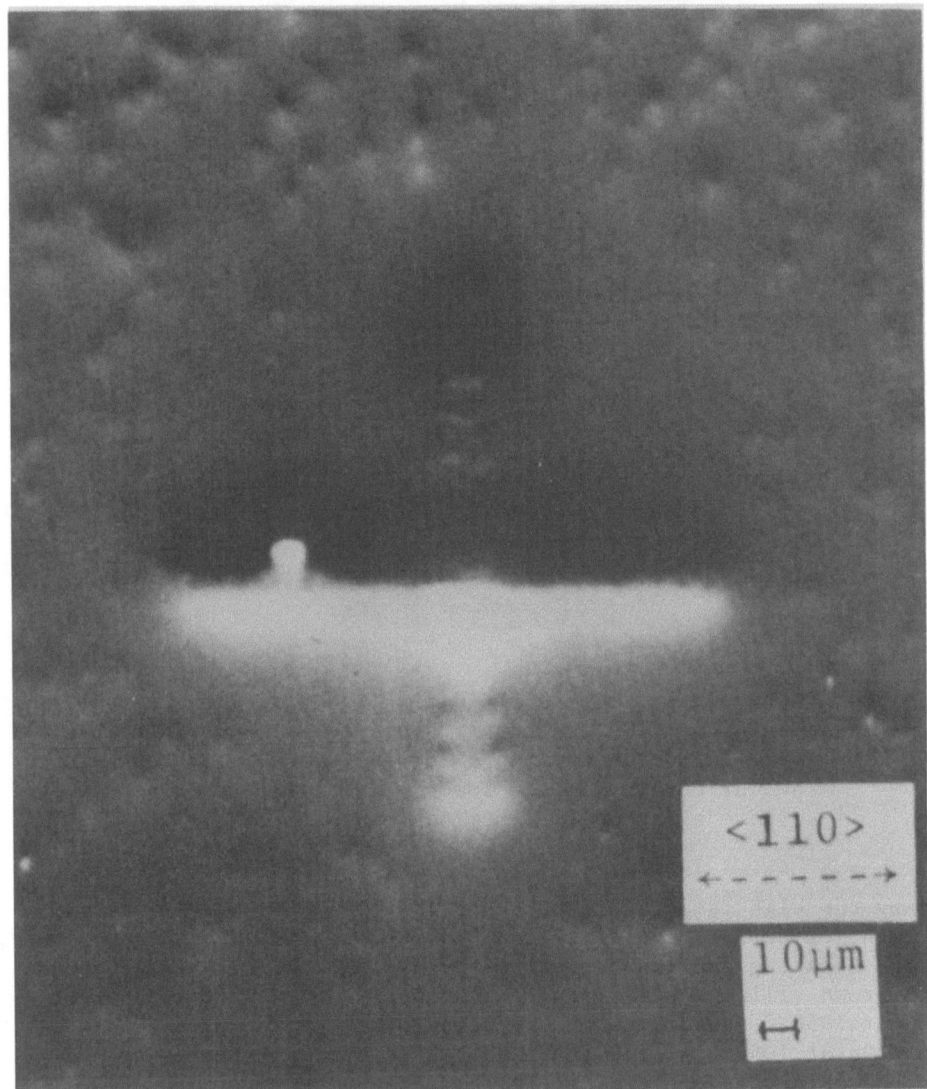

Fig. 25. Photovoltage image of a dislocation rosette in (100) silicon. The rosette is an array of dislocations which emerge from the surface along the <110> directions, radiating out from the point of indentation. (From Ref. 50)

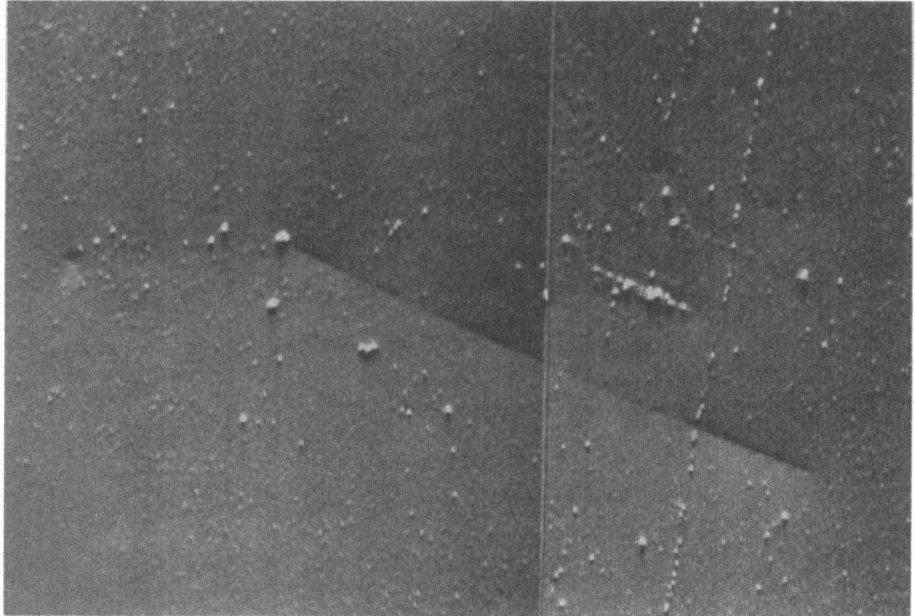

Fig. 26. Scratches and an emergent dislocation seen in the photo-voltage image of a device grade silicon wafer. The major scratch, about 1000 μm long, is residual damage from the polishing process.

Because the ends of the stacking faults appear similar, with no major variations in intensity, the recombination seen in the image is very likely intrinsic rather than due to an impurity decorating the defect.

The stacking faults visible in Fig. 25 measure approximately 30 μm in length [99], and the width of the image of the bounding dislocation is about 2 μm. The image was scanned from top to bottom, producing a light to dark contrast reversal in the image of the stacking faults due to the differentiation of the photovoltage signal. It is interesting to note that the stacking faults in the image do not lie directly under the sensing electrode, which lies in the lower right hand corner. The photovoltage signal generated by scanning over the stacking faults is coupled to the electrode by a conductive surface channel which forms when the slightly depleted surface is illuminated by the laser spot.

A defect of rather interesting geometrical shape is the dislocation rosette found in the photovoltage image shown in Fig. 25. The rosette, which does not normally occur in silicon device

fabrication, was formed [97] by indenting the (100) surface with a
diamond tip at a load of 25g for 30 sec. During indentation, the
sample was held at 600°C, and then annealed for 30 min. at 970°C.
The indentation produces an array of dislocation loops radiating
out in the <110> directions from the indentation [98]. Each loop
penetrates the silicon surface at a pair of points on opposite sides
of the arm. These paired emergent dislocations can be seen along
the arms of the rosette in the photovoltage image. Additionally,
a high density of small defects is found in the region around the
rosette but, curiously, the immediate vicinity of the rosette is
denuded of the visible spots. These defects could be stacking faults
or precipitates which may or may not be decorated by impurities.
Since it is a sink for impurities and vacancies, the rosette may
have either reduced the size of the adjacent defects or depleted
them of decorating impurities. Through the use of a sequence of
phtotovoltage images at progressive stages of annealing, it should
be possible to study the gathering action of various surface
treatments in the production of silicon devices.

Several silicon defects commonly occuring in the production
of integrated circuits are found in the photovoltage images of
Figs. 26-28 which were measured on nominally defect free, p-type
(100) silicon. Defects visible in the image are carrier generation/
recombination sites which can be detected, but not imaged, by methods
such as the Zerpst technique [100]. These defects may cause mal-
function of devices fabricated on them. One of the most common
defects on silicon surfaces is seen as small spots in Fig. 28.
These spots are thought to be precipitates in the silicon crystal
which may be decorated with impurities and which may have generated
associated small dislocation loops or stacking faults. Precipitates
in silicon were found by Ravi and Varker [102,103] to nucleate the
formation of stacking faults during high temperature processing.
It has not been determined whether bare precipitates, without
associated impurity decoration or crystallographic defects, would
be visible in photovoltage images. Scans of many crystalline surfaces show the spots to form a pattern on the surfaces, and to
occur over a wide range of areal density.

The surface defects shown in Figs. 26 and 27 are residual damage
left from the cutting and polishing operations on (100) silicon
wafers. Before measurement, the wafers were oxidized to form a
thickness of 1000 Å of SiO_2. The scratches in Fig. 26, one of
which is more than 1000 μm long, lie in an array which is not
parallel to the <110> direction on the surface. The recombination
which forms the images of the scratch is due to a row of small
dislocation loops punched out along the length of the scratch. The
defect, lying along the <110> direction in Fig. 22, is identified
as a microsplit [104]. By comparison with TEM images shown by
Schwuttke, the structures lying in the <110> direction off of each
end of the microsplit are thought to be arrays of stacking faults.

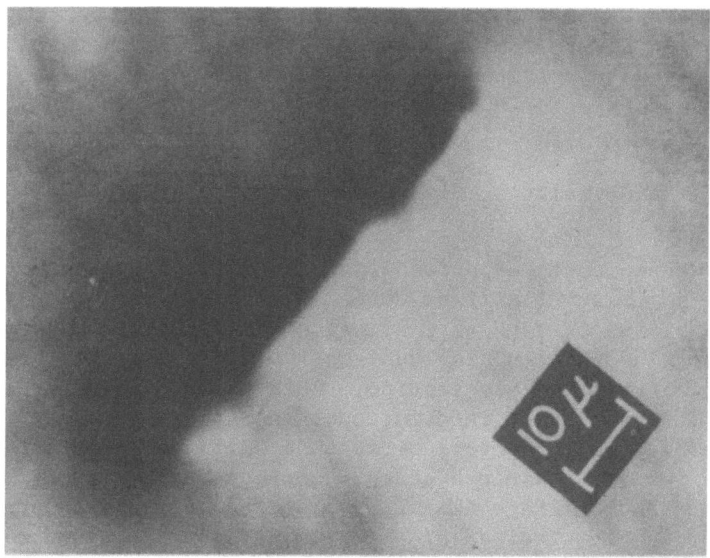

Fig. 27. A photovoltage image of a microsplit, with stacking faults on each end, seen on a device grade silicon (100) wafer. (From Ref. 52)

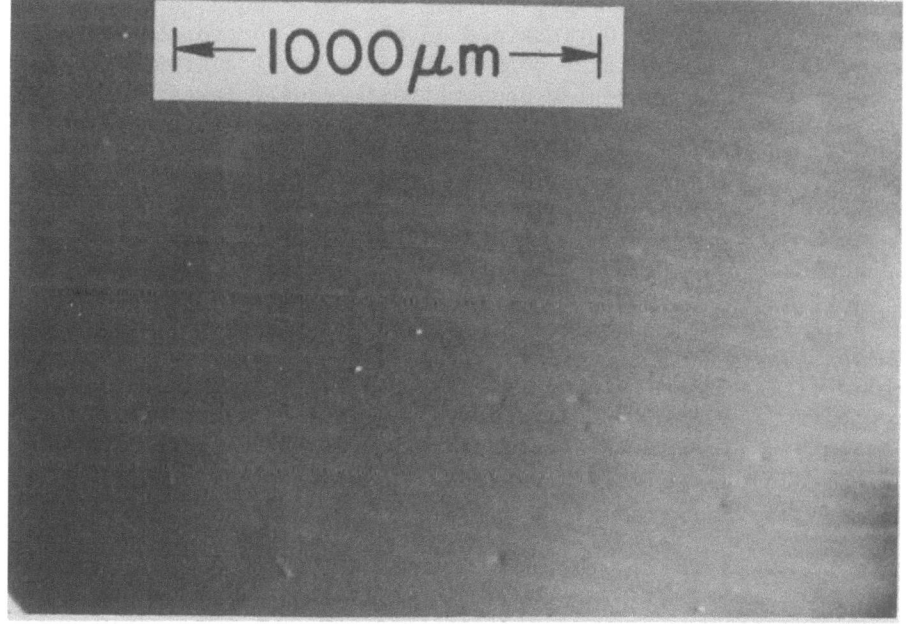

Fig. 28. Photovoltage image of a (100) surface of silicon showing small defects thought to be SiO_2 precipitates.

Schwuttke has shown that microcracks produced during sawing at silicon wafers penetrate so deeply that they may not be removed in the lapping and polishing operations. Overall, photovoltage images are seen to provide information on the surface perfection of semiconductors after cutting, lapping and polishing.

B. Correlation with Chemical Etch Patterns

The scanned surface photovoltage technique provides non-destructively an image of surface defects which are currently detected and imaged by any of several chemical etch pattern techniques. Chemical etching is of course destructive, so the technique is useful for batch or sample testing. Both photovoltage imaging and chemical etching provide information on surface defects at a similar spacial resolution. However, because of different contrast mechanisms, the images attained by each technique are not necessarily equivalent. Photovoltage images have been compared with etch patterns on the same sample obtained by selective etching [50] and by anodic etching [105] with the result that the etch patterns and the photovoltage images provide similar information for surface defect detection. Where the defects lie below the surface, there is a difference in the sensitivity of the several techniques.

Both photovoltage images and selective chemical etch patterns of an angle lapped sample of silicon are shown in Fig. 29. The photovoltage images (A´, B´, C´, and D´) correspond to the etch patterns (A, B, C, and D) obtained on the same portion of the surface. It can be seen that both techniques provide similar images, although the defects appear somewhat differently in each. The sample was a piece of saw cut silicon which has been lapped at and polished at $2°$ to remove the same damaged material to an increasing depth. Photovoltage images of the oxidized sample were obtained at several positions along the bevel before the sample was selectively etched and photomicrographs obtained at the same position along the bevel. Even at the most heavily damaged portion of the surface, D and D´, the photovoltage image provides information equivalent to the etch pattern, although the photovoltage image is somewhat blurred. The defects visible in the angle lapped sample are topologically connected to the surface, so that selective etching is expected to delineate them. In the case of defects, such as precipitates, that do not intersect the surface, the defect may not show up in an etch pattern although they may appear in a photovoltage image if they are within an absorption length of the surface.

A comparison between photovoltage images and anodic etch patterns shows that the two provide similar information [105]. Selective anodic etching of n-type silicon occurs at a rate proportional to

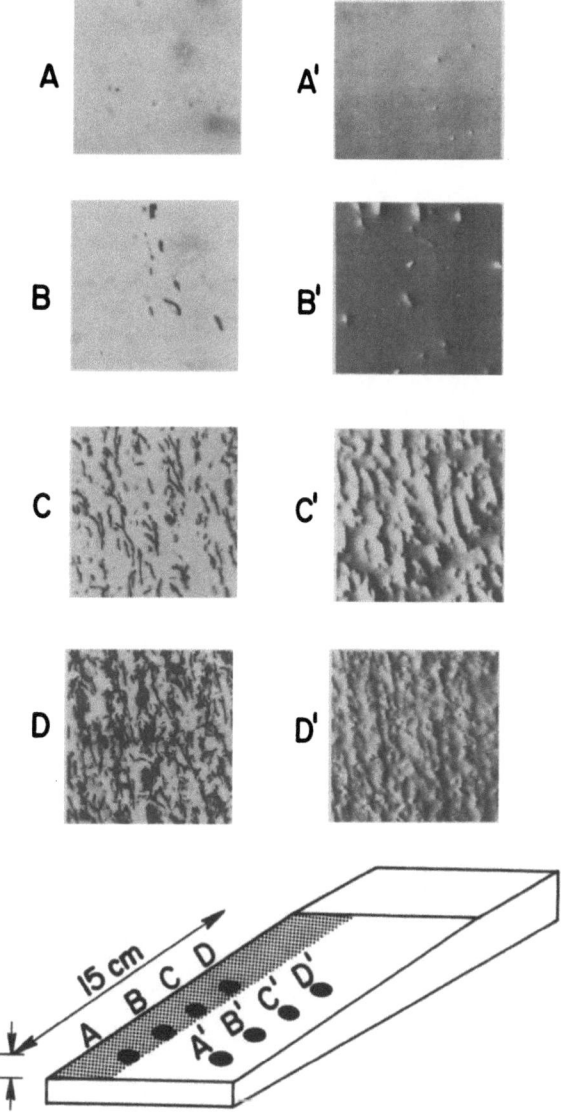

Fig. 29. A comparison between the photovoltage image and the etch pattern on a saw cut silicon wafer which has been bevel lapped. The photovoltage images A - D were obtained on areas from which saw damage has been removed to a depth determined by the 2° bevel angle. Subsequent etching of adjacent areas A - D yields the pattern shown. (From Ref. 50.)

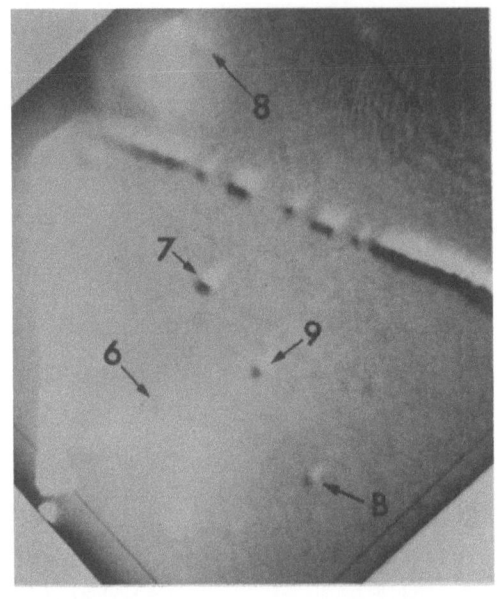

(a)

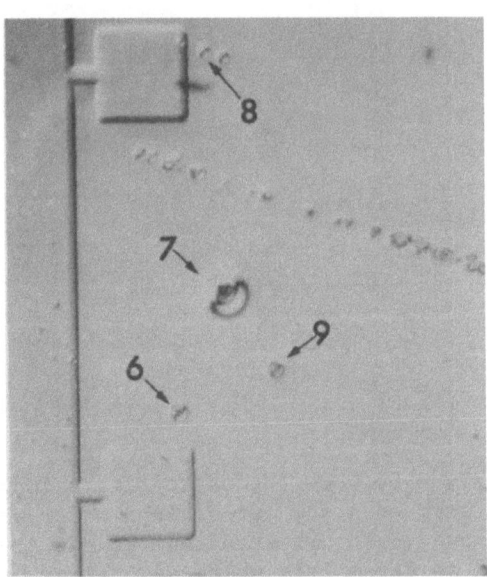

(b)

Fig. 30. Silicon wafer with patterning and defects shown in (a) a photovoltage image and (b) a photomicrograph of an anodic etch pattern, both on the same sample (From Ref. 105, courtesy of J. L. Deines, D. B. Dove, J. W. Philbrick and M. R. Poponiak).

the hole current to the surface [106], so that the etching proceeds
at a higher rate in the vicinity of a defect which generates holes.
A defect forms an etch pit, even though it is below the surface,
if the holes generated are able to reach the surface. Since defects
which are a locus of generation are also a locus of recombination,
it is expected that a photovoltage image and an anodic etch pattern
of the same surface would show the same defects. To a large extent,
this is true, as can be seen in Fig. 30, showing the results of
both techniques used sequentially on the same surface. The sample
is a n-type epitaxial layer on (100) silicon. Both show a scratch
and several small defects. However, the photovoltage image shows
defects A and B not seen in the etch pattern, while the etch pattern
shows many defects over the buried n^+ region (the lithographically
defined region) not seen in the photovoltage image. When anodized
at a higher voltage, the etch pattern shows additonal defects
which, presumably, lie deeper in the material. Overall anodic
etching and photovoltage imaging provide the same type of information,
but with somewhat different sensitivity.

C. Surface Charge

Certain types of surface charge inhomogeneities can be detected by scanned photovoltage techniques, although the contrast
mechanisms are neither well understood nor particularly reliable.
Features in photovoltages due to local surface accumulation and
due to charged ions on a surface have been observed and can be
understood in terms of a simple, non-quantitative model. An understanding of the contrast mechanisms due to surface charge are of
some value in interpreting photovoltage images. Additionally,
scanned photovoltage techniques may be useful as a diagnostic
tool for determining charge nonuniformities in semiconductor devices
such as various charge storage structures.

In the case where a local charge causes an accumulation on
an underlying semiconductor surface, minority carriers produced
under the charge are not captured at the surface as they are at
a slightly depleted surface. Instead, the carriers are injected
into the bulk where they may recombine before diffusing to a depleted
portion of the surface where they can be captured. By this mechanism, local accumulation appears as areas of enhanced recombination.
An example of this mechanism is seen in Fig. 31, showing a photovoltage image of an MNOS structure on p-type silicon. Positive
charge was stored at the SiO_2- Si_3N_4 interface by a voltage pulse,
and then an oblong spot of UV light was used to discharge a small
region of the interface. A bias voltage applied to the electrode
depleted the surface except for the discharged area, which is left
in accumulation. The oblong spot appears in the photovoltage image
as an area of enhanced recombination. This technique may be useful

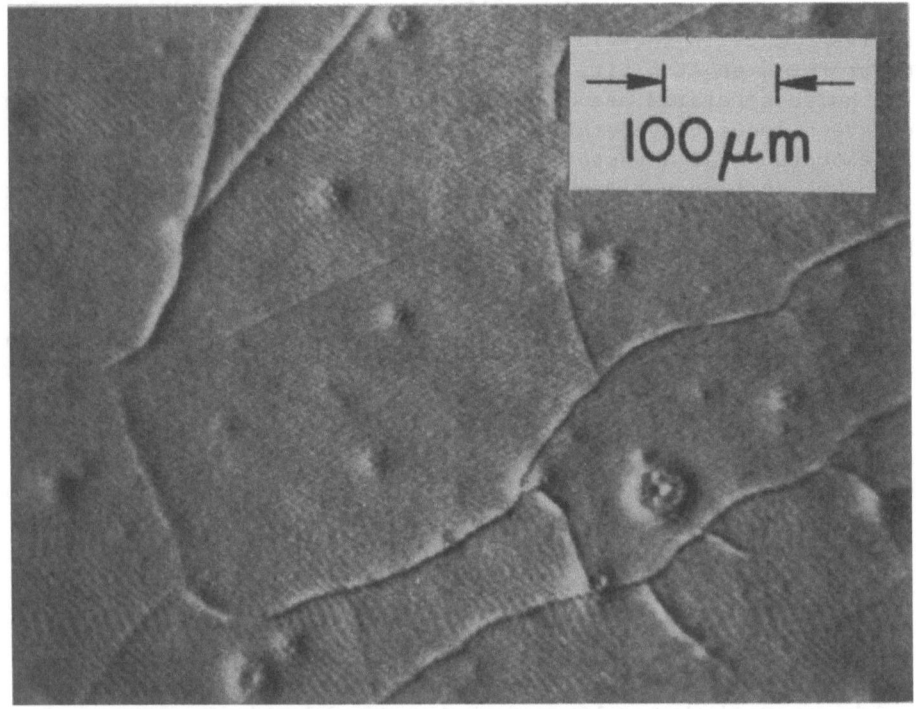

Fig. 31. Photovoltage image of an MNOS capacitor on p-type silicon, showing the image of charge trapped on the SiO_2-SiN_4 interface. The interface was first uniformly positively charged, and then discharged in a small oblong area by exposure to UV light.

in imaging non-uniformities that build up during the fatique of MNOS storage devices.

Another mechanism by which local surface charge produces detail in a photovoltage image involves enhanced surface recombination caused by the charge on a depleted surface. An example is the round spot in Fig. 32 due to a local region of sodium accumulation. During the measurement, the entire surface was in depletion. The sodium ions at the surface act as recombiantion sites, and as a result, appear in the photovoltage image. Similar images of sodium nonuniformities have been seen in the scanned electron beam analogs [89].

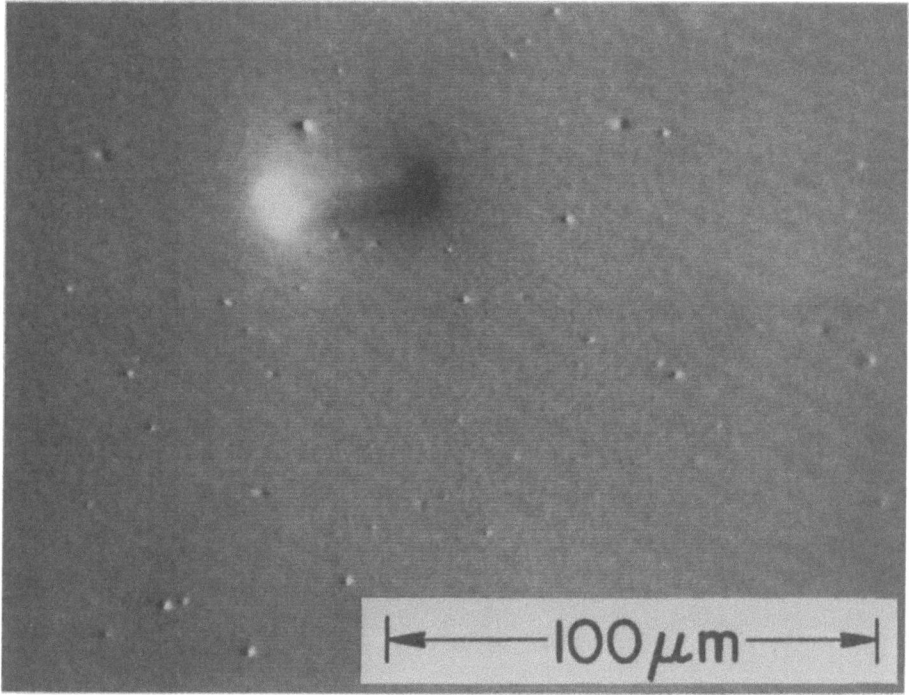

Fig. 32. Photovoltage image of a silicon surface showing a round region due to a sodium ion impurity.

D. Active Devices

Scanned photovoltage or photocurrent techniques have been used to prove active semiconductor devices including power devices [42,43], bipolar integrated circuits[44,48], C-MOS circuits [49], and photovoltaic cells [107-109]. The devices must, of necessity, be large relative to the wavelength of the proving light if detail is to be resolved in the device. Scanned electro-optical techniques have a somewhat limited potential for imaging LSI devices of the future. However, in probing large area power devices and solar cells, the technique may be quite useful.

The scanned photovoltage technique has been used to examine recombination at grain boundaries in a polycrystalline solar cell [108]. Fig. 33 shows a photovoltage image of a solar cell fabricated on large grain, a p-type silicon. The junction was formed by diffusion of a spin-on dopant. Grain boundary recombination is clearly visible in the image, which was obtained by differentiation of the

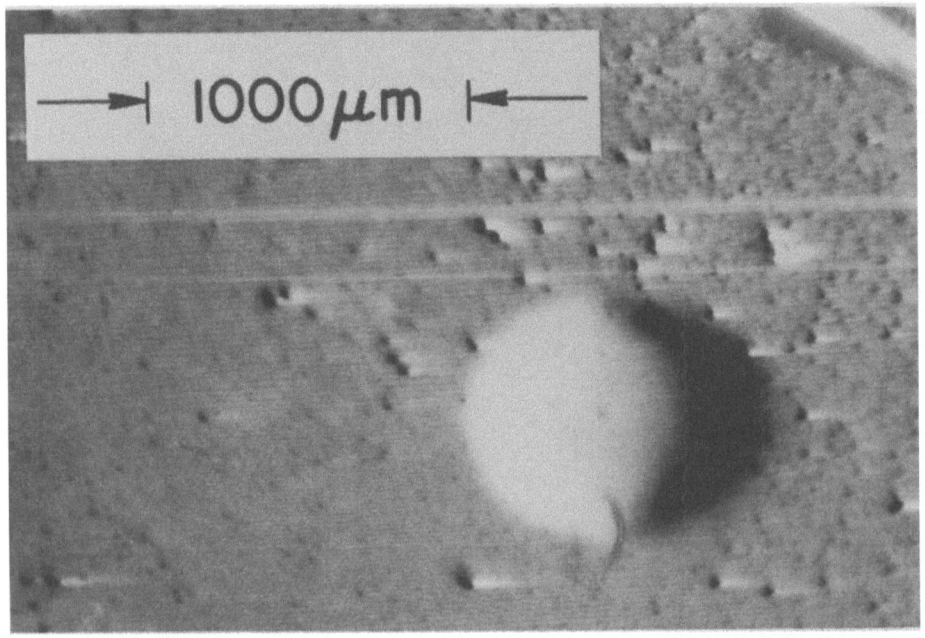

Fig. 33. Photovoltage image of polycrystalline silicon showing various grain boundaries. An n-type layer was formed on the surface of 1 Ω - cm p-type material by diffusion from a phospho-silicate glass (Emulsitone).

open circuit photovoltage produced by the scanned spot.

Scanned photocurrent methods have been used to examine active circuit elements, including the image of the bipolar transistor [44] shown in Fig. 34. Here the spot produces collector current when it scans the collector-base junction. More recently, Levy has used a scanned spot to probe a C-MOS circuit during sequential operation [49]. The spot produces a photocurrent which is detected in the power supply line when the probed device is activated. Overall electro-optical probing techniques may have some specialized application in probing LSI circuits, but as a general diagnostic tool it is somewhat limited, particularly in future sub-micron circuits.

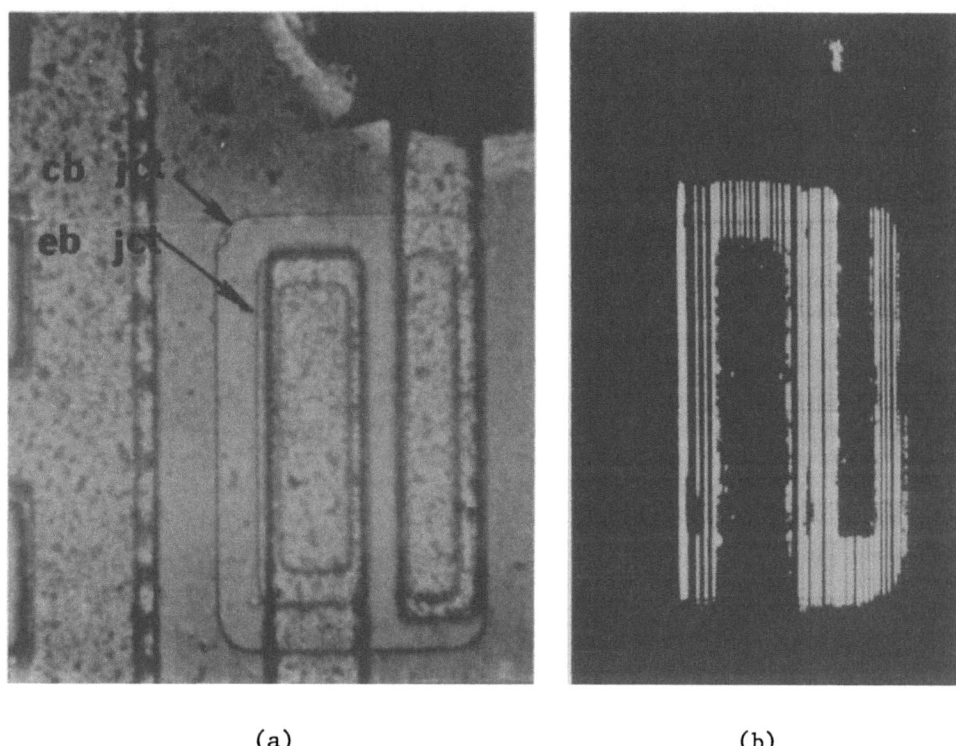

Fig. 34. A planar npn transistor shown in (a) an optical micrograph and (b) a scanned photovoltage image. (After Kasprzak, Ref. 44.)

VII. CONCLUSION

Each of the scanned electro-optical techniques has unique capabilities, advantages, and disadvantages for examining semiconductor surfaces and interfaces. Photoemission imaging is useful in determining the effects of an interface reaction or contamination. Few other techniques are capable of providing such information on an undisturbed interface. However, the photoemission technique is not easily amenable to use in process control or material evaluation because of the difficulty of the measurement and the requirement that the sample be partially transparent to light so that the beam can penetrate to the interface. Photoemission imaging may find application in researching and understanding specific problems such as Schottky barrier formation and degradation, interdiffusion in MOS structures, and the influence of contamination on insulator interfaces.

The photovoltaic imaging technique has greater potential for use in semiconductor material evaluation and in process control in the production of integrated circuits. Since the photovoltage may be detected capacitively, with no direct contact to the wafer, the wafers can be non-destructively examined at several stages of processing. Also, since no contact is required, silicon material can be inspected before being processed into integrated circuits. As a research tool, photovoltage imaging is somewhat less useful than EBIC techniques because of a lower resolution. However, the optical technique is useful in research on samples which are sensitive to ionizing radiation, such as MOS structures.

REFERENCES

1. A review of the electron beam techniques is given by C. J. Varker (in this volume).
2. Photoelectron Spectroscopy for the Examination of Surfaces and Interfaces is reviewed by W. E. Spicer (in this volume).
3. R. Williams in Semiconductors and Semimetals, 6, 97 (R. K. Willardson, Ed.), Academic (1970).
4. R. Williams and R. H. Bube, J. Appl. Phys. 31, 968 (1960).
5. R. Williams, Phys. Rev. 140, A569 (1965).
6. C. N. Berglund and W. E. Spicer, Phys. Rev. 136, A1030 (1964).
7. Evan. O. Kane, Phsy. Rev. 127, 131 (1962).
8. R. J. Powell and C. N. Berglund, J. Appl. Phys. 42, 4390 (1971).
9. Alvin M. Goodman, Phys. Rev. 144, 588 (1966).
10. C. A. Mead, Appl. Phys. Lett. 9, 53 (1966).
11. A. Rose, Concepts in Photoconductivity and Allied Problems, 121, Interscience (1963).

12. T. H. Ning and H. N. Yu, J. Appl. Phys. 45, 5373 (1974).
13. T. H. Ning, C. M. Osburn, and H. N. Yu, Appl. Phys. Lett. 26, 248 (1975).
14. B. E. Deal, E. H. Snow, and C. A. Mead, J. Phys. Chem. Solids 27, 1873 (1966).
15. A. M. Goodman, "Internal Photoemission is a Tool for the Study of Insulators," in Optical Properties of Dielectric Films (ECS Symposium, Boston) (1968).
16. C. G. Wang and T. H. DiStefano, Crit. Rev. Solid State Scie. 5, 327 (1975).
17. J. Andrews, J. Vac. Sci. Technol. 11, 972 (1974).
18. W. E. Spicer, RCA Rev. 19, 555 (1958).
19. J. J. Scheer, Phillips Res. Reports 15, 584 (1960).
20. W. E. Spicer, J. Appl. Phys. 31, 2077 (1960).
21. A. M. Goodman, Phys. Rev. 152, 785 (1965).
22. G. W. Gobeli and F. G. Allen, Phys. Rev. 127, 19 (1962).
23. E. H. Snow, A. S. Grove, B. E. Deal, and C. T. Sah, J. Appl. Phys. 36, 1664 (1965).
24. T. H. DiStefano and J. E. Lewis, J. Vac. Sci. Technol. 11, 1020 (1974).
25. R. Williams, J. Vac. Sci. Technol. 11, 1025 (1974).
26. C. R. Viswanathan and S. Ogura, Appl. Phys. Lett. 12, 220 (1968).
27. R. Williams, J. Appl Phys. 37, 1491 (1966).
28. C. R. Viswanathan and S. Oguran, Proc. IEEE 59, 1552 (1969).
29. B. E. Deal, A. S. Grove, E. H. Snow, and C. T. Sah, J. Electrochem. Soc. 112, 308 (1965).
30. N. J. Chou, Y. J. van der Neulen, R. Hammer, and J. Cahill, Appl Phys. Lett. 24, 200, (1974).
31. T. H. DiStefano and J. M. Viggiano, IBM J. Res. Dev. 18, 94 (1974).
32. J. S. Johannessen, W. E. Spicer, J. F. Gibbons, J. D. Plummer, N. J. Taylor, J. Appl. Phys. 49, 4453 (1978).
33. M. Avron, M. Shatzkes, P. J. Burkhardt, adn I. Cadoff, Appl. Phys. 47, 3159 (1976).
34. J. R. Haberer, "Photoresponse Mapping of Semiconductors," in Physics of Failure in Electronics, (T. S. Shilliday and J. Vaccaro, Eds.), 5, 51, Rome Air Developement Center-USAF, Rome, NY (1967).
35. R. A. Summers, Solid State Technol. 10, 12 (March 1967).
36. G. Adam, Physica 20, 1037 (1954).
37. J. Tihanyi and G. Pazstor, Solid State Electron 20, 235 (1967).
38. J. Oroshink and A. Many, J. Electrochem. Soc. 106, 360 (1959).
39. V. Stoyanov and R. Steranov, Phys. Stat. Sol. A7, K133 (1971).
40. T. H. DiStefano, Appl. Phys. Lett. 19, 280 (1971).
41. D. L. Lile and N. M. Davis, Solid State Electron. 18, 699 (1975).
42. C. N. Potter and D. E. Sawyer, "Optical Scanning Techniques for Semiconductor Device Screening and Identification of Surface and Junction Phenomena," in Physics of Failure in Electronics, (T. S. Shilliday and J. Vaccaro, Eds.) 5, 37, Rome Aire Devel. Center-USAF (1967); C. N. Potter and D. E. Sawyer, Rev. Sci. Instrum. 39, 180 (1968).

43. R. J. Phelan, Jr., and N. L. DeMeo, Jr., Appl. Optics 10, 858 (1971).
44. L. A. Kasprzak, Rev. Sci. Instrum. 46, 257 (1975).
45. H. W. Werlich and R. v. Pole, IEEE J. Quantum Elect. 13, D79 (1977).
46. R. Williams and M. H. Woods, J. Appl. Phys. 43, 4142 (1972).
47. D. L. Blackburn, H. A. Schafft, and L. J. Scwarzendruber, J. Electrochem. Soc. 119, 1776 (1972).
48. D. E. Sawyer, D. W. Berning and D. C. Lewis, Sol. St. Technol. 37 (1977).
49. M. E. Levy, "An Investigation of Flaws in CMOS Devices by a Scanning Photoexcitation Technique," (Proc. 15th Annual Reliability Physics Conf., Las Vegas, Nevada 1977), 44.
50. J. W. Philbrick and T. H. DiStefano, "Scanning Surface Photovoltage Study of Defects in Silicon," (Proc. 13th Annual Reliability Phys. Conf. Las Vegas, 1975). 159.
51. V. E. Kozhevin, Sov. Phys. Solid State 8, 1979 (1967).
52. T. H. DiStefano, "Photoemission and Photovoltaic Imaging of Semiconductor Surfaces," (NBS Pub. 400-23, NBS, 1976) 197.
53. C.J. Owen and W. R. Merwarth, Res./Dev. 66 (March 1970).
54. T. H. DiStefano, J. Appl. Phys. 44, 527 (1973).
55. R. Castagne and A. Vapaille, Electron. Lett. 6, 691 (1970).
56. G. H. Schwuttke, Tech. Rep. No. 1, ARPA Contract No. DAHC 15-72-C-0274 (1973).
57. T. H. DiStefano, Semi-Annual Tech. Rep. No. 2, ARPA Contract F19628-73-C-0006 (1973).
58. L. W. James, G. A. Antypas, J. J. Vebbing, T. O. Yep, and R. L. Bell, J. Appl. Phys. 42, 580 (1971).
59. C. Chr. Veerman, Matter Sci. Eng. 4, 329 (1969).
60. W. J. Baxter, J. Appl. Phys. 44, 608 (1973).
61. A. Gieroszynski and B. Sujak, Acta Phys. Pol. 20, 337 (1975).
62. W. J. Baxter and S. R. Rouze, J. Appl. Phys. 44, 4400 (1973).
63. W. J. Baxter and S. R. Rouze, Rev. Sci. Instrum. 44, 1628 (1973).
64. W. J. Baxter and S. R. Rouze, J. Appl. Phys. 46, 2429 (1975).
65. W. J. Baxter, J. Appl. Phys. 45, 4692 (1974).
66. Letokhov, in Laser Focus, (September 1975).
67. F. S. Goucher, Phys. Rev. 81, 475 (1951).
68. L. B. Valdes, Proc. IRE 40, 1420 (1952).
69. D. G. Avery and J. B. Gunn, Proc. Phys. Soc. B68, 918 (1955).
70. G. Schwab, Proc. 3rd Intl. Symp. on Silicon Materials Sci. and Technol., (Electrochem Soc. 1977), 481.
71. J. D. Kamm, in Proc. 3rd Intl. Symp. on Silicon Materials Sci. and Technol. (Electrochem. Soc., 1977), 491.
72. Chemning, Hu, IEEE Trans. Electron Devices ED-15, 822 (1978).
73. M. Watanabe, G. Actor, and H. C. Gatos, IEEE Trans. Electron Devices ED-24, 1172 (1977).
74. R. W. Bicknell, J. Vac. Sci. Technol. 14, 1012 (1977).
75. T. S. Moss, J. Electron, Cont. 1, 126 (1955).
76. A. Quilliet and P. Gosar, J. Phys. Rad. 21, 575 (1960).

77. A. M. Goodman, J. Appl. Phys. 32, 2550 (1961).
78. F. Bergman, C. Fritzche, and H. D. Riccius, Telefunkun Zig. 37, 186 (1964).
79. D. R. Frankl and E. A. Ulmer, Surface Sci. 6, 115 (1966).
80. D. L. Lile, Surface Sci. 34, 337 (1973).
81. W. H. Brattain and J. Bardeen, BSTJ 32, 1 (1953); C. G. B. Garrett and W. H. Brattain, Phys. Rev. 99, 376 (1955); W. H. Brattain and C. G. B. Garrett, BSTJ 35, 1019 (1956); C. G. B. Garrett and W. H. Brattain, BSTJ 35, 1041 (1956).
82. E. O. Johnson, Phys. Rev. 111, 153 (1958).
83. H. C. Gatos and J. Lagowski, J. Vac. Sci. Technol. 10, 130 (1973), and references cited therein.
84. J. J. Loferski and J. J. Wysocki, Bull. Am. Phys. Soc. 115, 265 (1960).
85. W. E. Phillips, Solid St. Elect. 15, 1097 (1972).
86. H. Reichl and H. Bernt, Solid St. Electron 18, 453 (1975).
87. R. S. Nakhmanson, Solid St. Electron 18, 617 (1975); R. S. Nakhmanson, Z. S. Ovsyuk, and L. K. Popov, Solid St. Electron 18, 627 (1975).
88. T. H. DiStefano, IEEE Trans. Electron. 22, 1055 (1975).
89. W. R. Bottoms and D. Guterman, J. Vac. Sci. Technol. 11, 965 (1975); W. R. Bottoms, D. Guterman, and P. Roitman, J. Vac. Sci. Technol. 12, 134 (1975).
90. J. Tauc, Czeck. J. Phys. 5, 178 (1955).
91. I. A. Baev and E. G. Valyashko, Soviet Phys.-Solid State 6, 1357 (1964); 7, 2090 (1966).
92. J. Tauc, Photo and Thermoelectric Effects in Semiconductors, (Int'l. Series of Monographs on Semiconductors, Ed. by H. K Henisch, Pergamon, 1962), 98.
93. H. F. Matare, Solid State Technol. 20, 56 (1977).
94. D. J. Dumin and W. N. Henry, Metallurgical Trans. 2, 677 (1971).
95. K. V. Ravi, Phil. Mag. 30, 1081 (1974); e1, 405 (1975).
96. S. M. Hu, J. Appl. Phys. 45, 1567 (1974).
97. S. M. Hu, J. Appl. Phys. 46, 1465 (1975).
98. S. M. Hu, J. Appl. Phys. 46, 1470 (1975).
99. A previously published photovoltage image of thiese stacking faults (Ref. 50) includes an incorrect dimensional marker.
100. M. Zerpst, Z. Agnew. Phys. 22, 30 (1966).
101. D. K. Schroder adn J. Guldberg, Solid State Electron. 14, 1285 (1971).
102. K. V. Ravi and C. J. Varker, J. Appl. Phys. 45, 263 (1974).
103. C. J. Varker and K. V. Ravi, J. Appl. Phys. 45, 272 (1974).
104. G. H. Schwuttke, "Damage Profiles in Silicon and Their Impact on Device Reliability," Tech. Rep. 1, Jan. 1973, and Tech. Rep. 2, July 1973, ARPA Constract DAH C 15-72-C-0274.
105. J. L. Deines, D. B. Dove, J. W. Philbrick, and M. R. Poponiak, to be published (1979).
106. D. R. Turner, J. Electrochem. Soc. 105, 402 (1958).
107. V. Stoyanov and R. Strfanov, Phys. Stat. Sol. A7, K133 (1971).

108. T. H. DiStefano and J. J. Cuomo, Appl. Phys. Lett. 30, 351 (1977).
109. T. Warabisako, T. Saitoh, H. Otoh, N. Nakamura, and T. Tokuyama, Jap. J. Appl. Phys. 17, 309 (1978).

Chapter 10

SEM METHODS FOR THE CHARACTERIZATION OF SEMICONDUCTOR MATERIALS AND DEVICES

C. J. Varker

Semiconductor Research and Development Lab

Motorola Inc.

5005 E. McDowell Road

Phoenix, Arizona 85062

I. INTRODUCTION TO SCANNING ELECTRON MICROSCOPY

The SEM has been used extensively as an analytical tool for high resolution surface microscopy [1,2]. When it is applied in this conventional mode, using secondary electron emission, it offers great depth of focus with high surface resolution and generally requires a minimum of sample preparation [3]. In recent years, the SEM has undergone major improvements in instrument design to accommodate a wide variety of newly developed analytical methods and imaging techniques [4]. Analytical methods such as energy dispersive x-ray spectroscopy [5] (EDS) and Auger electron spectroscopy [6,7] (AES) and the utilization of transmitted electrons in thin samples using the scanning transmission electron microscope [8,9] (STEM) represent a few examples of these developments.

As an analytical tool for the investigation of solid state materials and microelectronic devices [10], the SEM offers distinct advantages over conventional optical and transmission electron microscopy methods. It can be used to record and process a wide range of information (non-destructively) which results from the electron beam interaction with the sample. This information is contained in the energy released by electron emission and backscattering, photon radiation, electron absorption, and charge carrier generation processes in the solid. The characterization of

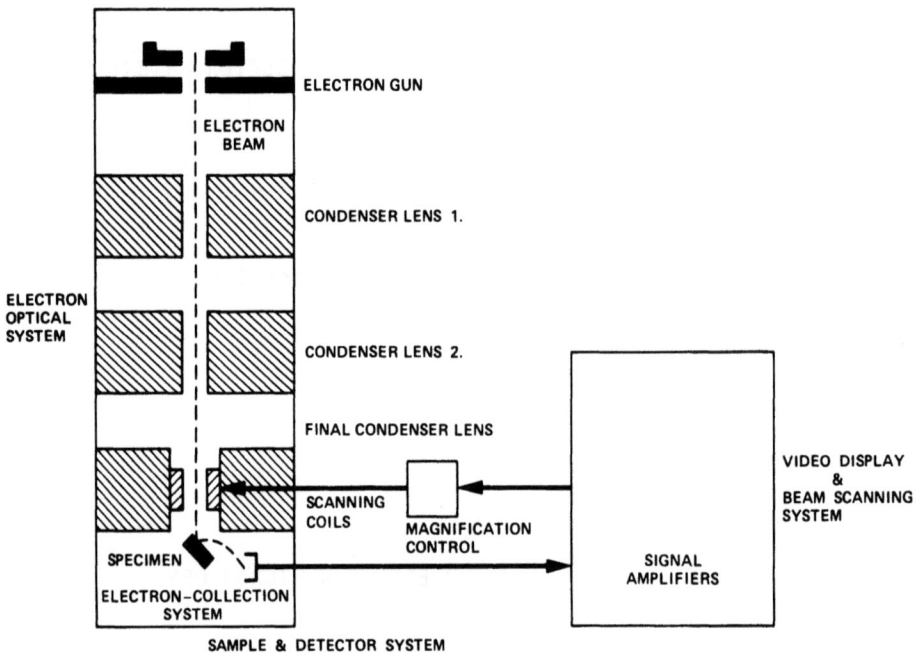

Fig. 1. Schematic representation of SEM.

electrical phenomena in semiconductor materials and devices is often best exploited by utilizing the low energy secondary electrons emitted at the surface and the charge carriers generated within the material by energy dissipation mechanisms. In these applications the low energy secondary electrons can be used to detect the presence of electric and magnetic fields and to measure potentials [11] at the surface of the device. The low energy charge carriers which are generated within the semiconductor can be collected at internal electric fields in the device to detect crystal defects and impurity inhomogeneities within the material [10,12,13].

A. General Principles of Operation

Figure 1 is a schematic representation with illustrates the basic operating principle of the SEM. The principle components of the SEM are: the electron optical system, the beam scanning and video display system, and the sample and detector system. The electron optical system consists of an electron gun which provides a beam of high energy electrons and a series of 2 or 3 electromagnetic lenses which demagnify this beam to produce a finely focused electron probe. A typical electron gun with a heated tungsten cathode produces a beam of electrons with a diameter of 25-50 µm, at a current density of 1-10 amps/cm². In a high quality SEM the electromagnetic lenses will produce a focused beam with a current of $\sim 10^{-11}$ amps into a spot diameter of 10 nm. The magnitude of the current in the focused beam is directly proportional to the current density available at the cathode of the electron gun. In recent years high current density electron sources such as the lanthanum hexaboride cathode [14] and the field emission cathode [15,16] have been developed. These newer gun designs provide even greater current densities at the cathode and thus provide a higher current density in the focused beam.

In the SEM the sample is scanned in a raster pattern by the electron beam and the image is generated in a manner analogous to a TV scan system with the exception that it generally operates at lower scan rates ($\sim$30 ns/line), and at greater scan line densities ($\sim$1000 lines/frame). The beam deflection coils in the SEM and the video display CRT are driven in synchronism by a common scan generator which provides the horizontal and vertical deflection waveforms. The resulting scan raster defines the area covered by the electron beam at the surface of the sample and on the fluorescent screen of the CRT. The current in the deflection coils of the SEM is adjustable to control the size of the scan raster at the sample and thus determine the magnification of the image displayed on the CRT.

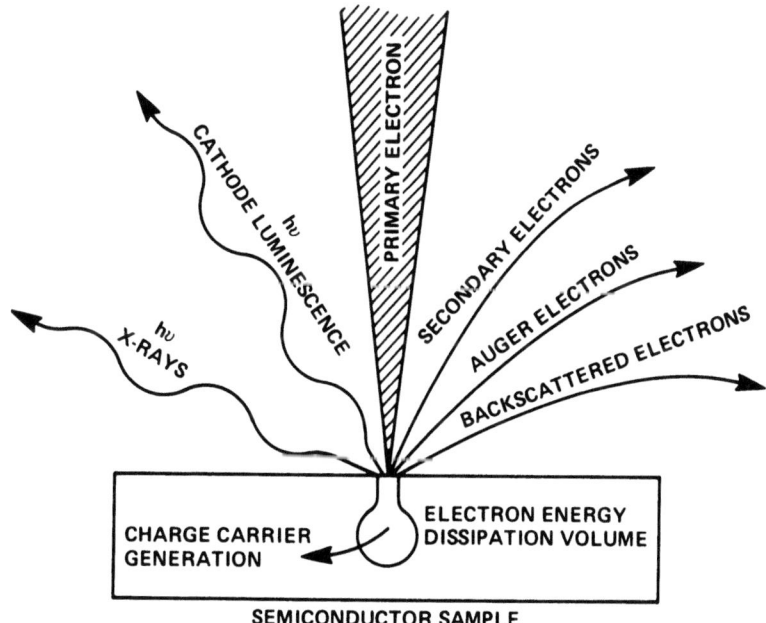

Fig. 2. Electron beam interaction with a solid.

When the primary beam strikes a thick sample the electrons interact with the solid giving rise to secondary and backscattered electrons, photon radiation and the generation of charge carriers in the solid. These interactions are illustrated schematically in Figure 2. In the conventional imaging mode the secondary electrons are collected with a detector adjacent to the sample. The image is generated by modulating the current in the beam of the display CRT with a signal obtained from the secondary electron collector. This produces an image in which the degree of brightness corresponding to each picture point on the screen of the CRT is related to the number of secondary electrons collected at each point on the surface of the sample.

The information content of the image displayed on the CRT depends on the type of signal generated at the sample. For example, with secondary electron microscopy the low energy secondary electrons, which emerge from the surface near the point of entry of the primary beam, are capable of generating a high surface resolution display because they possess a very short mean free path in the solid (1.0 - 5.0 nm). In this case the surface resolution is limited primarily by the spot diameter of the electron beam on the surface of the sample.

When the high energy backscattered electrons are collected, the resolution is limted by the penetration range in the solid of the elastically scattered electrons. At a typical beam energy ∿15 keV this range is several microns. In multi-element samples the contrast in the backscattered electron image is relative to the atomic number of the elements involved and therefore provides compositional information.

Equations which relate the focused spot diameter d to spreading effects within the beam and the lens parameters in the SEM are [17]:

$$d_s = \frac{1}{2} C_s \theta^3 \qquad (1)$$

$$d_c = C_c \frac{\Delta V}{V_o} \theta \qquad (2)$$

$$d_d = \frac{7.4}{\sqrt{V_o}\,\theta} \qquad (3)$$

$$d_g = \frac{1}{\theta} \frac{I_o}{\sqrt{\frac{\pi}{4} F J_c (\frac{11600}{T_e}) V_o}} \qquad (4)$$

where C_s and C_c are the spherical and chromatic abberation coefficients, θ the half angle of convergence for the beam at the target, and I_o and V_o are the beam current and voltage respectively. For a stationary beam with a current in the range of 10^{-11} amps the spot diameter as shown in Figure 3 [18] is limted to a first approximation by the spherical abberations of the electromagnetic lenses (equation (1)) and the transverse thermal velocity of the electrons in the beam (equation (4)). Although the resolution that can be realized is determined by the effective diameter of the electron spot on the target, there are other limitations on resolution [3] which involve the mode of operation of the SEM and the signal to noise ratio imposed by the detector and display system. In many applications for semiconductor materials and devices using electron beam induced current (EBIC) microscopy, x-ray microanalysis and cathode luminescence techniques, the resolution is limited primarily by electron penetration and energy dissipation in the sample.

B. Electron Penetration and Energy Dissipation

Electron penetration in solids have been studied extensively [19,20] and a variety of theoretical models have been used to calculate the electron penetration and energy loss [21-23]. Range and energy relationships have been derived by measuring transmitted and backscattered electrons in thin films [19,24] and from composite layers of material having different atomic numbers [25]. Measurements of the visible glow generated by penetrating electrons in low pressure gases and in phosphorescent material have also been used to establish these relationships [26-27]. In recent years, the electron dissipation in solids has been investigated using electron sensitive materials and plastics [28,29] to determine the shapes of the energy dissipation profiles.

When a monoenergetic beam of kilovolt electrons penetrates a solid, the electrons lose energy and change direction as a result of interactions with the atoms in the target. The energy loss occurs primarily during the inelastic collisions between the primary electrons and electrons in the material. The elastic collisions which occur with the nuclei of the atoms in the material produce large angle scattering which tends to spread the beam in the solid. The net result of these scattering events is shown in Figure 4 which illustrates the energy dissipation profiles for a material of low atomic numbers. Note that the lateral spread of the beam increases with penetration depth. In high atomic number materials the elastic scattering events occur more frequently and the energy dissipation contours approach a hemispherical geometry.

The energy dissipated in the solid can be defined by a universal depth-dose curve [26] as shown in Figure 5. For a given material the shape of this curve is virtually independent of beam energy.

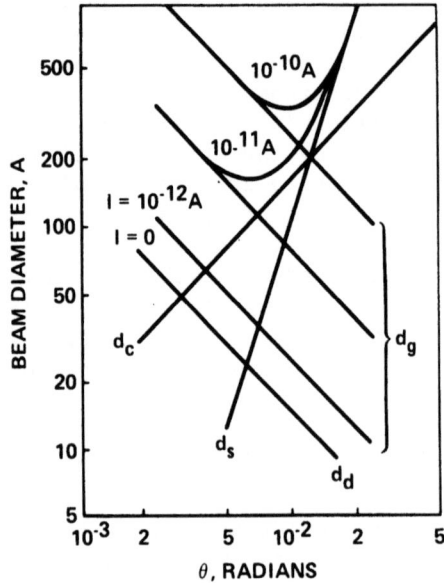

Fig. 3. Calculated spot diameter vs. convergence angle
($C = 2$ cm, $C_C = 1.6$ cm, $J_C = 5$ A/cm, $\Delta V/V = 10^{-4}$, $V = 10$ kV . . . see Equation 1-4 in text). Ref. [18]

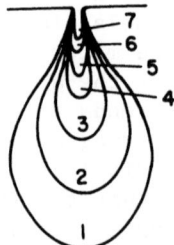

Fig. 4. Energy dissipation contours in a solid of low atomic number.

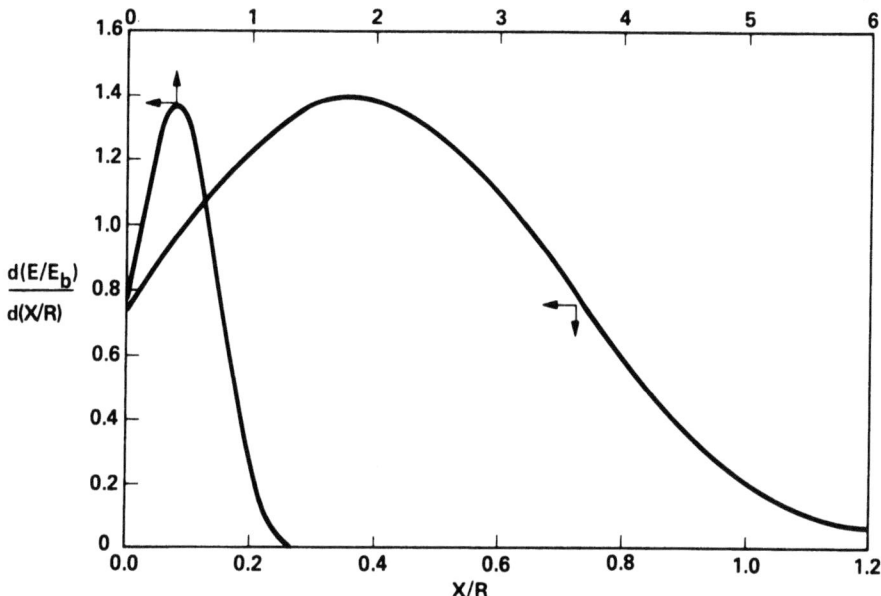

Fig. 5. Universal depth dose curve. Ref. [26 & 30].

The validity of this function has been verified experimentally on MOS capacitor structures for the aluminum-silicon dioxide-silicon system [30]. The experimentally determined range-energy relationship for this material is given by the following expression:

$$R_G = 4.0 \ E_b \ (keV)^{1.75} \ mg \ cm^{-2} \qquad (5)$$

This value defines the relative size of the energy dissipation volume as shown in Figure 4. The significance of the depth dose function and the range energy relationship is that the energy dissipation can be calculated at any depth and the corresponding charge carrier generation rate and the spacial resolution can be estimated.

C. Resolution and Contrast

The resultant image resolution obtained in the conventional mode with the SEM depends on the diameter of the focused electron spot, the dimensions of the electron emission region at the surface, and the signal to noise ratio in the recorded image. The electron beam current required to produce an acceptable image depends on the number of contrast levels M desired in the image, the number of picture elements N, and the exposure time T. An expression for the beam current I is given by [3]:

$$I = \frac{25 \, M^2 N^2}{4T} q \qquad (6)$$

The numerical values in equation (6) are based on a signal to noise ratio of 5 to 1 and the gray level allocations in the image. For example, if we utilize a 1000 line scan with 10 gray levels with a recording time of 60 sec, the minimum beam current $I = 1.7 \times 10^{-12}$ amp.

Additional considerations are required when contrast conditions are imposed, and a conversion efficiency factor must be included to calculate the required beam current. The mechanisms which determine the conversion efficiency depend on the escape depth and the range of the secondary electrons which contribute to the collected current. In general, these factors are dependent on the electron backscattering coefficient of the material.

When the SEM is used for material and device studies, utilizing EBIC microscopy, the resolution is limited primarily by electron penetration and energy dissipation in the solid. In this mode, there are also additional limitations on resolution and contrast which involve the charge carrier transport properties of the material. These considerations will be discussed in more detail in Section III.

II. PROPERTIES OF SEMICONDUCTOR MATERIALS AND DEVICES

A. Electrical Properties of Semiconductor Crystals

The basic electrical properties of crystalline semiconductor materials can be defined by the charge carrier concentration and transport of free electrons in the solid [31-34]. The energy scheme for visualizing electrons in a solid crystal is illustrated with an energy band diagram as shown in Figure 6. The electrical properties which characterize the material as a metal, semiconductor or insulator, are determined by the state of occupancy of the upper band (conduction band) in the diagram. The energy separation between the lowest conduction band state and the highest valence band state is defined as the band gap energy, E_g. Electrons in a solid can absorb energy from external sources in the form of heat, light, x-rays, energetic electrons, electric fields, etc. Energy absorption occurs when the amount of energy supplied by these interactions is equal to the difference in energy between two allowed states.

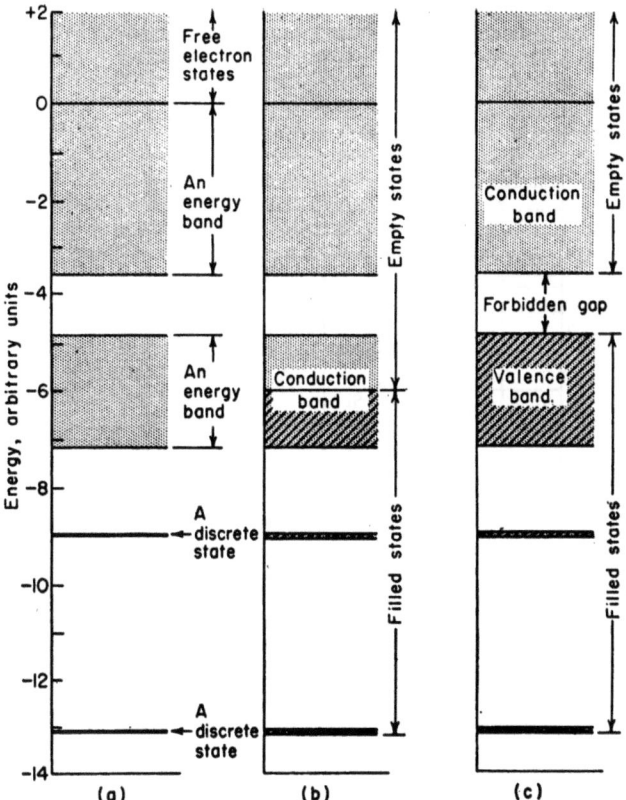

Fig. 6. Energy level diagram. (a) allowed states in hypothetical solid, (b) electron occupation in an ideal metal, (c) electron occupation in an ideal semiconductor or insulator.

1. <u>Charge Carrier Concentrations in Thermal Equilibrium</u>. In semiconductors and insulators, when the electron enters the conduction band, the vacant energy state left behind is defined as a hole. In pure materials of this type electrons in the conduction band and holes in the valence band will occur in pairs. When the crystal is in thermal equilibrium at a finite temperature the electron-hole pair generation rate is balanced by the carrier pair recombination rate. In the process of direct band gap transitions the recombination rate is proportional to the product of the electron concentration n and the hole concentration p, i.e., $np = n_i^2$, where n_i is the concentration of electrons or holes in the pure material.

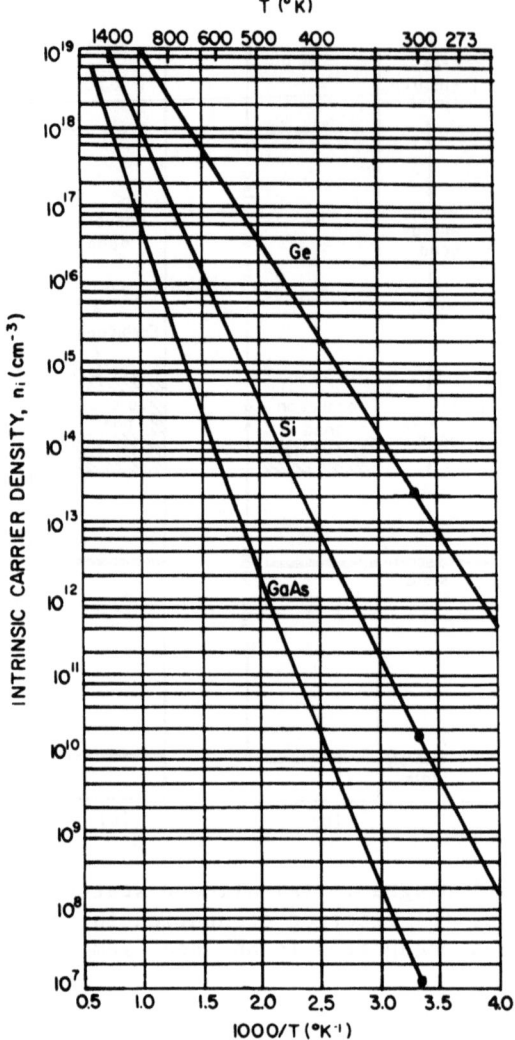

Fig. 7. Intrinsic carrier density vs. reciprocal of temperature. Ref. [31].

The intrinsic carrier concentration n_i depends on the band gap energy E_g. Figure 7 presents experimental data of the intrinsic carrier concentration as a function of temperature T, for Ge, Si and GaAs. The data clearly indicates that n_i is dependent on the band gap energy E_q, i.e., $n_i \sim \exp[-E_g/2kT]$.

When a semiconductor such as silicon is doped with the appropriate donor or acceptor impurity, shallow energy levels are introduced in the band gap. Since typical donors and acceptors (group V and group III elements) have small ionization energies they are

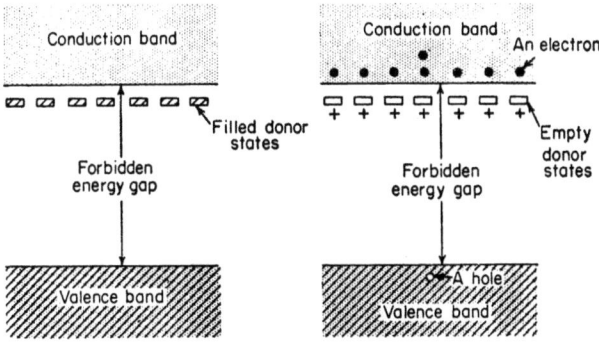

Fig. 3a

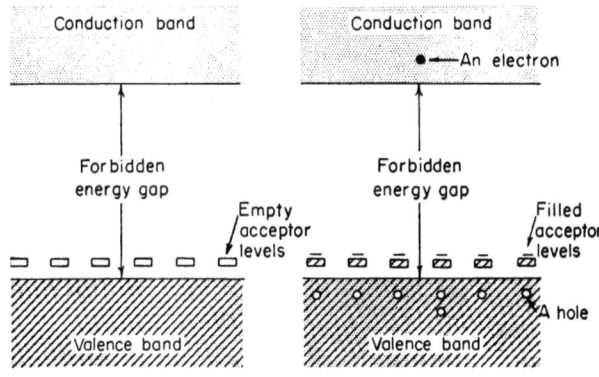

Fig. 3b

Fig. 8. Energy level scheme for a crystal containing donor impurites. (a) absolute zero, and (b) normal temperature.

thermally ionized at normal temperatures. The state of occupancy of these levels is shown in Figure 8. At normal temperatures the impurity atom can either donate an electron to the conduction band or accept an electron from the valence band. The impurity atom which has donated an electron has acquired an immobile positive charge. The impurity atom which has accepted an electron has acquired an immobile negative charge.

For n type silicon the impurity ionization is essentially complete at room temperature and the concentration of free electrons n is approximately equal to the concentration of the donor impurity N_D, i.e., $n \simeq N_D$. A similar argument is presented for p type material in which $p \simeq N_A$. Furthermore, when $(N_D - N_A) \gg n_i$ or $(N_A - N_D) \gg n_i$ the majority carrier concentration is relatively independent of temperature in the extrinsic range as

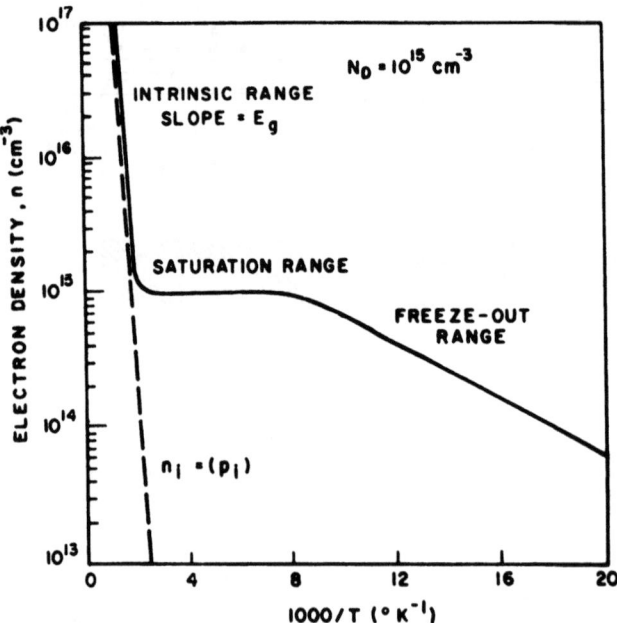

Fig. 9. Electron density vs. reciprocal of temperature. Ref. [31].

in Figure 9. At relatively high temperature the number of thermally generated carriers $n_i \gg |N_D - N_A|$ and the crystal exhibits its intrinsic carrier properties as shown in Figure 7. Whereas, at low temperature the majority carriers begin to freeze out and the free carrier concentration decreases with temperature.

The energy distribution of electrons in the solid are governed by Fermi-Dirac statistics. The probability F(E) that an energy state E is occupied by an electron is given by

$$F(E) = \frac{1}{1 + \exp\left(\frac{E-E_F}{kT}\right)} \tag{7}$$

where E_F is the Fermi level. When the occupation probabilities for energy states are considered, and the density of states are calculated from quantum mechancics, the electron and hole concentrations are given by the following equations:

$$n = N_C \exp\left(-\frac{E_C-E_F}{kT}\right) = n_i \exp\left(\frac{E_F-E_i}{kT}\right) \tag{8}$$

$$p = N_V \exp\left(-\frac{E_F-E_V}{kT}\right) = n_i \exp\left(\frac{E_i-E_F}{kT}\right) \quad (9)$$

These equations are valid for non-degenerate semiconductors where $N_A, N_D < 10^{19}$ cm^{-3}. The Fermi level for an intrinsic semiconductor is obtained from equations (8) and (9) as:

$$E_F = E_i = \frac{E_C+E_V}{2} + \frac{kT}{2} \ln \frac{N_V}{N_C} \quad (10)$$

From the above equations the product of the electron and hole concentration in equilibrium is obtained as:

$$pn = n_i^2 = N_C N_V \exp\left(-\frac{E_g}{kT}\right) \quad (11)$$

Under conditions of equilibrium in which impurites are homogeneously distributed, conservation of charge demands the net charge density is the crystal will be zero. This implies that

$$p - n = N_A - N_D \quad (12)$$

By combining equations (11) and (12), the concentration of free electrons n_{no} is an n type semiconductor in thermal equilibrium is given as:

$$n_{no} = \frac{1}{2}(N_D - N_A) + \sqrt{(N_D - N_A)^2 + 4 n_i^2} \quad (13)$$

If we further assume that $|N_D - N_A| \gg n_i$ then equation (7) reduces to:

$$n_{no} = N_D - N_A \quad (14)$$

Using equations (11) and (12) for holes in p type material a similar set of equations is obtained. If we then combine these results in the equilibrium product relationship, the equilibrium concentration of minority carrier for n and p type material respectively are given in the following two equations:

$$p_{no} = \frac{n_i^2}{N_D-N_A} \quad (15)$$

$$n_{po} = \frac{n_i^2}{N_A \, N_D} \quad (16)$$

We have considered the mechanism whereby free carriers are generated in a semiconductor under conditions of thermal equilibrium. The motion or transport of charge carriers in the solid can be described in terms of carrier drift and carrier diffusion. Thermally generated electrons and holes are in random motion in the crystal during thermal equilibrium. The energy transfer between electrons and atoms in the crystal during collision is balanced by the thermal energy absorbed by the electrons with the result that no net transfer of energy occurs during thermal equilibrium.

2. <u>Charge Carrier Transport</u>. When an electric field is present in the crystal, energy is gained by the electrons and holes and a drift velocity is impressed on their random motion. For moderate electric fields ($<10^4/cm^{-2}$) the drift velocity v_D is proportional to the electric field E.

$$v_D = \mu E \quad (17)$$

where μ is the proportionality constant, defined as the carrier mobility. The carrier mobility μ is dependent on the effective mass of the particle and the absolute temperature. The current density J which results from carrier drift is given by the product of the conductivity σ and the electric field E.

$$J = \sigma E \quad (18)$$

where $\sigma = (q \mu_n n + q \mu_p p)$

The electrical resistivity $\rho = 1/\sigma$. Thus the resistivity of an n type semiconductor $n \gg p$ is determined by the temperature characteristics of the electron mobility μ_n. In Ge and Si the mobility

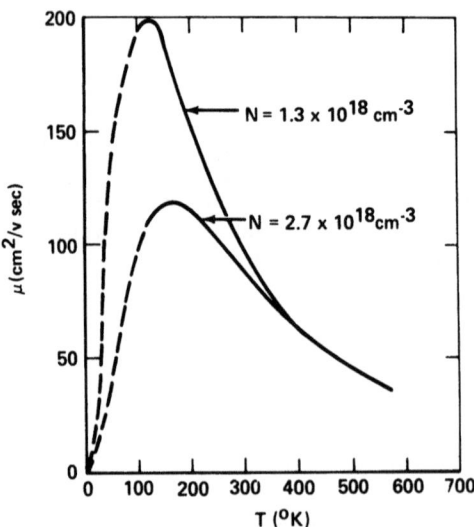

Fig. 10. Hole mobility μ vs. temperature. Ref. [32].

is dominated by two mechanisms: acoustic phonon scattering and
ionized impurity scattering. The influence of temperature on the
mobility of holes in silicon is shown in Figure 10. At sufficiently
large electric field strength the drift velocities for Ge and Si
approach a scattering limited velocity $v_d \sim 10^7$ cm sec^{-1}.

If the electric field in the semiconductor is further increase,
such that, $E \geq 10^5$ V/cm, electrons and holes can gain sufficient
energy during their mean free time between collisions with atoms
to create additional electron-hole pairs. This carrier pair multiplication process can lead to avalanche breakdown in the semiconductor. The field strength required for avalanche breakdown is a
function of the band gap energy and temperature.

In the absence of an electric field, the random motion of
free carriers is governed by local charge carrier concentration
gradients in the crystal. Since diffusion processes are a result of
random motion, the diffusion constants depend on the mean velocity
of the free carriers between collisions with atoms in the crystal.
The relationship between the diffusion constant the the free carrier
mobility is given by the Einstein relationship:

$$D = \frac{kT}{q} \mu \qquad (19)$$

3. <u>Non-Equilibrium Conditions</u>. Thus far we have discussed the
electrical properties of the semiconductor in a state of thermal
equilibrium. However, since most semiconductor devices operate
under non-equilibrium conditions, i.e., $pn \neq n_i^2$, the conditions
whereby the excess carriers return to a state of equilibrium will
be considered.

When a n type semiconductor is uniformly illuminated with an
external source of energy, such as optical radiation, the time
rate of increase the minority carrier concentration dP_n/dt can be
expressed as:

$$\frac{dP_n}{dt} = G - U \qquad (20)$$

Where G is the minority carrier generation rate and U the net
recombination rate given by $U = R - G_{th}$, where R is the total recombination rate and G_{th} is the dark thermal generation rate. If we
assume a simple linear relationship between the net recombination
rate U and the excess or injected minority carrier concentration,
Δp then the following expression is obtained.

$$U = \frac{1}{\tau_p} \Delta p = \frac{1}{\tau_p} (P_n - P_{no}) \qquad (21)$$

where τ_p represents the lifetime of the excess minority carriers. Conbinbing these expressions we obtain the following equation:

$$\frac{dp_n}{dt} = G - \frac{P_n - P_{no}}{\tau_p} \qquad (22)$$

In the solution to a differential equation of this form the rate constant which characterizes the recombination process is $1/\tau_p$, and the solution is given by

$$P_n = P_{no} + (\tau_p G) e^{-t/\tau_p}$$

Minority carrier lifetimes in Si and Ge are strongly dependent on imperfections and impurites in the crystal. The recombination-generation-trapping processes in these materials occur through intermediate centers in the band gap region. Therefore, in addition to the shallow level donor and acceptor impurities which are used to control the concentration of electrons and holes in the semiconductor, there are other impurities which introduce discrete energy levels deep within the energy band gap. These deep level centers tend to increase the carrier generation and recombination rates and result in a reduction of carrier lifetime.

The theory for the generation-recombination-process through-intermediate energy levels was developed by Hall, Schockley and Reed [35]. A complete discussion of this theory and its derivations is beyond the scope of this lecture. The essential features of the generation-recombination process for a single level center with 2 charge states is presented in Figure 11. The four basic processes involve electron capture and emission, and hole capture and emission.

An equation for the steady state net recombination rate U can be developed by substituting the appropriate expression for the electron occupation probability into the following rate equation:

$$U = R_{nc} - R_{ne} = R_{pc} - R_{pe}$$

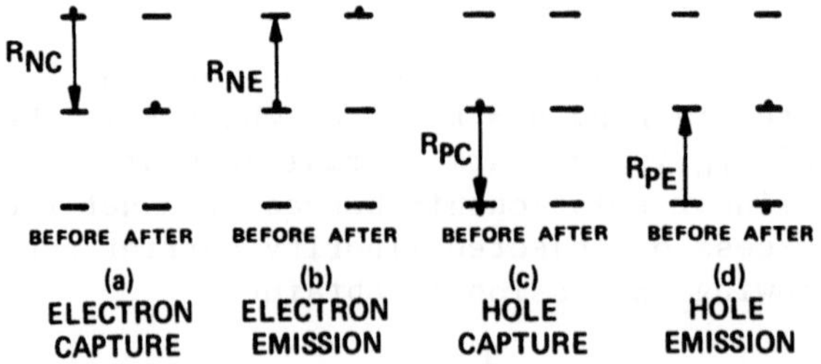

Fig. 11. Emission and capture processes through intermediate centers.

where the subscripts refer to the corresponding processes shown in Figure 11.

The resulting equation is given as:

$$U = \frac{\sigma_p \sigma_n v_{th} N_t (np - n_i^2)}{\sigma_n \left[n + n_i \exp\left(\frac{E_t - E_i}{kT}\right) \right] + \sigma_p \left[p + n_i \exp\left(\frac{E_i - E_t}{kT}\right) \right]} \quad (23)$$

For the case of low level infection in n type material $n_n \gg p_n$, and $n_n \gg n_i \exp\left(\frac{E_t - E_i}{kT}\right)$, and for $\sigma_p = \sigma_n = \sigma$ can be approximated by:

$$U = \sigma_p v_{th} N_t (p_n - p_{no}) \quad (24)$$

From equation (21) the lifetime for holes τ_p, for low level injection is given by:

$$\tau_p = \frac{1}{\sigma_p v_{th} N_t} \quad (25)$$

A similar expression can be obtained for electrons in p type material.

B. Physics of PN Junctions and Schottky Barrier Devices in Thermal Equilibrium

At the junction between a p and n type semiconductor, Figure 12, the diffusion of free carriers from each side of the junction sets up an electric field due to the space charge corresponding to the ionized donor and acceptors. The magnitude of the electric field is just sufficient to balance the diffusion flow of electrons from the n to the p side of the junction with an equal and opposite flow of electrons from the p to the n region. Similarly, the field must also be sufficient to balance the corresponding hole flows in both directions.

The diffusion currents in the junction are determined by the concentration gradients of the free carriers, and the drift currents result from the build in electric field of the space charge region. Thus, at thermal equilibrium, the electron and hole currents J_n and J_p are given by the following equations respectively:

$$J_n = q\mu_n nE + qD_n \frac{\partial n}{\partial x} = q\mu_n \left(nE + \frac{kT}{q} \frac{\partial n}{\partial x} \right) \quad (26)$$

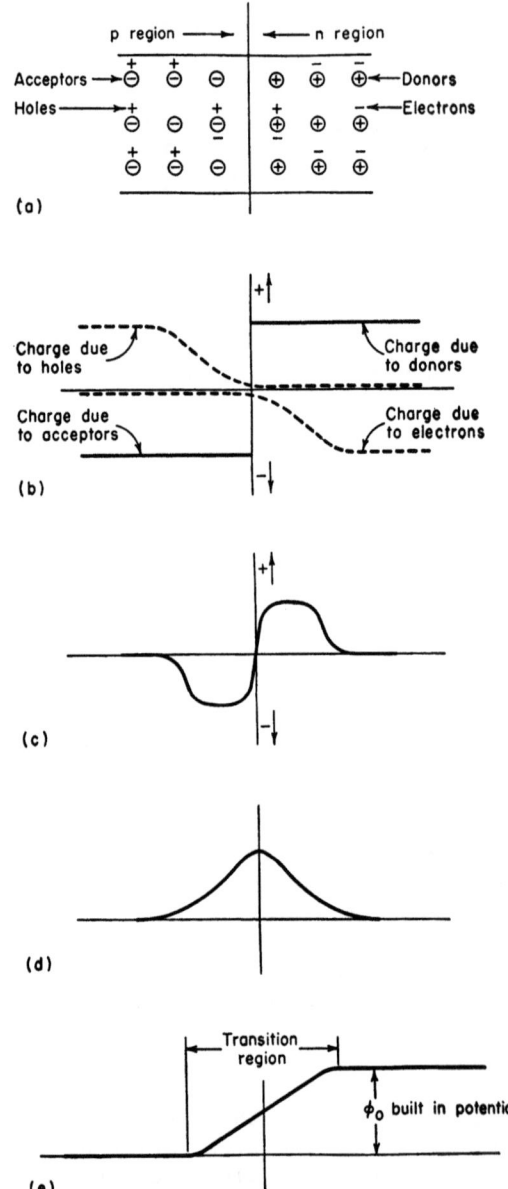

Fig. 12. Charge and field distribution at a P-N junction. (a) fixed and free charges, (b) individual charge carrier contributions, (c) net charge, (d) electric field, and (e) potential distribution.

$$J_p = q\mu_p pE - qD_p \frac{\partial p}{\partial x} = q\mu_p\left(pE - \frac{kT}{q}\frac{\partial p}{\partial x}\right) \qquad (27)$$

The first term in these equations represents the drift current and the second term the diffusion current. From equations (26) and (27) and the thermal equilibrium product relationship which is given by:

$$n_{no}p_{no} = n_{po}p_{po} = n_i^2$$

the concentration of carriers outside the space charge region on both sides of the pn junction is dependent on the built in junction potential ϕ_B and can be expressed by the following pair of equations

$$n_p = n_n \exp\left(-\frac{q\phi_B}{kT}\right) \qquad (28)$$

$$p_n = p_p \exp\left(-\frac{q\phi_B}{kT}\right) \qquad (29)$$

where, the built in junction potential or diffusion potential is given by

$$\phi_B = \frac{kT}{q} \ln \frac{n_{no}p_{po}}{n_i^2} \simeq \frac{kT}{q} \ln \frac{N_D N_A}{n_i^2} \qquad (30)$$

For the simple case of a one sided step junction as shown in Figure 13, the following important relationships can be derived. The built in junction potential ϕ_B is give as

$$\phi_B \simeq \frac{2kT}{q} \ln \frac{N}{n_i} \qquad (31)$$

where N equals the net impurity concentration $|N_D - N_A|$. The width W of the depletion region at the pn junction is given by

$$W = \frac{2\varepsilon_s(\phi_B + V)}{qN} \qquad (32)$$

where B is the applied voltage. And the maximum electric field E_{max} is expressed as

$$E_{max} = \frac{2\phi_B \pm V}{W} \qquad (33)$$

These equations apply equally well to both pn junction devices and to Schottky barrier devices where a one sided step junction approximation is valid. In the Schottky device, the barrier height is essentially independent of the dopant level and is approximately equal to 2/3 E_g. However, in Schottky devices the dominant current

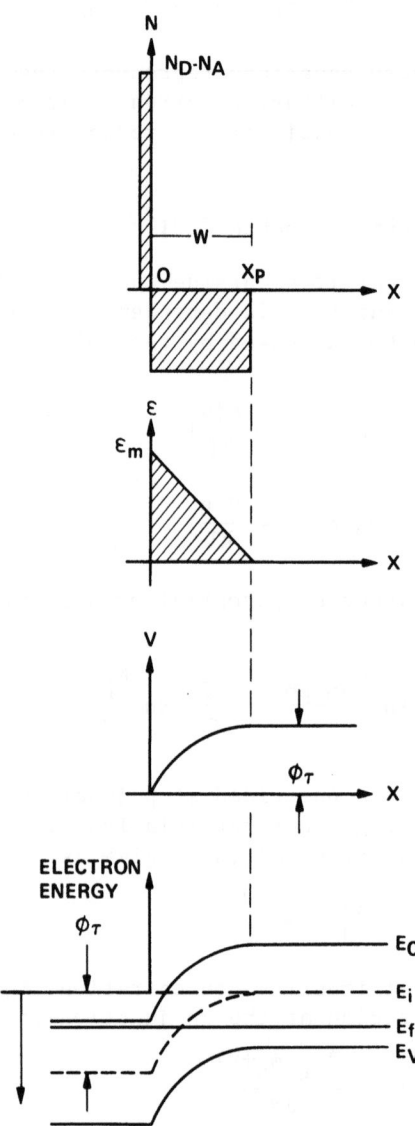

Fig. 13. One sided step junction approximation showing distribution net charge (N_D-N_A) Electric Field E, potential and band diagram.

is carried by majority carriers, whereas, in pn junction diodes, it is carried by minority carriers.

1. <u>Current-Voltage Characteristics</u>. The current-voltage characteristics of a pn junction diode are developed on the basis of carrier generation-recombination mechanisms [36,37]. The two principle components are generally referred to as generation or recombination current and the diffusion current. The analysis generally proceeds by considering two regions in the vicinity of the pn junction, i.e., the space charge, and the neutral or field free region outside the space charge region.

For the reverse biased, pn junction diode the space charge region is depleted of free carriers when $V_R >> kT/q$ i.e., p, $n >> n_i$ and the current I_g is due to carrier generation in the space charge region. With no external energy sources this current is determined by the net rate of generation U of electron-hole pairs is given by

$$I_g = q|U|WA = q \frac{n_i}{2\tau_o} WA \tag{34}$$

where A is the area of the pn junction, τ_o the effective lifetime in the depletion region, and U is the generation rate with no external energy sources applied other than thermal energy. In the neutral region, the carriers transport is by diffusion. As in the previous case, with no external energy sources, the diffusion component I_D is given by:

$$I_D = q \frac{Dn_i^2}{NL} A \tag{35}$$

The forward current in the space charge region depends on the number of carriers lost by recombination processes in the space charge region. This current can be approximated by considering the maximum recombination rate U in the space charge region which is given by:

$$U_{max} = \frac{n_i}{2\tau_o} \exp\left(\frac{qV_F}{2kT}\right) \tag{36}$$

Consequently, the recombination current I_R is given by

$$I_R = - \frac{qn_i}{2\tau_o} WA \exp\left(\frac{qV_F}{2kT}\right) \tag{37}$$

and, the diffusion current I_D for the forward biased diode is given by:

$$I_d = -qD \frac{n_i^2}{NL} A \exp\left(\frac{qV_F}{kT}\right) \tag{38}$$

where the diffusion length $L = \sqrt{D\tau}$.

The forward currents in the diode consist of the same diffusion component and generation term used in the reverse bias case, multiplied by an exponential factor $\exp\left(\frac{qV_F}{nkT}\right)$.

By arranging the total diffusion and the generation-recombination components separately, the pn diode equation for a one sided step junction can be expressed as

$$I_D = -\frac{qDn_i^2}{NL} A \left[\exp\left(\frac{qV_F}{kT}\right) - 1\right] \tag{39}$$

$$I_{g-r} = -\frac{qn_i WA}{2\tau} \left[\exp\left(\frac{qV_F}{2kT}\right) - 1\right] \tag{40}$$

for direct comparison, the equations corresponding to a Schottky barrier diode are given as:

$$I_T = (RAT)^2 \exp\left(-\frac{\phi}{kT}\right) \left[\exp\frac{qV_F}{kT} - 1\right] \tag{41}$$

$$I_D = \frac{qDn_i^2 A}{NL} \left[\exp\left(\frac{qV_F}{kT}\right) - 1\right]$$

Equation (41) represents the majority carrier flow by thermionic emission for the Schottky barrier, where R is the Richardson constant. For Schottky barrier diodes on silicon, i.e., Al on silicon, $N_D = 10^{16} \text{cm}^{-3}$, the majority carrier current expressed in equation (41) is dominant and the minority carrier diffusion is negligible.

For silicon pn diodes at room temperature the generation current component, is normally dominant for reverse biased diodes and the forward current exhibits both a generation and a diffusion component depending on the bias voltage. In either case the total diode current is the sum of the two components expressed separately in equations (39) and (40) for the pn diode and in equations (41) and (42) for the Schottky barrier diode.

C. Fabrication Technology for Material and Devices

Conventional single crystal silicon manufacturing starts with the production of high purity polysilicon ingots by chemical vapor deposition from purified $SiHCl_3$ or $SiCl_4$. The bulk polysilicon material is used as the charge for the silicon melt utilizing the Czochralski crystal growth method [38]. Typical ingots produced with this growth technique have a diameter of 76 mm or more with a (100) or (111) growth orientation and range from 0.5 - 1.0 meters in length. These ingots are often specified as 'dislocation free' and have radial dopant uniformities within ± 10% depending on the crystal orientation and dopant type.

In recent years the major concern with crystal quality has centered on the oxygen [39] and carbon [40] content of the crystals and the inhomogeneities which result from impurity-point defect complexes [41] as well as the axial and radial uniformity of the donor and acceptor impurities [42]. Single crystal ingots which meet the material quality specifications are machine ground to produce a cylindrical form with a surface orientation flat. The ingots are then saw cut to produce silicon slices measuring $\sim$ .020" in thickness. Extensive wafer shaping and surface preparation methods are used to produce a flat wafer of the desired thickness and surface finish. For integrated circuit manufacture, the incoming material is typically a wafer with an 'optical' quality surface finish with tight specifications on the surface flatness and edge quality.

In some manufacturing processes the pn junction devices are fabricated directly in the wafers produced from the single crystal ingot. Whereas, in other manufacturing operations a thin single crystallin layer is deposited onto these substrates using CVD technology [43]. In this case, the devices are normally fabricated in the epitaxial layer and the substrate wafers essentially provides a mechanical support for the epitaxial film with some added advantages in terms of device design.

The subsequent wafer processing operations which are used to produce the device structures have been described extensively in the literature [44]. No attempt will be made here to cover these processing details. However, the geometry of the resulting device structure and the major processing sequence which includes the thermal oxidations and diffusions are illustrated schematically in Figure 14.

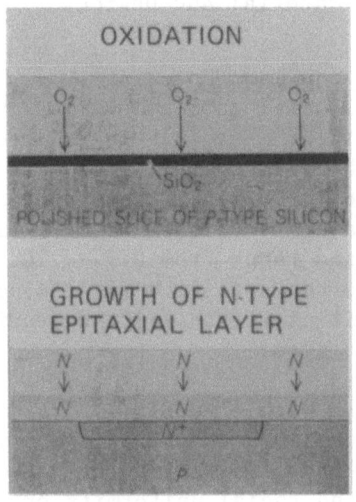
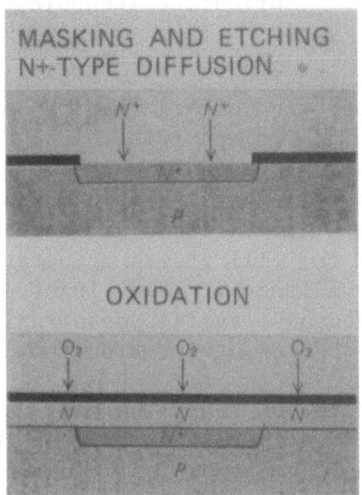

Fig. 14(a)

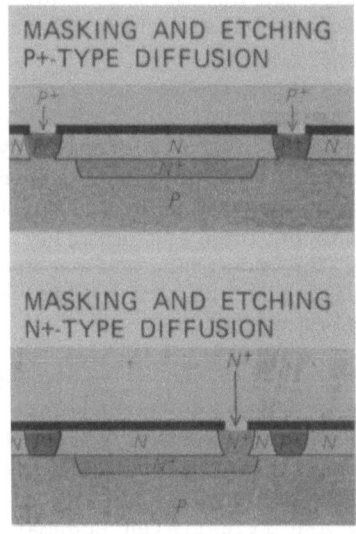

Fig. 14(b)

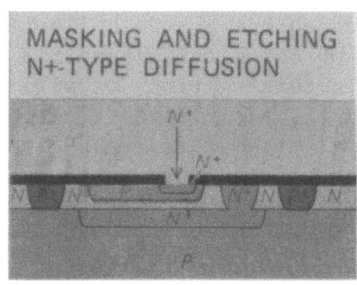 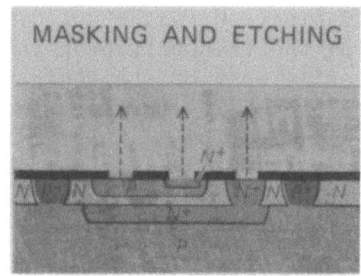
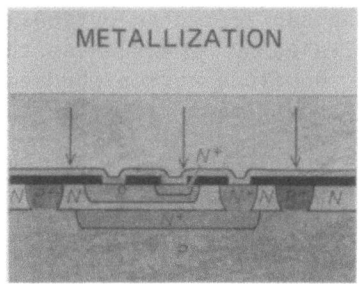 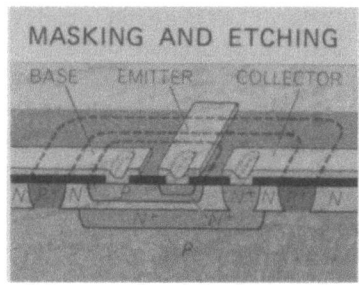

Fig. 14(c)

Fig. 14. Schematic diagrams showing process sequence used during manufacturing.

III. PRINCIPLE SEM METHODS AND TECHNIQUES

A. Electron Beam Induced Current Microscopy

Early investigations of semiconductor material and devices utilizing EBIC microscopy have been well documented in published literature [45-48]. The display mechanism is basically dependent on the generation of charge carriers by electron penetration and energy dissipation in the solid. Subsequent charge collection depends on the transport properties of the material and the collection efficiency of the device. In this section, we will develop the concept of EBIC microscopy using the basic diode equations developed in Section II. The material parameters which give rise to contrast in EBIC microscopy have been discussed in recent publications [49,50]. A rather extensive bibliography has recently been published on the applications of EBIC microscopy [51].

Typical pn diodes and Schottky barrier devices, when used for material characterization studies are fabricated on optically flat, chemical slurry polished wafers. Polished surfaces such as these, totally eliminate any modulation of the diode current by secondary emission variations and electron backscattering at the surface. The diffused pn junctions are planar and parallel to the polished surface with a junction depth $x_j < 1$ µm. The diode is normally designed with a junction area which is compatible with the anticipated defect density. With current silicon material having low defect densities, junction areas in the range of $10\text{-}100 \times 10^{-3} cm^2$ are required for effective sampling of the material.

The scan raster for EBIC displays generally lies within the vertical boundary of the pn diode. If we consider a diode with an active area A defined by a uniform electron irradiation field and neglect edge effects due to the vertical depletion field at the perimeter of the diode, then the diode equations are one dimensional for a plane junction parallel to the surface.

The minority carriers generated in the neutral or field free region of the pn diode above and below the space charge region must diffuse to the edge of the depletion field for collection. The magnitude of the diffusion current generated during electron irradiation, can be determined by solving the diffusion equation for minority carrier flow under steady state generation conditions. For electrons in the p region of the diode the equation is given as:

$$D_n \frac{d^2 n_p}{dx^2} + G(x) - \frac{n_p - n_{po}}{\tau_n} = 0 \qquad (43)$$

where G(x) is the generation rate function for minority carriers due to electron energy dissipation in the solid 30. If we assume that the minority carrier concentration $n_p = 0$ in the depletion field, and use the depletion edge as a boundary in the solution $n_p(0) = 0$ the solution to equation (43) is given as:*

$$n_p(x) = (n_{po} + \tau_n G(x))(1-e^{-x/L_n}) \qquad (44)$$

where $L_n = \sqrt{D_n \tau_n}$

The electron diffusion current at the edge of the depletion field is given by

$$I_{D_n} = \frac{qD_n(n_{po} + \tau_n G(x))A}{L_n} = \frac{qD_n(n_i^2 + \tau_n G(x)N_a)A}{N_A L_n} \qquad (45)$$

and the hole current by

$$I_{D_p} = \frac{qD_p(p_{no} + \tau_p G(x))A}{L_p} = \frac{qD_p(n_i^2 + \tau_p G(x)N_d)A}{N_d L_p} \qquad (46)$$

In these equations A represents the active area of the diode, neglecting edge effects resulting from irradiation and surface leakage. Now for the condition

$$G(x) >> \frac{n_{po}}{\tau_n}, \frac{p_{no}}{\tau_p}$$

the diffusion component of the diode current for both electrons and holes can be expressed as

$$I_D = \frac{qD\tau G(x)A}{L} = q\, G_D(x) LA \qquad (47)$$

Thus the diffusion current is proportional to the diffusion length L of the minority carrier on both sides of the junction. Although the generation function $G_D(x)$ has not yet been explicitly defined it represents the generation of excess carriers in the neutral region of the diode due to the total energy dissipated in these regions.

The generation component of the diode current will be the contribution due to generation of carriers in the space charge region by the electron beam. For the condition where

---

* In this solution G(x) is not explicityly defined. The solution in equation (44) assumes that G(x) is a slow varying function of x.

$$G(x) \gg \frac{n_i}{\tau_o},$$

the generation current in the depletion region is give by

$$I_G = q|U|WA = q\, G_W(x)WA \qquad (48)$$

where $G_W(x)$ is the generation rate between the edges of the depletion field. The total diode current is given as

$$I = qG_D(x)LA + qG_W(x)WA \qquad (49)$$

$$= qG_D(x)(L_n+L_p) + qG_W(x)WA \qquad (50)$$

For typical conditions $L_n, L_p > W$, and therefore the total diode current is essentially independent of voltage for reverse bias conditions, i.e.,

$$V_R \gg \frac{kT}{q}$$

Equation (50) is analogous with the diode equation for a photovoltaic cell. The first term relates the total diode current to the minority carrier diffusion length L which is a basic electrical parameter of semiconductor materials as discussed previously. And, it includes the minority carrier lifetime and the carrier mobility which together determine the transport characteristics of the material. The interrelationship is given as

$$L \equiv \sqrt{D\tau} = \left(\frac{kT}{q}\right)^{\frac{1}{2}} \sqrt{\mu\tau} \qquad (51)$$

The diffusion component of the total diode current which is represented by this term consists of the minority carriers generated in the neutral regions which have reached the edge of the depletion field and are swept across the depletion region, and collected.

The second term in equation (50) represents the component of current arising from the direct generation of minority carrier in the depletion region by the electron beam. However, since the depletion width W is dependent on the impurity dopant concentration and the applied voltage V as shown in equation (32) the carrier collection volume will then determine the magnitude of the second term in equation (50).

When a finely focused electron beam is scanned over the surface of a pn diode as shown in Figure 15, the maximum spacial resolution will depend on the shape and effective cross sectional area of the

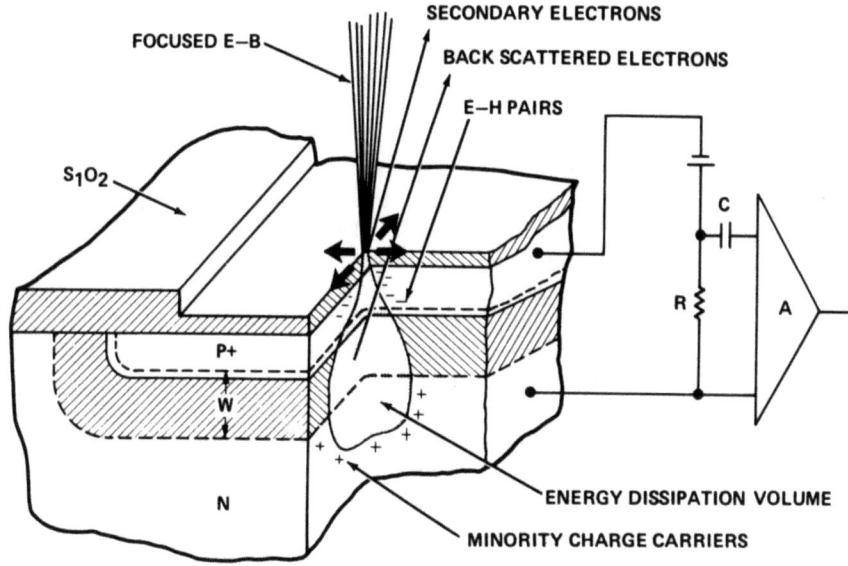

Fig. 15. Schematic illustrating EBIC microscopy.

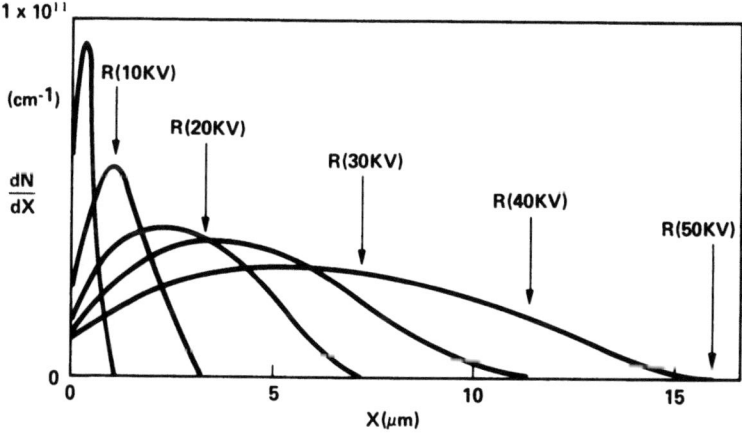

Fig. 16. Hypothetical steady state minority carrier distribution in silicon, resulting from electron penetration and energy dissipation.

generation rate function G(x). This function defines the excess
steady state carrier density per unit depth below the surface for
electron of a given energy E_b. An exact solution to equation (50)
would require an explicit definition of the G(x) term and the
limits over which the function applies in both the neutral regions
of the diode and the depletion field. The curves describing the
generation rate G(x) follow a universal form when normalized to the
electron range R.

In Figure 16 the steady state minority carrier distribution is
plotted as a function of depth below the Si surface for electron
beam energies in the range 10-50 keV. These hypothetial charge
carrier distributions represent the conditions where I_b = 1 na
and τ = 1 μs. These values were chosen as typical electron beam
and material parameters. However, the distribution plots represent
a degenerate solution corresponding to a carrier mobility μ = 0,
and therefore they do not represent the effects of carrier diffusion.

The actual minority carrier distribution for a typical material
with a lifetime τ = 1 μs would appear as a nearly homogeneous
distribution on a range scale as shown in Figure 16 if surface
recombination were neglected.

The expression used to calculate the steady state density is
given as:

$$\frac{dN}{dx} = \frac{kE_b I_b}{E_a q} \cdot \frac{\tau}{R} \qquad (52)$$

where E_b and I_b are the beam voltage and current respectively, τ
is the minority carrier lifetime, E_a is the carrier pair ionization
energy ≈ 3.6 eV for Si, and R is the electron range given by the
following expression:

$$R = 0.0171 \, E_b^{1.75} \; (\mu m) \qquad (53)$$

The curves in Figure 16 also indicate the effect of the beam
energy E_b on spacial resolution. This is of particular significance when spacial resolution is limited primarily by the electron
range R. However, there are additional factors which relate to
charge carrier transport during steady state generation which
impose additional limits on resolution via the carrier diffusion
length term.

Figures 17, 18, and 19 are used to illustrate the normalized
steady state carrier concentration (cm^{-1}) for a diffused n^+p
diode having a one sided step junction approximation. The diode

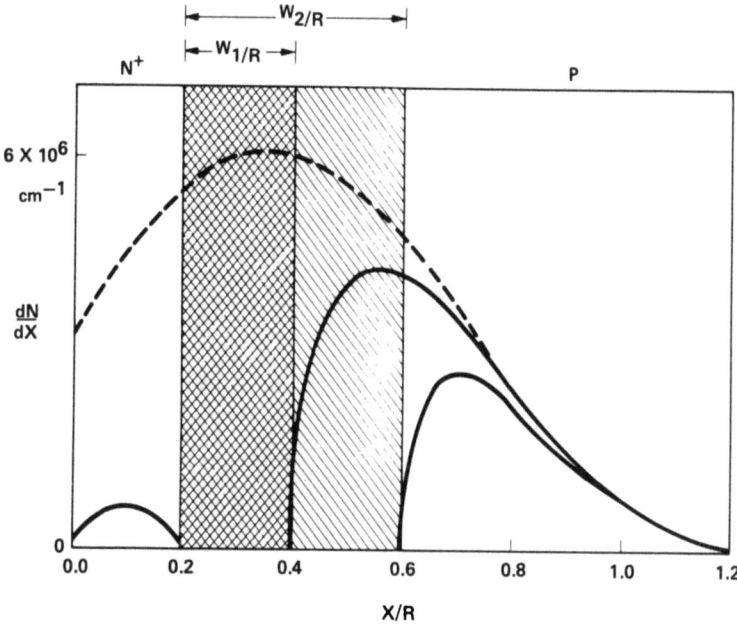

Fig. 17. Normalized steady state carrier concentration. cm^{-1} vs. normalized depth. R/L = 6 (See text for details).

is shown with a junction depth x_j = 0.5 μm on 1 ohm-cm silicon with two values of depletion width W, defined by external bias conditions. The <u>dashed</u> <u>curves</u> are used to represent the hypothetical steady state minority carrier distribution generated by a 15 keV electron beam with a 1 na beam current. The material is assumed to have a carrier mobility μ = 0 with a finite carrier lifetime to establish a carrier generation rate.

The distribution plots corresponding to the dashed curves shown in all figures, exclude the effects of the depletion field and therefore represents a zero charge collection condition in a homogeneous medium with a carrier mobility μ = 0. The main purpose of showing this hypothetical solution is to provide a visual reference for discussing the spacial resolution of the technique.

In Figure 17 a minority carrier lifetime of 40 ps was selected to satisfy the condition R/L = 6. For this case, the range of the generated carriers is less than the energy dissipation range and the resolution is limited primarily by the electron range R. The normalized carrier density dN/ds (cm^{-1}) as shown on the y axis

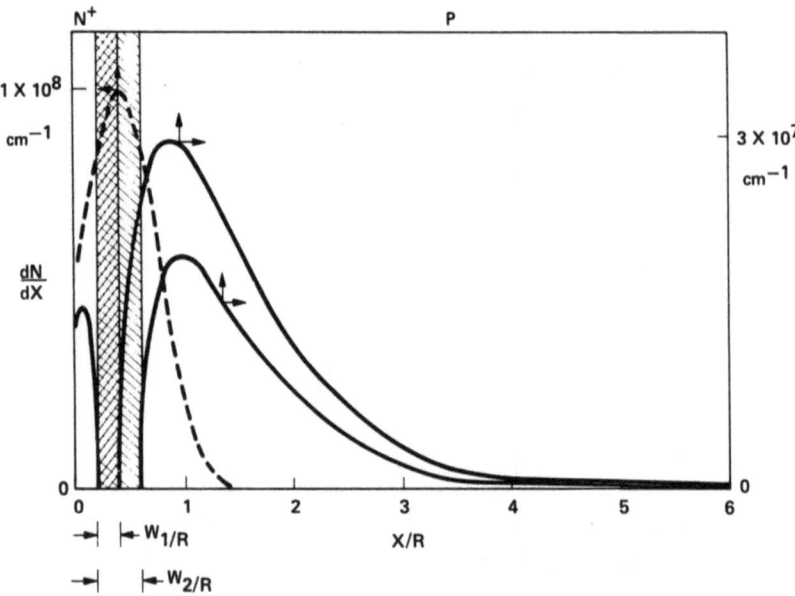

Fig. 18. Normalized steady state carrier concentration. cm^{-1} vs. normalized depth. $R/L = 1$. (See text).

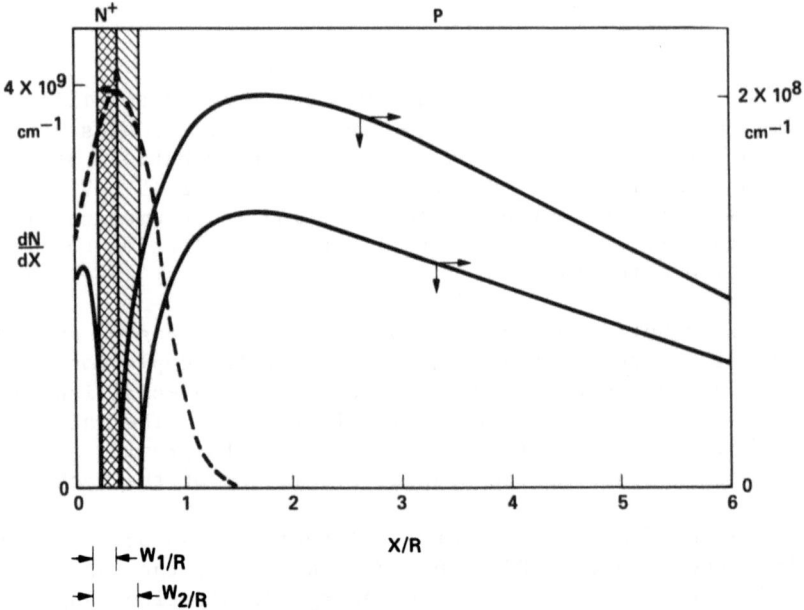

Fig. 19. Normalized steady state carrier concentration. cm^{-1} vs. normalized depth. R/L 1/6. (See text).

can be used to estimate the steady state minority carrier density for a given depth X/R. In this case, the dN/dx values for both curves are in the same range and the ordinate is common to both distributions. The collection efficiency η in the depletion region is assumed to be 100%.

If the carrier lifetime τ in the material is increased to 1 ns, R/L = 1, the steady state carrier density at one diffusion length exceeds the carrier density defined by the energy dissipation range for $\mu = 0$ as shown in Figure 18. Hence, the spacial resolution in the neutral region below the depletion field becomes dominated by the exponential tail $e^{-R/L}$ resulting from carrier diffusion in this region. Note, the different ordinates corresponding to the carrier density for the dashed and solid curves. The normalized x axis has also been extended to include the added range effects due to carrier diffusion.

If we further increase the carrier lifetime to 25 nS, where R/L = 1/6 (figure 19) with all other parameter unchanged, carrier diffusion in the neutral region is completely dominant and the resultant resolution is limited by the diffusion length L. The steady state carrier distribution for the <u>solid curves</u> in Figures 17 and 19 are calculated using an electron mobility μ_n = 1500 cm^2/v-sec at room temperature. The resultant distribution shows the effects of both the depletion field where it is assumed that carriers are collected with 100% efficiency and the diffusion length L. The peak carrier distribution in the n$^+$ region is reduced to compensate for a smaller diffusion length in this region. This would be expected in a one side step junction where the dopant concentration is normally large. In addition, one would expect an improved spacial resolution in the neutral n$^+$ region due to the reduced carrier diffusion length in this region.

By comparing the steady state distribution for the generated carriers shown in Figure 19 the dominant effect of the carrier diffusion length is readily apparent. Note the small values of carrier lifetime $\tau \sim 40$ ps which was required to satisfy the condition where R/L = 6 (Figure 17) where maximum spacial resolution is obtained. In typical silicon devices, the carrier lifetime ≈ 1 µs. This corresponds to a carrier diffusion length of ≈ 100 µm which would result in extremely poor resolution. Thus, to realize maximum spacial resolution, the local carrier generation range L must be smaller than R. The local lifetime τ corresponding to a local inhomogeneity or defect must be many orders magnitude smaller than the lifetime in the surrounding field material for optimum detection.

Calculation of the collection efficiency for gaussian diffused structures have been obtained by Hoff and Everhart [53] on the basis of the energy-range relationship in silicon. These calculations show the effects of both surface and bulk recombination on the collection efficiency and excess carrier concentration for steady state conditions.

In practical pn junction device structures, the dopant concentration and junction depth are specified by device design rules and processing conditions which are optimized for device characteristics. Collection efficiencies and device geometries are rarely optimum and the display conditions must be determined experimentally by selecting the appropriate electron beam energy and current and by adjusting the depletion width W with external voltage bias.

In typical semiconductor materials the electrical properties are not homogeneously distributed in the crystal as assumed in the basic equations of the previous section. The crystal generally contains localized and extended crystal defects such as dislocations, stacking faults and impurity precipitates.

The ability to detect dopant inhomogeneities and crystal defects depends on the size and shape of the excess carrier generation volume dN/dx which is defined by local charge carrier transport properties of the material being sampled, i.e., $L \alpha \sqrt{\mu \tau}$. In the field free region of the diode the local recombination as defined by equation (25) for low level injection at the defect, will reduce the local steady state concentration of minority carriers which are transported by diffusion to the edge of the depletion field. The contrast at the localized defect would be expected to depend on the ratio of the local carrier lifetime to the average lifetime in the material.

In the depletion region, one would expect the collection efficiency to be reduced by trapping in a similar manner if equation (25) is valid thus reducing the steady state concentration of carrier which are separated by drift in the depletion field region.

In addition to local recombination contrast at defects, dopant inhomogeneities in the semiconductor can be detected by changes in the depletion field width W as expressed in equation (32) since $W \alpha\ 1/\sqrt{N}$. The net result is a change in the relative magnitudes of the first and second terms in equation (50).

Image contrast will depend on the steady state collection efficiency η. This can be defined as the ratio of the number of charge carriers collected by the depletion field to the total number generated. One can visualize the dependence of the collection efficiency on depletion width W by comparing the areas under the curves shown in Figure 17, where R/L = 6. When the depletion width is doubled, i.e., by applying an external voltage V, see equation (32) Section II, the number of additional carriers collected is given approximately by the difference in area under the two solid curves. However, when the carrier lifetime is increased to 25 ns, R/L = 1/6, as shown in Figure 19 the relative increase in the number of collected carriers is smaller for the same doubling of the depletion width due to the increased diffusion length L.

The excess minority carrier concentration increases with carrier lifetime as shown by the magnitude of the peak carrier concentrations in each figure. Consequently, if we define efficiency in terms of the diode short circuit current it is clear that the efficiency will steadily increase with carrier diffusion length as indicated by equation (50) until the limiting efficiency is reached when the diffusion length L >> R.

Figure 20 illustrates the dependence of the collection efficiency η on W/L and R/L for a Schottky barrier diode structure [52]. This curve will represent the limiting case for a n^+p junction where $x_j \to 0$, in shallow diffused junction structures. Note that the depletion width W has a relatively small effect on collection efficiency where R/L is small. Whereas, a relatively large change of η occurs where R/L is large. The effect is more clearly illustrated in Figure 21 where the depletion widths corresponding to Figure 17 through 19 are indicated on these respective R/L curves. Thus, the spacial resolution of the technique is determined by the range and shape of the steady state carrier density dN/dx. When L R, the resolution is limited by the volume defined by the range energy relationship. Image contrast is determined by the magnitude of G(x) and the collection efficiency η of the diode. These parameters are dependent on the electron beam current and voltage, the ionization energy for charge carrier pair generation, the charge carrier transport properties of the material and the depletion width W of the diode. The resulting change in the diode current during the scan will produce a contrast change which is related to the local dopant concentration.

Charge carrier multiplication or generation can also occur in local high field regions of the diode when an inhomogeneity or defect results in a concentrated field. In this case, the energy gained by the charge carriers in the field is sufficient to produce additional carriers by impact ionization. Consequently, the local steady state carrier concentration may exceed the excess carrier concentration generated by the beam, producing a local increase in the diode current at the generation site.

In Section IV experimental results will be presented which illustrate the application of the EBIC technique for the characterization of electrically active defects in silicon devices.

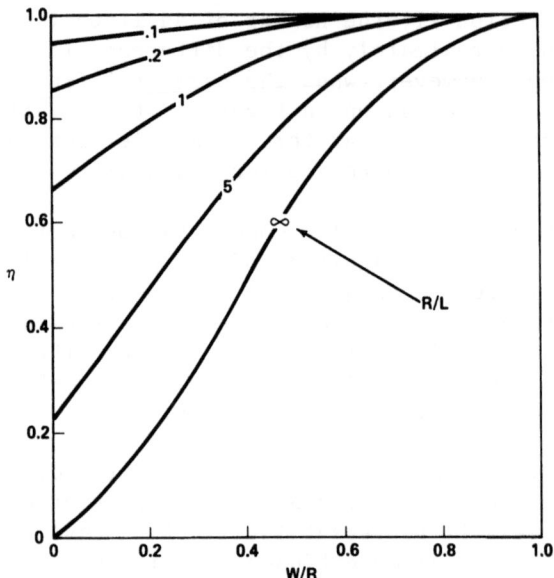

Fig. 20. Collection efficiency vs. normalized depletion width. Ref. [52].

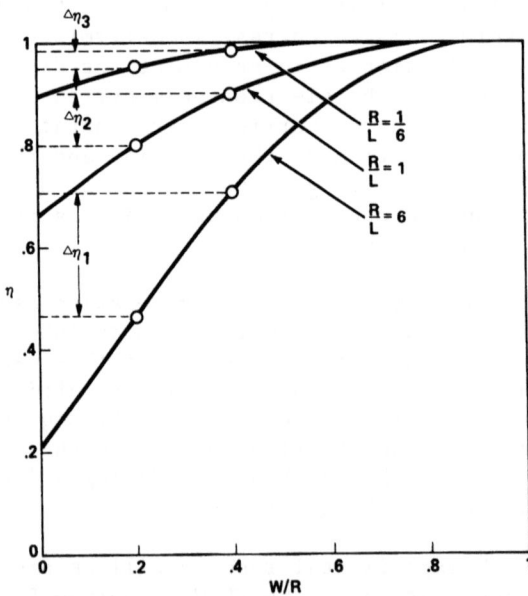

Fig. 21. Collection efficiency vs. normalized depletion width showing conditions corresponding with Figures 17, 18 and 19.

B. Secondary Electron Microscopy

Secondary electron (SE) microscopy provides the most effective means for examining the surface of solids with high special resolution. The escape depth for the low energy electrons is in the range of 1 to 5 nm for most conductive materials. Thus the resolution is limited by the spot diameter of the focused electron beam as shown in Figure 3. Additional secondary electrons which are generated by backscattering emerge at greater distances (typically 1-5 µm) from the primary beam. This latter group contributes to background noise in the display and does not enhance image quality.

The energy distribution for secondary electrons is virtually independent of the incident electron energy [54]. For most metals the energy distribution peaks at ~ 2 eV and is approximately 5 eV wide at half peak height. The total yield of secondary electrons δ can be expressed as the sum of the secondary electron emission δ_p at the entrance point of the primary beam, and, the secondary electron emission at the point where the backscattered electrons emerge

$$\delta = \delta_p + \eta \delta_B$$

The relative efficiency of the primary to backscattered electrons for generating secondaries is $\delta_p / \eta \delta_B$. Thus conditions which minimize backscattering, i.e., low atomic number materials and low beam energies, will produce a higher relative efficiency for generating secondary δ_p and thus produce a higher image resolution.

In SE microscopy the sample is normally tilted at an angle of $30°$ to $60°$ with respect to the primary beam. This increases the efficiency for the electron collector which is placed in close proximity to the sample. The SE collector generally consists of a positive biased electrode assembly which establishes an electric field at the surface of the sample. The secondary electrons are attracted to the collector assembly where they are accelerated to higher energies (~ 10 keV) before striking a phosphor or scintillator material. Subsequent amplification is obtained with a photomultiplier assembly.

In the absence of surface potentials, other than those generated by the collector, image contrast in SE microscopy is dependent on the angle between the electron beam and a line extended normal to the surface at the point of beam incidence. Consequently, when the beam is scanned over a moderately rough surface the secondary electron image consists of a relatively high contrast display of the surface topography with high spacial resolution.

For topographical detail, the image contrast arises from the variation of the local angle θ between the direction of the primary beam and the normal to the surface. For efficient collection of secondary electrons, with a positive biased collector, the collection efficiency γ approaches unity. The collected current I_C can be expressed in terms of the beam current I_b as:

$$I_C = \gamma \delta I_b$$

where δ is the secondary electron emission coefficient. The change in the collected current I_C which determines local image contrast is expressed as:

$$\Delta T_C = k \left(\frac{d\delta}{d\theta}\right) \Delta\theta$$

Typically, θ is set to ~45° so that

$$\Delta I_C = k \frac{d\delta}{d\theta}$$

The secondary electrons emitted at the point of the incidence between the primary beam and the surface constitute approximately 65% of the total video signal [55] if we consider the electrons with an energy <50 eV. If only the very low energy electrons are considered, i.e., <5 eV, this figure is further reduced to ~12% of the video signal. This represents approximately 5% of the total beam current. It is these low energy secondaries (<5 eV) that gain or lose energy depending on the potential of the surface at the emission point. Typical voltage contrast results when the surface voltage of the emission point alters the trajectories and energy distribution of these low energy electrons.

When local surface potentials exist on the sample surface a voltage dependent contrast is obtained in the SE display, the local contrast depends on the geometry of the local field and its effectiveness in suppressing or enhancing the secondary electrons emitted in this region. In general, the resultant voltage contrast is determined by the local electric field at the surface of the sample and the design of the electron collector. The SE collector can be designed for either directional sensitivity or for energy selectivity to optimize the range of contrast for the voltage sensitive detector.

C. X-ray Microprobe Analysis and Other Supporting Techniques

EBIC microscopy provides an effective non-destructive method for identifying crystal defects with modest resolution (~1 μm) and for characterizing their electrical effects in pn junction devices. However, the crystallographic identification of the specific defect type and the identification of the impurity elements which often decorate electrically active defects require other techniques such as transmission electron microscopy [56] and x-ray methods [57] to complete the analysis.

In transmission electron microscopy, the primary electrons must be transmitted with minimum energy loss in the solid. Thus, the sample thickness must be less than the mean range for elastically scattered electrons typically 0.5 µm for silicon at 100 keV. The technique is basically destructive in nature because of the thickness requirement. Sample preparation is often tedious and unreliable due to the unpredictable quality of chemically etched surfaces. Conventional sample preparation techniques utilize chemical jet etching and other chemical dissolution techniques for thinning the sample.

X-ray microprobe analysis is based on the emission of characteristic x-rays from a solid which is irradiated with energetic electrons. Figure 22 illustrates the generation of K_α x-rays from a copper target. The incident electron beam ejects an electron vacancy which is subsequently filled by an L_3 level electron. The quantum energy released in the process emerges as an x-ray photon with an energy E_x given by the difference between these two levels $E_K - K_{L3} = 8979-931 = 8.048$ eV. This characteristic energy is used to identify the element involved, which in this case is Cu.

The two commonly used methods for the measurement of this energy are: wavelength dispersive spectroscopy and energy dispersive analysis. The former technique utilizes x-ray diffraction in a crystal spectrometer where the diffraction condition is satisfied by Bragg's law, $N\lambda = 2d \sin \theta$ where λ is the wavelength of the x-ray photon.

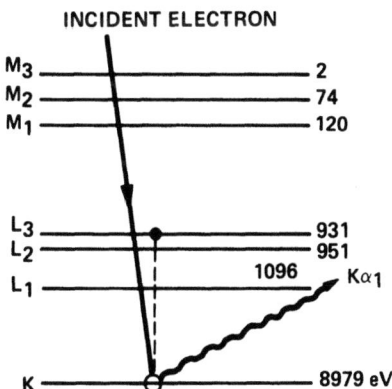

Fig. 22. Generation of copper k-alpha x-rays by energetic photons.

Energy dispersive analysis is the most commonly used technique for SEM studies. It utilizes a lithium drifted silicon detector placed adjacent to the SEM sample which generates charge carrier pairs when x-rays are absorbed. This gives rise to a current pulse in which the amplitude is proportional to the energy of the x-ray photon.

The spacial resolution of this technique is limited by the same mechanism discussed previously for EBIC microscopy, i.e., electron penetration and energy dissipation in the solid. The technique has a minimum detection limit for $\sim 10^{19}$ atoms cm^{-3} but is less sensitive to elements with low atomic number. In semiconductor devices impurity precipitates often develop in the near surface region of the device. The energy dispersive x-ray method [58] has been used successfully for identifying the elemental impurities in precipitates which often decorate near surface defects such as dislocations and stacking faults.

IV. CHARACTERIZATION OF 'ELECTRICALLY ACTIVE' DEFECTS IN SILICON MATERIALS AND DEVICES USING EBIC MICROSCOPY

A. Experimental Procedures and Device Design Considerations

When the objective of the experiment is to develop quantitative correlations between the electrical characteristics of the device and specific crystal defects, the selection and characterization of the silicon ingot is of utmost importance. The experimental matrix will usually consist of wafers drawn from a single ingot. Preliminary resistivity measurements are obtained to determine the net dopant concentration and its uniformity in the ingot. These usually consist of bulk resistivity measurements on the ingot and on wafers subsequently sliced from the ingot, using 2 and 4 point probe resistivity measurement techniques [59,60]. Radial resistivity homogeneity measurements are obtained on selected final polished wafers using a spreading resistance probe [61]. The final characterization of polished wafers is completed using optical microscopy on wafers chemically etched, using chemical reagents which preferentially decorate crystal defects such as dislocations [62], stacking faults [62], and impurity inhomogeneities referred to as impurity striations and 'swirl'. It is currently a standard practice to 'thermally activate' the material defects prior to chemical decoration--etching by subjecting the polished wafers to a thermal anneal or oxidation prior to etching.

After final polish, further characterization of the substrate material, is feasible using EBIC microscopy by fabricating a Schottky barrier device [64-66], using Au-Pd or Ti electrodes. If the barrier metal is sufficiently thin, 20-40 nm, electron energy loss in the film is negligible and the electrode is virtually electron transparent. The advantages of this technique will be illustrated in a subsequent section on single crystal characterization.

Careful selection and characterization of the material reduces the uncertainties in statistical trends caused by crystal quality variations, and provides a common thermal history for the wafers. The net result is that the credibility of the data is enhanced particularly for repeated experiments.

A typical diffused pn junction test device for material and process characterization is generally a diode or transistor structure with an area compatible with the anticipated material defect densities and electrical sampling methods. It is also convenient to select the test device area for compatibility with the maximum scan area that can be used effectively in the SEM. A convenient diode area for EBIC displays is 10-20 x 10^{-3} cm^2 which permits a single scan survey of the material at a magnification between 50 and 100.

For effective characterization the test device must be subjected to the same thermal processing history as the product or material being investigated. Thus, the design of the test structure must be compatible with the process mask set when complex devices or processes are being studied.

The test vehicle mask design should provide sufficient flexibility to terminate the processed wafer at major process steps, i.e., subsequent to epitaxial growth, isolation diffusion, base diffusion, and emitter diffusion. In this manner a completed test device with pn junctions can be fabricated which has been subjected to the cumulative thermal history of the process at the termination point in the process sequence. However, since the topology of the metal layer contacts and the geometry of the diffused junctions are normally optimized for EBIC microscopy, some tradeoffs in electrical testing are required. Fortunately, DC parametric measurements are normally sufficient to establish the necessary correlation for trends developed during wafer processing.

Test structures designed for this type of analyses are not optimized for dynamic and functional testing. They provide little insight into the more subtle material related electrical effects arising during high speed transients because of the excessive capacitance and spreading resistance associated with the non-optimum device geomtery. Consequently, the test vehicle should include companion test structures to supplement the DC parametric testing.

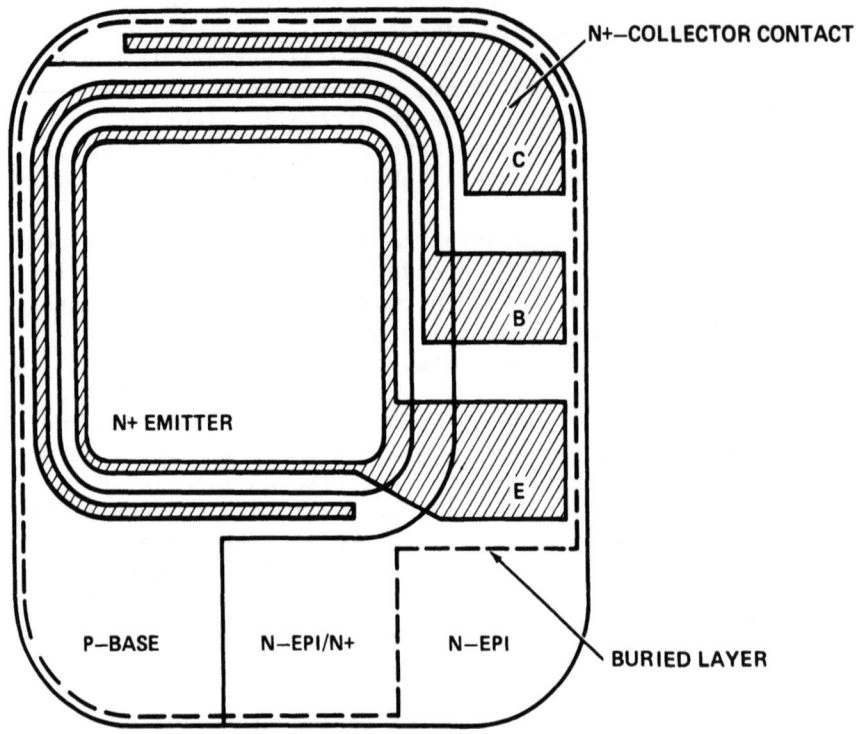

Fig. 23. Bipolar transistor test structure design for EBIC microscopy.

Figure 23 shows a test structure designed for a bipolar process which contains a device for EBIC microscopy as well as other test structures in which the geometries are optimized for dynamic testing.

It should be emphasized that conventional device structures used in complex integrated circuits are normally unacceptable for material and device studies using EBIC microscopy. These devices are typically characterized by large perimeter to area ratios, producing relatively large beam induced currents at vertical junction boundaries. In addition, they often contain complex metal layer interconnect patterns which mask and reduce the useful junction area leading to poor display conditions for material studies.

B. Methods of Establishing Correlations and Trends

Preliminary correlation and trends are established by selecting a set of DC test conditions such as forcing current or voltage which permits differentiation of test results displayed in graphic

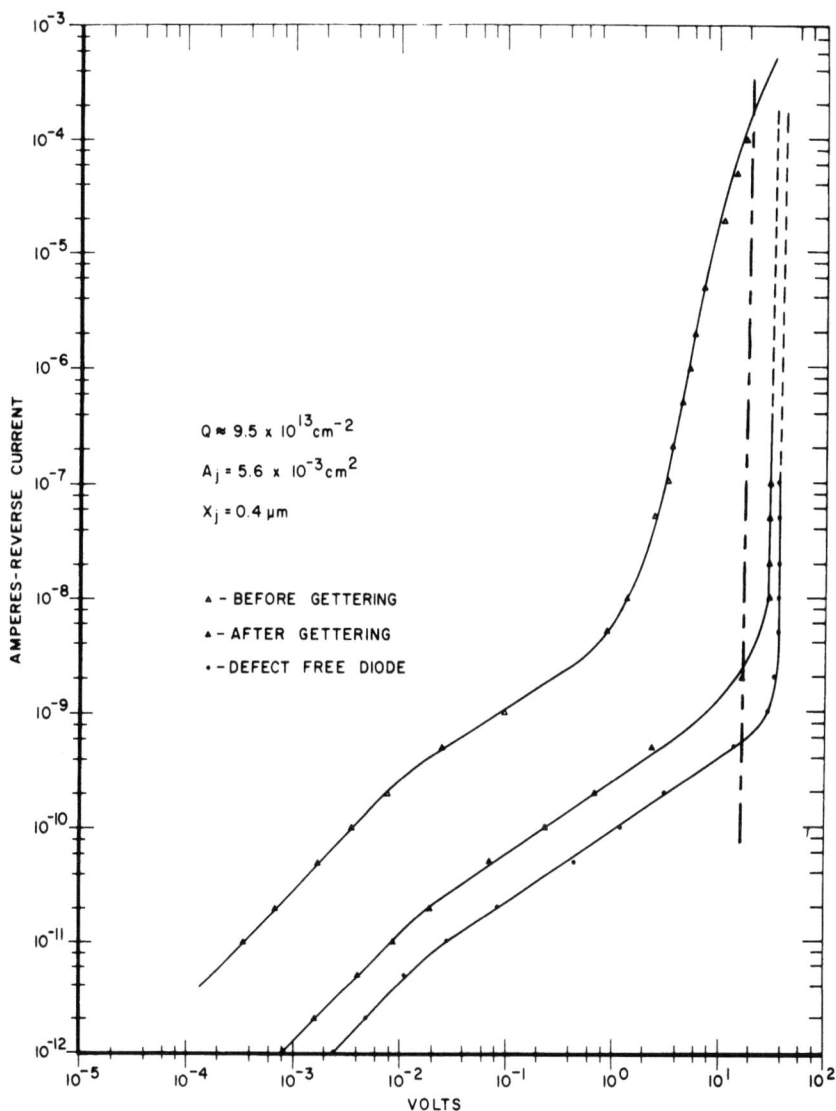

Fig. 24. Test conditions for optimum partitioning of test data.

form. The test results are normally presented as a data histogram in which the sample frequency is plotted against the desired DC parameter. When die position correspondence is desired a wafer test map is generated.

Manual test procedures can be used effectively when the sample size is small. However, when full wafer maps are desired, an automatic probe station with a computer controlled DC parametric test system and full complement of graphics software is required. Defining a DC parametric test program requires a prior knowledge of the I-V characteristics corresponding to the particular defect mode. Thus, when electrically active stacking faults (EASF) are present in a p^+n diode, the reverse I-V characteristics can be partitioned into 2 distinct regions which are separated by the threshhold voltage V_{th} of the stacking fault as shown in Figure 24 [67]. When the reverse voltage $V_R < V_{th}$ the diode exhibits the normal thermal generation current expected for a silicon diode at room temperature as given by equation (34) in Section II. However, when $V_R > V_{th}$ a voltage power law dependence is obtained where $I_R \propto V^n$, n ~ 5.0.

Now if we carefully select a test voltage V_R which is less than the avalanche voltage BV_R for the diode such that $V_{th} < V_R < BV_R$, the desired separation can be obtained for the I_R readings, corresponding to diodes containing EASF's and diodes with normal I-V characteristics. The net result for this data when it is plotted as a histogram is a bimodal distribution which reveals the relative diode test yield in the two categories as shown in Figure 25. The degree of skewing will depend on the magnitude of the test voltage. If this data is subsequently used in a wafer mapping routine, the distribution of the diodes containing EASF's can be presented in their relative locations on a wafer map. The graphic presentation

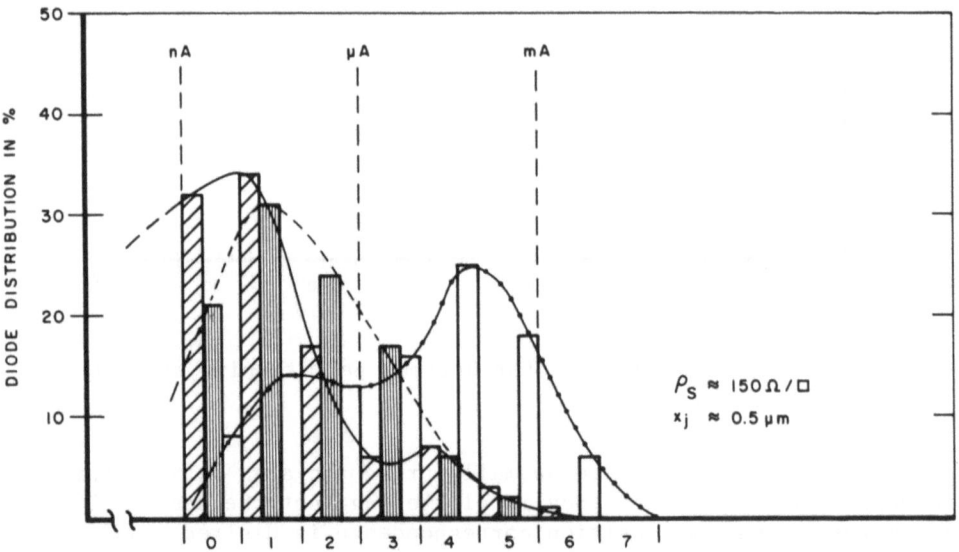

Fig. 25. Examples of bimodal distributions resulting from electrically active stacking faults.

of the data in this manner, can be used to determine the relationship between the distribution of these defects and the homogeneity of the silicon ingot. Moreover, subsequent shifts in the distribution can be monitored and related to specific wafer processing operations to identify the areas where process optimization is required.

DC parametric analysis as described above often provides sufficient spacial resolution in the wafer maps to identify macroscopic correlations with known crystal defects. An example of this is shown in Figure 26. The test device is a small area transistor in which a specific DC current parameter correlates directly with thermally induced dislocations. In this case the device which falls in the selected category is inked automatically following the test. The corresponding test wafers thus reveal the distribution of these transistors and the locations of the thermally generated dislocations.

Distribution plots and wafer maps are also used to select typical devices which represent specific test categories. When device selection is made on the basis of manual testing, rather than with an automatic test system, the selection procedure is extremely tedious, if not futile because of poor sampling statistics. Whereas, devices selected from full wafer maps can be subjected to further testing such as I-V characterization before the final selection is made with the test system. The main advantages of the computerized test system are: precision testing with large sampling statistics at high test rates, and the supporting software for test routines, data reduction, and graphics.

Representative devices are then selected for more detailed electrical characterization on the basis of these preliminary data. In some cases, detailed I-V characteristics are obtained from the devices in wafer form. Whereas, in other cases, the detailed testing is done after the selected die have been bonded, wire and packaged. In the latter case, the device is subsequently examined in the SEM using the EBIC mode to identify the specific electrically active defects.

Establishing the correlation between the I-V characteristics of pn diodes and the electrically active defects is fairly straightforward when the preliminary test and select approach described previously is used. Particularly, if the electrical test routine is defined on the basis of established electrical characteristics for the defect, as in the case of electrically active stacking faults.

In most cases, the preliminary testing is a necessary procedure for die selection. But, it is not sufficient to identify the electrically active defect or to determine the relative importance of the material and process related factors which give rise to the electrical effects observed. This type of information requires a

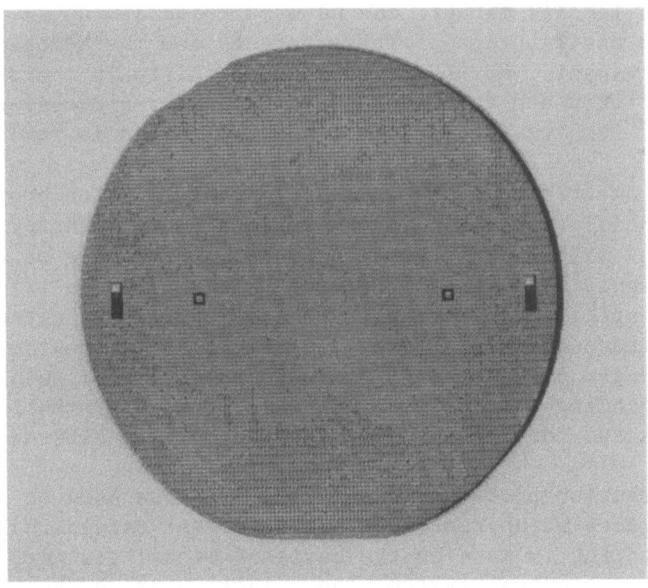

Fig. 26(a)

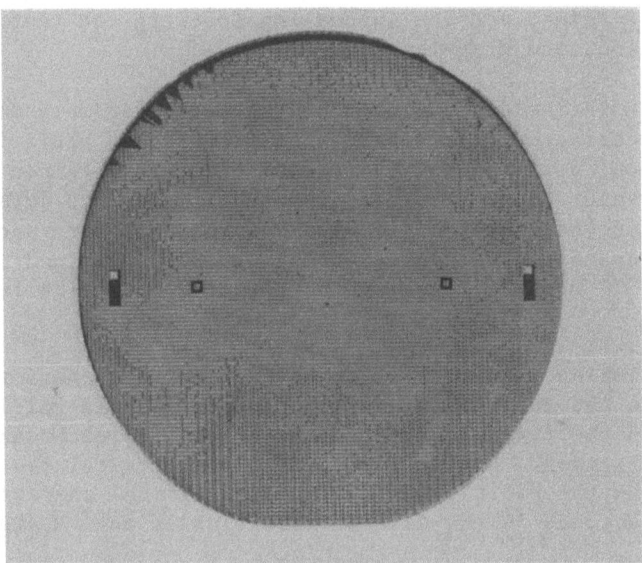

Fig. 26(b)

Fig. 26. Electrical test reject map showing distribution of thermally induced dislocations.

more direct approach. EBIC microscopy provides a nondestructive method which can be used effectively to investigate the interaction between the active defect and the depletion field of the diode. In addition, it can produce a medium resolution (∼1 μm) defect map for subsequent work using x-ray methods and transmission electron microscopy.

C. DC Parametric Analysis for Correlations With EBIC Displays

The current voltage characteristics of the device are normally examined with a curve tracer such as Tektronix 576 prior to EBIC microscopy. The purpose is to define the test conditions for the device and to establish the mode of display for effective characterization. Since excess junction currents are normally the result of electrically active defects, we prefer to use capacitance coupling between the device and the operational amplifiers. This avoids the electrical instabilities in the display which are normally associated with excess junction currents. Moreover, it eliminates the need for balancing the DC offset current resulting from power supply adjustments. When ambiguity exists in the display contrast, the DC coupled mode can be used to establish the relative contrast levels.

In cases where the device has not been characterized using preliminary test routines or where the defect mechanism is unknown, the I-V curve tracer and the SEM provide the primary diagnostic tools. The normal procedure is to photograph the I-V curve for reference during testing at the SEM. The device is then biased to the corresponding I-V operating point on the oscillograph where the electrical anomalies exist. Subsequently, the beam current and voltage are adjusted to optimize the display conditions for detection of the electrical effect.

The specific procedures which are used to investigate the active defects are dependent on the effects desired by the operator. However, there are sound general rules which can be applied, that are based on the physics of electron beam interactions with the solid and the geometry of the pn junction structure including the depletion field as discussed in the previous section. This can best be illustrated with several specific examples of EBIC microscopy.

D. Single Crystal Characterization

When dislocation free silicon crystals are used in complex semiconductor device processing, the crystal defects generated during the fabrication of the device are often the end result of

inhomogeneities formed during the crystal growth process. These
defects (impurity complexes and dislocation loops) are often
transformed during subsequent wafer processing by thermal-diffusional mechanisms to produce the final defect structure (complex
dislocation colonies and precipitates) which bear little resemblance
to the initial defect. The progressive development of native defects during thermal oxidation and diffusion has been investigated
in silicon pn junction devices using EBIC microscopy [68]. The
results clearly indicated that a complex interrelationship can
exist between major processing steps to produce the final defect
structure. Typically, the end result, is the progressive degradation of the electrical characteristics of pn junction devices.

Inhomogeneities in dislocation free silicon crystals have
been detected with a variety of analytical techniques [69]. The
most common manifestations of these are: non-uniform axial dopant
distributions which result from normal segregation during freezing,
radial dopant micro-inhomogeneities due to microsegregation and
rotation, and the formation of impurity point defect complexes.
The micro-inhomogeneites and impurity complexes are attributed to
crystal interface growth instabilities such as: fluctuation of the
growth velocity, thermal assymmetry of the melt, non-steady state
melt convection and mechanical and electrical instabilities in the
crystal puller.

Nondestructive characterization of silicon crystals with EBIC
microscopy can be achieved by utilizing pn junction devices [70,71]
which are incorporated into the manufacturing process or by fabricating Schottky barrier devices on 'unprocessed' wafers [66].
Both techniques offer advantages and limitations.

The characterization of unprocessed silicon wafers with Schottky
barrier devices is useful for starting material analysis and for
the investigation of crystal growth technology. In addition, this
technique can be used in carefully designed experiments to isolate
the thermally related material factors which control defect nucleation and growth.

Single crystal characterization with EBIC microscopy on processed pn junction devices includes the thermal history of the
material and the process related factors such as oxidation surface
reactions, diffusion related stresses and the precipitation processes which result from metallic impurity complexing. Although
both structures can be subjected to DC parametric testing the pn
junction device contains the total processing history of the material.
Moreover, the standard bipolar and MOS process design rules can
be applied to the diffused junction structures to produce an
end product that can be electrically tested as a process compatible
operational device.

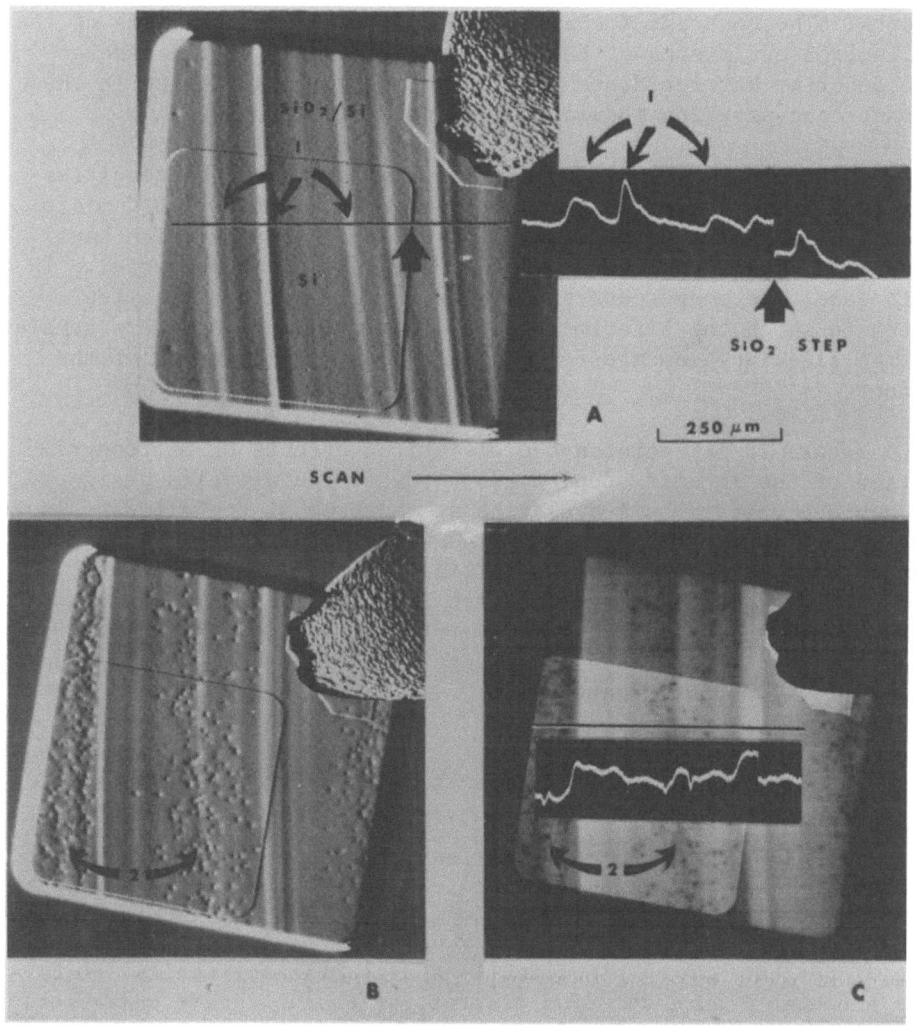

Fig. 27. Crystal inhomogeneities revealed by EBIC microscopy (a) dopant striations due to microsegregation, (b,c) microdefects resulting from impurity point defect complexes (swirl).

Figure 27 illustrates the application of a pn junction device for the detection of dopant inhomogeneities in the crystal [72]. The contrast striations shown in (a) correspond to regions where micro-segregation of the phosphorus dopant has occurred in the crystal as a result of interface instabilities during crystal growth. The contrast mechanism can be interpreted in terms of the collection efficiency of the diode as discussed for the case of the Schottky barrier diode [55] in which the local change in the dopant concentration modulates the diode depletion width W.

In Figure 27 (b) and (c) the crystal growth inhomogeneities referred to as 'swirl' are shown in a similar pair of EBIC micrographs. The small dark sites represent microdefects which have developed from impurity point defect complexes in the crystal. In this case, the image contrast arises from local recombination sites in which the lifetime of the minority charge carriers generated by the electron beam are reduced locally by the presence of the defect.

Figure 28 illustrates the electrical effects of microsegregation of the impurity dopant [73]. Here, the diode is designed to achieve uniform bulk avalanche at the base of the diffused layer. The local impurity dopant striations which result from microsegregation clearly establish the initial avalanche regions for the diode. The enhanced image contrast (brighter regions) results from the more efficient charge carrier generation in regions where the local depletion field is stronger than the surrounding areas due to local dopant enhancement.

E. Electrically Active Defects in PN Junction Devices

To demonstrate the characterization of electrically active defects in silicon pn junction devices we will utilize experimental results obtained from a study of oxidation induced stacking faults in silicon p^+n diodes [67]. The reverse I-V characteristics of p^+n diodes containing electrically active stacking faults (EASF) are shown in Figure 29. Two distict regions which are characteristic of most diodes containing EASF's are observed in the I-V characteristics shown in Figure 29. The two current-voltage domains are partitioned by the effective threshold voltage of the stacking fault. When the reverse voltage $V_R < V_{th}$, the diodes exhibit the normal I-V characteristics for a silicon diode at room temperature. When $V_R > V_{th}$ a voltage power low dependence is observed where $I_R \alpha V^n$, n = 4.75.

Using the experimental test procedure described previously, we would expect a bimodal frequency distribution for the reverse current histogram corresponding to a large number of devices tested.

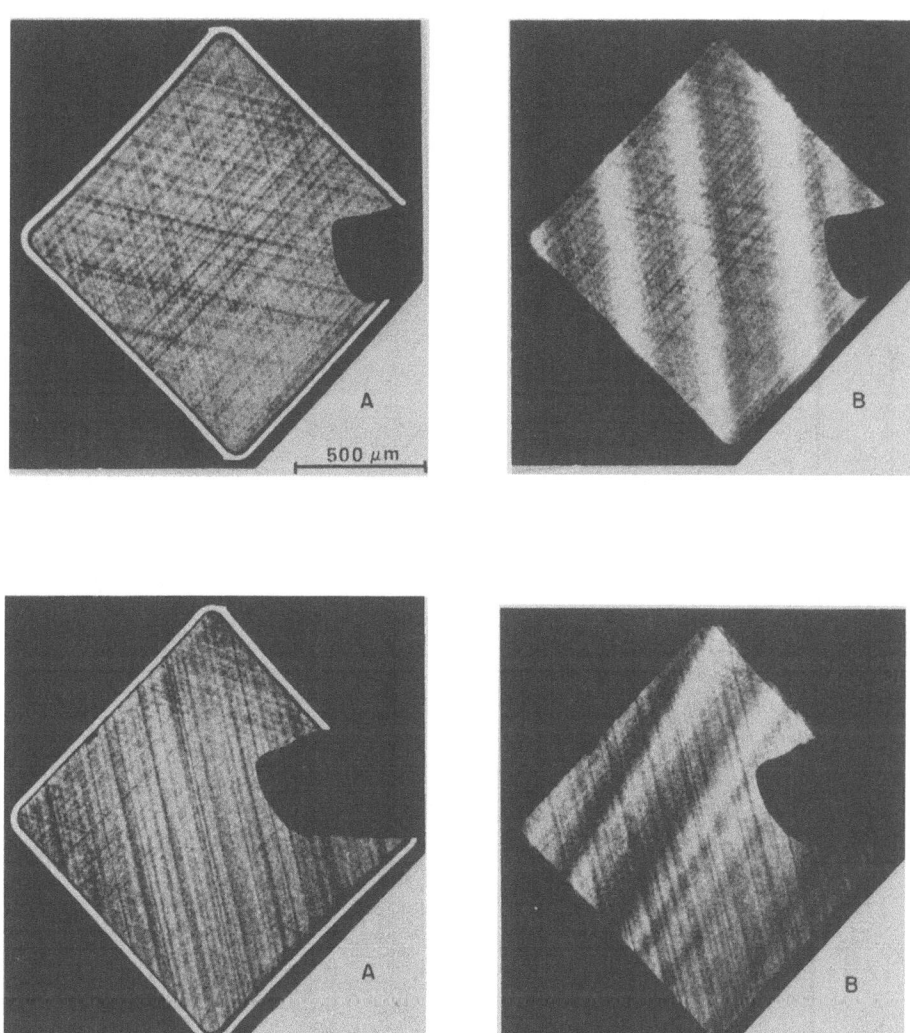

Fig. 28. Avalanche breakdown bands resulting from impurity dopant striations. (a) zero volt bias, and (b) 14 V BV_R

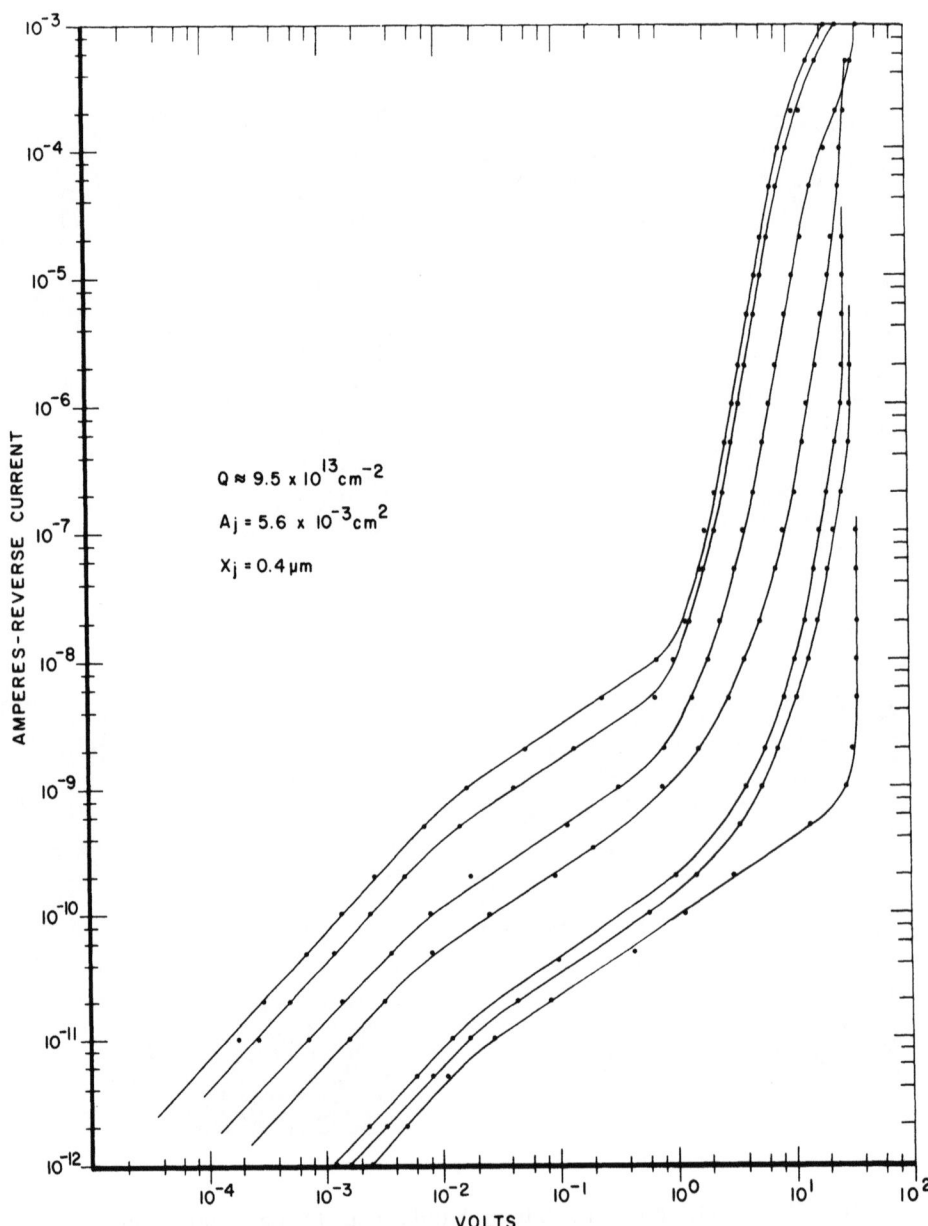

Fig. 29. Room temperature I vs. V characteristics of diodes containing electrically active stacking faults.

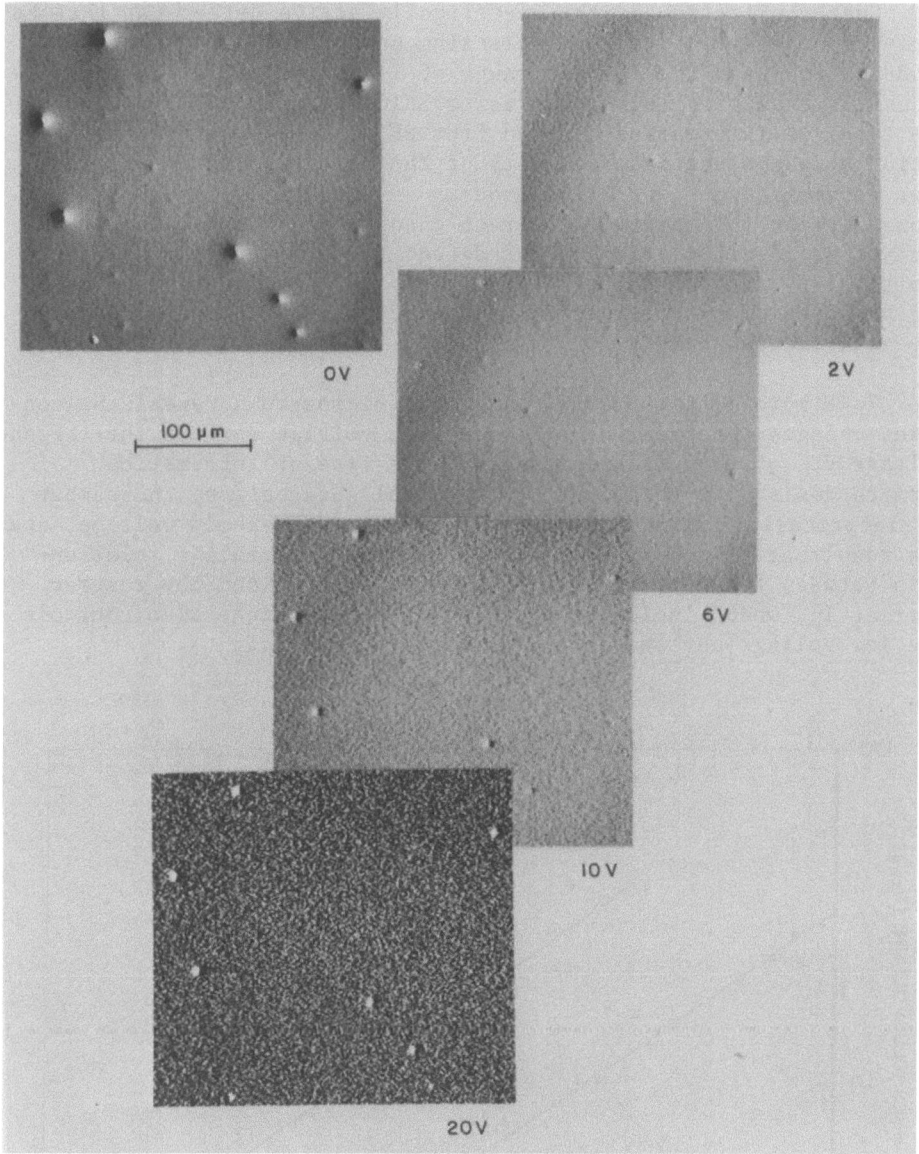

Fig. 30. Selected area EBIC display showing current enhancement effects at electrically active stacking faults.

This is particularly true if the test voltage is selected for maximum dispersion of the reverse current.

Characterization of the electrically active defect can then be achieved nondestructively by selecting an appropriate number of diodes which span the desired range of I-V characteristics and subsequently examining each diode using EBIC microscopy. Typically, the types of information desired from EBIC analysis are: verification of the electrical activity of the defect in terms of carrier recombination-generation under reverse voltage bias, correlations between the observed contrast changes and the I-V characteristics, identification of the defect type if within resolution limits, its size and distribution. This information is then used for the development of material and process defect models based on the experimental results.

In Figure 30 the charge collection micrographs reveal the contrast changes observed when the reversed voltage exceeds the threshold voltage V_{th} of the stacking fault. This type of information permits desired correlations to be established between the number of electrically dominant active sites, their threshold voltage, and the resultant I-V characteristics. Figure 31 reveals a relationship between the effective threshold voltage V_{th} and the reverse current I_R, which indicates that the defect g-r zone also controls the low voltage current in the diode where $V_R < V_{th}$.

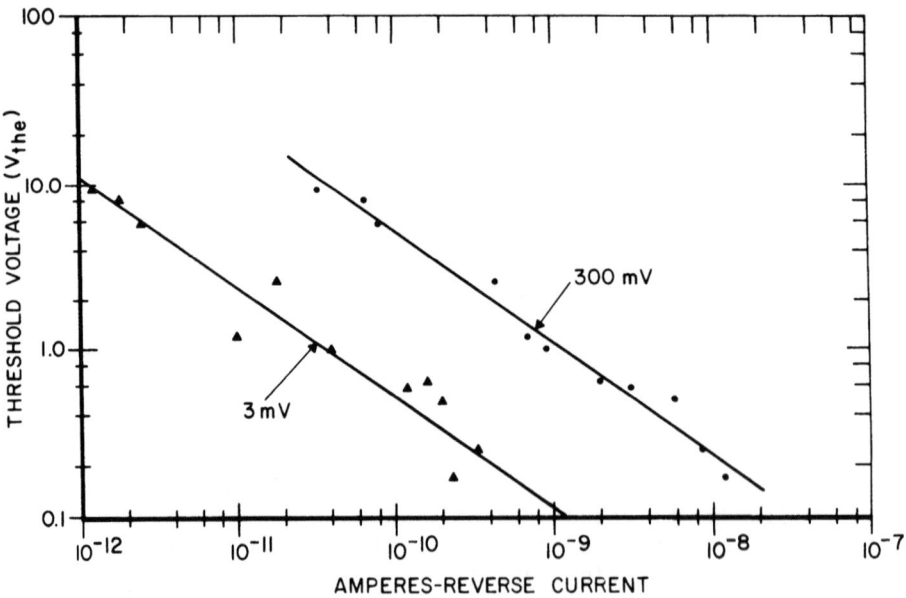

Fig. 31. Effective threshold voltage vs. reverse current.

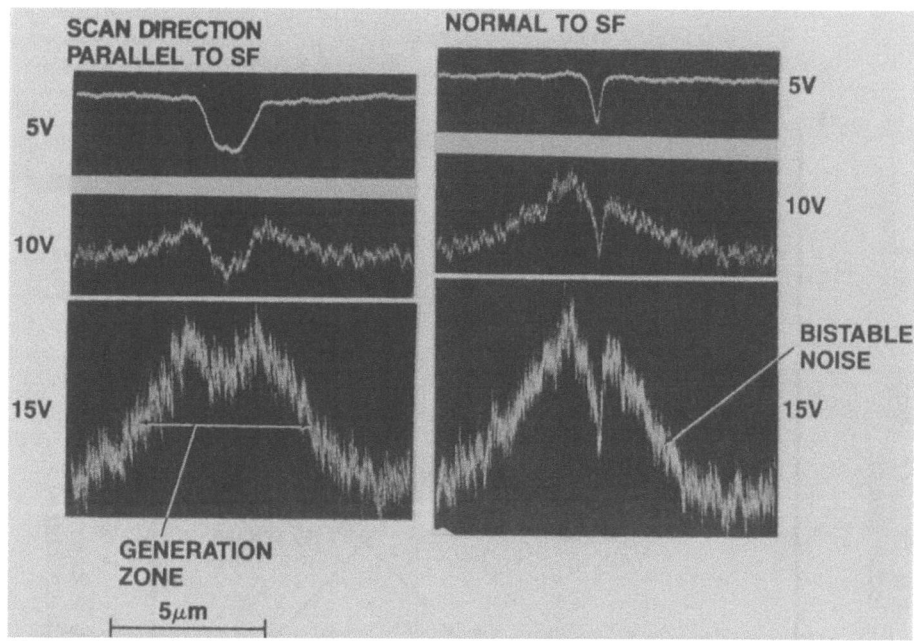

Fig. 32. Single line EBIC scan showing dimensions of stacking fault and generation zone.

More detailed examination of selected typical defects can produce information on the size of the electrically active zone surrounding the defect as shown in Figure 32. Figure 33 shows calculations based on experimental results obtained with EBIC microscopy which show the relationship between the effective lifetime or density of g-r centers surrounding the EASF and the reverse current I_R. When the data obtained with EBIC microscopy is combined with transmission electron microscopy a composite defect model emerges as shown in Figure 34.

F. Investigation of the Growth and Development of Defects During Processing and Their Subsequent Modeling for Process Control

The growth and development of defects during processing can in principle be monitored nondestructively on an appropriately designed test device. Ideally the test structure can be reinserted into the process sequence after EBIC microscopy for subsequent analysis at a later point in the processing sequence [74]. In a typical manufacturing operation this procedure is not feasible because of the lag time required for the analysis and its

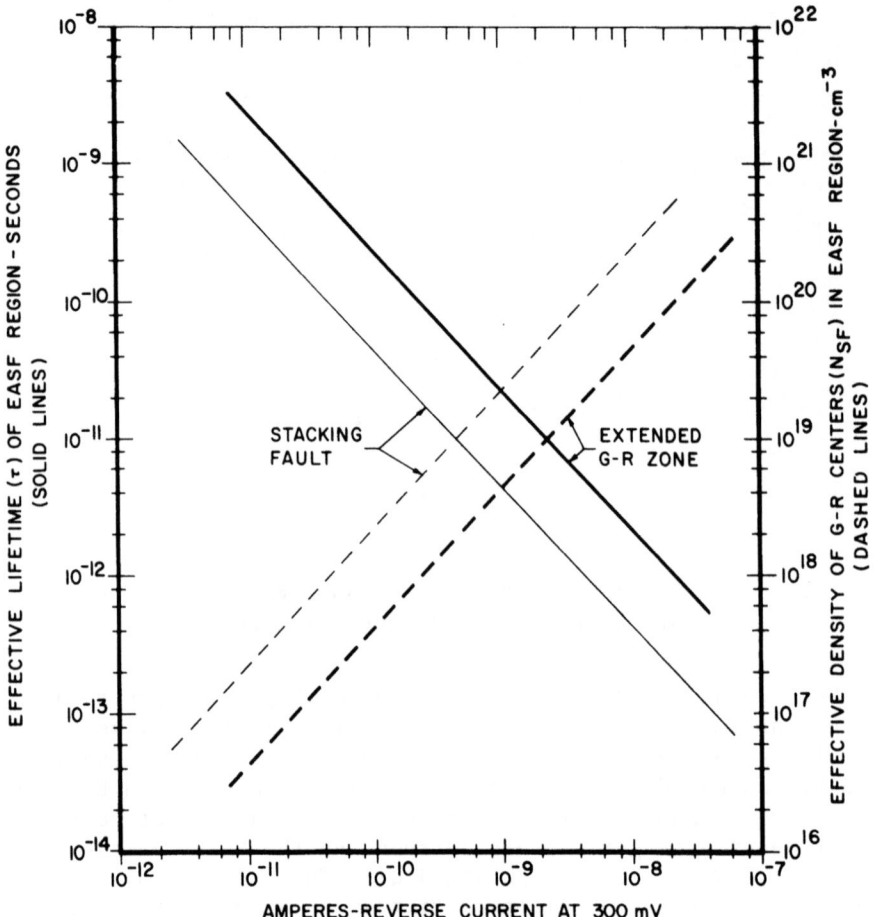

Fig. 33. Calculations showing relationship between effective lifetime, τ, and the density of G-R centers (associated with an electrically active stacking fault) vs. the reverse current at 300 mV.

incompatibility with manufacturing methods. However, the principles can be applied to a variety of experimental conditions.

The application of these concepts can be illustrated with experimental results obtained from an investigation of the growth

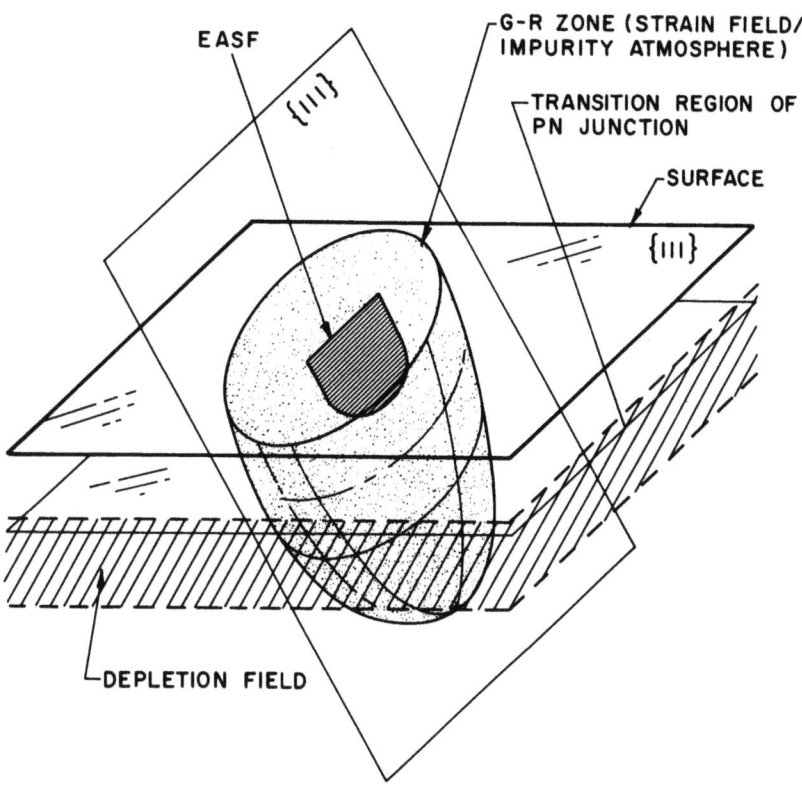

Fig. 34. Defect model of electrically active stacking fault.

and development of emitter pipes in bipolar transistors [68]. Emitter pipes represent a major mode of electrical degradation in shallow diffused bipolar transistors [75], and many techniques have been developed [76-78] to locate these defects although they are basically destructive techniques.

The electrical effects of emitter pipes are commonly detected electrically in the finished product as an excess component of current between the collector and the emitter in three terminal I-V measurements as shown in Figure 35. Using EBIC microscopy transistors containing emitter pipes were examined following preliminary electrical testing. Electrically active sites were located as shown in Figure 36, and characterized electrically to verify the three terminal measurements. Using information obtained

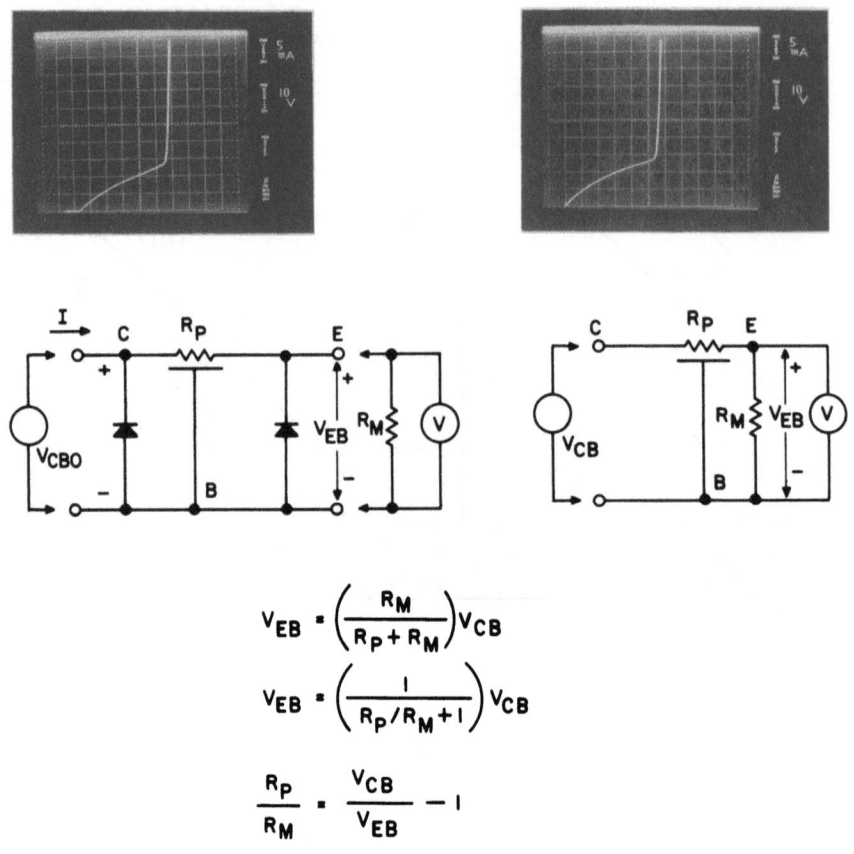

Fig. 35. I vs. V characteristics and equivalent circuit showing current limiting effects of emitter pipes.

from the EBIC displays and the I-V characteristics, three terminal DC equivalent circuits were synthesized as shown, for electrical modeling of the defect structure.

Selected areas of the emitter containing pipes were mapped and recorded on the EBIC micrographs. Subsequently using transmission electron microscopy, detailed analysis at high resolution was obtained which identified the defect structure and its crystallographic orientation in the device. The defect site was found to consist of radial dislocation arrays surrounding a small central core which is the remnant of earlier defects as shown in Figure 37.

SEM METHODS FOR CHARACTERIZATION OF MATERIALS

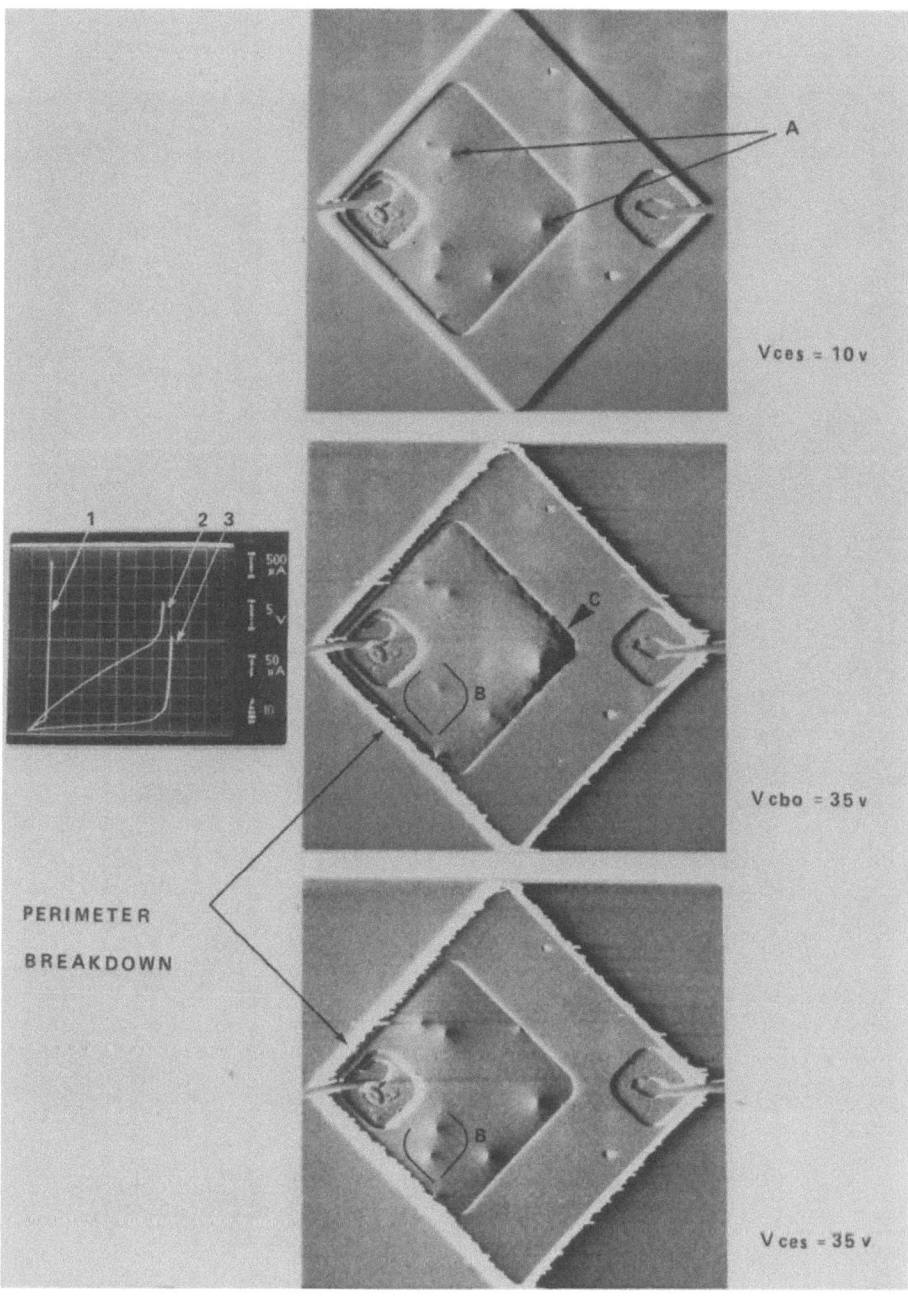

Fig. 36. EBIC micrograph showing the locations of typical emitter pipes.

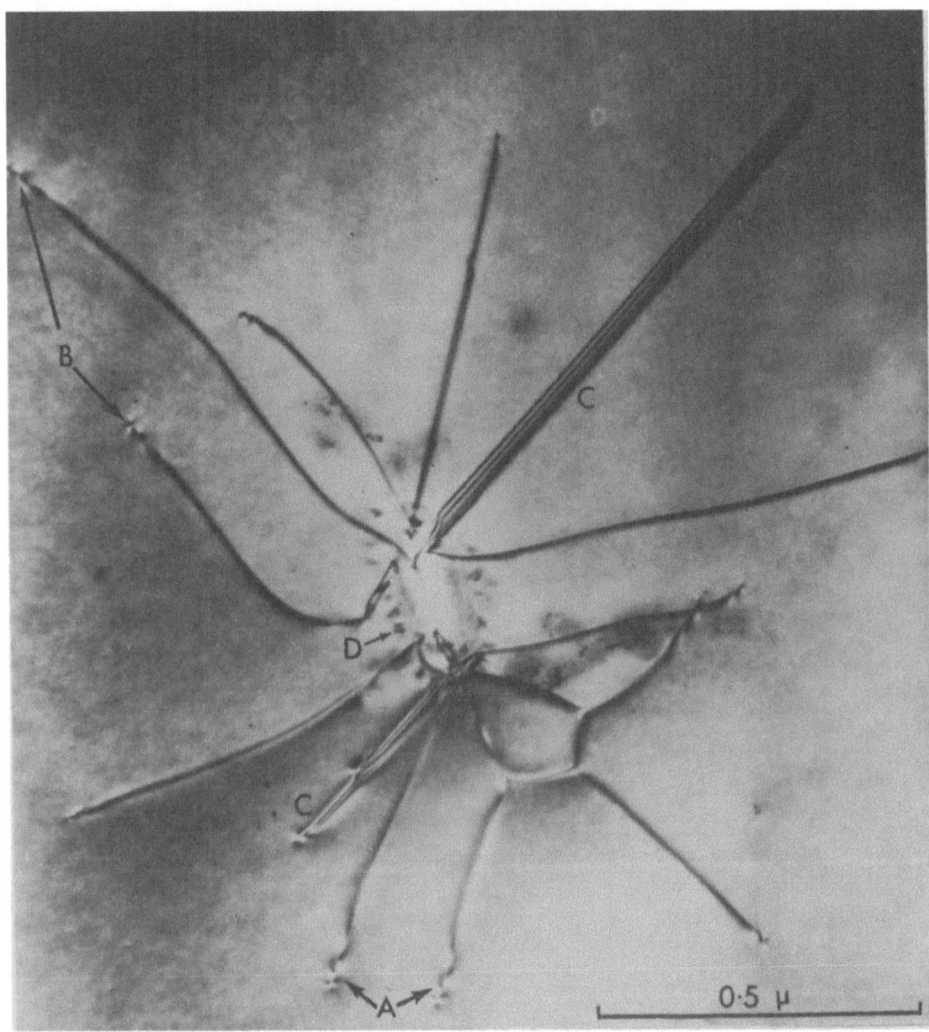

Fig. 37. Transmission electron micrograph of typical emitter pipe showing (a) and (b) dislocations, (c) extrinsic stacking fault, and (d) site of original precipitate.

To establish the origins of this defect, an experimental method was developed whereby partially processed wafers were examined with EBIC microscopy and reinserted into the process [74]. The information obtained using this investigative procedure produced direct correlation between emitter pipes and pre-existing base defects. Utilizing energy dispersive x-ray analysis on the base defect structures, Cu precipitates were identified. The precipitates were found to decorate partially dissociated stacking faults formed prior to the base diffusion.

The results of earlier investigations on defect nucleation and growth provided additional information to supplement the experimental results described above. The defect structure which constitutes an emitter pipe evolves by the progressive development during thermal processing of native defects formed during crystal growth. These mechanisms can be described as follows: the native defects postulated to be point defect complexes [79,80] and/or dislocation loops are converted into extrinsic stacking faults a result of oxidation [67]. The faults grow and/or function as dislocation sources as a result of heat treatment during the base diffusion process [81,82]. Impurities are introduced during the base diffusion forming precipitates which nucleate in the surface region of the device. The emitter diffusion further modifies the base defect which is now partially leached by chemical processes during photochemical processing prior to the emitter diffusion.

From these results a self consistent defect model emerges whereby latent crystal defects formed during crystal growth give rise to a terminal emitter defect which dominates the electrical characteristics of the bipolar transistor.

V. ELECTRICAL ANALYSIS OF INTEGRATED CIRCUITS

A. Electrical Analysis of IC's Using Voltage Contrast Microscopy

The information obtained from test devices using EBIC microscopy can be combined with conventional DC parametric and functional testing of complex integrated circuits to establish statistical relationships between product yields and defect mechanisms. Conventional electrical testing of complex IC's in wafer form is normally done with a computer controlled test system. These systems are designed for efficent testing using well designed test and data reduction software programs. Many of the systems now provide effective data graphic displays for the presentative of complex data, and provide a terminal control mode for interactive communication between the system and the test engineer.

Under normal circumstances, these test systems will produce sufficient information for thorough diagnostics of complex circuits. However, when additional information is required for functional analysis of complex IC's which cannot be obtained with conventional electrical test methods, the SEM can be used for nondestructive voltage measurements using SE voltage contrast microscopy.

Voltage contrast microscopy is currently used in many laboratories for 'qualitative' measurements of devices and integrated circuits. In recent years, a substantial effort has been made to develop methods for linearizing the voltage dependent secondary emission signal. These aspects of voltage contrast microscopy have been discussed in a recent review paper [11]. The dynamic measurement of voltage on operating devices and integrated circuits requires the application of time resolved measurement methods. There are basically two methods of detecting periodically varying surface voltages using a time resolved detection technique: stroboscopic scanning [83] and time sampling [84-86]. In stroboscopic scanning, the electron beam is strobed during the desired portion of the voltage waveform producing a voltage contrast display which corresponds to the surface voltage in phase with the gating period. The technique requires that the voltage waveform frequency be larger than the line scan frequency. The voltage contrast display can be obtained for any point along the fundamental waveform by changing the phase angle between the voltage waveform and the gating pulse.

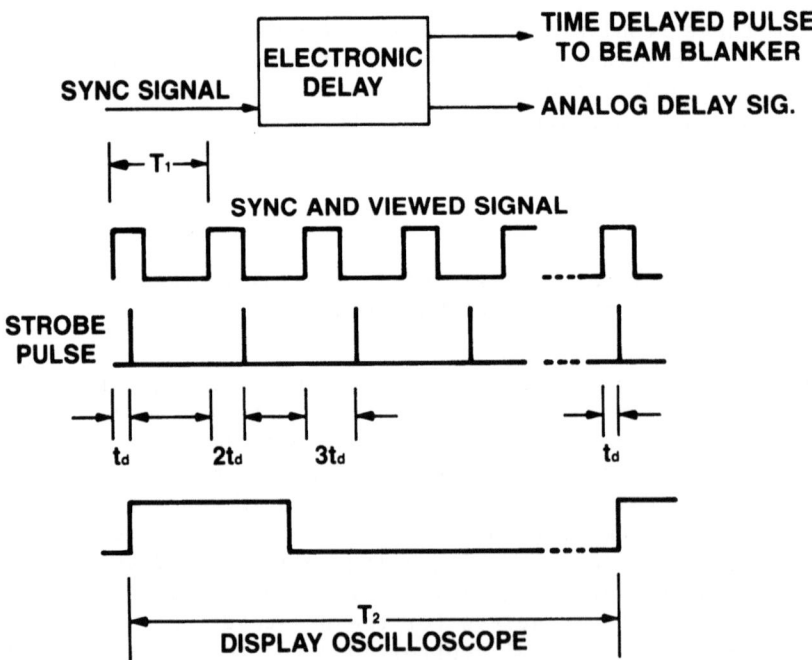

Fig. 38. Timing sequence corresponding to time sampling mode in SEM.

In the second method, the electron beam is held stationary at some point on the circuit and the beam is blanked with a timing waveform in a manner analogous with a sampling oscilloscope. The principle of this technique is shown in Figure 38. A sync signal is inserted into an electronic delay system which produces a delayed sync pulse and an analog ramp proportional to t_d. The delay period of the sync pulse is increased linearly with time until it reaches its maximum delay period and is then reset. The secondary electron signal provides the vertical waveform and the analog delay provides the horizontal ramp. The final display waveform represents the real time pulse T_1 on an expanded time scale T_2 to produce on time integrated noise free signal.

The secondary electrons emitted at the sample surface have an average energy ~ 10 eV. Consequently relatively small electric fields will influence their trajectories. The design of the secondary electron collector normally determines the voltage resolution obtained, neglecting ambient factor such as surface charging, stray electric and magnetic fields, etc. However, the surface voltage to collector current ratio is normally only linear over a small range of voltages for a conventional collector unless specially designed velocity analyzers are used.

Quantitative voltage contrast displays with the SEM require that the voltage dependent contrast effects are linearly dependent on the energy of the electrons leaving the surface. This assumes that the emitted electrons gain an amount of energy equal to the bias potential at the emission point. Ideally, for a constant secondary electron spectrum a change in the emission point voltage shifts the electron energy curve linearly along its energy axis.

Energy analyzers for secondary electron detection are of two general types: cylindrical mirror analyzers [87] and the retarding potential type [88]. Theoretical analysis [89] indicates that the retarding potential hemispherical analyzer should be capable of the greatest voltage resolution. Recent results obtained with a time sampling SEM using a beam current 10^{-11} amp have demonstrated a voltage resolution better than 100 mV with a measurement risetime of 100 ps using the hemispherical retarding potential analyzer design.

REFERENCES

1. V. E. Johnson, *SEM 1975, 763, IITRI, Chicago.
2. O. C. Wells, SEM 1972, 375, IITRI.
3. O. C. Wells, Scanning Electron Micrsocopy, MacGraw-Hill (1974).
4. H. Yakowitz, SEM 1975, 2 IITRI

---

*Scanning Electron Microscopy - Annual Conf. Proceeding edited by O. M. Johari.

5. R. Fitzgerald, Microprobe Analysis, (C.A. Anderson, Ed.), 1, Wiley (1973).
6. R. K. Hart, Scanning Electron Microscopy (M.A. Hayat, Ed.), IV, VI, Van Nostrand-Reinhold (1975).
7. N. C. McDonald, SEM 1971, 89, IITRI.
8. P. M. Petroff, et al, SEM 1978, 325.
9. C. E. Lyman, SEM 1978, 529.
10. P. R. Thornton, Scanning Electron Microscopy, Chapman and Hall, LTD (1968).
11. A. Gopinath, SEM 1978, 375.
12. D. B. Holt, Quantitative Scanning Electron Microscopy, 213-286, (D.B. Holt, et al, Eds.) Academic Press (1974).
13. D. W. Wittry, Microprobe Analsyis (C. A. Anderson,ed.), 123, Wiley (1973).
14. A. N. Broers, SEM 1974, 9, IITRI.
15. W. P. Dyke adn W. W. Dolan, Advances in Electronics and Electron Physics 8, 89-185, Academic Press (1956).
16. A. V. Crews, Quarterly Reviews of Biophysics, 3, 137-175, (1970).
17. P. Grivet, Electron Optics, 424, Pergamon Press, (1965).
18. T. E. Everhart, et al, Report ERL-66-11, Univ. of Calif. (May 1966).
19. V. E. Cosslett and R. N. Thomas, Brit. J. Appl. Phys. 15, 1283, (1964).
20. B. E. Schumacher, 1st Intern. Conf. on Electron & Ion Beam Science and Technology, 5-70, (R. BAkish, Ed.) Wiley (1965).
21. H. A. Bethe, M. E. Rose, and L. P. Smith, Proc. Amer. Phil. Soc. 78, 573 (1938).
22. T. E. Everhart, J. Appl. Phys. 31, 1483 (1960).
23. R. Shimizo and K. Murata, J. Appl. Phys. 42, 387 (1971).
24. H. Kanter, Ann. Phys. Lpz. 20, 144 (1957). Phys. Rev 121, 461(1961).
25. S. Thomas and E. B. Pattinson, J. Phs. D: Appl. Phys. 2, 2539, (1969).
26. A. E. Gruen, Z. Naturforsch, 12A, 89 (1957).
27. W. Ehrenberg and B. E. N. King, Proc. Phys. Soc., Londong, 81, 751 (1963).
28. M. Hatzakis, J. Electrochem. Soc. 116, 1923 (1969).
29. R. Shimizo, et al, J. Appl. Phys. 46, 4, 1581 (1975).
30. T. E. Everhart and P. H. Hoff, J. Appl. Phys. 42, 13, 5837 (1971).
31. S. M. Sze, Physics of Semiconductor Devices, Wiley-Interscience (1969).
32. A. S. Grove, Physics and Technology of Semiconductor Devices, John Wiley & Sons, (1967).
33. A. K. Johnscher, Principles of Semiconductor Device Operation, Wiley (1960).
34. J. L. Moll, Physics of Semiconductors, McGraw Hill (1964).
35. R. N. Hall, Phys. Rev. 87, 387 (1952); W. Schockley and W. T. Read, Phys. Rev. 87, 835 (1952).
36. W. Schockley, Bell Sys. Tech. J. 28, 435 (1949).
38. R. A. Laudise, The Growth of Single Crystals, Prentice Hall (1970).
39. L. E. Katz and D. W. Hill, J. of Electrochem. Soc. 125, 1151 (1978).

40. H. Graff, J. Hilgarth and H. Neobrand, Semiconductor Silicon 1977, 575, H. R. Huff and E. Sirtl, Electrochem. Soc.
41. J. Burtscher, Proc. of European Summer School, 63, Germany (July 1974).
42. T. Abe, V. Abe and J. Chikawa, Semiconductor Silicon 95, (H. R. Huff and R. R. Burgess, Eds.), Electrochem Soc. Inc. (1973).
43. J. Bloem, Semiconductor Silicon1977, 201, (H. R. Huff and E. Sirtl Eds.) Electrochem Soc.
44. W. R. Runyan, Silicon Semiconductor Technology, McGraw Hill (1965).
45. J. J. Lander, H. Schreiber, Jr., and T. M. Buck, Appl. Phys. Lett. 3, 206 (1963).
46. W. Czaja and G. H. Wheatley, J. Appl. Phys. 35, 2782 (Sept. 1974).
47. T. E. Everhart, O. C. Wells, and R. K. Matta, Proc. IEEE 1642 (Dec. 1964).
48. C. J. Varker, T. E. Everhart and A.J. Gonzales, 2nd Intern. Conf. on Ion Beam Science and Tech., New York (1966).
49. L. C. Kimmerling, et al, Semiconductor Silicon 1977, 468, Electrochem. Soc.
50. A. J. Gonzales, SEM 1974, 942, IITRI.
51. K. O. Leedy, Solid State Technology, 45, (February 1977).
52. H. J. Leamy, L. C. Kimmerling and S. D. Gerris, SEM 1978, 717.
53. P. Hoff and T. E. Everhart, IEEE Trans. on Electron Devices ED-17, 6, (June 1970).
54. R. Kullath, Ann. Phys. (Leipzig) 1, 357 (1947).
55. T. E. Everhart, I. C. Wells, and C. W. Oatley, J. Electronics and Control 7, 97 (1959).
56. G. T. Thomas, Transmission Electron Microscopy of Metals, John Wiley & Sons (1962).
57. K. F. J. Henrich, NBS Special Public. 298, U.S. Govt. Printing Office (1968).
58. J. I. Godsteing, H. Yakowitz, D. E. Newbury, E. Lifshin, J. Colby and J. Coleman, Practical Scanning Electron Microscopy, Plenum Press (1975).
59. L. Baldes, Proc. IRE 42, 420 (1954).
60. R. Holm, Electric Contacts Handbook, Springer, Berlin (1967).
61. J. R. Ehrstein, Spreading Resistance Symposium, NBS Maryland (June 1974).
62. E. Sirtl and A. Adler, Z. Metallkd, 52, 529 (1961); W. C. Dash, J. Appl. Phys. 27, 1193 (1956); F. Secco D'aragona, J. Electrochem. Soc., 119, 948 (1972).
63. M. W. Jenkins, J. Electrochem. Soc. 124, 757 (1977).
64. T. R. Cass and R. A. Burmeister, J. Elec. Mater. 3, 497 (1974).
65. A. J. R. deKock, S. D. Ferris, L. C. Kimmerling and H. J. Leamy, Appl. Phys. Lett. 27, 313 (1975).
66. H. J. Leamy, L.C. Kimmerling and S.D. Ferris, SEM 1976, 529.
67. C. J. Varker and K.V. Ravi, J. Appl. Phys. 45, 1, 272 (1974).
68. C. J. Varker and K. V. Ravi, Semiconductor Silicon 1977, 785, (H.R. Huff and E. Sirtl, Eds.) Electrochem. Soc.
69. Semiconductor Silicon 1973, (H.R. Huff and R.R. Burgess, Eds.) Part II Material Prep. and Characterization.

70. K.V. Ravi and C.F. Varker, Semiconductor Silicon 1973, 136, (H.R. Huff and R. Burgess, Eds.) ECS. Inc.
71. C.J. Varker and D.V. Ravi, Semiconductor Silicon 1973, 670 (H.R. Huff and R. Burgess Eds.) ECS, Inc.
72. K.V. Ravi and C.J. Varker, Appl. Phys. Lett. 25, 1, 69 (1974).
73. C. J. Varker, 9th Annual Proc. Reliability Physics, 55, (1971).
74. C.J. Varker and K.V. Ravi, Met. Trans. 4, 367 (January 1973).
75. J.E. Lawrence, Semiconductor Silicon 1973, (H.R. Huff and R.R. Burgess, Eds.)Electrochem Soc.
76. G. H. Plantinga, IEEE Trans. on Elect. Dev. ED-16, 394 (1969).
77. M. V. Kulkarni, et al, IEEE Trans on Elect. Dev. ED-19, 1098 (1972).
78. D. K. Seto, et al, 651, also W.K. Tice, et al, 639, Semiconductor Silicon 1973 (H.R. Huff and R.R. Burgess, Eds.) Electrochem. Soc.
79. A.J.R. deKock, J. Electrochem. Soc. 118, 1851 (1971)
80. P.M. Petroff and A.J.R. deKock, J. Crystal Growth 35,4 (1976).
81. K.V. Ravi, Trans. AIME 4, 681 (1973).
82. K.V. Ravi. Phil. Mag. 30, 1081 (1974).
83. G.S. Plows and W.C. Nixon, J. of Sci. Instr., (J. of Phys. E) 2, 1, 595 (1968).
84. G.S. Plows and W.C. Nixon, Microelectronics and Reliability 10, 317 (1971).
85. A. J. Gaonzales and M. W. Powell, IEDM Tech. Digest, 119, IEEE (1975).
86. K.G. Gopinathan, P.R. Thomas, A. Gopinath, and A.R. Owerns, Electr. Lett. 12, 501 (1976).
87. O.C. Wells and C.G. Bremer, J. of Phys. E. Sci. Instru. 1, 902 (1968).
88. W.J. Tee and A. Gopinath, SEM 1976, 595 IITRI.
89. A. Gopinath and W.J. Tee, SEM 1976, 603, IITRI.

Chapter 11

BACKSCATTERING SPECTROMETRY AND RELATED ANALYTICAL TECHNIQUES

M.-A. Nicolet

California Institute of Technology

Pasadena, California 91125

I. INTRODUCTION

Analytical techniques based on the interaction of collimated beams with solids can be classified by the nature of the incident beam and of the detected radiation. Many possibilities can be conceived, particularly in the realm of particles. For the analysis of solids in particular, three types of radiations have gained notable significance: electromagnetic radiation, electrons, and light ions (mainly hydrogen and helium). Table 1 shows the nine distinct types of analytical techniques which one can devise with these three types of radiations. The table lists in the appropriate combination the names or acronyms of those analytical techniques which presently exist and are used routinely or on a research basis in the solution of problems in solid-state physics. It is apparent from the table that the frame of techniques based on electromagnetic and electron radiation, or cross-combinations of the two, are most common. Some of those methods (SEM, photoemission spectroscopy) are the subject of topical sessions in this Institute (see Chapters 8 and 11).

Backscattering spectrometry (BS) makes use of particle beams both for the incident and for the detected ion and thus falls outside of this frame of traditional analytical tools. The different nature of this radiation as compared to electron or electromagnetic radiation is at the origin of the distinct and unique analytical capabilities of BS. Because of their mass, swift atomic projectiles react weakly with the very much lighter electrons. The reaction is much more energetic with the nuclei of atoms where the mass and the electric fields are strong enough to highly perturb the path of an atomic projectile. Consequently, BS is a tool which primarily senses the presench of atoms (nuclei). In contrast to neutrons, hydrogen and helium ions do interact with electrons also, albeit weakly. Consequently, the fast particles continuously lose energy as they penetrate into a solid. This loss limits the range of the particle in the solid to the order of micrometers at 1 MeV. But the fact that over the range of the projectile, a specific energy can be associated with each point in depth, provides BS with another useful property: depth perception. The combined sensitivity to atoms and to depth gives BS its very particular position among analytical techniques.

The presence of BS in a treatise on nondestructive techniques deserves comment. The collision of two atoms is inherently a violent process, and such violent collisions are the basic ingredient of BS. After the irradiation, the sample also contains within it additional hydrogen or helium atoms which are introduced during the exposure to the analyzing beam. To what extent the radiation damage and the presence of the foreign atoms are tolerable or intolerable modifications cannot be answered in general. For example, the leakage current of a helium-irradiated pn junction can increase manyfold, while the doping profile or the thin metal film used as a contact hardly change. The severity of damage must be discussed in connection with specific properties of a sample in mind. This issue arises with all analytical tools. Particular to BS is the high energy involved in an individual collision process. Questions of damage thus tend to arise sooner here than with low-energy tools such as LEED or optical reflectance, where the energy level of the radiation is 10^4 or 10^6 time smaller.

Part II of this Chapter is devoted to a review of BS. The treatment is cursory. Interested readers are referred to the book, Backscattering Spectrometry, W. K. Chu, J. W. Mayer and M-A. Nicolet (Academic Press, New York 1978) for details and references. The present treatment relies heavily on that text for figures and examples. Part III discusses briefly some related analytical techniques. Details on those techniques may be found in the book, Ion Beam Handbook for Material Analysis (J. W. Mayer and E. Rimini, Eds., Academic Press, New York 1977), which served as the major reference source for the present summary.

TABLE 1

The nine possible analytical methods combining a collimated incident beam and a detected beam of electromagnetic radiation, electron radiation, and atomic radiation. (A much more complete compilation of a similar kind is given by C.J. Powell in Surface Science, Vol. 1, p. 143-169 (1978).)

INCIDENT RADIATION

	Electromagnetic Waves	Electrons	Neutrons	Ions (hydrogen and helium)	Increasing Energy
Electromagnetic Waves	Reflectance IR Fluorescence Optical Ellipsometry UV Diffraction x-rays	E Probe Characteristics x-rays	Neutron Activation Analysis	PIX 2keV-2MeV Nuclear reactions ~ MeV	→

Electrons	Photoemission Optical
	                UV
	ESCA x-rays
		LEED 200 eV
		RED ≥2 keV
		AUGER ≥2 keV
		SEM ≥2 keV
		TEM ≥10 keV
		          ≥100 keV
			Neutron Activation Analysis	Nuclear reactions ~ MeV	→

Neutrons			Th. neutron scattering .2-2 keV		
Ions	Lu Probe		Neutron Activation Nuclear reactions Analysis ~MeV	LEIS (=ISS) .2-2 keV SPUTTERING 2-200 keV IμP SIMS BS 5-3 MeV	→

II. BACKSCATTERING SPECTROMETRY (BS)

A. Concept of the Method of Backscattering Spectrometry

BS rests on two basic processes: the large-angle scattering of two colliding nuclei and the energy loss incurred by swift particles in a medium. This section discusses these two processes briefly.

1. Large-Angle Scattering of Two Nuclei.

1a. Kinematic Factor K_M. Consider a light particle of mass m and energy E impinging on a stationary heavy particle of mass M (see Figure 1). If the two particles undergo an elastic collision and the light mass is deflected by an angle θ from its original direction of motion, the light mass will have a residual energy $K_M E$ which is less than its original energy, i.e., $K_M < 1$. The value of K_M follows from the conservation of energy and momentum:

$$K_M = \left(\frac{M^2 - m^2 \sin\theta + m\cos\theta}{M + m}\right)^2 \tag{1}$$

Note that K_M does not depend on the energy E before the collision;

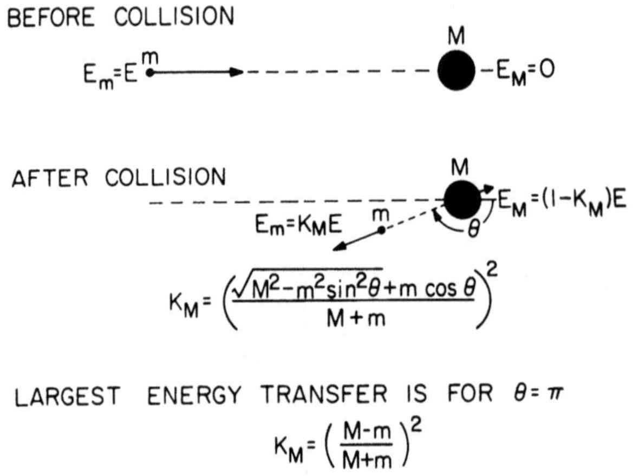

BEFORE COLLISION

$E_m = E \xrightarrow{m} \quad\quad\quad -E_M = 0$ (M)

AFTER COLLISION

$E_m = K_M E \xrightarrow{m} \quad\quad E_M = (1-K_M)E$ (M), angle θ

$$K_M = \left(\frac{\sqrt{M^2 - m^2\sin^2\theta} + m\cos\theta}{M+m}\right)^2$$

LARGEST ENERGY TRANSFER IS FOR $\theta = \pi$

$$K_M = \left(\frac{M-m}{M+m}\right)^2$$

Fig. 1. Schematic representation of an elastic large-angle collision of a swift light particle of mass m with a stationary heavy mass M. The kinematic factor K_M gives the fraction of the inital energy that resides in the light particle after the collision.

the fraction of the energy retained by the light particle after
collision is the same for any initial energy value E. The fraction
depends explicitly on the angle of deflection θ. For $\theta = \pi$, the
value of K_M has its lowest value, which is

$$K_M = \left(\frac{M-m}{M+m}\right)^2, \quad \theta = \pi \tag{2}$$

For $\theta = 0$, the value of K_M has its highest value, which is unity.
This case corresponds to no collision at all. The ratio K_M is a
function of the mass ratio m/M only. In an experiment, the projectile mass m is fixed and M varies with the sample, so that the
factor is viewed as a function of M, as the subscript indicates.
In backscattering spectrometry, as the name indicates, θ is
almost always close to π. As θ departs from π, say $\theta = \pi-\delta$, K_M
increases only quadratically in δ from its minimum value given
above, and can be approximated by

$$K_M \simeq 1 - 4\frac{m}{M} + \delta^2 \frac{m}{M} \tag{3}$$

for $\delta \ll \pi$ and $m \ll M$. A table of values for ^{4}He as a projectile is given in the Appendix.

In BS, the light mass in this model stands for the particle
of the analyzing beam (usually a proton or a ^{4}He$^+$ ion). The heavy
mass M represents an atom in the sample to be analyzed. The model
of an elastic collision between two isolated masses just described
is, of course, only an approximation. In reality, the heavy mass
is chemically bound to adjacent atoms in the solid, and these surrounding atoms affect the collision as well. Chemical binding
energies are of the order of magnitude of 1 to 10 eV. If the energy of the analyzing particle were of the same order of magnitude,
the chemical binding in the sample would have a major effect on the
energy and momentum balance of a collision. On the other hand,
these binding forces will be of lesser significance the higher is
the energy E of the analyzing particle. In BS, this energy is of
the order of 0.1 to 1 MeV. At these energies, the chemical binding
of one atom to other atoms is of vanishing consequence to the
scattering process and the simple model of two isolated masses
becomes an accurate description. By raising the level of the energy
in the collision process far above the level of chemical binding
energies, the kinematics becomes simple and can be accurately
described quantitatively. The drawback in so doing is that all information on the chemical state of the heavy mass M in the sample
is lost. Hence, <u>backscattering spectrometry provides information
only on the atomic constituents of a sample, not its chemical composition</u>.

The kinetic aspect of the collision answers the question "What is the energy after a collision?". The next topic answers the question "How frequent is a collision?".

1b. <u>Rutherford Scattering Cross Section</u>. To obtain an answer, the laws of conservation of energy and momentum must be supplemented by a description of the forces at work during the collision. Once more, there exists a simple model: the electrostatic repulsion of two bare nuclei (<u>Coulomb scattering</u>). This model is correct as long as the light projectile has sufficient energy to penetrate deeply into the first s electron orbit of the heavy atom M. If this atom has a nucleus of charge $+Ze$ and the projectile has a charge $+ze$, this condition requires that $E \gg (zZ) \cdot 13.58$ eV. For He ($z=2$) and the heaviest atoms ($Z \gtrsim 100$) this limit is at about 400 keV. A primary energy of several hundred keV is thus required to assure the validity of this model (in the worst case) and He as the projectile. On the other hand, the primary energy should not be as high as to permit the onset of nuclear reactions or resonances. These processes have threshold energies which may be as low as several hundred keV for protons and light nuclei, and are at about 2 or 3 MeV for ^{4}He. In the energy window in which the nuclear effects are absent and the full penetration of the electron clouds is assured, the model of Coulomb scattering describes the probability of a collision quantitatively.

This probability is expressed in terms of a differential scattering cross section $d\sigma_Z/d\Omega$, where $d\Omega$ is the differential solid angle of detection and $d\sigma_Z$ is the differential value of the cross section in that differential solid angle (Fig. 2). $d\sigma_Z/d\Omega$ has the dimension of an area per steradian whose meaning is based on a geometrical interpretation of the probability that the scattering will result in a signal at the detector. One imagines that each nucleus of an atom presents an area $d\sigma_z$ to the beam of incident particles. It is also assumed that this area is quite small and that the atoms within the target are randomly distributed in such a way that the small areas $d\sigma_Z$ of the nuclei do not overlap. Let S be the surface area of the sample illuminated uniformly by the beam. Then the total number of atoms eligible for a scattering collision in the sample and t is its thickness, assumed to be much smaller than the range of the projectile atom. The ratio of the total cross sectional area of all eligible atoms, $SNt \cdot d\sigma_Z/d\Omega$, to the area S actually exposed, is then interpreted as the probability that the scattering event will be recorded by the detector; that is, this ratio is set equal to $(dQ/Q)/d\Omega$. Here Q is the total number of particles which have hit the sample and dQ is the number of these particles which are scattered into the differential solid angle $d\Omega$. The division by $d\Omega$ of both $d\sigma_Z$ and dQ and $d\sigma$ with $d\Omega$, this geometrical contribution to the number of counts dQ is eliminated. The cross section defined in this way thus becomes a value per unit of solid angle, hence the name <u>differential</u> scattering cross section,

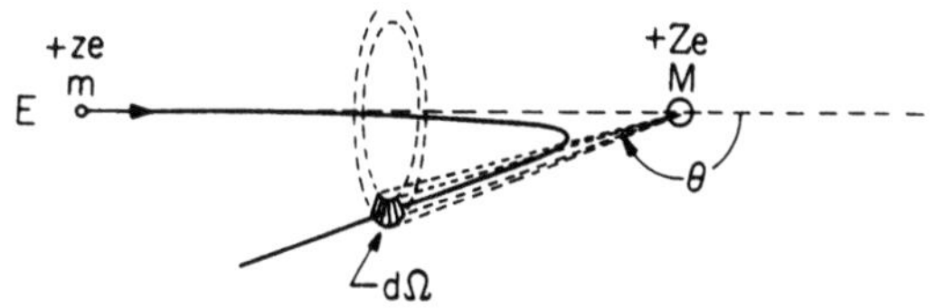

$$\frac{d\sigma_z}{d\Omega} = \left(\frac{zZe^2}{2E\sin^2\theta}\right)^2 \frac{\left[\cos\theta + \sqrt{1-\left(\frac{m}{M}\sin\theta\right)^2}\right]^2}{\sqrt{1-\left(\frac{m}{M}\sin\theta\right)^2}}$$

for $\theta = \pi$

$$\frac{d\sigma_z}{d\Omega} = \left(\frac{zZe^2}{2E}\right)^2 \left[\frac{1-(m/M)^2}{2}\right]^2 \propto \left(\frac{zZ}{E}\right)^2$$

Fig. 2. Schematic representation of an elastic large-angle collision of a swift light particle of mass m and charge ze with a stationary heavy mass M and charge Ze. The differential scattering cross section $d\sigma_Z/d\Omega$ gives the likelihood that the scattering results in the detection of the light particle in the differential solid angle $d\Omega$.

therefore the notation $d\sigma_Z/d\Omega$. For the model of simple Coulomb scattering, one finds (Darwin, 1914)

$$\frac{d\sigma_Z}{d\Omega} = \left(\frac{zZe^2}{4E}\right)^2 \cdot \frac{4(\cos\theta + \sqrt{1-(\frac{m}{M}\sin\theta)^2})^2}{\sin^4\theta \sqrt{1-(\frac{m}{M}\sin\theta)^2}} \qquad (4)$$

which is a value known as the <u>Rutherford (differential) scattering cross section</u>. It is a function of the scattering angle θ and of the mass ratio m/M. For any m/M ratio, the Rutherford differential scattering cross section has its lowest value at $\theta = \pi$, which is

$$\frac{d\sigma_Z}{d\Omega} = \left(\frac{zZe^2}{4E}\right)^2 [1 - (\frac{m}{M})^2]^2, \quad \theta = \pi \qquad (5)$$

In the vicinity of $\theta = \pi$ and for a small projectile mass (m << M) the cross section is approximated by

$$\frac{d\sigma_Z}{d\Omega} = \left(\frac{zZe^2}{4E}\right)^2 \left[\frac{1}{\sin^4\frac{\theta}{2}} - 2\left(\frac{m}{M}\right)^2\right] \tag{6}$$

The magnitude of the Rutherford scattering cross section is predominantly determined by the first factor $(zZe^2/4E)^2$.* For example, the differential cross section of 1 MeV ^{4}He ($z = 2$) in Si($Z = 14$) is 1.02 b for a unit steradian (1 b $\equiv 10^{-24}$cm^2). This is a typical order of magnitude for Rutherford differential scattering cross sections of ^{4}He in the MeV range. A monolayer of a solid contains about 10^{15} atoms/cm^2. A 1 MeV particle thus typically traverses many thousands of monolayers before being scattered out of its path by a large-angle Rutherford scattering collision. For a BS experiment, this result has the fortunate consequence that the intensity of the incident beam is virtually constant at any depth accessible to the analysis. The number of particles scattered out of the beam by large-angle collisions is so small as to be negligible. It is therefore unnecessary to correct for the loss of particles in the incident beam as well as in the outward flux of backscattered particles. This fact much simplifies the analysis of a backscattering spectrum.

In a backscattering experiment, z is known so that the cross section depends only on the atomic number Z of the atoms in the sample. The subscript in $d\sigma_Z/d\Omega$ stresses that fact. In the literature on backscattering spectrometry, it is customary to consider the average value of the differential scattering cross section over the solid angle of detection, i.e., $(1/\Omega) \int (d\sigma_Z/d\Omega) d\Omega$ where the integral extends over the solid angle of the detector. This average value is customarily abbreviated as σ and is referred to as a scattering cross section, which is both convenient and concise but formally an unfortunate usage.

2. <u>Energy Loss of Swift Particles in a Medium (dE/dx and ε)</u>.
Large-angle scattering collisions being highly unlikely, the fate of an average particles moving swiftly through a medium is controlled by other energy-loss interactions, termed <u>dE/dx losses</u>. The name refers to the transmission experiment sketched in Fig. 3 to measure this loss. As the thickness Δx of the target approaches sufficiently small values, the energy ΔE lost by the particles in transit through the target is proportional to Δx and one defines

---

* In formulae (4) - (6), the unit charge e is measured in cgs units, as is customary in the nuclear physics literature, and has the value e = 4.803 x 10^{-10} statC (1statC $\equiv$ (cm^{2}dyn)$^{1/2}$). A convenient constant to remember is that e^2 = 1.440 x 10^{-13}MeVcm, with which values of differential cross sections can be estimated quickly when E is given in MeV, as is usual.

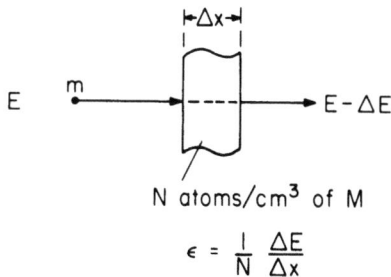

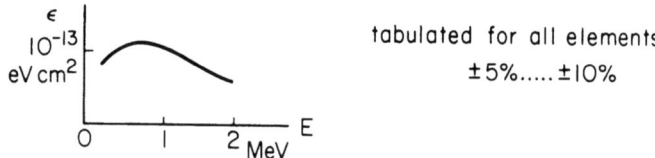

Fig. 3. Schematic representation of a transmission experiment to determine the dE/dx loss of a swift particle moving through a medium (top). The thickness Δx is assumed to be quite small, and in the limit, one defines

$$\varepsilon \equiv (1/N) \lim_{\Delta x \to 0} \frac{\Delta E}{\Delta x}$$

as the stopping cross section. The general dependence of ε on the energy is sketched also (bottom). Theoretical calculations of stopping cross sections in the energy range of interest to BS are difficult. ε is best considered as an experimentally determined quantity. Typical accuracies are 5%, and sometimes worse.

$$\lim_{\Delta x \to 0} \frac{\Delta E}{\Delta X} = \frac{dE}{dx}(E) \qquad (7)$$

for that particle and energy in that medium. For hydrogen and helium of MeV energies, the interaction responsible for the energy loss is that between the fast projectile and the electrons in the medium. The simplest model (Bohr, 1913) ignores all bonding and movement of the electrons in the sample, these binding energies being again very small compared to the kinetic energy of the particle. One then expects that the energy lost is proportional to the total number of electrons in the path of the particle, or $\Delta E \propto NZ\Delta x \cdot \pi b^2$, where N is the atom density in the sample, and b is an average distance between the moving particle and an interacting electron. In each individual encounter the electrostatic force ze^2/b^2 will act for a typical time b/v between the moving particle and a stationary electron, where v is the velocity of the particle. The momentum $(ze^2/b^2)b/v$ is thereby imparted to the electron, which picks up an energy of $(ze^2/bv)^2/2m_e$, where m_e is the free electron mass. From these elementary considerations one predicts an energy loss of

$$\Delta E = \Delta x \cdot NZ \frac{\pi}{2} \frac{(ze^2)^2}{m_e v^2} \qquad (8)$$

A quantum-mechanical treatment of Bethe (1930) gives

$$\frac{dE}{dx} = NZ \frac{4\pi(ze^2)^2}{m_e v^2} \ln(\frac{2m_e v^2}{I}) \qquad (9)$$

where the energy I is an average over various excitations and ionizations of the electrons in a target atoms. This equation suggests that dE/dx scales as

$$\frac{dE}{dx} = NZ(ze^2)^2 \, f(E/m) \qquad (10)$$

which is a useful rule to estimate how the dE/dx loss changes with the atomic number of the elements and with the energy and mass of the particle. For instance, the dE/dx of He (z = 2, m = 4) at 4 MeV is four times dE/dx of hydrogen (z = 1, m = 1) at 1 MeV for the same sample. Also, the dE/dx of Ni (Z = 28) is about twice that of Si (z = 14) for the same particle. Neither statement is quite accurate, but both are very useful. Typical values of dE/dx for 1 MeV ^{4}He range from 10 to 50 eV/Å.

Since dE/dx depends on the number of electrons that each atom contributes to the medium through which a particle passes, it is advantageous to introduce a quantity which measures the effectiveness of each atom in stopping the particle. From the measurement shown in Fig. 3, if ΔE is measured on an elemental sample of thickness Δx which has N atoms per unit volume, one defines

$$\varepsilon(E) = \frac{1}{N} \frac{dE}{dx}(E) \qquad (11)$$

which is called the stopping cross section of that particular element for that particle. The conventional units are eVcm^2/atom, usually abbreviated eVcm^2.* Typical values for 1 MeV ^{4}He go from 20 to 150 eV/(10^{15}atoms/cm^2), i.e. eV per monolayer traversed by the particle. An advantage of the stopping cross section ε over

---

*Another definition (used predominantly in the nuclear physics literature) sets $\varepsilon = \frac{1}{\delta} \frac{dE}{dx}$, where δ is the mass density (g/cm^3) of the element and ε is usually given in keVcm^2/g.

dE/dx is that the variations in dE/dx due to the mere change in atom density of the elements is removed and ε becomes a smoothly varying function of Z, making interpolations easy. Another benefit is that the stopping of compound media is easy to compute if one can assume that atomic stopping cross sections are additive; then one molecule of the compound A_mB_n contributes

$$\varepsilon^{A_mB_n} = m\varepsilon^A + n\varepsilon^B \qquad (12)$$

so that the compound has a dE/dx loss of $N^{A_mB_n}\varepsilon^{A_mB_n}$, where $N^{A_mB_n}$ is the volume density of the molecule in the compound. This statement of additivity of elemental stopping cross sections is known as Bragg's rule (Bragg and Kleeman, 1905). Minor violations have recently been reported in gaseous organic compounds and in compounds where one element is a gas in its elemental form (e.g. oxides or nitrides). In the energy range of 1-2 MeV which is typical for BS, the rule is usually applicable. The main uncertainty stems from errors in the elemental stopping cross sections. Theoretical calculations are difficult to perform and in the energy range of interest to BS, the elemental stopping cross sections are best considered as experimentally determined phenomenological quantities whose values are found in tables. The two books cited at the end of the Introduction (Section I) of this review list references and give tables. Their accuracy is rarely much better than 5%. In general, $\varepsilon(E)$ has a maximum around 0.5 to 1 MeV, as indicated in Figure 3.

At very much lower energies, the interaction between the moving particle and the stationary atoms in the sample changes in character and resembles the collision of two partially ionized atoms (so-called <u>nuclear stopping</u>, as opposed to the <u>electronic stopping</u> discussed so far). Nuclear stopping is of major significance to dE/dx of heavy particles in the MeV range or of light particles at much lower energies. The phenomenon of sputtering is closely connected with this mode of energy loss in media because the kinetic energy is transferred from the moving particle to the atoms (nuclei) of the samples rather than to the electrons. These phenomena are of minor consequence in the energy range typically used with H or ^{4}He in BS. Sputtering is the basis for those (destructive) analytical tools which rely on layer removal for depth profiling.

3. <u>Energy Straggling</u>. A second-order effect worth mentioning is that of energy straggling, because it places an ultimate limit to the accuracy with which energies can be determined in a backscattering spectrum. The statistical nature of the energy loss mechanism leads to a broadening of the energy distribution of

monoenergetic particles as they traverse a layer of thickness Δx. The variance in ΔE clearly will increase proportionally with the average number of the collisions over the length Δx, so that $<(\delta\Delta E)^2> \propto \Delta x$. The simple model of free stationary electrons adopted to calculate ΔE also yields an estimate for the variance of ΔE; one obtains

$$<(\delta\Delta E)^2> = \Delta x \cdot 4\pi NZ(ze^2)^2 \tag{14}$$

which is referred to as the Bohr value (Ω_B) of energy straggling. For ^{4}He a useful rule is that

$$\frac{\Omega_B}{\Delta E} \simeq \frac{\bar{E}}{\Delta E} \times 10^{-2} \tag{15}$$

where $\bar{E}$ is an average energy of the particle along the track through Δx. ^{4}He ions of 2 MeV undergoing a loss of 200 keV thus have a standard deviation of energy straggling which is about $\sqrt{10\%} \simeq 3\%$.

In addition to energy straggling, fluctuations of independent origins have to be considered as well (e.g., beam energy, noise in the detector and in the electronics). Their variances add quadratically to that of energy straggling, and together they form the actual system resolution of a BS system.

B. Concept of a Backscattering Spectrum

This section describes how the three basic concepts introduced so far (kinematic factor K_M, scattering cross section $d\sigma_Z/d\Omega$ and stopping cross section ε, or dE/dx) enter into a backscattering spectrum. We consider the ordinate of the spectrum first, which plots counts, and the abscissa last, which plots energy.

1. <u>Backscattering Yield (Counts) From a Slice Δx.</u> Consider a very thin slice of thickness Δx at some depth within a sample, and consider only atoms of a particular kind therein (mass M, charge Z, see Fig. 4). If Q is the total number of particles incident on the sample, a number ΔQ of the particles that emanate from the layer are scattered back into the solid angle Ω of the detector and produce a count in a channel of the multichannel analyzer. By the definition of the differential scattering cross section, we have

$$\Delta Q = Q\Omega \frac{\overline{\left(\frac{d\sigma_Z}{d\Omega}\right)}}{} N\Delta x = Q\Omega \cdot \sigma \cdot N\Delta x \tag{16}$$

where N now is the volume density of atoms of mass M and atomic number Z. If one channel of a multichannel analyzer is adjusted in its energy position and its energy width to record just the

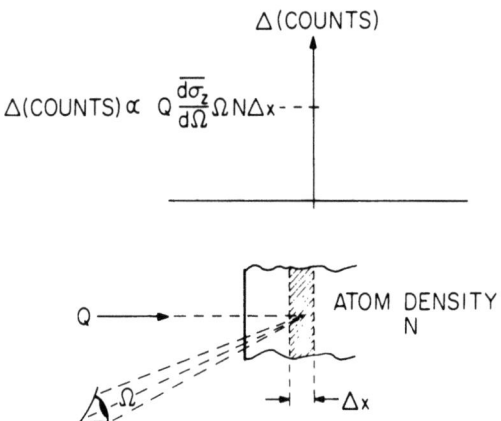

Fig. 4. Schematic representation of the connection that exists between backscattering events in a slice of thickness Δx at some depth in a sample, and the number of counts ΔQ in a channel of the backscattering spectrum.

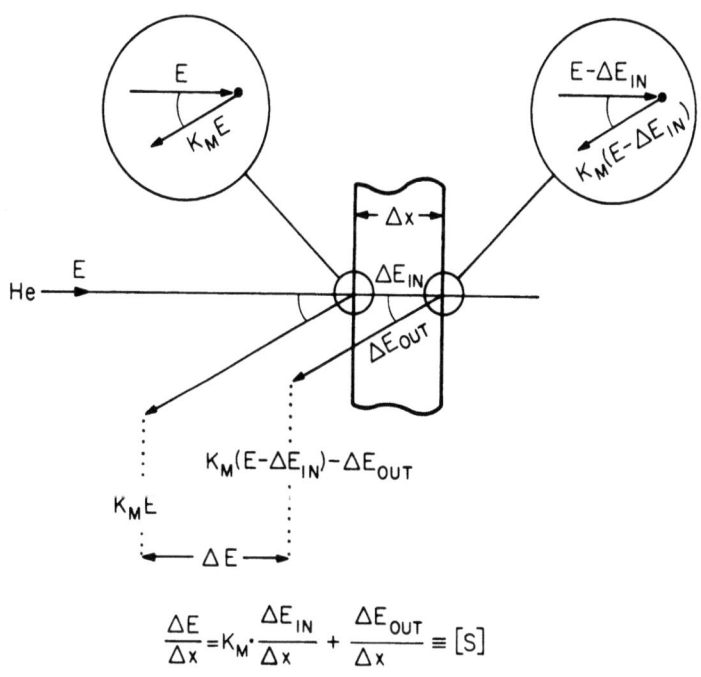

Fig. 5. Schematic representation of the energy spread ΔE of the particles backscattered at the front and back of a slice of thickness Δx, and the resulting definition of the energy loss factor $[S]$.

counts generated in the detector by those particles, that channel will register ΔQ counts and display that number as an ordinate. The essential point here is that ΔQ is proportional to $N\Delta x$, the number of atoms M per unit area in the slice Δx.

2. Energy Width of Particles Backscattered From a Slice Δx. Because the slice Δx has a finite width, the particles scattered back from atoms of mass M and atomic number Z in the slice have a finite energy width ΔE as well. From Fig. 5, one has

$$\Delta E = [S]\Delta x \qquad (17)$$

where

$$[S] \equiv K_M \frac{\Delta E_{in}}{\Delta x} + \frac{\Delta E_{out}}{\Delta x} \qquad (18)$$

can be calculated by looking up the energy loss in the medium in question for the analyzing particle at the energy E before the collision, for ΔE_{in}, and at the energy $K_M E$ after the collision, for ΔE_{out}. [S] is called the energy loss factor. The factor is tabulated for most elements and some fixed scattering angle (such as $170°$), and can be calculated with Bragg's rule for any material of known composition and a given scattering angle. In a real experiment, it is ΔE which is fixed, not Δx, because a multichannel analyzer subdivides the energy axis (x-axis) into equal increments of value ε, so that to each channel number, there is a corresponding slice of thickness

$$\Delta x = \varepsilon/[S] \qquad (19)$$

from which the backscattered particles produce the recorded counts. A detailed analysis reveals that the value of [S] changes slightly if the layer Δx varies in its position within the sample, but to a good approximation the effect can be neglected.

3. Average Energy of Particles Backscattered From a Slice Δx at Depth x. The average value of the particles scattered from the slice of thickness Δx depends on the depth x of that slice. This average energy will determine at what place in the energy scale (x-scale) the number ΔQ of counts will be displayed (see Fig. 6). To find this energy, the argument contained in Fig. 4 can be reapplied to the layer of finite thickness x, and one finds that now the energy ΔE is given by

$$\Delta E = [\bar{S}]x \qquad (20)$$

Because this ΔE is now finite and much larger than the ΔE of equation 17, the values of ΔE_{in} and ΔE_{out} have to be evaluated at some intermediate energy along the inward and outward track of the

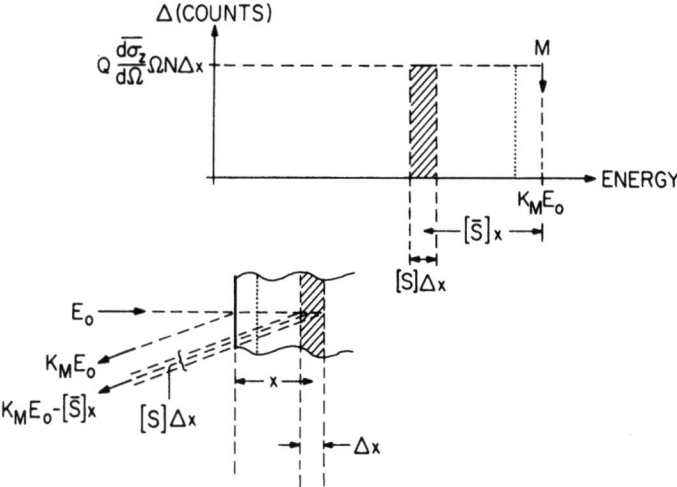

Fig. 6. The position of a thin slice of thickness Δx at a depth x in the sample is related to the position of the corresponding channel in the energy spectrum by the energy loss factor [$\bar{S}$].

particle, rather than at the fixed energies E and K_ME immediately before and after the collision, hence the notation [$\bar{S}$] instead of [S]. If the stopping cross sections were independent of energy, [$\bar{S}$] and [S] would be the same. In practical situations, the two quantities can differ by 10% or 20%.

Equation (20) states that the energy at which the counts ΔQ are displayed decreases linearly in first approximation with the depth x at which the slice is located. Since each channel contains a number of counts ΔQ which is proportional to the number of atoms NΔx in the corresponding slice, the trace of the yields ΔQ in successive channels along decreasing energies (decreasing x-axis) images the concentration of the atom of mass M and atomic number Z in the successive slices along increasing depth in the sample. This argument assumes that the coefficient

$$Q\Omega \left(\frac{d\sigma_Z}{d\Omega}\right)$$

in equation (16) is the same for all slices. Q and Ω are so to a high degree, but the differential scattering cross section depends on the energy E of the particle before scattering, and this energy decreased with depth in the sample. As a result, the spectrum of a uniform distribution of M in depth displays an increasing yield with decreasing energies (e.g., Fig. 11 and 12).

The surface of the sample ($x = 0$) coincides with the position $K_M E_0$ on the energy axis. This position is marked by an arrow labeled M in Figure 6 and is usually referred to as the edge of atom M on the energy axis. It corresponds to the energy of a particle impinging on the sample with the incident energy E_0 and scattered at the surface of the sample by an atom of mass M. Since E_0 is known, the position $K_M E_0$ can serve as an identification of the mass M of an atom producing a backscattering event at the surface, if M is not known.

C. Backscattering Spectrum of a Thin Compound Film

A thin compound film is a good example to demonstrate how the concepts discussed in the preceeding sections combine in a backscattering spectrum.

Consider a thin film composed of a uniform mixture of a two elements of mass M and m, such as a binary compound or a solid solution. To reduce the example to its simplest form, the substrate will be ignored. Assume for simplicity that the atomic concentrations of both elements are the same. The situation is represented graphically in Figures 7 (a) and (b). Figure 7(c) gives a schematic representation of the backscattering spectrum of this thin film. The signal of the heavy element of mass M has its highest energy at the edge $K_M E_0$ which is larger than the highest energy signal of the light element m, which is at the edge $K_m E_0$, because $K_m < K_M$ (eq. (3)). The counts at these energy edges come from particles which are scattered at the surface of the sample by atoms of mass M or m. Similarly, the low-energy edges of each signal corresponds to backscattering events at the rear (left) interface of the sample. Particles scattered there twice traverse the sample over its full thickness. Upon reaching the detector, their energy is lower by an amount ΔE_M and ΔE_m, respectively, than the energy of particles scattered at the surface. Particles scattered at intermediate depths will have intermediate energies, so that each signal is an image of the elemental distribution of the atoms M and m in the film. The signal widths ΔE_M and ΔE_m are generally unequal because after scattering, the energy of the scattered particle is larger after a collision with M than with m, and the dE/dx energy loss is generally a function of the particle energy. Along the outward tracks, the energy losses in the two cases are therefore different. One writes

$$\Delta E_M = [\bar{S}]_M^{comp} t, \quad \Delta E_m = [\bar{S}]_m^{comp} t \qquad (21)$$

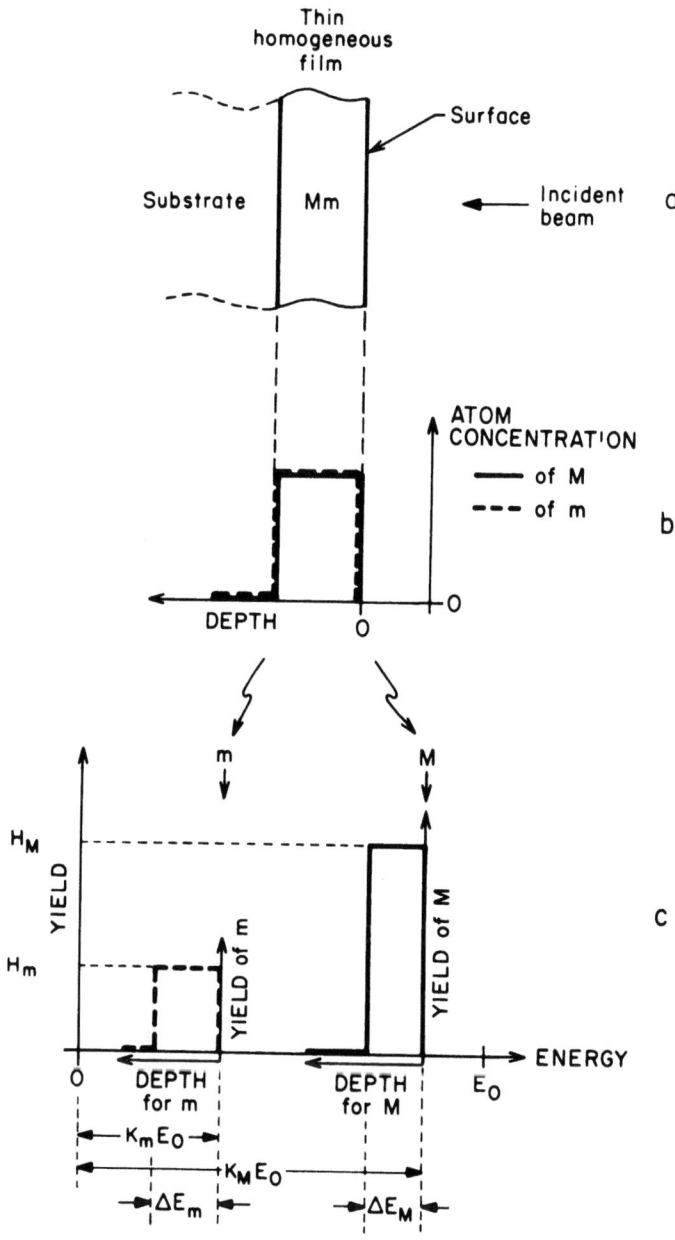

Fig. 7. A thin film of homogeneous composition Mm as sketched in (a) and (b) has a backscattering spectrum as sketched in (c) (see text).

where t is the film thickness and the two energy loss factors are given by equation (18) with the same ΔE_{in} but two different ΔE_{out}. Usually ΔE_M and ΔE_m are of very similar magnitude, i.e.

$$[\bar{S}]_m^{comp} \simeq [\bar{S}]_m^{comp}.$$

The yield of the two signals also differ, that of the heavy mass being the larger. This is so because the scattering cross section varies with the square of the atomic number Z of the element (eq. 5) so that

$$H_A = Q \Omega \sigma_A N_A^{comp} \Delta x \qquad (22)$$

from (eq. 16), where ΔQ is replaced by H_A to conform with usual notation, A stands for either M or m, N_A^{comp} is the number of atoms of species A per unit volume in the compound, and Δx is the width of a slice corresponding to one channel, i.e. (see eq. 19)

$$\Delta x = \varepsilon/[S]_A^{comp} \qquad (23)$$

If the yield is normalized, i.e. divided by the scattering cross section of the element, the ratio of the normalized yields then gives

$$\frac{H_M/\sigma_M}{H_m/\sigma_m} = \frac{N_M^{comp}}{N_m^{comp}} \cdot \frac{[S]_m^{comp}}{[S]_m^{comp}} \qquad (24)$$

(We follow standard practice and use subscripts of mass, atomic number for chemical symbols interchangeably, although σ is primarily a function of the atomic number, and K one of the atomic mass.) The ratio of the energy loss factors is usually close to unity, so that the normalized yield ratio directly gives the proportions of the heavy and light atoms in the film. This atomic ratio, estimated by neglecting the difference in the stopping factors, can be used to estimate the ratio of the [S] factor with which one can then obtain a better value of N_A^{comp} and N_B^{comp}. A further iteration is rarely necessary. In practice, the height H_M of a spectrum is not determined on a single channel; rather, an average over a number is used. Graphical interpolation is usually adequate because of the significant statistical variations in the counts of adjacent channels.

There is a second way to extract the ratio N_M^{comp}/N_m^{comp} from the spectrum by considering the total number of counts, A_M and A_m, contained in the two signals. Assume that the width ΔE_M corresponds to n channels, that is $\Delta E_M = n\varepsilon$. If each channel has the same height H_M as sketched in Fig. 7(c), the total A_M is

$$A_M = Q\Omega\sigma_M N_M^{comp} \frac{\varepsilon}{[\bar{S}]_M^{comp}} \cdot \frac{\Delta E_M}{\varepsilon} \tag{25}$$

but ΔE_M is $[\bar{S}]_M^{comp} t$ (eq. 21), so that

$$A_M + Q\Omega\sigma_M N_M^{comp} t \frac{[\bar{S}]_m^{comp}}{[\bar{S}]_M^{comp}} \tag{26}$$

The last ratio of [S] factors is unity if one neglects the energy dependence of dE/dx, so that to first order, that factor can be ignored; hence

$$\frac{N_M^{comp}}{N_m^{comp}} = \frac{A_M/\sigma_m}{A_m/\sigma_m} \tag{27}$$

To neglect the ratio of [S] factors in eq (26) usually introduces smaller error than in eq. (24) and eq. (27) tends to give the better results.

If the backscattering system is characterized in absolute terms meaning that the solid angle Ω and the total incident dose Q are known absolutely, it is also possible to obtain the number of atoms M per unit area, $N_M^{comp}t$, in the compound. To calibrate a backscattering system absolutely to a few percent is tedious. Note also that the measurement of Nt does not provide the thickness of the film, but the number of atoms per unit area. The thickness follows from the measurement only if the atom density is known independently.

A similar remark applies to eq. (21). The [S] factor there is given by dE/dx along the incident and the outward tracks (eq. 18), and dE/dx depends on $N\varepsilon(E)$ on each track (eq. 11). ΔE_M thus depends on the product $N_M^{comp}t$, which is independent of the spatial density of the atoms M and depends only on the number of atoms per unit area. To obtain the physical thickness of the film from ΔE_M requires an independent knowledge of the atom density N_M^{comp}.

D. Backscattering Spectrum of a Thick Compound Sample

If the film in Fig. 7(a) increases in thickness (or, more correctly stated, if the number of atoms per unit area in the film increases), the widths ΔE_M and ΔE_m in the spectrum of Fig. 7(c) increase also toward decreasing energies. If the sample becomes too

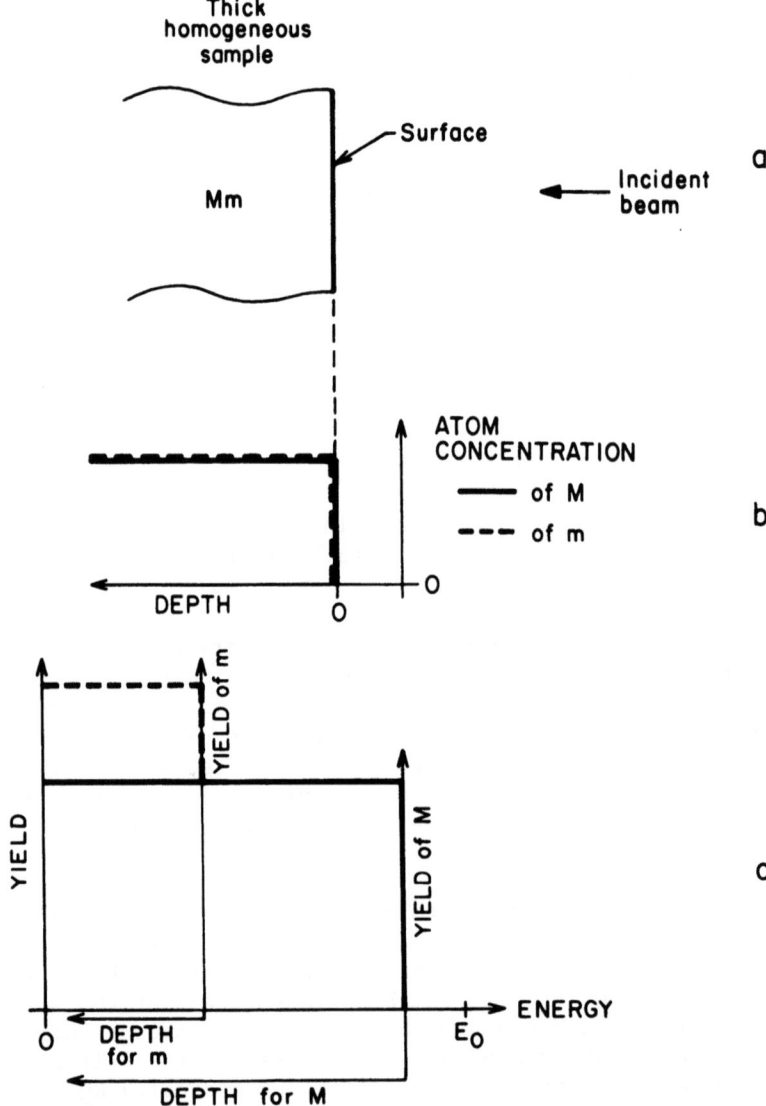

Fig. 8. A bulk material of homogeneous composition Mm as sketched in (a) and (b) has a backscattering spectrum as sketched in (c). (See text)

thick, particles may lose all their energy within the sample and the spectrum will extend to zero energies. (In practice, the first few 100 keV of the spectrum are dominated by spurious signals and must be discarded.) A schematic representation of this situation is shown in Figure 8.

Because the signal of each element now extends to zero energy, it is no longer possible to identify corresponding areas A in each signal, and atomic ratios can be derived from the signal heights only (eq. 24). Another observation is that in spectra of thick samples the signals of individual elements tend to overlap and add. Since light elements give lower yields recorded at lower energies than heavy elements, the sensitivity to light elements is quite poor when heavy elements are present as well. For example, oxygen can be detected well in Si only if it is present in chemical amounts (e.g. SiO_2 or SiO). The sensitivity to heavy impurities in light hosts is much better. The practical limits lie at about 10^{18} atoms/cm^3. For metallurgical-type studies, this sensitivity is most useful. Unfortunately, the range of interest to dopants in semiconductors is out of reach for BS.

E. Examples and Applications

1. **Surface Impurities**. As a first example, we present in Fig. 9 a schematic energy spectrum of ^{4}He backscattered from a light substrate with Cu, Ag and Au on the surface, each in the amount of about 10^{15} atoms/cm^2. This is of the order of one monolayer of surface coverage. The spectrum was taken with a ^{4}He$^+$ beam of incident energy E_0 = 2.5 MeV. The upper abscissa gives the energy scale of the backscattered ^{4}He particles. The lower abscissa gives the mass M associated with the positions $K_M E_0$ for the three elements (eq. 1) Note that the mass-to-energy conversions established via K_M is unique, but nonlinear. Au is the only element in this example which has only one stable isotope and produces only one signal in the spectrum. The two signals of the Ag isotopes cannot be distinguished, because the resolution is too coarse. The signals of the two Cu isotopes are just barely resolved. The area under each peak is proportional to the number of atoms per unit area and the scattering cross section of the element (which is known). Since the surface coverage is about equal for all three impurities, the size of the signals approximately reflects the change in cross section (dashed line), which is $A^2_{Au}:Z^2_{Ag}:Z^2_{Cu} = 79^2:47^2:29^2 \simeq 62:22:8.4$. To determine the exact ratio of atoms/cm^2 between these elements, one divides the area of the signals through the respective scattering cross sections and obtains quantitative results without using standards of calibration. For example, the signals of the two Cu isotopes directly indicate their relative abundance, and which of the

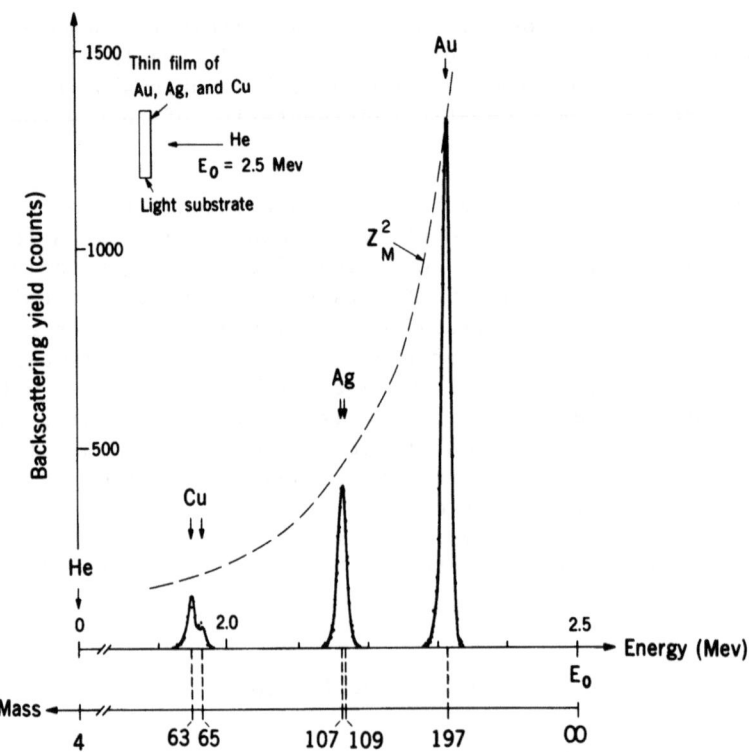

Fig. 9. Energy spectrum of 2.5 MeV ^{4}He$^+$ ions backscattered from Cu, Ag and Au atoms deposited on a light substrate (signal not shown) in approximately equal amounts of about 10^{15} atoms/cm^2 ($\sim$ one monolayer). The width of the signal of the monoisotopic Au is determined by the system resolution. The change in the scattering cross section follows the Z_M^2 dependence shown as a dashed line. The second abscissa gives the positions of $K_M E_0$ for the five isotopes (eq. (1) with $\theta = 170°$).

two is the more frequent one. (That ratio is known to be ^{63}Cu:^{65}Cu = 69%:31%.) As the figure shows, the heavy impurities on the surface of a light substrate can readily be detected in sub-monolayers amounts.

2. Impurity Distribution in Depth. Figure 10 shows an energy spectrum of 2 MeV ^{4}He backscattered from a wafer of Si($Z = 14$, $M = 28, 29$ and 30) implanted with As($Z = 33$, $M = 75$). The As signal appears at higher energies than the signals of the three Si isotopes and is thus clearly visible in spite of the relatively

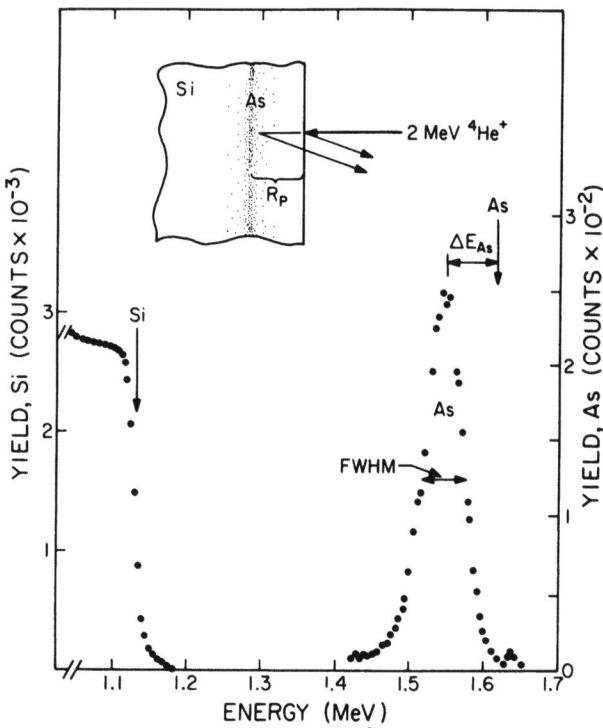

Fig. 10. Energy spectrum of 2.0 MeV ^{4}He$^+$ ions backscattered from a Si wafer implanted with 1.2×10^{15} As ions/cm^2 of 200 keV. The vertical arrows mark the energy position $K_{Si}E_0$ and $K_{As}E_0$ of ^{4}He particles scattered from surface atoms of ^{28}Si and ^{75}As ($\theta = 170°$, C = 5.0 keV/channel).

small concentration. That As atoms are located below the surface is evident from the fact that the As signal appears at energies below that of the As edge, which is at $K_{As} \cdot E \times 2$ MeV = 1.627 MeV for $\theta = 170°$. The signals ot the main ^{28}Si isotope, however, is located at the corresponding edge (E = 1.129 MeV for $\theta = 170°$). By assuming initially that the As concentration is relatively low, one can neglect the presence of the As to calculate the energy losses ΔE_{in} and ΔE_{out} and obtain the depth-to-energy conversion for the As signal, i.e. one uses the stopping cross section of pure Si. The shift of the As peak by ΔE_{As_0} = 68 keV below the As edge then corresponds to a range of 1430 Å if one assumes a bulk value for the atom density of Si in the implanted layer. Since the energy-to-depth conversion is closely linear, the As signal reveals without computation that the As profile in depth is nearly

Gaussian and that the full width half maximum is of the same order of magnitude as the average range of the As ions.

The concentration of the As at the maximum can be estimated by comparing the normalized height of the As signal to that of the Si signal; with eq. 24 one gets

$$\frac{N_{As}^{Si}}{N_{Si}^{Si}} = \frac{H_{As}/\sigma_{As}}{H_{Si}/\sigma_{Si}} \frac{[S]_{As}^{Si}}{[S]_{Si}^{Si}} \tag{28}$$

and obtains $N_{As}^{Si} = 8.3 \times 10^{19}$ atoms/cm^3 using $N_{Si}^{Si} = 4.98 \times 10^{22}$ atoms/cm^3 for the atom density of Si.

To find the total number of implanted As ions, the total number of counts in the As signal, as given by eq. (26) is compared with the height of the Si signal as given by Eqs. (22) and (23)

$$(Nt)_{As}^{Si} = \frac{A_{As}/\sigma_{As}}{H_{Si}/\sigma_{Si}} N_{Si}^{Si} \frac{\xi}{[S]_{Si}^{Si}} \tag{29}$$

where $[S]_{As}^{Si}$ and $[\bar{S}]_{As}^{Si}$ have been set equal (a very good approximation). One finds $(Nt)_{As}^{Si} = 1.2 \times 10^{15}$ As/cm^2.

Note that these quantitative results are derived without recourse to independent standards or an absolute calibration of the system. It is the signal of the Si wafer, whose properties are known, which serves to calibrate the As signals. Since both signals are recorded simultaneously, most of the systems parameter (Ω, Q, electronics settings, etc.) cancel out in the comparative analysis.

The spectrum of Fig. 10 conveys an idea about the sensitivity of backscattering spectrometry. Neutron activation analysis or secondary ion mass spectrometry can be (much) more sensitive than backscattering spectrometry, but backscattering spectra can be interpreted quantitatively without recourse to standards. The sensitivity is much reduced for impurities lighter than Si because the impurity signal is then superimposed on the Si signal. The stattistical variations in the counts of the Si signal will drown a superimposed signal which is small in amplitude. As a rule of thumb, the limit of detectability for a bulk impurity which is heavier than the host lattice is

$$\frac{N_{impurity}}{N_{host}} \geq \frac{Z_{host}^2}{Z_{impurity}^2} \times 10^{-3} \tag{30}$$

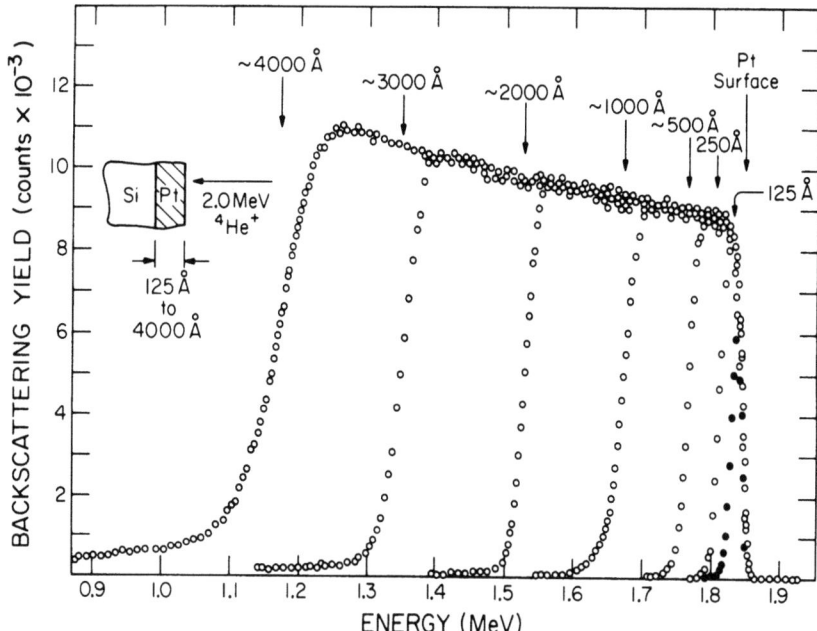

Fig. 11. Superimposed energy spectra of 2.0 MeV ^{4}He$^+$ ions backscattered from Pt films of various thicknesses to show the closely linear relationship between the width of the Pt signal and its thickness. The signals of the Si substrates are not shown.

3. <u>Thickness Measurements</u>. BS is excellently suited for the measurement of the thickness (actually: of the number of atoms per unit area) of films in the range of 200 to 5000 Å (with ^{4}He), or to 10,000 Å (with ^{1}H). Figure 11 shows spectra of 2.0 MeV ^{4}He ions backscattered from Pt films of various thicknesses. Several spectra are plotted in the same figure to illustrate the relationship between the energy width of the signal and film thickness. The two are closely proportional (eq. 21) and follow the relationship ΔE_{Pt} = 149 eV/Å ± 5% over the whole thickness range shown. The inaccuracy is determined mainly by the errors in the stopping cross sections of Pt used for the evaluation of the [S] factor.

A second example is given in Figure 12, where the backscattering spectra of three films of different materials and roughly comparable thickness are shown in superposition. The spectra were taken with the same total dose Q of incident ions, so that the difference in the yield of the signals reflects the difference in the scattering cross sections of Ti(Z = 22), Pd(Z = 46) and Pt(Z = 78). The vertical arrows give the positions $K_M E_0$ of the three elements and the

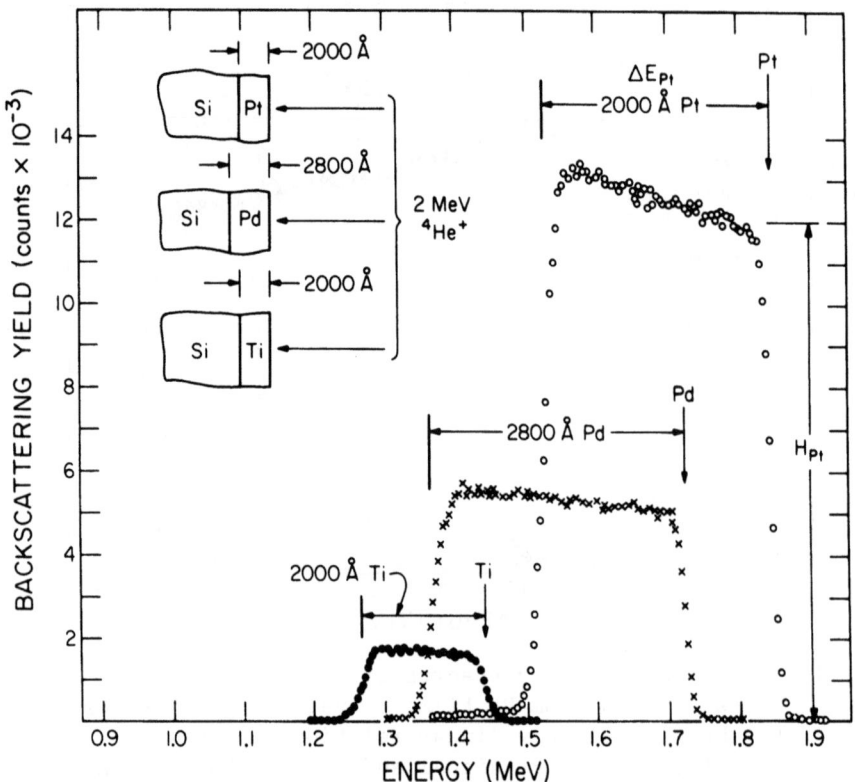

Fig. 12. Superimposed energy spectra of 2.0 MeV ^{4}He$^+$ ions backscattered from three different thin films. The spectra were recorded under unchanged systems conditions, so that the differences in the three spectra reflect those in the scattering cross sections (yields), in the mass (positions of energy edge given by the downward vertical arrows), and in the energy loss factors [S] (signal widths). The substrate signals are not shown.

width ΔE of the three signals is proportional to the [S] factors of the element as given by the dE/dx losses along the incident and backward track across the films.

4. <u>Bulk Composition</u>. If the composition of the first few thousand angstroms of a material are representative of the bulk composition, a backscattering analysis will yield quantitative ratios of the atomic composition of the bulk. Figure 13 gives the spectrum of a garnet whose composition was known to consist of X_8O_{12} molecules, where X was a combination of Fe, Ga, Y and Eu. The normalized yields of these four elements (dashed steps) provided the

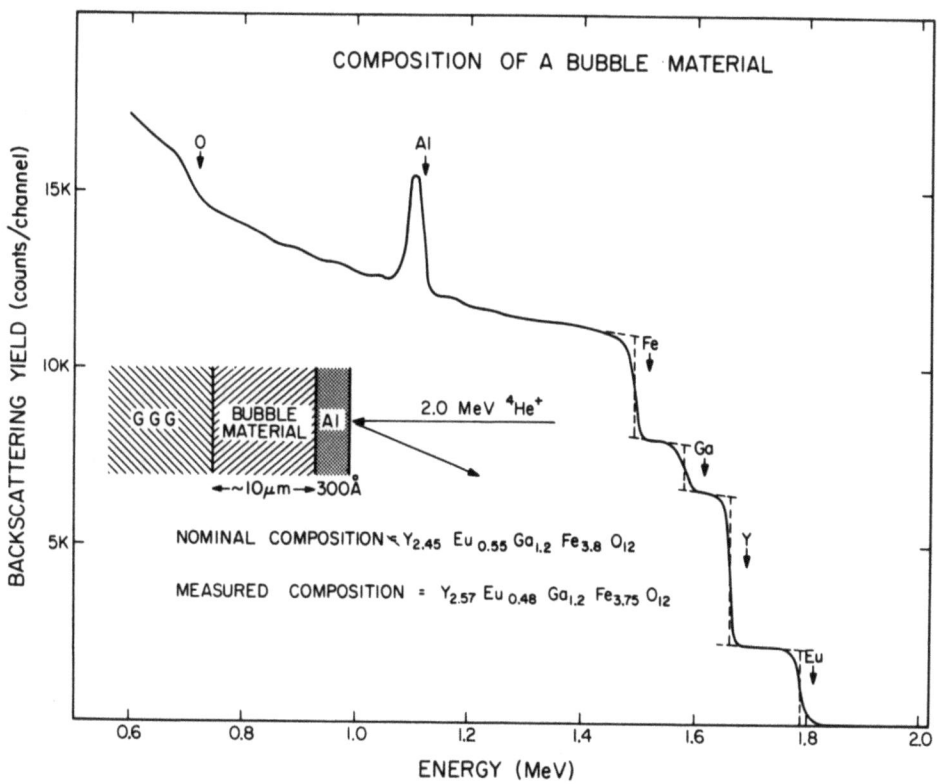

Fig. 13. Energy spectrum of 2.0 MeV ^{4}He$^+$ ions backscattered from a garnet covered with a thin layer of Al to provide a return path for the current during exposure to the beam. The normalized step heights of the individual signals (dashed) give the relative atomic ratios of the metal atoms in the garnet.

relative atomic concentrations directly as Fe:Ga:Y:Eu = 3.75:1.2: 2.57:0.48. Note that these ratios do not establish that the material is a garnet, because a backscattering spectrum contains no information on the chemical state of the atoms. Note further that the system resolution would make a positive identification of the individual elements difficult, particularly because the sample is insulating and had to be covered with a thin layer of Al to close the current path during irradiation of the sample. Note finally the weak signal generated by the light oxygen atoms; the step height would be hard to evaluate exactly, although oxygen atoms are plentiful in the garnet.

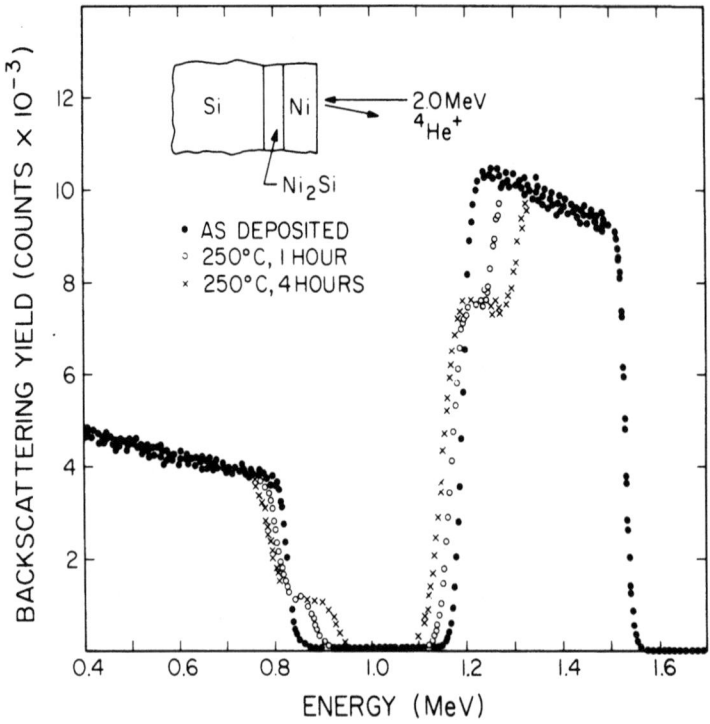

Fig. 14. Energy spectra of 2.0 MeV ^{4}He$^+$ ions backscattered from a 2000 Å film of Ni vacuum-evaporated on a Si substrate before and after thermal annealing (from Tu et al., 1975).

5. <u>Interface Reactions</u>. Changes in the near-surface region of a sample can be monitored quantitatively with backscattering analyses, although it is sometimes necessary to establish by tests that the exposure to the beam does not affect the outcome. Numerous investigations on the reaction of thin metal films with Si (and other semiconductors as well) have been performed with BS in recent years because of the significance of these phenomena in semiconductor device technology. Figure 14 is an example, showing the spectrum of a Ni film on a Si substrate as deposited, and after thermal annealing in an inert He ambient or in vacuum for 1 and 4 hr at 250 C. The step which develops in both the Si and the Ni signals reveals the formation of a new phase whose atomic composition is Ni:Si = 2:1, as derived from the normalized step heights. Glancing x-ray diffraction patterns confirmed the presence of Ni_2Si in the samples. The growth rate is parabolic in time, indicating a transport-limited process.

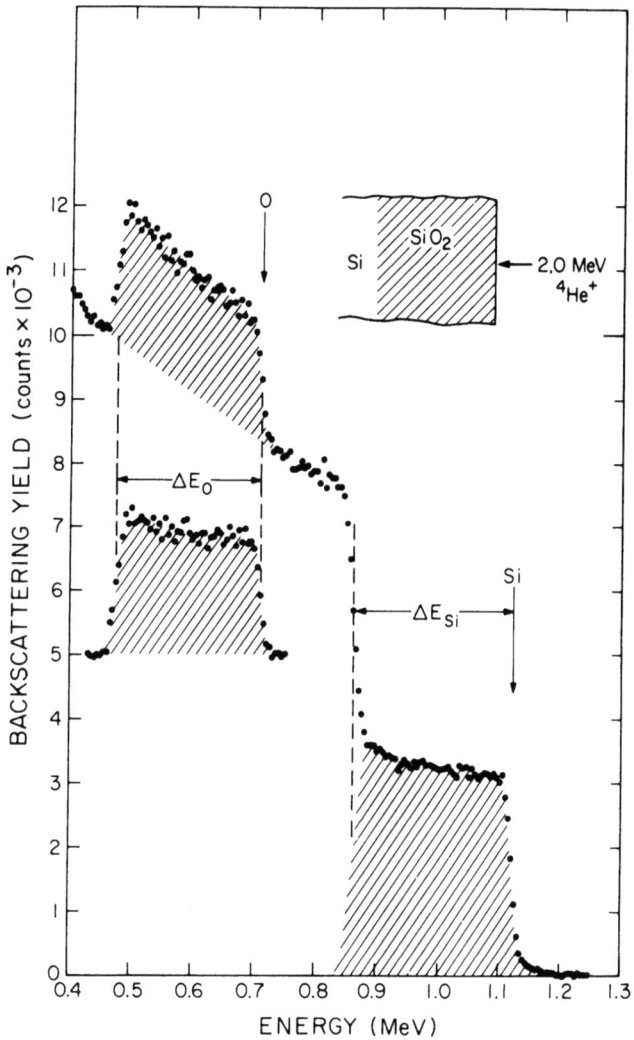

Fig. 15. Energy spectrum of 2.0 MeV $_4$He$^+$ ions backscattered from a thermally oxidized Si wafer.

Another example is the passivation of semiconductor surfaces by oxides and nitrides. Figure 15 gives the spectrum of a thermally oxidized Si wafer from which the thickness and the atomic composition ratio can be determined. The analysis of this spectrum based on the equations derived in sections II-B and II-C, gives a thickness of 5000 Å of SiO$_2$ if a bulk density of 2.28 × 10^{22} SiO$_2$ molecules/cm^3 is assumed for the oxide film. Note again that the estimate of the height and area of the oxygen signal is limited in accuracy by the background of the Si signal, and note also that this spectrum alone only ascertains that a layer of atomic ratio Si:O = 1:2 is present, not that this layer is SiO_2.

F. Channeling

When the sample analyzed by BS is a single crystal with sufficiently limited number of defects and imperfections, the phenomenon of channeling is observed. Figure 16 shows a model of a diamond lattice in three perspectives: a view in a "random" direction, a "planar" direction, and an "axial" direction. Because the wave length of a 1 MeV photon or ^{4}He ion is much shorter than the interatomic distances in solids (typically 1 Å), these views are suggestive representations of a single crystal target as it presents itself to a fast particle approaching the crystal from a random, planar or axial direction. It is evident that if the flux of the incident particles is exactly collimated and bundled within the planes and channels of the single crystal, the number of backscattering collisions per unit length along the path of a particle will diminish substantially as the beam moves from a random direction of incidence to a planar alignment, and that the backscattering yield will be even less for axial incidence. The

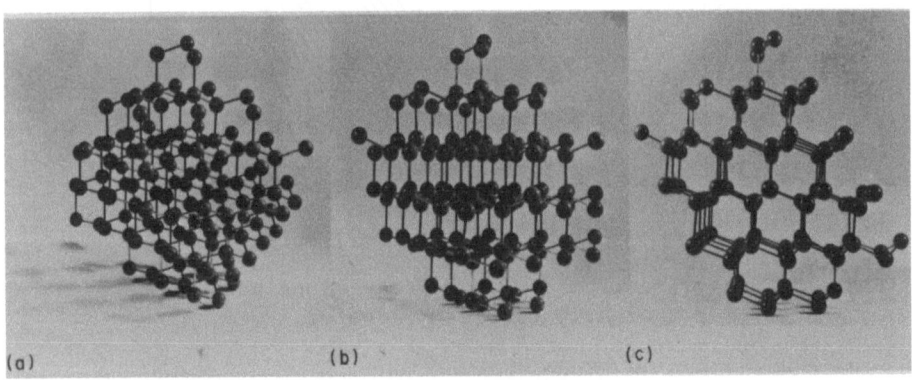

Fig. 16. Three views of a model of the diamond lattice, as seen from a "random" direction, a "planar" direction (a (111) plane) and a "axial" direction (a <110> direction). The bundling of the incident flux along the void spaces of the lattice where close-encounter collisions with atoms in the lattice are much less likely is responsible for the effects of channeling. To achieve these effects, a highly collimated beam is directed onto a single-crystal target in alignment with the planar or axial directions, as shown in the present views.

flux of the incident beam of particles approaching the crystal is not bundled, though, and particles colliding with the first few front atoms of planes or rows of atoms will be scattered and eliminated. This process generates the finite backscattering yield observed under aligned conditions, called the (relative) minimum yield χ_{min} of a backscattering signal measured in planar or axial orientation of the incident beam (see Fig. 17(c)). Among the

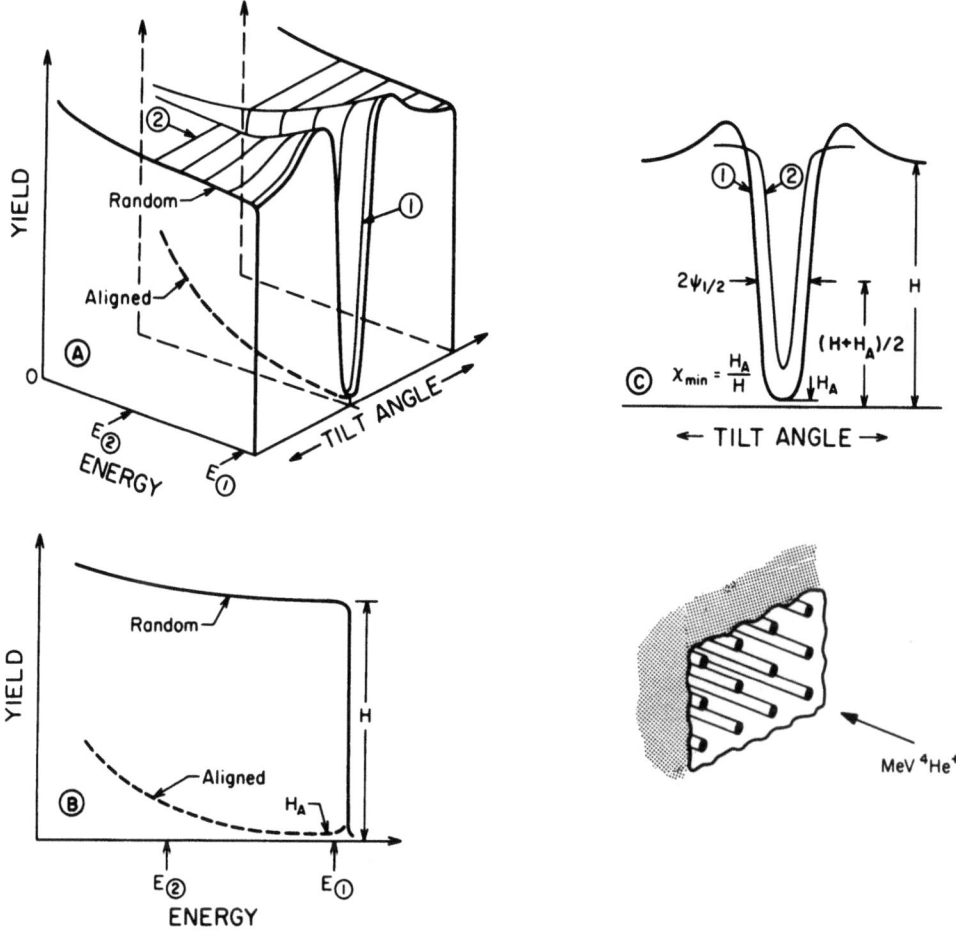

Fig 17. Schematic representation of a channeling experiment in an axial direction. If the single-crystal sample is tilted away from the direction of exact axial alignment with the incident beam, the backscattering yield increases to its "random" value as shown in Fig. (a). Fig. (b) gives the energy spectrum of backscattered particles under exact axial alignment and for random direction of incidence. Fig. (c) gives the change in backscattering yield as a function of tilt angle for two energies.

particles that penetrate the lattice along voids or channels between planes or rows, only those that are exactly equidistant from the neighboring planes or rows are not subjected to such forces and follow oscillatory trajectories within the voids formed by lattice planes or rows, and ultimately undergo a large-angle collision with a nucleus. As a result, the backscattering yield of a spectrum measured with an aligned beam rises with decreasing energy toward the yield of an unaligned beam (see Fig. 17(b), dashed line).

If the single crystal target is tilted slightly away from alignment with the beam, the backscattering yield increases (Fig. 17(c)). The phenomenon is experimentally characterized by the <u>critical angle</u> or the <u>half-angle</u> $\Psi_{\frac{1}{2}}$ at which the yield is halfway between the minimum yield and the yield for random incidence, H. Usually, the minimum yield is determined at an energy slightly less than the edge of the random signal (at energy E_1 in Fig. 17) to avoid the surface peak generated by backscattering from surface imperfections. Figure 17(a) is a composite drawing of the backscattering yield as a function of the energy and the tilt angle away from an axial alignment, as it were measured with a finely collimated beam.

An estimate of the minimum yield for axial channeling is given by

$$\chi_{min} = Nd\pi r^2_{min} \tag{31}$$

where d is the atom spacing along the row and r^2_{min} can be evaluated as the mean square vibrational amplitude measured perpendicularly to the row for an atom near the surface. The critical angle of a monoatomic row of atoms can be estimated by the expression

$$\Psi_{\frac{1}{2}} = \frac{2zZe^2}{Ed} \tag{32}$$

For atomic rows continuing more than one type of atoms or uneven spacings, Z and d are interpreted as an average nuclear charge or an average atomic spacing. For 1 MeV ^{4}He incident in the <110> channeling direction of a Si single crystal, χ_{min} is about about 0.03 and $\Psi_{\frac{1}{2}}$ is about .75°.

Channeling is useful for investigating a number of specific features in single crystals. An example is the determination of the fraction of heavy impurity atoms which is incorporated substitutionally in a row of the host lattice, as shown in Fig. 18. Only if all impurity atoms are located within the atomic rows of the channeling direction will the impurity signal decrease as completely as the signal of the host lattice (black squares) upon channeling.

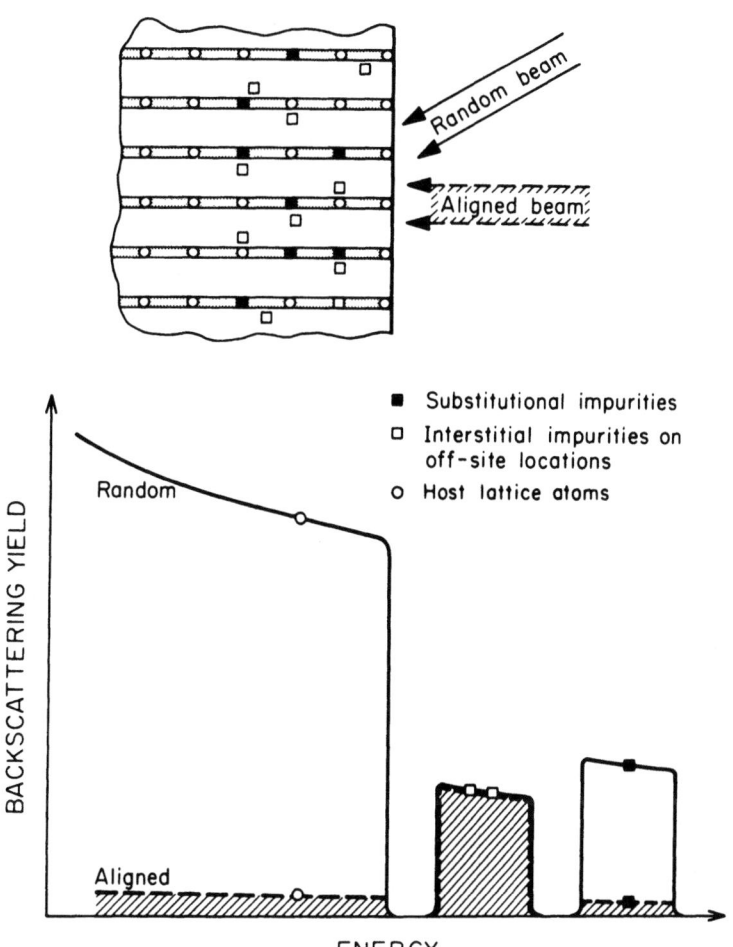

Fig. 18. The comparison of backscattering spectra taken with a collimated beam in a random and an axially aligned direction of incidence reveals if a heavy impurity atom is located in the row of atoms of the host lattice aligned with the beam (full squares) or not (open squares).

Locations elsewhere in the lattice give no reduction of the yield when the beam is aligned with the row (open squares). Other applications are the determination of the thickness of amorphous layers produced after ion implantation in single crystal substrates, the study of epitaxial layers, and the detection of lattice disorder. As a rule, channeling responds indiscriminantly to any departures from perfect crystallinity, because most departures produce additional scattering sites which are readily detected when present in sufficient numbers. These crystalline imperfections produce first-order modifications in the spectrum, in contrast to X-ray diffraction methods, where crystalline imperfections enter only as second-order effects. On the other hand, channeling techniques are nonspecific and the sensitivity to the number of defects per unit volume is no better than that for heavy impurities. Channeling techniques are particularly useful in the study of epitaxial growth where the simulataneous response to defects and to depth are most valuable.

G. Summary of Characteristics, Capabilities and Limitations of BS

Table 2 lists some of the most relevant experimental parameters of a backscattering experiment and typical choices for their values. ^{4}He is the preferred particle because it gives higher mass and depth resolution than H. For higher masses than ^{4}He, the energy resolution and the lifetime of a Si surface barrier detector decreases; only Li has characteristics which compare favorably with ^{4}He. The vacuum requirements in the backscattering chamber are very modest when compared with the 10^{-9} to 10^{-10} Torr range typical of surface-analytical techniques. The reason is that in BS samples, the distribution of atoms is in depth, not just on the surface. To be decipherable, a spectrum has to have reasonably smooth signals, which requires adequately large number of counts in each channel. Typically, such a spectrum is obtained in 10-20 minutes and requires of the order of 10^{14} incident ^{4}He$^+$ ions. The area of the sample illuminated by the beam can vary. Note, however, that it is always assumed that a sample is quite accurately uniform over the whole illuminated area. If the atomic distribution in depth changes from point to point, or if the sample is not geometrically flat, severe distortions of the spectrum can occur. Because of the modest vacuum requirements and the relatively short exposure times normally needed, BS is fast compared with other depth profiling techniques.

A backscattering spectrum contains information of three different kinds

1. the atomic mass of the target consitutents
2. the distribution of these atoms in depth
3. the structural arrangement of the atoms in space (with channeling, when the target is single crystalline, or nearly so).

Table 3 gives approximate capabilities obtained with a typical set-up, assuming ^{4}He of 1-3 MeV as projectile. The mass and depth resolution are determined primarily by the solid-state detector.

TABLE 2

Relevant Parameters of a Backscattering Experiment and Typical Values

Particle	$^1H^+$, $^4H^+$, $^4He^{++}$, $^7Li^+$, $^{14}N^+$, ...	
Beam Energy E	1 - 3	MeV
Accessible Depth	~ 5000	Å
Beam Cross Section	~ 1	mm^2
Beam Current	$\lesssim 50$	nA
Power in Target	$\lesssim 100$	mW
Total Dose	~ 10	μC
Exposure Time	~ 10	min
Vacuum	$\sim 10^{-6}$	Torr

TABLE 3

Summary of Backscattering Capabilities

Mass Perception	1 amu up to 40 amu 10 amu for heavy elements	
Depth Perception	Range Resolution Accuracy No external standards required	$\lesssim 1$ μm ~ 200 Å $\sim 5\%$
Composition Analysis	Atomic ratios or atoms per unit area Accuracy No external standards required	$\sim 5\%$ or better
Sensitivity	Heavy elements in light matrix Light elements in heavy matrix Similar masses - poor	$\gtrsim 10^{18}/cm^2$ $\gtrsim 10^{-1}$
Crystal Structure	Lattice location Epitaxy Point defects sensitivity Line defects sensitivity	$\gtrsim 5 \times 10^{20}/cm^3$ $\gtrsim 10$ cm/cm^2

Special techniques (cooled detector, grazing angles for the incident and/or backscattered particles) can improve the resolution several fold, but it must be kept in mind that to fully benefit from such an improved detection resolution, the vacuum in the chamber must be improved as well. The accuracy in depth perception also is limited by the uncertainties in tabulated values of stopping cross sections and additionally, by the fact that the volume density of atoms in the target must be known if the depth is indicated in a unit of length (e.g. Å) rather than in an areal density of atoms.

Table 4 lists the main deficiencies of BS, and those analytical techniques which can bring remedy. Part III of the review is dedicated to a brief outline of the methods devised to overcome the inadequate sensitivity of BS to light elements by using nuclear reactions.

TABLE 4

Deficiencies of BS and Remedies

Deficiency	Remedy
Only atomic composition, no chemical information	x-rays AUGER ESCA
No microscopic lateral resolution	SEM TEM µ probe
Poor sensitivity to light elements	MS SIMS µ probe PIX nuclear reactions
Radiation damage	

III. NUCLEAR REACTIONS AND ION-BEAM INDUCED X-RAYS

In the energy range below a few MeV, nuclear reactions can be initiated in light nuclei by using H, D or He as projectiles. These reactions are element-specific, so that they can be used to detect the presence of specific elements in a sample. Nuclear reactions analysis is complementary to BS, which has a poor sensitivity to light elements. Elastically backscattered particles are of course present in a nuclear reaction experiment as well, so that the detection system must be able to differentiate between the reaction products and the backscattering events. Radiation usually detected in nuclear reaction analysis are proton, α particles or γ rays.

The characteristic x-rays induced by fast light ions in a target can also be used for the analysis of a target. When atoms are bombarded with ions, there is a high probability that inner shell electrons are ejected and, by subsequent filling of a vacancy by an outer electron, the energy of the transition may be radiated as a characteristic x-ray. These x-rays are characteristic of each atom, so that the method is element specific too.

Common to both nuclear reaction and induced x-ray analysis is the necessity of possessing standards of reference to reliably translate the measured intensities into bulk concentrations. Profiles of atomic distributions with depth are difficult to obtain by induced x-rays, because the radiation emanates from all depths reached by the incident ion. The same can be true for nuclear reaction whose cross sections depend weakly on the energy of the penetrating beam. In other cases, the nuclear reaction can exhibit strong, sharp resonances. These cases are of particular interest because of the possibility they offer to obtain depth profiles of the light elements involved in the reaction. The following discussion gives brief account of that particular technique.

A. Nuclear Reaction Analysis of Hydrogen

Hydrogen offers a good example of elemental detection and profiling by the technique of nuclear reaction. Hydrogen is a common contaminant in solids, but its accurate detection is an outgrowth of relatively recent practical materials problems mainly related to energy. For example, H irradiation of the fuel cladding of fast-breeder reactors, or of the first wall of proposed thermonuclear reactors, produces degradation of materials properties. When the hydrogen concentration exceeds a certain limit, bubbles can form below the surface of the material resulting in blister formation and ultimately in flaking. The phenomenon appears to be basically a mechanical failure process on a microscopic scale, and recent results tend to support the view that extreme stresses are the main cause of blistering. Containment of hydrogen is a key concept in the proposed hydrogen-oxygen fuel economy. But failures in certain high-strength steels have been attributed to high hydrogen concentrations and the consequent hydrogen embrittlement. The discovery that amorphous Si films deposited under a pressure of H can have resistivities as high as 10^{10} Ω cm and more as compared to the typical value of a few hundred Ω cm in the absence of H has generated much excitement in the field of photovoltaic conversion. Amorphous Si of this type can be doped p and n; p-n junctions can be made. Amorphous Si absorbs sunlight more strongly than crystalline material, so that films of the order of 1 μm may be envisaged for efficient cells. The addition of hydrogen has relatively little effect on the absorption. Conversion efficiencies of 6% have been reported for laboratory versions of Schottky barrier solar cells, as compared to 15% for good crystalline cells and a maximum theoretical efficiency of about 15-20% for this type of solar cell. Economical operation is claimed to require at least 10% efficiency.

Hydrogen has also been found to be relevant in problems relating to solar wind, the hydration of glass surfaces and the proposed dating of man-made glass, the superconducting transition temperature of Nb_3Ge and the retaining quality of containers of ultra cold neutrons ("neutron bottles").

The concept of nuclear reaction profiling is described schematically in Figure 19. It is assumed that a nuclear reaction exists between the incident particle and the atom to be detected (e.g., hydrogen). The excitation curve is assumed to be very sharp and large in the vicinity of an energy E_r, and practically zero elsewhere, i.e., the nuclear reaction takes place only if the incident particle impinges on the reacting nucleus (hydrogen) with the resonance energy E_r (Fig. 19(a)). For simplicity, assume further that the incident particles do not initiate other nuclear reactions with other isotopes or elements that happen to be in the target also. If the beam energy is less than E_r, no reaction will take place because the incident particles cannot impinge on a hydrogen atoms with the energy required for the reaction. If the beam energy is raised to E_r exactly, only the hydrogen atoms at the uppermost layers of the sample can undergo a reaction. No reactions take place below the surface, because the beam particles reaches hydrogen atoms below the surface with energies less than E_r, the detector signal will be proportional to the hydrogen concentration at that position below the surface at which the particles have been slowed down to the energy value E_r (See Fig. 19(b)). The raw data of the experiment consists of a plot giving the detector response versus incident beam energy E_r. If the energy loss of the incident particles in the solid is known, and if the detected

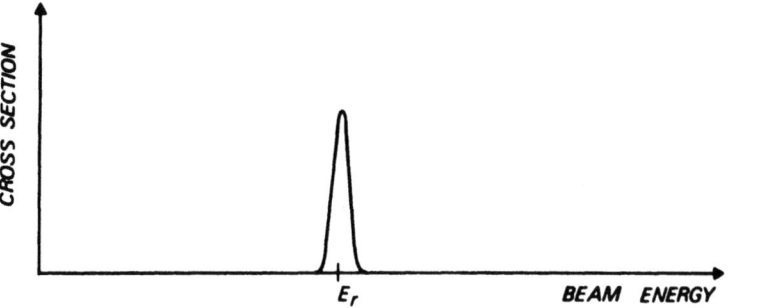
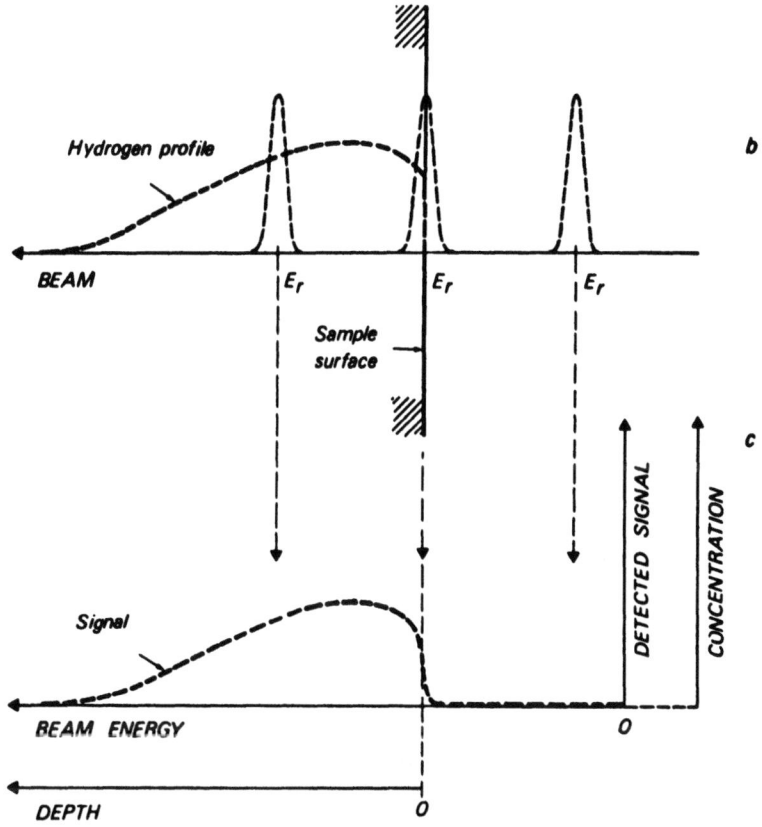

Fig. 19. Schematic diagram explaining how a narrow resonance reaction can be used to profile a specific atom in a sample. (a) Reaction cross section as a function of the incident beam particle. (b) If the beam energy exceeds E_r, the resonance energy falls below the surface of the sample and the signal is proportional to the concentration of the reacting element there (here hydrogen). (c) The resulting signal vs. beam energy reflects the concentration profile in the sample.

radiation does not suffer attenuation on its path from the reacting
nucleus to the detector, the plot can be recalibrated easily to
give a distribution profile of hydrogen atoms versus depth (Fig.
19(c)). To calibrate the ordinate in atoms/cm^3 requires an absolute reaction cross section. One usually prefers to compare the
measurement against that of a standard target whose hydrogen content
is known independently and accurately. The whole experimental set-up
is then actually used as a bridge to compare the two samples and
the conditions imposed on the systems are merely those of stability
over the two experiments rather than that of absolute calibration.
In practice, a number of second-order effects enter the extraction
of the real profile from the raw data (deconvolution for the finite
width of the resonance, for nonvanishing values of the cross section outside of the resonance, and for energy straggling in the
beam, correction for background, etc.).

The nuclear reaction that takes place when a hydrogen atom is
bombarded with ^{15}N of 6.385 MeV is excellently suited for hydrogen
profiling. The reaction has a narrow, isolated resonance of the
type $^1H(^{15}N,\alpha\gamma)^{12}C$, that is

$$^1H + ^{15}N(6.385 \text{ MeV}) \rightarrow \alpha + ^{12}C^* \rightarrow \alpha + \gamma(4.43 \text{ MeV}) + ^{12}C$$

The full width of the resonance at half maximum is theoretically
$\Gamma = 6$ keV, as calculated from the known width of 0.4 keV in the
center-of-mass reference system. Outside of the resonance, the
excitation curve falls by about 3 orders of magnitude. Finally,
the reaction is specific to hydrogen, because the Coulomb barrier
for ^{15}N is above 6.4 MeV for all other elements. Figure 20 shows
the resonance measured on a hydrogen surface layer of an equivalent
thickness of 12 keV with a γ ray detector. This resonance allows
near-surface depth resolutions of about 50 Å in materials such as
Si.

The advantage of γ-ray detectors (NaI(Tl) detector with photomultiplier) is that it can be placed outside of the vacuum chamber,
yet close to the sample inside so as to span a large solid angle of
detection. There is no interference by the α particles which are
stopped by the chamber walls and the background at those high γ-ray
energies is small with proper shielding. The relatively poor energy
resolution NaI detectors is of no consequence since the depth resolution is determined by the reaction resonance, not the detector.
Typical beam currents are in the tens of nA, and an individual datum
point requires about 5 µC, or 100 s for a H concentration higher
than about 1 at %. A spectrum with 40 data points thus requires a
total dose of 200 µC, which is large. The power dissipated on the
beam spot is also significant. Beam-induced modifications in the
hydrogen profile and in the sample itself must therefore be rec-
koned with. The sample must be mounted with good heat sinking and
the beam spot must be shifted at regular intervals. The spot size

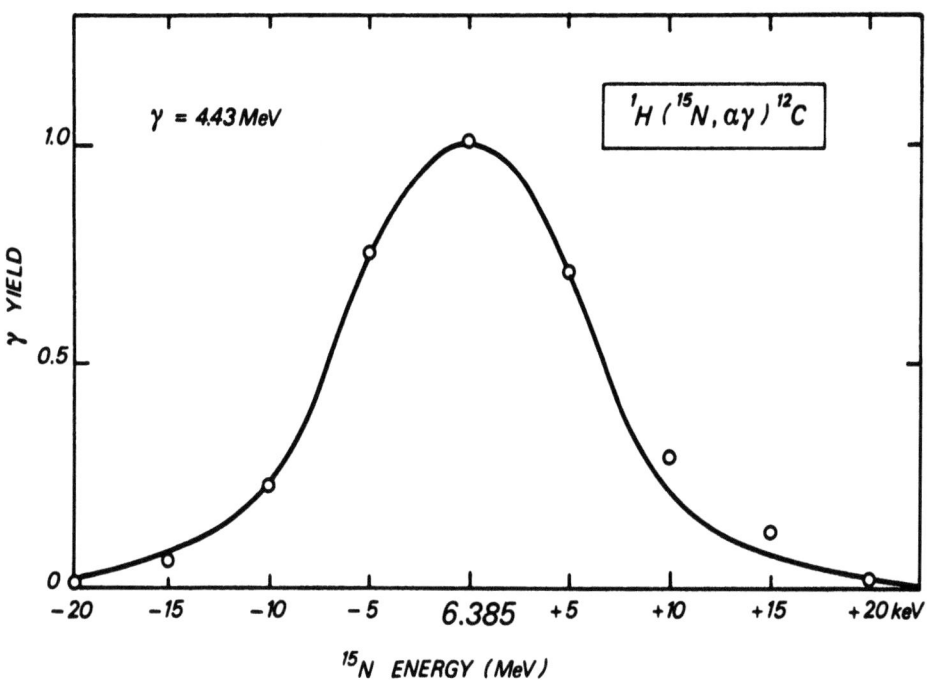

Fig. 20. Measured profile of hydrogen at the surface of a sample along with a fit obtained by assuming an equivalent uniform hydrogen thickness of 12 keV and a resonance width of Γ = 6 keV full width at half maximum (from W. A. Lanford, Nucl. Instr. Meth., 149, (1978), 1).

varies with operating conditions and may reach 10 mm^2, so that the size of the sample required is a bit larger than for BS analyses.

Figures 21 and 22 show H profiles measured with the ^{1}H(^{15}N,$\alpha\gamma$)^{12}C resonant reaction. Figure 21 gives the distributions of H implanted into polycristalline Si at 3 and 20 keV to a dose of 3.10^{16}H atoms/cm^2. Both profiles are skewed, and both samples have large surface concentrations of H which are not explained from range theory of ion implantation. Figure 22 gives the H profile of an amorphous Si-H film. It can be seen that the H atomic concentration is uniform in the bulk of the film and has a value of about 12%. In this case also, there is a surface accumulation of H, as well as an indication of a depletion in the first 100 to 200 Å below the sample surface.

Table 5 summarizes the characteristics and capabilities of this profiling technique, and lists three other resonant reactions which have also been used to detect H. The choice of the reaction

Table 5. List of the essential characteristics and the capabilities of four resonant nuclear reactions used for H detection and profiling (from Thomas et. al., Revue de Physique Appliquee, in press).

Nuclear Reaction		$^1H(^{11}B,\alpha)^8Be$	$^1H(^7Li,\gamma)^8Be$	$^1H(^{15}N,\alpha\gamma)^{12}C$	$^1H(^{19}F,\alpha\gamma)^{16}O$
Resonance Energy E_r	(MeV)	1.793	3.070	6.385	6.418
Resonance width Γ	(MeV)	~66	~70	~6	~45
Energy range*	(MeV)	1.8 – 2.4	3–6	6.4–7.5	6.4–7.5
Detected radiation		α	γ	γ	γ
Detected energy	(MeV)	≤ 3	14.7 17.6	4.43	6.13 6.92 7.12
Analyzable range*	(Å)	~6000	~70.000	~10.000	~6.000
Depth resolution on Si surface	(Å)	~400	~1700	~50	~170
	(H at/cm^2)	10^{13-14}	10^{12}	10^{14}	10^{13-14}
Sensitivity	(H ppm)	10^{2-3}	1–10	10^3	10^{2-3}

* Determined by other resonances or increased non-resonant yields at higher energies.

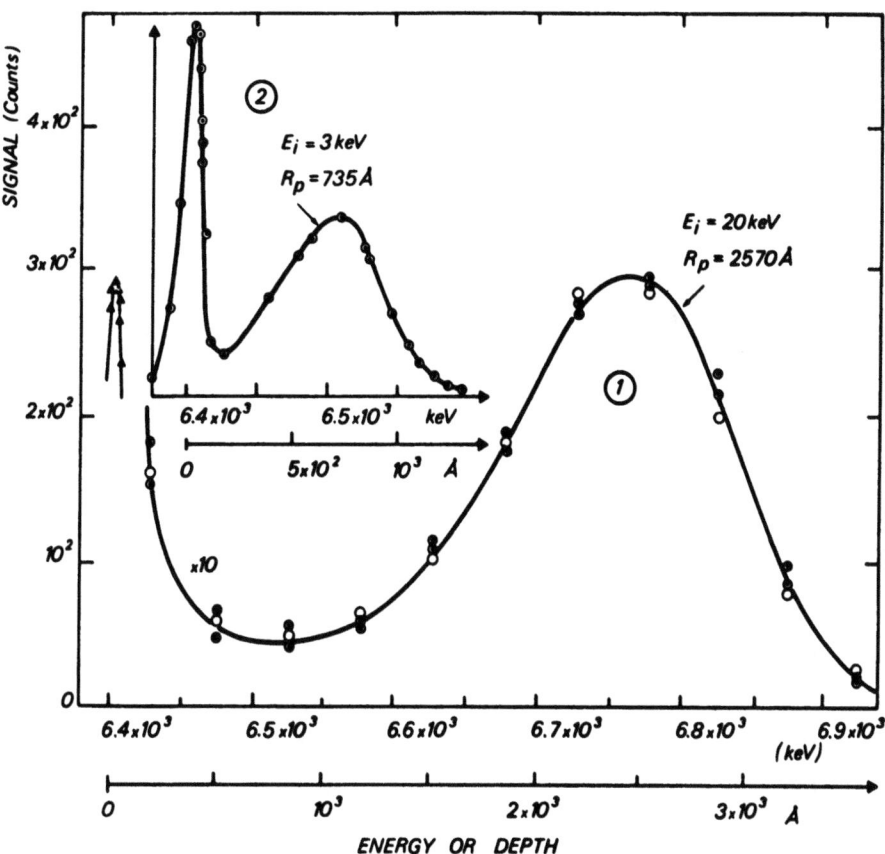

Fig. 21. Hydrogen profiles measured by the nuclear reaction $^1H(^{15}N,\alpha\gamma)^{12}C$. The hydrogen was ion implanted into polycristalline Si samples at a dose of 3.10^{16} H atoms/cm^2 with (1) 20 keV and (2) 3 keV (from J. P. Thomas et. al., Revue de Physique Appliquee, in press).

used is usually determined by the available equipment. A recent comparison of the results obtained by various methods on identical samples of Si implanted with H concludes that accuracies of better than 8% are presently possible[*].

---

[*] J. F. Ziegler et. al., Nucl. Instr. Meth., 149, 19-39, (1978).

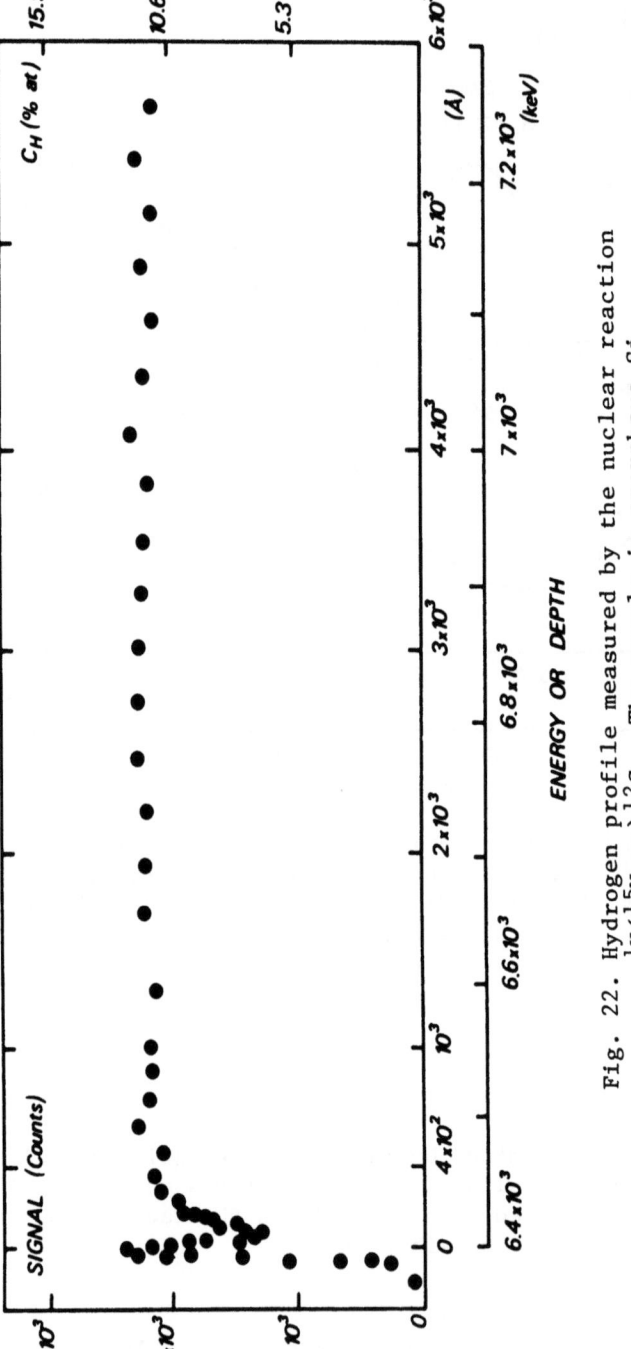

Fig. 22. Hydrogen profile measured by the nuclear reaction $^1H(^{15}N,\alpha\gamma)^{12}C$. The sample is amorphous Si deposited by glow discharge on a glass substrate (from J. P. Thomas, unpublished).

Table 6

Some nuclear reactions used for C, N and O detection and profiling.

Detected Element	Nuclear Reaction	Beam (MeV)	Detected	
			Radiation	Energy (MeV)
C	C(d,p) C	1.20	p	3.1
	C(p,) N	0.46		2.36
	C(d,p) C	0.64	p	5.8
	C(p,) N	1.75		9.17
				6.43
				2.74
N	N(d,) C	1.2		6.7
		1.5		9.9
	N(p,) C	0.8		3.9
	N(p,) C	0.43		4.4
O	O(d,p) O	0.9	p	1.6
		0.9	p	2.4
	O(d,) N	0.93		2.6
	O(p,) N	0.73		3.4

B. Nuclear Reaction Analysis of Light Elements

Any nuclear reaction with a cross section that varies smoothly with energy can be applied to depth profiling provided a charged particle is emitted in the reaction. The depth information is obtained from an energy analysis of the emitted charged particles. If $(dE/dx)_1$ is the energy loss of the incident particle (normal incidence), the reduction $\Delta x (dE/dx)_1$ over the distance Δx in the energy E_1 of the incident particle immediately before the reaction will lower the energy E_2 of the emitted particle immediately after the reaction by

$$\Delta x (dE/dx)_1 \; (\partial E_2 / \partial E_1)$$

where the last factor is given by the kinematics of the nuclear reaction. Over the distance Δx, the emitted particle undergoes an additional energy loss of $(\Delta x / \cos \theta)(dE/dx)_2$, where $(dE/dx)_2$ is the energy loss per unit length for the emitted particle. The difference ΔE observed in the emitted particle at angle θ for reactions occuring at depths Δx apart is therefore

$$\Delta E = - [(dE/dx)_1 \; (\partial E_2 / \partial E_1) + (dE/dx)_2 / \cos \theta] \Delta x$$

which shows that the energy separation corresponding to a depth interval depends both on the reaction kinematics (1st term) and on the energy loss of the incident and emitted particles. The depth resolution is thus a function of stopping powers, incident energy, detector angle, and will vary with depth in the target. The maximum probing depth is limited by a number of factors, such as the presence of other reactions, the acceptable combined energy straggling of incident and emitted particles, the highest accessible energy, or the energy dependence of the reaction cross section. The elastically backscattered particles usually constitute a major background and limit the sensitivity unless they are filtered out. In that case, the background is low and the sensitivity is high, but with same loss in energy resolution.

The number of reactions available for detection and profiling of light elements are numerous. They have been listed by several authors

Table 6 lists some reactions used for C, N and O detection and profiling, these being among the most ubiquitous impurities in solids.

REFERENCES

1. W. K. Chu, J. W. Mayer and M-A. Nicolet, Backscattering Spectrometry, Academic Press (1978).
2. J. W. Mayer and E. Rimini, Eds., Ion Beam Handbook for Material Analysis, Academic Press, (1977).
3. H. A. Bethe, Ann. Physik $\underline{5}$, 325 (1930).
4. N. Bohr, Phil. Mag. $\underline{25}$, 10 (1913).
5. W. H. Bragg and R. Kleeman, Phil. Mag. $\underline{10}$, S318 (1905).
6. C. G. Darwin, Phil. Mag. $\underline{28}$, 499 (1914).
7. K. N. Tu, W. K. Chu and J. W. Mayer, Thin Solid Films $\underline{25}$, 403 (1975).
8. J. R. Bird, B. L. Campbell, P. B. Brice, Atomic Energy Review $\underline{12}$, 275-342 (1974).
9. J. Bottiger, S. T. Picraux and N. Rud, in Ion Beam Surface Layer Analysis, O. Meyer, G. Linker and F. Kappeler (Eds.), Plenum Press $\underline{2}$, 811-819 (1976).
10. A. Cachard, J. P. Thomas and E. Ligeon, in Material Characterization Using Ion Beams, J. P. Thomas and A. Cachard (Eds.), Plenum Press, 367-397 (1978).

APPENDIX

Table A-1. This table gives the kinematic factor, K_M defined by equation (1), for ^{4}He as projectile (m = 4.002603 amu) and integral atomic masses for the target atom M = 6 to 216 amu). The parameter is the scattering angle θ (180°-90°) measured in the laboratory frame of reference, as shown in Fig. 1.

Table A-2. This table gives the mean kinematic factor $\bar{K}$, which is the weighted average of the kinematic factors K_M for ^{4}He as projectile and the stable isotopes of an element. The kinematic factor K_M of an individual isotope is calculated by eq. (1), using 4.002603 amu for m (^{4}He) and the actual masses (in amu) for M of each isotope of an element (e.g., 27.976929 amu for ^{28}Si, 28.976496 amu for ^{29}Si, and 29.973763 amu for ^{30}Si). The kinematic factors calculated in this fashion for each isotope are then weighted by the relatibe abundance of each isotope to obtain $\bar{K}$ (e.g. 0.9221, 0.0470, and 0.0309 for the abundances of ^{28}Si, ^{29}Si and ^{30}Si).
The atomic mass and relative abundance of the stable isotopes were taken from the Handbook of Physics, (E. U. Condon and H. Odishaw, Eds., McGraw-Hill 1967). The parameter is the scattering angle θ (in degrees) measured in the laboratory frame of reference as shown in Fig. 1.

Table A-1

ATOMIC MASS	180.	170.	160.	150.	140.	130.	120.	110.	100.	90.
6	0.0399	0.0407	0.0433	0.0479	0.0554	0.0668	0.0838	0.1092	0.1465	0.1997
7	0.0742	0.0755	0.0796	0.0868	0.0980	0.1143	0.1375	0.1699	0.2140	0.2724
8	0.1109	0.1126	0.1179	0.1271	0.1411	0.1610	0.1883	0.2249	0.2726	0.3330
9	0.1477	0.1497	0.1559	0.1667	0.1827	0.2051	0.2351	0.2741	0.3235	0.3843
10	0.1834	0.1857	0.1926	0.2044	0.2220	0.2460	0.2777	0.3180	0.3681	0.4283
11	0.2175	0.2200	0.2273	0.2400	0.2586	0.2837	0.3164	0.3573	0.4073	0.4664
12	0.2498	0.2523	0.2600	0.2733	0.2925	0.3183	0.3515	0.3926	0.4420	0.4998
13	0.2800	0.2827	0.2907	0.3043	0.3239	0.3501	0.3834	0.4243	0.4730	0.5292
14	0.3084	0.3111	0.3192	0.3331	0.3530	0.3793	0.4125	0.4530	0.5007	0.5553
15	0.3349	0.3377	0.3459	0.3599	0.3798	0.4062	0.4391	0.4790	0.5257	0.5787
16	0.3598	0.3625	0.3708	0.3848	0.4047	0.4306	0.4636	0.5027	0.5483	0.5998
17	0.3830	0.3857	0.3940	0.4080	0.4279	0.4538	0.4860	0.5244	0.5685	0.6188
18	0.4047	0.4075	0.4157	0.4296	0.4493	0.4750	0.5067	0.5444	0.5877	0.6362
19	0.4251	0.4278	0.4360	0.4498	0.4693	0.4947	0.5258	0.5627	0.6050	0.6520
20	0.4442	0.4469	0.4551	0.4687	0.4880	0.5129	0.5435	0.5796	0.6209	0.6665
21	0.4622	0.4648	0.4729	0.4864	0.5055	0.5300	0.5600	0.5953	0.6355	0.6798
22	0.4791	0.4817	0.4897	0.5030	0.5218	0.5459	0.5754	0.6099	0.6491	0.6921
23	0.4950	0.4976	0.5055	0.5186	0.5371	0.5608	0.5897	0.6235	0.6617	0.7035
24	0.5100	0.5126	0.5203	0.5333	0.5515	0.5748	0.6033	0.6361	0.6734	0.7141
25	0.5242	0.5267	0.5344	0.5472	0.5650	0.5879	0.6157	0.6480	0.6843	0.7240
26	0.5376	0.5401	0.5476	0.5602	0.5778	0.6003	0.6275	0.6591	0.6946	0.7332
27	0.5503	0.5527	0.5602	0.5726	0.5899	0.6120	0.6386	0.6695	0.7042	0.7418
28	0.5623	0.5647	0.5721	0.5843	0.6013	0.6230	0.6491	0.6793	0.7132	0.7499
29	0.5737	0.5761	0.5833	0.5954	0.6121	0.6334	0.6590	0.6886	0.7217	0.7574
30	0.5846	0.5869	0.5941	0.6059	0.6223	0.6432	0.6683	0.6973	0.7296	0.7646
31	0.5949	0.5972	0.6042	0.6159	0.6320	0.6525	0.6772	0.7056	0.7372	0.7713
32	0.6047	0.6070	0.6139	0.6254	0.6413	0.6614	0.6856	0.7134	0.7443	0.7776
33	0.6141	0.6164	0.6232	0.6344	0.6500	0.6698	0.6936	0.7208	0.7511	0.7837
34	0.6231	0.6253	0.6320	0.6431	0.6584	0.6779	0.7012	0.7279	0.7575	0.7894
35	0.6316	0.6338	0.6404	0.6513	0.6664	0.6855	0.7084	0.7346	0.7636	0.7948
36	0.6398	0.6420	0.6485	0.6592	0.6740	0.6928	0.7153	0.7410	0.7694	0.7999
37	0.6476	0.6498	0.6562	0.6667	0.6813	0.6998	0.7218	0.7470	0.7749	0.8048
38	0.6551	0.6572	0.6635	0.6739	0.6883	0.7064	0.7281	0.7529	0.7802	0.8094
39	0.6623	0.6644	0.6706	0.6808	0.6950	0.7128	0.7341	0.7584	0.7852	0.8138
40	0.6692	0.6713	0.6774	0.6874	0.7014	0.7189	0.7398	0.7637	0.7900	0.8181
41	0.6759	0.6779	0.6839	0.6938	0.7075	0.7248	0.7453	0.7688	0.7946	0.8221
42	0.6822	0.6842	0.6901	0.6995	0.7134	0.7304	0.7506	0.7736	0.7990	0.8260
43	0.6884	0.6903	0.6962	0.7058	0.7191	0.7358	0.7557	0.7783	0.8032	0.8297
44	0.6943	0.6962	0.7019	0.7114	0.7245	0.7410	0.7605	0.7828	0.8072	0.8332
45	0.7000	0.7019	0.7075	0.7169	0.7297	0.7459	0.7652	0.7871	0.8111	0.8366
46	0.7054	0.7073	0.7129	0.7221	0.7348	0.7507	0.7697	0.7912	0.8148	0.8399
47	0.7107	0.7126	0.7181	0.7271	0.7396	0.7554	0.7740	0.7952	0.8184	0.8430
48	0.7158	0.7176	0.7231	0.7320	0.7443	0.7598	0.7782	0.7990	0.8218	0.8461
49	0.7207	0.7225	0.7279	0.7367	0.7488	0.7641	0.7822	0.8027	0.8251	0.8490
50	0.7255	0.7273	0.7325	0.7412	0.7532	0.7682	0.7860	0.8062	0.8283	0.8518
51	0.7301	0.7318	0.7370	0.7456	0.7574	0.7722	0.7898	0.8097	0.8314	0.8545
52	0.7345	0.7363	0.7414	0.7499	0.7615	0.7761	0.7934	0.8130	0.8344	0.8571
53	0.7389	0.7405	0.7456	0.7540	0.7654	0.7798	0.7969	0.8162	0.8373	0.8596
54	0.7430	0.7447	0.7497	0.7579	0.7693	0.7835	0.8003	0.8193	0.8400	0.8620
55	0.7471	0.7487	0.7536	0.7618	0.7729	0.7870	0.8035	0.8222	0.8427	0.8643
56	0.7510	0.7526	0.7575	0.7655	0.7765	0.7903	0.8067	0.8251	0.8453	0.8666
57	0.7548	0.7564	0.7612	0.7691	0.7800	0.7936	0.8097	0.8279	0.8478	0.8688
58	0.7584	0.7600	0.7648	0.7726	0.7834	0.7968	0.8127	0.8306	0.8502	0.8709
59	0.7620	0.7636	0.7683	0.7760	0.7866	0.7999	0.8156	0.8333	0.8526	0.8729
60	0.7655	0.7670	0.7717	0.7793	0.7898	0.8029	0.8184	0.8358	0.8549	0.8749
61	0.7689	0.7704	0.7750	0.7825	0.7928	0.8058	0.8211	0.8383	0.8571	0.8768
62	0.7721	0.7737	0.7782	0.7856	0.7958	0.8086	0.8237	0.8407	0.8592	0.8787
63	0.7753	0.7768	0.7813	0.7886	0.7987	0.8113	0.8262	0.8430	0.8613	0.8805
64	0.7784	0.7799	0.7843	0.7916	0.8015	0.8140	0.8287	0.8453	0.8633	0.8823
65	0.7814	0.7829	0.7873	0.7944	0.8043	0.8166	0.8311	0.8475	0.8652	0.8840
66	0.7844	0.7858	0.7901	0.7972	0.8070	0.8191	0.8334	0.8496	0.8672	0.8856
67	0.7872	0.7887	0.7929	0.7999	0.8095	0.8216	0.8357	0.8517	0.8690	0.8873
68	0.7900	0.7914	0.7956	0.8026	0.8121	0.8240	0.8379	0.8537	0.8708	0.8888
69	0.7927	0.7941	0.7983	0.8051	0.8145	0.8263	0.8401	0.8557	0.8726	0.8903
70	0.7954	0.7967	0.8009	0.8076	0.8169	0.8285	0.8422	0.8576	0.8743	0.8918
71	0.7979	0.7993	0.8034	0.8101	0.8193	0.8307	0.8442	0.8594	0.8759	0.8933
72	0.8004	0.8018	0.8058	0.8125	0.8215	0.8329	0.8462	0.8612	0.8775	0.8947
73	0.8029	0.8042	0.8082	0.8148	0.8238	0.8350	0.8482	0.8630	0.8791	0.8960
74	0.8053	0.8066	0.8105	0.8170	0.8259	0.8370	0.8501	0.8647	0.8806	0.8974
75	0.8076	0.8089	0.8128	0.8192	0.8280	0.8390	0.8519	0.8664	0.8821	0.8987
76	0.8099	0.8112	0.8151	0.8214	0.8301	0.8410	0.8537	0.8681	0.8836	0.8999
77	0.8121	0.8134	0.8172	0.8235	0.8321	0.8429	0.8555	0.8697	0.8850	0.9012
78	0.8143	0.8156	0.8193	0.8256	0.8341	0.8447	0.8572	0.8712	0.8864	0.9024
79	0.8164	0.8177	0.8214	0.8276	0.8360	0.8465	0.8589	0.8727	0.8878	0.9036
80	0.8185	0.8197	0.8234	0.8295	0.8379	0.8483	0.8605	0.8742	0.8891	0.9047
81	0.8205	0.8217	0.8254	0.8315	0.8397	0.8500	0.8621	0.8757	0.8904	0.9058
82	0.8225	0.8237	0.8274	0.8333	0.8415	0.8517	0.8637	0.8771	0.8917	0.9069
83	0.8244	0.8257	0.8293	0.8352	0.8433	0.8534	0.8652	0.8785	0.8929	0.9080
84	0.8263	0.8275	0.8311	0.8370	0.8450	0.8550	0.8667	0.8799	0.8941	0.9090
85	0.8282	0.8294	0.8329	0.8387	0.8467	0.8565	0.8682	0.8812	0.8953	0.9101
86	0.8300	0.8312	0.8347	0.8404	0.8483	0.8581	0.8696	0.8825	0.8964	0.9111
87	0.8318	0.8330	0.8364	0.8421	0.8499	0.8596	0.8710	0.8837	0.8976	0.9120
88	0.8335	0.8347	0.8381	0.8438	0.8515	0.8611	0.8724	0.8850	0.8987	0.9130
89	0.8353	0.8364	0.8398	0.8454	0.8530	0.8625	0.8737	0.8862	0.8997	0.9139
90	0.8369	0.8381	0.8414	0.8470	0.8545	0.8640	0.8750	0.8874	0.9008	0.9148
91	0.8386	0.8397	0.8430	0.8485	0.8560	0.8654	0.8763	0.8886	0.9018	0.9157
92	0.8402	0.8413	0.8446	0.8500	0.8575	0.8667	0.8776	0.8897	0.9029	0.9166
93	0.8418	0.8429	0.8461	0.8515	0.8589	0.8681	0.8788	0.8908	0.9039	0.9175
94	0.8433	0.8444	0.8476	0.8530	0.8603	0.8694	0.8800	0.8919	0.9048	0.9183
95	0.8448	0.8459	0.8491	0.8545	0.8616	0.8706	0.8812	0.8930	0.9058	0.9191
96	0.8463	0.8474	0.8506	0.8558	0.8630	0.8719	0.8824	0.8941	0.9067	0.9200
97	0.8478	0.8488	0.8520	0.8572	0.8643	0.8731	0.8835	0.8951	0.9076	0.9207
98	0.8492	0.8503	0.8534	0.8585	0.8656	0.8743	0.8846	0.8961	0.9085	0.9215
99	0.8506	0.8516	0.8548	0.8599	0.8669	0.8755	0.8857	0.8971	0.9094	0.9223
100	0.8520	0.8530	0.8561	0.8612	0.8681	0.8767	0.8868	0.8981	0.9103	0.9230
101	0.8533	0.8544	0.8574	0.8624	0.8693	0.8778	0.8878	0.8990	0.9111	0.9238
102	0.8547	0.8557	0.8587	0.8637	0.8705	0.8790	0.8889	0.9000	0.9120	0.9245
103	0.8560	0.8570	0.8600	0.8649	0.8717	0.8801	0.8899	0.9009	0.9128	0.9252
104	0.8573	0.8583	0.8612	0.8661	0.8728	0.8812	0.8909	0.9018	0.9136	0.9259
105	0.8585	0.8595	0.8625	0.8673	0.8740	0.8822	0.8919	0.9027	0.9144	0.9266
106	0.8598	0.8607	0.8637	0.8685	0.8751	0.8833	0.8928	0.9036	0.9151	0.9272
107	0.8610	0.8619	0.8649	0.8696	0.8762	0.8843	0.8938	0.9044	0.9159	0.9279
108	0.8622	0.8631	0.8660	0.8708	0.8772	0.8853	0.8947	0.9053	0.9166	0.9285
109	0.8633	0.8643	0.8672	0.8719	0.8783	0.8863	0.8956	0.9061	0.9174	0.9292
110	0.8645	0.8654	0.8683	0.8730	0.8793	0.8873	0.8965	0.9069	0.9181	0.9298
111	0.8656	0.8666	0.8694	0.8740	0.8804	0.8882	0.8974	0.9077	0.9188	0.9304
112	0.8667	0.8677	0.8705	0.8751	0.8814	0.8892	0.8983	0.9085	0.9195	0.9310
113	0.8678	0.8688	0.8716	0.8761	0.8823	0.8901	0.8991	0.9093	0.9202	0.9316
114	0.8689	0.8699	0.8726	0.8771	0.8833	0.8910	0.9000	0.9100	0.9209	0.9322
115	0.8700	0.8709	0.8736	0.8781	0.8843	0.8919	0.9008	0.9108	0.9215	0.9327
116	0.8710	0.8719	0.8747	0.8791	0.8852	0.8928	0.9016	0.9115	0.9222	0.9333
117	0.8721	0.8730	0.8757	0.8801	0.8861	0.8936	0.9024	0.9122	0.9228	0.9338
118	0.8731	0.8740	0.8767	0.8810	0.8870	0.8945	0.9032	0.9129	0.9234	0.9344
119	0.8741	0.8750	0.8776	0.8820	0.8875	0.8953	0.9040	0.9136	0.9241	0.9349
120	0.8751	0.8759	0.8786	0.8829	0.8888	0.8962	0.9047	0.9143	0.9247	0.9354
121	0.8760	0.8769	0.8795	0.8838	0.8897	0.8970	0.9055	0.9150	0.9253	0.9360
122	0.8770	0.8778	0.8804	0.8847	0.8905	0.8978	0.9062	0.9157	0.9259	0.9365
123	0.8779	0.8788	0.8814	0.8856	0.8914	0.8986	0.9070	0.9163	0.9264	0.9370
124	0.8788	0.8797	0.8823	0.8865	0.8922	0.8993	0.9077	0.9170	0.9270	0.9375

Table A-1 (cont.)

125	0.8797	0.8806	0.8831	0.8873	0.8930	0.9001	0.9084	0.9176	0.9276	0.9379
126	0.8806	0.8815	0.8840	0.8882	0.8938	0.9009	0.9091	0.9182	0.9281	0.9384
127	0.8815	0.8824	0.8849	0.8890	0.8946	0.9016	0.9098	0.9189	0.9287	0.9389
128	0.8824	0.8832	0.8857	0.8898	0.8954	0.9023	0.9104	0.9195	0.9292	0.9394
129	0.8832	0.8841	0.8866	0.8906	0.8962	0.9030	0.9111	0.9201	0.9297	0.9398
130	0.8841	0.8849	0.8874	0.8914	0.8969	0.9038	0.9117	0.9207	0.9303	0.9403
131	0.8849	0.8857	0.8882	0.8922	0.8977	0.9045	0.9124	0.9212	0.9308	0.9407
132	0.8857	0.8866	0.8890	0.8930	0.8984	0.9051	0.9130	0.9218	0.9313	0.9411
133	0.8866	0.8874	0.8898	0.8937	0.8991	0.9058	0.9136	0.9224	0.9318	0.9416
134	0.8873	0.8882	0.8906	0.8945	0.8998	0.9065	0.9143	0.9229	0.9323	0.9420
135	0.8881	0.8889	0.8913	0.8952	0.9005	0.9072	0.9149	0.9235	0.9328	0.9424
136	0.8889	0.8897	0.8921	0.8959	0.9012	0.9078	0.9155	0.9240	0.9332	0.9428
137	0.8897	0.8905	0.8928	0.8967	0.9019	0.9084	0.9161	0.9246	0.9337	0.9432
138	0.8904	0.8912	0.8936	0.8974	0.9026	0.9091	0.9166	0.9251	0.9342	0.9436
139	0.8912	0.8920	0.8943	0.8981	0.9033	0.9097	0.9172	0.9256	0.9346	0.9440
140	0.8919	0.8927	0.8950	0.8988	0.9039	0.9103	0.9178	0.9261	0.9351	0.9444
141	0.8926	0.8934	0.8957	0.8994	0.9046	0.9109	0.9183	0.9266	0.9355	0.9448
142	0.8933	0.8941	0.8964	0.9001	0.9052	0.9115	0.9189	0.9271	0.9360	0.9452
143	0.8941	0.8948	0.8971	0.9008	0.9058	0.9121	0.9194	0.9276	0.9364	0.9455
144	0.8947	0.8955	0.8978	0.9014	0.9065	0.9127	0.9200	0.9281	0.9368	0.9459
145	0.8954	0.8962	0.8984	0.9021	0.9071	0.9133	0.9205	0.9286	0.9372	0.9463
146	0.8961	0.8969	0.8991	0.9027	0.9077	0.9138	0.9210	0.9290	0.9377	0.9466
147	0.8968	0.8975	0.8997	0.9034	0.9083	0.9144	0.9215	0.9295	0.9381	0.9470
148	0.8974	0.8982	0.9004	0.9040	0.9089	0.9150	0.9220	0.9300	0.9385	0.9473
149	0.8981	0.8988	0.9010	0.9046	0.9095	0.9155	0.9226	0.9304	0.9389	0.9477
150	0.8987	0.8995	0.9016	0.9052	0.9100	0.9160	0.9230	0.9309	0.9393	0.9480
151	0.8994	0.9001	0.9023	0.9058	0.9106	0.9166	0.9235	0.9313	0.9397	0.9484
152	0.9000	0.9007	0.9029	0.9064	0.9112	0.9171	0.9240	0.9317	0.9400	0.9487
153	0.9006	0.9013	0.9035	0.9070	0.9117	0.9176	0.9245	0.9322	0.9404	0.9490
154	0.9012	0.9019	0.9041	0.9075	0.9123	0.9181	0.9250	0.9326	0.9408	0.9493
155	0.9018	0.9025	0.9047	0.9081	0.9128	0.9186	0.9254	0.9330	0.9412	0.9497
156	0.9024	0.9031	0.9052	0.9087	0.9133	0.9191	0.9259	0.9334	0.9415	0.9500
157	0.9030	0.9037	0.9058	0.9092	0.9139	0.9196	0.9264	0.9338	0.9419	0.9503
158	0.9036	0.9043	0.9064	0.9098	0.9144	0.9201	0.9268	0.9343	0.9423	0.9506
159	0.9042	0.9049	0.9069	0.9103	0.9149	0.9206	0.9272	0.9347	0.9426	0.9509
160	0.9048	0.9054	0.9075	0.9108	0.9154	0.9211	0.9277	0.9350	0.9430	0.9512
161	0.9053	0.9060	0.9080	0.9114	0.9159	0.9215	0.9281	0.9354	0.9433	0.9515
162	0.9059	0.9066	0.9086	0.9119	0.9164	0.9220	0.9285	0.9358	0.9436	0.9518
163	0.9064	0.9071	0.9091	0.9124	0.9169	0.9225	0.9290	0.9362	0.9440	0.9521
164	0.9070	0.9076	0.9096	0.9129	0.9174	0.9229	0.9294	0.9366	0.9443	0.9524
165	0.9075	0.9082	0.9102	0.9134	0.9179	0.9234	0.9298	0.9370	0.9446	0.9526
166	0.9080	0.9087	0.9107	0.9139	0.9183	0.9238	0.9302	0.9373	0.9450	0.9529
167	0.9086	0.9092	0.9112	0.9144	0.9188	0.9243	0.9306	0.9377	0.9453	0.9532
168	0.9091	0.9097	0.9117	0.9149	0.9193	0.9247	0.9310	0.9380	0.9456	0.9535
169	0.9096	0.9103	0.9122	0.9154	0.9197	0.9251	0.9314	0.9384	0.9459	0.9537
170	0.9101	0.9108	0.9127	0.9159	0.9202	0.9255	0.9318	0.9387	0.9462	0.9540
171	0.9106	0.9113	0.9132	0.9163	0.9206	0.9260	0.9322	0.9391	0.9465	0.9543
172	0.9111	0.9117	0.9137	0.9168	0.9211	0.9264	0.9326	0.9394	0.9468	0.9545
173	0.9116	0.9122	0.9141	0.9173	0.9215	0.9268	0.9329	0.9398	0.9471	0.9548
174	0.9121	0.9127	0.9146	0.9177	0.9219	0.9272	0.9333	0.9401	0.9474	0.9550
175	0.9126	0.9132	0.9151	0.9182	0.9224	0.9276	0.9337	0.9404	0.9477	0.9553
176	0.9130	0.9137	0.9155	0.9186	0.9228	0.9280	0.9340	0.9408	0.9480	0.9555
177	0.9135	0.9141	0.9160	0.9191	0.9232	0.9284	0.9344	0.9411	0.9483	0.9558
178	0.9140	0.9146	0.9164	0.9195	0.9236	0.9288	0.9348	0.9414	0.9486	0.9560
179	0.9144	0.9150	0.9169	0.9199	0.9240	0.9292	0.9351	0.9417	0.9489	0.9563
180	0.9149	0.9155	0.9173	0.9203	0.9245	0.9295	0.9355	0.9421	0.9491	0.9565
181	0.9153	0.9159	0.9178	0.9208	0.9249	0.9299	0.9358	0.9424	0.9494	0.9567
182	0.9158	0.9164	0.9182	0.9212	0.9252	0.9303	0.9361	0.9427	0.9497	0.9570
183	0.9162	0.9168	0.9186	0.9216	0.9256	0.9306	0.9365	0.9430	0.9499	0.9572
184	0.9167	0.9173	0.9191	0.9220	0.9260	0.9310	0.9368	0.9433	0.9502	0.9574
185	0.9171	0.9177	0.9195	0.9224	0.9264	0.9314	0.9371	0.9436	0.9505	0.9576
186	0.9175	0.9181	0.9199	0.9228	0.9268	0.9317	0.9375	0.9439	0.9507	0.9579
187	0.9179	0.9185	0.9203	0.9232	0.9272	0.9321	0.9378	0.9442	0.9510	0.9581
188	0.9184	0.9189	0.9207	0.9236	0.9275	0.9324	0.9381	0.9444	0.9512	0.9583
189	0.9188	0.9194	0.9211	0.9240	0.9279	0.9328	0.9384	0.9447	0.9515	0.9585
190	0.9192	0.9198	0.9215	0.9244	0.9283	0.9331	0.9387	0.9450	0.9517	0.9587
191	0.9196	0.9202	0.9219	0.9248	0.9286	0.9335	0.9391	0.9453	0.9520	0.9589
192	0.9200	0.9206	0.9223	0.9251	0.9290	0.9338	0.9394	0.9456	0.9522	0.9592
193	0.9204	0.9210	0.9227	0.9255	0.9294	0.9341	0.9397	0.9458	0.9525	0.9594
194	0.9208	0.9214	0.9231	0.9259	0.9297	0.9344	0.9400	0.9461	0.9527	0.9596
195	0.9212	0.9217	0.9234	0.9262	0.9301	0.9348	0.9403	0.9464	0.9530	0.9598
196	0.9216	0.9221	0.9238	0.9266	0.9304	0.9351	0.9406	0.9467	0.9532	0.9600
197	0.9219	0.9225	0.9242	0.9270	0.9307	0.9354	0.9409	0.9469	0.9534	0.9602
198	0.9223	0.9229	0.9246	0.9273	0.9311	0.9357	0.9411	0.9472	0.9537	0.9604
199	0.9227	0.9233	0.9249	0.9277	0.9314	0.9360	0.9414	0.9474	0.9539	0.9606
200	0.9231	0.9236	0.9253	0.9280	0.9317	0.9364	0.9417	0.9477	0.9541	0.9608
201	0.9234	0.9240	0.9256	0.9284	0.9321	0.9367	0.9420	0.9479	0.9543	0.9610
202	0.9238	0.9243	0.9260	0.9287	0.9324	0.9370	0.9423	0.9482	0.9545	0.9611
203	0.9242	0.9247	0.9264	0.9290	0.9327	0.9373	0.9426	0.9484	0.9548	0.9613
204	0.9245	0.9251	0.9267	0.9294	0.9330	0.9376	0.9428	0.9487	0.9550	0.9615
205	0.9249	0.9254	0.9270	0.9297	0.9334	0.9379	0.9431	0.9489	0.9552	0.9617
206	0.9252	0.9258	0.9274	0.9300	0.9337	0.9381	0.9434	0.9492	0.9554	0.9619
207	0.9256	0.9261	0.9277	0.9304	0.9340	0.9384	0.9436	0.9494	0.9556	0.9621
208	0.9259	0.9264	0.9281	0.9307	0.9343	0.9387	0.9439	0.9497	0.9558	0.9622
209	0.9262	0.9268	0.9284	0.9310	0.9346	0.9390	0.9442	0.9499	0.9560	0.9624
210	0.9266	0.9271	0.9287	0.9313	0.9349	0.9393	0.9444	0.9501	0.9562	0.9626
211	0.9269	0.9275	0.9290	0.9316	0.9352	0.9396	0.9447	0.9504	0.9564	0.9628
212	0.9273	0.9278	0.9294	0.9320	0.9355	0.9398	0.9449	0.9506	0.9566	0.9629
213	0.9276	0.9281	0.9297	0.9323	0.9358	0.9401	0.9452	0.9508	0.9568	0.9631
214	0.9279	0.9284	0.9300	0.9326	0.9361	0.9404	0.9454	0.9510	0.9570	0.9633
215	0.9282	0.9288	0.9303	0.9329	0.9364	0.9407	0.9457	0.9513	0.9572	0.9634
216	0.9286	0.9291	0.9306	0.9332	0.9366	0.9409	0.9459	0.9515	0.9574	0.9636

Table A-2

ATOM	Z	180.	170.	160.	150.	140.	130.	120.	110.	100.	90.
B	5	0.2112	0.2136	0.2209	0.2334	0.2517	0.2767	0.3091	0.3500	0.4090	0.4593
C	6	0.2501	0.2526	0.2604	0.2736	0.2929	0.3187	0.3519	0.3929	0.4423	0.5001
N	7	0.3086	0.3113	0.3194	0.3333	0.3531	0.3795	0.4127	0.4532	0.5009	0.5555
O	8	0.3597	0.3625	0.3708	0.3848	0.4047	0.4309	0.4635	0.5027	0.5433	0.5398
F	9	0.4251	0.4278	0.4360	0.4498	0.4693	0.4946	0.5258	0.5627	0.6050	0.6523
NE	10	0.4472	0.4499	0.4580	0.4717	0.4909	0.5158	0.5463	0.5822	0.6233	0.6687
NA	11	0.4948	0.4974	0.5053	0.5185	0.5370	0.5637	0.5896	0.6233	0.6615	0.7034
MG	12	0.5143	0.5169	0.5246	0.5375	0.5556	0.5789	0.6069	0.6397	0.6767	0.7171
AL	13	0.5500	0.5525	0.5600	0.5724	0.5897	0.6117	0.6384	0.6693	0.7040	0.7416
SI	14	0.5632	0.5657	0.5730	0.5852	0.6022	0.6238	0.6499	0.6801	0.7139	0.7505
P	15	0.5946	0.5970	0.6040	0.6156	0.6318	0.6523	0.6770	0.7054	0.7370	0.7711
S	16	0.6053	0.6075	0.6144	0.6259	0.6417	0.6619	0.6860	0.7138	0.7447	0.7779
CL	17	0.6352	0.6374	0.6440	0.6548	0.6698	0.6887	0.7114	0.7374	0.7662	0.7970
AR	18	0.6689	0.6709	0.6770	0.6871	0.7010	0.7186	0.7395	0.7634	0.7897	0.8179
K	19	0.6630	0.6650	0.6712	0.6814	0.6955	0.7133	0.7344	0.7588	0.7856	0.8142
CA	20	0.6698	0.6718	0.6779	0.6880	0.7019	0.7194	0.7403	0.7641	0.7904	0.8184
SC	21	0.6897	0.7016	0.7073	0.7166	0.7295	0.7457	0.7650	0.7869	0.8139	0.8365
TI	22	0.7152	0.7170	0.7224	0.7314	0.7437	0.7592	0.7776	0.7985	0.8214	0.8457
V	23	0.7298	0.7316	0.7368	0.7454	0.7572	0.7720	0.7896	0.8095	0.8312	0.8543
CR	24	0.7345	0.7362	0.7414	0.7498	0.7615	0.7761	0.7934	0.8129	0.8344	0.8570
MN	25	0.7468	0.7485	0.7534	0.7615	0.7727	0.7867	0.8033	0.8221	0.8425	0.8642
FE	26	0.7504	0.7520	0.7569	0.7649	0.7760	0.7898	0.8062	0.8247	0.8449	0.8662
CO	27	0.7618	0.7634	0.7681	0.7758	0.7864	0.7997	0.8154	0.8331	0.8524	0.8728
NI	28	0.7612	0.7628	0.7675	0.7752	0.7859	0.7992	0.8152	0.8329	0.8522	0.8726
CU	29	0.7770	0.7785	0.7829	0.7902	0.8002	0.8129	0.8276	0.8442	0.8624	0.8815
ZN	30	0.7825	0.7839	0.7883	0.7954	0.8052	0.8175	0.8319	0.8482	0.8659	0.8846
GA	31	0.7946	0.7960	0.8001	0.8069	0.8162	0.8279	0.8415	0.8570	0.8738	0.8914
GE	32	0.8020	0.8033	0.8073	0.8139	0.8229	0.8342	0.8475	0.8624	0.8786	0.8956
AS	33	0.8074	0.8087	0.8126	0.8191	0.8279	0.8389	0.8519	0.8663	0.8820	0.8986
SE	34	0.8163	0.8176	0.8213	0.8275	0.8359	0.8464	0.8588	0.8727	0.8877	0.9035
BR	35	0.8183	0.8195	0.8232	0.8293	0.8377	0.8481	0.8603	0.8741	0.8890	0.9046
KR	36	0.8259	0.8271	0.8307	0.8366	0.8446	0.8546	0.8664	0.8796	0.8938	0.9088
RB	37	0.8290	0.8302	0.8337	0.8395	0.8474	0.8573	0.8689	0.8818	0.8958	0.9105
SR	38	0.8329	0.8340	0.8375	0.8431	0.8509	0.8605	0.8719	0.8845	0.8982	0.9126
Y	39	0.8351	0.8362	0.8396	0.8452	0.8529	0.8624	0.8736	0.8861	0.8996	0.9138
ZR	40	0.8389	0.8400	0.8433	0.8488	0.8563	0.8656	0.8765	0.8888	0.9020	0.9159
NB	41	0.8416	0.8427	0.8460	0.8514	0.8588	0.8679	0.8787	0.8907	0.9038	0.9174
MO	42	0.8461	0.8471	0.8503	0.8556	0.8628	0.8717	0.8822	0.8939	0.9066	0.9198
TC	43	0.0	0.0	0.0	0.0	0.0	0.0	0.0	0.0	0.0	0.0
RU	44	0.8532	0.8542	0.8573	0.8623	0.8692	0.8777	0.8877	0.8989	0.9110	0.9236
RH	45	0.8558	0.8569	0.8599	0.8648	0.8716	0.8800	0.8899	0.9008	0.9127	0.9251
PD	46	0.8603	0.8613	0.8642	0.8690	0.8756	0.8838	0.8933	0.9040	0.9155	0.9276
AG	47	0.8620	0.8630	0.8659	0.8706	0.8771	0.8852	0.8946	0.9052	0.9165	0.9284
CD	48	0.8673	0.8682	0.8710	0.8756	0.8818	0.8896	0.8987	0.9089	0.9199	0.9313
IN	49	0.8698	0.8707	0.8735	0.8780	0.8841	0.8917	0.9007	0.9106	0.9214	0.9326
SN	50	0.8738	0.8747	0.8773	0.8817	0.8877	0.8951	0.9037	0.9134	0.9239	0.9348
SB	51	0.8767	0.8776	0.8802	0.8845	0.8903	0.8976	0.9060	0.9155	0.9257	0.9363
TE	52	0.8820	0.8829	0.8854	0.8895	0.8951	0.9020	0.9102	0.9192	0.9290	0.9392
I	53	0.8814	0.8823	0.8848	0.8889	0.8945	0.9015	0.9097	0.9188	0.9286	0.9388
XE	54	0.8852	0.8860	0.8885	0.8925	0.8979	0.9047	0.9126	0.9215	0.9310	0.9409
CS	55	0.8865	0.8873	0.8897	0.8937	0.8991	0.9058	0.9136	0.9223	0.9317	0.9415
BA	56	0.8899	0.8907	0.8931	0.8969	0.9021	0.9086	0.9162	0.9247	0.9338	0.9434
LA	57	0.8911	0.8919	0.8942	0.8980	0.9032	0.9096	0.9172	0.9256	0.9346	0.9440
CE	58	0.8919	0.8927	0.8950	0.8988	0.9039	0.9103	0.9178	0.9261	0.9350	0.9444
PR	59	0.8926	0.8933	0.8956	0.8994	0.9045	0.9109	0.9183	0.9266	0.9355	0.9448
ND	60	0.8949	0.8956	0.8979	0.9016	0.9066	0.9128	0.9201	0.9282	0.9369	0.9460
PM	61	0.0	0.0	0.0	0.0	0.0	0.0	0.0	0.0	0.0	0.0
SM	62	0.8989	0.8997	0.9018	0.9054	0.9102	0.9162	0.9232	0.9310	0.9394	0.9481
EU	63	0.9000	0.9007	0.9028	0.9064	0.9111	0.9171	0.9240	0.9317	0.9400	0.9487
GD	64	0.9032	0.9039	0.9059	0.9094	0.9140	0.9197	0.9265	0.9339	0.9420	0.9504
TB	65	0.9041	0.9048	0.9069	0.9103	0.9149	0.9206	0.9272	0.9346	0.9426	0.9509
DY	66	0.9061	0.9067	0.9088	0.9121	0.9166	0.9221	0.9287	0.9359	0.9437	0.9518
HO	67	0.9075	0.9081	0.9101	0.9134	0.9178	0.9233	0.9299	0.9369	0.9446	0.9526
ER	68	0.9087	0.9094	0.9113	0.9145	0.9189	0.9244	0.9307	0.9378	0.9454	0.9533
TM	69	0.9096	0.9102	0.9122	0.9154	0.9197	0.9251	0.9314	0.9384	0.9459	0.9537
YB	70	0.9116	0.9122	0.9142	0.9173	0.9215	0.9268	0.9329	0.9398	0.9471	0.9548
LU	71	0.9125	0.9132	0.9151	0.9182	0.9224	0.9276	0.9337	0.9404	0.9477	0.9553
HF	72	0.9143	0.9149	0.9168	0.9198	0.9239	0.9290	0.9350	0.9417	0.9488	0.9562
TA	73	0.9153	0.9159	0.9178	0.9207	0.9249	0.9299	0.9358	0.9423	0.9494	0.9567
W	74	0.9166	0.9172	0.9190	0.9219	0.9260	0.9309	0.9369	0.9432	0.9502	0.9574
RE	75	0.9176	0.9182	0.9200	0.9229	0.9269	0.9318	0.9375	0.9439	0.9508	0.9579
OS	76	0.9197	0.9203	0.9221	0.9249	0.9288	0.9337	0.9393	0.9456	0.9523	0.9593
IR	77	0.9201	0.9207	0.9224	0.9252	0.9291	0.9339	0.9394	0.9456	0.9523	0.9592
PT	78	0.9212	0.9218	0.9235	0.9263	0.9301	0.9348	0.9403	0.9464	0.9530	0.9598
AU	79	0.9219	0.9225	0.9242	0.9270	0.9307	0.9354	0.9408	0.9469	0.9534	0.9602
HG	80	0.9234	0.9239	0.9256	0.9283	0.9320	0.9366	0.9420	0.9479	0.9543	0.9610
TL	81	0.9246	0.9252	0.9268	0.9295	0.9332	0.9377	0.9429	0.9488	0.9551	0.9616
PB	82	0.9255	0.9260	0.9276	0.9303	0.9339	0.9383	0.9435	0.9493	0.9555	0.9619
BI	83	0.9262	0.9268	0.9284	0.9310	0.9346	0.9390	0.9442	0.9499	0.9560	0.9624

Chapter 12

THE ACOUSTIC MICROSCOPE: A TOOL FOR NONDESTRUCTIVE TESTING

J. Attal

Centre d'Etudes d'Electroniques des Solides

Universite des Sciences et Techniques du Languedoc

Place de Bataillon

34060-Montpelier-Cedex, France

I. INTRODUCTION

Historically the microscope has proved to be one of the most powerful scientific tools. This has been especially true in the biological sciences where many of the most significant advances have been founded on microscopic observations. Moreover, each time a microscope based on a new class of radiation has been developed, our understanding of the microscopic structure in nature has been extended. The introduction of acoustic radiation to microscopy can be expected to have a similar impact. This is the motivation underlying the development of the acoustic microscope.

It can be surprising that this instrument has required such a long time to become operational since the analogy between acoustics and optics had been demonstrated by the late nineteenth century. The difficulties arose in designing an instrument of this type since it is not obvious how one might visualize the pattern of acoustic energy that reproduces the essential features of the specimen that is to be viewed. This problem does not exist with the optical microscope but with acoustic waves there is nothing equivalent to either the human eye or photographic film.

Unlike other forms of radiation, an acoustic wave interacts directly with the elastic properties of the material through which it propagates. By using the acoustic field to form an image we can study the spatial variations in these properties directly. This ability makes the following two classes of objects particularly interesting candidates for acoustic microscopy:

(a) The first group concerns samples obtained from living systems. Cells or tissues characteristically show very little intrinsic optical or electron contrast. Accordingly, the biologists have devoted considerable effort toward developing specific stains to provide the necessary contrast. Unfortunately these techniques are not universally applicable nor do they leave the specimen in its natural state. Beyond improved contrast, an acoustic micrograph can give fundamental information about the elastic properties of a biological specimen without damaging the natural structural relationships. This is the acoustic microscope's unique capability.

(b) The second class of samples of particular interest in acoustic microscopy is comprised of materials which are completely opaque optically. In most cases these materials can be readily penetrated by an acoustic wave. With an acoustic microscope, detail lying beneath the surface of such a sample could be revealed. This capability could find application to the problems of non destructive testing on a microscopic scale.

The challenge in designing an acoustic microscope is to devise a means of visualizing the acoustic properties of a specimen with the highest possible resolution. In any microscope the finest detail which can be resolved is determined by the wavelength of the radiation that is used. For example, the light of the visible spectrum is centered about a wavelength of 0.5 µm. Thus the light microscope can give us information about a sample down to this level of size. In many materials the sound velocity is five orders of magnitude less than the velocity of light. Acoustic waves with frequencies in the gigahertz range therefore have wavelengths comparable to those of visible light. During the past ten years, a technology has become available which allows acoustic waves in this range to be readily generated. With this advance the way was cleared for the development of an acoustic microscope with resolution rivaling that of the best optical microscope.

II. SCANNING ACOUSTIC MICROSCOPE DESIGN IN TRANSMISSION
OPERATING MODE

A. Introduction

The acoustic microscope is an instrument designed to exploit the intrinsic ultrasonic properties of matter. According to the size of the object to be acoustically viewed and the advance in technology many systems have been developed. The resolution capability of this

instrument ranges between 1 mm when operating at 1 MHz to 1 μm at
1 GHz. Low frequency acoustic waves have been used for a number of
years to look at relatively gross biological objects i.e. bone
tumors, and other objects of that size. Waves of this frequency
range do have the advantage of being able to penetrate a considerable
distance into the human body, but they can distinguish only the
larger structures. This technique is also used in aeronautics to
detect eventual cracks inside the wings of aircrafts.

To achieve resolution comparable with that of the optical
microscope it is necessary to operate in the frequency range from
1000 to 3000 MHz, the so-called microwave acoustics region. Only,
in recent years there have been advances in microwaves acoustics
technology to work at that frequencies. Indeed, it was necessary
to await the development of transducers which could serve to effic-
iently convert electromagnetic energy into acoustic energy. It
was also necessary to await for creative suggestions as to how the
acoustic patterns could be rendered visible. This occurred in the
last ten years at three different locations in USA and England.

The first technique developed at University College London
consisted of reading the acoustic image with a scanning focused
laser beam employing light scattering processes. At Stanford
University they developed a non-scanning system that utilized the
acoustic radiation pressure on one micron latex spheres immersed
in liquid. The radiation pressure was sufficient to condense the
latex spheres into a pattern that reproduced on a one-to-one scale
the acoustic pattern. The image as developed by the latex spheres
was viewed with a conventional optical microscope. The resolution
of the instrument was close to 10 microns. These two systems
suffered from two disadvantes since they were dependent upon some
form of light in the final readout of the image, the ultimate
resolution must always be less than the optical micrsocope, and
the required level of acoustic power was rather excessive (1 mW/cm^2).
Furthermore, the required intensity increased when the acoustic
frequency was raised to improve the resolution.

For these reasons, Quate and Lemons at Stanford University [1]
turned to a system which used piezoelectric films for generating
and detecting the acoustic waves. They combined this with mechanical
scanning in order to record the entire image. In this system the
acoustic intensity is reduced to a minimum because of the high
efficiency of the piezoelectric detector. Optical waves are not
used in any part of the acoustic system and the wavelength of light
does not, therefore, influence the ultimate resolution that might
be attainable. The resolution of the acoustic microscope now
stands at a value slightly better than the optical instrument.
This is somewhat ironic since the physical principles which establish the resolution in the two cases are entirely different.

B. Outline of System

The primary problem that has to be faced with any acoustic imaging device is the absence of a medium equivalent to the emulsion of a photographic film. One must devise a different method for recording the image. A scanning technique using acoustic lenses has been adopted wherein the acoustic absorption is recorded point by point across the specimen. This information modulates the intensity of the electron beam of a TV monitor.

C. Lens Element Design

The basic confocal lens geometry used in the acoustic microscope is shown schematically in Fig. 1. This system consists of a symmetric pair of lens elements connected by a small volume of liquid. In practice this liquid is water and is held in place by surface tension. Each lens element is formed by a polished concave spherical surface in the end face of a crystal rod. At the opposite end of the rod a thin film piezoelectric transducer is centered on the axis of the lens surface. The input transducer converts an oscillatory electromagnetic field into an acoustic wave of the same frequency. In the crystal the acoustic energy propagates as an approximately collimated beam until it is refracted at the lens surface. Since the propagation velocity in the liquid is generally much less than it is in the crystal, the beam in the liquid is sharply focused. The function of the transmitter element is to generate an acoustic probing beam of micrometer size at its focus. When the object is scanned through this focus the amplitude and phase of the transmitted acoustic wave will be modulated by the structure.

An identical receiver element faces the transmitter in a confocal position. With this arrangement the receiver lens will collect and collimate the transmitted acoustic energy. Finally, at the output transducer the oscillatory strain of the acoustic field is converted back into an electromagnetic signal. In addition to the resolution advantage discussed in chapter II the confocal geometry greatly reduces background interference. The signal that is detected represents information which only comes from the focus.

D. The Mechanical Scanning System

1. <u>General Description</u>. Figure 2 shows a generalized diagram of the mechanical assembly used in the microscope. For clarity the lenses have been drawn apart as they are when a specimen is mounted. There are three design requirements for the mechanical scanning system used in the microscope. First, the specimen must be moved with a precision and repeatability that exceeds the limits set by the resolution of the lenses. Secondly, the scan should provide enough

THE ACOUSTIC MICROSCOPE

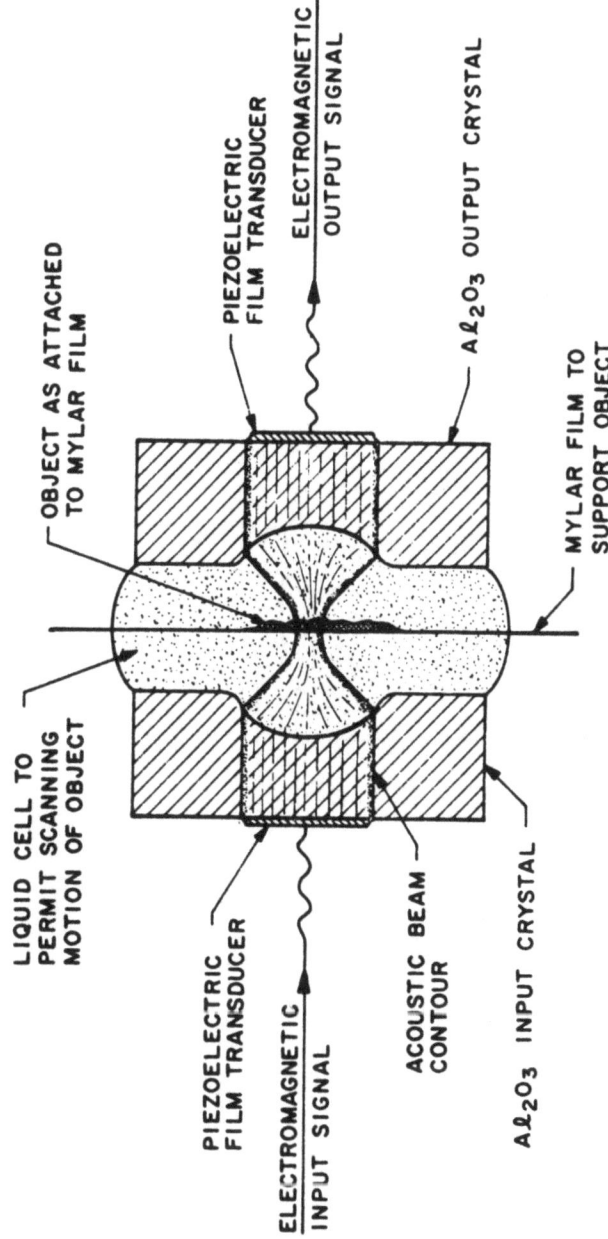

Fig. 1. Schematic diagram of the acoustic microscope lens configuration

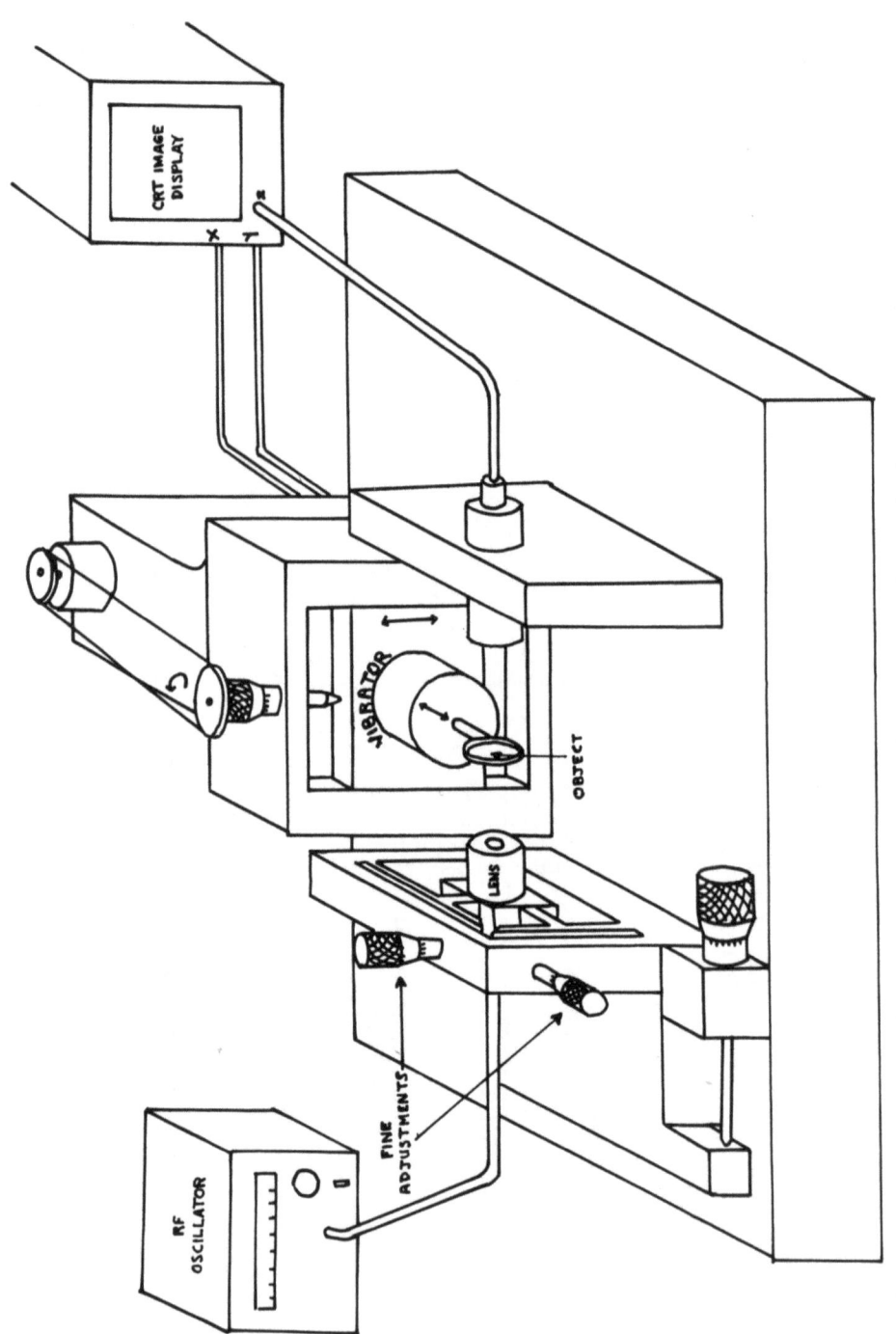

Fig. 2. Generalized diagram of the scanning acoustic microscope.

field of view to include 10^5 to 10^6 resolution elements. For a 500 MMz microscope this would require approximately 5 mm of motion in both the vertical and horizontal directions. Lastly, the scan should be carried out rapidly enough to complete an image in a few seconds or less.

For convenience the mechanical scanning is carried out in a raster pattern. The fast line scan is obtained from the movement of a vibrator. To constrain the motion to one dimension, centering springs are used to guide the specimen ring mounted on the vibrating rod. This mechanism provides the required accuracy of motion. The total mass of the specimen ring, springs and the steel rod is approximately 10 g. With this loading at the frequency of 30 Hz, the vibrator can easily produce 5 mm peak to peak amplitude. In practice the vibration is driven sinusoidally to minimize the deterioration of the resolution produced by transients. The line scan on the displays is, of course, also driven sinusoidally. Although this produces a variation in the velocity of the display beam, it is not very apparent as a brightness variation across the screen. The scan from line to line is accomplished by mounting the entire vibrator assembly on a precision translation stage. This stage is mounted vertically and is moved with the aid of a micromotor acting as a micrometric head. An inductive pick-up is used to transduce the position of the micrometer.

The entire scanning assembly is mounted on a crossed pair of translation stages as shown in Fig. 2. One of these stages serves to make coarse adjustments of the horizontal position of the object. The other one provides course and fine focussing adjustments for the object.

2. <u>The Specimen Support</u>. In order to image with the acoustic microscope, the object must be scanned through the focused beam. During the scan it is essential for the object to remain within the depth of focus of the acoustic lenses. This is achieved by mounting the specimen on a planar support which in turn is attached to the mechanical scan system. In transmission microscopy this support consists of a 2 μm thick mylar membrane stretched across a metal snap ring. For its thickness the mylar is amazingly strong. This simplifies the sample preparation considerably. In addition it has good acoustic properties such as low sound velocity and low acoustic impedance.

3. <u>Confocal Alignment</u>. The most sensitive adjustment on the acoustic microscope is the confocal alignment of the transmitter and receiver lenses. A relative displacement of the lenses by a fraction of the wavelength either laterally or axially can reduce the output drastically. When operating at a frequency of 1 GHz this requires the lenses to be aligned in three dimensions with submicron precision. To achieve the lateral alignment one lens is

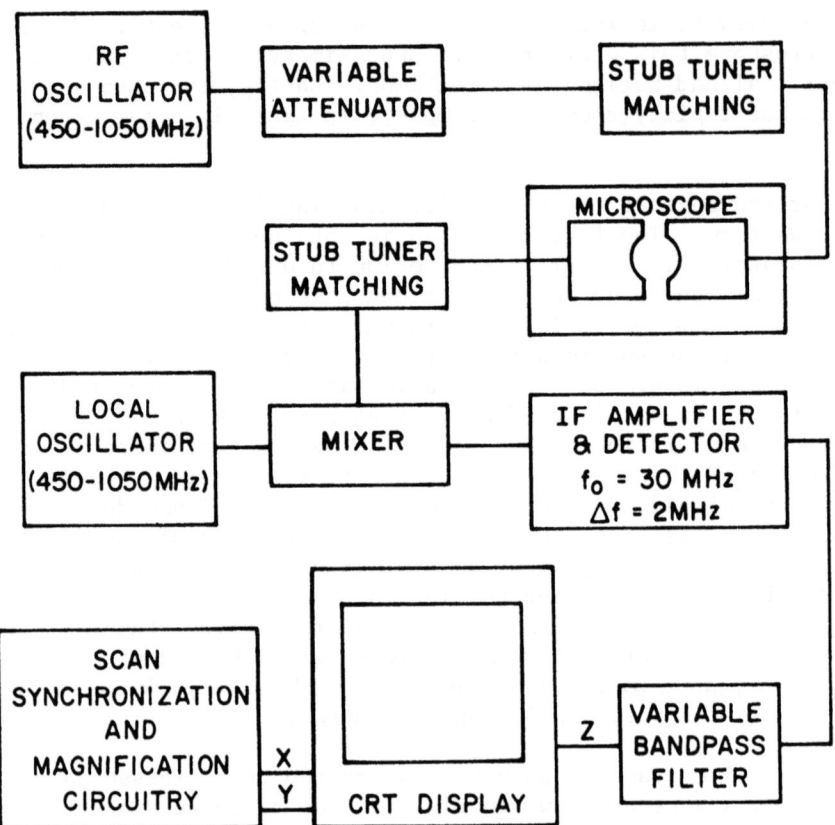

Fig. 3. Block diagram of the electronics associated with transmission imaging.

fixed while the second is attached to a pair of orthogonal elastic levers as shown in Fig. 2. A micrometer is used to position the free end of each of the levers. At the lens, a mechanical reduction by approximately a factor 20 is achieved by a combination of the lever ratio and flexure. The axial alignment of the lenses is achieved by mounting this entire assembly on a precision translation stage.

E. Instruments and Electronics

A block diagram of the electronics associated with the transmission mode of operation is shown in Fig. 3. In practice, the system is operated on a continuous wave basis. Conventional heterodyne detection is used in conjunction with a low noise IF amplifier. The IF amplifier contains an internal envelope detector. Thus, the output can be used to modulate the CRT display directly; however, a variable bandpass audio amplifier is usually included in

THE ACOUSTIC MICROSCOPE

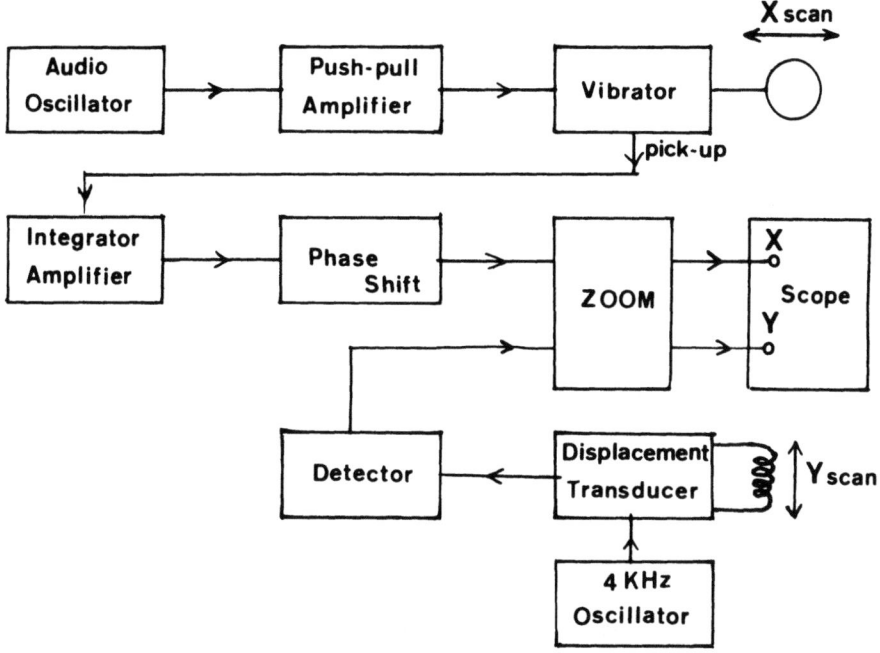

Fig. 4. Schematic diagram of the scan control circuit.

the circuit. Matching circuits are used at both the input and the output of the microscope to match the electrical impedance of the transducer to the impedance of the electronics.

The X-Y scan circuitry produces the electrical signals necessary to synchronize the beam on the display with the position of the object. A schematic diagram of this circuit is shown in Fig. 4. The oscillator which drives the vibrator also provides the voltage for the horizontal deflection on the display. Since there is a phase shift between the driving voltage and the mechanical motion of the vibrator a compensating, phase shift network is used in the X axis output circuit. This signal also triggers a pulser which blanks the return scan line. The vertical deflection signal is produced by a pick-up coil connected to the micrometer drive. Since the vertical and horizontal scans are independent, a provision is made to equalize the respective magnifications. Finally, the vertical and horizontal scan circuits are connected by a ganged potentiometer enabling a continuous variation of the overall magnification.

III. RESOLUTION PERFORMANCE

A. Introduction

The resolving power of the acoustic microscope is primarily determined by six factors:

- Transducer efficiency
- Quality of the lens: spherical aberrations, diffraction effects
- Absorption loss in the liquid cell
- Loss from acoustic impedance mismatch
- Power which misses the lens
- Electronic bandwidth and noise figures

In this chapter, we will look at all of them and discuss their influence on the final resolution of the microscope.

B. Transducer Efficiency

1. <u>Transducer Geometry</u>. The transducer structure consists of three layers (Fig. 5). On the face of the sapphire rod opposite the lens surface, a thin counter electrode of gold is deposited. This layer is generally 0.2 to 0.3 µm thick. Growing on a Z-cut face of the sapphire, the Au take a predominently [1 1 1̄] orientation. The piezoelectric ZnO layer is deposited by RF sputtering onto this Au substrate. The orientation of the Au layer facilitates the growth of a ZnO layer with maximum coupling to the longitudinal acoustic mode. This thickness of the ZnO film is usually chosen to be a quarter of the acoustic wavelength at a center frequency of approximately 750 MHz. With this design a bandwidth approaching 100% can be achieved. This enables us to work over the frequency range of 400 to 1100 MHz. Finally a circular top dot electrode of Al is deposited on the ZnO. The diameter of this top dot electrode determines the diameter of the collimated acoustic beam in the transmitter element of the microscope. It is chosen as we shall see in part II-D to be equal to the lens aperture.

It is also possible to make good efficiency transducers with lithium niobate thin plates. The technology is different. The face of the plate and the flat face of the sapphire are first coated with an indium layer at liquid nitrogen temperature to prevent the indium from oxidizing after being withdrawn from the bell jar. Then the lithium niobate and sapphire are bonded by pressing them together at room temperature without other precaution. The exact thickness is obtained by grinding and polishing the plate, for example, down

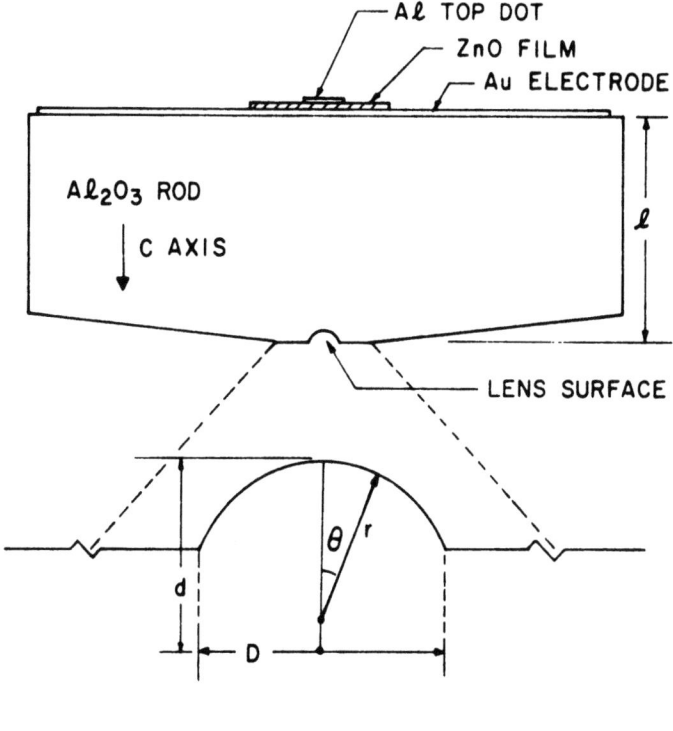

r = 0.135 mm
d = 0.156 mm
D = 0.207 mm
θ_{max} = 50°
F/ = 0.75
ℓ = 2.00 mm

Fig. 5. Design parameter for the high frequency sapphire-water acoustic lens.

to a few microns for operation at 1 gigahertz. A top dot of aluminium is then evaporated.

2. <u>Efficiency Measurements</u>. The ZnO transducers used on the microscope typically have an optimum two way insertion loss of ∼ 10 dB around 600 MHz when conjugately matched to the impedance of the electronics. Independent measurement of the transducer insertion loss of the lens elements is difficult because the lens surface disperses the reflected power. For this reason a test rod with parallel faces is mounted along with the lens element during transducer deposition. This test rod is then used to evaluate the

efficiency of the transducer by a pulse echo technique. For example Figure 6 shows a typical two way insertion loss for a transducer optimized at 2.2 GHz. Probably due to mechanical loss or quality of the ZnO film this optimized value increases with frequency.

C. Quality of the Lens

1. <u>Introduction</u>. As in optics, it is evident that the quality of the lens is the prime parameter which determined the resolving power of the acoustic microscope. The quality of the lens means the diameter of the acoustic beam at the focus. The better the quality, the less this diameter must be. Limitations are imposed by diffraction effects arising from a finite lens aperture and by the spherical aberration which is inherent to a spherical surface. In this part we will calculate the magnitude of this geometrical aberration and show that the large velocity difference at the interface solid-liquid makes spherical aberration a negligible factor in limiting the resolution of the single surface acoustic lens.

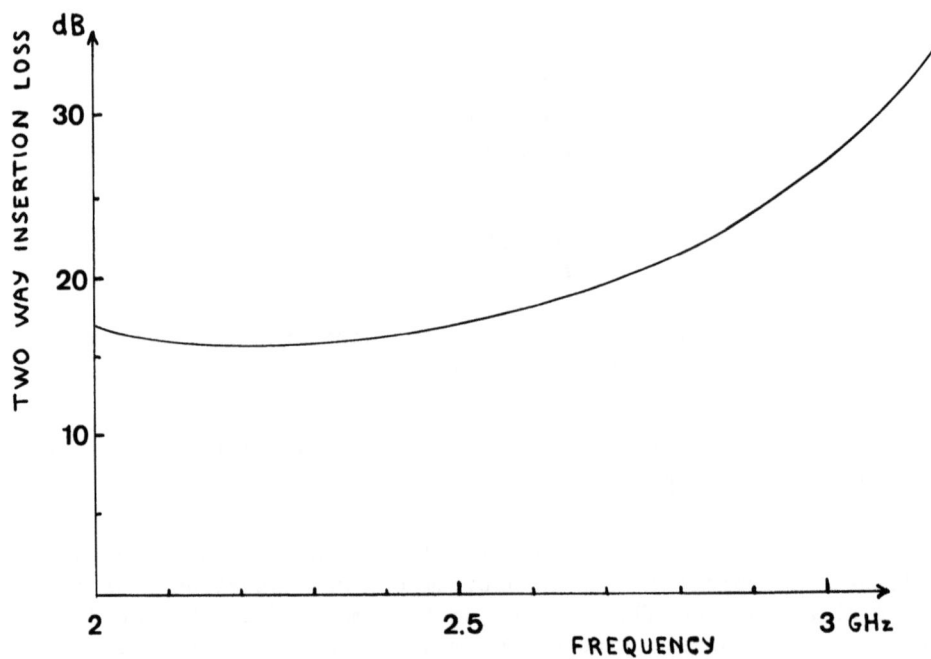

Fig. 6. Typical two-way insertion loss for a ZnO transducer optimized at 2.2 GHz.

2. Spherical Aberration. The lenses of the acoustic microscope as in most imaging systems consist of spherical interfaces between media with different propagation velocities. The use of a spherical surface is a practical one owing to the relative simplicity of grinding and polishing a spherical shape. This is particularly important for the acoustic microscope lenses since they must be made with very small radii of curvature due to the very high absorption of the liquid (see part III-4).

In most cases spherical surfaces do not produce an ideal image. The actual image is subject to a number of defects known as aberrations. Basically this type of aberration prevents a collimated beam from being focussed to a point even in the absence of diffraction. In order to evaluate the importance of spherical aberration to the imaging performance of the acoustic lens, it is useful to model it from two points of view.

First, using the techniques of ray tracing, the deviation of incoming rays from ideal focus can be calculated directly. The advantage of this approach is that it gives a somewhat intuitive feeling for the spreading of the beam focus due to the spherical aberration.

Consider the geometry of Fig. 7 in which a cross section of the acoustic lens is represented as a circle with its center at the origin of the coordinate system. The lens properties of this interface are specified by the ratio of C of the wave velocity in the liquid (c_2) to the wave velocity in the solid (c_1):

$$C = \frac{c_2}{c_1} \qquad (3\text{-}1)$$

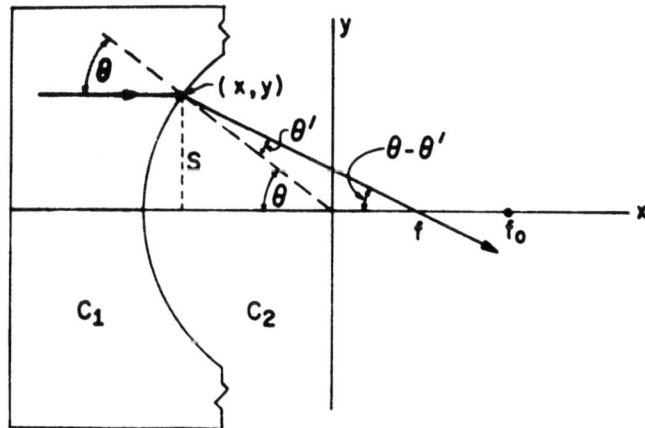

Fig. 7. Geometry for the ray tracing analysis of spherical aberration.

If the generating transducer is normal to the symmetry axis of the lens, the incoming rays will make an angle θ with the normal to the surface at the point (x,y). Because the incoming ray is parallel to the symmetry axis, θ is also the aperture angle of the point (x,y). For a value $C < 1$, such a ray will be refracted toward the axis, crossing it at a point f. The refracted angle θ' will, of course, be determined by Snell's law [2]:

$$\sin \theta' = C \sin \theta \tag{3-2}$$

For this geometry the crossover distance is given by the expression:

$$f = y/\tan(\theta - \theta') + x \tag{3-3}$$

Expanding $\tan(\theta - \theta')$ yields the result:

$$f = y \frac{1 + \tan\theta \tan\theta'}{\tan\theta - \tan\theta'} + x \tag{3-4}$$

In evaluating this expression, the algebra is considerably simplified if the dimensions are normalized by the radius of curvature of the spherical interface ($r = 1$) and if the substitution $\sin \theta = s$ is made. With this simplification, the following substitutions can be made:

$$y = s \; ; \; x = -\sqrt{1 - s^2} \quad \tan\theta = \frac{s}{\sqrt{1 - s^2}} \; ; \; \tan\theta' = \frac{Cs}{\sqrt{1-c^2 s^2}} \tag{3-5}$$

yielding the expression:

$$f = \frac{C}{\sqrt{1 - C^2 s^2} - C\sqrt{1 - s^2}} \tag{3-6}$$

For small values of s the denominator simplifies to give the familiar paraxial formula

$$f_0 = \frac{C}{1 - C} \tag{3-7}$$

The distance f_0, as measured from the center of curvature of the lens, will be designated the focal length of the single surface lens. To this degree of approximation, the focal length is independent of the incident angle and the lens provides ideal focussing. The exact expression (3-6), however, shows that rays incident on the lens at a finite distance from the axis cross it short of the paraxial focus (see Fig. 7). Moreover, the magnitude of this aberration increases rapidly as the aperture angle of the

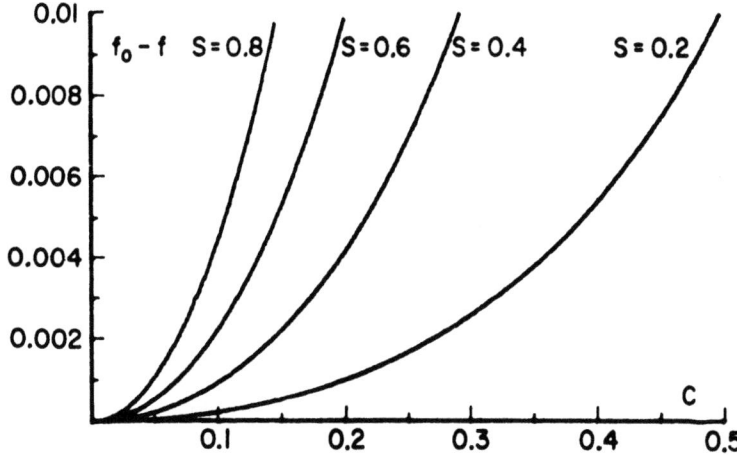

Fig. 8. Spherical aberration ($f_0 - f$ in units of r) as a function of the velocity ratio. Each curve is designated by the aperture angle of the incident ray ($s = \sin\theta$).

lens is expanded. This can be a grave difficulty for a high resolution system since large lens aperture angles are required to minimize the limitations of diffraction.

Inspection of equation (3-6) shows that the spherical aberration is also dependent upon the relative propagation velocity. It is this dependance which is exploited in the acoustic microscope. In Fig. 8, the difference between the paraxial focus and the crossover for a given ray ($f_0 - f$) is plotted as a function of C. Each curve thus represents a ray incident at a different ordinate y. These curves show that the spherical aberration can be greatly reduced if C is minimized. By chosing the materials for the acoustic lenses judiciously, extraordinarily small values of C can be achieved. That is the reason that sapphire was chosen as the material in which to grind the lens surface. Not only is sapphire an excellent transducer substrate and propagating medium, but the longitudinal acoustic velocity is 11.1×10^5 cm/s. Using water as the liquid component of the lens ($c_2 = 1.5 \times 10^5$ cm/s) gives a velocity ratio of $C = 0.135$. To appreciate how much reduction in spherical aberration this value of C provides, Figure 9 compares the ray tracing in the vicinity of the paraxial focus for a sapphire water acoustic lens with that for an analogous air-glass optical lens. The only difference between these two cases is the value of C used to generate the ray tracing. In optics with the ratio of light velocity equal to 0.667 the effects of shperical aberration are clearly evident. The foreshortened focusing of the outer rays has blurred the focus into a broad distribution of radiation. The point at which the diameter of this distribution is a minimum is known as the circle of least confusion. The size of this circle is approximately the limit on resolution imposed by spherical aberration. For an air-glass lens of Fig. 9, this resolution would be aberration limited to 40 μm. This is vastly worse than the diffraction limitation imposed by 0.5 μm wavelength.

In contrast, the spherical aberration for the acoustic sapphire water lens is negligible. Using a velocity ratio of $C = 0.135$ the diameter of the circle at least confusion for the acoustic lens of Fig. 9 is approximately 0.5 μm. For this reason the resolution of the lens is essentially limited by diffraction only. Accordingly, an optical microscope made analogously to the scanning acoustic microscope could not employ the simple single surface lens. One final point is that spherical aberration is geometrical in character. It therefore scales with the size of the lens. The lenses used at high frequencies have radii of 0.1 mm which is eight times smaller than those used in the calculation. The magnitude of the spherical aberration is accordingly eight times smaller.

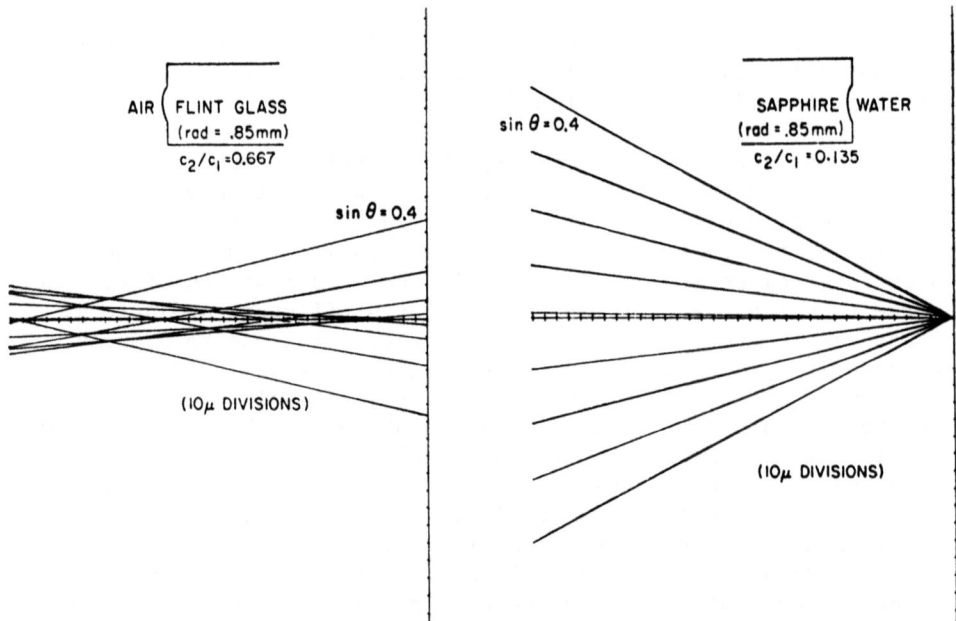

Fig. 9. Ray tracing comparison of the performance of a single surface lens in a light optical system (left) and an acoustic system (right). The paraxial focus lies at the ordinate.

As an alternative to the ray tracing approach, spherical aberration can be considered as a defect in the shape of the wavefront which emerges from the lens. Ideally the phase distribution at the exit pupil of the lens would be that of spherical wave converging to a single focal point. In the presence of spherical aberration the actual wavefront becomes distorted. The magnitude of the aberration can therefore be fully described by the discrepancy between the phase of the actual wavefront and that of the ideal spherical wave. It is possible to find an analytical expression for this phase error invoking Fermat's principle. Taking for illustration a sapphire water lens with an 0.4 mm radius of curvature we can show that the path error at an aperture angle of 50° would be approximately 0.4 μm. For a 500 MHz acoustic wave this is slightly more than an eighth of a wavelength.

3. <u>Diffraction Effects</u>. Now we have clearly established that spherical aberrations for a 0.2 mm diameter lens give at the focal point a circle of approximately 0.07 μm diameter. The diffraction pattern can be calculated as we do in optics. The only difference is the acoustic absorption in the liquid medium and the high refractive index of the lens. But this does not affect the classical result which is given by:

$$w = \frac{1.22 \lambda}{2 n \sin \theta_m} \qquad (3\text{-}8)$$

where n is the refractive index equal to 1/C and θ_m is the half angle aperture of the lens. For example, at 1 gigahertz, λ/n is equal to 1.5 μm and with θ_m equal to 60 degrees w is around 1 μm. Therefore, the limit in the resolution is imposed by the diffraction.

D. Absorption Loss in the Liquid Cell

1. <u>Generalities on Absorption in Liquids</u>. One of the primary objectives in designing lenses of the microscope is to achieve the highest possible resolution. Ultimately the resolution of the instrument is limited by the acoustic absorption in the liquid cell. Even though the attenuation of sound in water is less than that of most liquids (about 200 dB/mm at 1 GHz at 24°), it now appears that in this liquid, using the smallest lenses we can make, the maximum operating frequency will be 1500 MHz.

Prior to discussing the properties of specific liquids let us review the more important features of liquid absorption. The principal source of acoustic attenuation is the viscosity of the liquid and this is divided into two parts, a viscosity, η_s, arising from shear motion and a viscosity, η_v, arising from compression and rarefaction of the liquid. In addition there is a term associated with the conducting of heat which results from the cyclic variation of temperature created by the excess pressure at the crests of the wave. In most liquids, except metals, the heat conductivity loss due to variation in temperature is small compared to viscous losses. It can be neglected in most cases of interest here.

The waves propagate in the form $e^{j\omega t - jkz} e^{-\alpha z}$ where $k = \omega/c$ and α is the attenuation constant. The value of α is given by the equation:

$$\alpha = \frac{\omega^2}{2\rho c^3} (\eta_v + \frac{4}{3} \eta_s) \qquad (3-9)$$

or

$$\frac{\alpha}{f^2} = \frac{2\pi^2}{\rho c^3} (\eta_v + \frac{4}{3} \eta_s) \qquad (3-10)$$

For many liquids, α/f^2 is a function of frequency but for water cryogenic and metal liquids α/f^2 is a constant over a wide range of frequencies. Typically at 25°C, α/f^2 is equal to 22×10^{-17} cm^{-1} s^2. It is a point of reference which can be used to compare other liquids.

In the case of organic liquids this square law dependance of α with frequency is not valid. Other phenomena associated with motion of the molecule such as rotation and vibration give an important contribution to α at frequencies near 300 MHz.

2. <u>Coefficient of Merit for a Liquid</u>. In the microscope, the resolution is determined by the wavelength and, therefore, we want to characterize those liquids that will permit us to achieve the shortest wavelength λ. We can write this in the form:

$$\lambda^2 = \frac{c^2}{f^2} = \frac{c^2}{\alpha L} \frac{\alpha}{f^2} L \qquad (3-11)$$

L is the total attenuation through a cell of length L. The value of L is fixed by the geometry of the lens and we use the smallest lens so as to minimize this parameter. The value of αL is determined by the electronics of the system and we strive to maximize this value through a careful design of that system. The liquid properties are contained in the term $c^2(\alpha/f^2)$ and we search for the liquid which will give us the minimum value for this particular combination. Since water at 25°C is so convenient we choose this as a reference for comparing it with all other liquids. We, therefore, assign to each liquid a coefficient of merit M defined by:

$$M = \lambda_w/\lambda \qquad (3-12)$$

where λ and λ_w are respectively the shortest wavelength that we can achieve in the liquid and in water corresponding to the frequencies f and f_w. Equations (3-11) and (3-12) lead to the following expression of the coefficient of merit:

$$M = \frac{c_w}{c} \frac{(\alpha_w(f_w)/f_w^2)^{1/2}}{(\alpha(f)/f^2)^{1/2}} \qquad (3-13)$$

with the condition

$$\alpha(f) = \alpha_w(f_w) \qquad (3-14)$$

For some liquids where α/f^2 is independent of the frequency the coefficient of merit can be written:

$$M = \frac{c_w}{c} \frac{(\alpha/f^2)_w^{1/2}}{(\alpha/f^2)^{1/2}} \qquad (3-15)$$

For most of organic liquids we must know the frequency dependance of α and in such a case the coefficient of merit becomes:

$$M = \frac{c_w}{c} \frac{f}{f_w} \qquad (3-16)$$

with

$$f = \alpha^{-1} \alpha_w(f_w) \qquad (3-17)$$

With these definitions we see that high values of M will permit shorter wavelength and improved resolution. We have reported in Table 1 the acoustic properties of some liquids which have a coefficient of merit greater than one. We see that near room temperature only liquid metals on the list are substantially better than water. In the next chapter we shall see how it is possible to use these liquids when imaging solid state objects. However, with biological specimens, water remains a better choice.

E. Loss From Acoustic Impedance Mismatch

The benefits obtained by making the microscope lenses from materials which provide a large velocity ratio between the solid and liquid components have been discussed in section III-C.

TABLE I
ACOUSTIC PROPERTIES OF SOME LIQUIDS

LIQUID	TEMPERATURE	$\alpha/f^2 \times 10^{17} cm^{-1} sec^2$	Velocity $\times 10^5 cm\ sec^{-1}$	Coefficient of merit
Water	25° C	22	1.500	1
Gallium	30° C	1.58	2.870	1.82
Mercury	23.8° C	5.8	1.449	1.89
Nitrogen	73.9° K	10.6	0.962	2.10
Helium	4.22° K	260	0.183	2.23
Helium	0.45° K	5.8*	0.238	24
Hydrogen	17° K	5.6	1.187	2.34
Oxygen	87° K	8.6	0.952	2.36
Argon	85.2° K	10.1	0.853	2.43
Neon	27.09° K	23.2	0.595	2.45
Xenon	164° K	22	0.630	2.38

* This value has been extrapolated at 1 GHz from the measurements of Roach et al (Phys. Rev. Lett. 25, 1002 and Phys. Rev. A 5, 2205, (1972). In this range of temperature, the absorption α follows a linear law with frequency.

Unfortunately, maximizing the ratio of velocities also results in
a large acoustic impedance mismatch at the boundary. This acoustic
impedance Z is defined by the product of the density times the
velocity. This important acoustic parameter provides the reflection
and transmission coefficients at the interface of two media. For
instance with normal incidence these two coefficients are given by:

$$R = \frac{Z_1 - Z_2}{Z_1 + Z_2}^2 \qquad \text{reflection coefficient} \qquad (3\text{-}18)$$

$$T = \frac{4 Z_1 Z_2}{(Z_1 + Z_2)^2} \qquad \text{transmission coefficient} \qquad (3\text{-}19)$$

For the lenses employed in the acoustic microscope, the mismatch is
such that less than 15% of the power incident on the lens surface
will be transmitted. For the confocal geometry, the total power
transmitted through both lenses is reduced approximately 17 dB
by this effect.

It is possible to reduce this reflection loss. One may apply
thin layers of material to the lens surface. By choosing the
acoustic impedance and thickness of these layers properly, the
reflection at the interface can be greatly reduced. For example,
one solution to the problem of normally incident waves indicates
that an excellent single layer antireflection coating can be made
by applying a quarter wave thickness of a material whose acoustic
impedance is the geometric mean of those composing the interface.
For a sapphire-water interface, this would require a material with
an acoustic impedance of 8.1×10^5 g/cm^{2}s. Unfortunately, few
materials have an impedance in this range and in addition are not
damaged in contact with water. A compromise has been found using
SiO_2 which has an impedance of 13.1×10^5 g/cm^{2}s.

F. Power Which Misses the Lens

We have seen that the diameter of the aluminum top dot
electrode deposited on the ZnO transducer determines the
diameter of the collimated acoustic beam in the sapphire rod ended
with the lens. It is desirable to adjust the size of this top dot
to minimize the acoustic power which falls outside the aperture of
the lens. Such power is not only wasted but it can also interfere
coherently with the energy that is transmitted through the lens,
causing artifacts in the final image. In addition, it is desirable
to spread the input electrical power over as large an area as
possible to reduce the danger of burning the transducer. A compromise between these goals is obtained by making the top dot diameter

approximately equal to the lens aperture and by choosing the length of the sapphire rod to put the lens aperture at the Fresnel focus of the transducer. In Fig. 10, the axial distribution of normalized amplitude as a function of distance from a circular transducer is plotted. The distance to the final maximum is known as the Fresnel focal length. This length is given by the expression:

$$l_0 = \rho_0^2/\lambda_1 \qquad (3\text{-}20)$$

where ρ_0 is the radius of the transducer top dot and λ_1 is the wavelength in the sapphire. At this distance the diameter of the acoustic beam is approximately equal to the diameter of the transducer. The lateral distribution of amplitude at various planes parallel to the transducer is also shown schematically in Fig. 10. Between the transducer and the Fresnel focus there are large fluctuations in the lateral amplitude distribution. The phase distribution also fluctuates. Beyond the Fresnel focus both the amplitude and phase resemble the smoother Fraunhofer pattern of the far field. For best performance, it is usually desirable to illuminate the lens aperture with as uniform a phase and amplitude distribution as possible. To do this, the lens aperture should be placed at or beyond the Fresnel focus.

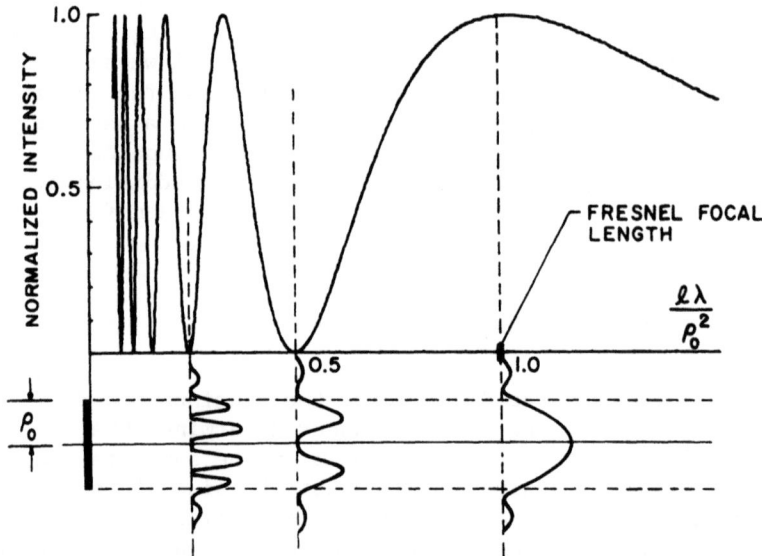

Fig. 10. The axial intensity distribution produced by a circular transducer of radius ρ_0 is plotted as a function of distance 1 from the transducer. λ is the wavelength in the medium. Below this curve approximate transverse intensity distributions are plotted for selected distances from the transducer.

THE ACOUSTIC MICROSCOPE

Unavoidably, some acoustic power will fall outside the lens aperture. To reduce the adverse effects, the surface of the sapphire around the lens is ground with a coarse abrasive to scatter this unwanted power. In addition, the area outside the lens is painted with an absorbing layer of plastic. Together these precautions greatly reduce troublesome interference effects from both incident radiation and multiple reflections in the crystal. One advantage of the large absorption in the liquid is that multiple reflections in the water cell are not a problem.

Power Required for Imaging

The electrical bandwidth that is required to display all of the information coming from the microscope is determined by the resolution, the scan rate and the field of view. For a typical scan rate of 160 lines per second and with 250 resolution elements per line, a 40 KHz bandwidth is required.

The least detectable signal power P_{min} is determined by the thermal noise level

$$P_{therm} = kT \, \Delta f \qquad (3-21)$$

where k is Boltzman's constant, T is the absolute temperature, and Δf is the effective noise bandwidth. For the detection system of Fig. 3, this bandwidth is given by:

$$\Delta f = (2 \, \Delta f_1 \, \Delta f_2)^{\frac{1}{2}} \qquad (3-22)$$

where Δf_1 is the bandwidth of the IF amplifier (2 MHz) and Δf_2 is the bandwidth of the final audio amplifier (40 KHz). With these parameters P_{therm} = -118 dBm. This noise level, along with the instrument loss and the noise figure for the electronics determines the sensitivity of the instrument. The noise figure for the mixer that is used is $\sim$ 7 dB, and the noise figure for the IF amplifier is $\sim$ dB. Thus, the minimum detectable power should be -49 dBm. The measured input power at 600 MHz for a signal to noise ratio of one is -47 dBm.

The dynamic range of the instrument is determined by the sensitivity level and the maximum allowable input power. The input power level is ultimately limited by the breakdown of the acoustic transducer. In practice the maximum power available from the RF oscillator that is used is +23 dBm. This gives a useful dynamic range of 70 dB.

Combining the losses which have been categorized above we would expect a total loss of 57 dB for the microscope operating at

600 MHz using 400 µm diameter lenses near room temperature. These losses can be decomposed as follows:

Transducer conversion losses (both	9 dB
Acoustic impedance mismatch (both lenses)	17 dB
Power missing the lens (both lenses)	4 dB
Absorption loss in water	28 dB
Total	58 dB

This corresponds closely with the measured loss of 59 dB.

Knowing the instrument loss and the sensitivity we can calculate the power density at the specimen. If we assume a Gaussian beam profile with a focal 3 dB width of 2 µm at 600 MHz, the peak power density at the sample is approximately 0.5 mW/cm^2 when the input power is equal to -47 dBm. Averaged over the scan, the power density is close to 1.25×10^{-5} mW/cm^2.

IV. ACOUSTIC IMAGES: NON DESTRUCTIVE EVALUATION OF SOLID STATE DEVICES

A. Introduction

The flexibility of a microscope derives from the variety of samples it can image and the kinds of information that can be extracted. In addition to the transmission mode described in the previous chapters, reflection, phase contrast, acoustic second harmonic imaging modes, dark field and stereo viewing have been demonstrated with the scanning acoustic microscope. Each of these techniques provides different information than is available from amplitude modulated transmission imaging. By comparing the acoustic micrographs obtained with these different modes of operation, a more thorough understanding of the elastic properties of the specimen is realized. A large part of this chapter will be devoted to reflection microscopy because of its direct applications to non-destructive evaluation in microelectronics.

B. Acoustic Transmission Image

1. <u>Resolution Check</u>. In Fig. 11 we have imaged a thin metal grid (2,3 millimeter diameter on top view) as immersed in a water cell. The operating frequency is 600 MHz and the scales from top to bottom are respectively 1 division equals 300 µm, 36 µm, and 12 µm. This is in agreement with the calculated predictions on the resolu-

Fig. 11. The acoustic photo of a metal grid immersed in water. Diameter of this grid is 2.3 mm. Scale from **left to right** is 1 division equal respectively to 300 µm, 36 µm, and 12 µm.

tion performance which also shows up the high accuracy of the scanning system.

2. Observation of Biological Samples. Prior to discussing the possibilities of the acoustic microscope in the field of non destructive evaluation of solid state devices, we will present a few acoustic micrographs of biological specimens to illustrate the potential of this instrument in that field. To date, it has been undoubtedly the widest area of application. The majority of these images were obtained with an acoustic frequency of 600 MHz (λ = 2.5 µm) and more recently at 900 MHz (λ = 1.7 µm) with a resulting resolution of approximately 1 µm. With this resolution, variations in the elastic properties within a single cell can be readily observed. The types of specimen which have been imaged can conveniently be divided into the following four categories: cell smears, sections of normal tissue, sections of tissue showing pathology and living cells. The acoustic micrographs of these samples as shown in Fig. 12 (a) and (b) demonstrate that the expectation of large intrinsic acoustic contrast has been borne out completely. Indeed, every specimen that has been examined has exhibited ample contrast to provide sharp, well-defined acoustic image without the staining so necessary to optical microscopy. More importantly, there are indications in these images that the acoustic microscope may have diagnostic applications. For example, collagen deposits are readily detected in an acoustic micrograph as regions of very high acoustic attenuation. Acoustic microscopy might therefore provide a fast and sensitive technique for analyzing abnormal distributions of collagen without the time consuming stains that are ordinarily required. The mechanical properties of each material in a sample are characterized by the density, the compliance tensor, and the viscosity tensor composed of both shear and volume components. These parameters will in turn determine the local

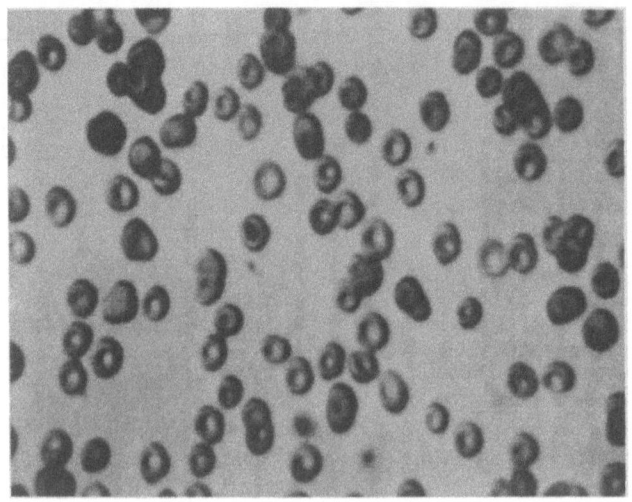

Fig. 12 (a). Normal red blood cell in the scanning acoustic microscope. X 225

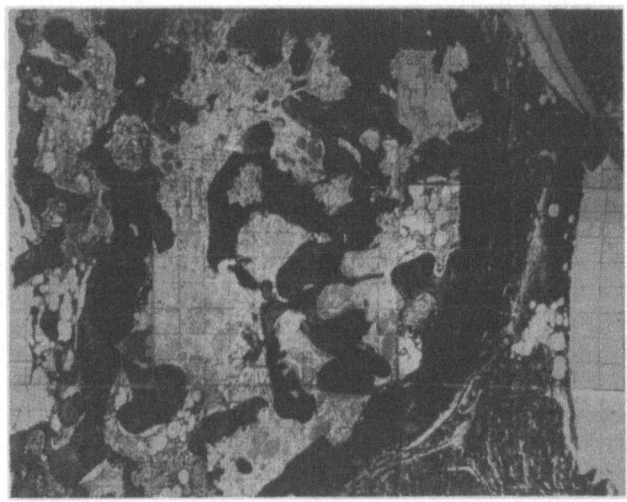

Fig. 12 (b). Section of bone tissue. X 100

acoustic velocity impedance and absorption. Thus, variations in the mechanical properties will cause changes in the amplitude and phase of the acoustic beam which passes through the sample. The changes in attenuation which are observed can be produced either directly by absorption or by scattering, arising from the acoustic impedance mismatch between a region of the specimen and the surrounding material. Using the detection circuitry described in Chapter II, only the amplitude information in the transmitted acoustic wave is displayed. In presenting the images, a convention was chosen so that points on the sample with greater acoustic transmission correspond to lighter regions of the image and dark areas correspond to regions of large acoustic attenuation.

C. Acoustic Reflection Operating Mode

1. <u>Selection of Samples</u>. For certain samples it is advantageous to form an image with reflected acoustic energy rather than transmitted energy. One important application of this mode of operation is the investigation of integrated circuits and solid state devices. Indeed, the silicon substrate for an integrated circuit is usually too thick to be accomodated in a transmission system if the lenses are kept in a confocal position. The higher acoustic velocity in most solids would also produce diffraction spreading and spherical aberration in transmission. The interesting details of an integrated circuit are, however, near the surface. This makes the integrated circuit particularly suitable for reflection imaging. However, it is not the only pole of interest in solid state physics and applications such as micrometallurgy, ion implanted surfaces, epitaxial layers of semiconductor have a large place in acoustic microscopy.

2. <u>Experimental Procedure</u>. Measurements in reflection are accomplished in the scanning acoustic microscope by using a single lens element. In this case, one transducer both generates and detects the acoustic signal. The generated acoustic wave is focussed into the water by the lens surface just as in the transmission mode. A portion of this energy will be reflected by a specimen held at the lens focus. The same lens will recollimate the reflected wave in the sapphire crystal before it is converted back into an electromagnetic signal. One advantage of using a single lens element is that it eliminates the need for the precise confocal alignment of two lenses. The common technique used in macroscopic acoustic reflection imaging relies on pulse-echo operation. Application of this technique in the acoustic microscope is somewhat more difficult because the time delays are so short. The total round trip delay in the high frequency lens elements is less than 0.5 µs. For efficient time gating, acoustic pulses of much shorter duration than this, would be required. In practice, stub timers which match the electrical impedance of the transducer to that of electronics, are not suitable. With these tuners it is difficult to obtain a sharp acoustic pulse of duration

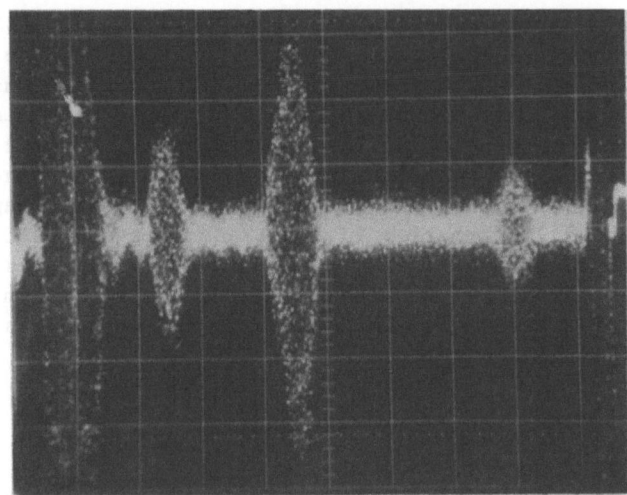

Fig. 13. Display on a sampling scope of the reflected signal. The second echo is the reflected signal. The others are echos in the sapphire. Time scale 100 ns per division.

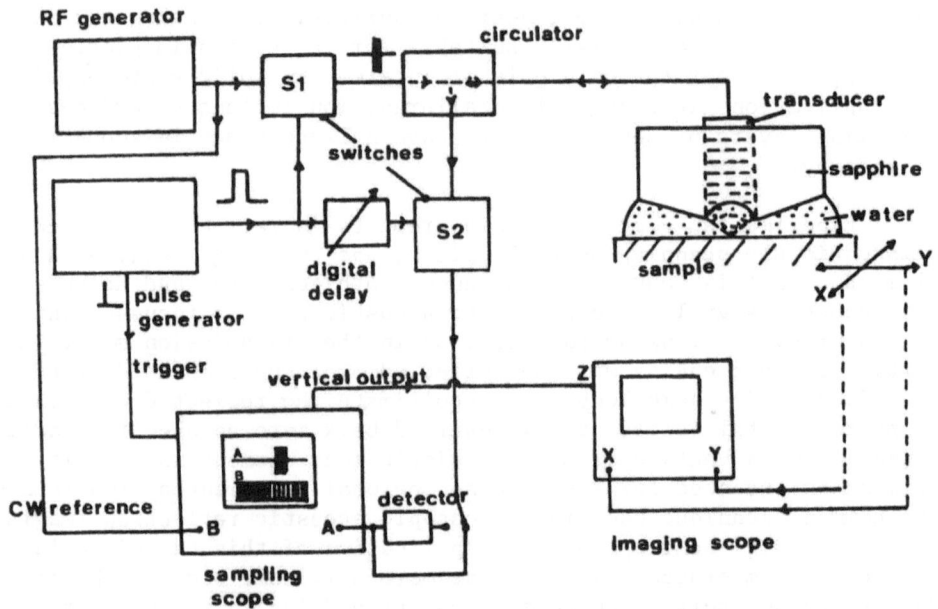

Fig. 14. Block diagram of the electronics associated with relection imaging.

less than 1 µs. This obstacle is eliminated when broad band matching networks are designed for the piezoelectric transducers. Such networks have been described in literature and ordinarily consist of strip lines on ceramic substrate. Finally, short pulses are used: typically 4 ns RF nonmodulated for Stanford's workers and 20 ns to 50 ns RF modulated for other researchers. A circulator separates the input from the output signal followed by a detection based upon a time gating system which selected the reflected signal from other spurious reflections (Fig. 13). A schematic diagram of the circuitry is shown in Fig. 14.

3. <u>Acoustic Reflection Micrograph</u>. In Fig. 15, we show an acoustic reflection micrograph of a part of an integrated circuit supplied by Sescosem compared with an SEM micrograph of the whole circuit. The acoustic photo is made at a frequency of 600 MHz. Because of the relatively large area of this circuit (1 mm^2), we cannot picture in one operation the entire circuit with all the desirable resolution. Actually, this resolution is neither limited by the acoustic system nor the scanning assembly but by the diameter of the spot beam of the imaging scope when looking at an area of the object larger than 0.25 x 0.2 mm^2. Nevertheless, in Fig. 15 the quality of the images is high enough to illustrate the possibility of this instrument. We note that the contrast of the acoustic picture is much higher than that of the electron picture. Even more interesting is the fact that the various defects are visible in the acoustic image. Naturally, it is difficult to give a good explanation of the complexity of such a circuit consisting of superimposed diffused layers and several coatings. To clarify, we propose to study some cases of particular interest to bring into evidence the originality of such an instrument. In general, the principal difference between acoustic and optical microscopic images turns on the inability of the optical instrument to reveal anything but surface features since in solid state devices most of the materials are opaque. In contrast, acoustic waves can penetrate into these materials and allow an exploration of the region beneath the surface. This is one of the most important features of the acoustic microscope since no other forms of radiation can be used so easily for microscopic observations. In addition, it does not require destructive sample preparation and does not modify their structure during acoustic irradiation.

4. <u>Exploration of Subsurface</u>. In order to show the microscope's unique capability, we have constructed a test object consisting of three different materials as shown in Figure 16. An electron microscopy grid is glued on a half mil thick sheet of mylar coated with a few thousand angstroms of aluminum so that the grid and the coating are on each side of the mylar. Viewed optically from the coated side, the object is completely opaque and the surface is perfectly smooth. We show in Fig. 17 the acoustic image when focussing on the aluminum coating (a) and on the grid (b) through

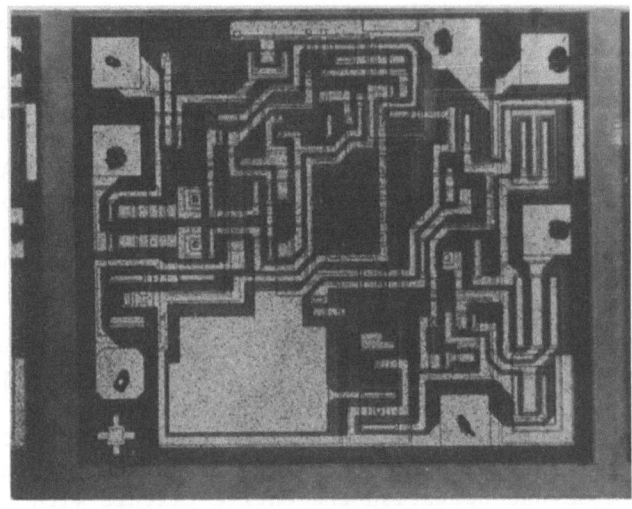

200 µm

Fig. 15 a

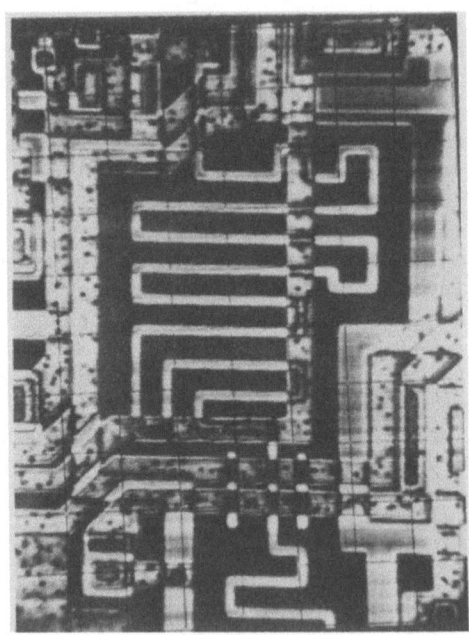

50 µm

Figure 15 b. Electron (a) and acoustic (b) micrographs of a part of the integrated circuit SFC 2741 supplied by Sescosem.

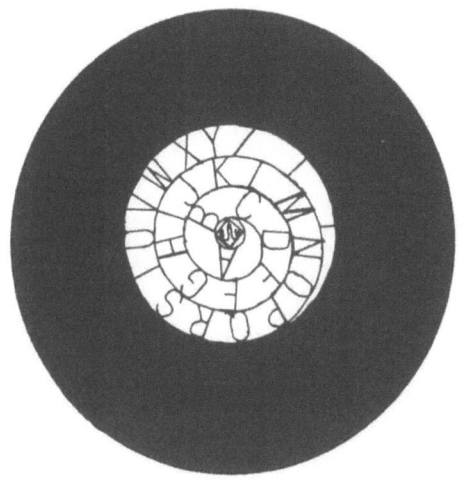

TOP VIEW

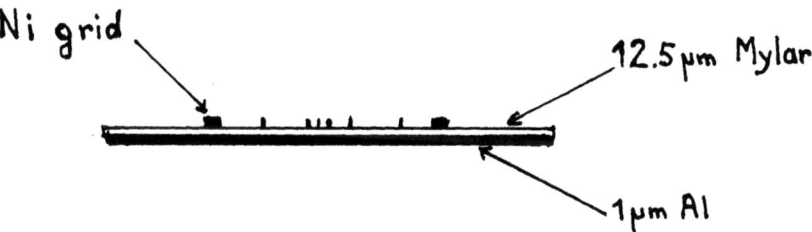

EDGE VIEW

Fig. 16. Test object for exploration of subsurface.

the aluminum and mylar. It is clearly evident that we are able to sharply focus inside the object in spite of the large acoustic reflection which occurs between water, aluminum and mylar. When we focus onto the aluminum coating the grid is blurred and large areas of interference patterns are superimposed on the image.

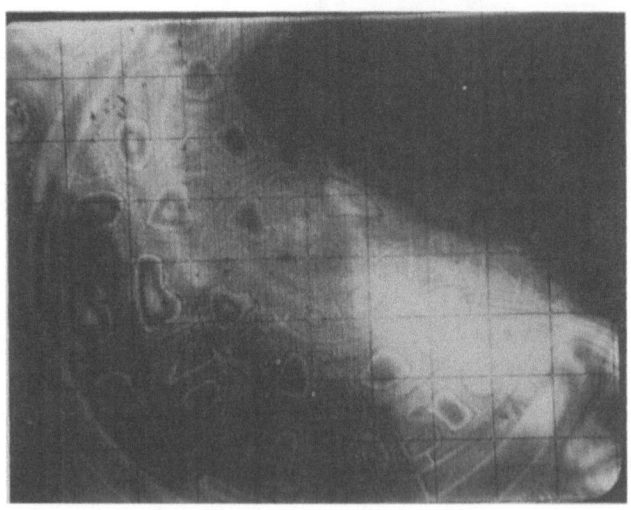

a

250 µm

b

Figure 17. Acoustic image of the test object for two different focussings inside the sample.
(a) focussing is onto the surface
(b) focussing is 12.5 µm inside the object.

5. **Depth of Field and Depth of Penetration of the Focussed Beam.** On this simple example we have demonstrated the important feature of the acoustic microscope. If we are interested in the problem of viewing the interior of an object we have to introduce two parameters characterizing this ability. The first one is the depth of field which is the maximum distance between two points on the axis which are simultaneously in focus. A simple calculation gives its expression p as a function of the wavelength and the aperture of the lens:

$$p = \frac{2\lambda}{\sin^2\theta_m} \tag{4-1}$$

taking $\sin \theta_m = 3/2$ p is of the order of 3 λ. For example at 600 MHz in water, p is about 7.5 µm. Inside this object, the determination of p is rather complicated because of the nature and position of the object with respect to the center of curvature of the lens. This leads us to introduce the second parameter which is the penetration of the focussed beam inside an object. As we see in Fig. 18, it is dependent on the distance between the position of the interface sample-liquid and is given in first approximation by the linear relation:

$$d' = \frac{d}{n} \tag{4-2}$$

where d is the distance between the focal point in water when the object is withdrawn and the interface liquid-object as shown in

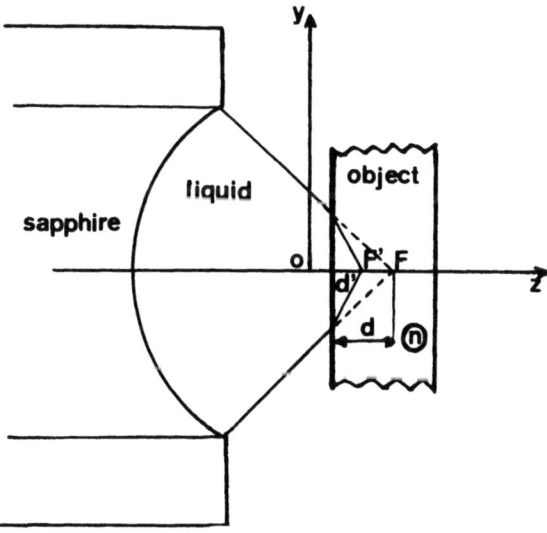

Fig. 18. Geometry for the ray tracing analysis inside an object.

Fig. 18, d' the distance between the actual focal point in this
object and the interface, n is the refractive acoustic index. For
high value of n (that is for most solids) the acoustic beam strongly
converges to a point near the interface and no significant penetration can be noticed. In order to improve this penetration, recent advances in liquid absorption and lenses design have substituted gallium or mercury for water. These two metallic liquids
offer the advantage of a rather solid-like acoustic impedance which
increases the acoustic transmission between the sapphire and liquid
and lowers the reflection due to impedance mismatching at the interface liquid-object. Contemplating the relation (4-2) gallium is
a better candidate than mercury for increasing the depth of penetration since its velocity is almost twice that of water. Naturally
the resolution get worse but for certain applications in the field
of non-destructive evaluation inside the object, it might be necessary to trade penetration for resolution. Besides, we have made
successful attempts with mercury having solved, in particular, the
delicate problem of wetting the surfaces. This liquid is adequate
when we want to improve the resolution of the microscope since its
velocity is almost equivalent to water and its acoustic absorption
approximately one quarter that of water. So, the depth of field
is almost the same as water and should be more appropriate for
subsurface examination down to a few microns.

6. <u>Some Typical Applications</u>. Acoustic microscopy by reflection
is particularly suitable for observation of surfaces and defects,
as well as structures or stresses inside semiconductors, metals and
generally opaque optical materials. These defects can not only be

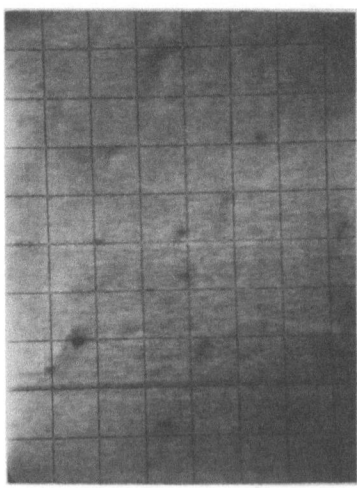

Fig. 19 a

THE ACOUSTIC MICROSCOPE   665

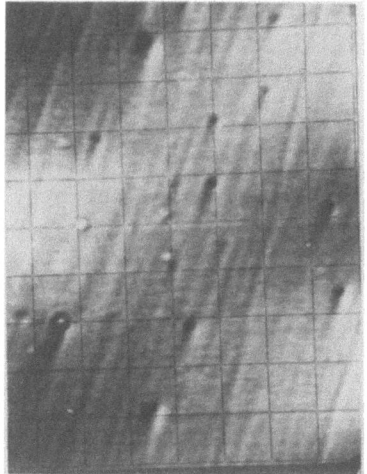

Fig. 19 b

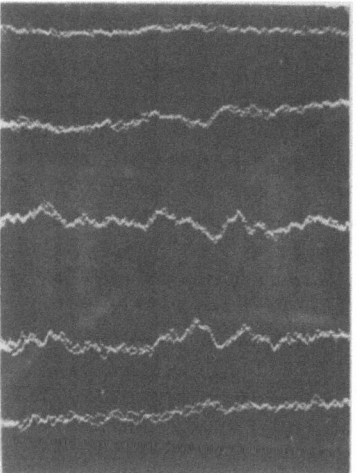

Fig. 19 c

Figure 19. Acoustic examination of an epitaxial layer of
$Ga_{1-x}Al_x$ Sb on a ZnTe substrate (x = 10%)
Scale 12 μm per division
(a) focussing on the surface
(b) focussing down to the substrate
(c) recordings of one line scan for different focussings from the surface (top trace) to the interior of the substrate (bottom trace).

microcracks, dislocation voids, joints, etc., but also changes of structure or crystalline phase of an element, compound or alloy. Defects under coatings or ion implanted regions are of prime importance in microelectronics. This list is, of course, not exhaustive and to illustrate one of these applications we have chosen an epitaxial layer of $Ga_{1-x}Al_xSb$ on a ZnTe substrate (x = 10 per cent). Fig. 19 (a) shows the acoustic image of the surface of such a device which is relatively smooth with some pin holes. When focussing down to the substrate at 3.5 μm from the surface, we see gallium inclusions (white spots) and trails due to polishing. To monitor this change we have recorded on Fig. 19 (c) the output signal for one line scan at different positions inside the object. This clearly shows up the transitions between the surface (top trace) and the interior of the substrate (bottom trace) with a maximum of detail at the interface.

In addition, the possibilities of the present instrument can be extended by phase measurements as discussed in the next section.

D. Phase Contrast Imaging

The basic objective of the phase contrast method is to transform variations in the phase delay introduced by the object into an intensity pattern in the final images. The design of the scanning acoustic microscope makes this particularly easy to achieve. Since the piezoelectric transducer is a linear device, the phase of the output electromagnetic signal is exactly the same as the phase of the acoustic wave incident on the output transducer. Thus, by comparing the phase of the output electromagnetic signal to that of the input, the acoustic delay through each point on the sample can be measured. To achieve this result, a directional coupler splits the incident power from the RF oscillator to provide the phase reference. Owing to the excellent temporal coherence of the oscillator, it is not necessary to compensate for the transit delay in the microscope. The output signal from the microscope can be directly compared with this reference. As the object is scanned through the focused acoustic beam, spatial variations in the acoustic phase delay appear as a phase modulation of the output electromagnetic signal. A doubly balanced mixer combines this signal with the reference and produces a voltage proportional to the difference in phase between them. After amplification, the output voltage from the mixer is used to modulate the beam intensity on the CRT display. The appearance of the image can be varied continuously between these two extremes by introducing a variable phase into either the reference circuit or the microscope circuit. In this case, the separation between the acoustic lenses was increased by approximately half of an acoustic wavelength in going from the bright image to the dark. Comparing the phase contrast images with the amplitude modulated image shows that there are

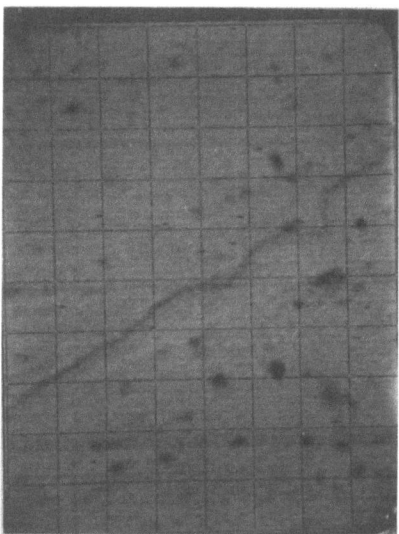

Fig. 20. Acoustic image of a quartz implanted with ions ^{4}He$^+$.
The dark line is the transition between implanted
(right) and non-implanted (left) regions.
Scale: one division equals 75 µm.

substantial variations in the acoustic phase delay introduced
by the sample. In addition to phase variations, the amplitude
modulation produced by the sample is also contained in the phase
contrast images. Such independent measurements of amplitude and
phase can have important applications. For example, in a sample
which has an approximately uniform thickness, it is important to
know how much of the attenuation is due to absorption and how much
to impedance mismatch. Since there is very little variation in
the density of biological material, large variations in acoustic
impedance must be due to changes in the acoustic velocity. More
recently, phase imaging in reflection mode has been investigated
to monitor the thickness of layers deposited onto a polished surface of a single crystal and the small scale variation of the
elastic parameter in these materials.

We have performed such an experiment on quartz whose surface
was partially implanted with ^{4}He$^+$. Fig. 20 shows the image of the
transition between the implanted and non-implanted region. As we
see in this image of amplitude, the contrast is not very strong.
For more convenience we have displayed in Fig. 21 the recording of
the amplitude and the cosine of the phase corresponding to one
line scan and brings out a jump between the two regions. This jump
clearly observed on the phase curve is equal to $\lambda/8$ and corresponds
to a step of 780 Å caused by an expansion of the implanted region.

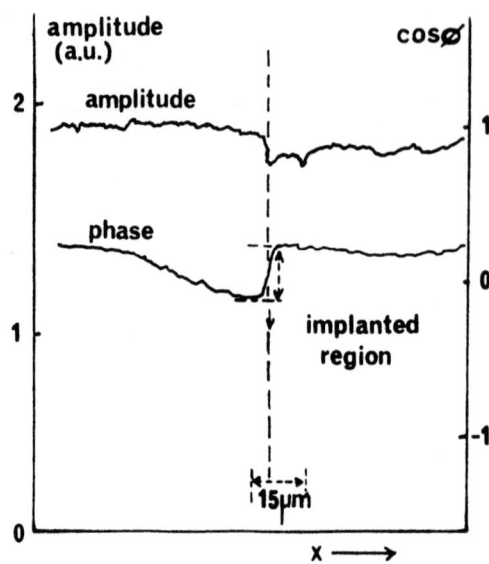

Fig. 21. Variations of amplitude and phase of the reflected signal for one line scan. The sample is the same as that in Fig. 20.

With this technique we should be able to estimate a variation of phase of about $\lambda/30$ which is much better than we can expect with amplitude.

E. Non Linear Acoustic Microscopy, Dark-Field and Stereo Viewing

The sharply convergent acoustic beam used in the microscope can provide sufficient intensity to produce strong nonlinear effects at microwave frequencies. Second harmonic acoustic radiation generated in the vicinity of the beam focus is readily detected and used to form an image of an object placed in the focal plane. Successful attempts have been carried out at a fundamental frequency of 450 MHz and a detector tuned at 900 MHz. In order to interpret the content of the second harmonic images, it will be necessary to understand in detail the interaction between the specimen and the generation of second harmonic radiation through the microscope. Whenever a new way of seeing things is discovered, problems of interpretation arise. Second harmonic acoustic imaging is no exception. Another type of application of the acoustic microscope is the dark-field and stereo viewing. Additional details can be obtained if we use an off-axis arrangement wherein the axis of the output lens is tilted with respect to the input lens. Observing the wide-angle scattering from an object in the ultrasonic beam is analogous to dark-field

THE ACOUSTIC MICROSCOPE

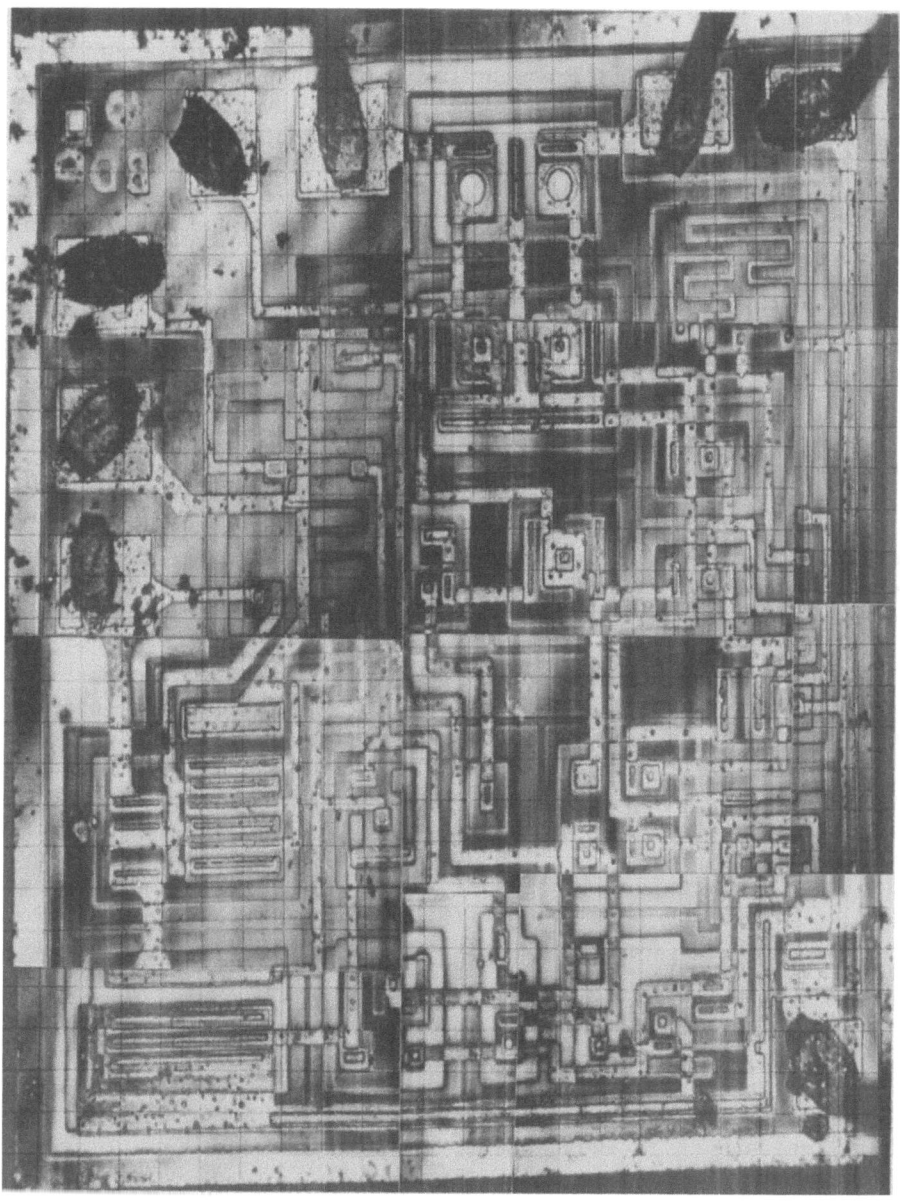

Fig. 22. Acoustic micrograph of an integrated circuit (1.2 mm x 1 mm) showing imaging of the contacts underneath the aluminum wires.

microscopy with an optical microscope. The plane of the object can, therefore, be conveniently turned to an angle farther from the usual position transverse to the incident beam and to the observing beam. Furthermore, since the source and the receiving beam are small in cross section, there is a limited depth of field in the noncollinear case which may be as small as the width of the beams when the viewing is at right angles. Indeed, one may advantageously observe the inside of an object in this geometry without interference from the nearby out-of-focus structure. Acoustic stereograms can be also recorded by canting the plane of the object. With a stereoscopic viewer, depth information becomes apparent and texture can be seen.

F. Recent Advances in Acoustic Microscopy

1. <u>Non-Destructive Test of the Bonding Contact of an Integrated Circuit</u>. One of the most delicate problems in integrated devices turns on the bonding contacts which often fail. To date, there is no convenient way to investigate the bonding contact without a destructive technique. Acoustic microscopy seems to be one of the best candidates for such an examination since its resolution and focussing penetration are suitable. Fig. 22 gives an example of what we can achieve in that field. The object which is imaged is an ordinary integrated circuit supplied by SESCOSEM where the cover of the package has been carefully removed. We see very clearly on the photograph that the aluminum wires are still bonded to their pads. The acoustic beam is focussed onto the surface in such a way that we can generate a picture of the contact through the aluminum wire. The size of these contact wires is of the order of 50 µm in diameter and we see underneath the wire, structures corresponding to local changes of the acoustic impedance of the bonding. This is of the greatest importance since we know that these changes could be correlated to a nonuniformity of the current distribution at the interface. At this early stage of this study we are not able to do this correlation since we do not yet know what types of defects are revealed by the acoustic microscope.

2. <u>Imaging with Mercury</u>. As has been shown in Section III-D, the strong attenuation in water limits the resolution of the acoustic microscope. However, water has the following advantages over other liquids: it is inexpensive, nonhazardous and does not chemically react with most objects. Besides this, it is easy to put in the microscope (one drop fills the lenses) and gives good images at room temperature and at frequencies up to one gigahertz. However at that frequency, the acoustic absorption in water is close to 200 dB per millimeter and special spherical lenses are required to minimize the acoustic path in the liquid. If we want to improve resolution by increasing frequency, we have to face the drastic increase of acoustic absorption in water which follows a square law

dependence with frequency. Therefore, a first solution consists in designing smaller lenses. Technologically we feel there is a limitation to diameters near a hundred microns. A second way to improve the resolution is to search for liquids which are less lossy than pure water. After a long and systematic investigation (as seen in Table I) it turns out that except for cryogenic liquids, very few fluids exhibit lower absorption than pure water. Using additives in water such as RbI gives an improvement of only 33%. Carbon diSulfide seems to be a good candidate but only at frequencies above 3 GHz. At room temperature only two liquids should be suitable. Both are metallic. Liquid gallium is less chemically reactive than mercury but its sound velocity is almost twice as much as pure water. This is of the utmost importance since it affects the final quality of the image by increasing the spherical aberrations of the lens. Mercury seems to be a better choice with a velocity of v_g equal to 1450 ms^{-1} a bit less than that of pure water. Its acoustic absorption of α/f^2 is equal to 5.8×10^{-17} cm^{-1}s^2 which is four times less than pure water at room temperature. Therefore the resolution which is approximately a fraction of the wavelength will be improved by the ratio:

$$2 \times \frac{v_{H_2O}}{v_{Hg}} \simeq 2.06 \qquad (4-3)$$

This high value, scarce for most liquids, had led us to investigate mercury in acoustic microscopy. It is not the only reason because its acoustic impedance which is about 20×10^6 kgs^{-1}m^{-2} matches that of most solids.

However one problem arises when using mercury for imaging. It is well known that mercury does not oxidize at room temperature but will not wet most solids, or when it wets it gives an amalgam (with gold in particular). Many workers have been interested in this problem when measuring acoustic absorption and velocity in this liquid. Some of them have described an original technique to get acoustic contacts between two rods of quartz separated with mercury. They have placed a grid consisting of 25 micron diameter silver wires between the crystal to prevent the mercury from breaking up into droplets when translating the rods. In addition, in order to insure that mercury would wet the quartz crystal, the faces of each crystal were coated with an evaporated film of silver before placing them in the bath. These precautions, of course, cause some contaminations of mercury but the amount (estimated to less than 0.1 percent) does not affect the acoustic parameters to any extent. However if this solution is suitable for flat surfaces when making ultrasonic measurements, no successful attempt has been obtained on spherical lenses with this technique. Indeed, after a while, mercury breaks up the silver coating on the lens depending on the handling of the object and how vigorous is the fast scan. Thus the acoustic transmission with the object is no longer guaranteed. For all these reasons not much attention has been given to the use of mercury in acoustic microscopy. We have now solved the delicate problem of mercury wetting using a simple technique: as shown in Fig. 23 we use only a tiny drop of this liquid to fill up the spherical lens. Because of the surface tensions this drop balls up and fits exactly the wall of the lens. This lens is ground deeper than for water to allow a large surface contact with the liquid metal. Neither silver coating, nor a silver grid is required. Acoustic contact is insured by the pressure of the scanned object. In fact, the complex problem of wetting depends on the cleanliness of the surfaces and the electrical surface charges on the material to be wet. For instance, ion bombardment makes mercury wet many solids but on the contrary absorption of gas or other impurities does not favor wetting. In the scanning microscope it is not necessary to have a good wetting of the surface of the object to make acoustic images. Indeed the scanning process tolerates a very thin layer of adsorbed gas or an oxide a few tens of angstroms thick which prevents mercury from wetting. Except for gold, tin, lead, indium and some other easily amalgamated materials which require a protective layer (carbon for example) we have a large choice of solids for acoustic microscopy. We now compare acoustic imaging with water and mercury. The microscope is still working in

Fig. 23. Photograph of a 0.2 mm diameter lens filled with a drop of mercury.

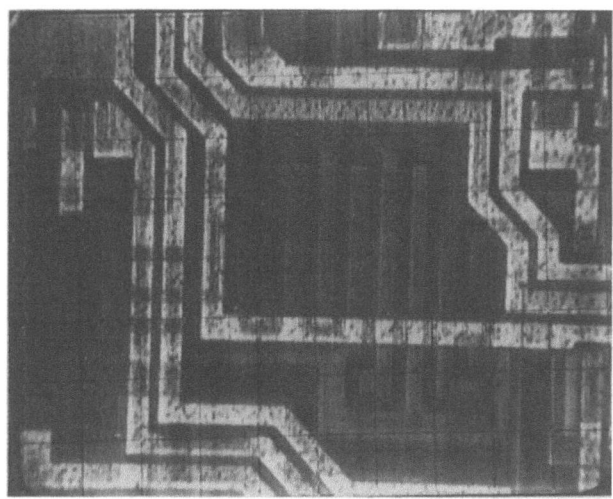

Fig. 24. Imaging with a sapphire water lens of a part of an integrated circuit at 600 MHz (50 μm per division). Focussing is on the surface.

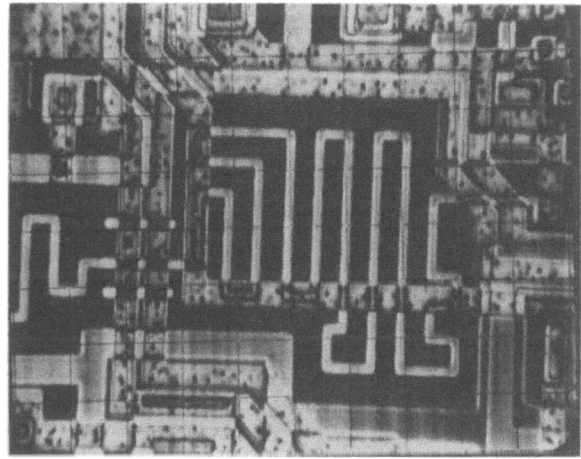

Fig. 25. Same as Fig. 24. Focussing is down to a few microns under the surface.

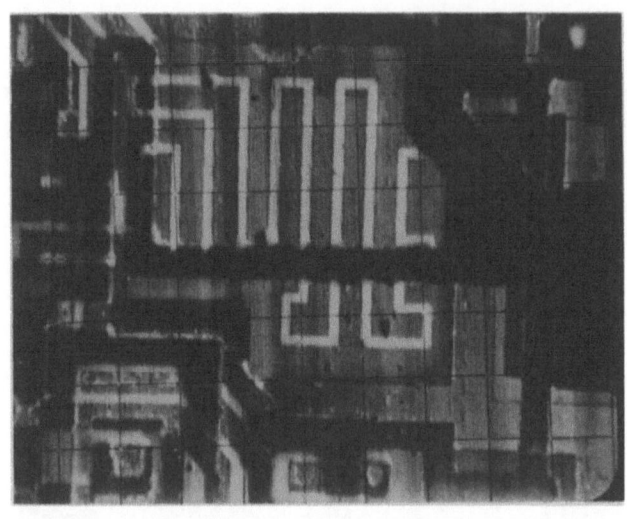

Fig. 26. Same as Fig. 24 but with sapphire mercury lens at 1.5 GHz. Focussing is on the surface.

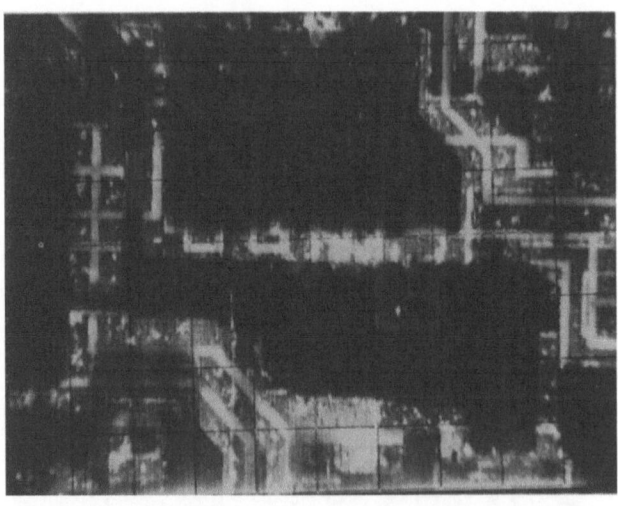

Fig. 27. Same as Figure 26. Focussing is down to a few microns under the surface.

reflection mode. The reference object is a part of the integrated circuit SFC 2741 previously imaged (see Fig. 15). Figures 24 and 25 are the images taken with water at 600 MHz for two different focussing of the acoustic beam inside the object. We note the strong contrast due to the large impedance mismatch at the interface water-object. Figures 26 and 27 are the corresponding images in field and focussing but with mercury at 1.5 GHz. Although we recognize the object they look different. First of all, the increased resolution makes the edges of the patterns sharper, secondly the better acoustic impedance matching gives less contrast but reveals structures when we focus inside the integrated circuit (see Fig. 27). Naturally, it is difficult to give a fair interpretation of the complexity of such a circuit. We shall note, for example, at the top part of the two mercury images the small depth of field of the microscope and its ability to show up details buried under coatings. This depth of field is of the order of 1.5 microns at 1.5 GHz in good agreement with our observations. With water at 600 MHz it is about 4 microns. We have not performed experiments at the same frequency to know whether the differences observed are due to frequency or impedance matching. At the present time with our 0.1 mm diameter lens, taking into account the 100 dB dynamic range of our electronic detection and 15 dB conversion losses of the transducer (two ways) we should be able to go up to 3.5 GHz. At that frequency the acoustic losses in mercury through the lens cavity are about 70 dB; that leaves 10 to 15 dB at our disposal to take an image which is suitable. The wavelength in mercury is 0.41 microns and the resolution will be less than 0.2 microns which is better than that of the best optical instruments.

Acknowledgments

I would like to thank Professor C. F. Quate of Stanford University, California, who introduced me to acoustic microscopy and also for his constant advice at all stages of this study.

REFERENCES

1. R. A. Lemons and C. F. Quate, Appl. Phys. Lett. 24, 163-165 (1974).
2. L. W. Kessler, J. Acoust. Soc. Am. 55, 909-918 (1974).
3. E. Bridoux, B. Nongaillard, J. M. Rouvaen, C. Bruneel, G. Thomin and R. Torguet, J. Appl. Phys. 49, 574-579 (1978).
4. J. Attal and C. F. Quate, J. Acoust. Soc. Am. 59, 69-73 (1976).
5. C. F. Quate, Trends in Biochemical Sciences, N 127-N 129 (June 1977).

6. R. A. Lemons and C. F. Quate, Appl. Phys. Lett. 25, 251-253 (1974).
7. C. S. Tsai, S. K. Wang and C. C. Lee, Appl. Phys. Lett. 31, 317-320 (1977).
8. J. Attal and G. Cambon, Electron. Lett. 14, 472-473 (1978).
9. H. K. Wickramasinghe and M. Hall, Electron. Lett. 12, 637-638 (1976).
10. R. D. Weglein and R. G. Wilson, Appl. Phys. Lett. 31, 793-796 (1977).
11. A. Atalar, C. F. Quate and H. K. Wickramasinghe, Appl. Phys. Lett. 31, 791-793 (1977).
12. J. Attal and G. Cambon, Submitted to Appl. Phys. Lett.
13. R. Kompfner and R. A. Lemons, Appl. Phys. Lett. 28, 295-297 (1976).
14. H. K. Wickramasinghe and C. Yeack, J. Appl. Phys. 48, 4951-4954 (1977).
15. W. L. Bond, and C. C. Cutler, R. A. Lemons and C. F. Quate, Appl. Phys. Lett. 27, 270-272 (1975).
16. J. Attal and G. Cambon, Application to the Non-Destructive Evaluation in Microelectronics, UltraSonic Symposium Proceedings IEEE #78 CH1344-15U (1978) (to be published).

Chapter 13

NON-DESTRUCTIVE TESTS USED TO INSURE THE INTEGRITY OF SEMICONDUCTOR DEVICES WITH EMPHASIS ON PASSIVE ACOUSTIC TECHNIQUES

George G. Harman

Electron Devices Division

National Bureau of Standards

Washington, DC 20234

I. INTRODUCTION

This paper reviews a number of important non-destructive tests used frequently in the semiconductor industry to test the mechanical integrity of semiconductor devices. Many of these tests are not rigorously quantitative, but rather, involve an element of human judgement or some empirical comparison for interpretation. As such, the usage of some of the tests is controversial even though they are specified in important military and other microelectronic standards. The scientist or engineer just entering the microelectronics field should look upon this as an opportunity to develop better tests rather than be discouraged by the lack of rigor.

The discussion is divided into two major sections. The first section begins with a brief review of device assembly techniques and problems. This serves as necessary background for all of the material that follows. Next follows a review of six important non-destructive tests that are used during and after device packaging to insure the mechanical integrity of completed electronic devices. Most of these tests are called out in MIL-STD-883 [1], a widely used guide for device testing, and are generally classified as screens. These tests, presented in the order that they are usually performed, are: (1) the non-destructive wire bond pull test, (2) internal visual inspection, (3) temperature cycling and shock, (4) package seal integrity (hermeticity), (5) burn-in (removing early failures), and (6) particle impact noise detection.

The first section concludes with a brief introduction to some factors that result in the choice of one screen over another and to production line statistical sampling appropriate to special high reliability device lots such as those used for space flight.

The second section begins with an introduction to acoustic emission (AE), and the status of its theory as it can be applied to microelectronics. Also, the published papers that have applied AE as a non-destructive test in electronics applications will be reviewed. Finally, acoustic emission measurement techniques developed at the NBS are applied to establishing the mechanical bond integrity of beam lead, flip chip, and tape bonded integrated circuits as well as components in hybrid microcircuits.

II. SOME CURRENT PRODUCTION LINE ASSEMBLY TESTS

A. Introduction to Microelectronics and Hybrid Packaging Methods

Microelectronics assembly starts after the scribe-and-break or sawing operation that cuts the individual die (chips) from the wafer. Once cut out they are usually die-attached to the package by gold-silicon eutectic, a solder, or an epoxy and then conventionally interconnected by wire bonding. Other attach technologies, such as flip chip, beam-lead, and tape-carrier bonding can essentially combine die attach and interconnection bonding into a single operation. The first, most familiar, and still overwhelmingly dominant method of interconnection is to use flying wires, an example of which is shown in Fig. 1. These wires, typically 25 μm diameter aluminum or gold, are welded on to the semiconductor bonding pads by thermocompression [2], ultrasonic [3], or thermosonic [4] (a combination of both) techniques.

The second method of interconnection uses gold beam leads, which are made during wafer processing in place of wire bonds. Fig. 2 is an example of such a device. These beams are welded to a package, usually a hybrid by thermocompression techniques in which the substrate is heated to ~250°C and a heated bonding tool compresses the leads against the substrate with forces >100 kg/cm^2. The flip-chip approach requires building up the bonding pads, usually with a solder bump, and then reflow soldering the bumps to the package face down, thereby obscuring the joints so that they cannot be visually inspected and discouraging the use of this technology. Several typical flip-chips are shown in Figure 3. Neither beam leads nor flip chips are used widely at present and their main applications are for large scale in-house production and consumption.

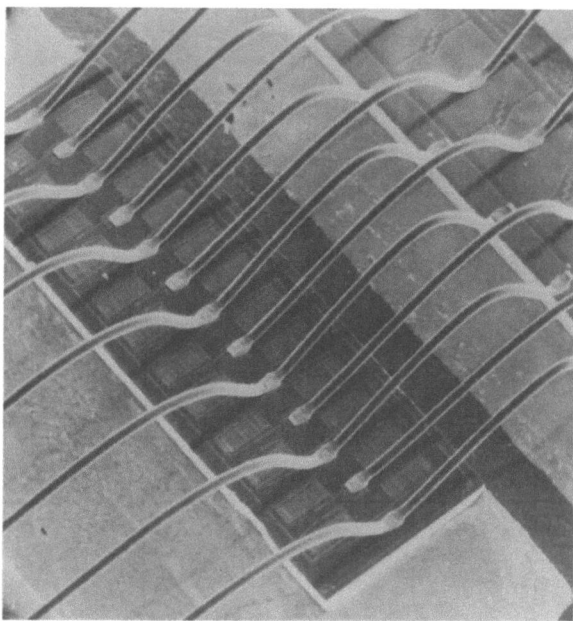

Fig. 1. An example of 50 m diameter aluminum ultrasonic wire bonds (flying wires) interconnecting a microwave power transistor.

Fig. 2. (A) is a gold beam lead device shown face up, and (B) bonded into its hybrid microcircuit. The contact and metallization system is platinum silicide, titanum, platinum, gold. The sealed junction is obtained by depositing silicon nitride over oxide layers.

(A)

(B)

Fig. 3. (A) flip chip device shown soldered onto a substrate (B). Several devices shown face up.

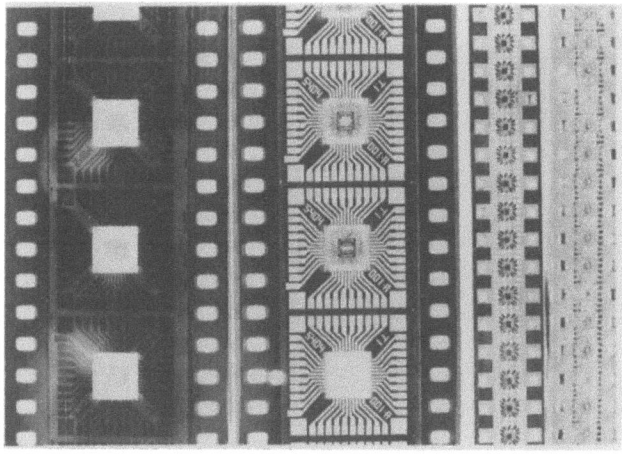

Fig. 4. (A) is a 35 mm three layer (metal, adhesive, polyamide) testable-tape bonded device.
(B) is an example of an 11 mm copper all-metal single layer tape with an IC chip inner-lead-bonded in the center. The leads are made of tin plated copper.

Fig. 5. A high technology 5 layer thick-film hybrid microcircuit. Arrows point to chip capacitors (A), and chip resistors (B). (Courtesy of General Dynamics Electronics Division).

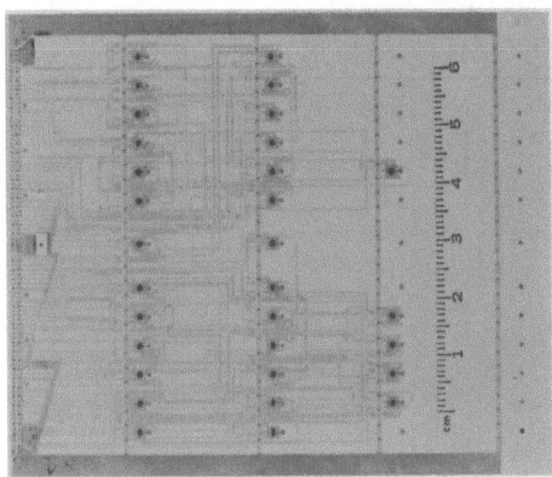

Fig. 6. A large thin-film hybrid substrate. The ICs are beam lead devices. The substrate is alumina ceramic and interconnections are Ti, Pd, and Au.

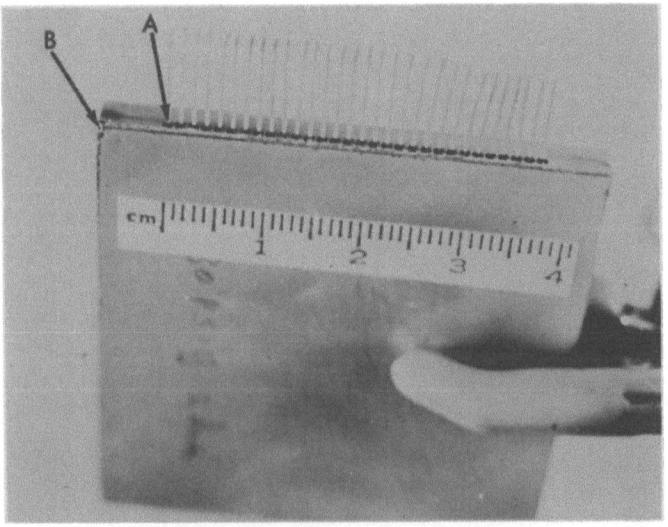

Fig. 7. A 5 x 5 cm sealed hermetic-type hybrid package with numerous glass-metal seals (A). A leak in the package lid is also indicated (B).

The final technique of interconnection which may be new to many is called tape bonding. For this, one normally builds up the semiconductor aluminum bonding pads into copper or gold "bumps" by plating techniques, then a special metallized tape with individual extended leads is bonded to the bumps by a thermocompression or reflow solder process. This is termed inner-lead bonding. Figure 4 gives two examples of chips bonded to such tape. The tape in 4A is "so called" testable tape. Here the individual insulated leads terminate on pads that may be probed and permit the device to be electrically tested after bonding. The tape in Figure 4B is of all metal construction which cannot be tested before packaging. Once the chips are attached to the tape, it is usually wound onto reels stored until final assembly into either an IC package or a hybrid. In the case of some calculators and watches, the leads are bonded onto a printed circuit board and the chip is protected by a drop of epoxy.

Figure 5 is a typical high technology thick film hybrid. In this figure, A is a chip capacitor, and B is a chip resistor. Some hybrids are very large (e.g., 5 x 10 cm) and the substrate may crack during certain assembly operations such as the thermocompression bonding. Some methods of detecting such cracks will be described in Section III B. Figure 6 is a photograph of such a large substrate. Once the hybrid components are assembled, the substrate is usually put in a package or, epoxy coated. In the case of packages, there may be moisture leaks if the lid seal and metal leads are damaged or improperly assembled. The seal integrity of packages that contain numerous glass-to-metal lead-throughs is a potential reliability problem. Such seals are shown in Figure 7. Thomas [5] has stated that the glass seals are the major source of hermetic moisture ingress and that these may open only at high temperature and subsequently reseal, avoiding detection later in a normal leak test.

B. Review of Typical Nondestructive Tests Used to Reveal Devices with Mechanical Defects

There are a number of mechanical screens that may be applied to devices that have specific reliability requirements. The decision to implement one test over another can be based on such considerations as cost, the specifying engineers' personal familiarity with one test, or the availability of test personnel. It is therefore appropriate to briefly review the most common non-destructive tests used on assembly lines to verify mechanical integrity. These will be described in the order that they are usually performed on a production line:

1. **The Non-destructive Wire Bond Pull Test.** The purpose of the non-destructive wire bond pull test (NDPT) is to remove weak wire bonds having pull forces less than a designated force value, while

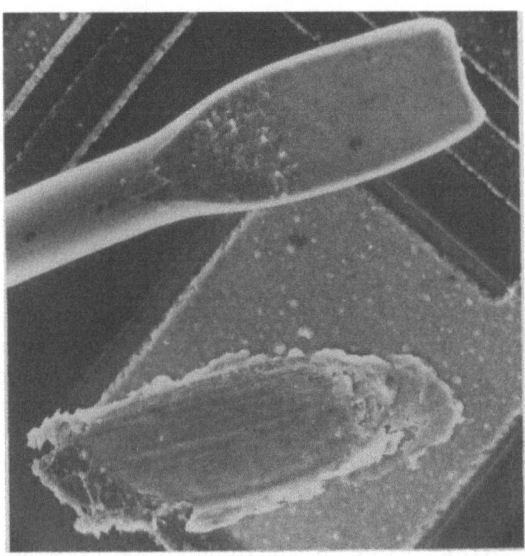

Fig. 8. SEM photograph of a weak bond resulting from poor process control. A 25 μm diameter ultrasonic wire bond failure resulting from a layer of glassivation or other contamination on the surface of the bonding pad. The visually perfect bond is shown in the upper part of the figure. The bond lifted at ∼ 0.2 gf.

avoiding damage to acceptable bonds. An example of a weak 25 μm diameter aluminum bond revealed by the NDPT is shown in Figure 8. This bond lifted at 0.2 grams force (gf). The bond would have passed any internal visual inspection (see Section II B 2) and its weakness could only be revealed by a pull test or possibly temperature cycling (See Section II B 3).

To perform the test, a hook, made of tungsten or steel wire, 2 to 3 times the diameter of the wire to be tested, is placed under the loop and a specified pull force is applied vertically to that loop. The bond and perhaps the device is rejected if the wire breaks. The pull force is usually specified for a given wire diameter, but it can also be derived from appropriate equations [6] based on the results of a sample destructive pull test and the metallurgical characteristics of the particular wire being tested. A typical equation is:

$$F = 0.9(\bar{X} - 3S_X)$$

where F is the non-destructive pull force (in gf), $\bar{X}$ and S_X are the mean and standard deviation of the destructive pull force respectively

PASSIVE ACOUSTIC TECHNIQUES

This equation is appropriate for small diameter (25-50 μm) aluminum wire used for ultrasonic bonding.

This test is used rather extensively in large hybrids and in integrated circuits for space and other applications requiring very high reliability. The NDPT has been reported to cost about half the price of the original wire bonding [7]. Although there is considerable evidence that the test is indeed non-destructive (it has been used on $>10^7$ wire bonds), it is nevertheless controversial. Some organizations require it on a 100% basis and others absolutely prohibit its use. This test can be both used and prohibited on different contracts for the same system.

2. <u>Internal Visual Inspection</u>. The purpose of this test is to check the internal construction, and workmanship of microcircuits for compliance with the requirements of the applicable specifications.

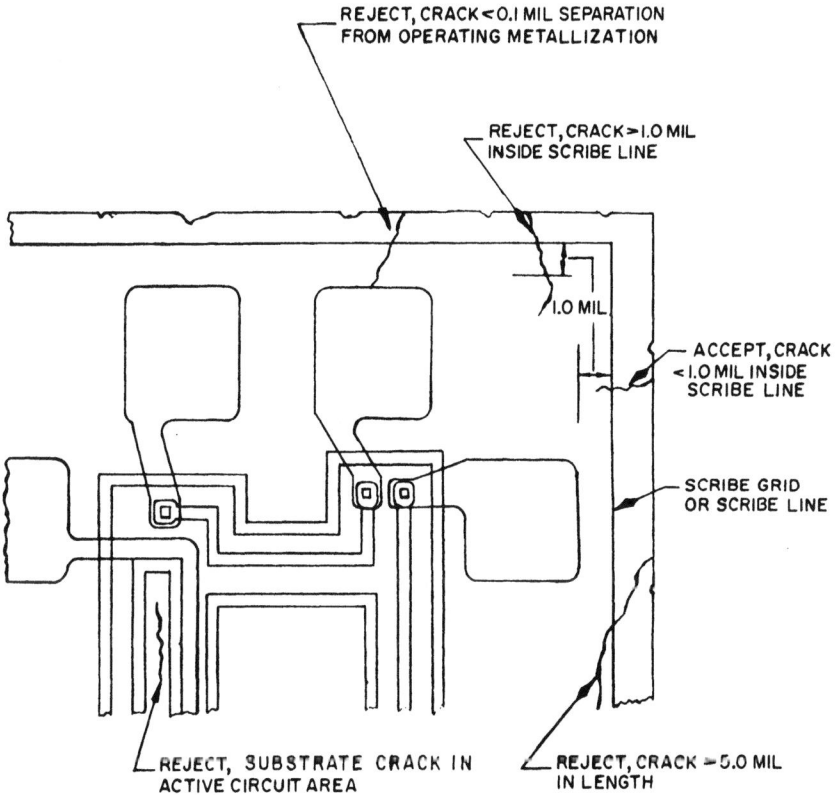

Figure 9. Figure 2010-24 from MIL-STD-883 [1] which is an example of visual inspection criteria for scribing and die defects.

This test will normally be used prior to capping or encapsulation on a 100% inspection basis to detect and eliminate devices with visually detectable internal defects that could lead to device failure in normal application. It may also be employed on a sampling basis prior to capping to determine the effectiveness of the manufacturer's quality control and handling procedures for microelectronic devices.

Most visual inspection criteria are accompanied by photographs or sketches so that the inspector has a basis for comparison. An example of one of these from MIL-STD-883B [1], Method 2010.3 is shown in Figure 9. Items generally covered by such inspection include placement of wire bonds with respect to the pad, lead dress, device metallization scratches, contamination, obvious passivation and diffusion faults, and die attach faults. Sometimes a scanning electron microscope (SEM) examination of devices on a sampling basis is specified to look for defects such as metallization coverage of oxide steps and diffusion window defects.

Visual inspection is a relatively expensive test in which much is left to the judgement of the inspector. At best, it is 80% effective in detecting faults and, at times, good product is rejected. If the same device is recycled through the same or a different inspector, different defects are frequently discovered and original ones may pass unobserved. The test is of limited usefulness for large scale integrated circuit chip integrity because of the high magnification and time required. Nevertheless, visual inspection is considered useful and important for removing gross chip and assembly induced defects.

3. <u>Temperature Cycling</u>. After the package is sealed additional non-destructive mechanical integrity tests are often performed. The purpose of this test is to determine the mechanical resistance of a part to exposure to various thermal environments that may be encountered during system life. Internal mechanical stresses result from the different thermal expansion coefficients of different parts of the device. This test is the only effective means of revealing poorly welded or soldered caps or damaged glass or ceramic to metal lead throughs.

The apparatus usually consists of two chambers, one cooled and one heated. Many devices are placed in a holder which is rotated or otherwise moved from one chamber to another. Heat transfer takes place by forced air. Typically, the device is tested for 10 to 20 cycles of from -65° to 125°C with 5 minutes equilibrium time at each temperature; but these test conditions may vary depending on anticipated device usage. A hundred or more cycles is usually considered destructive. Temperature cycling is considered an effective mechanical test and is inexpensive to perform. It is usually followed by an electrical test to reveal failures arising

Fig. 10. Aluminum chloride corrosion product on a glassivat IC chip. (After Ebel [8]).

during cycling. Temperature shock is a variation of this test. The intent is similar to that of temperature cycling, but shock is a more severe test. The temperature shock test consists of liquid to liquid transfer. Although a typical temperature range would be -55° to 125°C, the most severe specified condition in MIL-STD-883 is -195° to 200°C, where the low temperature liquid is liquid nitrogen and the high temperature liquid is a fluorocarbon. The transfer time from one liquid to the other is less than 10 seconds. The test is usually carried out for 15 cycles, where one cycle includes one high and one low temperature immersion.

Most operating devices will experience some temperature cycling during system life due to system turn on and off (e.g., automobile engine compartment electronics). However, it is not reasonable to expect a device or system to undergo liquid to liquid thermal shock in any anticipated usage. Therefore, it a choice is given, temperature cycling is usually the preferred test.

4. **Package Seal Leak Tests (Hermeticity)**. Moisture ingress in electronic packages is the cause of many device failures and, thus, the explanations of leak testing given here will be longer than other non-destructive production line screens. In addition, some of the moisture induced failure mechanisms are worth reviewing since moisture is a major, long-term reliability problem for semiconductor devices. The usual point of corrosion attack is the

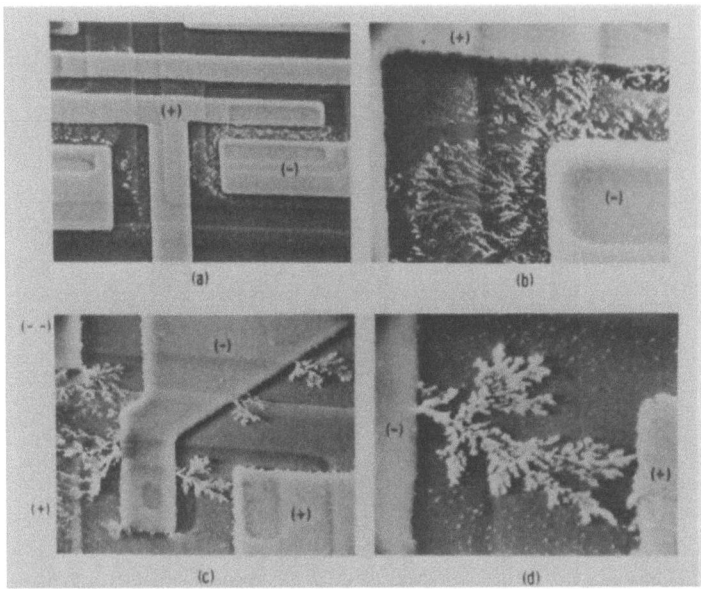

Fig. 11. Gold migrative resistive shorts. The (+) symbol is the anode and the (-) symbol the cathode. (a) is an SEM view of a portion of a microcircuit showing migrated gold between a T-shaped stripe and two neighboring stripes below the T bar. (b) is a close-up of the right half portion of the T-shaped stripe. Dendritic or fern-like features between the stripes have been identified to be gold by energy dispersive analysis of x-rays. (c) shows the center stripe as the cathode. Dendrites are shown growing from both sides of the cathode (center stripe) and proceeding toward the neighboring anodes. (d) a close-up of (c) illustrates the migrated-gold dendritic features typically observed. (After Schumka [9]).

exposed bonding pad aluminum, and subtle metallurgical and contamination induced interactions are common. An example of chlorine-moisture induced corrosion on an IC [8] is shown in Figure 10, and an example of metal migration [9] in Figure 11. One of the less obvious sources of contamination is spittle which may be deposited in micron sized droplets as a production line operator speaks to an associate [8]. The droplets contain contaminants such as sodium, phosphorus, sulphur, and chlorine. The initial water dries quickly, but later after package seal, the deliquescent contaminants may be reactivated by moisture ingress through a leaking package.

The purpose of the various leak tests is to determine the hermeticity of the seal of microelectronic and semiconductor devices with designed internal cavities. There are a number of methods such as bubble, dye, weight gain, halogen leak detection, helium leak detection, and radioisotope leak detection. These have been critically reviewed by Ruthberg [10] and the following treatment is derived from his work. There are variations on the methods of performing each test. Because of the small volumes and package constructions, most of the test methods require back pressurization, a process of driving a tracer gas or fluid into the interior by pressurization, and then detection of the tracer on reemission. The bubble, dye, and weight gain methods are appropriate for the gross leak range, which is taken to be $\geq 10^{-6}$ Pa·m^3/s.* The leak detector and radioisotope methods are essentially intended for the fine leak range ($\leq 10^{-6}$ Pa·m^3/s), but can be used for detection into the gross leak range depending on the size of the package internal volume.

Dye penetration techniques are more appropriate to the destructive testing of individual components for diagnostic purposes where decapping or other physical alteration of the package occurs. Dye techniques are used non-destructively, however, for devices with transparent walls. The halogen leak detector method is not frequently used for semiconductor components because of its corrosion potential. Bubble and the helium leak detection are the most widely used of all of the leak tests and will be described in some detail.

a. <u>Helium Mass Spectrometer Leak Detector Test</u>. The use of the helium leak detector is well documented. This instrument has the widest leak rate application range for general use. For this procedure, the packages to be tested are pressurized in a simple pressure bomb with helium gas, removed, transferred to the helium leak detector, evacuated, and tested for effusing helium. A numerical indication is obtained from the leak detector which can be related to true leak rate if an appropriate theoretical relationship is available to relate these two quantities. The correlation depends upon the regime of the gas flow into the test object, the pressurization parameters, internal free volume, the delay time between pressurization and readout, and the flow mechanism for helium effusion from the test part. Since enough of the helium must first be driven into the part to give discernible effusion, the pressurization times can be quite long for packages of large internal volume when tested to package leak rates $\gtrsim 1 \times 10^{-8}$ Pa·m^3/s. In most standards for helium leak detector use, the package leak rate is determined from an expression based upon the molecular flow regime. The

---

* Units of flow rate are conventionally atm·cm^3/s or torr·l/s, but in the International System (SI) of metric units the unit of of flow rate is the Pa·m^3/s. 1 Pa·m^3/s = 9.86926 atm·cm^3/s and 7.50064 torr·l/s.

equation is:

$$R = P_b \cdot L \left[\frac{1}{P_0} \left\{ 1 - \exp\left(-\frac{L}{P_0 V}T\right) \right\} \exp\left(-\frac{L}{P_0 V}t\right) \right]$$

where R is the machine reading for helium, L is the package leak rate under conditions of one atmosphere of helium pressure upstream and zero pressure downstream, P_0 is the pressure of one atmosphere, P_b is the bombing pressure, V is the internal free volume, T is the pressurization time, and t is the delay or dwell time between pressurization and readout. The first exponential term describes the pressure rise of helium within the package due to pressurization, while the second exponential term describes the fall off of pressure due to effusion. Since this expression is based upon the molecular flow regime, it is in principle only applicable to fine leaks; whereas in practice, it is applied to the whole leak range. The leak rate (L) is double valued as a function of machine reading (R) and package volume (V) as indicated by calculations which predict that a gross leaker may not be distinguishable from a fine leaker without further manipulation of test variables. In practice it is assumed that a minimum and maximum detectable leak rate exists and the range of leak rates between these two limits will be detected for any given bombing pressure P_b (in atmospheres), internal free volume (V) and minimum detectable machine reading (R_{min}).

The failure criteria from MIL-STD-883 is as follows: "devices with an internal cavity volume of 0.01 cc or less shall be rejected if the equivalent standard leak rate (L) exceeds 5×10^{-8} atm cc/secHe. Devices with an internal cavity volume greater than 0.01 cc and equal to or less than 0.4 cc shall be rejected if the equivalent standard leak rate (L) exceeds 1×10^{-7} atm cc/secHe. Devices with an internal cavity volume greater than 0.4 cc shall be rejected if the equivalent standard leak rate (L) exceeds 1×10^{-6} atm cc/secHe.

b. **Bubble Emission Tests.** There are two classes of bubble tests. One is simply direct immersion of the test object into a hot, clear, inert fluorocarbon liquid of low surface tension. If a leak is present, bubbles will appear as the gas in the device expands on heating. The leak test range is narrow. The second class is superior in test range and detection of leakers. For this, the component is exposed first to vacuum and then back pressurized with a high vapor pressure fluorocarbon liquid so that if a leak is present the fluorocarbon is driven into the component. On immersion in a hot, low surface tension, indicator fluid, the fluorocarbon bubbles out of a leaky device.

Although bubble size and frequency have been related to leak rate under ideal conditions, the interpretation of bubble tests are subjective in practice. They are tedious, results are very dependent upon the geometry of the package, and a gross leak comprised of several fine leaks can be missed. The use of liquids requires that this test be performed <u>after</u> fine leak tests have been completed to avoid the plugging of leaks; it also provides the possibility of residual contamination in the accepted components which have undetected leaks.

5. <u>Burn-In Test</u>. The burn-in test is not considered to be a mechanical integrity test (although it may reveal such problems), but it is included in the present discussion because of its importance in assuring device reliability. The burn-in test is performed for the purpose of screening or eliminating marginal devices, those with inherent defects or defects resulting from manufacturing aberrations which cause time and stress dependent failures. These are generally referred to as freaks. In the absence of burn-in, these defective devices would be expected to result in "infant mortality" or early lifetime failures under use conditions. Therefore, it is the intent of this screen to stress microcircuits above maximum rated operating conditions in order to reveal time and stress dependent failure modes in a practical length of time. The procedure involves placing the device in an oven at a fixed temperature for a specified number of hours (e.g., 125°C for 168 hours) with an applied bias, forward or reverse, depending on the stated

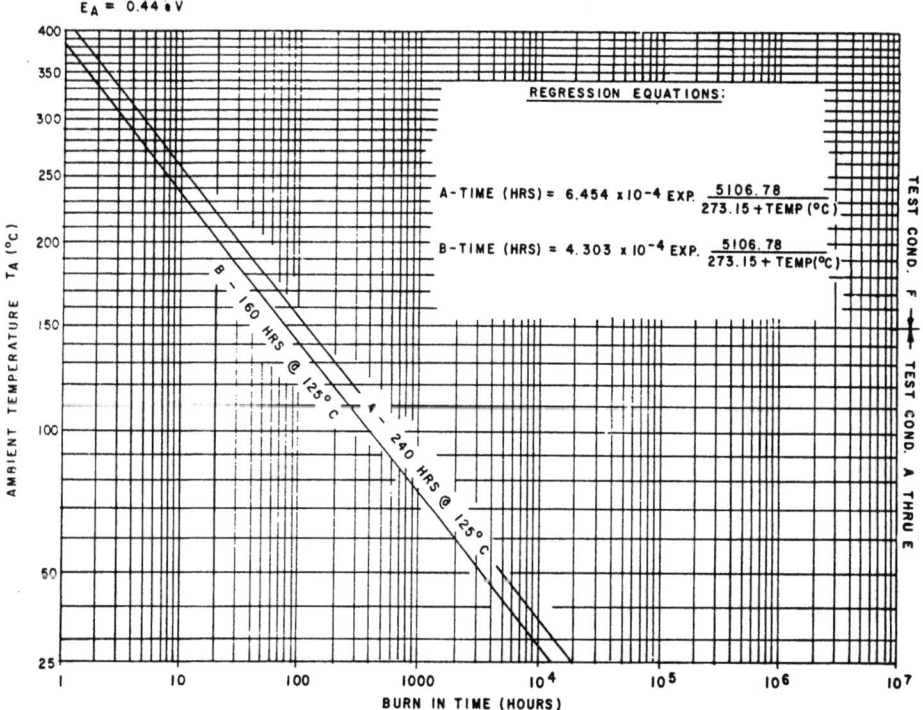

Fig. 12. Burn-in regression curve from MIL-STD-883B [1], Figure 1015-1.

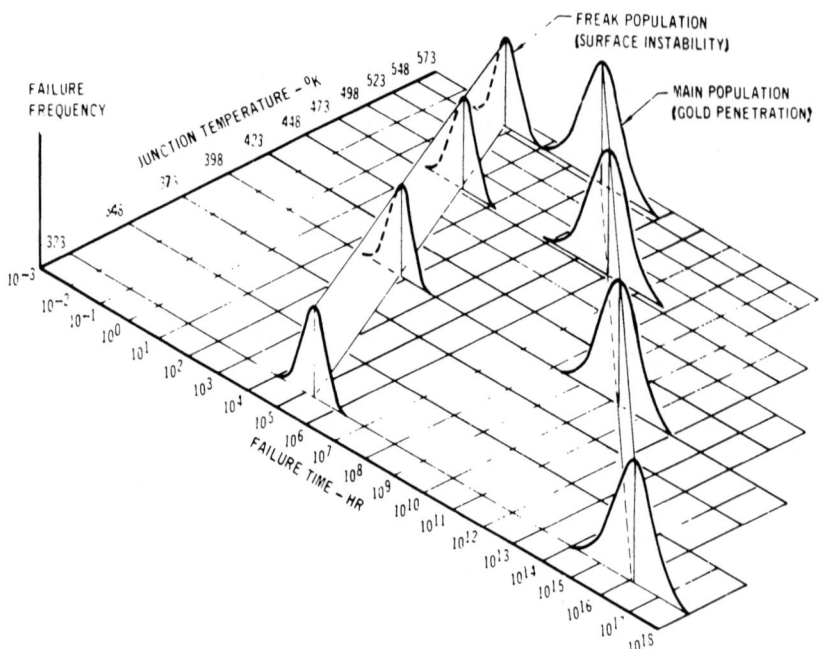

Fig. 13. Isometric graph of accelerated test failure distribution for a device with Au-Ti-W metallization showing freak and main population distributions. (After Johnson [15]).

test conditions. The regression equations are based on a simple Arrhenius reaction rate equation where the burn-in time (t) is given as:

$$t = A \exp \frac{E_A}{kT}$$

where A is a constant, E_A is the apparent activation energy in eV (in MIL-STD-883B it is 0.44 eV) chosen for freak population removal, T is the absolute junction temperature, and k is Boltzmann's constant. Figure 12 gives the burn-in regression curve from MIL-STD-883B [1]. This curve is applied to the overwhelming number of devices that are burn-in tested. However the temperature selected must not be high enough to create failure modes not related to freak removal. One such failure mode, known as purple plague (Au-Al intermetallic compound formation [11]), may occur when gold wire bonds are made to aluminum bonding pads. If burn-in is carried out for one hour at 400°C, the interconnects will fail due to this new failure mode.

Various activation energies have been reported for freak populations, 0.25 to 0.7 eV and main populations, 0.5 to 1.7 eV. These latter vary according to the technology (bipolar $\tilde{} 1.1$ eV, C-MOS $\tilde{}$ 1.3 eV, beam-lead silicon-nitride sealed-junction $\tilde{} 1.7$ eV). However, the variations between manufacturers or even wafer lots are often larger than this indicated range of values [12,13,14,15,16].

The freak population produces "infant mortality" and represents devices with relatively gross manufacturing defects. Many different defects may be involved such as cracked chips, nearly open interconnections, oxide pinholes and other oxide defects, mask alignment-induced coverage (oxide or metal) defects, gross contamination and failures that occur earlier than expected from main population failure mechanisms. Main population life may be limited by some of the above causes. It is more often related to ionic drift in oxides, but it may include such process related mechanisms as gold penetration through a barrier metal (for a complex metallization system). Figure 13 vividly demonstrates the different temperature-time dependence of freak and main populations [15]. For this case both freak population activation energy (0.8 eV) and the main population activation energy (2.0 eV) are quite high, indicative of what is expected from the beam lead type metallization-passivation system.

It has been pointed out by Stitch, et al., [16] that in addition to temperature, bias voltage is an important parameter in accelerating failures. In this case the Eyring [16] modification of the Arrhenius reaction rate is most applicable. It is:

$$t = \frac{G}{T} \exp\left(\frac{E_A}{kT} - V\left[C + \frac{D}{kT}\right]\right)$$

where G, C, and D are positive constants, and V is the bias voltage. This equation indicates that for a fixed temperature, as the voltage increases, the median life decreases. In one case on increasing the applied voltage from 5 to 15 V, the main population failure mode activation energy decreased by approximately 0.1 eV (apparent) [16]. The problem of directly applying this model results from the fact that in a complex integrated circuit it may be impossible to have the same voltage across each junction, thus one can only speak in terms of voltage applied to the overall device. Interpretation is further complicated by some tests that are carried out at temperatures $> 200\,°C$.

The burn-in test is generally considered to be the only test that can reduce electrical "infant mortality" in electronic systems that incorporate many devices. It is relatively inexpensive to perform and is required on essentially all hermetic military microelectronic devices. The use of this test on nonhermetic (plastic encapsulated) devices can be undertaken only after consideration of such additional factors as the plastic glass-transition temperature and decomposition temperature. Most plastic devices are burned-in at lower temperatures than hermetic devices. A comprehensive review

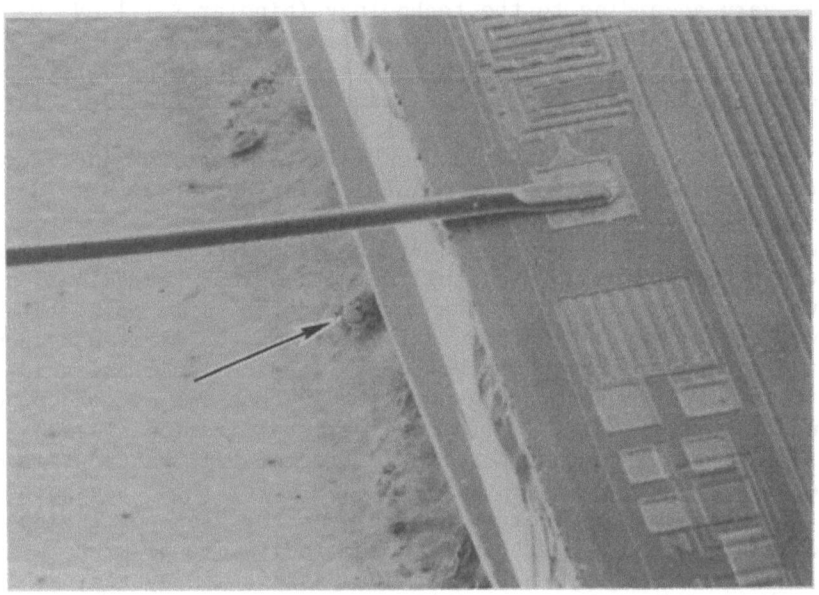

Fig. 14.(A) An IC having low lead dress and gold-silicon eutectic particles formed during die attach. A typical particle is indicated by arrow.

Fig. 14.(B) Shorting gold/silicon eutectic particle under different lead on same IC. (After Ebel [8]).

of accelerated testing has been given recently by Reynolds [17].

6. **Particle Impact Noise Detection (PIND) Test**. The purpose of this test is to detect loose particles inside a device package. To perform this test, the device is attached to a piezoelectric transducer with an ultrasonic coupling material. The transducer is attached to a shaker which vibrates the device and transducer, typically at 60 Hz, at accelerations in the 10 to 20 G level. Various specifications may require different frequencies for different package sizes in order to allow for the variation in time of flight of loose particles with different characteristics. The transducer and ultrasonic amplifier are usually peak tuned to 140 kHz. Most small particles of 25 µm diameter or more will produce ample acoustic signals at that frequency upon impacting with the walls of the device package. Unfortunately, the hissing of compressed air escaping, clapping of the hands and many other common noises may also excite the transducer leading to rejection of good, particle-free devices. Also, some particles do not break loose from the package so that they can be detected during the test. After the device is installed, these undetected particles can cause failure. Figure 14 (A) shows a wire bond with low lead dress and an arrow points to die attach eutectic particles which could break loose and possibly be detected by a PIND test. Figure 14 (B) shows a loose eutectic particle short between the wire bond and the chip.

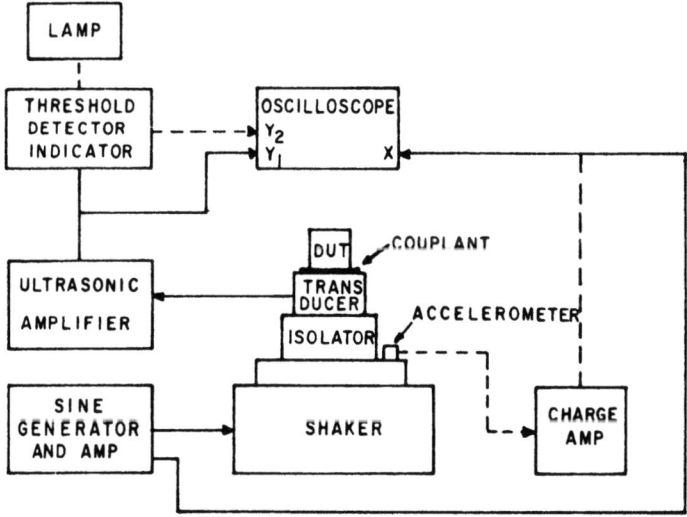

Fig. 15. Schematic of typical loose particle detection apparatus from MIL-STD-883B [1], Figure 2020-1.

The detection systems usually consist of an oscilloscope and headphones or a speaker which are connected to the apparatus through a hetrodyne system to reduce the frequency to the audible range. Fig. 15 is a sketch of a typical equipment setup for this test.

Before and/or during the test, the device is tapped by a "15 cm solid copper rod 2.5 mm in diameter with rounded end or other apparatus capable of imparting shock pulses of 200 to 1500 G to the device under test" (Method 2020 MIL-STD-883B) to loosen particles adhering to the device or package. Both mechanical and electromechanical apparatus, are available to shock the parts, but because of their differences they may contribute to variability in the test results.

From this brief description, it should be apparent that this test is neither quantitative nor particularly accurate. In fact, it is controversial. David considers it at best 50% effective [18]*. The test is costly to perform and costly because of the ambiguous results. It does not distinguish between metal particles which can cause shorts, and insulator particles that are usually benign. Small particles of aluminum are frequently undetectable. The limit of detection is typically ~ 0.02 µg, which represents an aluminum particle of approximately 25 µm in diameter.

Even though this test is of limited usefulness, loose metal particles have caused numerous field failures, and a high confidence level test to detect them is still needed. Polymer conformal coatings are used to immobilize the particles as an alternate solution to the problem adopted by many organizations. This is highly effective, but at times, these coatings can produce their own reliability problems.

7. **The Statistics of Sampling for Special Production Lots.** On a production line, all tests, whether destructive or nondestructive, are expensive and the decision to use one must be a compromise between cost and the reliability confidence level required by the end use of the devices. The costs of applying any individual test are not always apparent to those who must make decisions concerning their use. Implementation of a test is sometimes based on intuition rather than on the physics or statistics involved.

Once implemented, the obvious cost factors of a quality assurance test are the initial expense of test equipment, operator time, and the actual loss of product that results from the test. Some

---

*John Hilton, National Bureau of Standards (private communication), has done a study of the PIND test using seeded packages and his results generally agree with those of David.

TABLE 1. LTPD SAMPLING PLANS [1,2]

Minimum size of sample to be tested to assure, with a 90% confidence, that a lot having percent-defective equal to the specified LTPD will not be accepted (single sample).

Max. % Defective (LTPD)	50	30	20	15	10	7	5	3	2	1.5	1	0.7	0.5	0.3	0.2	0.15	0.1
Acceptance No. (c) (r=c+1)					Minimum Sample Sizes												
0	5 (1.03)	8 (0.64)	11 (0.46)	15 (0.34)	22 (0.23)	32 (0.16)	45 (0.11)	76 (0.07)	116 (0.04)	153 (0.03)	231 (0.02)	328 (0.02)	461 (0.01)	767 (0.007)	1152 (0.005)	1534 (0.003)	2303 (0.002)
1	8 (4.4)	13 (2.7)	18 (2.0)	25 (1.4)	38 (0.94)	55 (0.65)	77 (0.46)	129 (0.28)	195 (0.18)	258 (0.14)	390 (0.09)	555 (0.06)	778 (0.045)	1296 (0.027)	1946 (0.018)	2592 (0.013)	3891 (0.009)
2	11 (7.4)	18 (4.5)	25 (3.4)	34 (2.24)	52 (1.6)	75 (1.1)	105 (0.78)	176 (0.47)	266 (0.31)	354 (0.23)	533 (0.15)	759 (0.11)	1065 (0.080)	1773 (0.045)	2662 (0.031)	3547 (0.022)	5323 (0.015)

[1] Sample sizes are based upon the Poisson exponential binomial limit.

[2] The minimum quality (approximate AQL) required to accept (on the average) 19 of 20 lots is shown in parenthesis for information only.

TABLE 1. First three lines from LTPD Table B-1 in MIL-M-38510D [19].

of the less obvious costs are the test throughput time that may
cause delays in shipment (such as burn-in or lot-sample life-tests);
training of operators; maintenance of equipment; data evaluation
costs; record keeping; and the electrical and other costs of
running the test equipment (burn-in, lifetests, temperature cycle,
etc.). Time lost in making decisions on border line cases (such
as in visual inspection of expensive devices), cost of rejecting
good product due to incorrect decisions or to such problems as
faulty test equipment (e.g., electrical tester with high electrical
transients which damage devices or faulty probe adjustments that
damage pads).

Because the cost of 100% testing is often prohibitive, various
statistical sampling plans have become a normal part of semiconductor
device quality assurance. In the past, such plans were generally
based on AQL (acceptable quality level), but recently, perhaps
because they describe more directly the protection to the consumer
for individual lots, LTPD (lot tolerance percent defective plans)
are usually specified. MIL-M-38510D [19] contains LTPD tables that
are extensively used by the semiconductor industry. This sampling
method is valid for special product made in single lots. Parts for
space application or other high reliability usage often fall into
this category. Also, many hybrids are made and sold in unique
lots of 10, 50, or 100 units. The statistics used in controlling
the production of parts such as these are not, in general, the same
as are appropriate for month after month identical production of
a given product, where mean plus standard deviation ($\bar{x}$, σ) control
charts are used to build up confidence. LTPD sampling, as it is
practiced in the electronic device industry usually requires
segregating production over some period (e.g., 2 hours a day) from
one machine performing a particular assembly or test operation.
A number of devices (determined by the LTPD number) are then randomly selected from the lot and tested. Usually, but not necessarily, if one unit fails (acceptance number, C = 0) the entire
lot is rejected or must be 100% tested (if possible) before acceptance. Table 1 presents a portion of the LTPD table from MIL-M-
385D [19]. It should be noted that LTPD is based upon the number
of parts tested and not upon the percentage of the lot tested.
(If the lot is small, relative to the number to be tested, special
calculations are required.)

Figure 16 is a plot of LTPD at defect levels that can be
applied to high reliability parts. The values are calculated from
a paper by Schilling [20] in which an appendix derives the LTPD
equations. It can be seen that a very large percentage of units
must be tested to arrive at a low defect level. In fact if the
test is destructive, (such as a bond pull test), then it becomes
prohibitively expensive and an appropriate nondestructive test is
often substituted on a 100% basis (such as a nondestructive bond
pull test).

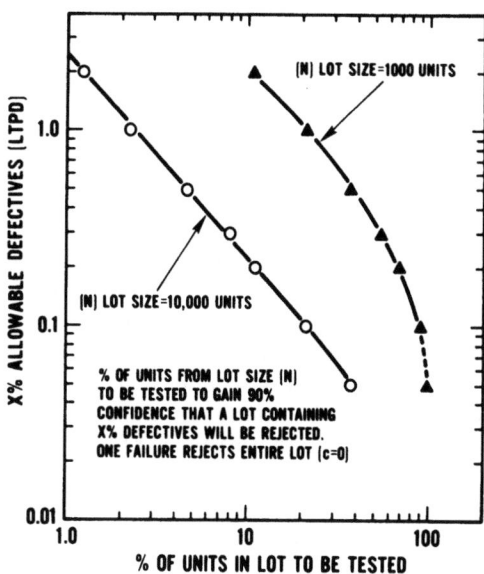

Fig. 16. Plot of LTPD vs. sample size to be tested at specified defect levels applicable to space quality parts.

As indicated, LTPD sampling is widely used in industry specifications, and it is almost as widely abused. The statistical purpose of LTPD plans is for protection with regard to single lots, and yet specifications call out the same LTPD values for both single lot and normal, continuous production. Typically in the electronics industry LTPD values of 5 to 10 (% defectives) are specified for all cases. This may result in product of much higher quality than indicated by the specified LTPD when applied to continuous production where a high confidence level, based on $\bar{X}$, S_X charts, is built up over a period of time for the production equipment and personnel. However, the same LTPD values are also used indiscriminately for small single lots as well and the user apparently expects lots which pass to have defect levels similar to space parts (~0.1% defectives).

8. **Conclusions of Section II.** Several nondestructive tests (screens) called out in MIL-STD-883B that are applied to establishing the mechanical integrity of electronic devices have been reviewed. Some of these (e.g., the PIND test) result in such low test confidence level that they should be rarely used, while others (e.g., burn-in) offer considerable assurance against infant mortality or other failures and should be increased in usage. The statistical sampling basis (LTPD) often used in applying these tests was also reviewed.

III. PASSIVE ACOUSTIC TECHNIQUES

A. Introduction to Acoustic Emission

Acoustic emission (AE) is generally defined as being a transient elastic wave or stress wave generated by the rapid release of energy within a material when that material undergoes fracture or deformation. The classic example of this has been known for years to the metallurgy industry as "tin cry" where merely bending a piece of tin will result in an audible sound. The first scientific report of sounds being emitted from metals during deformation was given by Joffe [21]. In 1950, Kaiser did the first comprehensive study of the phenomena [22]. His name is associated with the generally irreversible nature of AE, in which little or no acoustic emission occurs until previously applied stress levels are exceeded. Literally, a deformation or crack will only produce more AE if it is enlarged. The emitted stress waves may have frequencies ranging from the audible into the megahertz region, but the maximum energy is usually concentrated in the mechanical resonance modes of the test specimen. Detection of these waves usually takes place with ceramic piezoelectric transducers that are acoustically coupled to the specimen. However, wide band optical [23,24] and capacitive detection [25] have recently been used.

Many of the early materials studies were carried out on metals and correlations were made between various metallurgical properties and the AE released after the elastic limits were exceeded in stress-strain type of studies. Dunegan [26] for instance obtained an excellent fit of a mobile dislocation model for 7075-T6 aluminum, but little correlation was obtained for other metals. The various sources of acoustic emission that have been observed include: crack nucleation and propagation, twinning, grain boundary sliding, multiple dislocation slip, creation of multiple dislocations, solid-solid, solid-liquid, and liquid-solid phase transformations, and the Barkhausen effect (realignment of magnetic domains).

In general, most microelectronic uses of AE, with the exception of certain melt type welding applications, are more concerned with crack initiation and propagation than with structural dislocations or defects. Crack propagation in brittle materials was first explained by Griffith [27]. He postulated that elliptical cracks exist on the surface of brittle materials such as glass, lowering the tensile strength. Crack propagation occurs when the stress at the ends of the crack exceeds a theoretical value, σ_{th}. The stress (σ_e)* at the ends of such a crack as modified by Orowan [26] is given as:

$$\sigma_e = 2\left(\frac{c}{\rho}\right)^{\frac{1}{2}}$$

where c is the half length of an interior crack or the length of a surface crack and ρ is the radius of cruvature of the ends of the crack ellipse. For the crack to spread, the stress at the point of the crack must exceed the theoretical breaking stress (σ_b) of the material and

$$\sigma_b = \frac{\gamma Y}{4c} \left(\frac{\rho}{a}\right)^{1/2}$$

where γ is the specific surface energy and E is the Young's modulus. It can be seen that as the crack length increases, the stress necessary to keep the crack growing decreases. Thus, once started, Thus, once started, complete fracture can occur. The maximum or limiting crack propagation velocity is 0.38 the velocity of sound in the material.

For brittle cracks in polycrystalline materials, including metals, γ is replaced by γ_p which is the "effective" specific surface energy and c is replaced by d, the average grain diameter. The equation now predicts that the strength of a polycrystalline metal which fails by brittle cleavage should vary as the reciprocal of the square root of the grain size. The energy (E_g) released during microcrack growth, assuming all energy in the grain is released as the crack passes through the grain, is $E_g = \alpha^2 \sigma^2$, where α is a constant depending on grain size. Presumably much of this energy is released in the form of stress waves (AE), although there can be other mechanisms of energy loss. As the crack continues to propagate, stress waves (AE) are propagated in all directions from the crack tip. They include a white spectrum of frequency components up to many megahertz.

Rapid propagation as predicted by the Griffith crack theory, or its various modifications, is assumed to be valid for cracks in silicon chips as well as for glass- and ceramic-to-metal seals as found in semiconductor packages. However, for ductile-metal bonds made under nonoptimized or contaminated welding conditions, this theory is not necessarily correct. As reported in Section III-C bonds and various interconnections made under contaminated conditions usually consist of numerous individual microwelds (see Figures 22 and 37) which, in principal, can break individually without propagating a Griffith-type crack. Still, the crack will always propagate along the bond interface without major deformation of the joined pieces. This tends to make crack propagation at least somewhat "Griffith like" rather than what one might expect in a pure ductile fracture. One might calculate breaking stress of such weak "welds" but the microweld dimensions and number, when made under contaminated conditions, are unknown and this could lead to

---

* Units for stress could be grams force/cm^2.

unrealistic conclusions. Perhaps the closest description of this type of crack propagation has been given for adhesive bond (e.g., epoxy) fracture. Some workers [28] have modified the Griffith equations to include terms for plastic energy dissipation, and others have used a thermodynamic approach. Most are complex and require computer based computations. Anderson, et al., [28] have included an excellent review of the theories of adhesive fracture along with the computer techniques used to solve them.

Most aspects of acoustic emission theory are either in a state of controversy or are incompletely developed. Several papers have attempted to develop models that correlate the amount of acoustic emission to the size of a dislocation source, the grain size (for stress wave scattering), and the distance from the AE detector. However, as pointed out by Green [24,29], it is usually necessary to oversimplify the model for purposes of calculation as illustrated in Figure 17. Typically, special point sources and uniform propagation

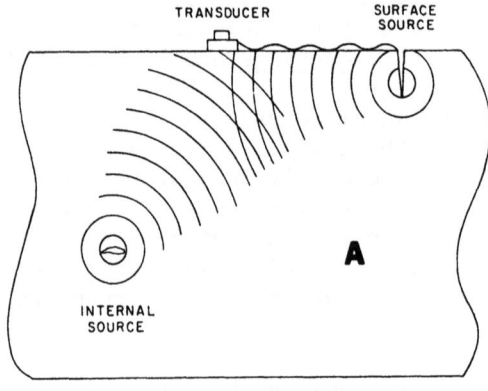

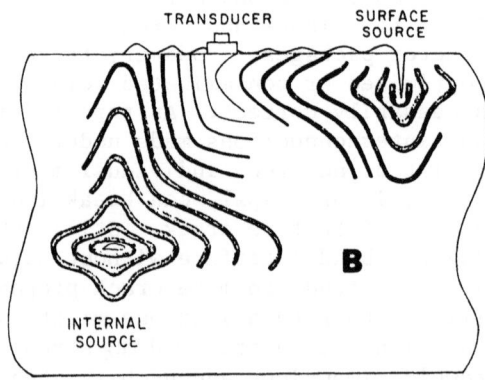

Fig. 17. (a) Oversimplified model of acoustic emission sources. (b) More realistic simplified model of acoustic emission sources. (After Green and Pond [29])

velocities are assumed; but in fact, the sources may have various nonsymmetrical shapes and the emitted wavefront continues to change due to directional variation in wave velocities associated with linear elastic wave propagation in anisotropic solids. Also, the wave amplitude decreases as it travels from the source because of the expanding wavefront and preferential attenuation caused by thermoelastic effects, grain boundary scattering, acoustic diffraction, and scattering effects are frequency sensitive. The signal may be further complicated by internal reflections and interferences that depend upon the specimen geometry. Thus, the detected signal amplitude, phase, and frequency may not be characteristic of the source. Therefore, most theory can only be verified on large, simple geometry samples. In addition, if the piezoelectric detector has high sensitivity, it usually has a high Q (narrow frequency) response. The amplifiers used for detecting threshold signals usually have a limited band pass to reduce noise so that all signals are somewhat similar in appearance and ringing is normal. Green [24] has given comparisons between acoustic emission signals from a typical piezoelectric-transducer tuned-amplifier combination and signals from a laser interferometer and optical detector - broad-band amplifier system. The preciseness of the optical detector output is striking as compared to the ringing output of the piezoelectric ceramic transducer system.

Hsu, et al., [25] have developed a theory to combine as much information as is known about a simple AE source, its propagation and detection by a wide band (capacitive) transducer. The theory is based on a Fourier inversion technique. Experiments to verify this theory were carried out on a well characterized system consisting of a 2.5 x 23 x 44 cm aluminum plate with a simulated AE source directly across the smallest dimension from the detector. This system simplified the waveform interpretation and excellent agreement with theory resulted.

It is not even clear how a piezoelectric detector responds to an AE wave front. Harris, et al., [30] demonstrated that the detector responds to the square root of the energy released during a given deformation process; whereas, Jon, et al., [31] show that the voltage generated by the transducer is related to the rate of energy released by the source. Thus, with all of the variables and uncertainties, it is extremely difficult to quantitatively relate a microscopic theory of stress wave emission with experimentally measured frequency and waveforms. In addition, it is not clear on a microscopic basis how cracks or other mechanisms actually generate "white" AE.

Many of the concerns of Green [24] and others are not of practical importance to AE detection in samples of dimensions encountered in microelectronics. The various mechanical resonant modes of a specimen may be as high as several megahertz and multiple

internal reflections are inevitable. Investigators are usually
constrained to attach a transducer, which may be larger than the
specimen, in whatever manner is possible (such as using tapered
acoustic waveguides) and work with whatever signal is received.
Waveform signatures, of frequency and amplitude, are recorded and
empirically correlated with appropriate mechanical stress tests
(e.g., destructive pull tests) for interpretation. It is obvious
that the interpretation of AE signals from typical electronics
applications is undeveloped and may never be amenable to clear
mathematical solution.

Two books are avilable that give theory, equipment and applications of acoustic emission to a variety of nonelectronic problems
and these should be read for more detail than is given above [32,33].
Also, a recent critical review of the status of the AE field has
been given by Lord [34].

B. Review of Acoustic Emission Applications to the Real Time Nondestructive Testing of the Mechanical Integrity of Electronic Components

The application of passive acoustic measurement techniques
are in their infancy in the electronics industry. Although the potential is great, it is yet to be exploited. Acoustic emission has
been used as a tool to study materials and monitor the physical
condition of large structures such as nuclear pressure vessels and
bridges but it has only recently been used to evaluate electronic
materials and assembly processes. The largest effort is perhaps
at the Western Electric Engineering Research Center, Princeton,
New Jersey and a majority of the published papers have come from
there. Most of the emphasis has been on real-time in-process
evaluation of some electronic production processes and these will
be reviewed in this section. These real-time applications will be
treated first and since the present author [35] was responsible
for the only study of post-production integrity screening, this
subject will be given in greater detail at the end.

The first published use of AE in electronics was by Vahaviolos
[36]. He used AE to reveal substrate cracking during the thermocompression bonding of beam lead devices to gold metallization on
hybrid ceramic substrates. During this bonding process the substrate
is subjected to a temperature of approximately 300° C and forces
greater than 100 kg/cm^2. Any ceramic flaw or warping located under
the bonding tool may initiate a microcrack which, if undetected,
can ultimately lead to failure of the entire device from moisture
related causes; or if the crack propagates during later qualification
screening or during device life, it may open up a metallization

interconnection. A beam leaded substrate of the type and vintage used during that work was shown in Fig. 6. The large size and completed cost of the substrate made it mandatory to stop assembly and reject the unit as soon as any crack was initiated (this problem is even more acute on sophisticated multilayer hybrids).

Since the AE detection work of Vahaviolos required placement of a transducer on the heated bonding stage, he had to make his own high temperature detector mounts and chose piezoelectric sensors according to his special needs. He used a modified lead zirconate-titanate transducer which could detect both longitudinal and shear waves and had a Curie temperature of 360°C. He also used a lead metaniobate transducer which only detected longitudinal waves but has a Curie temperature of $>600°C$ so that it can be used in many thermal tests.

As in most real-time AE detection systems, the many extraneous production noises that are present must be discriminated against. These may have the same frequency and amplitude characteristics as the AE. For example, during beam lead bonding, there is first the impact force of the hot bonding tool against the beams and substrate which generates noise similar to AE. As the beams and the substrate metallization undergo plastic deformation, they also emit AE. If the substrate cracks during this time, a rather sophisticated electronic system is required to distinguish crack generated AE from the extraneous signals. Therefore, most real-time tests are relatively insensitive to cracks 50 to 100 μm in length in brittle materials, whereas a less sophisticated electronics system used to screen completed devices can easily detect brittle cracks 5 to 10 μm long.

Saifi and Vahaviolos [37], have reported the use of AE for real-time nondestructive evaluation of laser spot welding of small insulated wires to electronic terminal posts. For this a pulsed YAG Laser was used (40 pulses per second, pulse length ∿3.5 ms, energy per pulse ∿20 J). They determined the conditions required for insulation vaporization as well as for the ideal combination of wire and terminal-post metallurgy (copper wires, monel terminal posts). Production noise was minimized by gating the detection system only during the process period when welding occurred, and after all insulation was vaporized. Although this work is not closely related to microelectronic devices, it could be the beginning of reliable laser welding in smaller structures such as electronic device packages.

Carlos and Jon [38] reported the detection of cracking in high reliability, high voltage ceramic capacitors that were intended for submarine cable use. In this paper, AE is used to detect cracks created in the ceramic casings due to the thermal shock of soldering. Cracks so generated could not be observed in a visual inspection because the susceptable area was covered with solder, however such cracks could later cause reliability problems. The electronic system was designed to operate continuously and to process each burst of AE from the sensor, using pattern recognition to determine whether any particular burst resulted from cracking the ceramic or was from such extraneous noise sources as the soldering iron scraping against the ceramic. The system could detect brittle cracks of the order of 25 µm but most actual cracks were nearer 1 mm in length. This AE system was described as the only possible real-time method of 100% nondestructive checking for cracks in such capacitors.

Jon et al., [31,39] have described the use of AE for the nondestructive evaluation of the quality of several types of resistance welds. In one case of tantalum capacitor leads, the material combination in the vicinity of the weld was unique (Ta, Ta_2O_5, Cu, solder, and steel) and it presented considerable AE interpretational problems. For materials of complicated geometry and metallurgical composition, it is not easy to identify any individual intrinsic processes from the total AE generated during the welding process. This identification difficulty is due to the frequent overlap of the generated AE due to different causes. For example, one element of the composition such as copper could begin melting while the already-liquid solder could be going through expulsion from the compressed joint at the same time. Signals generated in such a manner are usually very difficult to distinguish electronically. In such cases, the best way to analyze the problem is to monitor the generated AE during a time period where only clearly understood events are occurring. From a metallurgical point of view, the cooling period is easy to identify experimentally. During this period, weld nuggets start to solidify and generate AE signals because of both the plastic deformation and the build-up of the residual stresses. Both of these processes can be used to indicate the weld strength because large nugget volume, a good physical indicator of weld strength, will give rise to more plastic deformation as well as a higher residual stress build-up. The result is to generate more AE signals during this time frame. In order to avoid extraneous AE signals that bore no relationship to weld quality the authors gated their AE detection system to be responsive from 10 to 40 ms after the peak of the weld current, during the nugget solidification. This period produced AE signals that were directly correlated with weld quality.

Another evauulation of spot welding quality by the AE monitoring of 500 µm diameter nickel wire welded in electronic components was carried out by Knollman and Weaver [50]. The all nickel system was simpler and the wires larger than that described by Carlos and Jon [38]. In this case, a commercial AE weld quality control system was found to be adequate. The authors found that weak welds could be detected at a confidence level of better than 97%.

In 1976, Ikoma, et al., [41] used acoustic emission to study
dislocations in semiconductor materials. Many failures or shortened
useful life in GaAs light emitting devices have been shown to result
from dislocations propagating into active regions of the device.
Also it had been shown by Kotani, et al., [42] that improperly
controlled thermocompression bonding can create defects in GaAs
devices and reduce their reliability. Previously Sedgwick [42]
had shown that AE was emitted from dislocation loops in KCl and
LiF. Thus, Ikoma designed experiments to explore the possibility
of observing similar AE in GaAs by intentionally creating defects.
His experimental procedure was as follows. A quartz rod with a
smooth 1 mm hemispherical radius on its end and an AE transducer
on the other was used as a pressure probe. The polished GaAs
sample was placed on a heat stage. A load of 100 gf was applied in
all cases to force the probe against the GaAs sample and the
substrate was heated to various temperatures from 25° to 420°C.
A study of the dislocation patterns of samples was carried out on
the area under the probe by varying the temperature. At each
temperature, the sample was etched and the dislocation density and
pattern determined. No dislocations were observed to occur below
150°C. At temperatures between 300° and 400°C clear "rosette"
patterns were observed with "arms" stretching radially along the
<1$\bar{1}$0>, <01$\bar{1}$>, and <$\bar{1}$01> directions on a <$\bar{1}\bar{1}\bar{1}$> surface. AE was only
observed at and above 300°C. The AE signal increased linearly
with temperature from 300° to 400°C and then decreased above 400°C.
The reason for the decrease was not understood, since dislocations
were still generated. Since the AE-active temperatures and pressures
were the same as those used in thermocompression bonding, it would
be logical to apply the results of this investigation to controlling
the thermocompression bonding process. Without giving any details, the
the authors states, "This technique is not being applied to the
detection of dark-line-defects in GaAs-laser diodes and to the
real-time inspection of a thermocompression bonding process of GaAs
device fabrications."

Very recently Ikoma, et al., [44] presented evidence that AE
is emitted from GaP light emitting diodes during electrical over
stressing. The devices were operated at currents up to three times
their rated values. The measurement apparatus consisted of a 4.7
mHz transducer and high gain (69 db), low noise, tuned amplifier.
The typical AE signals were very small ($\sim$100 nv at the transducer).
The authors correlated increased AE with decreased light output.
Diodes that showed no decrease in light output produced no AE.
After the tests, the devices were etched and those that had produced the most AE displayed the greatest dislocation density. There
were large unpredictable variations in degradation from diode to
diode. At present, there is no obvious application of the AE
correlation with decreased LED performance since the decrease in
light output can be more easily measured. However, the authors
have clearly shown that electrical and thermal stresses which

produce crystal dislocations in semiconductor materials can be
revealed by AE. Applications of these techniques for screening out
failure prone devices may be possible in the future.

C. Acoustic Emission as a Post-Production Screen for Bond Integrity in Microelectronics

1. <u>Introduction</u>. The previous section reviewed acoustic
emission tests that could detect a weak weld or a cracked substrate
during the actual production process (real-time). From the
standpoint of production economics, this is the ideal time to dis-
cover a defect so that no additional production effort and money
is lost in further assembly. However, there are numerous cases
where such real-time detection is not practical, but instead a
screen applied at a later time is necessary to remove production
defects, as with tests in Section I. For instance, in cases of
simultaneous multiple bonding such as the 40 lead tape bonding of
integrated circuits, it is not possible to assure that all leads
are well bonded in real-time. Some stress tests must be applied
later. Likewise, when multiple devices are simultaneously soldered,
such as in wave soldering, acoustic emission signals are not inter-
pretable. Epoxy bonded components yield little or no acoustic
emission during curing, and thus, cannot be tested in real-time.
Also, tests are often required as screens for incoming inspection
for individual parts such as packages. Thus, the ability to test
the component at a later time may be the only way to assure bond
or package integrity. The next section describes acoustic emission tests on completed units and will start with tests developed
to assure beam lead bond integrity. Then the techniques will be
applied to tape bonded devices and hybrid components [35].

Beam lead and other gold-gold thermocompression bonding is
generally reliable, once an optimum bonding schedule is achieved.
However, as with any bonding system, contamination in the bond
interface may inhibit welding [45,46] on one or more beams in an
unpredictable manner. In addition, if the hardness of the gold
varies for devices from different wafer lots of different manufacturers, a bonding schedule optimized for one lot may produce
erratic bond reliability for another. Thickness irregularities in
thick-film bonding metallization may also reduce bond adherence for
one or more beams on a multibeam device. Thus, it is desirable
to have a simple 100% nondestructive test that will detect one or
two poorly bonded leads out of a large number of well bonded ones
and not require a subsequent visual inspection or electrical
test to reveal the results.

AE has been studied in a variety of materials by many workers.
However, there is only one known study of such emission from gold,

the material used in the beam lead bond system. Schofield [47] reported that "the occurrence and behavior of AE in gold was undoubtedly the most consistent and certainly the most active of any of the face centered cubic metals previously studied." He verified the Kaiser effect [22] for the high frequency emission. However, he found that certain "burst emissions" were reproducible without annealing. Schofield's work was on gold single crystals in two orientations; however, some specimens that had been elongated during a first test were then annealed and developed a relatively coarse grain structure. These polycrystalline samples also emitted AE upon further testing. Thus it was concluded from the NBS investigation that the gold in beam leads should be capable of AE if poorly welded beams lead devices could be stressed adequately to strain or break some of the microwelds in the few bonded areas.

2. <u>Methods of Applying a Mechanical Stress to the Beam-Lead System</u>. The problem of mechanically stressing a bonded beam-lead device in a nondestructive manner is a formidable one. The most desirable method is to lift the device upward. This would stress both the beam anchors (the attachment to the chip) as well as the beam bonds. The most obvious method of doing this would be to slip hooks under the corners of the bonded beam-lead device and pull upward. However, since the bugging height* can be less than 25 µm and may vary from corner to corner, such a grappling hook could crack the thinned silicon at the edge of the chip, causing extraneous AE unless extreme care was exercised by the operator. Another method of pulling the device upward would be to epoxy tiny hook-shaped wires to the top of the chip and pull the device with a wire bond puller. A more convenient version of this technique is to use a hot-melt-glue pull-test, applying a force well below the pull-off level. However, the glue used for this purpose would remain on the chip and that particular organic material is not a desirable additive to a hybrid package. Also, the brittle hot melt glue can develop cracks during pulling and emit extraneous, misleading AE.

One simple alternative to the glue method is to apply a "dab" of a silicone rubber (SR) to the top of the beam lead chip and let it cure overnight at a temperature of about 50°C. Then a sharply pointed hook can easily pierce the rubber parallel to the chip as shown in Figure 18. The hook is then pulled upward; about 40 gf

---

* Defined as the vertical distance from the substrate metallization to the bottom of the chip.

**If the chip is not clean, this resin may pull free with about 20 to 30 gf. An adequate cleaning procedure is to immerse the bonded substrate in a fluorocarbon solvent and then air blow off the remainder of the solvent.

can be applied to a 1 x 1 mm chip before the rubber breaks.** A tweezer type of device or a flatter shaped hook could be used in place of the present hook if it is desired to apply greater pull-forces. To verify that this method produced no extraneous AE, similar sized "dabs" of SR were bonded directly to the substrate and pulled with the hook. The SR emitted no measurable AE until the force that ruptured the rubber was reached. The silicone rubber used for this purpose was usually the stiff version of the methanol-based resin that some organizations use to protectively coat beam-lead devices. When the device is subsequently encapsulated, the new resin will seal the punctured rubber and fill the cavity under the chip. Therefore, this pull method and its residue can be considered nondestructive to both the device and the substrate. Either a 100% test or a sample quantity of devices could be tested on each substrate as a control. Alternately, a high strength silicone rubber can be applied to the chip and this hook method can then be used as a destructive pull test. Pull forces of up to 100 gf have been applied to 1 x 1 mm chips.

Another method of pulling a chip is to mold a vacuum cup out of silicone rubber or other elastomer into shape of the chip. When vacuum is applied to a device, an upward force of about 0.5 gf per beam can be obtained for a 14 or 16 beam device having an Electronic Industries Association registered chip outline. A third method of forcing the chip upward and the beams outward is to inject the silicone protective coating resin under the chip and allow it to cure. The material has a much higher coefficient of expansion than the gold beams. Preferential heat (such as infrared) applied to the chip will expand the rubber enough to stress very weak bonds. This method works best with high bugging heights which allow more resin under the chip.

An alternative method of stressing the bonds is to push downward on top of the chip. Some of the force will be applied to the bond heels in the shear direction. This is also the simplest method of applying force to the beam lead system. Depending on the angle at which the beams project from the chip, the uniformity of the bugging height around the chip, variation in beam dimensions, and the gold hardness, a 16 lead device will collapse to the substrate, accompanied by large bursts of AE, with the application of about 30 to 60 gf. This collapse force may be less if it is applied off-center or if the bugging height is not uniform. An improved variation on the simple push-down technique is to simultaneously push downward on top of the chip and, with an equivalent force, push horizontal force is applied to the bonds in a peel direction. Care must still be exercised to prevent collapse of the bugging height.

It has been found that roughly three times as much force in a downward direction can be safely used if it is only applied to the beams, leaving the chip free. A simple resolution of forces analysis based on typical bonded beam dimensions indicated that with all

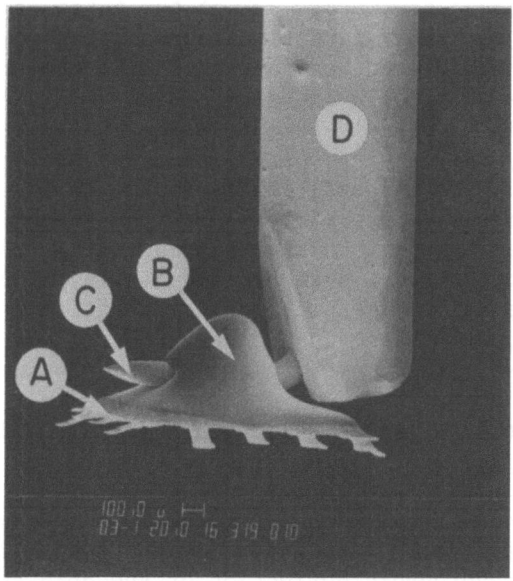

Fig. 18. Scanning electron micrograph of a weakly bonded beam lead device (A) with a silicone rubber "dab" (B) on top, that has been pulled up by an electrolytically etched, 150 μm diameter tungsten hook (C). (It is important that the hook be very smooth and have a sharp point so that the rubber is not torn while it is being pierced.) The hook carrier (D) is a section of a No. 22 hypodermic needle in which the hook had been inserted and rigidly epoxied in place.

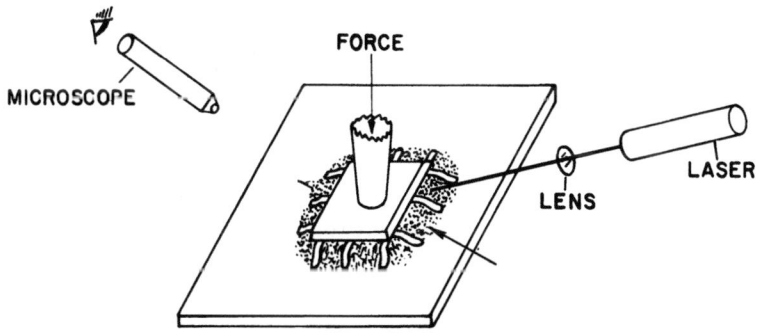

Fig. 19. A simple method of measuring the threshold of downward motion of a beam lead chip. The arrow ponts to the static interference pattern (the dots between the beams). When force is applied, the pattern changes.

beams uniformly stressed downward, 90% of the force is applied at the bond heel in the shear direction and no net torque is applied to the beam anchors (the beam attachment to the chip). However, when the beams along only one side of the chip are stressed at one time, the chip is rigidly held in place by the other beams and then essentially all of the applied force appears as a torque on the anchors. Thus, it is possible to provide an AE test for both beams and anchors by appropriately applying the stress. This assumes that the beams project horizontally outward from the chip for about 50 μm before curving downward and bonding occurs about 100 μm away from the chip. In some bonding situations, the beams leave the chip at about 45 deg. in a downward direction and the bond occurs only 25-50 μm outward from the chip. No anchor torque is possible in this case. Thus, in order for this test method to be applicable, great care in beam alignment during bonding is essential to obtain uniform bugging height.

To determine the effectiveness of the chip push-down method, a simple technique was developed to determine the downward force necessary to produce threshold deflection of the chip as well as of the bonded beams. A 1 mw He-Ne laser with a focused spot diameter of approximately 25 μm or less was directed under the chip at a low angle between two of the beam leads as shown in the sketch of Fig. 19. The force was applied by the apparatus described in Section III-C-4 that is normally used for such purposes in the AE test. The laser light was multiply-reflected back and forth between the substrate and the chip and established a complex static interference pattern that could be seen extending outward from the edges of the chip for 125 to 250 μm. Downward or upward deflection of the chip by only a small fraction of a wavelength produced changes in the interference patterns that could be easily seen through a 40X binocular microscope, even though no direct motion of the chip was discernible. The threshold of observable motion of an individual unbonded beam could also be seen by this method.* Such motion of the chip occurs with an applied force as low as 3 to 5 gf to the chip before movement is observed, and this requirement varies according to the angle at which the beam lead approaches the substrate.

---

*Although qualitative in nature, this simple technique may be useful in other types of visual inspection and should be a valuable aid in the observation of the relative thermal expansion of components as well as for studying creep phenomena. In order to be effective, the substrate must be coated with metal. A ceramic substrate under the chip results in a confusing pattern of internal reflections and interference patterns.

3. <u>Preparation of Controlled Bondability Substrates</u>. Various unpredictable bonding conditions can result in one or more of the beam leads not having a strong weld. However, it is difficult to deliberately obtain weakly bonded leads. The usual method for obtaining weakly bonded gold to gold leads, by lowering the bonding temperature, is unreliable. The first beam leads of such an intentionally weak bonding series may increase in bond strength while other devices are being bonded. Even substrate temperatures as low as 85° to 150°C for one hour can significantly increase gold to gold bond strength on uncontaminated bonding surfaces [45,48], and higher temperatures require even less time for bond improvement. Such strengthening of the bonds has been verified in these studies. Therefore, low temperature methods of producing weak bonds are not desirable for use in developing new measurement methods.

In order to obtain weak bonds specified as to both number and position, an effect that is normally avoided was used. It is well known that chromium-locked-gold metallization must be kept at relatively low temperatures or the chromium will diffuse to the

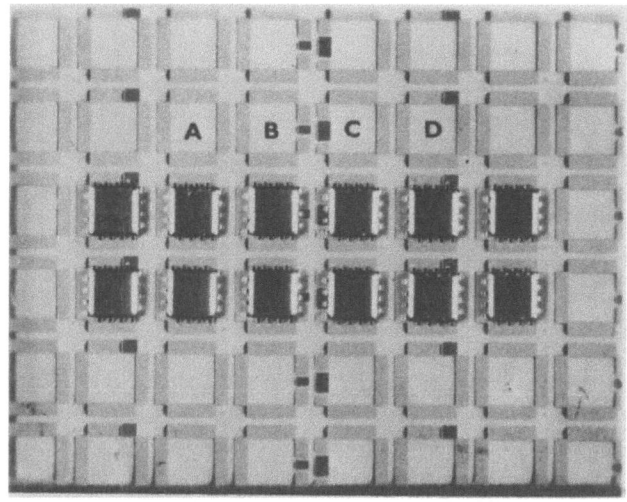

Fig. 20. A patterned substrate showing bonded beam lead devices. Chromium oxide covered areas for bond strength inhibition are stained black to increase visibility. The vertical row patterns are: (A) bonding controls (all good bonds), (B) pattern containing one weak bond in the center of the span, (C) pattern containing two weak bonds in the ceneter, (D) pattern containing one weak bond on a corner. The chip dimensions are 1 mm on a side.

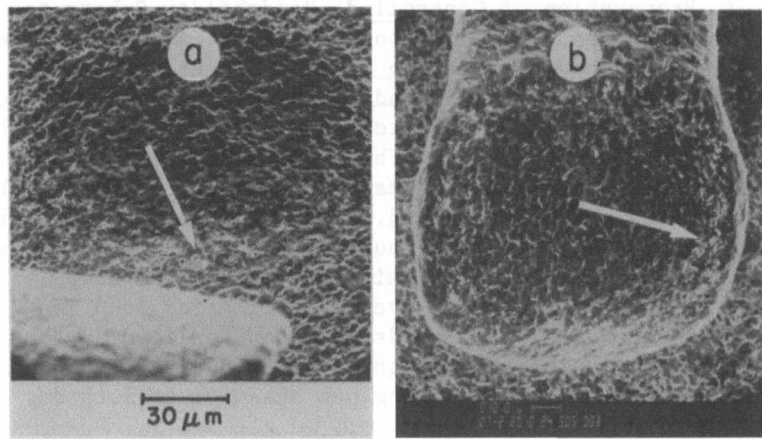

Fig. 21. (a) is an SEM photograph of a chromium diffused, gold coated ceramic substrate with a peeled up beam lead in the foreground. An arrow reveals some of the tiny broken substrate welded areas. (b) is an SEM photograph of a peeled up beam lead revealing the tiny white appearing dots near the perimeter that were the welded ones.

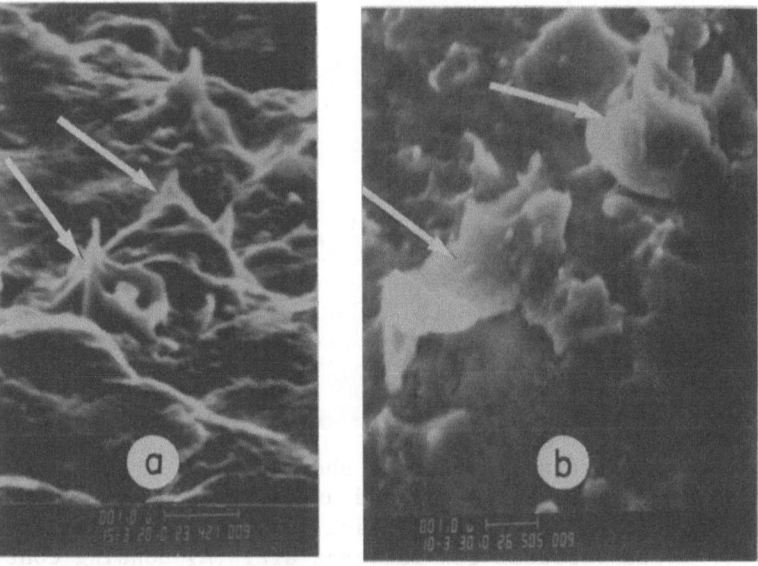

Fig. 22. (a) is the SEM photograph showing pieces of the substrate pulled off and sticking to the beam lead. (b) is a high magnification SEM photograph of substrate weld breaks. (These are indicated by the arrows.)

surface, oxidize, and severely decrease the thermal compression bondability [49]. Therefore, tantalum nitride-chromium-gold* substrates were heated to 310°C for two hours to diffuse the chromium to the surface. A special photomask set was used to pattern the metallization, and a cerric ammonium nitride etch [50] was used to preferentially remove the chromium oxide in all but specifically designated areas. Figure 20 is a photomicrograph of such a substrate bonded with beam lead devices. For clarity of presentation, the chromium oxide covered areas have been darkened (normally they are only slightly darker than the rest of the metallization). The four different patterns can be clearly seen. They include, from left to right, a single poorly bonded beam on a corner location, (D), a single weak beam in the center, (B), two weak beams in the center, and a control pattern for making all well bonded beams.

It should be pointed out that when using the chrome-diffused gold bonding pads, some degree of control over the beam-lead bond-peel-strength (essentially all the bonds so prepared, peel) can still be exercised by varying the bonding parameters (force and temperature). In this manner the peel force for an individual beam can be varied from less than 0.5 to approximately 3 gf.

It was established that the chrome oxide method of producing controlled weak bonds still left a few areas that were welded and therefore could produce valid AE signals when the lead was stressed. In addition, it was desirable to determine the minimum AE signals that could be detected with the available equipment. Devices were bonded following various bonding schedules to substrates that had chromium oxide on the surface. The devices were stressed and acoustic emission signals were recorded on digital pulse-capturing equipment. The poorly bonded beams were then peeled back and examined for evidence of torn welded areas representing AE point sources. Fig. 21a is an SEM photograph of the chromium-gold coated ceramic substrate with a peeled-up beam in the foreground. Examination of the beam lead bond depression in the substrate revealed tiny broken welded areas around the perimeter where deformation is greatest. The largest of these are indicated by the arrow. The beam is shown in Fig. 21b. The tiny white dots near the perimeter are the broken welded areas. The fact that the welded areas lie around the perimeter is in agreement with the deformation theory of thermal compression bonding by Tylecote [51]. A higher magnification view of substrate weld breaks is given in Figure 22. It should be noted that the bonds made under such contaminated conditions consist of a large number of individual microwelds. Each

---

*Efforts to use chromium-gold metallized substrates resulted in rapid etching of the undiffused chromium, thus undercutting the gold. The reason for this is unknown.

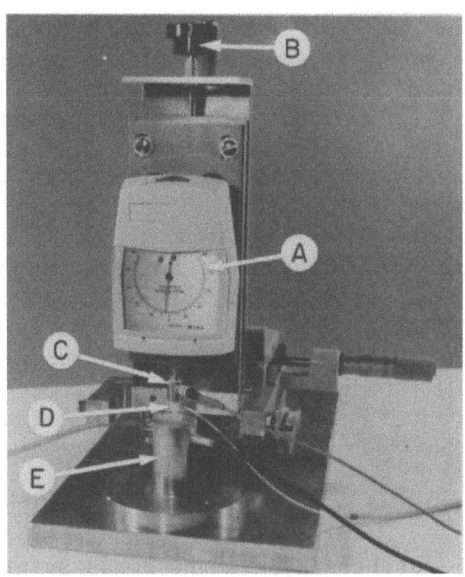

Fig. 23. Apparatus used to apply upward or downward force on beam lead devices and to detect any resulting AE. (A) force gage, (B) force angle control, (C) acoustic emission detector, (D) acoustic waveguide and force probe, (E) substrate holder with vacuum hold down and substrate AE detector.

one of these may be broken relatively independently of distant ones, and thus, this type of bond does not necessarily follow the crack propagation velocity as described in III-A. The arrows in the figure designate broken gold welds of one to two micrometers width. Figure 22b shows part of the beam-lead, clearly revealing pulled-off metal pieces from the substrate. In this case, their size is two to three micrometers in width.

4. Experimental Apparatus. Most mechanical stressing experiments and AE measurements were made using the apparatus of Fig. 23. The force gage (A) can measure either an upward or downward force. The entire gage-probe apparatus can be rocked in front-to-back and side-to-side directions by the knob (B) enabling a downward or upward force to be applied at an angle to a single row of leads at a time. The top AE detector-force probe (C-D) is screwed into the gage, facilitating rapid probe changes. The bonded test substrate is held on a chuck (E) which contains the substrate AE detector.

Close-up photographs of several top detector probes are shown in Fig. 24. Figure 24a shows a ceramic probe designed to

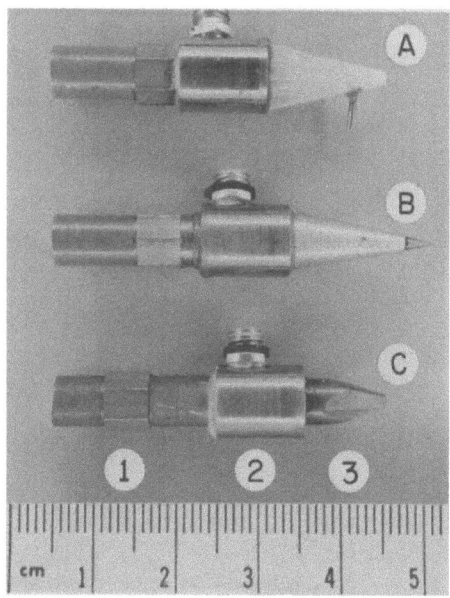

Fig. 24. AE probe-detectors. (1) Adaptor for attachment to the force gage, (2) AE lead zirconate titanate type detector, (3) acoustic waveguide and probe. Probe (A) is designed to apply uniform pressure on SR encapsulated lead chips. The waveguide portion is ceramic and its tip is coated with SR or polyamide. Probe (B) is designed to probe individual beams. The center tip is of tungsten carbide and has a 75 μm flat portion which is coated with SR. Probe (C) is a modified tungsten carbide beam-lead bonding tool in which the inner walls are about 25 μm larger than the silicon chip on all sides.

apply a uniform pressure on beam lead chips. Figure 24b shows a tungsten carbide conical probe with a 75 μm diameter flat on the bottom. This probe is used to stress individual leads. The probe in Figure 24c is made of tungsten carbide and is essentially a beam lead bonding tool with small dimensions so that it only contacts the horizontal projection of the beam near the silicon. The tips of all probes are coated with from 25 to 75 μm of silicon rubber (SR) both to increase the acoustical coupling to the leads or chips and to avoid metal to metal or metal to silicon contact since such scraping can result in extraneous AE-type noise.

A substrate detector fixture is shown in Figure 25. In this particular fixture, the test substrate is held against the AE detector by a cylindrical weight. (Other substrate detector

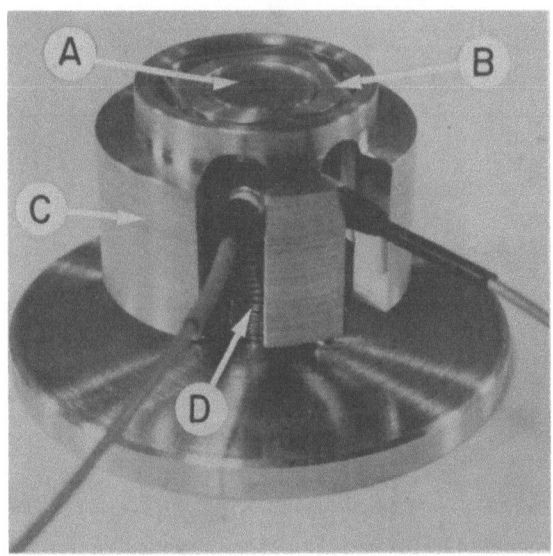

Fig. 25. Substrate holder using a weight to force the substrate against the detector: (A) AE detector, (B) simulated substrate (glass), (C) removable brass weight, (D) spring to force the detector against the substrate.

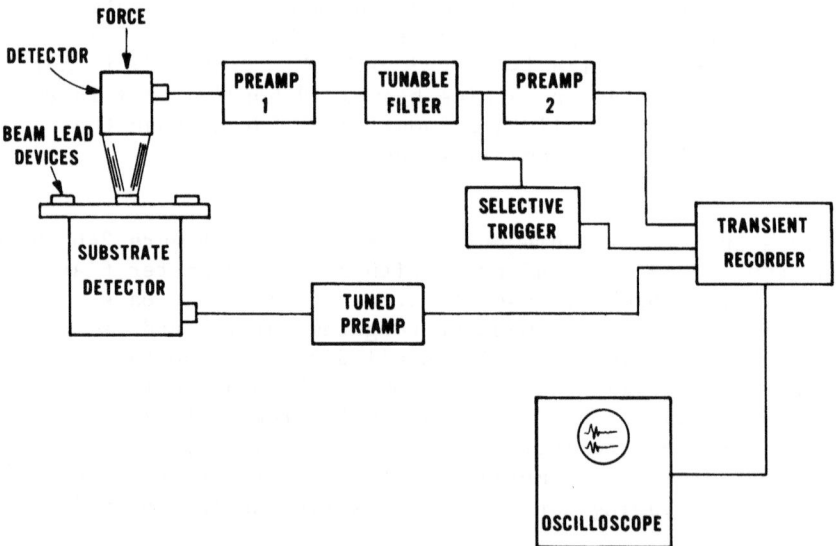

Fig. 26. Block diagram of the acoustic emission test apparatus.

fixtures, such as the one that was shown in Fig. 23 use a vacuum hold down.) The detectors are forced upward against the substrate by a spring. The surface of the substrate detectors are coated with a thin film of very compliant SR to facilitate acoustical mating with the ceramic substrate, in order to avoid the use of various sticky organic coupling materials which must be removed later. A textured SR surface is preferable. This is obtained by pressing ground glass, treated with a mold release agent, against the detector while the resin cures.

Several arrangements of signal preamplifiers, filters, and the digital pulse-capturing equipment have been employed. However, the block diagram of the most frequently used system is given in Figure 26. The total gain in each amplifier channel is 80 dB. Each channel has a 24 dB/octave band pass filter, a tunable filter, or both. The special digital trigger circuit [52] requires that a given number of cycles, selectable from 1 to 10, of a separately specified positive and negative amplitude signal occur within a total specified time frame for triggering the dual-channel transient recorder. The overall system is capable of detecting AE signals barely above the average noise level of the preamplifiers and considerably below various system and line transients. Most AE detector output signals produced in the present experiments were in the range of about 10 to 100 µV and were easily captured by the above equipment. Because of the variety of gain adjustments possible (preamplifiers, pulse capturing equipment, and oscilloscope), the vertical scale of most AE oscillograms will not be specified. The only important consideration is the signal to noise ratio and this can be easily observed from the traces.

Experiments were run using various AE substrate detectors with the tunable filter. It was found that the maximum AE output was obtained from the thin, 2.5 cm square ceramic substrates at 350 to 400 kHz, and from bonded 16-lead, beam-lead chips at approximately 1 MHz. The actual boundary conditions were unknown and could not be included in calculations of the mechanical resonances of the substrate and chip, and were off by more than a factor of two. As a result of the measurements, the substrate preamplifiers and detectors were chosen to peak at 375 kHz and the chip probe equipment at 1.1 MHz. The AE experiments described here will all assume such frequency responses unless otherwise stated. The digital trigger has been used in either the substrate or probe circuit; but in most cases, it was used in the probe circuit.

When the probe was in contact with the chip, there was essentially no mutual response from the probe and substrate detectors resulting from random AE sources remote to the beam lead device regardless of the operating frequencies of each detector. A crack in the substrate would yield a strong signal in the substrate detector, but not in the probe. A ceramic scribe-scratch on the tapered probe, which saturated its detector preamplifier, registered negligibly on the substrate detector. However, stress waves generated by a failure within the beam bond-anchor system resulted in a substantial signal in both channels. Therefore, in experiments where both substrate and probe detectors were employed, some AE output was required from each detector in order to define a failure although the relative amplitude and number of bursts recorded from each detector often varied considerably when the detectors were operated at different frequencies. The purpose of the present study was to develop a specific test method; however, monitoring a single AE source in two or more frequency bands, as in the present experiments, may be a fruitful approach to understanding the nature and mechanism of stress wave emission.

5. Experimental Results

5a. AE Results from Pulling Beam Leaded Devices. The silicone rubber-hook method of pulling beam-leaded devices, as described in Section III-C-2, was used to obtain quantitative information of various beam failure modes, once the pulling force is equally distributed between all beams. For this work the apparatus of Figure 23 was used. The hook shown in Figure 18 was substituted for the probe detector and all AE was picked up by the substrate detector.

In order to demonstrate the sensitivity of the AE method, all beams except one were cut and the remaining one was pulled to destruction. It broke at the bond heel. Figure 27 gives its AE pattern. Clipped waveform peaks indicate that the substrate detector output was significantly greater than one millivolt peak-to-peak during the initial part of the break. AE from such breaks generally continue erratically for several times the 200 µs shown in the figure. Most AE signals in this work are much smaller and of shorter duration. For comparison, a well bonded device with weak anchors had all but one of its beams cut in a similar manner to the above. Figure 28 gives the AE pattern of the single anchor failure. The peak to peak detector output in this case was approximately 0.3 mV.

Pull tests were conducted on several well bonded devices that had weak anchors (peel strength ∿3 gf/anchor), a series of short bursts was observed starting at ∿1.5 gf/beam. Figure 29 gives a typical AE signal from peeling anchors. Similar well-bonded devices with strong anchors produced no AE until a force per beam of approximately 2.5 gf was applied and the bursts in this case were longer and higher in amplitude.

Pull tests were conducted on strongly bonded beam-lead devices to serve as controls for weakly bonded-beam experiments. A number of devices from four different manufacturers were tested. The devices from three of these manufacturers produced no detectable

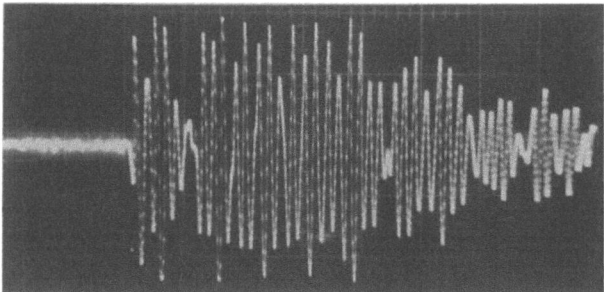

Fig. 27. Oscillogram of the AE obtained from pulling off a single well bonded beam lead with a force of ∿3.5 gf. Horizontal scale is 20 µs/div. The peak to peak output of the detector was greater than 1 mV.

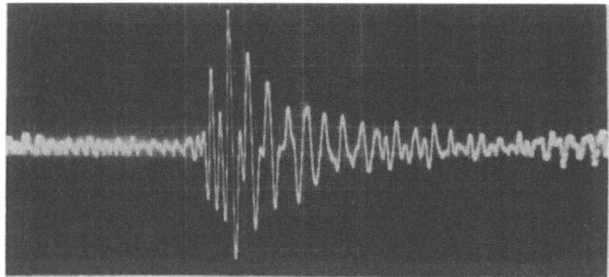

Fig. 28. Oscillogram of the AE obtained form a single anchor peeling off under a load of ∿3 gf. All other leads on the device were cut. Horizontal scale is 16 µs/div.

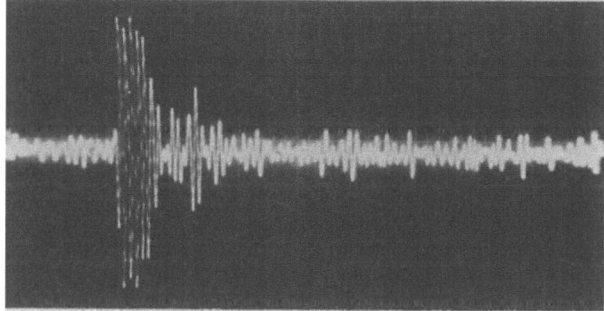

Fig. 29. A typical AE burst from a weak anchor. A pull-force of only 1.5 gf/beam was applied. The entire sweep is 200 µs in duration and the main AE burst is about 15 µs long.

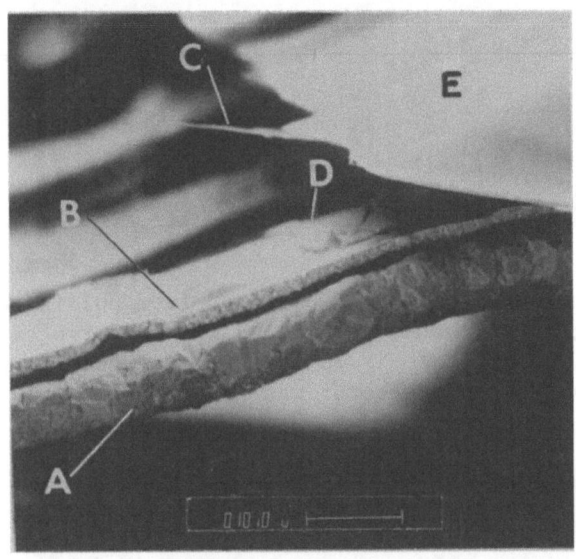

Fig. 30. SEM photograph of a beam lead from a device having poor mechanical integrity. The device had been subjected to a pull force of approximately 2.5 gf/beam. (A) gold beam, (B) separated titanium layer, (C) silicon nitride, (D) broken piece of silicon, and (E) silicon chip.

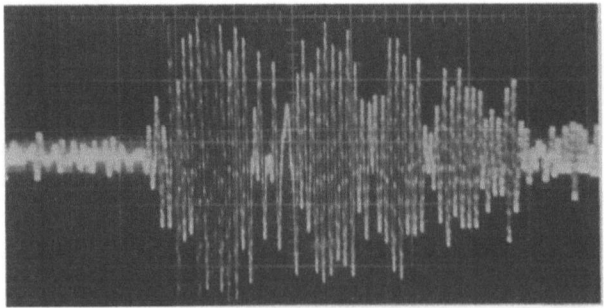

Fig. 31. An AE burst from a device having poor mechanical integrity as shown in Fig. 29. The pull force to produce this burst was only 1.2 gf/beam. Horizontal scale is 20 μs/div.

AE until stressed to about 2.5 gf/beam, at which point the beam and the anchor system began to deteriorate. However, devices from the fourth manufacturer were quite different. Large bursts of AE were emitted when the devices were stressed to only 1 gf/beam and these bursts increased with increasing stress. This result was verified on three different device types and on lots purchased 18 months apart. Examination of these devices after they had been stressed to the 1 gf/beam level revealed no obvious problems; however, examination of them after stressing to the 2.5 gf/beam level (the point where well bonded devices from other sources generally emited their first AE bursts) revealed elongation of the beams, separation of the relatively thick titanium layer, anchor peeling, nitride separation from the beams or silicon, and chips of silicon broken off at the anchor location (see Figure 30). Figure 31 gives a typical AE burst obtained by pulling a similar device to 1.2 gf/beam. Any of the mechanisms of beam system degradation shown in Gifure 30 could be responsible for bursts such as that of Figure 31. It should be noted that a normal destructive pull-off or push test would not have revealed any problem since these beams ultimately broke with forces similar to those from other sources.

Poor mechanical integrity could possibly lead to premature electrical problems resulting from thermal cycling if the device is encapsulated in silicone rubber (SR). Dias [53] has calculated the forces on the beam lead system during bonding, and although not explicit in his calculations, it appears that forces high enough to produce beam system degradation may occur during bonding. If so, the devices with poor mechanical integrity could be damaged during the bonding process and predisposed to relatively early field failure. It should be emphasized, however, that there is no experimental proof of this possible result of poor beam-anchor mechanical integrity, since no electrical tests have been performed in this study.

In a large series of SR pull-tests on devices bonded to the chrome diffused substrates shown in Figure 20, it has found that a pull force of between 1.0 and 1.5 gf/beam was required to produce AE from one or two weakly bonded beams on an otherwise well bonded device. A lower force, of between 0.5 and 1.0 gf/beam, was often sufficient to produce AE when all of the beams were poorly bonded (i.e., the device would pull off at a force of 1.0 to 1.5 gf/beam).

5b. <u>Tests That Apply Force Only to the Beams</u>. The two silicone rubber tipped probes, designed to avoid contact with the chip, have been used to apply a downward force on the horizontal portion of the beam extending outward from the chip. As previously stated, the simple resolution of forces analysis of a single beam indicates that approximately all of that force is applied as a torque tending to peel the anchor. Thus, probing a single beam or a single row

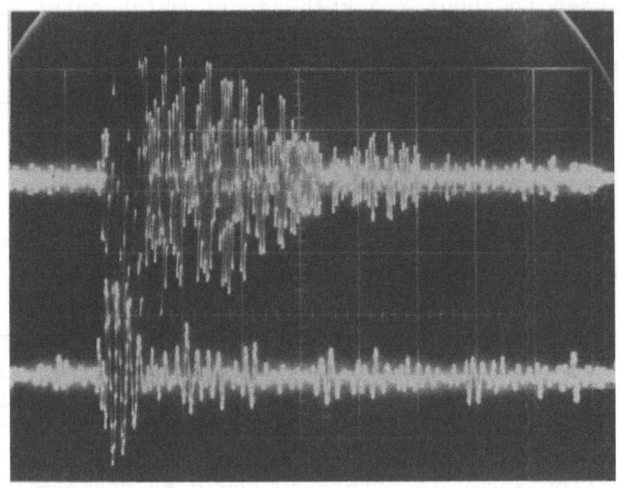

Fig. 32. AE output from applying a downward force of ~3.5 gf to a single well binded beam that had a weak anchor. The horizontal scale is 20 μs/div. The upper trace is from the probe of Fig. 23(B) (peak response is 1.1 MHz). The lower trace is from the substrate detector (peak response is 375 kHz).

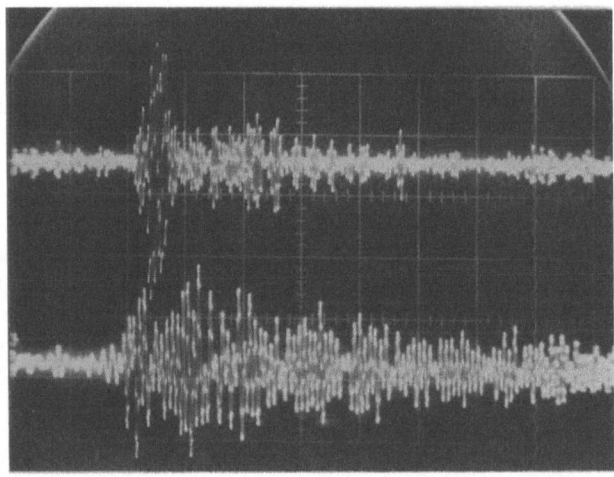

Fig. 33. AE output oscillogram resulting from a downward force of ~4 gf with the probe of Fig. 24(B) on a single very weakly bonded beam (~1 gf pull force). The scales are the same as in Fig. 32.

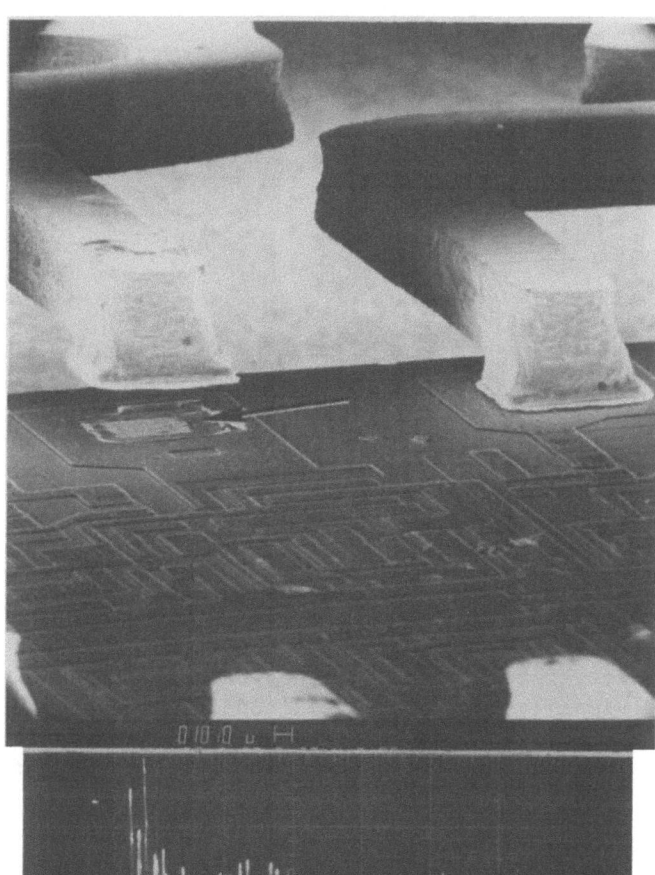
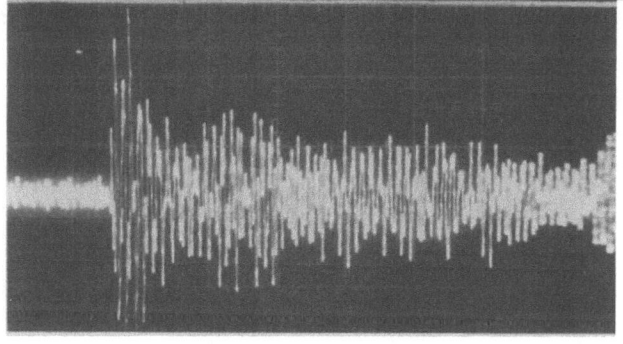

Fig. 34. The top illustration is an SEM photograph of a portion of an automated gang-bonded integrated circuit. The bond on the left lifted up during minimal bending of the lead-frame. The lower illustration is the AE waveform resulting from the lift up. The signal was picked up by the substrate detector tuned to 375 kHz. The horizontal scale is 20 μs/div.

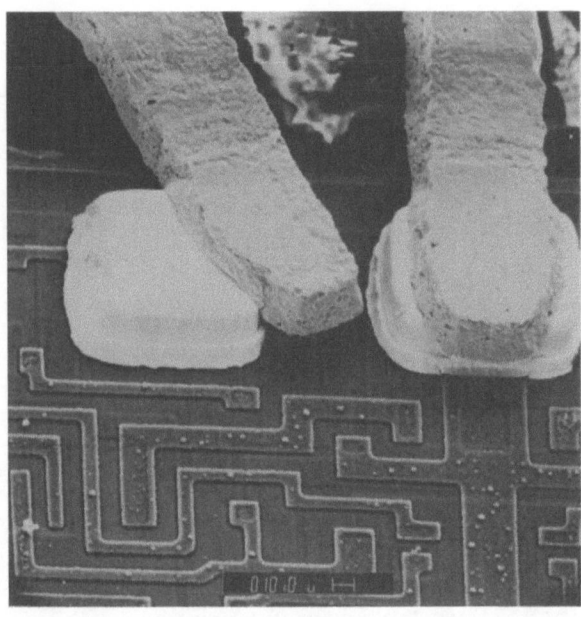

Fig. 35. An SEM photograph of two bonds from a tape bonded integrated circuit. This device, including the visually poor bond, remained intact and produced no AE even though the bonds were stressed to approximately four times the value required to produce the bond break in Fig. 34.

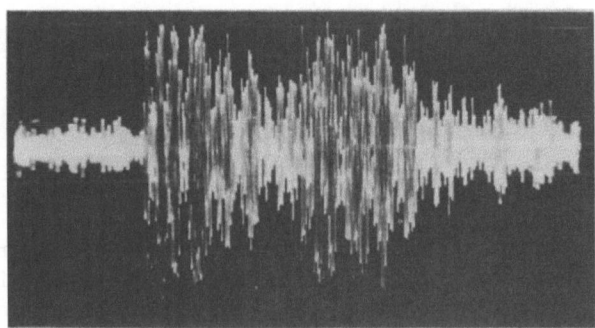

Fig. 36. The acoustic emission burst from a tape bonded lead that partially lifted at a stress of 6 gms. It gave 219 counts on an AE counter. The lead later completely lifted at 12 gf.

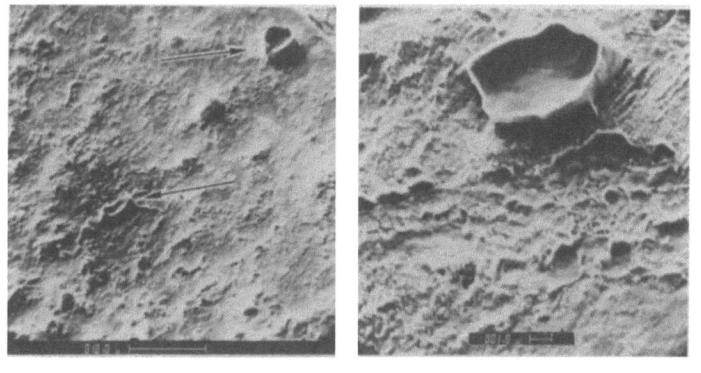

Fig. 37. Examples of microwelds left on the gold plated bump from the copper tape indicated by arrows in (A). Some of these failed by ductile "cup fracture." (B) Apparently a grain from the tape lead was left welded on the bump.

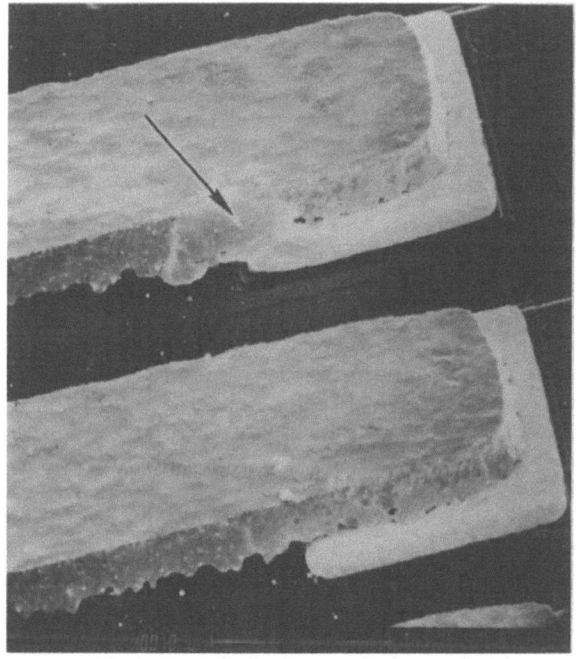

Fig. 38. SEM photograph of original unstressed tape bonded leads. The arrow on the upper lead points to the large tin-gold intermetallic compound lump. Part of the gold bump has partially disolved in the compound. Only minimal intermetallic compound is ovservable on the lower lead.

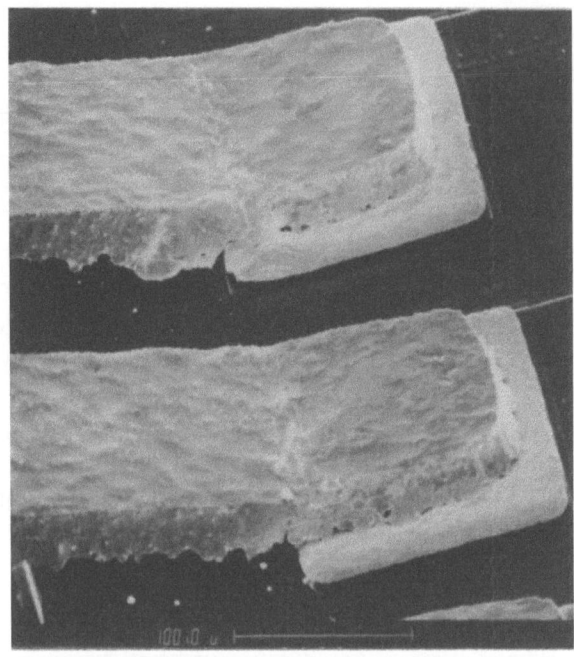

Fig. 39. SEM photograph of the same two leads in Fig. 38 after stressing the upper lead to 10 gf and the lower one to 7 gf. Cracks at the heel of the bond in both are obvious.

of beams along one side of the chip provides an anchor adherence test. If all beams are probed at the same time, the torque cancels out, and force is applied only to the bond system.

Individual beams were probed with the silicone rubber tipped tungsten carbide probe shown in Fig. 24b to establish AE patterns for both anchor and beam failures. Figure 32 gives the twin AE oscilloscope traces resulting from applying a downward force of approximately 2 gf to a beam with a weak anchor. The anchor failed at an applied force of 3.5 gf. A well bonded beam having a strong anchor would typically collapse (curve downward until it touched the substrate) with a downward force of from 6 to 10 gf depending on the beam curvature and the bugging height. The beam-probe AE detector usually produces a larger signal than the substrate detector for anchor failures. When a very weak bond (failing at $\lesssim 1$ gf/beam pull force) is probed in a similar manner to about 3 or 4 gf, the AE signal intensities are generally reversed, as shown in Figure 33. However, it should be emphasized that while these are typical AE patterns for their respective failure modes, these same failure modes may at times produce entirely different patterns.

PASSIVE ACOUSTIC TECHNIQUES   729

Some experiments were performed using the single probe to try to detect AE from silicon nitride breaks. In general, breaks in the thin ($\sim$2000 Å) nitride skirt were not detected under normal circumstances. This was believed to result from higher frequency emission as well as poor stress-wave coupling into the chip and substrate. This was verified by coating the single beam probe with a viscous acoustic mating compound and moving it sideways into an extended nitride skirt. A small AE burst was recorded in the probe detector circuit (1.1 MHz) but not in the substrate detector.

6. <u>Application of Acoustic Emission to Determine the Integrity of Tape Bonded Devices and Hybrid Components</u>. There are a number of semiconductor device areas in which AE can be used to insure mechanical integrity. One of the most straightforward uses is in testing to assure the bond integrity of automated tape carrier systems. To do this automatically, the mechanical stressing of bonds may be accomplished as the carriers bend during winding on a reel. A rubber coated detector could be pressed agains part of the inner bonded lead frame as the frame undergoes some maximum allowable flexing or bending during or before being wound onto the reel.

To show the feasibility of this procedure, some AE tests were performed on different types of automated gang-bonded integrated circuits. The first was of solder bump-Kovar inner-lead construction. A slight bending applied 2 gf per lead to the uncut inner-lead frame and produced the lifted lead shown in the upper illustration of Fig. 34. The lower illustration of Fig. 34 shows the substrate detector response to the AE resulting from that single bond lifting. Examination of this device and others from this lot revealed a tendency for the solder bump and its interfacial plating to separate from the aluminum bonding pad. A second type of gang-bonded device having an aluminum inner-lead construction was tested in a manner similar to that used for the solder bump unit. One of these devices had several weakly bonded leads which emitted bursts larger than that shown in Fig. 34 when they separated. Similar tests on several modern tape bonded devices reveal no failures even though the visual appearance of one was quite poor (see Fig. 35). Destructive pull tests verified that both of these leads were well bonded. Thus, it appears that an AE test can be used to assure bond integrity on such gang-bonded systems. In addition, it showed that a lead that would have been rejected in · a visual inspection was adequately bonded.

Failure was indicated by AE in the above cases when the lead completely lifted up. Low stresses applied to weak bonds on similar device structures often gave preliminary warnings of peel failures that would later occur at higher forces. Figure 36 is the AE burst from a lead that partially lifted up at a stress of 6 gf. Later the lead completely lifted at 14 gf. This is a demonstration that a catastrophic Griffith-type break (see III-A) is not necessarily valid for typical tape bonded leads used in integrated circuits and that the microwelds are capable of breaking, perhaps in groups; but a peel (crack) may be arrested part way into the bond and not propagate further until a higher stress is applied. An investigation into the fracture mechanism of this weakly bonded system revealed that the copper lead to gold plated bump weld contained numerous individual microwelds similar to those of Fig. 22 for the gold-gold system. The only difference was the mode of fracture which was frequently the "cup fracture" type, characteristic of OFHC* copper [54]. Figure 37 shows two such copper cups remaining on a gold plated bump after the lead had lifted at 14 grams.

A different metallurgical tape bonded system was shown to give AE results that were not correlated with the bond integrity. This system consisted of gold bumps on the chip and tin plated copper leads on the tape. During the process of thermocompression bonding the tin melts and may form intermetallic compounds with the gold bump, as is evident in Fig. 38. These compounds are relatively brittle, and if the lead is stressed, they can crack at comparatively low forces. Such cracks are shown in Figure 39. The upper lead cracked at 5 gf and was further stressed to 10 gf with no additional AE. The lower lead was stressed at 4 gf and cracked. It was further stressed to 7 gf with no additional AE. Figure 40 is a close-up of the lower lead with the AE burst associated with the small double crack. Upon further stressing, these two leads (and other similar ones) required 38 to 45 gf to break. Thus, for this alloyed metallurgical system, it appears that the low-stress AE, while indicative of brittle intermetallic compound formation, is not related to the ultimate bond strength of the weld.

Flip-chip devices are another area where AE may be used to verify the bond integrity. In this case visual inspection of the solder joints is almost impossible as may be seen from the SEM photograph showing a side-on view of such a flip chip in Fig. 41. Therefore, it was decided to investigate the use of AE for this application. The usual method of testing for bond strength is to measure the shear strength of the chip. This is typically of the order of 50 to 75 gf/bump for good solder joints but decreases to 0 for bad ones. When an AE shear test was applied, it was found that, with rare exceptions, there were no pre-break AE bursts. Although a large burst was detected when the chip broke away, this could hardly be considered a nondestructive test. Further investigations using the stressing apparatus of Fig. 23 showed that poorly soldered flip chips did emit AE when the probe was pressed down on top of the chip and rotated about the vertex

---

* A pure grade of copper "oxygen free high conductivity."

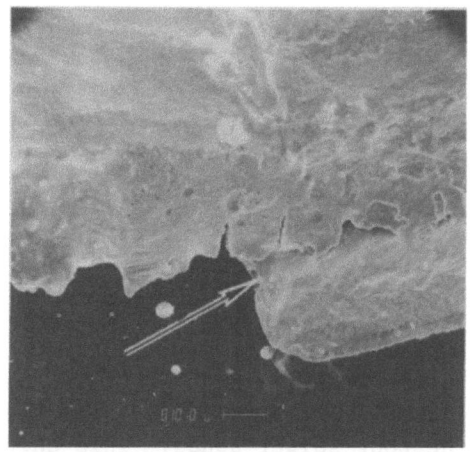
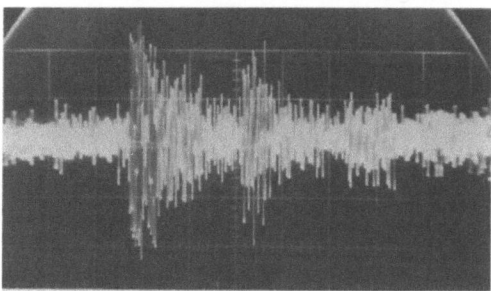

Fig. 40. A close-up of the lower cracked lead and bump of Figure 39. A small amount of intermetallic compound is evident filling the area as the lead leaves the bump as shown by arrow. The double acoustic emission burst that occurred when the lead was stressed to 4 gf is shown below. The horizontal scale is 130 μs/div.

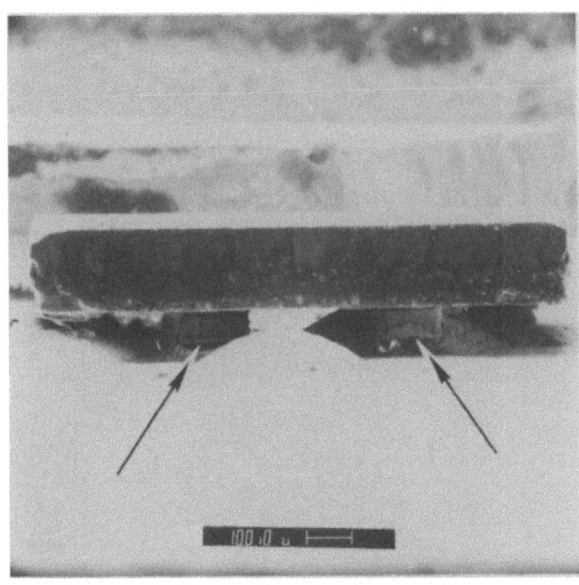

Fig. 41. The SEM photograph of a side-on view of a flip chip poorly bonded into its hybrid circuit. Arrows point to the poor solder bonds. This chip was detected by AE probing. It is almost impossible to inspect the bonds on flip chips.

Fig. 42. A poorly soldered flip-chip shown face-up beside its normal position in a hybrid microcircuit. The poor bond qulity was revealed by AE and after removal it is evident that the bumps were only partially soldered.

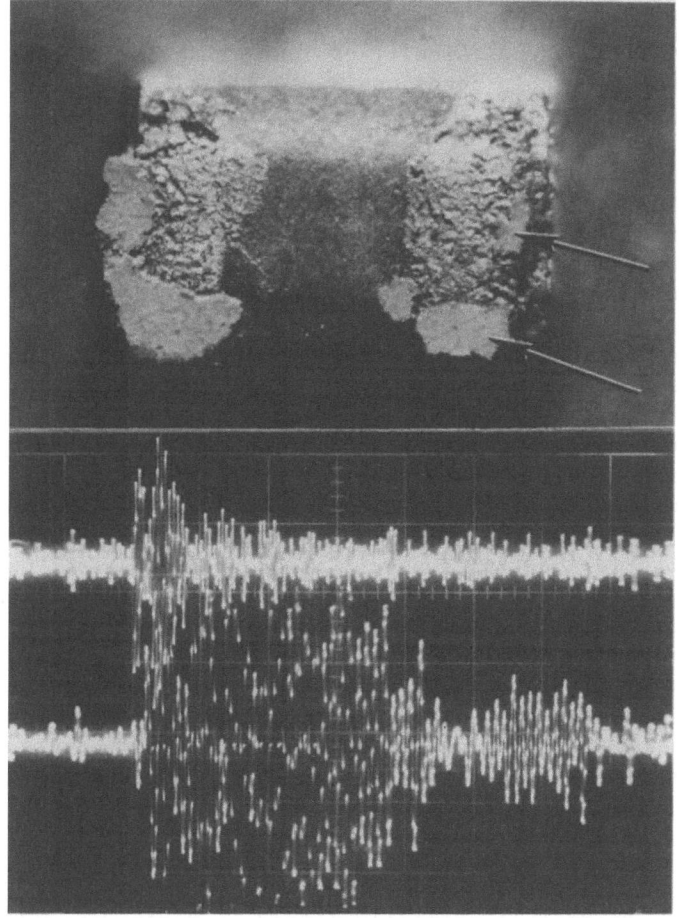

Fig. 43. The upper illustration is a photomicrograph of a 2 mm long chip capacitor removed from its hybrid circuit. The arrows point to the very small areas that had been conductive epoxy bonded to the circuit. Below is the AE waveform resulting from applying 200 gf with an AE probe to the top on the capacitor before it was removed from the circuit. Both 1.1 MHz probe detector and 375 kHz substrate detector waveforms are shown. The horizontal scale is 20 µs/div.

of the probe (orbited). An example of a very poorly soldered chip is given in Figure 42. The weak joints were revealed by AE during downward rotating force application. The chip was then sheared off and photographed. The AE burst resulting from applying a downward force of 42 gf is shown in the lower sector of Figure 42. In numerous tests, the AE technique revealed all <u>known</u> poorly soldered bonds. However, the difficulty of inspecting and verifying the condition of flip chip bonds would require a long expensive study to verify that the test indeed is a relible screen.

Various discrete components such as chip capacitors bonded into hybrids can be stressed by applying a small downward or shear force. If weakly bonded, AE should be detectable. One such capacitor was subjected to a downward orbiting force of 200 gf. It emitted the AE signals shown in the lower portion of Fig. 43. The capacitor was then broken free and photographed. Only about 15% of the intended area was actually epoxy bonded as shown by the arrows on the right side.

Since propagating cracks emit stress waves, cracks in power device chips should be detectable by current pulsing the device. Nonuniform heating of the chip during such pulses should expand the crack and cause the emission of stress waves. Cracks and flaws in hybrid substrates should also be detectable. One such cracked substrate was detected in the course of the present work while pressure testing silicone encapsulated devices. General package integrity should be assessable with AE by stressing the package under pressure or with rapid heating. Such conditions have been observed to destroy the hermeticity of potentially defective packages [5]. AE detection equipment could be used in conjunction with the nondestructive wire-bond pull-test to assist in determining the maximum nondestructive force to be applied. It could then be used to monitor that test to give ultimate assurance of its nondestructive nature. A limited evaluation of this was carried out as a preliminary to the present work and it appeared promising. Bonds that partially lifted up during a nondestructive pull were easily detected.

D. Conclusions of Section 3

Previous studies of acoustic emission applied to real-time control or evaluation of electronics assembly processes have been reviewed. In addition, work directed towards developing test AE based tests to establish the mechanical integrity of electronics devices as screens after production are described. These studies have revealed considerable differences in the mechanical integrity of beam lead bond-anchor systems and demonstrated that AE testing offers a unique method of assessing new beam lead-nitride-anchor designs and of maintaining quality control on normal production. General deterioration of the beam-anchor system begins at pull forces of from 1.0 to 2.5 gf/beam, depending on the manufacturer. Thus, no test can be considered nondestructive that applies forces higher than about 2.0 gf/beam to the mechanically strong beam systems and perhaps 0.8 gf/beam to the weak ones. The maximum safe force for each separate manufacturing procedure must be obtained experimentally. It was found that a pull force from about 1.0 to 1.5 gf/beam was required to reveal a few poorly bonded beams in otherwise well bonded devices; however, this force is equal to the beam-system deterioration force for devices with poor mechanical integrity. For such devices, no meaningful nondestructive pull test is possible. The forces applied to the beam-anchor system for all methods of stressing except the pull test, are dependent upon the shape of the individual beams as they extend from the chip, as well as upon the uniformity of the bugging height. Thus, to effectively use these tests, more operator care is required than is usually achieved in typical production line environments. The SR pull test is simple to employ and can be considered nondestructive if the user does not object to leaving cured SR in the package. The same material is, after all, often used as a conformal coating. The silicone resin could be applied to chips with modified epoxy die-attach equipment at either a 100% or some lower percentage sampling basis. Of the methods studied, only the SR pull test could reliably reveal weak bonds having equivalent strengths greater than 1 gf.

The main difficulty in the work with beam lead devices was encountered in the development of means of nondestructively stressing delicate, irregularly extending beam leads. However, many other uses of AE in electronics offer no such problems. Any system whose bond strength normally is destructively tested by shearing or probing, such as flip chips or capacitor chips in hybrids, can be nondestructibely tested by that same method at a lower force using AE as the failure indicator. Both the inner and outer lead bonds on automated tape-bonded integrated circuits can be flexed or probed, while monitoring for failures with AE equipment, to gain assurance that they are well bonded. The mechanical integrity of large packages can likewise be assessed by rapid heating, high or low pressure, or other means of stressing. Thus, it appears that AE will have an increasing role in assuring reliability in microelectronics.

Acknowledgment

The author gratefully acknowledges valuable discussions on hermeticity testing with S. Ruthberg and on statistical sampling with Dr. M. Natrella. The manuscript was prepared by Mrs. Kaye Dodson.

REFERENCES

1. Test Methods and Procedures for Microelectronics MIL-STD-883B, (November 1974)
2. H. Christensen, Bell Laboratories Record 36, 127-130 (April 1958).
3. G. G. Harman, Ed., Semiconductor Measurement Technology: Micro-Electronic Ultrasonic Bonding, NBS Spec. Publ. 400-2(January 1974).
4. D. R. Johnson and E. L. Chavez, Proc. Intl. Microelec. Symp. (ISHM), 88-94, (October 1976).
5. R. W. Thomas, IEEE Trans. Parts, Hybrids, and Packaging PHP-12, 3, 167-171 (1976).
6. G. G. Harman, Proc. 12th Annual IEEE Reliability Physics Symposium, 205-210, (April 1974).
7. J. Roddy, N. Spann, and P. Seese, IEEE Trans. Comp. Hybrids and Manufacturing Tech. CHMT-1, 3, 228-236 (1978).
8. G. H. Ebel, Proc. 15th Annual IEEE Reliability Phsyics Symposium, 70-81, (April 1977).
9. A. Shumka, and R. R. Piety, Proc. 13th Annual IEEE Reliability Physics Symposium, 93098, (April 1975).
10. S. Ruthberg, Nondes. Testing Standards - A Review, ASTM STP 624, 246-259 (April 1977).
11. G. G. Harman, Proc. 12th Annual IEEE Reliability Phys. Symp., 253-254 (April 1975).
12. D. S. Peck, Proc. 13th Annual IEEE Reliability Physics Symposium, 253-254 (April 1975).
13. B. Maximow, E. M. Reiss, and S. Kutunaris, Proc. 15th Annual IEEE Reliability Physics Symposium, 212-216, (April 1977).
14. D. S. Peck, Proc. 16th Annual IEEE Reliability Phys. Symp., 1-6, (April 1978).
15. G. M. Johnson, Accelerated Test Techniques for Microcircuits, Final Tech. Report MDC-E 1208, January 1975, McDonnel Douglas Astronautics Co., East, Contract Report for NASA Goddard Space Flight Center.
16. M. Stitch, G. Johnson, B. Kirk, and J. Brauer, IEEE Trans. on Reliability R-24, 4, 238-250 (1950).
17. F. H. Reynolds, Proc. 15th Annual IEEE Reliability Physics Symposium, 166-178, (April 1977).
18. R. F. S. David, Proc. 28th IEEE Electronics Components Conf., 281-285, (April 1978).
19. Military Specifications, Microcircuits, General Specification for MIL-M-38510D, (August 1977).
20. E. G. Schilling, J. Quality Technolgy 10, 2, 47-51 (1978).
21. A. Joffe, The Physics of Crystals, McGraw Hill, (1928).
22. J. Kaiser, Untersuchungen uber das Auftreten von Geraschen Beim Zugversuch, Arkiv fur das Eisenhuttenwesen 24, 1/2 43-45.
23. C. H. Palmer and R. E. Green, Appl. Opt. 16, 9, 2333-2334 (1977).
24. R. E. Green, Acoustic Emission: A Critical Comparison Between Theory and Experiment, Proc. Ultrasonics International Conf. Brighton, England (June 1977).

25. N. N. Hsu, J. A. Simmons, and S. C. Hardy, Materials Evaluation 35, 100-106 (October 1977).
26. H. L. Dunegan and D. O. Harris, Ultrasonics, 160-166 (July 1969).
27. O. L. Anderson, The Griffith Criterion for Glass Fracture, Wiley & Sons, Inc. (1959), and R. E. Reed-Hill, Physical Metallurgy Principles, 2nd Ed. Nostrand, (1973).
28. G. P. Anderson, S. J. Bennett and K. L. DeVries, Analysis and Testing of Adhesive Bonds, Academic Press, (1977).
29. R. E. Green and R. B. Pond, The Ultrasonic Detection of Tatigue Damage in Aircraft Components, Air Force Office fo Scientific Res., Contract F44620-76-C-0081, (March 1977).
30. D. O. Harris, A. S. Tetelman, and F. A. Darwistt, Acoustic Emission, ASTM STP 505, 238-249.
31. M. C. Jon, H. A. Duncan, S. J. Vahaviolos, Analysis of Stress Wave Emission in Resistance Welding of Tantalium Capacitor, Materials Evaluation, to be published.
32. Jack C. Spanner, Acoustic Emission Techniques and Applications, Soc. for Nondestructive Testing, (1974).
33. Acoustic Emission, ASTM Tech. Publ. STP 505, Amer. Soc. of Testing and Materials, (1972).
34. A. E. Lord, Acoustic Emission, Physical Acoustics, W. P. Mason and R. N. Thurston, Eds., 289-353, Academic Press (1975).
35. G. G. Harman, Proc. 14th Annual IEEE Reliability Phsyics Symp. 86-97, (April 1976), also see IEEE Trans. Parts, Hybrids, and Packaging PHP-10, 3, 152-159 (September 1974).
36. S. J. Vahaviolos, IEEE Trans. Parts, Hybrids, and Packaging PHP-10, 3, 152-159, (September 1974).
37. M. A. Siafi andS. J. Vahoviolos, IEEE J. of Quantum Elec. QE-12, 2, 129-136 (1976).
38. M. F. Carlos and M. C. Jon, Proc. 28th IEEE Elec. Components Conference, 336-339 (April 1978).
39. M. C. Jon, C. A. Keskimaki and S. J. Vahoviolos, Materials Evaluation 36, 4, 41-51 (1978).
40. G. C. Knollman and J. L. Weaver, Proc 3rd Acoustic Emission Symposium, 413-427 (September 1976).
41. T. Ikoma, M. Ogura and Y. Adachi, Ibid, 329-341.
42. M. Kotani, S. Mitsui, and K. Shirahata, Trans. of IECE of Japan, 58-C, 583-590 (1975).
43. T. Sedgwick, J. Appl. Phys. 39, 1728-1740 (1968).
44. T. Ikoma, M. Ogura and Y. Adachi, Acoustic Emission Study of Defects in GAP-LEDS, and unpublished talk presented at the 20th Elec. Materials Conference. Univ. Calif. at Santa Barbara, (June 1978).
45. J. L. Jellison, IEEE Trans. Parts, Hybrids, and Packaging PHP-11 3, 206-211 (1975).
46. P. H. Holloway and D. W. Bushmire, Proc. 12th Annual IEEE Reliability Physics Symposium, 180-186 (April 1974).
47. B. H. Schofield, Acoustic Emission Under Applied Stress, Technical Document Report No. ASD-TDR-63-509, Part II, 1-29, (May 1964), Af. Materials Lab., Wright Patterson AFB, Ohio.

48. C. E. Wirsing, Solid State Tech. 16, 48-50 (1973). Also see J. L. Jellison, Proc. 26th IEEE Electronics Components Conference, 92-97 (April 1976).
49. N. T. Panousis and H. B. Bonham, Proc. 10th Annual IEEE Reliability Physics Symposium, 21-25 (April 1973).
50. P. H. Holloway and R. L. Long, Jr., IEEE Trans. Parts, Hybrids, and Packaging PHP-11, 2, 83-88 (1975).
51. R. F. Tylecote, Trans. Institute of Welding, 153-178 (November 1945).
52. This circuit was designed and built by T. F. Leedy.
53. J. L. Dais, Proc. 25th IEEE Electronics Components Conference, 43-51, (May 1975).
54. H. C. Rogers, Trans. Met. Soc. of AIME 218, 498-506 (1960).

Chapter 14

LIFETIME DATA ANALYSIS

F. H. Reynolds

Post Office Research Centre

Martlesham Heath

Ipswhich Ip5 7RE

England

I. INTRODUCTION

The outcome of electronic-component reliability studies is usually a sample of <u>lifetimes</u> and the task is then one of drawing inferences about the conceptual population from which the sample has been withdrawn. Quite often "lifetime" will refer to the interval between the real time at which a component enters service and the later time at which it fails to satisfy its performance specification for any reason. The term may also be applied however to a lifespan determined by specified causes only. In the laboratory, lifetimes may arise from simulated service operation or operation under conditions deliberately different from service, usually more onerous. Still further, lifetimes may be measured on only an elementary part of a component.

Whatever the origin of the lifetimes, they will need to be analyzed; the objective is the characterization of the parent population about which a variety of descriptive and predictive statements can be made.

II. LIFETIME FUNCTIONS

Much of the discussion to follow is concerned with statistical estimation but for the present, it is necessary to conceive a

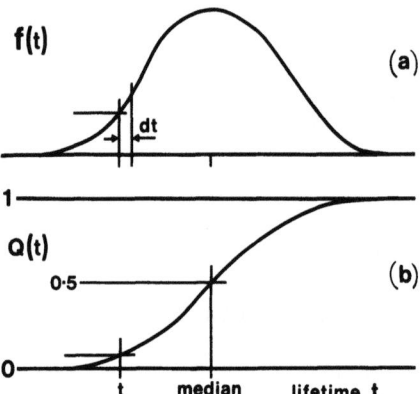

Fig. 1. Distribution of lifetimes (a) frequency density function (fdf), f(t), (b) cumulative distribution function (cdf), Q(t).

large lifetime population the particulars of which are fully known and describable by a <u>frequency density function</u> (fdf) f(t) shown in Fig. 1 where f(t) is such that the fraction of the population having lifetimes in the range t to t + dt is f(t)dt [1,2,3]. The fraction of the population having a lifetime less than that of a selected ordinate is called the <u>cumulative distribution function</u> (cdf) Q(t), also shown on Fig. 1 where

$$Q(t) = \int_0^t f(t) \, dt \qquad (1)$$

The fraction of the population having a lifetime greater than t is called the <u>reliability function</u> R(t) and so

$$R(t) = \int_t^\infty f(t) \, dt \qquad (2)$$

and of course

$$R(t) + Q(t) = 1 \qquad (3)$$

Another lifetime function needed is the <u>hazard function</u> defined as

$$\lambda(t) = \frac{f(t)}{R(t)} \qquad (4)$$

LIFETIME DATA ANALYSIS

and is most easily interpreted by imaginging that all the members of the population of size N say, are put to service simultaneously at time zero. The <u>hazard</u> in a real time interval dt is then defined as,

$$\text{hazard in dt} = \frac{\text{number of members failing between t and t + dt}}{\text{number of survivors at time t}}$$

$$= \frac{Nf(t)\, dt}{NR(t)} \qquad (5)$$

$$= \lambda(t)\, dt$$

Dividing by dt then gives the hazard rate as $\lambda(t)$, the rate at which surviving members fail. The hazard function is thus a lifetime function which gives the hazard rate at a real time t when all the population members start life (are "born") at a time $t = 0$.

Another lifetime function which will be needed is the cumulative hazard, $H(t)$, given by

$$H(t) = \int_0^t \lambda(t)\, dt \qquad (6)$$

which is easily shown to relate to $Q(t)$ according to

$$H(t) = \ln[1/(1 - Q(t))] \qquad (7)$$

If, again, all the members of the population are born at time $t = 0$, the average hazard rate between times t_1 and t_2 is

$$\lambda_{t1,t2} = \frac{H(t_2) - H(t_1)}{t_2 - t_1} \qquad (8)$$

III. LIFETIME DISTRIBUTIONS

Although many different forms of the fdf have been proposed for lifetime distributions, only two, the lognormal and Weibull distributions [4,5,6] have been applied consistently to electronic parts. The expressions for their fdfs are both characterized by three parameters, of location, scale and shape, which for the lognormal distribution are, respectively γ, ψ and σ whence

$$f(t) = \frac{1}{\sigma(t - \gamma)\sqrt{2\pi}} \exp\left[-\tfrac{1}{2}\left(\frac{\ln\frac{t-\gamma}{\psi}}{\sigma}\right)^2\right] \qquad (9)$$
$$t > \gamma$$

For the Weibull distribution the parameters are (again respectively) γ, η and β, and

$$f(t) = \frac{\beta}{\eta}\left(\frac{t-\gamma}{\eta}\right)^{\beta-1} \exp-\left(\frac{t-\gamma}{\eta}\right)^{\beta} \quad t > \gamma \qquad (10)$$

For $t \le \gamma$, $f(t) = 0$ for both distributions.

The determination of the cdf is simplified by transforming the lifetime t to τ by

$$\tau = \ln(t - \gamma) \qquad (11)$$

and then introducing z, called the <u>probit</u> [7], for the lognormal distribution where

$$z = \frac{\tau - \mu}{\sigma} \qquad (12)$$

and the scale factor has, for convenience, been replaced by μ where

$$\mu = \ln \psi \qquad (13)$$

For the Weibull distribution, $\mathfrak{z}$, which by analogy may be called the <u>Weibit</u>, is correspondingly introduced by

$$\mathfrak{z} = \frac{\tau - \xi}{1/\beta} \qquad (14)$$

with

$$\xi = \ln \eta \qquad (15)$$

Having thus transformed the three parameters into a single, dependent, parameter (z or $\mathfrak{z}$) the cdf is then easily found as

$$Q(t) = \Phi(z) \qquad (16)$$

for the lognormal distribution, where $\Phi(z)$ is the Gaussian function given in most reference works on statistics and can be found from approximating formulae [8].

For the Weibull case, the cdf is simply

$$Q(t) = 1 - \exp(-\exp\mathfrak{z}) \qquad (17)$$

Equations (12) and (14) show that plots of z or $\mathfrak{z}$ against the transformed lifetime will be linear, the slope and intercept (on the abscissa) being 1/σ and μ for the lognormal distribution, and β and ξ for the Weibull as shown by Fig. 2. Because the cdf is a function of the probit or Weibit, the ordinate axis can be scaled with Q(t) according to equations (16) or (17) which is, in practice, easily effected using normal- or Weibull-scale graph paper.

LIFETIME DATA ANALYSIS

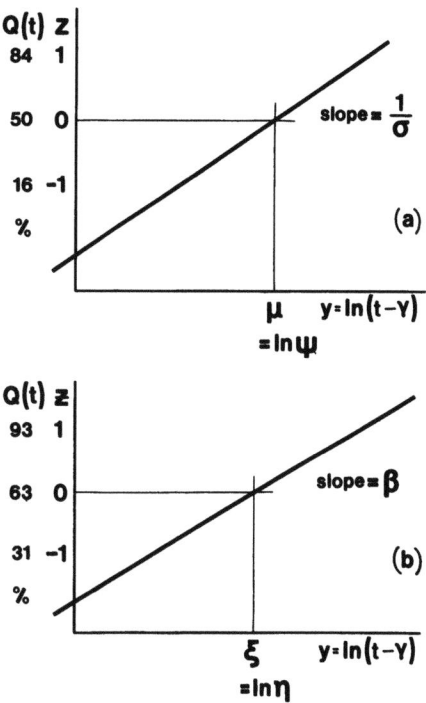

Fig. 2. The cdf, Q(t) plotted against the lifetime variable (a) lognormal scale probit: z, (b) Weibull scale Weibit:Z.

When $z = 0$ on the normal plot, the corresponding lifetime is $\psi + \gamma$ which is thus the <u>median</u> lifetime. On the Weibull diagram $Z = 0$ when the cdf has the value 0.63, the corresponding lifetime being $\eta + \gamma$ which is called the <u>characteristic</u> lifetime. Very often, γ is so small relative to ψ and η that the median lifetime for the normal distribution is taken to be ψ, and η is regarded as the characteristic lifetime for the Weibull distribution.

With the cdf known, equations (4), (9) and (10) can be manipulated to express the hazard function in terms of z or Z in a standardised manner as shown by Figs. 3 and 4, where a cdf abscissa scale is also added. The hazard function is multiplied by 10^9 before plotting on the ordinate scale. If then a population of components were all born at a time $t = 0$, the hazard rate at time t is given in fits, or failures per 10^9 component hours. These plots are often useful for the purpose of examining the trend of the hazard rate when a given fraction of failures has occurred.

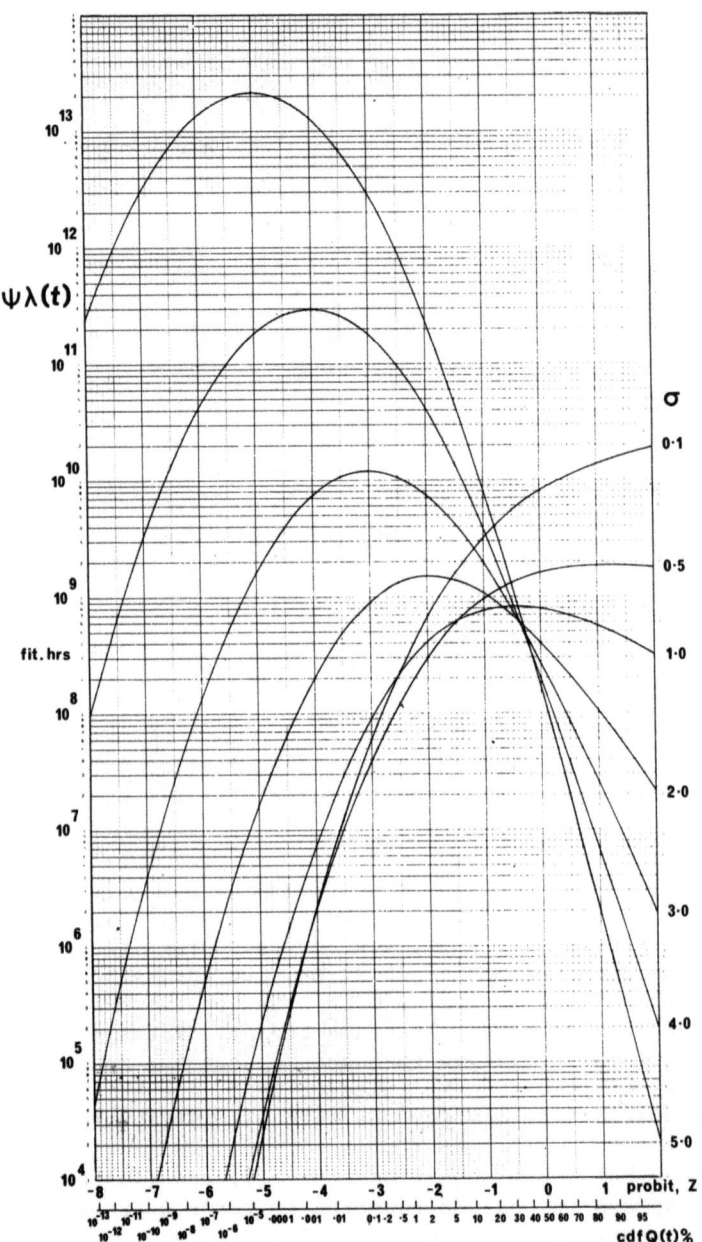

Fig. 3. Standardised plot of the hazard rate $\lambda(t)$ against the probit and the cdf for a lognormal lifetime distribution.
scale factor: ψ shape factor: σ

LIFETIME DATA ANALYSIS

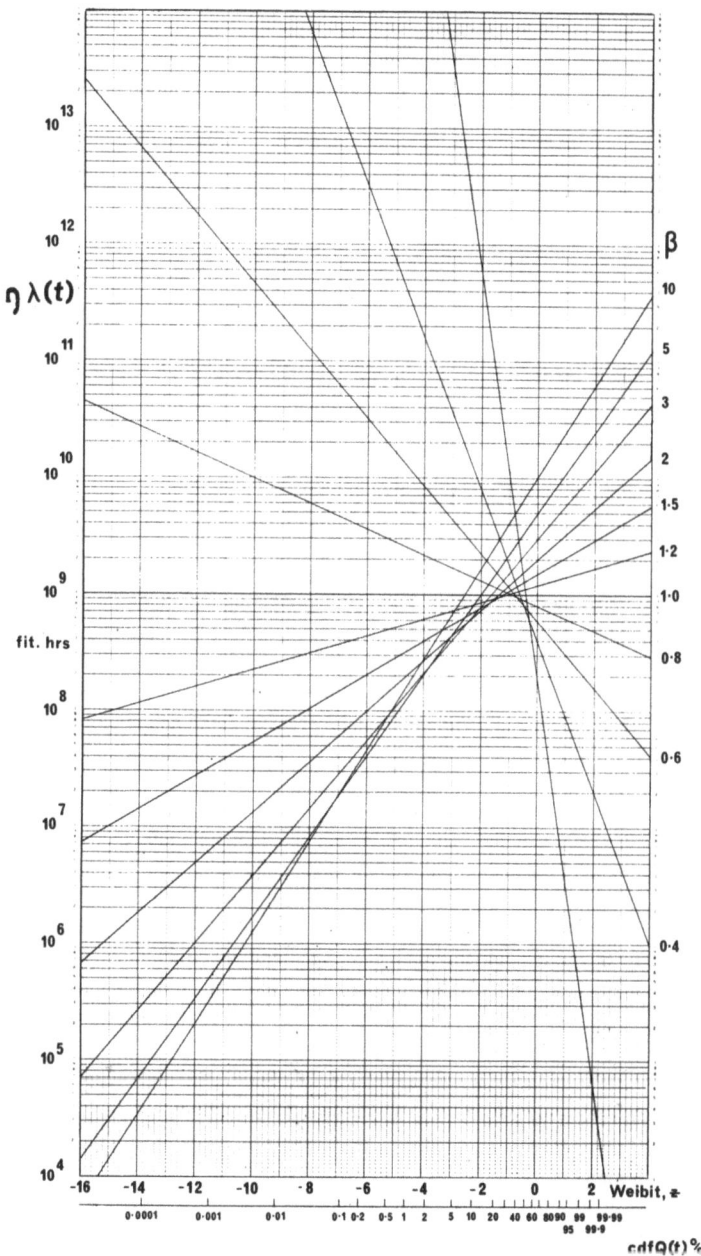

Fig. 4. Standardised plot of the hazard rate, $\lambda(t)$, against the Weibit and the cdf for a Weibull lifetime distribution.
scale factor: η shape factor: β

For the Weibull distribution, the special case $\beta = 1$, $\gamma = 0$ is important because, as seen from Fig. 4, the hazard rate is constant, and equation 10 reduces to

$$f(t) = \frac{1}{\eta} \exp(-\frac{t}{\eta}) \qquad (18)$$

with

$$Q(t) = 1 - \exp(-\frac{t}{\eta}) \qquad (19)$$

and

$$\lambda(t) = 1/\eta$$

and $f(t)$ in equation (18) is said to define the <u>exponential</u> distribution.

IV. PARAMETER SENSITIVITIES

Component lifetimes depend upon the environment and mode of usage and so therefore will the parameters of their distribution. Some very complicated dependencies can be envisaged in which each parameter, say, might depend upon the ambient temperature. From accumulated experimental evidence however, one simple relationship often occurs in which the scale parameter alone depends on the applied conditions according to

$$(\mu, \xi) = b_0 + b_1 x \qquad (21)$$

where x is a measure of the stress and the b's are constants. This <u>stress model</u>, the entire basis of which is the linear regression [9] of the logarithm of the scale parameter on the stress variable, causes the df plot to depend on x in the manner of Fig. 5, and some of the ways in which the stress can be applied are:

$$x = \frac{1}{T}, \qquad x = \ln\frac{1}{V}, \qquad x = \ln\frac{1}{J} \qquad (22)$$

where T is temperature, V is a voltage and J a current density. The combination of the lognormal distribution and temperature stress gives, for example,

$$\psi = \text{constant} \times \exp(\frac{b_1}{T}) \qquad (23)$$

which is a version of Arrhenius' Law [3,4,10,12,15,16]. In this form, b_1 is usually expressed after multiplying by the ratio Boltzmann's constant/electronic charge (k/e) when it is called the activation energy with the dimensions of electrons-volts and can sometimes have a physical significance. As another example, the

LIFETIME DATA ANALYSIS

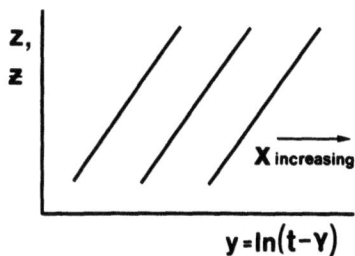

Fig. 5. Dependence of cdf, plot on the stress variable, x. (schematic)

Weibull distribution taken with voltage stress, gives

$$\eta = \text{constant} \times \frac{1}{V^{b_1}} \qquad (24)$$

which is an inverse power law [4,17,18] relationship. For the exponential distribution and temperature stress, the (constant) hazard rate can be expressed by

$$\lambda = \text{constant} \times \exp\left(-\frac{b_1}{T}\right) \qquad (25)$$

V. GRAPHICAL PARAMETER ESTIMATION

Attention can now be turned to the practical task wherein a population of components exists and its lifetime distribution is sought. Two broad approaches are possible, either graphical or numerical, and though a graphical approach is most widely used, it will be argued that a numerical procedure is advantageous. The working data is a sample set of lifetimes - quite often a very small sample - obtained from service or from laboratory lifetests, the latter including accelerated life tests.

The data may come in various forms; the simplest is a random sample in which each lifetime t_j is known, and the data is then said to be complete, thus

$$t_1 \quad t_2 \quad t_3 \ldots \ldots t_j \ldots \ldots t_n$$

where n is the sample size and the lifetimes are assumed to be ordered from smallest to largest. Quite often, however, and certainly on tests of components under simulated normal working conditions, only some members of the sample will fail in the time available for observation. The data then comprises lifetimes together with operating times without failure which are called <u>censoring times</u>, θ_j, so the data has the form

$$t \quad t \quad t \ldots \ldots t_j \ldots \ldots t_r \quad \theta_{r+1} \cdots \theta_j \cdots \theta_n$$

and the data is said to be <u>censored</u>.

In accelerated testing, the sample will generally be divided into sub-samples each of which is subjected to a different level of stress. A complication should be recognised in this case because it is usually necessary to restore the components to a normal level of stress in order to ascertain whether they have failed or not. The sample is thus observed at discrete times and the lifetiems will therefore be known only within limits*.

If the population parameters are to be estimated graphically, the sample or sub-sample data is used to estimate the cdf at each lifetime using

$$Q(t_j) = \frac{j - 0.5}{n} \qquad (26)$$

The shortest lifetime in a sample of 10 for example would give a cdf value of 0.05.

When service data is collected, the members of the sample may enter service at different real times so the available data, when ordered, will comprise lifetimes and censoring times intermingled. The data is then said to be <u>multiply-censored</u> and the above formula for the cdf can no longer be used. Instead, the lifetimes and censoring times must be commonly ordered [19] as in Fig. 6. The top line shows the number of survivors up to each lifetime and censoring time given in the second line. The hazard in the preceding interval is then known (the fraction failed in that interval) from Section II and hence the cumulative hazard is obtained at each time enabling the cdf to be found from equation (7) rewritten as

$$Q(t_j) = 1 - \exp[-H(t_j)] \qquad (27)$$

Multiply-censored data can arise in another way. If the causes of failure of the members of a sample are known, or perhaps just one

---

* If the components fail by a gradual deterioration, it may be possible to estimate the lifetimes by interpolation as used for the work of reference No. 45.

LIFETIME DATA ANALYSIS

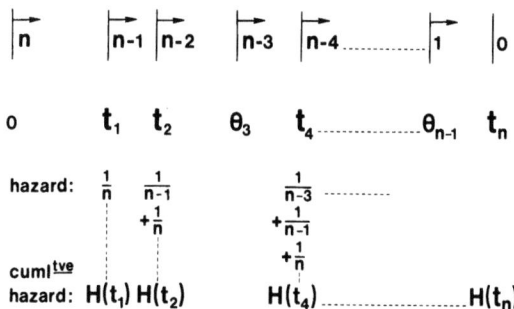

Fig. 6. Hazard calculations from multiply-censored data
lifetimes: t_1, t_2, . . . censoring times: θ_3. .
samples size: n

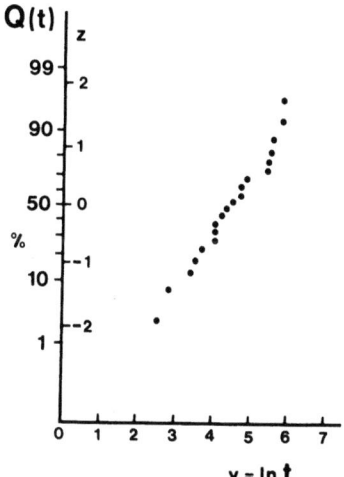

Fig. 7. Illustrative cdf plot for sample of complete data log-normal distribution.

of the causes, then it may be desired to know how the lifetime distribution would be altered if that cause were eliminated. It is then only necessary to convert the lifetimes for the failure cause being suppressed to censoring times [20,21].

Having thus obtained a value of the cdf for each lifetime, these values can be plotted in the style of Fig. 2 with $\gamma = 0$ with a results which might look like the (real data) plot of Fig. 7.

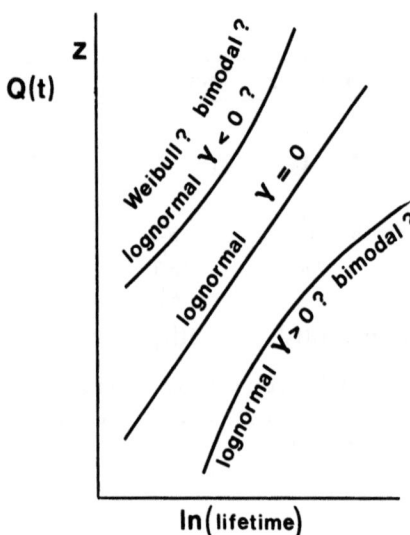

Fig. 8. Interpretation of 2-parameter lognormal cdf plots.

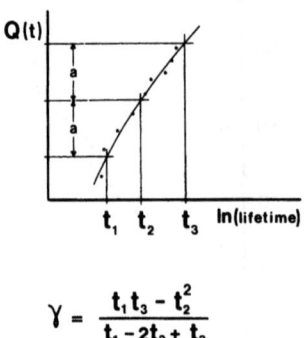

$$\gamma = \frac{t_1 t_3 - t_2^2}{t_1 - 2t_2 + t_3}$$

Fig. 9. Graphical estimation of location parameter, γ, from curved cdf plot. lognormal or Weibull distribution.

It has thus been effectively postulated that the population lifetime distribution is 2-parameter lognormal, the truth of which has to be judged by the linearity of the plot. If any curvature is rather obvious, Fig. 8 shows some possible interpretations. For downward concavity, a positive location parameter may be indicated and its value can be estimated by the method of Fig. 9 [22]. The data can then be replotted with the revised values of the lifetime variable and again examined for linearity. If the original curve were concave upwards, the location factor may be less than zero but the curve is more likely to indicate a Weibull distribution calling for the data to be replotted on a Weibit scale. If the curve appears to comprise two segments, the population is most likely <u>bimodal</u>, the discussion of which is deferred until section IV.

When a linear plot has been obtained, the parameter values of the parent population have effectively been estimated and for a single sample, or for each sub-sample, the graphical procedure is complete.

For accelerated test data, it is additionally necessary only to extract μ or ξ from each cdf plot in the manner of Fig. 2 and plot it against x as on Fig. 10, in order to find b_0 and b_1 of equation (21). The (constant) shape factor can be taken as the average of the sub-sample values.

VI. MODEL VALIDATION

It may, in practice be quite difficult to decide when a linear cdf plot has been obtained so it would be useful to have a quantitative method of judging the validity of the model thus derived; the question is whether the errors from the straight line can be dismissed as due merely to change. An answer can sometimes be obtained by means of a Kolmogorov-Smirnov [4,23] test. The method compares the observed cdf points with the expected values deduced from the drawn line in the way shown by Fig. 11. If the discrepancy, as measured by the statistic D, exceeds a critical value available from tables [23,24,25] and represented on Fig. 11 by the broken lines, then the postulated model has to be rejected.

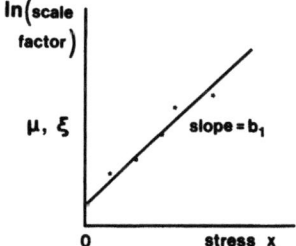

Fig. 10. Graphical estimation of scale parameter stress dependence
scale factors: ψ = exp μ, lognormal
η = exp ξ, Weibull

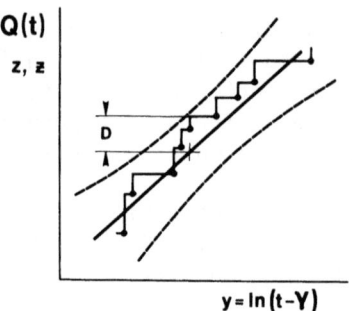

Fig. 11. Graphical illustration of Kolmogorov-Smirnov D-statistic computation from a lifetime sample cdf plot.

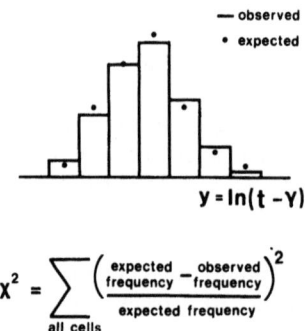

Fig. 12. Determination of x^2-statistic from classified sample lifetimes.

The tables require however, that a prior decision be made of the level of significance. A suitable value to take is 10% which means that if the model is rejected by the test, there is still up to a 10% chance that it is, after all, valid. When rejection occurs, a lower percentage significance (which broadens the acceptance band on Fig. 11) can be chosen so that the test only just rejects. That percentage is then the precise risk of rejecting the model incorrectly.

If the number of lifetimes in the sample is sufficient, an alternative statistic, x^2, can be calculated [26] by dividing the range of lifetimes into cells as in Fig. 12 and observing the number of lifetimes in each. From the estimated parameters, the cumulative hazard is calculated at each cell boundary giving the

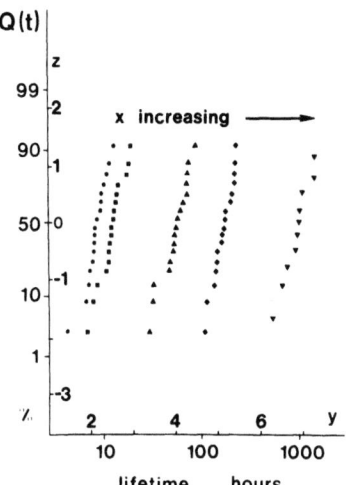

Fig. 13. Illustrative cdf plots for sub-sample data obtained in an accelerated life test (5 sub-samples).
lognormal distribution
stress variable: x

hazard in each cell. Multiplying by the average number of components exposed to failure in the cell interval, then yields the expected total of failures in each cell allowing x^2 to be calculated. For again a stated significance level, tables are available [27] showing the value of x^2 which, if exceeded, necessitates rejection of the model.

For data obtained from accelerated testing, an example of which is given on Fig. 13, a similar validation procedure can be applied to each sub-sample plot but when the data is complete and the sub-samples are of equal size, the additional assumptions, discussed in section IV can be tested, namely that the shape parameter is independent of the stress and the logarithm of the scale factor depends linearly on the stress variable, x. For 9 sub-samples, the parallelism of the cdf plots is examined by calculating Cochran's C-statistic [27,28] from

$$C = \frac{S_i \text{ largest}}{\sum_{i=1}^{S} S_i} \qquad (28)$$

where S_i is the variance* of the ith sub-sample.

As for the preceding tests, the hypothesis is rejected if C exceeds the appropriate tabulated value [27,29] given in terms of the number of sub-samples and their size n.

The linearity of the stress dependence is tested [14] by finding the F-statistic from

$$F = \frac{\sum_{\text{all SS}} (\text{SSsize} - \text{S mean})^2 /(g - 2)}{\sum_{\text{all SS}} \sum_{\text{each S}} (\text{lifetime} - \text{SS mean})^2 /n(g - 1)} \qquad (29)$$

where S is the sample and SS represents the sub-sample. and again comparing with tabulated values [26,29] given now in terms of (g - 2) and n(g - 1).

VII. NUMERICAL PARAMETER ESTIMATION

The application of quantitative validation procedures using graphically-determined parameters is clearly unsatisfactory, providing one reason for adopting a numerical estimation procedure. Several methods are available but the preferred technique is called the <u>maximum likelihood method of estimation</u> (mle) [4,7,31]. The principle is easily grasped with the help of Fig. 14 which shows a set of sample lifetimes t_1 to t_7. Suppose that trial population parameter values are used to yield the distribution profile of the broken curve. Multiplying the ordinates together would then give the <u>likelihood</u> L as shown which would be very small due to the several lifetimes in the tail of the distribution. For another proposed set of parameters, however, yielding the full curve, the product would be much larger and their exists indeed a unique fdf curve which maximizes the likelihood. If the data is censored, the likelihood function is modified to

$$L = \prod_{j=1}^{r} f(t_j) \prod_{j=1}^{c} [1 - Q(\theta_j)] \qquad (30)$$

for r lifetimes and c censoring times. By an analogous simple argument to that of Fig. 14 (using the cdf plot) it can be seen that the censored terms in equation 30 tend to shift the lifetime distribution to higher values which nonfailure of course implies.

This method of adjusting the parameters is most easily explained by first considering one parameter only, say σ in a log-

---

*the "unbiased formula"; e.g. Ref. 44, p. 135.

LIFETIME DATA ANALYSIS

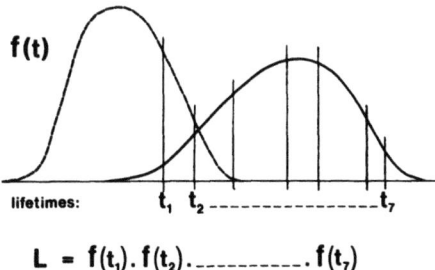

Fig. 14. Graphical illustration of the basis of the maximum
likelihood method of parameter estimation
fdf: f(t)   broken line: L_{low}
likelihood: L   full line, L_{high}

normal distribution. For a range of trial values of σ, the likelihood will pass through a maximum when $dL/d\sigma = 0$. It is virtually never possible to solve this equation for σ explicitly, so Newton's method [32] must be used which gives σ' as a better approximation than σ where

$$\sigma' = -\frac{\frac{dL}{d\sigma}}{\frac{d^2L}{d^2\sigma}} \qquad (31)$$

and the procedure can be iterated until σ has any desired accuracy.

The labor involved in the method is largely the inital task of obtaining the derivatives, which is however made much easier by taking the logarithm of the likelihood and maximising log L.

The second derivative has an additional significance because when the maximum in the likelihood function is sharp (say) implying that the uncertainty in the estimate is small, the second derivative will be highly negative. The reciprocal of the negated second derivative is thus a measure of the uncertainty and can actually be shown to be equal to the variance of the parameter estimate, which in turn can be used to place confidence limits on the expected value.

In the usual situation where several parameters have to be optimized, $dL/d\sigma$ must be replaced by the partial derivative of L with respect to each parameter and as many equations as there are parameters have to be solved simultaneously which is most easily carried out by standard matrix methods. The mle method

can quite readily be extended to include accelerated testing where
the stress model (Eq. (21)) has the effect of adding another variable to be optimized . The procedure is then equivalent to finding
the global set of parallel lines of best fit though all the points
of a sub-sample cdf plot as of Fig. 13.

VIII. ANALYSIS OF EXPERIMENTAL LIFETIME DATA

In the illustrations of the mle method which follow, a selection of the possible ways of presenting the results is made for
each example. Both lognormal and Weibull parameter estimations
were made but the D statistics always favored the lognormal distribution and no significant advantage was obtainable by resorting to a
non zero location parameter. All references to confidence levels
imply 90%.

The results will also show that although the median life is
the fundamental parameter of scale, its value is often so large
as to have a minimal practical impact. For descriptive purposes,
it is therefore more useful to quote a lower percentile, such as
the 2nd, and refer to the 2% reliable life. This percentile is
implied in all references to reliable life which follow.

A. An MOS Integrated Circuit in Service

The first illustration utilizes field data obtained on custom designed MOS integrated circuits in a telecommunications application [33]. The sample size was about 6500, and over a surveillance period of approximately 1000 days, 164 failures occurred
but 99 of them were due to a manufacturing fault affecting a short
period of integrated circuit production. As the lifetimes due to
this defect - corrosion of the metallization - could be identified by inspecting the failed components, the desired sample data for
fault-free production was obtained by converting all the lifetimes
caused by corrosion to censoring times, yielding multiply-censored
data. The parameters were then estimated by the mle method and the
solution validated by a x^2 test. From the resulting cdf plot of
Fig. 15 (shich also shows the observed, computer plotted, lifetimes)
the reliable life is about 10 years.

Further cdf plots could also be drawn representing, say,
th upper and lower confidence limits but in this instance they
lie very close to the expected line of Fig. 15 reflecting the large
sample size.

Of the other lifetime functions calculable from the solution,
the most important one is the hazard rate seen, on Fig. 16 to be
falling steadily as a function of operating time and dipping below

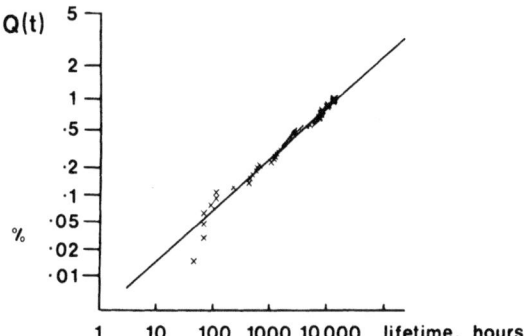

Fig. 15. Lifetime cdf plot for MOS integrated circuits in service
lognormal distribution
x - test significance level: 63%

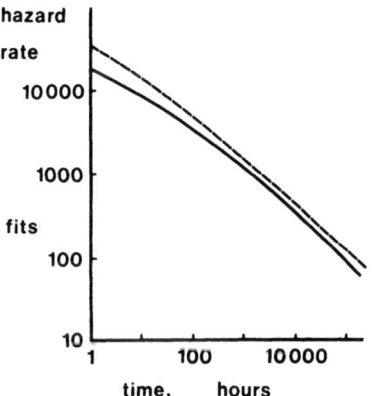

Fig. 16. Hazard plot for MOS integrated circuits in service.
lognoraml distribution
full line: expected hazard rate
broken line: upper 90% confidence limit.

100 fits at 20 years. This trend arises from the comparatively high shape factor, σ, of 5.7 as can be checked from Fig. 3.

It is important to appreciate that most of Figs. 15 and 16 constitute predictions because the observations extend only to somewhat over 2 years, so it is still possible that other long-term failure causes could exist adding another failure distribution. It is to this kind of risk that accelerated testing is directed.

B. Current-Stressed Metallization

The phenomenon of electromigration in very thin metal tracks, especially aluminum, carrying fairly heavy current densities, is accelerated both by current and temperature. The sub-sample cdf plots of Fig. 13 are an example of current stress at a constant temperature where a lifetime is determined by an open circuit. The mle solution for that data passed all the validating tests without question to yield the cdf lines of Fig. 17. These lines are not those which would have individually been estimated for each sub-sample. The dependence of the median life on the stress variable is shown by Fig. 18 (in the format of Fig. 10) together with the reliable-life line and 90% confidence limits. The slope of the lines gives b_1 (of Eqn. (21)) within the confidence limits 1.86 and 2.0 suggesting (eqn. (24)) that electromigration conforms to an inverse square law.

C. Temperature-Stressed Metallization

The lifetimes of the cdf plots of Fig. 19 are again due to the open circuit of metal tracks by electromigration but now the current density is constant and the temperature is the stress variable. The application of the Kolmogorov-Smirnov test to the sub-sample using individual lines of best fit (not shown) yielded the results of Fig. 20 showing that one of the plots is rejected at the 10% level although only by a small margin and hardly enough to justify halting the exercise. The Cochran Test of the parallelism of the three lines provided no basis for rejecting the hypothesis that the differences in the shape factors is merely a matter of chance. The test for the linearity of the stress plot (the validity of eq. 21) is however rather firmly rejected at the 10% level; the actual significance is 2.5% which really indicates that at least more measurements ought to be made. If this slim chance is accepted however, Fig. 19 shows the best global set of lines of best fit and Fig. 21 gives the resulting stress plot. By adding the experimental points, the confidence bands reveal how the uncertainty in the position of the true cdf lines is least in the region of the observations. The lower confidence limit

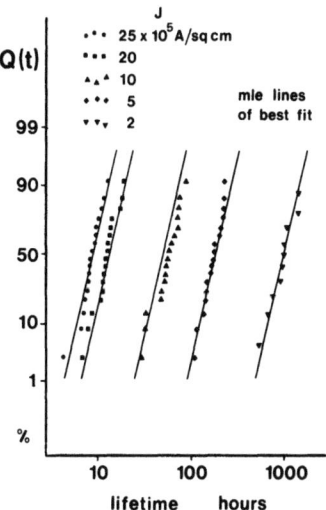

Fig. 17. Lifetime cdf plots for aluminum metallization tracks with
current density, J, as the stress variable
lognormal distribution   failure criterion: open circuit
track temperature: 180°C

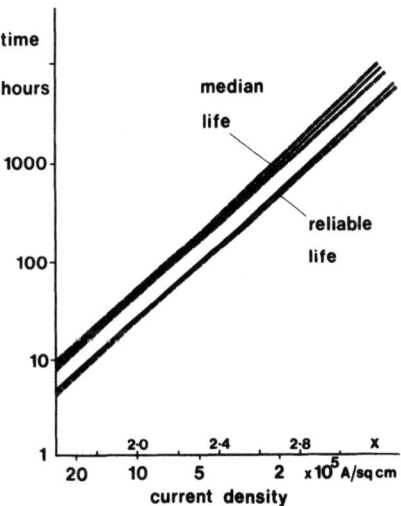

Fig. 18. Stress-model plot for aluminum metallization tracks with
current density as the stress variable
lognormal distribution   reliable-life fraction: 2%
confidence limits (broken lines): 90%
stress parameter: x = 8 - logJ   J in amp/sq cm

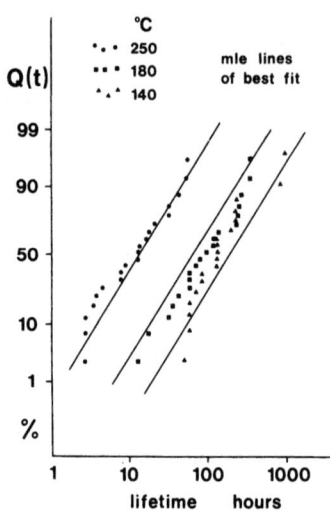

Fig. 19. Lifetime cdf plots for aluminum metallization tracks with temperature as the stress variable.
lognormal distribtuion   failure criterion: open circuit
current density = 2 x 10^{10} amp/sq cm

on the reliable life means, for example, that in 10 future samples out of 100, at 50°C, the time for 2% failures will be less than about 200 hours.

The slope of the lines yields the activation energy for electromigration as 0.44 eV with confidence limits of 0.35 and 0.52.

D. Operational Amplifiers

An example of the fairly scattered data often obtained on integrated circuits is shown by the cdf plot of Fig. 22, for temperature-stressed operational amplifiers with the mle lines already drawn. The lognormal model passed the Kolgomorov-Smirnov test on all the subsamples but because the data is censored, the stress model tests could no be applied. The stress plot of Fig. 23 now reveals rather wide differences between expected values and confidence limits. At a temperature of 50°C for example, the expected reliable life is about 20 years but in 10 samples out of every 1000, the time for 2% failures would be a mere 6 months. These deductions can also be made from the cdf plot at 50°C of Fig. 24 showing also the time dependence of the hazard rate where the discrepancy between expected and upper-limit values is very marked. After 1

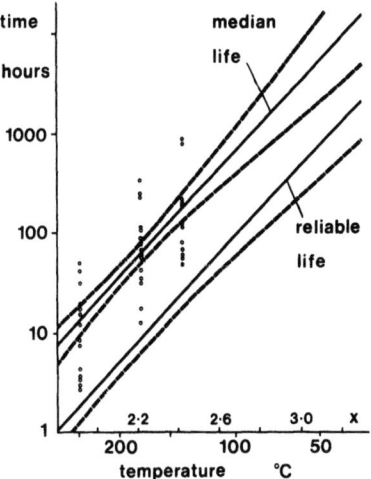

Fig. 21. Stress-model plot for aluminum metallization tracks with temperature as the stress variable.
lognormal distribution   reliable life fraction: 2%
confidence limits (broken lines): 90%
stress parameter: x = 1000/temperature (K)

year, the expected failure rate is about 50 fits but in 10 samples out of every 100 it exceeds 5000 fits. These results convey the general message that a simple graphical analysis would provide a very optimistic impression of the reliabiltiy of these components.

E. Reed Relays

As an example of a non-semiconductor component Fig. 25 shows a lognormal cdf plot for a sample of 20 reed relays operating at 10 Hz and switching currents simulating its normal usage. The median lifetime is very low at about 100 hours. Unlike solid states components however, the lifetime of a reed relay is proportional to its operating frequency so, without resorting to applied stresses, Fig. 25 can be used to predict, say, the reliable life at any other operating frequency. At 1 Hz, for example, the time for 2% failures is 14 hours but in 10 samples out of every 100 the reliable life is less than 9 hours. One could also say that at 10 Hz, the median number of operatins before failure is about 3 million.

IX. ALTERNATIVE APPROACHES

Amongst other semiconductor components which have been tested on the basis of the foregoing stress model are germanium transis-

Hypothesis			Statistic	Critical Value	Hypothesis Accepted or Rejected
lognormal subsample distributions	subsample	250	D = 0.12	0.17	A
	stress	180	D = 0.14	0.17	A
	temperature: °C	140	D = 0.21	0.18	R
shape parameter constant			C = 0.38	0.57	A
linear stress model			F = 5.4	2.8	R

Fig. 20. Model-validation statistics for metallization tracks with temperature as stress variable.
Significance level: 10%

tors [10,11], plastic-encapsulated transistors [34,35], silicon submarine-cable transistors [36], MOS transistors [37,38], gallium arsenide impact diodes[39] and TTL and CMOS integrated circuits [45]. Simple graphical cdf plots for a single sample are frequently made in reliability studies.

The procedures outlined often run into difficulties however, as when there are insufficient failures. A meaningful cdf plot cannot really be drawn with less than 10 lifetimes and the mle method really needs at least 20. Sometimes it is possible to

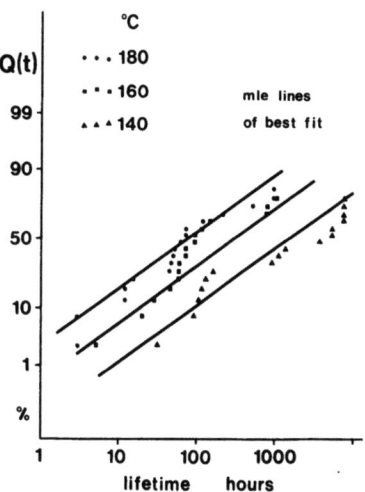

Fig. 22. Lifetime cdf plots for operational amplifiers with ambient temperature as the stress variable.
lognormal distribution   operating conditions: dc bias
failure criterion: manufacturers data sheet limits.

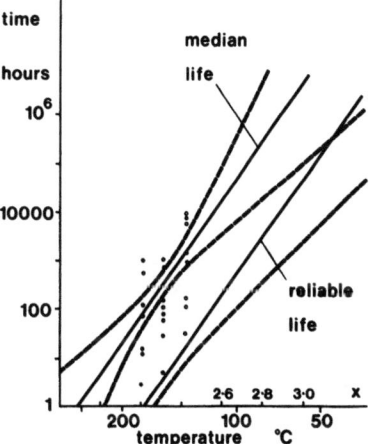

Fig. 23. Stress-model plot for operational amplifiers with ambient temperature as the stress variable.
lognormal distribution reliable fraction: 2%
confidence limits (broken lines): 90%
stress parameter: x = 1000/temperature (°K)

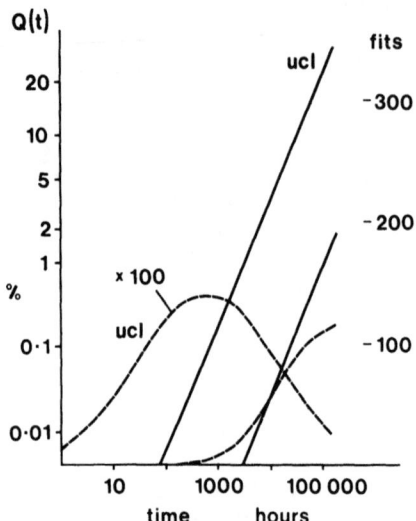

Fig. 24. Lifetime cdf and hazard-rate plot for operational amplifiers.
lognormal distribution   ambient temperature: 50 C (x=3.1)
full line: cdf with 90% ucl
broken line: hazard rate with 90% ucl referred to fit scale x 100.

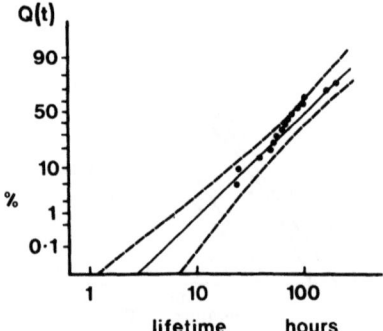

Fig. 25. Lifetime cdf plot for reed relays.
lognormal distribution
failure criterion: contact resistance > 1 ohm
operating conditions: 10 Hz, 50v from discharging cable.

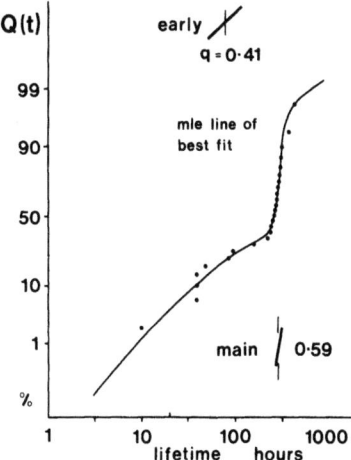

Fig. 26. Bimodal lifetime cdf plot for plastic encapsulated transistors. Lognormal distribution
$\sigma_{early} = 1.1$ $\sigma_{main} = 0.11$
$\sigma_{main} - \sigma_{early} = 1.3$

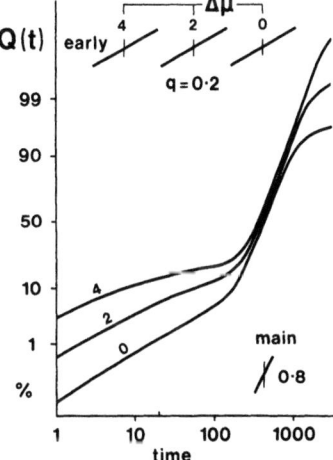

Fig. 27. Theoretical bimodal lifetime cdf plots. lognormal distribution.
$\sigma_{early} = 2.0$ $\sigma_{main} = 0.5$ $\mu_{main} - \mu_{early} = \Delta\mu(0,2,4)$

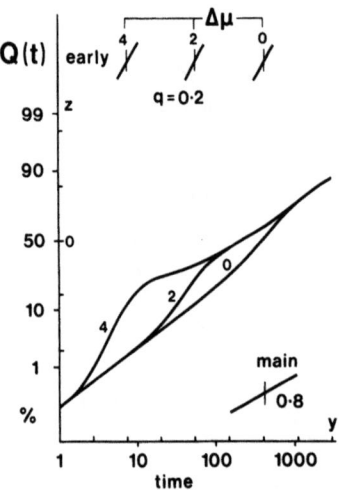

Fig. 28. Theoretical bimodal lognormal cdf plots.
lognormal distribution

$\sigma_{early} = 0.5 \qquad \sigma_{main} = 2.0$
$\mu_{main} - \mu_{early} = \Delta\mu \quad (0,2,4)$

adopt a more sensitive failure criterion - allowing, say, a smaller change in a specified logic level before the item is declared defective - and then correct the resulting analysis for the real criterion. As another option, widely taken, the exponential model (eq. 18) can be postulated although there is almost never any chance of proving its validity. Usually the expected (constant) failure rate λ is estimated rather than the characteristic lifetime. If then r failures occur after h hours of operation in a sample of size n,

$$\lambda = \frac{r}{nh} \qquad (32)$$

An upper 90% (say) confidence limit can be found from [40]

$$\lambda_{90\%} = \frac{x^2[0.9, 2(r + 1)]}{nh} \qquad (33)$$

where the value of x^2 is available from standard tables.

Another reason for the breakdown of the methods thus far described, is the withdrawal of the sample from a mix of two populations. If their fdfs are $f_1(t)$ and $f_2(t)$, and for every q lifetime belonging to the first population, 1 - q belong to the second, then the combined population fdf is

$$f(t) = qf_1(t) + (1 - q) f_2(t) \qquad (34)$$

from which the other lifetime functions can easily be determined using the relationships of section 2. An example of a bimodal cdf plot obtained on plastic-encapsulated transistors in a humid environment is shown by Fig. 26. As with a single population, it is possible to estimate the parameters of the two constituent distributions both graphically [41,42] and by the mle method but the data is seldom sufficient for other than postulated 2-parameter distributions (zero location parameter).

As an aid to a preliminary dissection of a bimodal cdf plot, Figs. 27 and 28 have been prepared for the usual situation where there is an early or "freak" distribution characterized by a low value of q. Each curve corresponds to a different interval, $\Delta\mu$, between the logarithms of the median lifetimes and the dispersions are represented schematically. Plots of this form are quite often derived from accelerated testing [43,45].

The line already drawn on Fig. 26 was obtained from the mle solution. The outcome provides confidence limits on all four parameters as well as on q which was found to be between 0.23 and 0.6.

X. CONCLUSION

A mathematical, as opposed to a graphical, analysis of lifetime data has specific advantages enabling the validity of a postulated lifetime model to be assessed quantitatively or at least permitting an objective comparison between alternatives. The maximum likelihood method also provides confidence limits on the estimated parameters which can help to place the results, and particularly predictions, in a better perspective. It is true that for many purposes, a graphical procedure will suffice but having made a once-only investment, mostly in computer programming, a mathematical analysis is actually easier to perform.

The practical application of the foregoing procedures is often compromised by the changing population problem; before further samples can be withdrawn for service use, the manufacturer will have changed his process, intentionally or otherwise. These reservations do not apply to very high reliability applications as in submerged telephone-cable repeaters, satellites an some military applications where it is feasible to impound a manufacturing batch until it has been characterized by testing. Even when manufacturing changes occur, some lifetime characteristics can be expected to endure, such as the form of the lifetime distribution and its stress sensitivities, particularly the activation energy in thermal acceleration. The best that a user can do to ensure manufacturing constancy is insist upon conformance to a suitable procurement specification.

ACKNOWLEDGEMENT

Acknowledgement is made to the Director of Research, British Post Office Telecommuncations, for permission to publish this paper.

REFERENCES

1. Reynolds, F. H., IEEE 15th Annual Proc. Reliability Physics, 166-178 (1977).
2. Bazovsky, I. Reliability Theory and Practice, Prentice-Hall (1961).
3. Reynolds, F. H., Proc. IEEE 62, 2, 212-222 (February 1974).
4. Mann, N. R., Schafer, R. E., Singpurwall, N. D., Methods for Statistical Analysis of Reliability and Life Data, Wiley (1974).
5. Hald, A. Statistical Theory with Engineering Applications, Wiley (1962).
6. Myers, Wong, K. L, Gordy, Reliability Engineering for Electronic Systems - Reliability Mathematics (Wong), Wiley (1964).
7. Finney, D. J. Probit Analysis, Cambridge Univ. Press, 3rd Edn., (1971).
8. Abramowitz, M. Stegun, I. A., Handbook of Mathematics Functions, (Eqns. 26.2.17 and 26.2.23), Dover Publications (Nov. 1970).
9. Draper, N. R., Smith H. Applied Regression Analysis, Wiley (1967).
10. Dodson, G. A., Howard, B.T., Proc. 7th Nat. Symp. on Reliability and Quality Control in Elec., IEEE, 262-272 (January 1961).
11. Peck, D. S. Semiconductor Reliability Predictions from Life Distribution Data, Semiconductor Reliability, Shwop, J. E., and Sullivan, H. J., Eds. Van Nostrand Reinhold (1961).
12. Nelson, W. IEEE Transactions on Electrical Insulation El-6, 4, 165-181 (December 1971).
13. Nelson, W. IEEE Transactions on Elec. Insulation El-7, 1 36-55 (March 1972).
14. Nelson, W. IEEE Transactions on Elec. Insulation El-7, 99-199, (June 1972).
15. Hahn, G. J., Nelson, W., Insulation/Circuits, 79-84 (Sept. 1971).
16. Hahn, G. J., Nelson W., IEEE Transactions on Reliability R-23, 1, 2-11 (April 1974).
17. Nelson, W. IEEE Transactions on Reliability R-21, 1, 2-11 (February 1972).
18. Nelson, W. IEEE Transactions on Reliability, R-24, 2, 103-107, (June 1975).
19. Nelson, W. Technometrics 14, 4, 945-966 (November 1972).
20. Nelson, W. J. of Quality Technology 2, 3, 126-149 (July 1970).
21. Nelson, W. IEEE Transactions on Reliability R-25, 4, 230-237, (October 1975).
22. Caplen, R. H. A Practical Approach to Reliability, Business Books (1972).
23. Massey, F. J., J. American Statistical Assoc., 68-78 (March 1951).

24. Barr, D. R., Davidson, T. Technometrics 15, 4, 739-757 (November 1973).
25. Lilliefors, H. W., J. American Statistical Assoc. 399-402 (June 1967).
26. Yule, Kendall, An Introduction to the Theory of Statistics, Griffin, 5th Imp. (1950).
27. Winer, B. Statistical Principles for Experimental Design, McGraw Hill (1962).
28. Eisenhart, C, Hartay, M. W. Wallis, W. A., Techniques of Statistical Analysis, McGraw Hill (1947).
29. Guenther, W. C., Analysis of Variance, Prentice Hall, (1964).
30. Cohen, A. C., Technometrics 7, 4, 579-588 (November 1965).
31. Nelson, W. Meeker, W. Q., IEEE Transactions on Reliability R-25, 1, 20-24 (April 1976).
32. Toft, L. McKay, A. D. D., Practical Mathematics, Pitman (1942).
33. Reynolds, F. H. Stevens, J. W., IEEE 16th Annual Proc. Reliability Physics (1978).
34. Peck, D. S. Zierdt, C. H. IEEE 11th Annual Proc. Reliability Physics 146-152 (1973).
35. Lawson, R. W. IEEE 12th Annual Proc. Reliabiltiy Physics, 243-247 (1974).
36. Rouhof, H. W. IEEE Transactions on Reliability R-24, 4, 226-229, (October 1975).
37. Reynolds, F. H. Parrott, R. W., Braithwaite, D., Proc. IEE, 118, 475-485 (March/April 1971).
38. Reynolds, F. H. IEEE 9th Annual Proc. Reliabiliity Physics, 46-56, (1971).
39. Staeker, P. Lindley, W. T. Murphy, R. A., Donnely, J. P. IEEE 12th Annual Proc. Reliability Physics, 293-297 (1974).
40. Epstein, B., IRE Transactions on Reliability and Control, 104-107 (April 1960).
41. Peck, D. S., IEEE 13th Annual Proc. Reliability Physics (1975).
42. Cran, G. W., Microelectronics and Reliability 5, 1, 47-52 (1976).
43. Peck, D. S., Zierdt, C. H., Proc. IEEE 62, 2, 185-211 (February 1974).
44. Byrkit, D. R., Elements of Statistics, Van Nostrand, Reinhold (1972).
45. Stitch, M. Johnson, F. M. Kirk, B. P., Brauer, J. B., IEEE Transactions on Reliability R-24, 4, 238-240 (October 1975).

INDEX

Adsorption
   In on Si, 449
   thermal spike, 448
AES, see Auger Electron Spectroscopy
Airborne dust, 10
$Al_xGa_{1-x}As$, 378, 380
Ambipolar diffusion coefficient, 355
Amorphous semiconductors
   GaP, 367
   Si, 360, 616
   thickness measurement, 611
   produced by implantation, 583
Anharmonic force constants, 324
Anodic oxides, GaAs, 366
Arrhenius law, 306 see also activation energy
Arsenic, 428
   oxide, 425, 426, 427
   surface atoms, 420, 421
Atomic depletion of surface, 443, 445
Atomic rearrangement, 434
Atomic stopping crossection, 589
Autocorrelation function, 260, 263, 265, 287, 290
   Fourier transform, 287
Auger electron spectroscopy (AES), 256, 398, 407, 515
   profiling, 398, 418
Autocovariance, 260

Backscatter spectroscopy, see Channelling, 560, 581
   composition measurement, 594, 597
   depth profile, 618
   films, 594, 597
   helium ions, 436
   limits of detectability, 602
   noise, 590
   spectrum, 590
   yield, 590
Band gap absorption, 376, 380

Band gap profile, 380
Bardeen Model of surface states, 440, 449
Barriers associated with Na^+, 479
   transmission of electrons over, 462
Barrier resistance effects, 585
Berglund integrals, 121, 126, 142
Beryllium, 364, 366, 385
Beveling, 54
Biological specimens, 656
$Bi-Nb_2O_5-Nb$, 480
Blister formation, 615
Bohr energy straggling, 590
Bond reliability
   beam leads, 718, 719
   brittle cleavage, 700
   bugging height, 706, 707
   collapse height, 707
   controlled weak bonds, 710, 712
   peel strength, 712
   pull test, 697, 722
   Griffith break, 728
Bonding
   automatic, 728, 734
   tape, 683, 718, 728
   thermocompression, 678, 703, 712, 717, 718
   thermosonic, 678
   ultrasonic, 678
Braggs rule, 589
Breakdown voltage, 226
Brewster's angle, 368
Brownian motion, 259
Bubbles
   gas test, 690
   H_2 in solids, 615
Bulk traps, see Traps
Burst noise, 293, 297
Burstein-Moss effect, 340, 375

Cadmium, Cd, 479
CdS, 203
CdTe, 340

Capacitance
  depletion layer, 487
  Gray-Brown measurement, 123
  space charge, 108, 206, 210
    211
  temperature variations, 247
Capacitance-Voltage measurements,
    (C-V), 50
  deep level transient spec-
    troscopy (DLTS), 133,
    251
  diode, 50
  edge effects, 218
  equivalent circuit, 114,
    222, 223
  errors, 121, 218
  frequency response, 110
  Hg probe, 50
  high frequency model, 117
  low frequency model, 118
  quasi-static, 118, 121
  Schottky barrier, 230, 336
  slow ramp, 118
  step capacitance (STEPCAP)
    method, 240
  thermally stimulated capaci-
    tance (TSCAP) method,
    242, 247
Cathodoluminescence, 375, 519
Censoring times, 750
Cesium, CS, 447
  on GaAs, 447, 450
    GaSb, 450
    InP, 450
    Si, 447
  Cs-O uniformity, 481
Channel shortening, 138
Channeling, 608
  critical angle, 610
  yield, 610
Characteristic device life-
    time, 743
Characteristic x-rays, 615
Charge
  compensation, 202
  neutrality, 202
  pumping technique, 130, 144
Charge coupled devices, 143
  fat zero level, 145
  transfer loss, 131

Chemical etching, 437
Chemical state
  binding energy, 583
  Si atoms, 436, 440
Chemical reactions, 202, 449
Chemically shifted energy
    levels, 342
Chemisorption, 425
  excited oxygen, 425, 426
  oxygen, 418, 422, 423, 432,
    434, 450
Chip downward bonding, 707
Clusters of emergent disloca-
    tions, 497
Conduction band, 410
Cochran C-statistic, 753, 758
Collector pipes, 102
Collinear four point (CFP)
    probe, 68, 79
Composition analysis, see
    Backscatter spectroscopy
    and Auger electron
    spectroscopy
Conchoidal fracture, see Cracks
Conductivity
  fluctuations, 289
  gradient, 494
  measurement, 68, 79, 331
Conformal mapping, 77
Constant final state (CFS)
    spectroscopy, 409, 410,
    440
Contacts
  arrangements, see Collinear
    and Square four point
    probes, and Spreading
    resistance measurements,
    and Hall Measurements
  barriers, 223, 457
  clover leaf, 82, 90
  composition of point con-
    tacts, 10
  edge of circular specimen,
    82
  finite radius, 3, 80, 90
  metal-semiconductor, 204, 398
  misalignment potential, 85
  noise, 85, 93
  non-linearity, 15
  position errors, 76, 80, 85,
    93

Contacts (cont'd.)
  pressure sensitivity, 13
Contact parameter, 42
  pressure sensitivity, 13
  resistivity dependence, 43
  Schottky barrier, 17
Contact potential, 421
Contamination, 457
  detergent residue, 477
Contrast
  acoustic microscopy, 660
  SEM, 521
Core levels, 416
Correction factors
  F, 69
  C, 70, 219
  conducting substrates, 4
  finite substrate thickness, 24
  geometric, 71, 92, 99
  insulating substrate, 4
  interpolation procedures, 33
  local slope approximation, 36
  sampling volume, 9, 25
Correlation function, 266
Cracks
  initiation, 699
  conchoidal fracture, 6
  propagation, 699, 703
Cross correlation function, 260, 277, 280
Cross Section
  capture, 134, 247
  cascade capture, 234
  coulomb scattering, 463, 584
  differential scattering, 584
  Rutherford scattering, 583, 584
  surface state capture, 125
  stopping, 588
Cubic spline functions, 33
Cumulative distribution function (CDF), 277, 278
Cumulative hazard, 741, 742, 750
Current streamlines, 91
Cyclotron resonance, 326, 375

Damage layer model, 334
  effect of mechanical damage on optical measurements, 332, 333, 364
Damping constant
  DC, 331
  free carrier, 324
  optical, 331
  phonon, 322
Dangling bonds, 105, 420
Dating of man-made glasses, 616
Debye length, 107, 208
Deep levels, see Deep level transient spectroscopy, Traps
  acceptor, 206
  bulk semiconductor, 105, 133, 201, 206
  donor, 206
  electron, 253, 255
  holes, 253, 255, 356
  multilevel traps, 247
  multiphonon emission, 234
Deep level transient spectroscopy (DLTS), 133, 251
Defects, 458, 660, 693, see Dislocations, Stacking faults
  crystallographic, 257, 258, 297, 457, 458, 491
  density, 297
  formation, 446, 448
  lattice disorder detection, 611
  stress associated with, 297, 481
Dember potential, 490
Dendrites, 465, 688
Density function, 260
Density of states, 405
Depletion, 109, 110, 202, 206, 226, 477
  capacitance, 487
  layer, 248, 336
  zero bias, 226
Dielectric constant, high frequency, 322
Differential sheet resistivity, 37

Diffusion
  constant, 484
  currents, 531
  dipole moment, 466
  drift, 466
  length, 470, 483, 487
  voltage, 212, 226
Dimensional resonances, 327,
  see Helicon
Direct transition, 320, 462,
  see Band edge absorption
Dislocations, 458, 555, 717, 718
  local stress, 297, 481
  loops, 497, 567
  rosettes, in Si, 498, 499
  voids, 661
Distribution function, 260
Dopants
  concentration, 211, 212,
    553, 599
  inhomogeneity, 493
  profiles, 9, 24, 46, 37, 213,
    214, 218
Drude model, 324, 328, 348

Edge effects, 218
Effective charge, 324
Effective mass, 324
Einstein relation, 529
Elastic properties, 631, 655
Electric sub-bands, 375, 380, see
  Franz-Keldysh effect
Electrically active stacking
  faults (EASF), 563, 569,
  576
Electromigration, 758
Electron beam induced current
  (EBIC) microscopy,
  519, 545, 556, 561
Electron escape depth, 407,
  414, 437, 560
Electron microscopy, see
  Scanning and transmission
  electron microscopy
Electron penetration, 519
  energy dissipation, 519,
  546, 588, 589
Electron transport, 522
Electro-reflectance, 380

Ellipsometry, 376
Emitter pipes, 576, 580
Energy dispersive x-ray spec-
  troscopy (EDS) (EDAX),
  515
Energy distribution, 401
  angle integrated, 411
  angle resolved, 411
Energy loss, 519, 546, 588, 589
Epitaxial films, 80, 612, see
  Films
  molecular beam, 419, 446
  resistance, 224
Equipotentials, 91
Excitation probability, 400
Excitonic transition, 425
  bulk, 409
  surface, 409, 410, 441
Exponential distribution, 746

F Statistic, 754
Faraday effect, 326
Faraday configuration, 326
Fermat's principle, 648
Fermi-Dirac statistics, 111
Fermi level pinning, 428, 479
Field effect transistor (FET),
  226, 304
  GaAs, 224
  junction, 304
Films, 69, 82 see Epitaxial
  films
  epitaxial, 80
  isolated graded layers, 37, 40
  molecular beam, 419, 446
  optical, 317, 319
  resistance, 224, 477
  semitransparent, 471
  thin conducting, 477
  thickness measurement, 597,
  603
Flaking, 615
Finite extinction, 319
Flat band condition, 108, 109,
  110
Flicker noise, 105, 134, 263,
  288, 471
Flip chips, 729, 733

INDEX

Four probe measurements, 37, 42, 74, 555
  collinear, 68, 79
  square, 71
  wafer geometry effects, 84
Franz-Keldysh effect, 380
Freaks, 691, 693
  population, 693, 767
Free carriers
  absorption, 324, 336
  collection efficiency, 553, 554, 559
  concentration, 67, 522, 550
  contour map, 382
  damping constant, 324
  generation-recombination, 134, 573
  graded plasmon, 359
  inhomogeneity, 339, 348
  interaction with oxide states, 290
  intrinsic, 524
  laser scan, 338
  lifetime, 473
  mobility, 67, 292
  Raman scattering, 374
  transfer inefficiencies, 145
  transport, 528, 532
  trapping, 105
Frequency density function, 740
Frequency spectrum, 287
Fresnel focus, 653

GaAlAs-GaAs heterojunctions, 419
GaAs, 203, 226, 228, 252, 328, 339, 340, 342, 344, 349, 350, 352, 355-57, 362, 364, 366, 385, 429, 445
  FET, 224
  free clean (110) surface, 419, 427
  optical parameters, 321, 355
GaAsSb, 378, 380, 381
$(GaAs)O_4$, 427
Gallium oxide, 426, 427
GaP, 252, 366
  ion implantation, 366
GaSb, 410, 428, 429, 440, 441, 443

GaSb (cont'd.)
  depletion due to Au, 443
Gamma ray detectors, 618
Garnet, 604
Gas phase diffusion, 102
Gate controlled diode, 133
Gaussian beam profile, 655
Gaussian function, 267, 742
Generation-recombination, 530, 536, 546
  generation rate, 69, 529, 546, 549, 553
  Hall-Shockley-Read, 112, 133, 234
  noise spectral density, 292
  surface recombination, 549, 553
Germanium, 328, 342, 374, 401
  filter, 374
Gibbs ensemble, 261
Glass substrates 319
Grain boundaries, 458, 465
Gray-Brown, technique, 123
Griffith break, 728
Growth rates, 606
Guided waves, 354
Gunn devices, 226

Hall effect
  lamellae, 89
  structures, 102
Hall generator, 87
Hall measurement, 67, 84-88, 93, 94, 102, 331
  asymmetric bridge, 85
  compensation circuit, 86
  double AC, 95
  electrostatic shorting of Hall field, 88
  magnets, 87, 88
Hazard
  function, 740
  rate, 743, 756
Heat of condensation, 447
Heat of formation, 446
Helicons, 326, 327
Helium, 479
Helium leak detection, 689
Hertzian stress analysis, 11

Heterojunctions
  devices, 378
  GaAlAs-GaAs, 419
  widths, 419
Hg acoustic properties, 665
HgCdTe, 380
Hot spots, 373
Hydration of glass surfaces, 616
Hydrogen-Oxygen fuel economy, 616
Hydrogenic impurities, 327

Image contrast
  SAM, 671
  SEM, 553, 560, 569
Implantation, 102, 668
Impurities
  C, 537
  complexes, 567
  hydrogenic, 327
  inhomogeneities, 213, 214
  ionization, 525
  multilevel, 247
  O, 537
  striations, 555
  surface, 599
  surface redistribution, 117
  swirl, 555, 569
Inclusions, 667
Index of refraction, 315, 359
Infant mortality, 691, 693, see Freaks
InGaAs, 361
Inhomogeneities, 457, 553
  axial, 67
  interfacial, 479
  localized, 67
  microsegregation, 67, 569
  one dimensional profiles, 98
  patchwork model, 113
  phosphorus deficit, 446
  precipitates, 567, 580
  random potential model, 116
  resistance, 372
  surface depletion, 477
  surface potential, 113, 122
  vertical lamellae, 27
InP, 417, 428, 429, 441, 445

Interdiffusion
  metal-semiconductor interfaces, 446, 449, 450
  segregation, 467
Interface, see Interdiffusion
  defect formation, 448
  diffuse Au-GaSb, 446, 450
  ideal metal semiconductor, 398, 442
  layers, 105
  semiconductor-insulator, 397, 419, 467
  SiO_2-Si, 105, 106, 366, 397, 439, 466, 468
  states, 105, 431, 436, 464
  III-V, 431
Interference
  acoustic, 653, 654
  dimensional resonances, 427
  optical, 406
Inversion, 109, 110, 202, 475
  surface channels, 137
Ion sputtering, 398, 418, see Sputter ion mass spectrometry and AES
  ion milling, 418
Ionic drift in oxides, 693

Johnson noise, 263, 280, 654
Josephson tunneling junction, 93
Joule heating, 93
  effect on resistivity measurements, 75
Junction breakdown voltage, 226
Junctions
  isolation, 44
  one sided, step, 533, 552
  p-i-n diodes, 299
  p-n devices, 293, 299, 533, 535-37, 545, 547, 553, 556, 562, 567
  Schottky, 243, 297, 336
  surface leakage, 105

Kinematic factor, 583
Kinetics of O_2 uptake, 425
Kolmogorov-Smirnov Test, 751, 758, 762
Kramers-Kronig analysis, 321

Laser spotweld, 704
Lateral resistivity profile, 45
Lattice damage
   ion implantation of GaAs, 362
   observations in the visible, 364
Leak tests, 687, 689
   moisture leaks, 683, 686, 703
Life tests
   long term reliability, 257
   lot sample, 697
   lot tolerance percent defective, 697, 698
   median lifetime, 743
   service laboratory, 747
Lifetime, 355, 530
   local fluctuations, 372, 486
   minority carriers, 484, 487
Light emitting diodes, 224
Likelihood function, 755
Lindhard, Scharff and Scratt (LSS) Theory, 50
Lithium niobate, 641
Local dipole layer, 264
Local mean free path, 264
Local modes
   B in Si, 374
   GaAs, 340
Loose particles, 695
Log normal distribution, 741, 754
LO phonon, 322
Lorentz model, 322
Low energy electron diffraction (LEED), 419, 421
Luminescence spectroscopy, 256

Magneto-Optics, 326 see Cyclotron resonance
   microwaves, 348
   Si MOSFETs, 375
Magnetoresistance, 86, 93
   non-linearities, 90
Mapping of properties, 96, 98, 99
Mask alignment, 693

Maximum likelihood method of estimation (MLE), 754
Metallization, 756
   thermal stress reliability, 758, 762
Mg, 366
Microcracks, 667, 700, 703
Microweld, 712
Microsegregation, 67 see Impurities, Inhomogeneities
Minority carriers, see Lifetime
   excitation, 488
   generation rate, 529
   injection, 67
Minimum detectable power, 654, 667 see Acoustic Microscope
Mobility
   Germanium and Silicon, 528
   surface channel, 138
   variation, 143
Monte Carlo calculations, 405
MOS, see C-V measurements
   capacitors, 102
   equivalent circuit with surface states, **111**
   gate controlled diodes, 133
   gated resistivity, 102
   integrated circuit reliability, 756
   magneto-optics, 375

Neutrons
   activation analysis, 602
   ultra-cold, containers for, 616
Newton's method, 755
NiAu, 448
NiCu, 448
Ni_2Si, 606
Niphow disc, 269
Nitrogen, 362
Noise
   bandwidth, 269
   density generator, 269
   figure, 269, 654
   1/f, see Flicker noise

Noise (cont'd.)
   power factor, 270
   power spectrum, 277
   root mean square, 277
   thermal, 280, 654   see Johnson noise
   time dependence, 287
   voltage spectral density, 305
Noise measurements
   acoustic emission, 714
   backscatter, 590
   digital, 277
   figure contours, 271
   flicker, 134, 288
   impedance matching, 266
   junction FET, 304
   lock-in technique, 280
   mechanical vibration, 94
   non-linear, 306
   popcorn, 293
   preamplifier, 273
   quadratic detectors, 277
   thermal methods, 280
   white noise, 269
Normalized yield ratio, 596
Nuclear reaction analysis, 615, 618, 622
   $^1H(^{15}N,\ )^{12}C$, 620, 621
   hydrogen, 615, 618
   light elements, 622
   narrow resonance reactions, 617
Nuclear resonance probes, 97
Nuclear stopping, 589 see LSS theory

Ohmic contacts, 230
Optical excitation matrix elements, 406
Optical samples, 316
Optimum specimen thickness, 87
Oscillator strength, 322
Oxidation induced stacking faults, 497
Oxides
   As, 425
   charge fluctuations, 106, 138, 143

Oxides (cont'd.)
   formation, 418, 427, 428, 567
   Ga, 425
   native, 336, 428
   thermal, 567, 607
   thick for devices on III-V, 431
Oxygen chemisorption, see Chemisorption

Particle impact noise detection test (PIND), 695, 696
Passivating layers, 425
Pattern recognition, 705
Pb, 342
PbSnTe, 297, 359
Penetration depth, optical, 356
Phase contrast imaging, 667
   reflection mode, 668
Photocapacitance, 251
Photoemission, 437
   energy distribution curves, 401
   microscopy, 465
   spectroscopy, 398, 399
   techniques
   "three step" model, 403
   UV, 399, 437
Photolithography, 85
Photoluminescence, 375, 385
Photovoltage, see Laser scan
   bulk, 475, 493
   capacitive coupling, 458
   chopped light, 490
   displacement current measurement, 477
   high frequency, 480
   low frequency, 483
   real time rastering, 469
   surface scan, 361, 488
Photovoltaic cell, 473, 547
Picture elements (PIXELS) 521
Piezoelectric transducers, 660, 667
   insertion loss, 642
   structure, 641
   thin film, 634
Pinholes, 667, 693

INDEX 779

Plasma
  dischargers, 428
  frequency, 324
  reflection, 326
  two dimensional, 375
Plastic deformation, 67
Poisson equation, 107, 207, 214, 239
Polishes
  aqueous, 10
  damaged layers, 333
  non aqueous, 10
Popcorn noise, 293
Potentiometric measurements, 93
Power spectral density, 289
Poyntings vector, 317
Precipitates, 567, 580
Probability density function, 267, 269
Probit, 742
Processing induced electrical changes, 67
Pull test, 683, 697, 706, 719, 722
  destructive, 722
  full force, 684, 734
  nondestructive, 683, 733
  silicone rubber, 706, 719, 733
Pulse echo technique, 643, 658
Purple plague, 692

Quasi-Fermi level, 131, 137
Quasi uniform model, 113

Recombination
  bulk, 69, 529, 546, 549, 553
  surface, 549, 533
Reflectivity, 465
Reliability, see Accelerated life test
  function, 739, 740
  life, 756
Residual Na^+ contamination, 477
Resistivity, see Spreading resistance, Four probe and Collinear measurements

Resistivity (cont'd.)
  contours of equal value, 99
  maps, 98, 99
  probing, 5, 555
  sheet, 82
  spatial average, 67
  substrate, 224
  thermal error voltage, 93, 228
Resolution
  acoustic microscopy, 649, 655, 675
  scanning electron microscopy, 547, 560
Restrahlen, 334

Sb on Au, 448
Scanned internal photoemission, 457, 471
  maps, 477
Scanned surface photovoltage, 488, see Photovoltage
  capacitor electrodes, 477
  image, 496
  inhomogeneities, 371, 372
  light spot, 490
  sampling function, 491, 492
Scanning acoustic microscopy (SAM), see Acoustic microscopy
Scanning electron microscope, 686
  raster pattern, 517
  resolution, 547, 560
  sample preparation, 562
  secondary backscattered electrons, 518
  secondary emission, 560, 561
Scanning laser, 368, see Photovoltage
  electro-optic, 469
  for LSI circuits, 475
  free camerabs, 338
  He-Cd laser, 479
  source, 469
Scattering
  acoustic phonon, 529
  electron-electron (el-el), 404

Scattering (cont'd.)
   electron-phonon (el-ph), 404
   electron-plasmon (el-pl), 404
   ionized impurity, 529
   length, 402
   multiple scattering, 406
   polar optical phonon, 338
   Rutherford cross section, 584, 585
Scattering length, 402
Schottky barriers, 17, 42, 201, 204, 218, 219, 222, 230, 237, 371, 397, 436, 440, 457, 458, 462, 465, 471, 487, 533, 536, 545, 554, 556, 567
Secondary emission coefficient, 561
Selection procedures for devices, 564
   screening of wafers, 458, 461
   statistics, 696
Semitransparent electrodes, 471
Shallow levels, 677
   acceptors, 682
   donors, 682
Si, 328, 362, 366, 374, 480, 679, 704
   boron implant, 374
   electron and hole traps, 255
   heavily doped, 356, 480
   surface analysis, 437
   surface states, 449
   surface transformation, 434
Silicon on sapphire, 102
Slip lines, 481
Snells law, 645
Sodium contamination, 466, 477
Solar wind, 616
Solid solutions analysis, 594
Sound propagation velocity, 634
Specimen geometry, see contacts, Van der Pauw
   circular, 82
   clover leaf, 82, 90
   square structures, 82

Spittle, 688
Spreading resistance, 2, 555
   corrections, 9, 25, see sampling volume
   differential sheet resistance, 37
   limitations, 6
   measurement conditions, 6, 10
   nonlinear contact, 15
   nonlinearity, 20
   optimum source resistance, 15
   recurrence relation for corrections, 33
   sensitivities to sample, 23
   single contact, 5, 555
   surface barrier, 15
Sputtering, 589
   Auger technique, 398, 418, 437
   etching, 437
   ion mass spectrometry (SIMS), 50, 602, 732
Square four probe (SFP) array, 71
Stacking faults, 458, 479, 555, see Electrically active stacking fault
Stochastic process
   "regular", 259
Stress model, 746
   simulation, 739
Stressing
   bias voltage, 693
   Hertzian analysis, 11
   mechanical 713
   model, 746
   thermal, 686, 687, 704, 758, 762
Subsurface changes, 606
   buried layers of Si_3N_4, 361
Surface chemistry, 416, 477
Surfaces
   atomic rearrangement, 422
   disordering phase change, 434
   electronic structure, 412, 421, 422
   leakage current, 67, 105
   strain, 425, 434
   III-V, 403

INDEX

Surface potential variation, 106, 560, see Inhomogeneity
  patchwork model, 113
  random potential model, 116
  statistical fluctuations, 113, 118, 125
Surface States, 105, 131, 202
  Bardeen model, 454, 463
  charge coupled device measurement, 143
  emission rate, 131
  emission time constant, 134
  empty, 421
  extrinsic, 106, 421
  fast states, 105
  fictitious peaks, 118
  filled, 421
  intrinsic, 410, 421, 441
  MOS capacitor equivalent circuit, 114
  oxide-free carrier interaction, 106, 290
  Si, 449
  trapping phenomena, 125, 464
  weak inversion channel current method, 137
Superconducting Nb_3Gc, 616
Swirl, 555, 569
Synchrotron emission pattern, 413, 414
  radiation, 407, 440
  storage ring and sources, 411

Tape bonding, see Bonding, tape
Te, 366
Test structures and systems, 563
  automatic, 231, 564
  bipolar process, 557
  bridge, 85, 231
  internal logic states, 370
  mapping of sheet resistivity, 96
Thermal imaging of microcircuits, 373
Thermal Stress, see Stressing
Thermionic emission, 17, 467
  microscope, 481

Thermocompression bonding, see Bonding, thermocompression
Threshold voltage, 563
Time averaging, 262
Time reversal symmetry, 473
Total internal reflectance of Ge prism, 367
Transfer function, 266
Transistors
  low injection, 300
Transmission electron microscopy, 562, 577
Transmission probability, see Tunnelling
Transverse optical (TO) phonon, 322
  Raman line of Si, 320
Traps
  concentration, 245, 248
  electron in GaAs, 253
  electron in Si, 255
  hole in GaAs, 253
  hole in Si, 255
  interfaces, 464
  profile, 249
  time constant, 251
Tunnelling, 17
  diodes, 400
  into oxides, 106
  through barriers, 223
  transmission probability, 290, 463
II-VI compounds, 448

Ultrasonic bonding 678
Ultraviolet photoemission spectroscopy (UPS), 399, 437

Valence band structure, 416, 432
  maximum, 417, 418
Van der Pauw measurement, 76, 80
  gate resistivity, 102, 350
Variance, 260, 263, 754
Viscosity, 648
Visual inspection, 685
Voigt effect, 326
Voltage noise figure, 270
  power factor, 270
Voltage spectral density, 290

Wave soldering, 718
Weibit, 742, 751
Weibull
   diagram, 743
   distribution, 741, 751
Weiner Khinchin theorem, 266, 290
Wetting of surfaces, 673
Work function differences, 202

X-ray microprobe analysis, 519, 561
X-ray photoemission spectroscopy (XPS), 399, 437

ZnO, 641
   transducers, 642, 643, 652

MIX
Papier aus verantwortungsvollen Quellen
Paper from responsible sources
FSC® C105338

If you have any concerns about our products,
you can contact us on
ProductSafety@springernature.com

In case Publisher is established outside the EU,
the EU authorized representative is:
**Springer Nature Customer Service Center GmbH
Europaplatz 3, 69115 Heidelberg, Germany**

Printed by Libri Plureos GmbH
in Hamburg, Germany